9th TSME-International Conference on Mechanical Engineering (TSME-ICoME 2018)

IOP Conference Series: Materials Science and Engineering
Volume 501

Phuket, Thailand
11 – 14 December 2018

ISBN: 978-1-5108-8539-4
ISSN: 1757-8981

Printed from e-media with permission by:

Curran Associates, Inc.
57 Morehouse Lane
Red Hook, NY 12571

Some format issues inherent in the e-media version may also appear in this print version.

Printed by Curran Associates, Inc. (2019)

Institute of Physics
Dirac House, Temple Back
Bristol BS1 6BE UK

Phone: 44 1 17 929 7481
Fax: 44 1 17 920 0979

techtracking@iop.org

Additional copies of this publication are available from:

Curran Associates, Inc.
57 Morehouse Lane
Red Hook, NY 12571 USA
Phone: 845-758-0400
Fax: 845-758-2633
Email: curran@proceedings.com
Web: www.proceedings.com

Table of contents

Volume 501

9th Thai Society of Mechanical Engineers, International Conference on Mechanical Engineering (TSME-ICoME 2018)

11–14 December 2018, Phuket, Thailand

Accepted papers received: 28 February 2019
Published online: 9 April 2019

Preface

Papers

Alternative Energy and Combustion

Biomechanics and Bioengineering

Computation and Simulation Techniques

Dynamic Systems, Robotics and Controls

Thermal System and Fluid Mechanics

IOP Conf. Series: Materials Science and Engineering **501** (2019) 011001 doi:10.1088/1757-899X/501/1/011001

Preface

The 9th TSME International Conference on Mechanical Engineering (TSME-ICoME 2018) was organized by the Department of Mechanical Engineering, Faculty of Engineering, Thammasat University, Phathumthani province, Thailand. The conference's venue was located at Tavorn Palm Beach Resort in Phuket and held from 11 to 14 December 2018.

Like other years the conference, being supported by Thai Society of Mechanical Engineering (TSME), is considered as a platform for innovators, researchers and industrial representatives around the world to discuss, collaborate and share ideas regarding mechanical-related topics of the century. For instance, the academic program included topics such as Alternative Energy and Combustion (AEC), Automotive, Aerospace and Marine Engineering (AME), Applied Mechanics, Materials and Manufacturing (AMM), Biomechanics and Bioengineering (BME), Computation and Simulation Techniques (CST), Dynamic Systems, Robotics and Controls (DRC), Energy Technology and Management (ETM), and Thermal System and Fluid Mechanics (TSF).

A total of 180 delegates from around the world including, Germany, Japan, Malaysia, Taiwan Slovakia, United States of America and Vietnam, contributed to initiate development of manufacturing technique, prototype machine, and automotive innovations. This volume of IOP Conference Series: Material Science and Engineering included 66 chosen papers whose authors presented at the conference. The aim of this publication is to inspire and spread novel ideas in the area of mechanical engineering.

On behalf of the editing team, we would like to acknowledge the conference organizing committee for their hard work resulting in successful collaborations during the conference. The team would like to express our gratitude to various sponsors, partners and industrial representatives for their supports. Ultimately, we would like to thank Thammasat University for financially supporting the conference TSME-ICoME 2018.

Asst. Prof. Dr. Monchai Prukvilailert

Conference Chair of TSME-ICoME 2018

IOP Conf. Series: Materials Science and Engineering **501** (2019) 011001 doi:10.1088/1757-899X/501/1/011001

Message from the President of TSME

TSME started to support the International Conference on Mechanical Engineering (ICoME) in 2010 to help inspire researchers in the of field mechanical engineering, aerospace engineering and other related engineering to reach their goal as an innovator. In my experience it is these innovations that have been an essential driven force for the development of our nation, especially in the industry sector.

My hope is that this conference will be a forum for academic researchers to collaborate and share information in their field of expertise in order to provide a clearer perspective of how technology will have an effect on this generation and the next. Furthermore, it is necessary for a conference like this to be held every year to help our society grow stronger. Therefore, it is my honor to congratulate participates who have devote their time to attend and present in this conference.

I also would like to express my appreciation to the Department of Mechanical Engineering, Thammasat University for their excellence work in organizing this year the 9th TSME International Conference on Mechanical Engineering (TSME-ICoME 2018), which will be held at Tavorn Palm Beach Resort, Phuket from 11 to 15 of December 2018.

Thank you for making an important contribution to our society and I look forward to being part of this collaborations again in the future.

Asst. Prof. Dr. Chinda Charoenphonphanich
President of Thai Society of Mechanical Engineers (TSME)

IOP Conf. Series: Materials Science and Engineering **501** (2019) 011002 doi:10.1088/1757-899X/501/1/011002

Keynote Speakers

Thermochemical Energy Storage for Utilization of Unused Heat

Dr. Keiko Fujioka
President of Functional Fluids Ltd., Japan
1-4-5, Utsubohonmachi, Nishi-ku, Osaka, 550-0004, Japan

Abstract: More than half of the annual primary energy consumption, about 20000 PJ, is disposed in Japan every year as waste heat, in spite that its energy consumption per GDP is the minimum level in the world. Utilization of waste heat and unused heat such as terrestrial heat and solar heat would contribute the reduction of greenhouse gas, however, the problem is the differences in time, location and quality between supply and demand sides of thermal energy. Thermochemical energy storage can recover, store and transport thermal energy, by using reversible endothermic and exothermic chemical reactions, to eliminate the above-mentioned mismatch. Among many systems for thermal energy storage, chemical heat storage/pump using reversible reactions of inorganic salts with gas substances (gas-solid CHP) have been expected to be an attractive way of effective use of thermal energy for their large energy density, long-term energy storage without heat loss, and wide range of working temperature. To develop highly efficient CHPs, improvement of heat transfer in reactor beds has been one of the most important subjects, since gas-solid reactions are often controlled by heat transfer. In this paper, heat transfer mechanism of porous materials is examined by taking account into the effect of variations of thermophysical properties with reaction progress, which is typical feature of gas-solid reaction. Then, technology development focusing on heat transfer enhancement is described. The effective thermal conductivity of the bed with inorganic salts or oxides is generally in the range of 0.1-0.2 W/mK. Various ways have been developed to increase it without decreasing gas permeability mainly by combining high thermal conductivity materials. Finally challenges in achieving social implementation of thermochemical energy storage, including thermal demand/supply data construction and waste heat analysis using remote sensing methods are presented.

Biography: Dr. Keiko Fujioka graduated with a Doctor of Engineering degree from the Department of Chemical Science and Engineering, Graduate School of Engineering Science, Osaka University in 1999. Dr. Keiko Fujioka research area included Heat-transfer in Porous Materials Energy Conversion, Energy Storage, Energy Transportation Chemical heat storage/heat pump system. In 1984 Dr. Keiko Fujioka Founded Shinsei Cooling Water System Ltd. And assumed position of the president (Company name changed to Functional Fluids in 2006). Additionally, Dr. Keiko Fujioka received many rewards and held many positions over the years including the SCEJ Award for Outstanding Women's Activity, President of Heat Transfer Society of Japan, Vice Chair of Japan Inter-Society Liaison Association Committee for Promoting Equal Participation of Men and Women in Science and Engineering (EPMEWSE) and Vice President of Society of Chemical Engineers.

IOP Conf. Series: Materials Science and Engineering **501** (2019) 011002 doi:10.1088/1757-899X/501/1/011002

Engineering Innovative Practice in Engineering Education: in case study of Rehabilitative and Assistive Innovation

Asst. Prof. Dr. Bunyong Rungroungdouyboon
Excellence Center in Creative Engineering and Development (CED[2])
Department of Mechanical Engineering Thammasat University, Thailand

Abstract: At present, the world's education has changed from being teacher-centered to learner-centered. Also, the 21st century learning skills have been introduced. The new set of skills is required for working in industry. Even, it's required for doing startup company. However, the education in Thailand, especially the engineering education, just start to prepare their students for this situation.Center of Creative Engineering Design and Development was established in 2005 at Faculty of Engineering, Thammasat University. The main goal of center is to use engineering innovation practice to improve the quality of engineering leaning and to solve the main country's problems by innovation simultaneously. The problems of rehabilitative and assistive technology in Thailand have been used for the target problem. Design thinking processes have been taught and practiced. After working more than ten years, the student's skills have been improved obviously. They show the improvement in ability to ask good questions, to categorize the data, to create conceptual solution, to find solution creatively, to design and prototype, to communicate and to work as a team. In addition, the newly designed innovations have been developed and can be commercialized.

Biography: Asst. Prof. Dr. Bunyong Rungroungdouyboon graduated in PhD level from Mechanical Engineering, Lehigh University, USA. He is currently lecturer at Department of Mechanical Engineering, Thammasat University, Thailand. His research work focused on Rehabilitative and Assistive Innovation. Asst. Prof. Dr. Bunyong Rungroungdouyboon is also the founder and head of the Excellence Center in Creative Engineering and Development (CED^2); a task force consisted of students who are interested in developing and prototyping equipments to support patients with servere fatigue conditions and patients who are doing physical therapy. Recently inventions included Manual Standing Wheelchair, Tiny Urban Scooter, Space Walker, and etcs. Over the years Asst. Prof. Dr. Bunyong Rungroungdouyboon also received many award including gold award iCREATe conventions (2017-2018), Gold Medal and Special Prize from Tiawan in the 42nd International Exhibition of Inventions of Geneva, Geneva Switzerland.

IOP Conf. Series: Materials Science and Engineering **501** (2019) 011002 doi:10.1088/1757-899X/501/1/011002

Invited speakers

Topic	Speakers
Comprehensive Conversion of Biomass and Waste to Super Clean Fuels	Assoc. Prof. Dr Kanit Wattanavichien Department of Mechanical Engineering Chulalongkorn University
Evolution of Science of Thermodynamics	Assoc. Prof. Dr. Anchasa Pramuanjaroenkij Department of Mechanical and Manufacturing Engineering Kasetsart University
Some Energy Saving Opportunites in Sugar Mill Cogeneration Plant	Prof. Dr. Chullaphong Chullabodhi School of Energy, Environment and Materials King Mongkut's University of Technology Thonburi

IOP Conf. Series: Materials Science and Engineering **501** (2019) 011003 doi:10.1088/1757-899X/501/1/011003

Conference Program Committee

Honorary Chair

Asst.Prof.Chinda Charoenphonphanich, D.Eng. (President, Thai Society of Mechanical Engineers) — King Mongkut's institute of technology Ladkrabang, Thailand

General Chair

Assoc.Prof. Thira Jearsiripongkul, Dr.-Ing. (Dean, Thammasat School of Engineering) — Thammasat University, Thailand

Prof. Oshima Marie, Ph.D. (Former President of Japan Society of Mechanical Engineers) — The University of Tokyo, Thailand

Conference Chair

Asst.Prof. Monchai Prukvilailert, D.Eng. (Head of Department of Mechanical Engineering) — Thammasat University, Thailand

General Conference Secretariat

Assoc.Prof. Watit Pakdee, Ph.D. — Thammasat University, Thailand

International Scientific Committee

Prof. Dr.-Ing. Heinz Peter Berg	Brandenburg University of Technology Cottbus- Senftenberg, Germany
Prof. Dr. Ashim K. Datta	Cornell University, USA
Prof. Dr. Ashwani K. Gupta	University of Maryland, USA
Prof. Dr. Katsunori Hanamura	Tokyo Institute of Technology, Japan
Prof. Dr.-Ing. Lothar Kroll	Chemnitz University of Technology , Germany
Prof. Dr. Ocktaeck Lim	University of Ulsan, Korea
Prof. Dr. Jau-Huai Lu	National Chung Hsing University, Taiwan
Prof. Dr. Naoki Maruyama	Mie University, Japan
Prof. Dr. Shinichi Nakasuka	University of Tokyo, Japan
Prof. Dr. Hironori Saitoh	Sojo University, Japan
Prof. Dr. Marco P. Schoen	Idaho State University, USA
Prof. Dr. Ning Zhu	Shizuoka Institute of Science and Technology, Japan
Assoc. Prof. Dr. Elmi Abu Bakar	Universiti Sains Malaysia, Malaysia
Assoc. Prof. Dr. Huynh Thanh Cong	Ho Chi Minh City University of Technology, Vietnam
Assoc. Prof. Dr. Seokwoo Jeon	Korea Advanced Institute of Science and Technology, Korea
Assoc. Prof. Dr. Iman K. Reksowardojo	Institute of Technology Bandung, Indonesia
Assoc. Prof. Dr. Hendri Syamsudin	Institute of Technology Bandung, Indonesia
Assoc. Prof. Dr. Le Anh Tuan	Hanoi University of Science and Technology, Vietnam
Assoc. Prof. Dr. Pham Huu Tuyen	Hanoi University of Science and Technology, Vietnam
Asst. Prof. Dr. Pham Tuan Anh	Ho Chi Minh City University of Technology, Vietnam

IOP Conf. Series: Materials Science and Engineering **501** (2019) 011003 doi:10.1088/1757-899X/501/1/011003

Local Scientific Committee

Prof. Dr. Sumrerng Jugjai	King Mongkut's University of Technology Thonburi
Gp.Capt. Prof. Dr. Noppon Hankra	Navaminda Kasatriyadhiraj Royal Thai Air Force Academy
Prof. Dr. Vitoon Uthaisangsuk	King Mongkut's University of Technology Thonburi
Asst. Prof. Dr. Jackrit Suthakorn	Mahidol University
Prof. Dr. Phadungsak Rattanadecho	Thammasat University
Prof. Dr. Suwat Kuntanapreeda	King Mongkut's Univ. of Technology North Bangkok
Assoc. Prof. Dr. Jarruwat Charoensuk	King Mongkut's Institute of Technology Ladkrabang
Prof. Dr. Somchai Wongwises	King Mongkut's University of Technology Thonburi
Assoc. Prof. Dr. Ratchaphon Suntivarakorn	Khon Kaen University

Host and Organizing Committee

Organizing Committee

Chairperson: Assoc. Prof. Dr. Thira Jearsiripongkul

IOP Conf. Series: Materials Science and Engineering **501** (2019) 011003 doi:10.1088/1757-899X/501/1/011003

Organizing chairperson Prof Dr Phadungsak Rattanadecho

Asst. Prof. Dr. Monchai Prukvilailert

Secretary: Assoc. Prof. Dr. Watit Pakdee

Asst. Prof. Dr. Charnnarong Assavatesanupap

Committee: Prof. Dr. Chaosuan Kanchanomai

Prof. Dr. Somchart Chantasiriwan

Assoc. Prof. Dr. Wiroj Limtrakarn

Assoc. Prof. Dr. Dulyachot Cholaseuk

Assoc. Prof. Dr. Chainarong Chaktranond

Assoc. Prof. Dr. Isares Dhuchakallaya

Asst. Prof. Dr. Chatchai Marnadee

Asst. Prof. Dr. Witawats Satasook

Asst. Prof. Dr. Bunyong Rungroungdouyboon

Asst. Prof. Dr. Krit Jiamjiroch

Asst. Prof. Dr. Sappinandana Akamphon

Asst. Prof. Dr. Somsak Vongpradubchai

Asst. Prof. Dr. Jakkrapun Chuanasa

Asst. Prof. Dr. Chonlada Luangarpa

Dr. Nattadon Pannucharoenwong

Dr. Snunkhaem Echaroj

IOP Conf. Series: Materials Science and Engineering **501** (2019) 011004 doi:10.1088/1757-899X/501/1/011004

Sponsorships

สภาวิศวกร
COUNCIL OF ENGINEERS

SCG
CEMENT-BUILDING MATERIALS

Café Amazon

BLCP POWER

GC

THAI TOHKEN THERMO CO.,LTD.

IOP Conf. Series: Materials Science and Engineering **501** (2019) 011004 doi:10.1088/1757-899X/501/1/011004

IOP Conf. Series: Materials Science and Engineering **501** (2019) 011004 doi:10.1088/1757-899X/501/1/011004

Conference Photographs

IOP Conf. Series: Materials Science and Engineering **501** (2019) 011005 doi:10.1088/1757-899X/501/1/011005

IOP Conf. Series: Materials Science and Engineering **501** (2019) 011005 doi:10.1088/1757-899X/501/1/011005

IOP Conf. Series: Materials Science and Engineering **501** (2019) 011005 doi:10.1088/1757-899X/501/1/011005

IOP Conf. Series: Materials Science and Engineering **501** (2019) 011005 doi:10.1088/1757-899X/501/1/011005

IOP Conf. Series: Materials Science and Engineering **501** (2019) 011006 doi:10.1088/1757-899X/501/1/011006

Peer review statement

All papers published in this volume of *IOP Conference Series: Materials Science and Engineering* have been peer reviewed through processes administered by the proceedings Editors. Reviews were conducted by expert referees to the professional and scientific standards expected of a proceedings journal published by IOP Publishing.

IOP Conf. Series: Materials Science and Engineering **501** (2019) 012001 doi:10.1088/1757-899X/501/1/012001

Numerical Prediction of Mixture Formation Process of an Ethanol Spray in a Rapid Compression and Expansion Machine

Akihiro UMENO[1], Koji UCHIDA[2], Norihiko WATANABE[2] and Hironori SAITOH[2,*]

[1]Graduate School of Mechanical Engineering, Sojo University, Ikeda 4-22-1, West Ward, Kumamoto City, 860-0082, Japan

[2] Department of Mechanical Engineering, Sojo University, Ikeda 4-22-1, West Ward, Kumamoto City, 860-0082, Japan

* Corresponding Author: E-mail: saitoh@mec.sojo-u.ac.jp, Telephone: +81-96-326-3729, Fax: +81-96-323-1351

Abstract. This study deals with the development of controlled-ignition technology for high performance compression ignition alcohol engines. Among the alcohol fuels, we focused on Ethanol as a promising candidate of alternative fuels replacing from petroleum. In our previous study, visualization tests of spray mixture formation process up to auto-ignition for Ethanol-Diethyl either blend fuels had been conducted by using a constant volume electrical heating combustion chamber. Quantitative evaluation of auto-ignition quality was summarized in a form of 3D map that indicates the effects of surrounding gas pressure, temperature and Oxygen concentration on ignition delay. However, maximum surrounding gas pressure in the experiments was 2.8MPa much less than that of real engine due to the structure of the combustion chamber. As the next step, we originally designed and manufactured a RCEM (Rapid Compression and Expansion Machine with variable compression ratio) in order to investigate the mixture formation process up to auto-ignition of an Ethanol spray under surrounding gas conditions similar to real engine operating conditions with the consideration of the interaction between fluid motion of surrounding gas and fuel injection. Prior to carrying out the experiments, we numerically predicted the auto-ignition quality of an Ethanol spray in the rapid compression and expansion machine. One of the commercial CFD codes; CONVERGE was used in the computational calculation with the considerations of turbulence, atomization, evaporation and detailed chemical reaction. Effect of fluid motion and surrounding gas pressure and temperature inside the combustion chamber of the RCEM on an Ethanol spray mixture formation is mainly discussed in this paper.

IOP Conf. Series: Materials Science and Engineering **501** (2019) 012001 doi:10.1088/1757-899X/501/1/012001

Nomenclature

T	: tempareture [K]
t	: time [ms]
λ	: excess air ratio [–]
ε	: compression ratio

Subscripts

ig	: ignition
spm	: Mean of a spray penetration mean
tip	: spray tip

Abbreviation

AMR	: Adaptive Mesh Refinement
ATDC	: After Top Dead Center
BDC	: Bottom Dead Center
BRICS	: Brazil, Russia, India, China, South Africa
BTDC	: Before Top Dead Center
FE	: Fixed Embedded
IPCC	: Intergovernmental Panel on Climate Change
KH	: Kelvin-Helmholtz
LES	: Large Eddy Simulation
RCEM	: Rapid Compression and Expansion Machine
RT	: Rayleigh-Taylor
TDC	: Top Dead Center
WMO	: World Meteorological Organization

1. Introduction

At present civil life and industrial activities are supported by mass consumption of fossil resources. Nowadays, energy consumption is increasing year by year with population growth and industrial development especially in BRICS. We have been faced the problems such as oil reserve depletion and global warming. Amount of recognized oil reserves is approximately 1071.1 billion barrels, and if we continue to consume oil at the present rate, available years is expected as 52.5 years [1]. Under the above introduced international movements with population, economic growth and industrial development depending on fossil fuel, global warming has been obviously appeared. IPCC(Intergovernmental Panel on Climate Change) founded by united nations and WMO(World Meteorological Organization) presented that carbon dioxide accounts for 60% of the total greenhouse gas emissions, and 80% of them are being emitted by consumption of fossil fuels by power plants and automobiles. Therefore, it is necessary for modern societies to pay an attention of decreasing the fossil fuel dependence and it is urgent to construct a recycling-oriented society. As one of the solutions, we focus on the development of high efficient energy conversion technology by using biomass-based alcohol fuel as an alternative fuel that is environmentally friendly and renewable.

2. Previous study and Objective

2.1. *Previous study*

The goal of our research is the establishment of controlled-ignition technology for high performance compression ignition alcohol engines. Among alcohol fuels, we focused on Ethanol as a promising candidate of alternative fuels replacing from petroleum. In our previous study, we conducted a theoretical calculation of mixture formation process based on the momentum theory of spray penetration proposed by Wakuri et.al [2]. In this theoretical approach, we employed two fundamental factors of mixture on auto-ignition in a spray. One was concentration factor; mixture have to be in its mixing ratio enough to advance chemical reactions. The other was temperature factor; temperature of mixture have to exceed a certain value that chemical reactions are promoted up to combustion. Focusing on this recognition, we defined concentration and temperature factors respectively as $\lambda_{tip} = 1$ and $T_{spm} \geqq T_{ig}$. λ_{tip} is excess air ratio of a spray tip, T_{spm} and T_{ig} are mean temperature of a spray and minimum ignition point of fuel, respectively. In our recognition, stable auto-ignition would be obtained by fuel injection if temperature factor is cleared when concentration factor is attained. Based on this fundamental mechanism of auto-ignition, prediction of mixture formation process of an Ethanol spray and a Gas oil spray was theoretically performed. Temporal histories of (λ_{tip}) and (T_{spm}) from fuel injection for Gas oil and Ethanol is shown in figure 1.

It can be understood from figure 1 that Gas oil seems to have stable auto-ignition because T_{spm} is much higher than T_{ig} when λ_{tip} is reached its concentration factor ($\lambda_{tip} = 1$). On the other hand, in the case of Ethanol, at the timing of $\lambda_{tip} = 1$, T_{spm} does not reach to its temperature factor (T_{ig}). In addition, leaner situation is easily expected when temperature factor is satisfied. These results allowed us to draw one conclusion on the reason of poor auto-ignition quality of Ethanol spray. That is the difficulty of simultaneous attainment of auto-ignition suit concentration and temperature during mixture formation process due to smaller stoichiometric air/fuel ratio and much larger latent heat of evaporation of Ethanol comparing to Gas oil. In other words, improvement of auto-ignition quality of Ethanol seems to be possible by retardation of leaner situation and faster temperature rising. Surrounding gas conditions such as pressure, temperature and Oxygen concentration, which corresponds to the entrained gas density, temperature and amount of heat supply into a spray, also influence the λ_{tip} and T_{spm} histories from fuel injection. Higher temperature of surrounding gas result faster T_{spm} rising and higher Oxygen concentration induce faster lean situation of mixture. Higher pressure of surrounding gas has the both effects of faster lean situation and temperature rising due to higher density of entrained gas into a spray.

Based on the above stated recognition, we conducted a spray visualization tests by using a constant volume electrical heating chamber under various kind of surrounding gas conditions. Ethanol-Diethyl ether blend fuel(changing blend ratio) was employed as test fuel in order to make clear the effect of fuel properties on their ignition delay. Value of fuel properties of Diethyl ether is close to Gas oil. As expected before experiments, shorter ignition delay was obtained with increasing of surrounding gas pressure, temperature and Oxygen concentration for Diethyl ether rich blend fuel. However, auto-ignition was not observed in case of Ethanol rich blend fuel even with retardation of leaner situation and faster temperature rising by surrounding gas conditions of low Oxygen concentration of 15% and 800 K in temperature, 2.8MPa in pressure. This result seemed to be attributed to insufficient surrounding gas density and temperature.

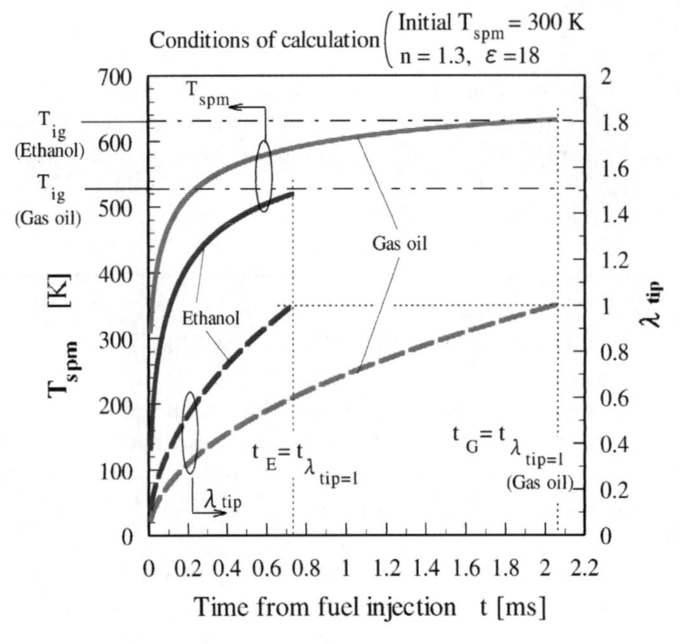

Figure 1. Theoretical evaluation of λ_{tip} and T_{spm} histories from fuel injection for Ethanol and Gas oil spray.

2.2. *Objective*

As the next step we originally designed and manufactured a RCEM (Rapid Compression and Expansion Machine with variable compression ratio) aiming to the observation of auto-ignition for an Ethanol spray. Figures 2(a) and (b) show the overall view of the RCEM and the cross section view of the device. RCEM consists of five parts. These are drive system, lubricating oil system, air intake/exhaust system, fuel injection system and compression parts by means of a long stroke crank-piston structure with cross-head mechanism. Compression ratio is adjustable with electronic control variable valve operating system. Maximum compression ratio is 22. Table 1 represents the principal particulars of the RCEM. Target values of compression pressure and temperature are 6.5MPa and 1000K, respectively for auto-ignition of an Ethanol spray. Prior to the experiments, in order to realize such surrounding gas conditions of high pressure and temperature we have to understand the RCEM operating conditions such as operating speed[rpm], initial intake gas condition, valve close/open timing and fuel injection timing synchronizing with crank angle. Then we performed numerical analysis of mixture formation process of an Ethanol spray with the consideration of flow behavior in combustion chamber in compression and expansion stroke. The objective of this study is to predict the spray mixture formation process of an Ethanol spray in the RCEM and to have a feasibility check for the attainment of surrounding gas conditions sufficient to occur auto-ignition.

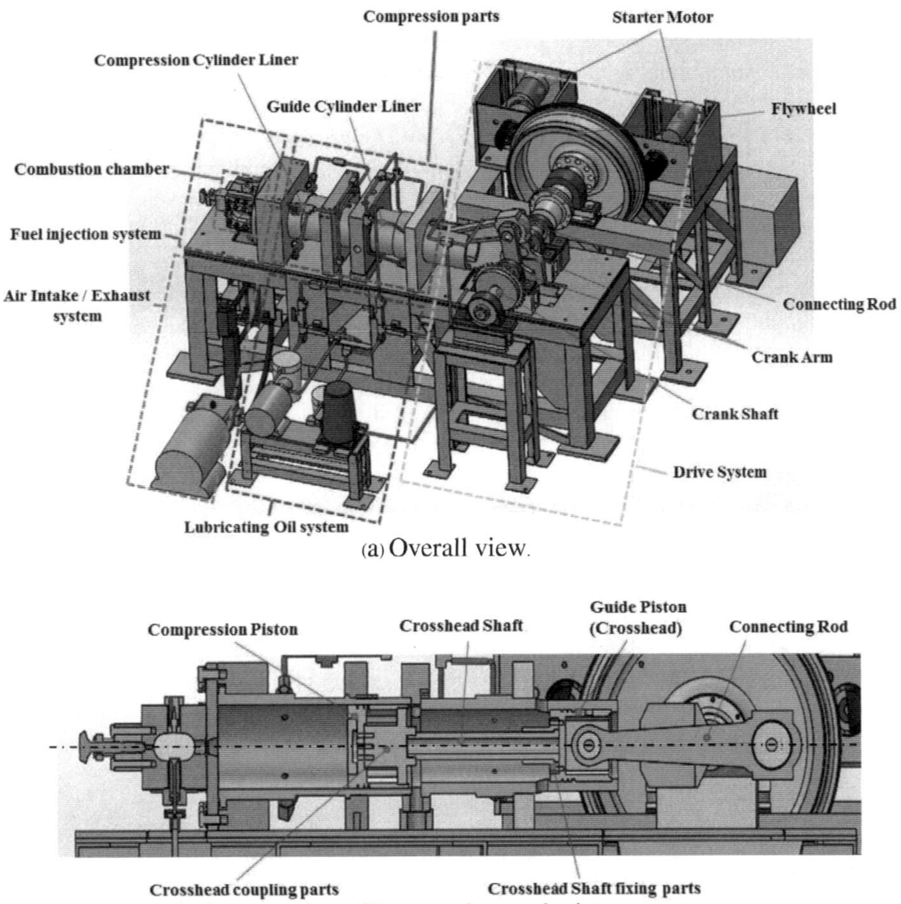

(a) Overall view.

(b) Cross section of long stroke crank-piston structure.

Figure 2. General view of the RCEM.

Table 1. Specification of the rapid compression and expansion machine with variable compression ratio.

Cylinder Bore [mm]	150
Stroke [mm]	280
Maximum Volume[mm^3]	5.18×10^6
Target Compression Temperature [K]	1000 (1200K pre-heating)
Target Compression Pressure [MPa]	6.5
Maximum Compression Ratio	22
Fuel Injection Pressure [MPa]	50
Fuel Injection Duration [ms]	4.5
Injector Nozzle (type, diameter:φ[mm]×number)	Hole type, 0.14×1

3. Numerical Analysis

3.1. *Computational Domain and Mesh Geometry*

Computational domain as shown in figure 3 consists of intake/exhaust port, combustion chamber and cylinder with changing its size by piston movement. Table 2 shows the mesh geometry employed in the numerical analysis. Base grid size was 4mm. Fine mesh of 0.5mm was set in the expected area where mixture formation is occurred. Entity of the fine mesh is cone shape as shown in figure 3 (light green region in the combustion chamber) and its geometry is presented as “Fixed Embedding Fine Mesh” in table 2. Commercial CFD code; CONVERGE used in the numerical analysis has an automatic mesh refinement function named “AMR”: Adaptive Mesh Refinement. We applied AMR for velocity, temperature and chemical species as listed in table 2. For each grid point, AMR activation is judged by the “sub-grid criterion” comparing to the calculated results of velocity, temperature and chemical species to their previous time step results at the same grid. Minimum mesh size of 0.0625mm was applied by AMR in the numerical analysis for correct calculation. Table 3 shows the analysis conditions. RCEM operating speed was set as 244 rpm calculated from the rotational speed of starter motor and gear ratio between starter motor and flywheel, intake air was set as suction from atmosphere with 0.1MPa in pressure, 300K in temperature and 21 volume % in Oxygen concentration. Large eddy simulation (LES) was used as turbulence model. Kelvin-Helmholtz and Rayleigh-Taylor models represented as KH-RT model was employed for atomization and evaporation of fuel droplets. SAGE detailed chemical elementary reaction model was also used in the case with the consideration of combustion after spray mixture formation.

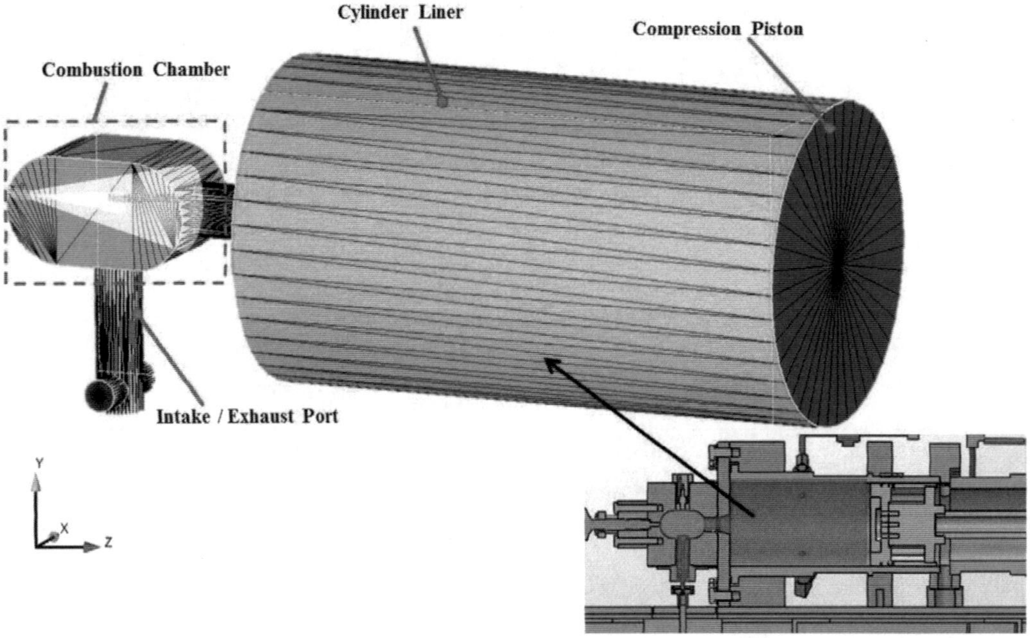

Figure 3. Computational domain.

IOP Conf. Series: Materials Science and Engineering **501** (2019) 012001 doi:10.1088/1757-899X/501/1/012001

Table 2. Mesh geometry in the numerical analysis.

Base grid	x		4 mm
	y		4 mm
	z		4 mm
FE (Fixed Embedded)	Entity type		Cone: INJECTOR
	Scale: Base grid × 2^(-scale)		3: 4mm × 2^{-3} =0.5 mm
	Radius 1		2 mm
	Radius 2		20 m
	Length		80 mm
AMR (Adaptive Mesh Refinement)	Velocity	Max embedded Scale	6: 4mm × 2^{-6} =0.0625 mm
		Sub-grid criterion	1.0 m/sec
	Temperature	Max embedded Scale	6: 4mm × 2^{-6} =0.0625 mm
		Sub-grid criterion	2.5 K
	Species	Max embedded Scale	5: 4 mm × 2^{-5} = 0.125 mm
		Sub-grid criterion	0.001 mole fraction

Table 3. Analysis conditions.

RCEM Operating Speed [rpm]	244
Fuel	Ethanol
Intake Gas Temperature [K]	300
Intake Gas Pressure [MPa]	0.1
Intake Gas Oxygen Concentration	21% (vol.)
Fuel Injection Pressure [MPa]	50
Fuel Injection Duration [ms]	4.5
Turbulence Model	LES
Spray Model	KH-RT Model
Chemical reaction Model	SAGE Detailed elementary reaction model

KH : Kelvin-Helmholtz, RT : Rayleigh-Taylor

3.2. *Governing Equations and Boundary Conditions*

Flow and temperature fields were solved under the conservation law of mass, momentum and energy. Three dimensional equation of continuity, Navier-Stokes equation and energy equation were employed with the consideration of fluid compressibility. Transport equation of chemical species was also applied for the calculation of combustion. Flow and temperature fields were numerically solved by finite volume method for the above introduced governing equations.

Non-slip condition was applied for flow field at the boundary of computational domain. Law of wall condition was also applied as thermal boundary condition.

4. Results and Discussion

Figure 4 shows the calculation result of flow field during intake and compression stroke. Although complicated flow with large-scale vortices was observed with valve motion during intake stroke, such complicated flow was disappeared and uniform velocity was formed in the combustion chamber at the crank angle around BTDC 40 degree in compression stroke. However, at TDC the end of compression stroke, asymmetric velocity distribution was observed again. In the region near the valve, higher velocity was obtained. In addition, temporal velocity fluctuation (acceleration and deceleration) was almost followed the piston speed.

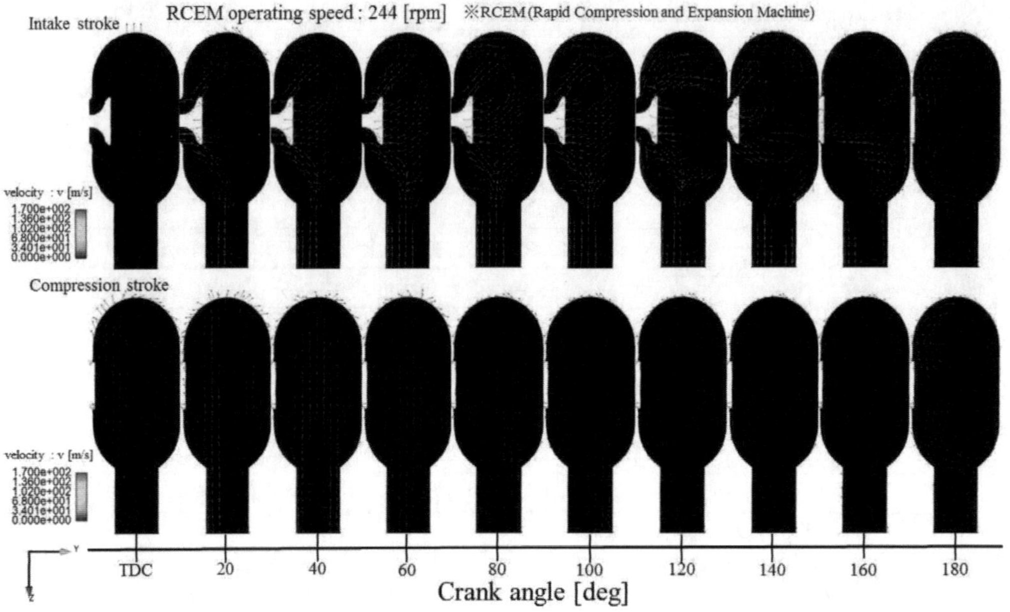

Figure 4. Numerical result of flow field in the combustion chamber during intake and compression stroke under the RCEM operating speed of 244rpm.

Figure 5 presents the temporal in-chamber gas pressure and temperature change during intake and compression stroke. Approximately 24% pumping loss was indicated in intake stroke. At the beginning of compression stroke at BDC, gas pressure and temperature was respectively indicated as 0.076MPa and 290K. Due to such pumping loss resulted compression pressure and temperature as 4.3MPa and 772K less than each target value. Therefore, decrease of pumping loss up to zero or device of supercharge is required to obtain the target compression pressure and temperature (6.5MPa and 1000K).

IOP Conf. Series: Materials Science and Engineering **501** (2019) 012001 doi:10.1088/1757-899X/501/1/012001

Table 2. Mesh geometry in the numerical analysis.

Base grid	x		4 mm
	y		4 mm
	z		4 mm
FE (Fixed Embedded)	Entity type		Cone: INJECTOR
	Scale: Base grid × 2^(-scale)		3: 4mm × 2^{-3} =0.5 mm
	Radius 1		2 mm
	Radius 2		20 m
	Length		80 mm
AMR (Adaptive Mesh Refinement)	Velocity	Max embedded Scale	6: 4mm × 2^{-6} =0.0625 mm
		Sub-grid criterion	1.0 m/sec
	Temperature	Max embedded Scale	6: 4mm × 2^{-6} =0.0625 mm
		Sub-grid criterion	2.5 K
	Species	Max embedded Scale	5: 4 mm × 2^{-5} = 0.125 mm
		Sub-grid criterion	0.001 mole fraction

Table 3. Analysis conditions.

RCEM Operating Speed [rpm]	244
Fuel	Ethanol
Intake Gas Temperature [K]	300
Intake Gas Pressure [MPa]	0.1
Intake Gas Oxygen Concentration	21% (vol.)
Fuel Injection Pressure [MPa]	50
Fuel Injection Duration [ms]	4.5
Turbulence Model	LES
Spray Model	KH-RT Model
Chemical reaction Model	SAGE Detailed elementary reaction model

KH : Kelvin-Helmholtz, RT : Rayleigh-Taylor

3.2. *Governing Equations and Boundary Conditions*

Flow and temperature fields were solved under the conservation law of mass, momentum and energy. Three dimensional equation of continuity, Navier-Stokes equation and energy equation were employed with the consideration of fluid compressibility. Transport equation of chemical species was also applied for the calculation of combustion. Flow and temperature fields were numerically solved by finite volume method for the above introduced governing equations.

Non-slip condition was applied for flow field at the boundary of computational domain. Law of wall condition was also applied as thermal boundary condition.

IOP Conf. Series: Materials Science and Engineering **501** (2019) 012001 doi:10.1088/1757-899X/501/1/012001

4. Results and Discussion

Figure 4 shows the calculation result of flow field during intake and compression stroke. Although complicated flow with large-scale vortices was observed with valve motion during intake stroke, such complicated flow was disappeared and uniform velocity was formed in the combustion chamber at the crank angle around BTDC 40 degree in compression stroke. However, at TDC the end of compression stroke, asymmetric velocity distribution was observed again. In the region near the valve, higher velocity was obtained. In addition, temporal velocity fluctuation (acceleration and deceleration) was almost followed the piston speed.

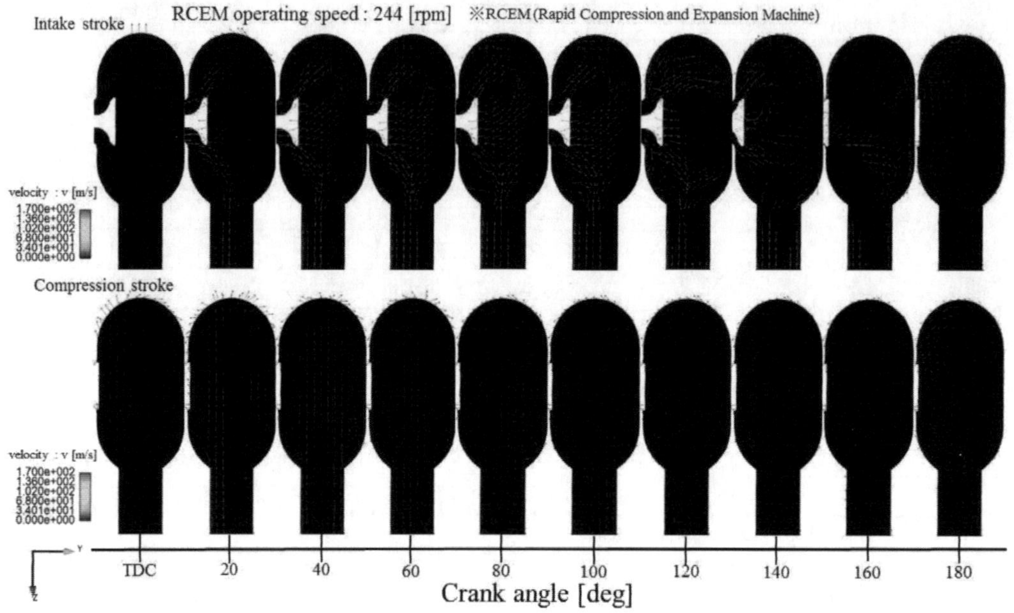

Figure 4. Numerical result of flow field in the combustion chamber during intake and compression stroke under the RCEM operating speed of 244rpm.

Figure 5 presents the temporal in-chamber gas pressure and temperature change during intake and compression stroke. Approximately 24% pumping loss was indicated in intake stroke. At the beginning of compression stroke at BDC, gas pressure and temperature was respectively indicated as 0.076MPa and 290K. Due to such pumping loss resulted compression pressure and temperature as 4.3MPa and 772K less than each target value. Therefore, decrease of pumping loss up to zero or device of supercharge is required to obtain the target compression pressure and temperature (6.5MPa and 1000K).

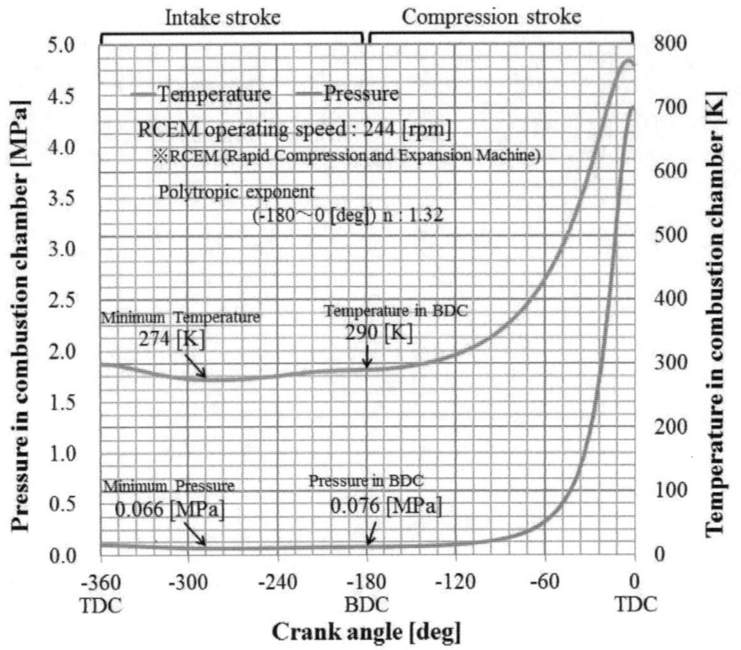

Figure 5. Pressure and temperature of in-chamber gas during intake and compression stroke.

Mixture formation of an Ethanol spray was simulated in expansion stroke from TDC to ATDC 7 degree in crank angle that corresponded to 4.78ms in time from fuel injection. Previously introduced flow (velocity distribution), pressure and temperature fields in the combustion chamber at the end of compression stroke was used as the initial gas condition at TDC before fuel injection. Result of calculation is shown in figure 6. Upper part of this figure indicates the history of spatial excess air ratio (λ) distribution along the axis of fuel injection (z direction), and lower part corresponds to those of temperature distribution. Although spray developed straight downward, both the excess air ratio and temperature inside a spray showed asymmetric distributions. Leaner and higher temperature mixture was observed in valve-side region. This can be resulted by promotion of evaporation due to stronger entrained gas flow, and this phenomenon can be caused by the asymmetric velocity distribution in the combustion chamber at TDC as represented in figure 4. In addition, auto-ignition was not confirmed under this surrounding gas conditions. Surrounding gas temperature was decreased at every crank angle due to increase of volume in expansion stroke. Figure 7 presents the excess air ratio (λ) and temperature (T) histories from fuel injection. They were calculated as the area-averaged λ and T of maximum x-y cross section in a spray. Both of the λ and T slightly increased with crank angle. This tendency is different from that previously introduced in figure 1 as theoretical analysis. This can be attributed to the decrease of entrained gas pressure and temperature at every crank angle in expansion stroke. Although temperature (T) exceeded the minimum ignition point of Ethanol around 0.5 degree in crank angle, at that timing excess air ratio (λ) showed the value greater than 1 defined as concentration factor for auto-ignition in figure 1.

In order to make clear the reason why auto-ignition was not occurred, we focused on instantaneous spatial λ and T distribution inside a spray from the viewpoint of temporal and spatial matching of auto-ignition suit mixture concentration and temperature.

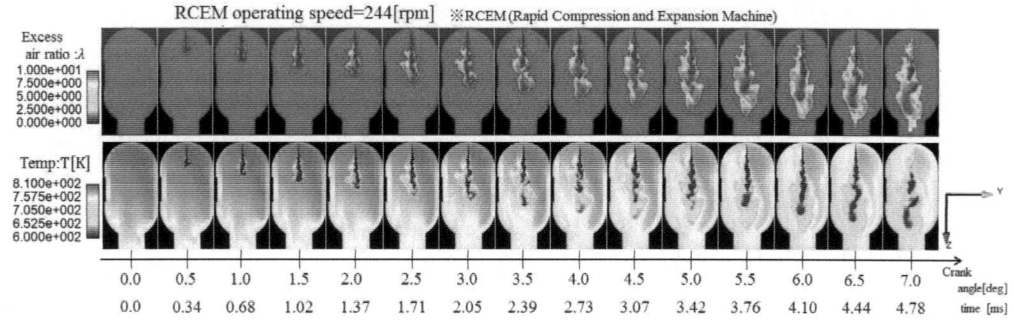

Figure 6. Prediction of the mixture formation process of an Ethanol spray in expansion stroke from TDC to ATDC 7.0 degree.

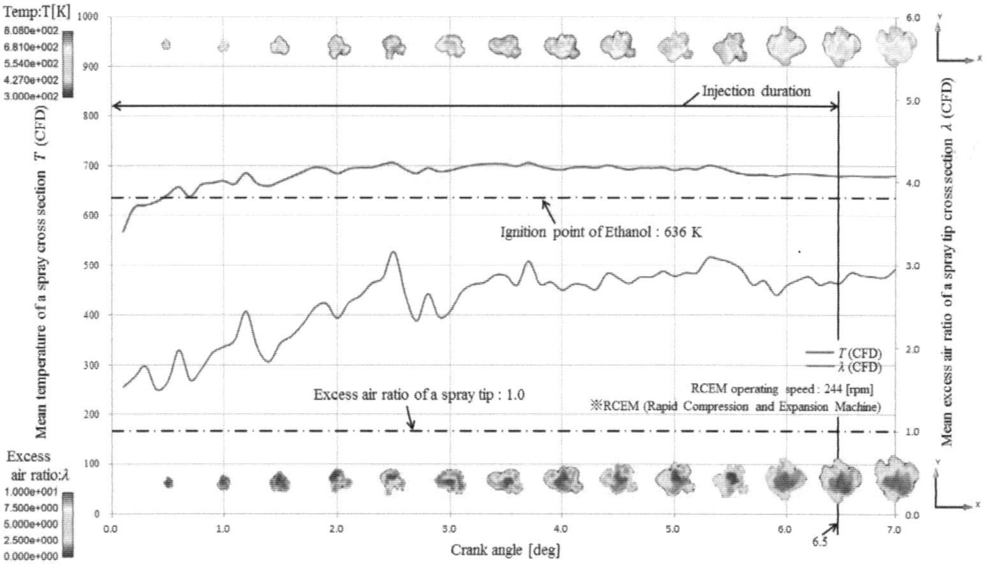

Figure 7. Numerically analyzed histories of mean excess air ration and temperature of an Ethanol spray cross section.

Figure 8 shows the local area of concentration and temperature factor attainment ($\lambda_{tip} = 1, T_{spm} \geqq T_{ig}$) and their change with crank angle. Upper part of this figure indicates λ distribution of *y-z* and *x-y* cross sections. Local area of concentration factor attainment for auto-ignition is coloured blue. Lower part corresponds to T distribution and local area of temperature factor attainment is coloured orange. Lowest part is the overlapped image of λ and T distributions at each crank angle. It is clearly confirmed from this figure that no overlapped area of two colours is found. Therefore, auto-ignition seemed not to be obtained.

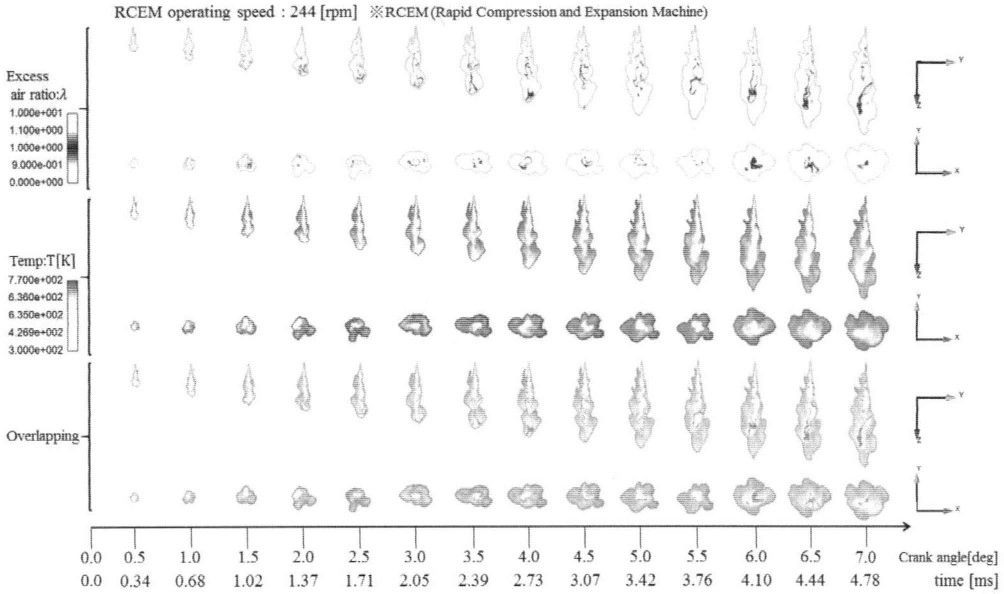

Figure 8. Temporal and spatial mismatch of auto-ignition suit concentration and temperature during the mixture formation process of an Ethanol spray.

5. Conclusions

1. Velocity distribution in combustion chamber at the end of compression stroke influences spray mixture formation process. Flow in the combustion chamber before fuel injection promote atomization and evaporation.
2. RCEM (Rapid Compression and Expansion Machine) has to be modified with negligible small pumping loss or supercharging device to realize target compression pressure and temperature (6.5MPa and 1000K).
3. As well as a spatial averaged discussion, in sort of local distributions of excess air ratio and temperature inside a spray, it is revealed that difficulty of simultaneous attainment of auto-ignition suit concentration and temperature during mixture formation process is the reason of poor auto-ignition quality of an Ethanol spray.
4. Decreasing of surrounding gas pressure and temperature with crank angle in expansion stroke prevents faster lean situation and faster temperature rising of an Ethanol spray, and based on this fact, we have to come up with new ideas to establish a new technology with stable auto-ignition in order to develop high performance alcohol engines.

Acknowledgments

The authors would like to express a special appreciation to the Japanese ministry of education, culture, sports, science and technology and Japan society for the promotion of science for their money support as grant-in-aid for scientific research.

References

[1] Narita T 2016 Resource Energy Yearbook Editorial Committee 2016 *Resource energy yearbook*, ed Resource energy yearbook editorial committee (Japan: Tsusan data publication association Ltd.) pp 215-218

[2] Wakuri Y, Fujii M, Amitani T and Tsuneya R 1960 Studies on the Phenomena of Fuel Spray of Diesel Engine. *Transactions of the JSME.* **25** 820-826

[3] Saitoh H and Uchida K 2015 Surrounding Gas Pressure and Temperature Dependence of the Spray Auto-ignition Phenomenon for Ethanol-Diethyl Ether Blended Fuels. *Transaction of the TSME 2016.* **4** 79-86

[4] Saitoh H, Tohjo Y and Uchida K 2016 Numerical Analysis on the Mixture Formation Process up to Auto-Ignition of an Ethanol Spray (AEC0025). *The 7th TSME Int. Conf. on Mechanical Engineering 13-16 Dec. 2016*

Investigation of Wood Sawdust Effect on Production and Final Quality of Composite Pellets Based on Sunflower Husks

M Matúš[1,*], P Križan[1], J Beniak[1], and Ľ Šooš[1]

[1]Slovak University of Technology in Bratislava, Faculty of Mechanical Engineering, Namestie slobody 17, 81231 Bratislava, Slovakia

* Corresponding Author: milos.matus@stuba.sk

Abstract. The pressure to diversify the raw material base of solid biofuels is steadily rising worldwide. The waste from agricultural production and its post-processing represents a huge but still little used energy potential. The use of such plant waste brings many technical problems. The paper is focused on research of increasing physical, mechanical and thermal quality indicators of composite pellets based on a mixture of sunflower husks and spruce sawdust. Tested mixtures varied by increasing the weight proportion of wood sawdust from 0% up to 40% in sunflower husks. The quality of this fuel pellets was evaluated using standardized parameters - particular density, bulk density, pellet moisture, mechanical durability, pellet hardness, calorific value, ash content etc.. Realized experiments have shown that it is not only possible to successfully transform this homogenized mixture into fuel pellets (pure sunflower husks represent a considerable problem with densification regard to their composition), but the qualitative parameters of the fuel also increase by increasing the weight proportion of wood in sunflower husks. The research results showed that the optimum ratio of wood component in the composite fuel is 30% based on the resulting mechanical quality parameters and economy of its production.

1. Introduction

Each type of densified material has its own specific properties. Processed materials differ in their structure and composition, what influences all the thermal and mechanical parameters of their densification process. The source of raw materials for the production of solid biofuels is mostly used soft and hard wood, bark, energy crops (e.g. fast-growing tree species) and phytomass. [1]

The field of phytomass densification for energy purposes is still not very well developed. Phytomass is used more often as raw material for gasification. The reason is the different chemical composition of the phytomass, which creates certain restrictions in combustion compared to wood. At present, the production of solid biofuels from phytomass used agricultural waste as a raw material, but also whole crops of cereals (straw together with grain). Phytomass has a much higher ash content compared to wood [2, 3]. Also, the concentration of nitrogen, sulphur and chlorine must be carefully monitored because they cause serious problems for combustion plants [1, 4]. Another critical aspect of the use of phytomass as a raw material for the production of solid biofuels is its low ash melting point [2, 5, 6] and increased fly ash and aerosol emissions [5, 7]. For these reasons, it is not appropriate to use such biofuels for small combustion systems. Mixing phytomass with woody raw material to produce solid biofuels with appropriate qualitative properties appears to be a solution. Suitable mixtures of these materials may reduce or even eliminate the above mentioned problems [8]. Another solution may be the addition of additives such as aluminium hydroxide, kaolin, calcium oxide or limestone. The result is an increase in the ash melting point and reduction of slag formation. At the same time, these additives also lead to an increase in mechanical quality indicators. Nevertheless, mixed biofuel from phytomass and dendromass are more widely used for large combustion plants. [2, 9, 10, 11, 12, 13, 14]

However, the phytomass has its advantages from the point of view of producing solid biofuels. High availability and low cost is its economic advantage. The results of scientific studies show that the densification process of phytomass needed significantly lower compacting pressures than dendromass to achieve the same pressings density. At the same time, it is possible to reduce the energy costs of drying the raw material, provided that the conditions of its collection and storage are kept in a dry state.

2. Materials and methods

2.1 Materials

The experiment is focused on the research of properties of composite pellets made from different ratios of the mixture of phytomass - sunflower husks (just a straw wrapper, not seed expeller) and dendromass – spruce sawdust (Figure 1). The problem of using pure sunflower husks as a biofuel in the form of pellets intended for automated combustion systems is a low content of binders (lignin etc.) what causes very low mechanical properties of the pellets with a high proportion of dust particles. The aim of the research is to define the effect of added spruce sawdust on the physical and mechanical quality parameters of the resulting composite pellets. The quality parameters of solid high-grade biofuels, which dependent on the ratio of raw materials has been investigated, are the particle density, bulk density, moisture content, mechanical durability and hardness.

Figure 1. Raw materials – sunflower husks and spruce sawdust.

The experiment comprises five proportions by weight of the mixtures listed in Table 1. It was produced 50 kg of pellets from each mixture for experimental purposes.

Table 1 Sample composition.

Mixture	Proportion by weight (%)		Moisture content of pelleting mixture (%)
	Sunflower husks	Spruce sawdust	
1	100	0	8.96
2	90	10	8.90
3	80	20	8.80
4	70	30	8.78
5	60	40	8.73

Table 2 Particle size distribution of sunflower husks and spruce sawdust

Particle size (mm)	Proportion by weight (%)	
	Sunflower husks	Spruce sawdust
≥ 4.00	21.04	2.56
2.00 - < 4.00	14.92	12.69
1.00 - < 2.00	37.53	35.92
0.50 - < 1.00	18.89	26.06
< 0.50	7.62	22.77

The preparation of the material consists in accurately weighing the ratios of the individual components of the mixtures and their significant homogenization. Since the fractional composition of both raw materials (Table 2) as well as their relative humidity (Table 1) are suitable for the used pelleting technology, the material was not required to be modified.

2.2 Densification process

Pellets were produced by an industrial scale process, by the Amandus KHAL 33-390 pelleting press. Pellets (diameter 6 mm) of each mixture sample were produced under the same conditions (pressing temperature, pressure, moisture, etc.). It must be mentioned that value of pressing temperature (measured on pelleting die) for each sample did not differ significantly, the value fluctuated 95±2 °C. It was caused by the same friction in pressing chambers which was controlled by pressure and moisture for each sample. Once the pelleting process had reached a stable state, a sample of 50 kg of each pellet mixture was used for testing (Figure 2).

Figure 2. Samples of pellets from 5 mixtures of sunflower husks and spruce sawdust.

2.3 Particle density

Generally, the particle density of pellets influences their combustion behaviour. Fuels with a higher particle density have a longer burnout time. [2]

The length, diameter and weight of pellets were measured using a digital calliper and an electronic balance. The particle density of the pellets was calculated by means of the ratio between pellet weight and its volume including pore volume. The volume of pellets was calculated as the volume of a cylinder with dimensions (length and diameter) measured as previously described, according to ISO 17829. The average particle density was calculated for each sample of 50 pellets at two stages – immediately after pelleting and 2 days after pelleting.

2.4 Bulk density
The bulk density is dependent on the particle density and the pore volume [2]. The higher the bulk density the higher is their energy density, and the lesser are the transport and storage costs. The bulk density of each sample of pellets was determined according to ISO 17828. The average value of bulk density for each sample was calculated from 10 measurements.

2.5 The pellet hardness
The hardness of pellets is not a biofuel qualitative parameter defined by ISO 17 225. Its importance is very similar to mechanical durability, but this test was carried out to compare the effect of papermaking sludge on mechanical properties of pellets and its positive ability to bind. [2]

The determination of pellet hardness is based on the comparative method, in which the tested pellet sample is placed between two spikes of the Hardness Tester Amandus KAHL and the maximum value of the resistance (hardness) of the pellet to the pressure is measured. The values of the pellet hardness are expressed in Newtons. The average value for each sample was calculated from 20 measurements.

2.6 Mechanical durability
When the mechanical durability of fuel pellets is low, a large amount of fines can cause bridging in the storage facility, blocking of the feeding screw, it can stop the fuel supply, increase particular matter emissions from combustion, as well as dust emissions during manipulation. Consequently, it can cause an explosion due to the dust. Therefore, the mechanical durability of pellets is one of the most important mechanical parameters of the fuel. [2]

Mechanical durability of each sample of pellets was determined according to ISO 17831. Measurements were performed by the Tumbler tester 1000 device, where the samples of tested pellets were subjected to controlled impacts of the mutual bombarding of the pellets and the pellets hitting the walls in the defined test chamber during a defined time period of 10 minutes. The mechanical durability of the pellets was calculated from the weight of the sample before insertion into the test chamber and the weight of the same sample after the impacts (the fines were separated).

2.7 Moisture content
The final moisture content of pellets is relevant to the net calorific value, the efficiency of the furnace and the combustion temperature [2]. The moisture content of the pellets was measured in 2 days after pelletizing. The determination of the moisture content was based on heating the pellets to 105±2°C until a constant weight was achieved according to the standard procedure described in ISO 18134-3.

3. Results and discussion
The physical and mechanical properties of the pellets significantly affect the quality of the fuel itself. These properties impact on the stability, dustiness and the space requirement during transport, handling and storage, and on the energy density of the fuel. The procedures for quantification of the individual parameters defined by standards were used to evaluate the quality of composite fuel.

3.1. Effect of wood content on particle density of composite pellets
Based on the research results, the particle density of pellets has an increasing tendency depending on the weight proportion of the spruce sawdust in pellets produced under the same technical and technological conditions. This fact is based on the higher density of the wood sawdust in comparison with sunflower husks. The results are shown in the Figure 3. The most significant increase can be found at mixture 2 in comparison with mixture 1. It is caused by absent of binder in pure sunflower husks, and after adding the wood containing lignin, the pellets become more compact.

Assubsequent experiments proved, the increase in pellet density has a direct effect on their mechanical durability and hardness. The higher density also favorably influences the shape and volume stability of the pellets.

IOP Conf. Series: Materials Science and Engineering **501** (2019) 012002 doi:10.1088/1757-899X/501/1/012002

Densification of the biomass causes the expansion of the pressings, therefore a certain period of stabilization after compacting has to be considered. It is possible to encounter two types of dilation for various materials and pressing conditions, namely dilation with decreasing or increasing in the pressing density. In the experiments, the dilatation was observed as a decrease in the pellets density at each mixtures. The results can be seen in the Figure 3. The maximum percentage change in pellet density after stabilization is only 0.5%, which means that the pellets are stable and the dilatation has only a slight effect on the change in density. Densification of the biomass causes the expansion of the pressings, therefore a certain period of stabilization after compacting has to be considered. It is possible to encounter two types of dilation for various materials and pressing conditions, namely dilation with decreasing or increasing in the pressing density. In the experiments, the dilatation was observed as a decrease in the pellets density at each mixtures. The results can be seen in the Figure 3. The maximum percentage change in pellet density after stabilization is only 0.5%, which means that the pellets are stable and the dilatation has only a slight effect on the change in density.

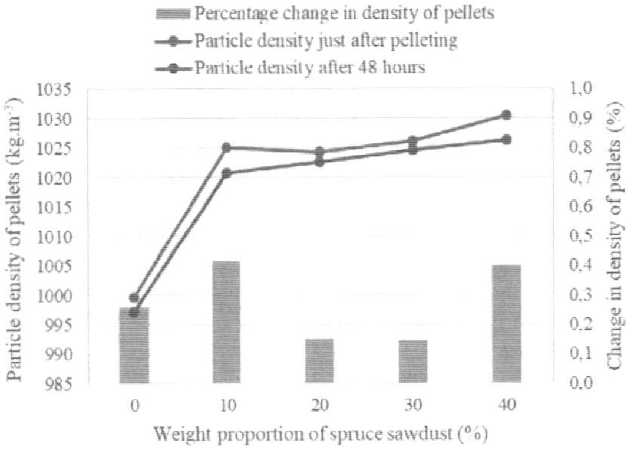

Figure 3. Variation of particle density and change in density of pellets with weight proportion of wood content in pellets.

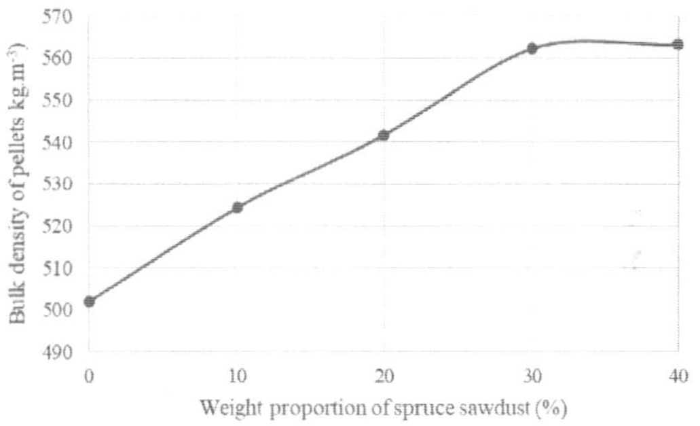

Figure 4. Variation of bulk density with weight proportion of wood content in pellets.

3.2. Effects of wood content on bulk density of pellets

Measurement results of bulk density are shown in the Figure 4. Experiments have shown that the bulk density of the pellets is increased by the addition of wood sawdust to sunflower husks. It can be seen from the results that the increase in the proportion of the wood component in the composite fuel over 30% by weight ratio is not significant. This is caused by increasing the length of the pellets in increasing the wood content. Therefore the pore volume in bulk fuel increases and the bulk density decreases, even despite the higher particle density. At this value of the wood content in the mixture 4, the bulk density of the pellets increased by 13% compared to the mixture 1. In view of the raw material costs it is economically advantageous to minimize the proportion of wood fuel component.

3.3. Effects of wood content on mechanical durability and hardness of pellets

Based on the experiment, it has been unequivocally demonstrated that increasing the proportion of wood sawdust in composite pellets significantly increases the value of mechanical durability. This is due to the effect of lignin as a bonding agent containing in sawdust. This significant increase can be observed till mixture 4 with wood content of 30% (Figure 5). The values of mechanical durability do not change significantly with a higher wood proportion (40%). The mechanical durability of pellets produced purely from sunflower husks has a value of 66.88%, so after the test there is a lot of dust. This leads to potential technical and security problems. However, the proportion of dust particles was significantly reduced by the addition of the wood component, thereby achieving the applicability and use of the energy potential of sunflower husks as raw material. The experiment results show the optimal proportion of wood 30% (mixture 4), when the mechanical durability of the pellets reaches 81.5%.

As Figure 5 shows, increasing the wood proportion in pellets causes a significant increase in their hardness. Information on the hardness of pellets is not subject to any standards, which is indicated by the size of the radial force that the pellet can resist. The force at which the pellet breaks down is the value of the highest hardness of the pellets. As expected, the highest hardness is achieved by the pellets of mixture 5 (wood content 40%) with an increase of 25% compared to the mixture 1.

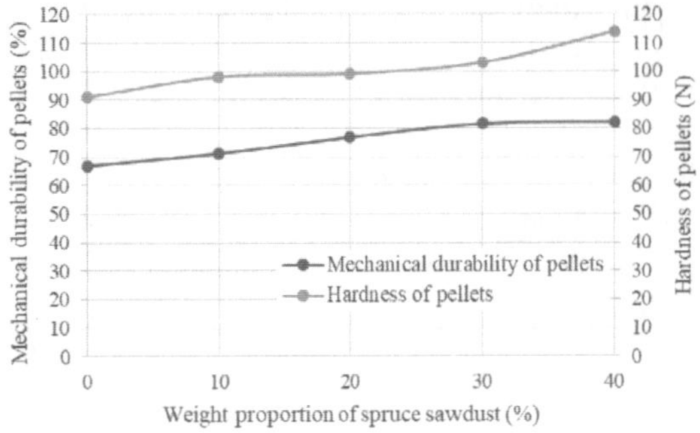

Figure 5. Variation of hardness and mechanical durability with weight proportion of wood in pellets.

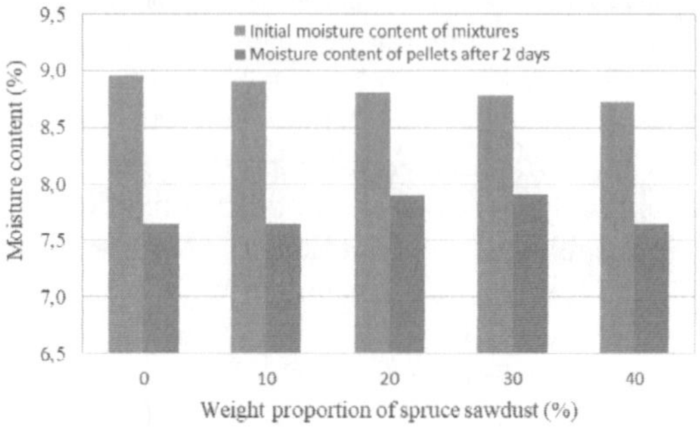

Figure 6. Comparison of moisture changes during the densification.

3.4. Effects of the change in moisture content

The initial moisture content of the prepared samples of the mixture was measured right before the entrance into the chamber of the pelleting press. The moisture content of pellets was measured 48 hours after the densification. The final pellets were stored at a room temperature 20°C, and with a stabile relative humidity of 55% in the laboratory.

Figure 6 shows the values of the optimum moisture content of each mixture of raw material entering the press chamber of the pelletizing press and the moisture content of the pellets after stabilization (48 hours). The graph shows that the increase in the proportion of wood causes a slight reduction of the optimum moisture content of the pellet feedstock and thus the finished pellets. The distribution of optimal moisture values of pelleted mixtures was very small, moisture values varied from 8.96% to 8.73%. The moisture values of the pellets of each mixture after stabilization are also very similar (from 7.64% to 7.91%). The average decrease in moisture content during the pelleting process was only about 1.1%.

Moisture content is one of the quality properties limited by biofuel standards. According to ISO 17225, biofuel pellets have to contain a moisture content lower than 10%. All samples of tested pellets had a moisture content lower than 10% what means that composite fuels meet the standard limits.

One more interesting conclusion can be drawn from the results of the minimum differences in moisture content of composite pellets. After performing the measurements it was found that the gross calorific value (GCV) of the sunflower husks and spruce sawdust has a very similar values (GCV of sunflower husks was 20.34 MJ.kg-1; GCV of spruce sawdust was 20.21 MJ.kg-1). On the basis of these findings, it was not worth exploring in more detail experimentally the effect of increasing the content of the wood component in pellets to the resulting net calorific value (NCV) of this composite fuel. Differences in the net calorific value are minimal, in the order of only 0.01 MJ.kg-1.

4. Conclusion

The results of the experiments clearly demonstrated that by increasing the weight of spruce sawdust added to the matrix of sunflower husks, the individual qualitative parameters of this biofuel also increase significantly. From the measured values, it can be stated that the optimal weight ratio of sunflower husks and wood sawdust is 70:30 (mixture 4). Further increase in the proportion of wood component does not significantly increase bulk density and mechanical durability. The values of moisture content and particle density are comparable for all composite mixtures. Only pure sunflower husks achieved different (lower) values of particular density. The hardness of the pellets was significantly increased by adding 30% by weight of the wood component. The results show that an increasing proportion of wood will continue to increase qualitative parameters of the biofuel, however, account must be taken to minimize the proportion of wood because the woody raw material is considerably more expensive than the raw material from the phytomass which otherwise would not be used its energy potential.

To improve the quality of this solid composite biofuel, it would also be desirable to change the compression conditions of the pressing channels in pelleting die to increase the pressing force. The application of these measures presupposes a positive increase of the individual indicators of the quality of solid composite fuel, which may be the subject of follow-up research.

Acknowledgments

The paper is a part of the research done within the project VEGA 1/0420/16 – "Determination and research of parameters influence on final briquettes quality at waste biomass densification". The authors would like to thank to the Ministry of Education of Slovak Republic and to the Slovak Academy of Sciences.

The paper is a part of the research done within the project KEGA 061STU-4/2017 – Increasing the level of the educational process in the field of production and environmental technologies by implementing of innovative tools. The authors would like to thank to the Ministry of Education of Slovak Republic.

References

[1] Obernberger I, Thek G. 2010. *The Pellet Handbook, The production and thermal utilisation of biomass pellets.* Earthscan Ltd., London.
[2] Matúš M, Križan P, Šooš Ľ, Beniak J 2018 *Fuel* **219:**196–204.
[3] Vassilev S, Baxter D, Vassileva C 2013 *Fuel* **112**:391–449.
[4] Vassilev S, Baxter D, Andersen L, Vassileva C 2013 *Fuel* **105**:19–39.
[5] Vassilev S, Baxter D, Vassileva C 2014 *Fuel* **117**:152–83.
[6] Vassilev S, Baxter D, Andersen L, Vassileva C 2013 *Fuel* **105**:40–76.
[7] Wang S, Jiang XM, Han XX, Wang H 2008 *Energy Fuel*, **22**:2229–35.
[8] Lu D, Tabil LG, Wang D, Wang G, Emami S 2014 *Biomass and Bioenergy* **69**: 287-296.
[9] Steenari BM, Lindqvist O 1998 *Biomass Bioenergy* **14**(1):67–76.
[10] Wilen C, Staahlberg P, Sipilä K, Ahokas J 1987 *Energy Biomass Wastes* **10**:469–84.
[11] Turn SO 1998 *J. Inst Energy* **71**:163–77.
[12] Llorente MJF, Arocas PD, Nebot LG, García JEC 2008 *Fuel* **87**:2651–8.
[13] Öhman M, Gilbe R, Lindstrom E, Bostrom D 2006 *Slagging characteristics during residential combustion of biomass pellets.* Proceedings of the 2nd World Conference on Pellets in Jönköping, Swedish Bioenergy Association, Stockholm, Sweden, pp. 93–100.
[14] Wang L, Skjevrak G, Hustad JE, Grønli M, Skreiberg Ø 2012 *Energy Proc,* **20**:30–9.

IOP Conf. Series: Materials Science and Engineering **501** (2019) 012003 doi:10.1088/1757-899X/501/1/012003

Optimization of Carbon Dioxide Reduction in Biohythane Using an Innovative Water Scrubber

S Chuayboon[1*], P Danugviluilux[1], S Prasertsan[2], and P Prasertsan[3]

[1]Department of Mechanical Engineering, Faculty of Engineering, King Mongkut's Institute of Technology Ladkrabang Prince of Chumphon Campus, Pathiu, Chumphon 86160, Thailand
[2]Department of Mechanical Engineering, Faculty of Engineering, Prince of Songkla University, Hatyai, Songkhla 90112, Thailand
[3]Department of Industrial Biotechnology, Faculty of Agro-Industry, Prince of Songkla University, Hatyai, Songkhla 90112, Thailand

* Corresponding Author: srirat.ch@kmitl.ac.th

Abstract. The major problem of biohythane production from palm oil mill effluent is the high carbon dioxide (CO_2) content. In this study, an innovative water scrubber system for upgrading biohythane has been experimentally investigated. Biohythane composing of ~53.34% of CH_4, ~39.12% of CO_2, and ~7.54% of H_2 was simulated regarding the composition of biohythane in the lab scale. Response surface methodology (RSM); a 5-level, 3- factor, central composite design (CCD), was employed to optimize three important parameters (biohythane flow-rate, water flow-rate, and operating time) in order to minimize the CO_2 content in the biohythane production. As a result, CO_2 concentration decreased with the increase of both the operating time and water flow-rate but inversely proportion to the biohythane flow-rate, which led to higher CO_2 absorption by water. The optimal condition regarding the maximum value of CO_2 reduction was found at: 3 Nl/min of biohythane flow-rate, 16 Nl/min of water flow-rate, and 9 min of operating time, thereby yielding 77.6% of CO_2 reduction.

1. Introduction

Currently, an interest in the eco-friendly energy from renewable sources seems to be one of the most important issues for any country because fossil fuels are being depleted. Biogas is one of the renewable energy sources used as a fuel to generate heat such as for cooking, for space heating, and for generating electricity. The biogas can be produced from renewable sources, e.g. cow dung, other animal waste, and palm oil mill effluent.

In palm oil milling industries, solid biomasses are used as materials to supply heat (steam) and electricity for milling processes. A study in 2017 revealed that Thailand had a total of 66 palm oil mills, of which 30 mills installed biogas plants. The total gas production was estimated at 315,000 m^3/day. Recent development in palm oil mill effluent (POME) biogas production was produced via a 2-stage digestion process in which hydrogen and methane gases are separately generated in the first and second stages, respectively. Thus, this is a good opportunity for using biohythane as power energy production. However, the raw composition of biohythane consists of approximately 35-38% of carbon , and the rests are 53-55% of methane (CH_4), and 5-10% of hydrogen (H_2) [1]. Thus, it is very necessary to remove CO_2 contained in the biohythane in order to upgrade its quality. Jeong et al [3] reported that the biogas power plant is characterized as relatively low efficiency, especially for internal combustion engine because the low flame speed of CH_4 was blamed for adverse effects [4]. Fortunately, hydrogen was used as compliment additive to improve the combustion of biogas and the mixture was named hythane [5]. For these reasons, the upgrading of biohythane is ultimately realized.

In order to enhance both CH_4 and H_2 of the biohythane, a water scrubbing absorption method was employed for CO_2 removal. According to this conventional concept, raw gases were fed into a packed-bed column from the bottom into the water scrubber tank. Then, pressurized water was sprayed from the top. The absorption is thus counter-current flow. The advantage of water scrubbing is that it is a simple technology [6,7]. However, the disadvantage is that the water contaminated by CO_2 is needed to be regenerated. In this study, an innovative water scrubbing concept was proposed. Both the pressurized raw gases and water were mixed before entering through the venturi device, which was installed at the top of the water scrubber tank, and they subsequently flowed into the tumbler area in side the tank in the downward direction, thereby absorbing CO_2 contained in the biohythane.

The objective of this present work is to study the optimization of three parameters; biohythane flow-rate, water flow-rate and operating time in order to reduce the CO_2 concentration that contained in the biohythane. Response surface methodology (RSM) with a 5-level and 3 factor central composite design (CCD) was employed to optimize these three parameters [8]. The goal is to prognosticate the proper response surface models of the relationship according to the reduction of CO_2 and these parameters.

2. Material and Methods

2.1. *Materials*

Biohythane was formed by the mixing of CH_4, H_2, and CO_2 via gas mixture. Impurities contained in the real biohythane, such as H_2S and ammonia (NH_3), were ignored due to its negligible amount. CH_4 was obtained from natural gas (CNG) for vehicles from the gas stations. Therefore, it was anticipated that the presence of some constituents such as propane and butane in the mixture would have slightly increased the heating value than the intended biohythane. Both the commercial H_2 (95% purity) and CO_2 (99.99% purity) were purchased form Linde, while the CNG used in this experiment was obtained from the gas station in Tasae District, Chumphon province, Thailand. Certificate of CNG compositions, listed in Table 1, was provided by the gas separation plant, and its composition mainly composed of 73.8 % of CH_4 and 17.1% CO_2. This data was also utilized in the calculations of simulated biohythane compositions.

Table 1 Compositions of natural gas (Tasae District, Chumphon province, Thailand) [9]

Gas composition (%) by volume	
Methane CH_4	73.836
Ethane C_2H_6	4.146
Propane C_3H_8	1.938
Isobutane i-C_4H_{10}	0.548
n-Butane n-C_4H_{10}	0.384
Isopentane i-C_5H_{12}	0.174
n-Pentane n-C_5H_{12}	0.104
Hexanes C_6H_{14}	0.229
Carbon dioxide CO_2	17.149
Nitrogen N_2	1.492
Specific gravity, (SG.)	0.791
Water H_2O (L_b/MMscf)	0.17

2.2. *Methods*

2.2.1. Procedure. Schematic diagram of experimental set up is illustrated in Figure 1. It mainly consists of the fuel supply system (H_2, CO_2, and CH_4(CNG)), water scrubber tank and water system.

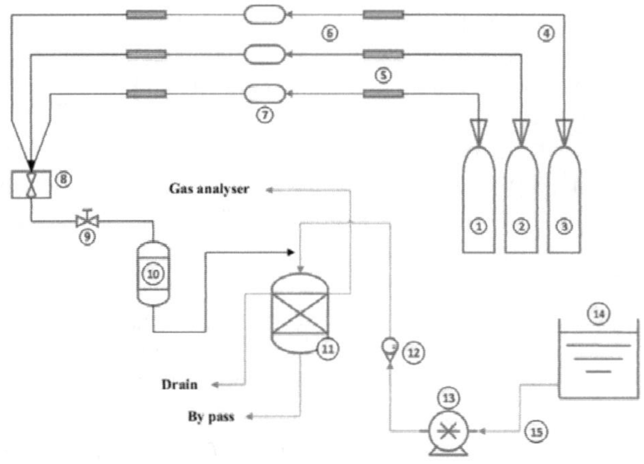

Figure 1. (a) Experimental set-up: 1 CNG tank; 2 H_2 tank; 3 CO_2 tank; 4 soft outlet tubes; 5 connectors; 6 copper inlet tubes; 7 mass flow controllers; 8 gas mixing unit; 9 ball valve; 10 receiver tank; 11 water scrubber tank; 12 rotameter and ultrasonic flow meter; 13 water pumping; 14 water tank, and 15 water inlet tube.

The key part of biohythane upgrading is the water scrubber tank. It composes of a vertical cylindrical tank (22 cm diameter) with two layers inside (total height: 29 cm and total volume:11 L) according to Figure 2.

The water scrubber tank is made of a metallic alloy (~2 mm wall thickness). It is pierced at the bottom to allow the passage of an outlet tube from which the water with CO_2 is drained after experiments. One pressure gage is positioned at the top of the tank for measuring the pressure inside the water scrubber tank, while the gas outlet port is placed at the top of the cylindrical part.

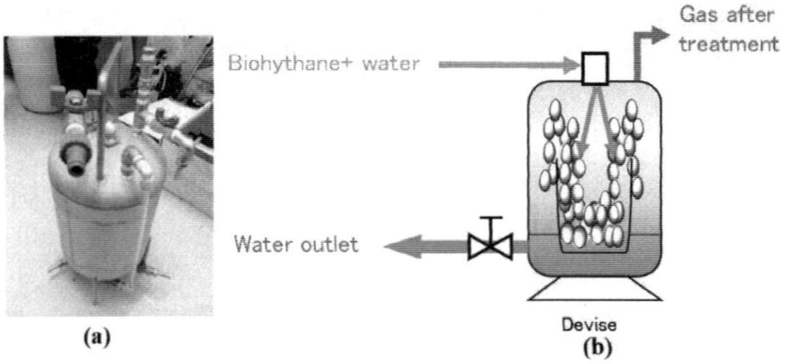

Figure 2. (a) water scrubber tank and (b) schematic diagram of the water scrubber tank.

H_2, CO_2, and CH_4 (CNG) were mixed via gas mixture under pressure (1.8 bar) to be simulated as biohythane. The simulated biohythane was then fed into the scrubber tank along with water. To control gas flow-rate, H_2, CO_2, and CH_4 mass flow controllers (Kofloc model 3660 ±1% full scale) were installed between the gases supply and water scrubber tank to control the composition of simulated biohythane. To complete mixing, a receiver tank was installed after a gas mixing unit. For safety concern, an anti-backfire valve was mounted at the H_2 fuel pipe line. An ultrasonic flow meter was used to measure the water flow-rate, which the reading was confirmed by the Rotameter. The operating temperature was controlled at ~20-25 °C (ambient temperature) throughout experiment. The mixing of biohythane and water was sprayed to the tumbler inside the scrubber tank (Figure 2b), thus producing bubbles and thereby absorbing CO_2 by water. The CO_2-rich water was drained through the outlet port after experiments. The experiments were carried out according to the experimental design matrix (Table

3). After experiments, the sampling biohythane was collected by gas sampling bags and then analysed by gas chromatography (model: GC-2014, Shimadzu).

2.2.2. Experimental design. Response surface methodology (RSM), with a 5-level and 3-factor central composite design (CCD), was employed in order to optimize the reduction of CO_2 in biothythane. Biohythane flow-rate ($\dot{q}_{gas}$), water flow-rate ($\dot{q}_{water}$), and time (t) are three independent variables that were used to find the optimization of the reduction of CO_2 in the biohythane after scrubbing process. Regarding the three independent variables, the axial parameter (α_x) is 1.68 (for rotatable CCD), which was calculated by equation (1) [10].

$$\alpha_x = \sqrt[4]{2^k} \tag{1}$$

where α_x is the axial parameter for rotatability, and k is the number of variables.

The experiments were considered for 5 levels of the independent variable identified as -1.68, -1, 0, 1, and 1.68 according to Table 2.

Table 2 Coding of independent variables

Independent variable	Coded level				
	-1.68	-1	0	1	1.68
$\dot{q}_{gas}$: biohythane flow-rate (Nl/min)	3	4.82	7.50	10.18	12
$\dot{q}_{water}$: water flow-rate (Nl/min)	8	9.62	12	14.38	16
t: time (min)	5	6.13	8	9.46	11

3. Results and Discussion

3.1. Experimental results

Table 3 summarizes the experimental design matrix generated by Multiple regression software according to the range of independent variables in Table 2 and experimental results for 17 experimental runs in a water scrubber.

Table 3 Operating conditions, simulated biohythane, and product gases concentrations after water scrubbing

No.	Independent variable			Simulated biohythane (%v)			product gas (%v)			CO_2 reduction (%v)
	$\dot{q}_{gas}$ (Nl/min)	$\dot{q}_{water}$ (Nl/min)	T (min)	H_2	CH_4	CO_2	H_2	CH_4	CO_2	
1	10.18	14.4	9.46	7.3	53.7	39.0	9.8±0.28	65.8±0.07	24.5±0.35	37.2±0.90
2	7.5	16	8	7.3	53.7	39.0	11.1±0.41	69.9±0.09	19.0±0.32	51.3±0.83
3	4.82	9.6	9.46	7.3	53.7	39.0	11.8±0.12	62.0±0.05	26.2±0.16	32.9±0.42
4	10.18	14.4	6.13	7.3	53.7	39.0	7.5±0.05	66.3±0.01	26.2±0.05	32.7±0.14
5	10.18	9.6	9.46	7.3	53.7	39.0	7.1±0.06	60.0±0.07	32.9±0.13	15.6±0.34
6	7.5	12	8	7.3	53.7	39.0	9.6±0.20	64.0±0.02	26.3±0.22	32.4±0.56
7	7.5	12	5	7.3	53.7	39.0	6.7±0.01	65.4±0.01	27.9±0.01	28.4±0.02
8	3	12	8	7.3	53.7	39.0	14.1±0.47	70.6±0.35	15.3±0.12	60.7±0.31
9	4.82	9.6	6.13	7.3	53.7	39.0	11.5±0.10	62.3±0.16	26.1±0.06	33.0±0.15
10	10.18	9.6	6.13	7.3	53.7	39.0	7.4±0.13	59.9±0.05	32.8±0.09	16.0±0.23
11	7.5	12	8	7.3	53.7	39.0	9.6±0.20	64.0±0.02	26.3±0.22	32.4±0.56
12	7.5	8	8	7.3	53.7	39.0	9.9±0.01	57.4±0.02	32.7±0.03	16.2±0.07
13	4.82	14.4	6.13	7.3	53.7	39.0	13.5±0.04	69.5±0.29	17.0±0.33	56.4±0.84
14	7.5	12	11	7.3	53.7	39.0	10.9±0.03	63.4±0.05	25.7±0.08	34.0±0.21
15	12	12	8	7.3	53.7	39.0	9.0±0.02	63.2±0.04	27.7±0.02	28.9±0.04
16	4.82	14.4	9.46	7.3	53.7	39.0	10.9±0.05	72.0±0.05	17.1±0.01	56.1±0.01
17	7.5	12	8	7.3	53.7	39.0	9.6±0.20	64.0±0.02	26.3±0.22	32.4±0.56

Experiments were carried to determine the reduction of CO_2 in the biohythane. The composition of simulated biohythane, by volume, was kept constant at 7.3% of H_2, 53.7% of CH_4, and 39.0% of CO_2. After completing experiments, it was reported that the reduction of CO_2 was in the range of 15.6-60.7 % by volume. Noted that no amount of the water contained in the product gases after scrubbing was detected.

3.2. Response surface models of results and statistical analyses

After considering the results of the effect of three independent variables regarding biohythane flow-rate, water flow-rate, and operating time on the reduction of CO_2, response surface models were applied to analyse the data in Table 3 by utilizing a multiple regression model to fit a second-order polynomial equation according to equation (2).

The equation (2) shows the relationship between the CO_2 reduction values and the other independent variables in the form of squared model.

$$CR = \beta_0 + \beta_1(\dot{q}_{gas}) + \beta_2(\dot{q}_{water}) + \beta_3(T) + \beta_4(\dot{q}_{gas}{}^2) + \beta_5(\dot{q}_{water}{}^2) + \beta_6(t^2) + \beta_7(\dot{q}_{gas}{}^3) + \beta_8(\dot{q}_{water}{}^3) + \beta_9(t^3) \quad (2)$$

Where CR is the reduction of CO_2 (%), $\dot{q}_{gas}$ is biohythane flow-rate, $\dot{q}_{water}$ is water flow-rate, t is time, and β is the coefficient value.

The model was considered by the probability of error value (p-value) in the detailed Table 4. The p-value was employed to exam the statistical significance of each regression coefficient. When their p-values are higher than 0.1, at 90% confidence level, the independent variables in a model were considered as negligible effects. In contrast, when their p-values are lower than 0.1, the independent variables were considered as significant parameters.

After considering the p-values, the negligible parameters in equation (2) were suppressed, thereby resulting in a new second-order polynomial equation according to equation (3).

$$CR = \beta_0 + \beta_1(\dot{q}_{gas}) + \beta_2(\dot{q}_{water}) + \beta_3(\dot{q}_{water}{}^2) + \beta_4(t^2) + \beta_5(\dot{q}_{water}{}^3) + \beta_6(t^3) \quad (3)$$

Table 4 shows coefficient values, p-values, R^2, and R^2 adjusted of equation (3) after suppressing the negligible parameters. According to this table, the operating time became to be the most significant term in the response models because of its p-value being lowest (0.04682). Therefore, it is very necessary to keep this parameter for calculating the reduction of CO_2 in biohythane by water scrubbing. Note that β_1 and β_5 were kept in the equation (3), although their P-values were higher than 0.1 because their coefficient values are important to be considered for the calculation in the reduction of CO_2.

Table 4 Coefficient and P values of response surface models.

Term	Coefficient value	P-value
β_0	-747.02	0.06725
β_1	-1.613	0.138
β_2	180.13	0.08451
β_3	-14.39	0.101
β_4	2.082	0.04682
β_5	0.387	0.110
β_6	-0.154	0.06508
R^2	0.809	-
R^2 adjusted	0.694	-

3.3. Response surface plots

Figure 3 shows contour plots of the influence of biohythane flow-rate, water flow-rate, and operating time on the reduction of CO_2. It was found that an increase in the operating time with the reduction of biohythane flow-rate has a positive effect on CO_2 reduction (Figure 3a). The percentage of CO_2 reduction reached ~60% at the operating time range 7-9 min while injecting biohythane in the range of 3-4 Nl/min. This implies that increasing operating time with decreasing biohythane flow-rate showed a positive impact on CO_2 reduction. As Figure 3b illustrated, the CO_2 reduction increased significantly when increasing water flow-rate, e.g. from 20% at ~9 Nl/min to 50% at 15 Nl/min, while injecting biohythane in the range of 5-11 Nl/min. This can be explained by the fact that higher water flow-rate leads to higher CO_2 absorption by water. Moreover, increasing water flow-rate while increasing operating time in the range of 4-8 min favours the reduction of CO_2 (40-70%), according to Figure 3c. For these variations, it can be summarised that increasing both water flow-rate and operating time while reducing the biohythane flow-rate inherently promoted the reduction of CO_2.

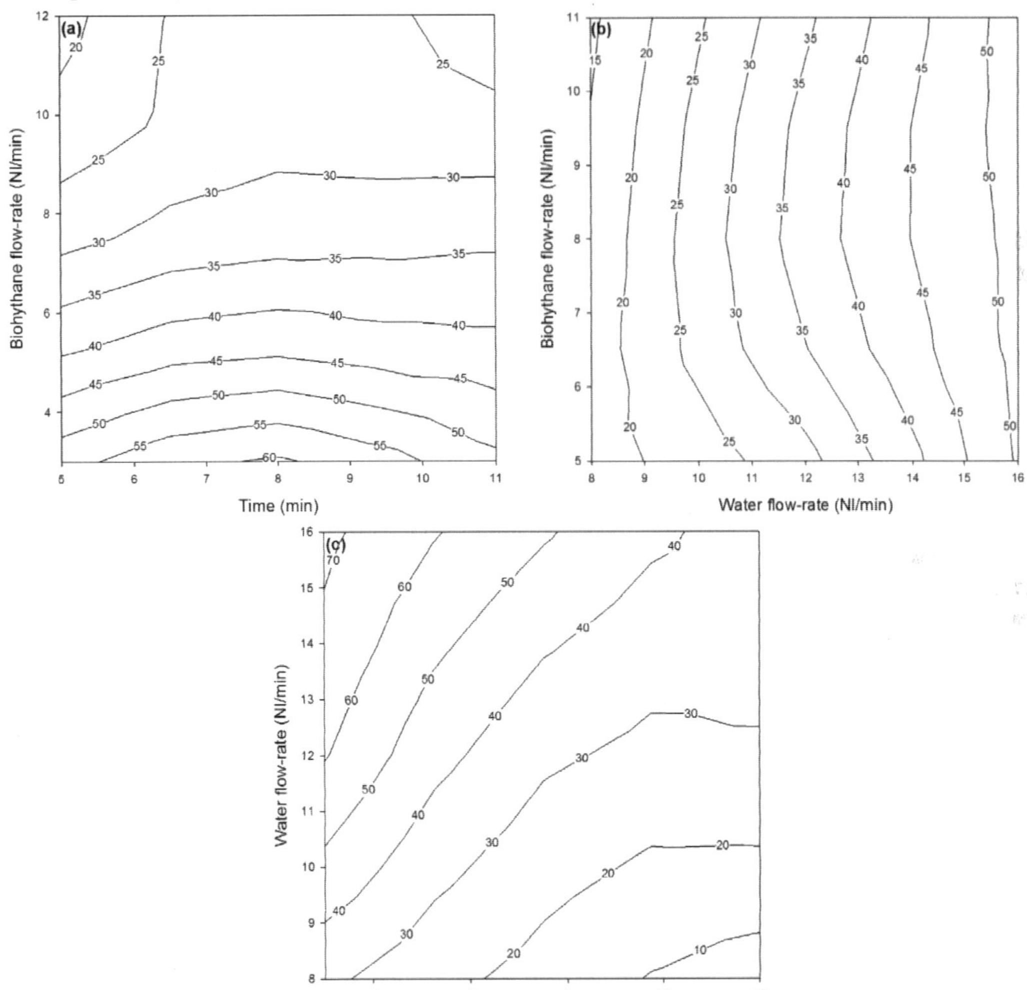

Figure 3. Percentage of the reduction of CO_2 (%v) as (a) a function of operating time and biohythane flow-rate, (b) a function of water flow-rate and biohythane flow-rate, and (c) a function of time and waterflow-rate.

3.4. Optimum conditions for reducing the CO_2 content in the biohythane

An optimal condition of the CO_2 reduction in biohythane via water scrubbing was performed by using Excel solver from Microsoft Excel add-in tool to calculate the optimal point from equation (3) (by taking the ranges of independent parameters in Table 2 into the calculation). The optimal condition for CO_2 reduction in biohythane using a water scrubber system in equation (3) is: biohythane flow-rate of 3 Nl/min, water flow-rate of 16 Nl/min, and operating time of 9 min, thus confirming that a reduction of the biohythane flow-rate with the growth of both water flow-rate and operating time is beneficial for removing CO_2 content. The experiments were conducted twice at the optimal condition to confirm and check the results of CO_2 reduction in the treated biohythane. It was found that the composition of biohythane after scrubbing is: 17.4±1.59% of H_2, 73.8±2.38% of CH_4, and 8.8±0.8% of CO_2, thereby yielding 77.6±2% of CO_2 reduction as compared to the raw gas. Moreover, such the value of CO_2 reduction (~77.6%) is slightly lower than that of the result obtained from the prediction of the equation (3), (85% of CO_2 reduction). Thus, this equation is both reliable for the prediction of the CO_2 reduction in biohythane performed in this water scrubber and beneficial for scaling-up the water scrubber tank.

4. Conclusions

The reduction of CO_2 in biohythane was performed in an innovative water scrubbing system. The optimization of three key parameters (biohythane flow-rate, water flow-rate, and operating time) with respect to the maximum of the CO_2 reduction was successfully demonstrated. CO_2 concentration decreased significantly with the increase of both water flow-rate and operating time while the flow-rate of biohythane was decreased. The optimal condition for CO_2 reduction in the biohythane using water scrubber was found at: biohythane flow-rate of 3 Nl/min, water flow-rate of 16 Nl/min, and operating time of 9 min, thereby yielding 77.6% of CO_2 reduction.

Acknowledgments

This study was financially supported by the Higher Education Research Promotion, and National Research University Project of Thailand, and King Mongkut's Institute of Technology Ladkrabang Prince of Chumphon Campus, Thailand. The study was also supported by the Faculty of Engineering and Faculty of Agro-Industry, Prince of Songkla University, Thailand. The authors also thank Dr.Tsuyoshi Imai from Yamaguchi University for supporting the water scrubber tank.

References

[1] O-Thong S, Prasertsan P, Karakashev D, Angelidaki I 2008 *Int J Hydrogen Energy* **33** 1204–14.
[2] Kapdi SS, Vijay VK, Rajesh SK, Prasad R 2005 *Renew Energy* **30** 1195–202.
[3] Jeong C, Kim T, Lee K, Song S, Chun KM 2009 *Int J Hydrogen Energy* **34** 9620–7.
[4] Chuayboon S, Prasertsan S, Theppaya T, Maliwan K, Prasertsan P *2014 Energy Procedia* **52** 659–65.
[5] Ortenzi F, Chiesa M, Scarcelli R, Pede G 2008 *Int J Hydrogen Energy* **33** 3225–9.
[6] Läntelä J, Rasi S, Lehtinen J, Rintala J 2012 *Appl Energy* **92** 307–14.
[7] Vo HT, Imai T, Teeka J, Sekine M, Kanno A, Van Le T 2013 *Water Res* **47** 4286–93.
[8] Somnuk K, Niseng S, Prateepchaikul G 2014 *Energy Convers Manag* **80** 374–81.
[9] Dscng.pttplc.com. (2018). PTT Distribution Service Center. [online] Available at: https://dscng.pttplc.com/(S(3maa3oidkkvxsurllgg1qxyi))/onlinegas/OnlineGasQuality.aspx [Accessed 16 Oct. 2018].
[10] Somnuk K, Smithmaitrie P, Prateepchaikul G 2013 *Energy Convers Manag* **68** 193–9.

Relationship between Raw Material Composition and Pellets Physical Properties

P Križan[1,*], M Matúš[1], J Bábics[1], Ľ Šooš[1], and J Beniak[1]

[1]Institute of Manufacturing Systems, Environmental Technology and Quality Management, Faculty of Mechanical Engineering, STU in Bratislava, Nám. Slobody 17, 812 31 Bratislava, Slovakia

* Corresponding Author: peter.krizan@stuba.sk

Abstract. The mutual interaction between pellets quality, which is represented by the pellets physical properties, and the input composition of raw material is very interesting and complicated itself. Therefore, the main aim of this research paper is to determine the relationship between the raw material composition, which influences the pellets quality and the pellets quality indicators - physical properties of pellets. During the biomass densification, also the raw material properties significantly influence the final solid biofuels quality. Therefore the raw material composition represented by different mixtures from spruce sawdust, wheat straw and spruce bark, in various ratios (90/10, 80/20, 50/50, 80/10/10 and 50/30/20) and in various raw material composition (sawdust/straw, straw/bark, straw/sawdust/bark) were investigated. During this research study in total 14 different material composition was used. Effect of mentioned raw material mixtures on pellets bulk density, mechanical durability, abrasion, particle density and hardness were determined. Research findings presented in this paper are based on a realized experimental research that was done on Faculty of Mechanical Engineering STU in Bratislava. This experimental research was a logical result of and based on authors practical experience and wide theoretical analyses of selected variables influence. Described experimental research was realized by real pelleting machine that is a part of semi-production pelleting plant on Faculty of Mechanical Engineering STU in Bratislava. The experimental findings presented here are showing the importance of mentioned variables and parameters during the pelletizing process.

1. Introduction

Mechanical treatment of biomass is necessary conditions before its final recovery. Densification process as a solid biofuels (pellets) production is included to the mechanical treatment and is a suitable option of biomass treatment to gain effective energy balance at energy recovery of biomass.

During the biomass pelletization, various technological variables and raw material parameters significantly affect the final solid biofuels (pellets) quality [1]. The quality of pellets is determined by the end user's requirements on the heating system and the handling properties [2, 3]. In general, the pellets quality depends on the properties of the raw material (feedstock) – in terms of biomass type, moisture content and particle size; and quality management of the manufacturing process – in terms of operating conditions, technological variables, pelletizer type and binding agent [4, 5]. Many parameters effect the pelletizing process and thus the quality of the final solid biofuels – pellets [5, 6]. The properties of the raw material, raw material treatment and composition as well as technological demands are very important during the solid biofuels production process [7, 8].

Among renewable resources, residual biomass is envisaged to play an important role in the new energy conception since agricultural and forest residues are a local energy resource with an important production throughout the world. With the development of solid biofuels market and biomass recovery technologies and with the gradual decrease of the quantities of usable wood biomass, increases the requirements for the processing of other types of biomass, e.g. agricultural - herbaceous biomass [5, 7]. Each different type of raw material requires a separate approach [6, 8]. As each type of raw material has its own specific properties and chemical composition, any small change in the properties of the raw material can influence the final quality of the pellets [6-8]. Different raw material properties produce different conditions during the densification process, and this causes the final quality of the pellets produced to be very different [2, 7]. This is the reason why it is so important to undertake experimental research to determine the relationship between raw material composition and pellets physical properties.

In general, solid biofuels production is very complicated process, because during solid biofuels manufacturing the technological and raw material variables influence the densification process and thus also the final solid biofuels quality (pellets). Quality of pellets is defined and given by EU standards [9], and is evaluated by physical-mechanical and thermo-chemical indicators [10]. In this study the pellets quality, as a final output of the pelletizing process, by its particle density, bulk density, mechanical durability, abrasion resistance and durability were evaluated. These physical-mechanical indicators of quality are evaluated from the storage, transport and dust emissions reasons [9, 10]. Low-quality pellets can cause operational problems during storage, in transport systems and in combustion systems, including undesired effects in the equipment, and may originate substantial amounts of and particulate matter (PM) emissions [2, 4, 8].

The main goal of presented experimental research is to determine the effect of raw material composition on pellets final quality. During this study, the spruce sawdust, spruce bark and wheat straw, as typical Slovakian biomass residues were used. The raw material composition represented by different mixtures from spruce sawdust, wheat straw and spruce bark, in various ratios (90/10, 80/20, 50/50, 80/10/10 and 50/30/20) and in various raw material composition (sawdust/straw, straw/bark, straw/sawdust/bark) were investigated. The experimental findings presented here are significant from pellets production point of view to make a solutions with regarding the economic aspects [5, 11].

2. Materials and Methods

2.1. *Raw material properties*

Spruce sawdust, wheat straw and spruce bark originating from Southwestern Slovakia were chosen for this experiment. Suitable spruce sawdust and spruce bark were obtained from wood processing company. Wheat straw was obtained from local agricultural company. Initially, Retsch Vibrating Sieve Equipment AS 200, according to the EN ISO 17827-1, for analyzing of the particle size distribution was used [12]. The raw material particle size distribution was determined, proportions by weight see in the Table 1.

Table 1 Raw materials particle size distribution.

	Proportion by weight (%)		
Particle size (mm)	Spruce sawdust	Wheat straw	Spruce bark
0 – 0.5	26.72	15.53	25.71
0.5 – 1.0	27.45	6.81	14.05
1.0 – 2.0	36.88	77.08	52.34
2.0 – 4.0	6.71	0.48	7.50
4.0 <	1.91	0.11	0.09

Moisture content of spruce sawdust, spruce bark and wheat straw before pelletizing was measured with the aid of a Kern MRS 120-3 balance. This measurement consisted in heating the raw sawdust (gravimetric method of moisture content measuring) [13] at 105 ± 2°C until a constant weight was achieved. Bulk density of spruce sawdust, spruce bark and wheat straw before pelletizing was determined according to the EN ISO 17828 [14]. This European Standard describes the determination of the bulk density of solid biofuels using a standardized measuring vessel. This method is applicable to all solid biofuels with an upper nominal size of 100 mm. Properties of raw materials before mixing and pelleting can be seen in the following Table 2.

Table 2 Properties of raw materials before mixing and pelleting.

Properties of raw material	Spruce sawdust	Wheat straw	Spruce bark
Bulk density ($kg.dm^{-3}$)	118.7	74.76	261.7
Moisture content (%)	9.8	7.49	12.5

2.2. *Mixtures material composition*

Main goal of this research was to determine the relationship between the raw material composition, which influences the pellets quality and the physical properties of pellets. The raw materials composition represented by 12 basic raw material mixtures were investigated and with 100% raw material of spruce sawdust and wheat straw were compared. Selected mixtures of raw material composition with its description can be seen in the Table 3.

Table 3 Mixtures material composition.

Sample number	Sample label	Description – material composition
1	100W	100 % of spruce sawdust
2	95W/5S	95 % of spruce sawdust + 5 % of wheat straw
3	90W/10S	90 % of spruce sawdust + 10 % of wheat straw
4	80W/20S	80 % of spruce sawdust + 20 % of wheat straw
5	20W/80S	20 % of spruce sawdust + 80 % of wheat straw
6	50W/50S	50 % of spruce sawdust + 50 % of wheat straw
7	100S	100 % of wheat straw
8	95W/5B	95 % of spruce sawdust + 5 % of spruce bark
9	90W/10B	90 % of spruce sawdust + 10 % of spruce bark
10	80W/20B	80 % of spruce sawdust + 20 % of spruce bark
11	80S/20B	80 % of wheat straw + 20 % of spruce bark
12	50S/50B	50% of wheat straw + 50 % of spruce bark
13	80S/10B/10W	80 % of wheat straw + 10 % of spruce bark + 10 % of spruce sawdust
14	50S/30B/20W	50 % of wheat straw + 30 % of spruce bark + 20 % of spruce sawdust

2.3. *Pellets production*

For the pellets production a vertical pelletizing machine KAHL 33-390 with flat round matrix was used. During the pelletizing process, a matrix with ø 6 mm holes was used. The constant operating parameters for the pelletizing press were the pressing pressure (12 MPa), the pressing temperature (100 °C) and the circumferential speed of rolls (2.2 m.s^{-1}). Semi-operational tests for pelletizing of mixtures were carried out without technical and technological complications. The pelletizing process was continuous. On following figures, you can see the produced samples of pellets from selected raw material composition.

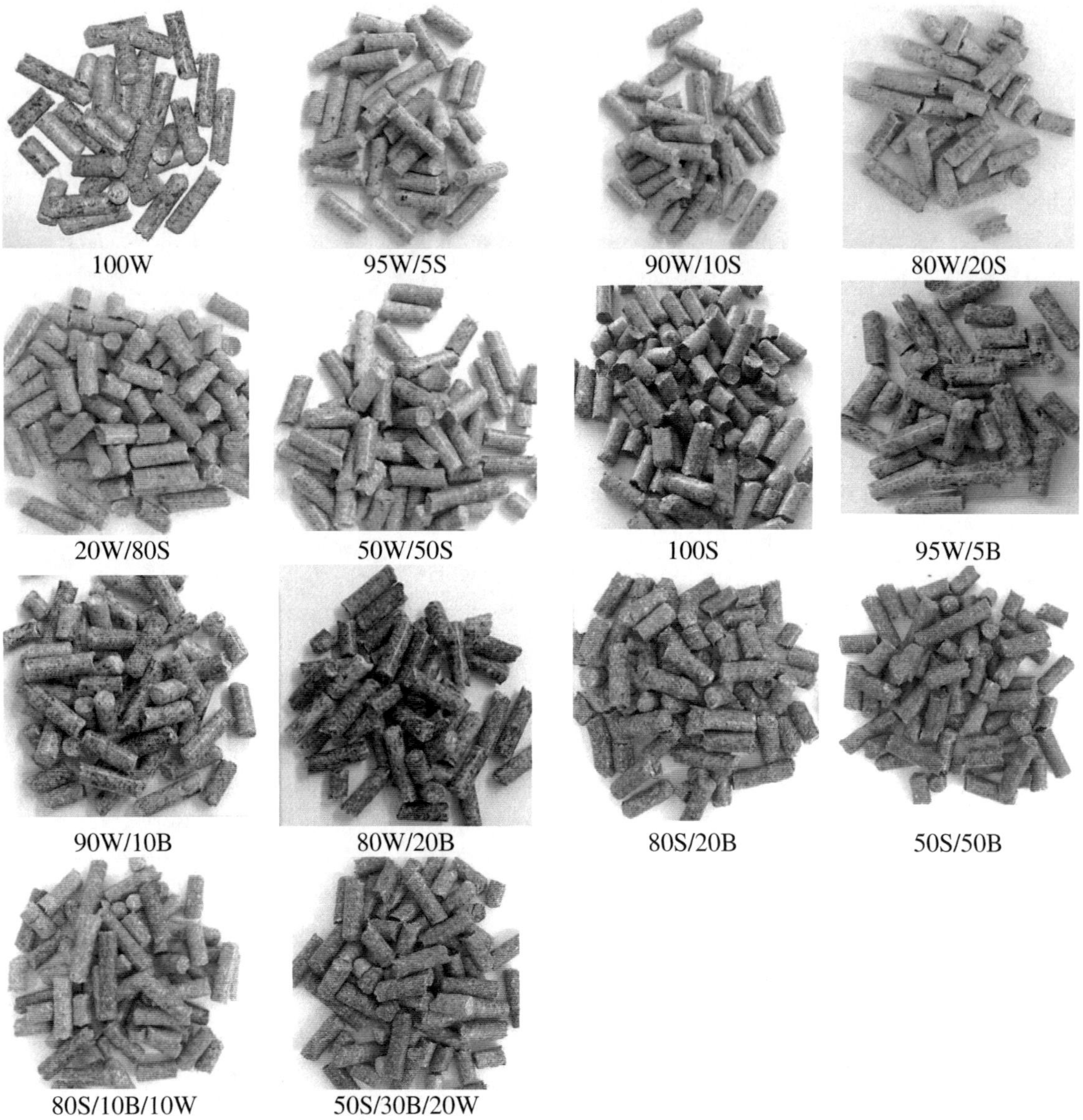

Figure 1. Samples of pellets from different material composition.

2.4. *Pellets evaluation*

Subsequently, pellets were subjected to laboratory testing and the determination of physical properties according to the EN technical standards [15]. Effect of mentioned raw material mixtures on pellets bulk density, mechanical durability, abrasion, particle density and hardness was determined. Pellets bulk density according to EN ISO 17828 [10], pellets mechanical durability according to EN ISO 17831-1 [16], particle density according to EN ISO 18847 [17], pellets abrasion according to DIN Standard [18] were determined. However, pellets hardness determination procedure is not supported by any Standard, the pellets hardness typically relates to the final pellets quality [19] and in this contribution was determined by specialized KAHL apparatus.

All results in this paper, showing the dependencies between lonely pellets physical quality indicators and input material mixtures, were depicted graphically, as seen in the figures below. Each of these displayed values is the average value from at least six measurements. It means at least six times each value was determined and measured. This value of repeatability was chosen according to requirements of standard mathematical and statistical methods for experiment evaluation.

3. Results and Discussion

The comparison of pellets particle densities can be seen in Figure 2. Wood/straw (samples 1-6) pellets particle density increases with increasing of straw amount in mixture. This was caused by straw better compressibility [20, 21] and material particles interconnection [22, 23]. Wood/bark pellets (samples 8-10) particle density decreases with increasing of bark amount in mixture, despite in straw/bark pellets (samples 11-12) the particle density increases with increasing of bark amount in mixture. Due to the properties and chemical composition of the bark [24, 25]. A similar trend was noted also at the comparison of pellets bulk densities, see Figure 3.

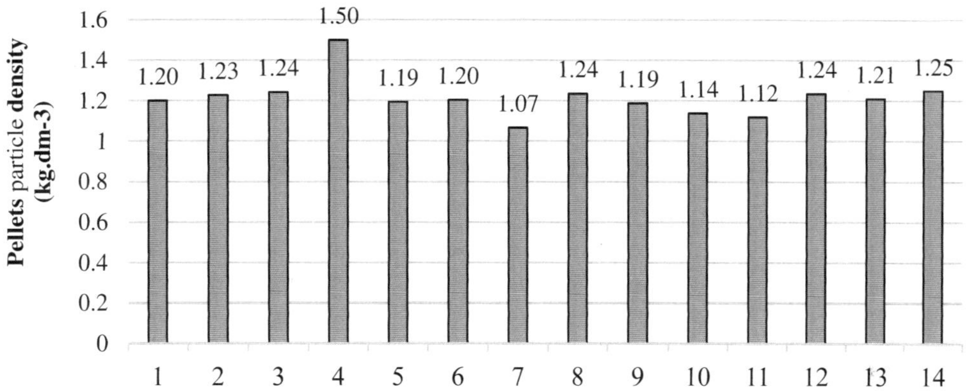

Figure 2. Comparison of particle density of pellets.

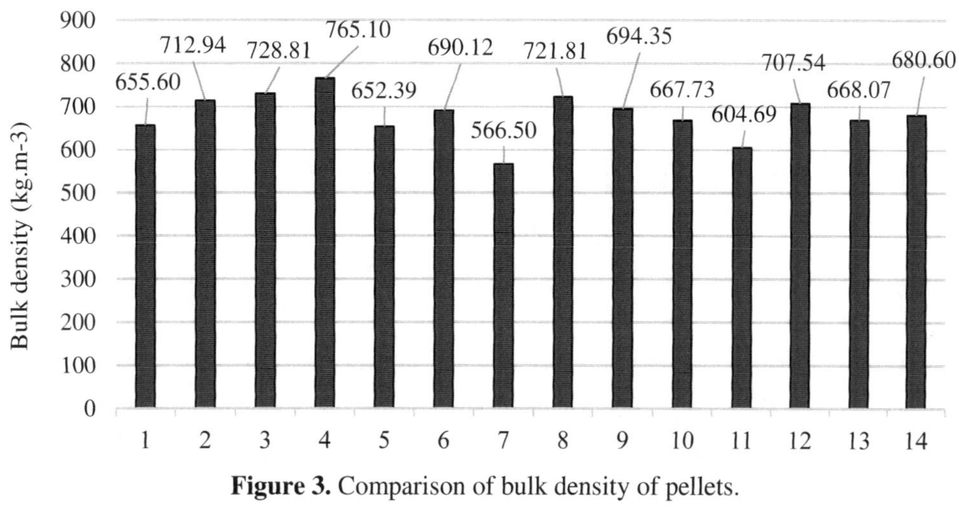

Figure 3. Comparison of bulk density of pellets.

The comparison of pellets mechanical durability can be seen in Figure 4. Wood/straw (samples 1-6) pellets mechanical durability decreases with increasing of straw amount in mixture, respectively with increasing of bark amount in mixture (samples 8-10). Vice-versa, straw/bark (samples 11-12) pellets mechanical durability increases with increasing of bark amount in mixture. Logically, wood/straw (samples 1-6) pellets abrasion resistance decreases with increasing of straw amount in mixture, respectively with increasing of bark amount in mixture (samples 8-10). Due to the structure and properties of the straw [24, 26]. A similar trend was noted also at straw/bark (samples 11-12) pellets where the pellets abrasion resistance decreases with increasing of bark amount in mixture. Comparison of pellets abrasion resistance is displayed on Figure 5.

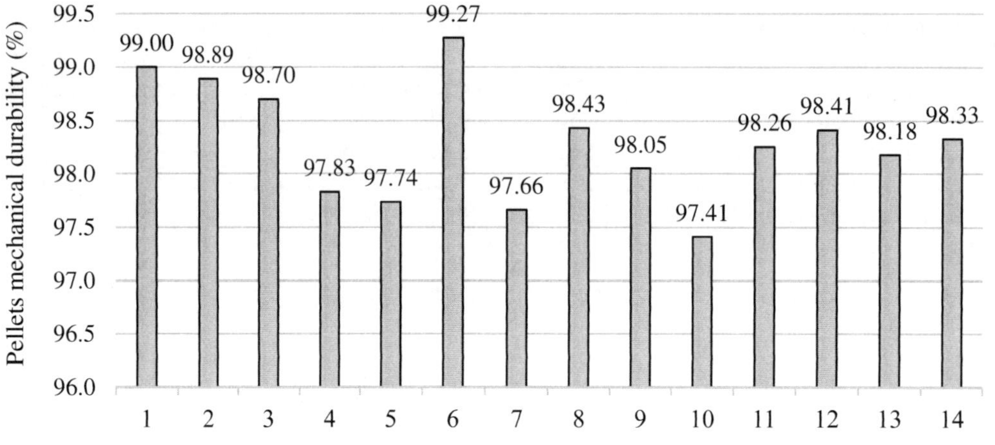

Figure 4. Comparison of mechanical durability of pellets.

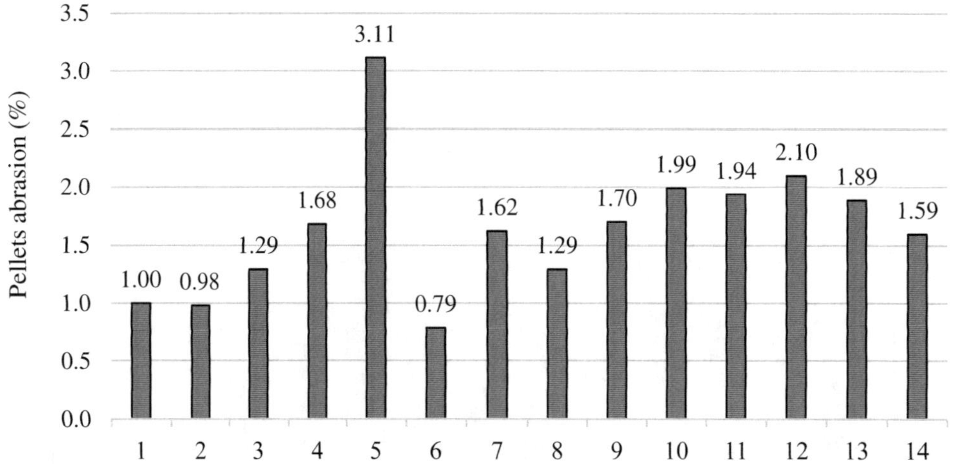

Figure 5. Comparison of abrasion of pellets.

Pellets hardness means pellets resistance to split by external force [11] and the strength of pellets in the direction perpendicular to the pellet axis is determined through the acting force value. Then the value of force that caused the pellets splitting is comparative criterion [21, 26].

The comparison of pellets hardness can be seen in Figure 6. Wood/straw (samples 1-6) pellets hardness very gently decreases with increasing of straw amount in mixture, respectively with increasing of bark amount in mixture (samples 8-10). Hardness determination showed that pellets with bark additions were more resistible against the splitting than pellets with straw additions. This was caused in order to better empty space filling by bark between raw material particles [21, 23] and thus better particles interconnection [5, 26].

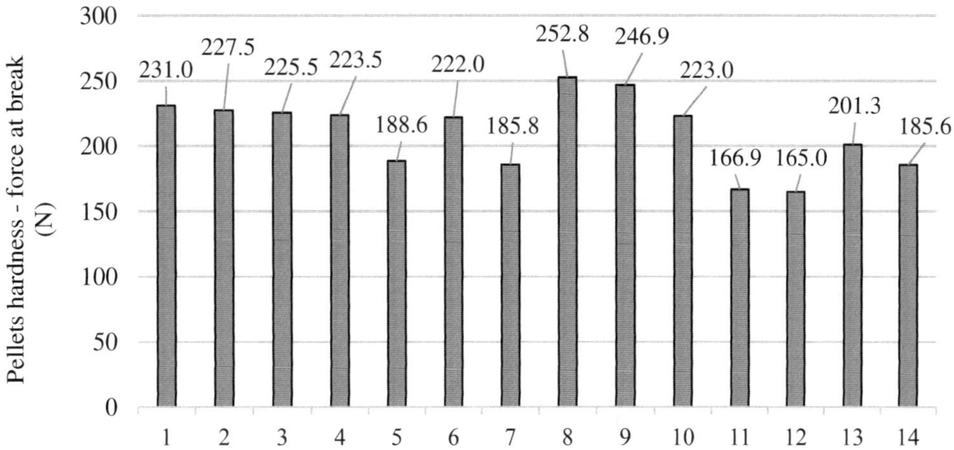

Figure 6. Comparison of hardness of pellets.

Conclusion
The main conclusions that can be withdrawn from this study are as follows:
- all pellets studied fulfilled the physical and mechanical requirements for non-industrial pellets **[15]**,
- raw material composition affected the pellets physical properties,
- increasing of straw amount in wood pellets positively influences the pellets particle density and pellets bulk density,
- increasing of straw amount in wood pellets negatively influences the pellets mechanical durability, pellets abrasion and pellets hardness,
- increasing of bark amount in wood pellets negatively influences the physical properties of pellets,
- increasing of bark amount in straw pellets positively influences the physical properties of pellets,
- both, the wheat straw and spruce bark can be effectively used for mixture pellets production in order to reduce the amount of clear spruce sawdust.

Acknowledgments
The paper is a part of the research done within the project VEGA 1/0420/16 – "Determination and research of parameters influence on final briquettes quality at waste biomass densification" and the project KEGA 061STU-4/2017 – "Increasing the level of the educational process in the field of production and environmental technologies by implementing of innovative tools". The authors would like to thank to the Ministry of Education of Slovak Republic and to the Slovak Academy of Sciences.

References

[1] Kaliyan N and Morey Vance R 2009 *Biomass and Bioenergy J.* **33** 3 337-359
[2] Nosek R, Holubcik M and Jandacka J 2016 *BioRes* **11** 44-53
[3] Lisý M, Baláš M, Špiláček M and Skála Z 2015 *A.Polytechnica* **55** 401-406
[4] Garcia Maraver A, Zamorano M, Fernandes U, Rabacal M and Costa M 2014 *Fuel* **119** 141-152
[5] Křižan P 2015 *The densification process of wood waste* (Berlin De Gruyter open)

[6] Miranda T, Montero I, Sepúlveda F J, Arranz J I, Rojas C V and Nogales S 2015 *Materials* **8** 1413-1427
[7] Castellano J M, Gómez M, Fernandéz M, Esteban L S and Carrasco J E 2015 *Fuel* **139** 629-636
[8] Gil M, Oulego P, Casal M D, Pevida C, Pis J J and Rubiera F 2010 *Bioresour. Technol.* **101** 8859-8867
[9] STN EN ISO 17225-1 2014 (European Committee for Standardization Brussels Belgium)
[10] Samuelsson R, Thyrel M, Sjostrom M and Lestander T A 2009 *Fuel Process. Technol.* **90** 1129-1134
[11] Križan P, Šooš Ľ, Matúš M, Beniak J and Svátek M 2015 *Wood research* **60** 2 301-316
[12] EN ISO 17827-1 2016 (European Committee for Standardization Brussels Belgium)
[13] STN EN ISO 18134-1 2015 (European Committee for Standardization Brussels Belgium)
[14] EN ISO 17828 2015 (European Committee for Standardization Brussels Belgium)
[15] EN ISO 17225-6 2014 (European Committee for Standardization Brussels Belgium)
[16] EN ISO 17831-1 2016 (European Committee for Standardization Brussels Belgium)
[17] EN ISO 18847 2017 (European Committee for Standardization Brussels Belgium)
[18] DIN 52182 1976 (Deutsches Institut für Normung Berlin Germany)
[19] Mani S, Tabil L G and Sokhansanj S 2006 *Biomass and Bioenergy J.* **30** 648–654
[20] Thoreson C P, Webster K E, Darr M J and Kapler E J 2014 *Energies* **7** 6 4019-4032
[21] Nielsen N P K, Gardner D J and Poulsen T 2009 *Wood and Fiber Science* **41** 4 414-415
[22] Serrano C, Monedero E, Lapuerta M and Portero H 2011 *Fuel Process. Technol.* **92** 699-706
[23] Križan P, Matuš M, Šooš Ľ and Beniak J 2015 *Energies* **8** 7 6382-6398
[24] Matúš M, Križan P, Šooš Ľ and Beniak J 2018 *Fuel* **219** 196-204
[25] Fernandes U and Costa M 2012 *Fuel Process. Technol.* **103** 51-56
[26] Lehtikangas P 2001 *Biomass and Bioenergy J.* **19** 351-360

IOP Conf. Series: Materials Science and Engineering **501** (2019) 012005 doi:10.1088/1757-899X/501/1/012005

Investigation of Solid Catalyst Poisoning Characteristics for Bio-diesel Fule(BDF) Synthesis by Using Ultrasonic Irradiation

N Zhu[1]*, M Kuo[2] and L Liu[2]

[1] Shizuoka Institute of Science and Technology, 2200-2 Toyosawa, Fukuroi, 4378555, Japan
[2] Graduate School of Shizuoka Institute of Science and Technology, 2200-2 Toyosawa, Fukuroi, 4378555, Japan
* Corresponding Author: zhu.ning@sist.ac.jp

Abstract. So far, a method of employing ultrasound irradiation and zeolites as solid catalyst for BDF synthesis has been suggested and experimentally verified. However, since catalyst poisoning will occur due to the repeated use of the same zeolite for BDF synthesis, hence at current paper, the catalyst poisoning characteristics of zeolite was investigated as the study target. During the study, firstly several zeolites of were selected and prepared to increase their basic characteristics. Secondly, the basic characteristics of the prepared zeolites were characterized by using SEM and X-ray diffractometer. Thirdly, BDF synthesis experiment was carried out by using the selected zeolite based on ultrasonic irradiation for several times. It was found that BDF yield ratio decreased as the number of the BDF synthesis experiments repeated while the same catalyst was used.

1. Background and purpose

It is extremely important to promote using alternative fuel such as alcohol or BDF for automobile in the era when mankind is facing the crisis of global warming while at the same time crude oil price is soaring so high. Recently many literatures[1-6] have being published focusing on the methods of synthesizing biomass fuels and their applications.

So far, a method of employing ultrasound irradiation and zeolites as solid catalyst for BDF synthesis has been suggested[7]. However, since catalyst poisoning will occur due to the repeated use of the same zeolite for BDF synthesis, hence at current paper, to investigate catalyst poisoning characteristics of zeolite is the study objective. During the study, several zeolites of were selected and were prepared to increase their basic characteristics. Then the basic characteristics was characterized by using SEM and X-ray diffractometer. Beside, BDF synthesis experiment was carried out by using the selected zeolite based on ultrasonic irradiation. It was found that BDF yield ratio decreased as the number of the BDF synthesis experiments repeated while the same catalyst was used.

2. BDF synthesis principle

The zeolite-catalyzed transesterification equation for producing BDF is shown in Figure 1. When vegetable oil(triglyceride) is mixed with alcohol(methanol) under the condition of ultrasonic irradiation and zeolite for some time, BDF and the by-product of glycerin are yielded.

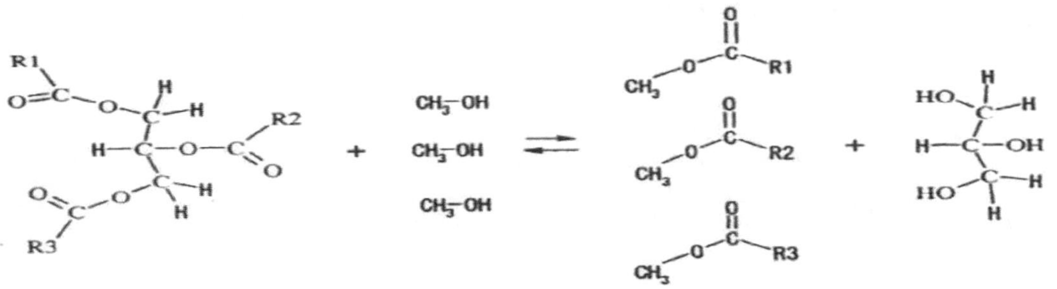

Vegetableoil Alcohol(methanol) BDF Glycerin (triglyceride)

Figure 1 BDF synthesis principle

3. BDF synthesis experiment

3.1. Catalyst preparation and evaluation

Since base strength could properly influence the BDF yield ratio, before BDF synthesis, zeolite was prepared as follows:

Step 1: Soak samples A3, A4, A5 and F9 with different NaOH concentration(3, 6 and 9 mol) respectively for 24 hours

Step 2: Wash the samples by using methanol and then dry the samples

Step 3: Sinter the samples by an electric furnace at the temperature of 400 ℃ for 2 hours

Before and after zeolites were prepared, SEM and X-ray diffractometer are used to character their base strength characteristics.

3.2. BDF synthesis

During batch type BDF synthesis(Figure 2), firstly, the mixture of the reactants of vegetable oil and methanol are prepared inside a glass tube. Secondly, the glass tube is put inside an ultrasonic reactor(Hondex frequency=28kHz) in which hot water is prepared. In order to decrease BDF synthesis time, the temperature of the hot water is set to be around 70 degree.

Condition of BDF synthesis is listed in Table.1. The zeolites of A3,A4,A5 and F9 as catalysts were used respectively. Volume ratio of vegetable oil to methanol is 5:1.

The sample of the BDF product is taken out after reaction time of 60 mins and BDF yield ratio is detected by using a GC(Shimazu GCMS-QP2020) system. The total BDF yield ratio is calculated based on the GC detection results.

Table.1 BDF synthesis condition

Physical quantities	Condition
Volume ratio of vegetable oil to methanol	5:1
Methanol purity	99.5%
Types of Zeolite	A3,A4,A5 and F9
Quantity of Zeolite	0.33wt%

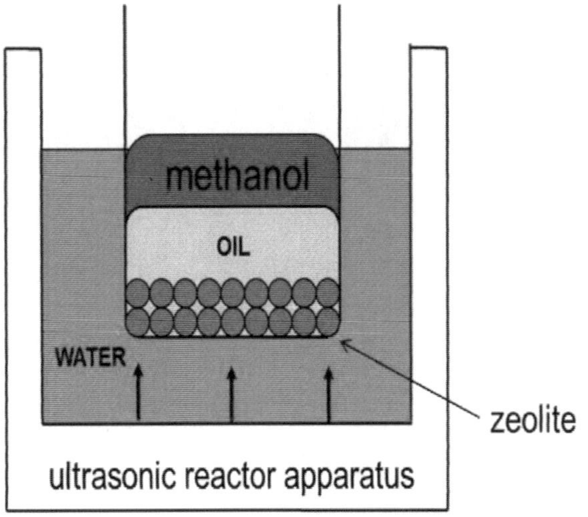

Figure 2 BDF synthesis by using batch method

3.3. Catalyst poisoning characteristics

Based on the BDF synthesis result, we selected A5 as the catalyst for catalyst poisoning investigation. The same amount of A5 was repeatedly used four times to produce BDF by the method described above. The BDF yield ratio was calculated based on the GC results.

4. Results and discussions

4.1. Catalyst evaluation

The solid catalysts used for BDF synthesis in this paper are 5 kinds of zeolites (A3, A4, A5 and F9) produced by Wako of Japan. Figure 3 shows the size and shape of zeolite A5 and its SEM analysis result. A5 has the round shape whose average external diameter is around 5-7 mm. According to the SEM analysis result, the surface area of A5 is very large which could absorb large amount of liquid.

Na+ content change after zeolite preparation is shown in Figure 4 based on SEM analysis. As for surface investigation, the Na+ content increased if the high NaOH concentration is used. In case of A4, A5 and F9, the Na+ content are over 20%. In regard of inner investigation, the Na+ content is lower than that of the surface because less NaOH is infiltrated in the inner of the zeolite

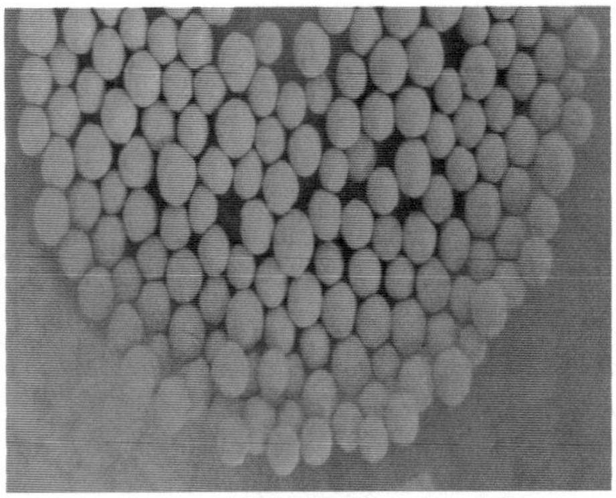

(a)Shape of zeolite A5

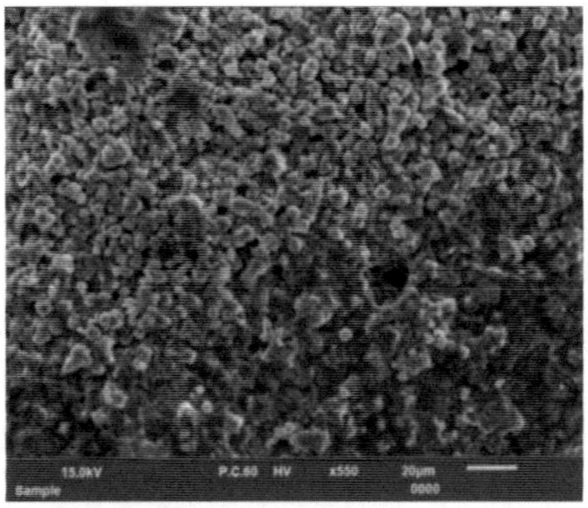

(b) SEM analysis result

Figure 3 Shape of zeolite A5 and its SEM analysis.

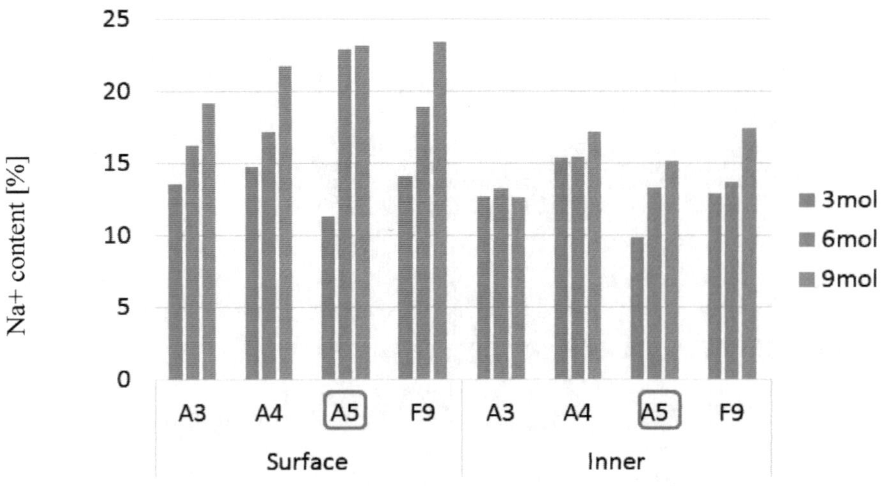

Figure 4 Na+ content for prepared zeolites

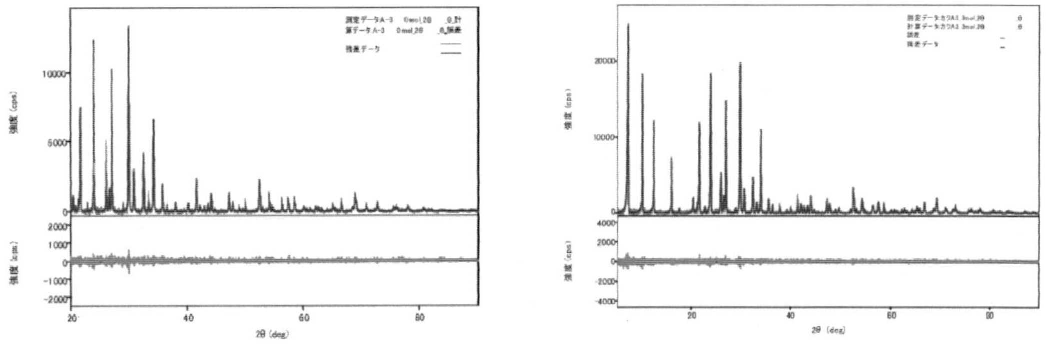

(a)Before preparation (b) After preparation for 3mol

Figure 5 X-ray diffractomete results for A5

Analysis result based on X-ray diffractomete are shown in Figure 5 for zeolite A5. It can been seen that the X-ray diffractomete results are different before preparation and after preparation. It seems that the X-ray diffractomete result changed because the soaked Na+ content changed its structure to some extent, which could likely effect the BDF synthesis efficiency.

4.2. BDF synthesis

In Figure 6, BDF yield ratio produced by different zeolites with different preparing method. It was found that by using A5 with 6mol NaOH in preparation, the BDF yield ratio reached the maximum value of 62%. Besides it could conclude that with the higher concentration of the NaOH for zeolite preparation, the more BDF yield ratio will be, which indicated that preparation was an effective way to increase BDF yield ratio. However, even 9 mol NaOH was used, the BDF yield ratio was not bigger than that of 6 mol. It appeared that there existed an optimum NaOH concentration for zeolite preparation.

4.3. Catalyst poisoning investigation

In Figure 7, the BDF yield ratio(right y axis) as well as change in Na+ (left y axis)content changed with the number of times of BDF synthesis reaction by using the same A5 are shown. At the beginning, the BDF ratio was 9%, however when it was used for the fourth time, the BDF ratio almost dropped to 0%, which means the effect of the catalyst poisoning. While at same time it is found that the Na+ content decreased from 12.7% to 10.8%. It appeared that the Na+ content would likely influence the BDF yield ration to some extent and how to keep the Na+ remain on the surface of the zeolite and how to reuse zeolites is a task to be challenged in the near future.

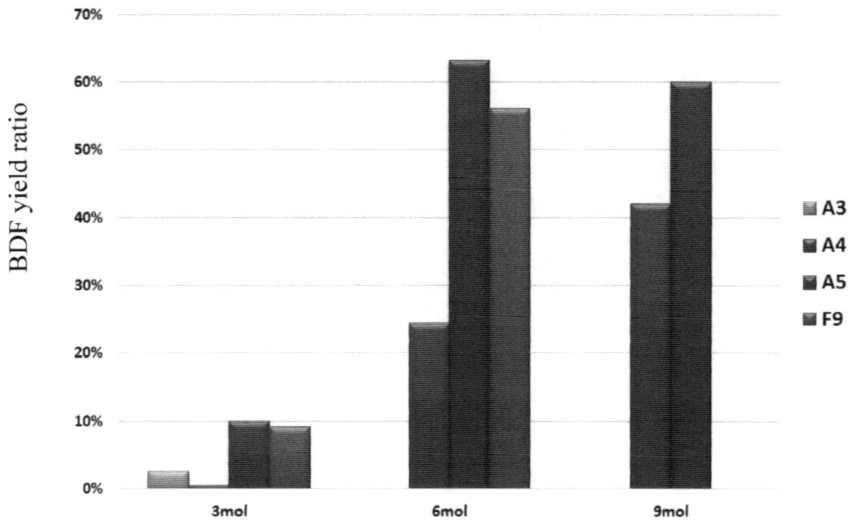

Different zeolites

Figure 6 BDF yield ratio

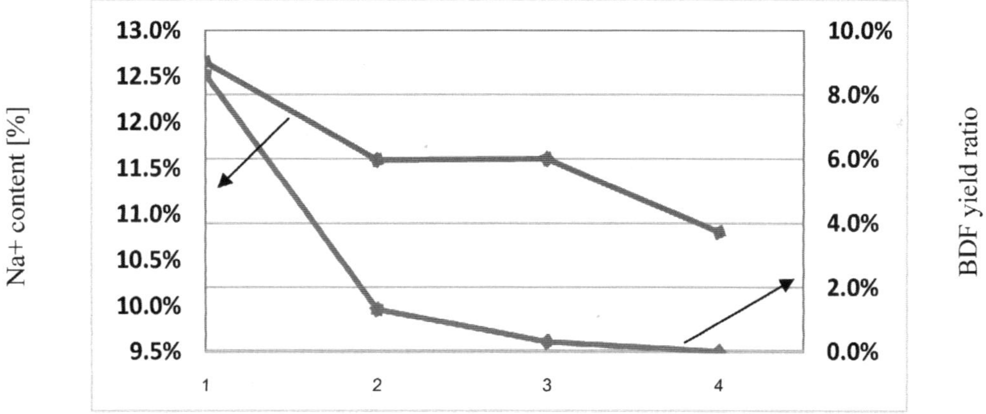

Number of times of BDF synthesis reaction by using the same A5

Figure 7 Change in BDF yield ratio and Na+ content after A5 was repeatedly used

5. Conclusion

In order to investigate the catalyst poisoning characteristics of the solid zeolite, a series of experiments including zeolite preparing, basic characteristics evaluation, BDF synthesis by repeatedly using the same prepared zeolite and some conclusions are obtained as follows:

(1)By using zeolite A5 with 6 mol NaOH preparation, the BDF yield ratio reached 62%;

(2)When zeolite A5 was used as catalyst for BDF synthesis at fourth time, the BDF could not be synthesized due to the influence of the catalyst poisoning.

Acknowledgement
This study is financially supported by Grant-in-Aid for Scientific Research(C17K06213) of JSPS, for which the author are very grateful.

References

[1] Furugawa S, Shimada, M, Okada M, Hirobashi L and Szuki Y 2004 *Trans. of JCS* **84-2** 1295-1296.
[2] Dszuki T, Hanehara H and Chita J 2005 *Proc. of JSME of Kans. Bran. Conf.* **80** 1247-1248.
[3] Yamane K, Shimamoto J and Ueda J 2001 *Trans. of JSA* **32**-2 25-30.
[4] Yamane K, Shimamoto J and Inoue H 2001 *Proc. of JSME of Kans. Bran. Conf.* **76** 1249-1250.
[5] Furuta S, Matsuhashi H and Arata K 2004 *J. of Catalysis Communications* **5**-12 721-723.
[6] Zhu N, Ttsuchiya T, Ito N and Kato S 2005 *Proc. of 3rd Int. Ener. Conv. Eng. Conf.* 2005-5543.
[7] Zhu N, Kuo M and Zhang T 2017 *Proceedings of 8th TSME-ICoME* AEC0017.

IOP Conf. Series: Materials Science and Engineering **501** (2019) 012006 doi:10.1088/1757-899X/501/1/012006

A Comparison of Temperature for Parabola and Sinusoidal Greenhouse Solar Dryer by CFD

A Srichat[1,*], P Vengsungnle[2], K Hongtong[1], W Kaewka[1] and J Jongpluempiti[2]

[1]Department of Mechanical Engineering, Faculty of Technology, Udon Thani Rajabhat University, Udon Thani, 41000, Thailand

[2]Department of Agricultural Machinery Engineering, Faculty of Engineering and Architecture, Rajamangala University of Technology Isan, Nakhon Ratchasima, 30000, Thailand

* Corresponding Author: saphichat@udru.ac.th

Abstract. This paper presents comparison for the thermal efficiency between a parabola greenhouse solar dryer and sinusoidal greenhouse solar dryer by computational fluid dynamics (CFD). The concept of design is to find ways to increase the solar area of the greenhouse solar drying. The greenhouse solar dryers used in the analysis are 2 types: 1) parabola greenhouse dryer has a one side solar roof structure for received solar energy area of 95.47 m^2 and 2) new design greenhouse is a sinusoidal greenhouse dryer has a two-sided solar roof structure for received solar energy area of 96.63 m^2, which increases the solar receiving area of 1.22%. The both volume of greenhouse dryers was about 94 m^3 (a parabola greenhouse dryer is a standard plant greenhouse dryer from department of alternative energy department and efficiency, Ministry of Energy, Thailand. The size: 8.2 of length and 6.0 of wide). The greenhouse dryer were received an average heat flux from the solar of 1,000 W/m^2 at 11.00am. - 02.00pm. in Udon Thani province, Thailand (in the absence of cloud cover) and the outside temperature is 28 °C. This experiment is an analysis of the effects of 6 hours of solar radiation at the same time for both greenhouse dryers to compare the temperatures using the computational fluid dynamic (CFD) program. The raw materials are drying at higher temperatures can be faster drying rate that resulting to drying time period is lower. From the results of the simulation, it was found that the average drying temperature of 6 hours for the sinusoidal and the parabola greenhouse is 322.26K (49.11°C) and 321.20K (48.05°C), respectively. The time of drying value of the sinusoidal greenhouse solar dryer is 3.40 hours at an average temperature value of 322.26K (49.11°C) and the parabola greenhouse solar dryer is 4.70 hours at an average temperature value of and 321.20K (48.05°C). In conclusion, a sinusoidal greenhouse has higher average drying temperatures with compared to a parabola greenhouse for the same drying time period. The sinusoidal greenhouse drying period takes less than the parabola greenhouse and higher drying efficiency.

1. Introduction

Thailand is mainly an agriculture country and a major food source in the world. Food production in Thailand is diverse such as rice, vegetables, fruits, fisheries and other crops. The food production is using for domestic consumption and export to foreign countries that is the major food exporter in Asia. Thailand's area of 230,850 km^2 is use for agricultural production and the agriculture employees about 60% of population [1]. The proportion of agricultural production exported to foreign countries accounted for 80% of all food production. However, some agricultural products for domestic are consume both fresh and preserved. The agricultural product preservation of Thailand is solar drying that Thailand has a relatively high proportion of solar energy. The preserving food by drying will keep the food can be stored for a long time. Solar drying is a good way to reduce the moisture content of the material.

Many countries use solar drying to conserve agricultural production [2]. In the Eastern Province of Saudi Arabia, the gable-even span greenhouse (solar greenhouse) of drying was experimented at King Faisal University that the daily average overall thermal efficiency was 57.2%. [3]. In India, a comparison of the properties of various solar greenhouse that research on the extent of drying and development to improve efficiency in terms of temperature and quality [4-5]. The types and factors of condition operating for solar greenhouses which natural convection type and forced convection type has been studied [6]. In addition, there are studies on the thermal modeling for the greenhouse drying of jaggery under force convection mode to predict the temperature and moisture evaporation [7]. As same in Bangladesh, the solar drying is studying for use in the future [8]. In China, there are studies on the solar collector use for the heat conducting oil can be heating to 230 °C and the average thermal efficiency is 66.5% [9]. Solar greenhouse is an enclosed structure having transparent walls and roofs that made up of polyethylene film. The working of solar greenhouse drying when the agricultural production are placed in trays receiving the solar radiations through the polyethylene film cover and moisture is removed by natural convection or forced convection condition. In Thailand, there are studies of the parabolic roof structure covered greenhouse type solar dryer for a small-scale and a large-scale of dried food industries with forced convection [10-11]. The parabolic solar greenhouses for a small-scale of 8.0 m in width, 20.0 m in length, 3.5 m in height and the parabolic solar greenhouses for a large-scale of 7.5 m in width, 20.0 m in length, 3.7 m in height which the both loading capacities about 1,000 kg of the agricultural production. From experiment, the drying air temperatures in the dryer varied from 35 - 65°C and 35 - 64°C, respectively. There is also the development and design of a PV-ventilated greenhouse dryer that is a black concrete floor with an area of $5.5\times 8.0\ m^2$ [12]. There are experiment drying with chili, banana, litchi and macadamia nuts [13-15]. The development of parabolic solar greenhouse has been experimented, designed and simulated for evaluate the parameters of drying. A present, the only parabolic solar greenhouse has been used for drying of the agricultural production in Thailand by Department of Alternative Energy Department and Efficiency, Ministry of Energy. In contrast, the other types of solar greenhouse have not been analysis for drying performance with compared to parabolic solar greenhouse. The type of sinusoidal solar greenhouse have been interested for drying performance analysis and expected to be more efficient than parabolic solar greenhouse. The concept of design is to find ways to increase the solar area of the greenhouse solar drying which sinusoidal roof structure covered greenhouse type solar dryer have an area of solar radiation more than parabolic roof structure covered greenhouse type solar dryer.

The studying of the temperature drying for the parabolic and sinusoidal solar greenhouse are interested which comparison the thermal efficiency of drying. The simple methods for predict to temperature drying of greenhouse dryer is the simulation by computational fluid dynamics (CFD) program with ANSYS. The simulation was showed that the temperature drying in the dry period for the agricultural production and can be compare between the parabolic and sinusoidal solar greenhouse. In this paper presents comparison for the thermal efficiency between a parabola greenhouse solar dryer and sinusoidal greenhouse solar dryer. The result of comparison will be using for building to experimental in next time.

2. The Solar Dryer Mathematical Model

Analysis of the behavior of the air to be transient with the energy equation [16]. The equation used for calculated from Eq. (1) – (4) [17]

Continuity
$$\frac{\partial \rho}{\partial t} + div(\rho u) = 0 \tag{1}$$

x-momentum $$\frac{\partial(\rho u)}{\partial t} + div(\rho uu) = -\frac{\rho p}{\partial x} + div(\mu \text{grad} u) + S_{Mx} \quad (2)$$

y-momentum $$\frac{\partial(\rho v)}{\partial t} + div(\rho vu) = -\frac{\rho p}{\partial y} + div(\mu \text{grad} v) + S_{My} \quad (3)$$

z-momentum $$\frac{\partial(\rho w)}{\partial t} + div(\rho wu) = -\frac{\rho p}{\partial z} + div(\mu \text{grad} w) + S_{Mz} \quad (4)$$

The computational fluid dynamics analysis is bases on the Navier-Stokes equation. The equation can be calculated from Eq. (5) is solves by the equation under the mass conservation equation, momentum conservation and energy conservation [18-19].

$$\left(\frac{\partial\rho\phi}{\partial t}\right) + \left(\frac{\partial\rho u\phi}{\partial x} + \frac{\partial\rho v\phi}{\partial y} + \frac{\partial\rho w\phi}{\partial z}\right) = \Gamma\left(\frac{\partial^2\phi}{\partial x^2} + \frac{\partial^2\phi}{\partial y^2} + \frac{\partial^2\phi}{\partial z^2}\right) + S(\phi) \quad (5)$$

When ϕ is the transport variable
V(u, v, w) is the Reynolds-averaged velocity vector
Γ is the diffusion coefficient
$S(\phi)$ is the source term

3. Models and method

3.1. *Models for analysis*

The comparison of the temperature between a parabola and sinusoidal greenhouse solar dryer by computational fluid dynamics (CFD) that design is to find ways to increase the solar area of the greenhouse solar drying. The greenhouse solar dryers used in the analysis are 2 types: 1) parabola greenhouse solar dryer has a one side solar roof structure for received solar energy with area surface of solar radiation of 95.47 m^2, as shown in Figure 1 and 2) new design greenhouse is a sinusoidal greenhouse dryer has a two-sided solar roof structure for received solar energy area of 96.63 m^2, as shown in Figure 2. The solar receiving area increasing of 1.22%. The both volume of greenhouse dryers was about 94 m^3 (a parabola greenhouse dryer is a standard plant greenhouse dryer from department of alternative energy department and efficiency, Ministry of Energy, Thailand. The size: 8.2 of length, 6.0 of wide and 3.0 of high). The greenhouse dryers were received an average heat flux from the solar of 1,000 W/m^2 at 11.00am. - 02.00pm. in Udon Thani province, Thailand (in the absence of cloud cover) and the outside temperature is 28 °C.

3.2. *Methods and hypothesis*

This experiment is an analysis of the effects of 6 hours of solar radiation at the same time for both greenhouse dryers to compare the temperatures using the computational fluid dynamic (CFD) program. The two equation k–ε turbulence models are widely applied in simulating turbulent flows. [20–26] The k–ε models have mathematical simplicity and need low computational demand [26]. The raw materials are drying at higher temperatures can be faster drying rate that resulting to drying time period is lower.

The hypothesis can be shown in Figure 3 and the conditions used for the analysis are given in Table 1 and Table 2.

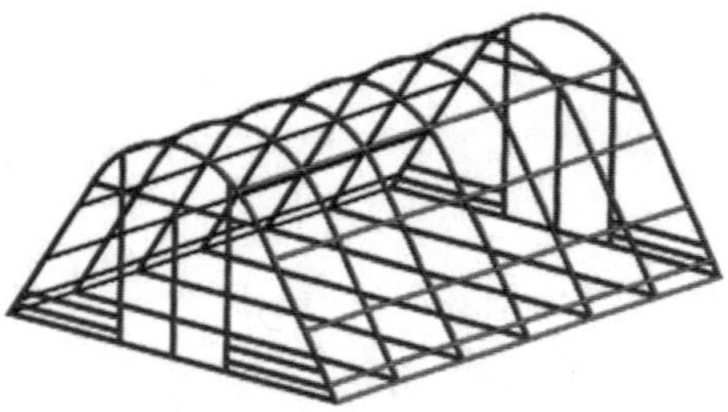

Figure 1. The parabola greenhouse solar dryer model

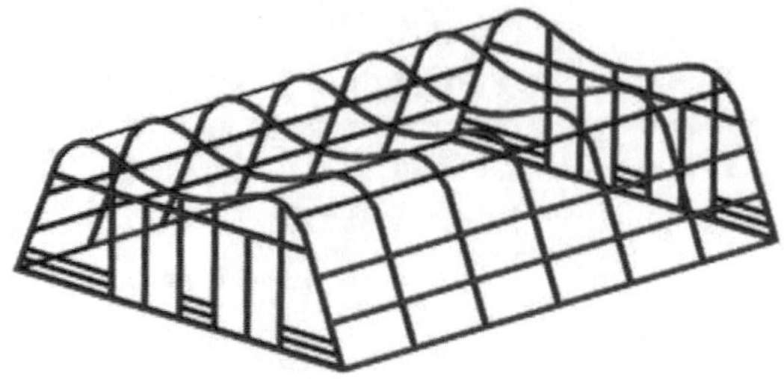

Figure 2. The sinusoidal greenhouse solar dryer model

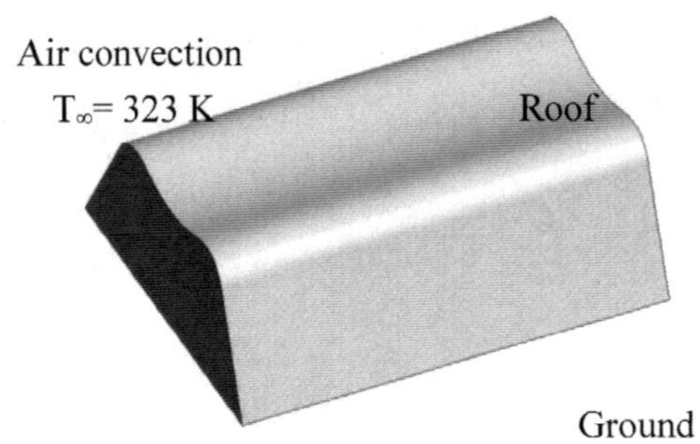

Figure 3. The hypothesis of sinusoidal greenhouse solar dryer

Table 1. Assumptions used in analysis

No.	Condition
1	Analysis of fluid properties depends on the coordinates and time (transience)
2	Use of air properties as a fluid property in analysis
3	Analysis in 3D
4	Analysis by considering the gravity
5	The viscosity model uses the Standard k-epsilon equation (2 eqn).
6	Solve the Pressure-Velocity Coupling SIMPLE Scheme

Table 2. Conditions used in analysis

No.	Parameter	Roof Type		Unit
		Parabola	Sinusoidal	
1	Air temperature starts in the greenhouse.	300.00	300.00	K
2	Outdoor air temperature	323.00	323.00	K
3	heat transfer coefficient	100.00	100.00	W/m^2K
4	simulation time	6	6	hr
5	time step size	60	60	s

3.3. *The Model for simulation*
Analyze the temperature inside the greenhouse solar model. The average temperature in the cross section XY and YZ plane at the middle of the greenhouse. The XZ plane is divide into each Y level starting from 0.0, 0.5, 1.0, 1.5, 2.0 and 2.5 m, respectively, as shown in Figure 4 and Figure 5.

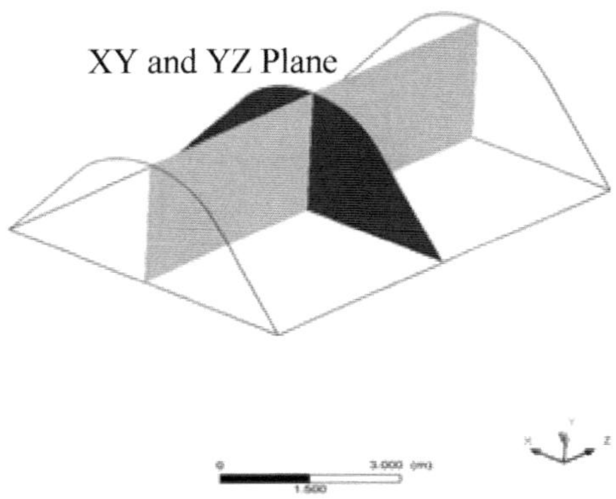

(A) parabola greenhouse solar dryer model

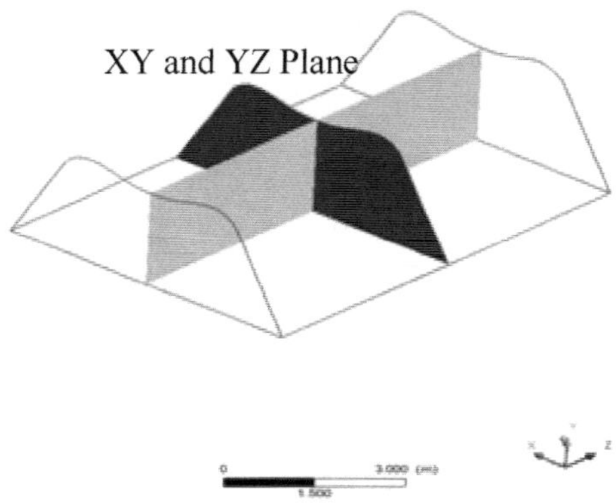

(B) sinusoidal greenhouse solar dryer model

Figure 4. XY and YZ plane of greenhouse solar dryer for simulation

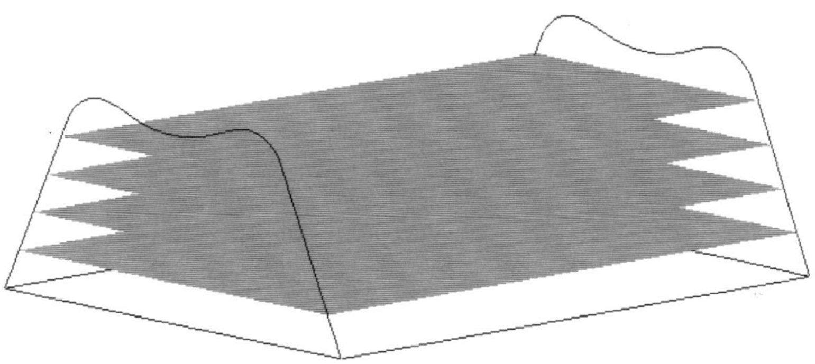

Figure 5. XZ plane (Y level layers) of greenhouse solar dryer for simulation

4. Results

4.1. *XY plane temperature of greenhouse solar dryer from simulation*
From the result of the simulation for temperature in the XY Plane of greenhouse solar dryers shown that the sinusoidal greenhouse solar dryer temperature value at the roof more than the parabola greenhouse solar dryer, as shown in Figure 6. When, the roof of greenhouses have heating from the sunshine that solar energy hits into the greenhouses. The roof will be the highest thermal energy and heat transfer down to the ground. From the result shown that the sinusoidal greenhouse can be receive, store and discharge heat energy better than the parabola greenhouse. Thus, the temperature stored in the sinusoidal greenhouse is higher than the parabola greenhouse. The roof and ground temperature of the sinusoidal greenhouse is 322.11K (48.96°C) and 322.20K (49.05°C), respectively. Likewise, the roof and ground temperature of the parabola greenhouse is 320.97K (47.82°C) and 321.29K (48.14°C), respectively. The different temperature value on the ground of greenhouse is 0.91K (°C).

IOP Conf. Series: Materials Science and Engineering **501** (2019) 012006 doi:10.1088/1757-899X/501/1/012006

4.2. *YZ plane temperature of greenhouse solar dryer from simulation*

From the result of the simulation for temperature in the YZ Plane of greenhouse solar dryers is similar with XY Plane that the sinusoidal greenhouse solar dryer temperature value at the roof more than the parabola greenhouse solar dryer, as shown in Figure 7. The roof and ground temperature of the sinusoidal and the parabola greenhouse is similar with XY Plane.

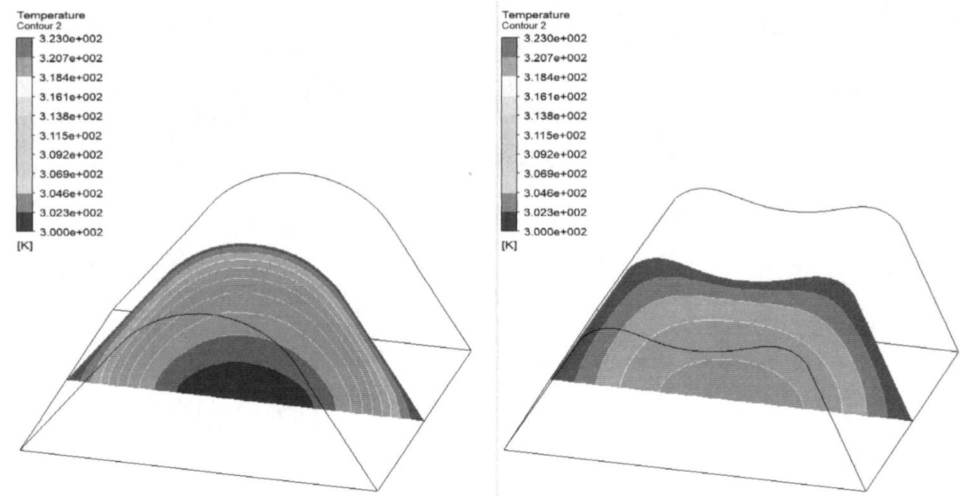

Figure 6. XY plane temperature of greenhouse solar dryer from simulation

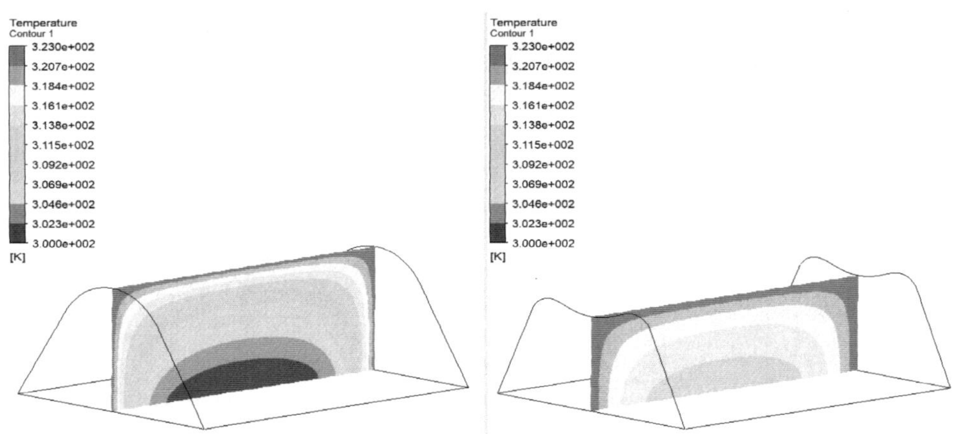

Figure 7. YZ plane temperature of greenhouse solar dryer from simulation

4.3. *XZ plane (Y level layers) temperature of greenhouse solar dryer from simulation*

From the result of the simulation for temperature in the XZ Plane of greenhouse solar dryers is similar with XY and YZ Plane that the sinusoidal greenhouse solar dryer temperature each layer value from Y level is 0.0 – 2.5 m is higher than the parabola greenhouse solar dryer temperature layer, as shown in Figure 8. Thus, the temperature stored in the sinusoidal greenhouse layer is higher than the parabola greenhouse layer. The comparison temperature layer of the greenhouse can be show in Table 3. The average temperature of the sinusoidal and the parabola greenhouse is 322.26K (49.11°C) and 321.20K (48.05°C), respectively. The average different temperature value of the both greenhouses is 1.06K (°C).

IOP Conf. Series: Materials Science and Engineering **501** (2019) 012006 doi:10.1088/1757-899X/501/1/012006

Table 3. Comparison temperature layer of the greenhouse

No.	Parameter	Roof Type		Unit
		Parabola	Sinusoidal	
1	Average temperature @ground	320.97	322.11	K
2	Average temperature @0.5 m	320.95	322.11	K
3	Average temperature @1.0 m	321.09	322.19	K
4	Average temperature @1.5 m	321.30	322.34	K
5	Average temperature @2.0 m	321.62	322.60	K
6	Average temperature at a roof	321.29	322.20	K
The average temperature of greenhouse		321.20	322.26	K
The average different temperature		1.06		K

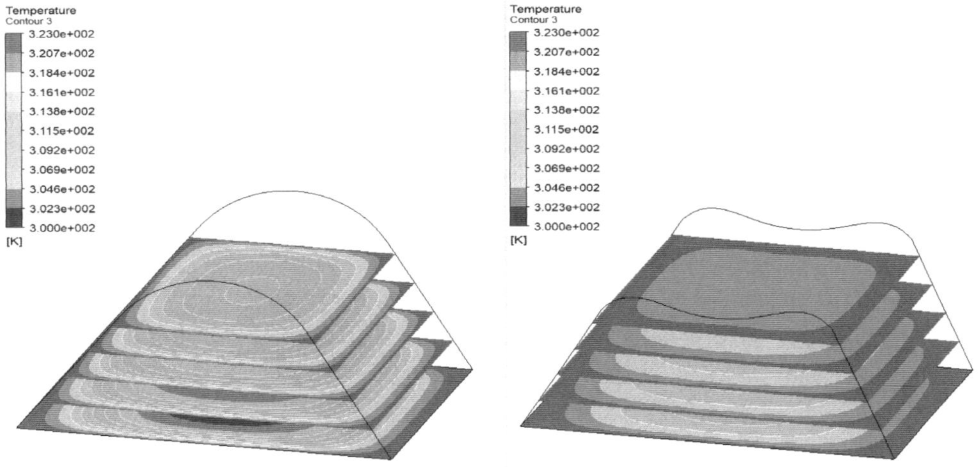

Figure 8. XZ plane (Y level layers) temperature of greenhouse solar dryer from simulation

4.4. *The transient of greenhouse solar dryer from simulation*

From the result of the simulation for the time of drying shown that the time of drying value of the sinusoidal greenhouse solar dryer is faster more than the parabola greenhouse solar dryer, as shown in Figure 9. When, the compare by the average highest temperature for drying shown that the time of drying value of the sinusoidal greenhouse solar dryer is 3.40 hours at an average temperature value of 322.26K (49.11°C) and the parabola greenhouse solar dryer is 4.70 hours at an average temperature value of and 321.20K (48.05°C). As a result, it was found that the drying time of the sinusoidal greenhouse was faster than the parabolic greenhouse.

5. Conclusion

The experiment was presents comparison for the temperature between a parabola greenhouse solar dryer and sinusoidal greenhouse solar dryer by computational fluid dynamics (CFD). The design is to find ways to increase the solar area of the greenhouse solar drying. The greenhouse used in the analysis are 2 types: 1) parabola greenhouse dryer has received solar energy area of 95.47 m^2 and 2) new design greenhouse is a sinusoidal greenhouse dryer has received solar energy area of 96.63 m^2, which increases the solar receiving area of 1.22%. The both volume of greenhouse is similar of 94 m^3 (a parabola greenhouse dryer is a standard plant greenhouse dryer from department of alternative energy department and efficiency, Ministry of Energy, Thailand. The size: 8.2 of length and 6.0 of wide). The greenhouse dryer were received an average heat flux from the solar of 1,000 W/m^2 at 11.00am. - 02.00pm. in Udon Thani province, Thailand (in the absence of cloud cover) and the outside temperature is 28 °C. This experiment is an analysis of the effects of 6 hours of solar radiation at the same time for both greenhouse dryers to compare the temperatures. The raw materials are drying at higher temperatures can be faster drying rate that resulting to drying time period is lower. From the results of the simulation, it was found that the sinusoidal greenhouse solar dryer temperature each layer value from Y level is 0.0 – 2.5 m is higher than the parabola greenhouse solar dryer temperature layer. The average different temperature value of the both greenhouses is 1.06K (°C) and the temperature stored in the sinusoidal greenhouse layer is higher than the parabola greenhouse layer. The average temperature of the sinusoidal and the parabola greenhouse is 322.26K (49.11°C) and 321.20K (48.05°C), respectively. The time of drying value of the sinusoidal greenhouse solar dryer is 3.40 hours at an average temperature value of 322.26K (49.11°C) and the parabola greenhouse solar dryer is 4.70 hours at an average temperature value of and 321.20K (48.05°C). The drying time of the sinusoidal greenhouse was faster than the parabolic greenhouse that it can be reduce the drying time of the raw materials about of 1.30 hours. From the results. In conclusion, a sinusoidal greenhouse has higher average drying temperatures with compared to a parabola greenhouse for the same drying time period. The sinusoidal greenhouse drying period takes less than the parabola greenhouse and higher drying efficiency.

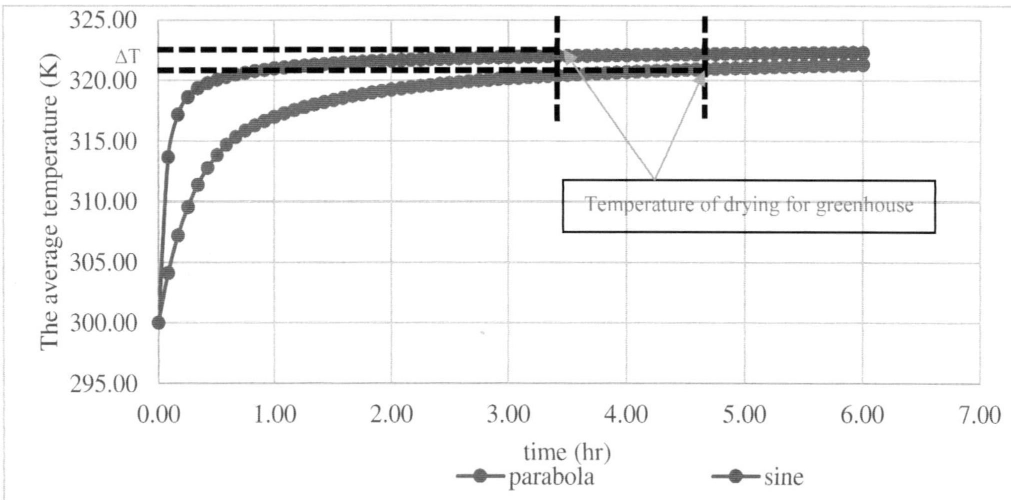

Figure 9. The transient of greenhouse solar dryer from simulation

Acknowledgments
The authors acknowledge the grant from Udon Thani Rajabhat University that all supported in this paper. The authors are also grateful to the Mechanical Design Research Unit, Department of Mechanical Engineering, Faculty of Engineering, Khon Kaen University for supporting the tools analyzer used in this research. The authors are also grateful to the Assoc. of Professor Dr. Paisan Naphon from Department of Mechanical Engineering, Faculty of Engineering, Srinakharinwirot University for supporting the tools analyzer used in this research.

References

[1] Sriwichailamphan T 2007 *Global Food Chains and Environment: Agro-Food Production and Processing in Thailand* (Ph.D-Thesis for Wageningen University) p 2
[2] Sharma A , Chen CR and Lan NV 2009 *Renewable and Sustainable Energy Reviews* **13(6–7)** 1185-1210

[3] Almuhanna Emad A. 2011 *Journal of Agricultural Science* **4(3)** 237-246
[4] Tiwari A 2016 *J Food Process Technol.* **7(9)** 1-12
[5] Kumar A and Shrivastava V 2017 *Indian Journal of Science and Technology,* **10(33)**1-16
[6] Nidhi and Verma P 2016 *IOSR Journal of Mechanical and Civil Engineering (IOSR-JMCE)* **Special Issue - AETM'16** 43-48
[7] Kumar A and Tiwari GN 2006 *International Journal of Agricultural Research* **1(3)** 265-279
[8] Mohsin ASM , Nasimul Islam Maruf Md., Sayem AHM, Mojumdar RR and Farhad Hossain MS 2011 P *IACSIT International Journal of Engineering and Technology* **3(2)** 165-170
[9] Ji X, Li M, Wang Y, Ling D and Luo X 2016 *J. of Bulgarian Chemical Communications* **48 Special Issue E** 120 – 125
[10] Janjai S 2012 *International Energy & Environment Foundation* **3(3)** 383-398
[11] Poolsak Intawee and Serm Janjai 2011 *International Energy Journal* **12** 39-52
[12] S.Janjai, V.Khamvongsa and B.K.Bala 2007 *International Energy Journal* **8** 249-258
[13] J.Kaewkiew, S.Nabnean and S.Janjai 2012 *Procedia Engineering* 32 433 – 439
[14] K.Tohsing, S.Janjai, N.Lamlert, T.Mundpookhier, W.Chanalert and B.K. Bala 2018 Experimental performance and artificial neural network modeling of solar drying of litchi in the parabolic greenhouse dryer *Journal of Renewable Energy and Smart Grid Technology* **13(1)** 1-12
[15] S.Janjai, C.Phusampao, W.Nilnont and P.Pankaew 2014 Experimental performance and modeling of a greenhouse solar dryer for drying macadamia nuts *International Journal of Scientific & Engineering Research* **5(6)** 1155-1161
[16] Cengel YA and Ghajar AJ 2014 *Heat and Mass Transfer 5th ed.* (United States: McGraw-Hill Education)
[17] Titus P, Damian T and Brezinski C 2005 *Basics of Fluid Mechanics and Introduction to Computational Fluid Dynamics* (Boston: Springer Science Business Media, Inc.)
[18] Bichkar P, Dandgaval O, Dalvi P, Godase R and Dey T 2018 *Procedia Manufacturing* **20** 195–200
[19] Baukal CE, Gershtein VY, and Li X Computational fluid dynamics in industrial combustion [Internet]. CRC Press; 2001. 630 p.
[20] Borchers O, Busch C, Sokolichin A, and Eigenberger G 1999 *Chem Eng Sci.* **54**(24) 5927–35.
[21] Mudde RF and Simonin O 1999 *Chem Eng Sci* **54**(21) 5061–9.
[22] Pfleger D, Gomes S, Gilbert N, and Wagner H-G 1999 *Chem Eng Sci.* **54**(21):5091–9.
[23] Sokolichin A, Eigenberger G. 1999 *Chem Eng Sci.* **54**(13) 2273–84.
[24] Behzadi A, Issa RI, and Rusche H 2004 *Chem Eng Sci.* **59**(4) 759–70.
[25] Selma B, Bannari R and Proulx P 2010 *Chem Eng Sci.* **65**(20) 5417–28.
[26] Liu Y and Hinrichsen O 2014 *Comput Fluids* **105** 91–100.

IOP Conf. Series: Materials Science and Engineering **501** (2019) 012007 doi:10.1088/1757-899X/501/1/012007

Heat balancing of direct reforming fuel cells in MGT-SOFC hybrid systems

H.P. Berg[1,*], M. Kleissl[2], A. Himmelberg[1], M. Lehmann[1], N. Prechavut[1], M. Vorpahl[2]

[1]BTU Cottbus-Senftenberg, Chair of Combustion Engines and Flight Propulsion, Siemens-Halske-Ring 14, 03046 Cottbus, Germany
[2]Prof. Dr. Berg und Kiessling GmbH, Burger Chaussee 25, 03046 Cottbus, Germany
* Corresponding Author: peter.berg@b-tu.de

Abstract. Solid oxide fuel cells convert chemical energy in electrical energy and are highly suitable for the conversion of hydrocarbon based fuels and products from PowerToGas conversions. Embedded in a micro gas turbine-cycle instead of the combustion chamber the heat energy of the injected fuel, released in the SOFC-Stack, can additionally converted into work and by a turbine-generator into electricity. In a compact system, SOFC stacks are designed to realize a direct steam-reforming process inside. In such direct-reforming fuel cell systems the SOFC operating temperature due to the heat demand for reforming, can be reduced to a value in the range of the outlet temperature of the downstream flow (650°C). If the cycle uses a recuperating system, the operating conditions of the SOFC can be realized without additional high-temperature heat exchanger systems. The system with an uncooled turbine, described in [1], just can reach the operating temperature with an additional high-temperature heat exchange. The system described in this publication does not require an additional high-temperature heat exchanger, because the X-value, as the ratio between the exchanged heat quantity and the required amount of heat to complete the MGT-cycle [2] due to the referring process, is nearly zero and can be realized via the inner container wall (of the MLC). The cycle is completed (between the SOFC and the turbine entry) by the heat input of an afterburner. Here the unreacted fuel of the SOFC fuel-stream is used to provide the necessary heat energy for the downstream turbo-generator-system.

1. Introduction

A MGT-SOFC hybrid system has been described in [1] and [2] in the planned "Turbo-Fuel-Cell (TFC)" design. A marketable system (developed by the B+K GmbH) [3] in cooperation with the BTU Cottbus-Senftenberg, LS VFA [4]) is scheduled to go into service in 2027. The TFC system [1] consists of a recuperated micro gas turbine process (MGT process) with an embedded SOFC subsystem ("Solid Oxide Fuel Cell"). This technology uses planar SOFC stacks because they are more economical and more compact than SOFC cylindrical cell stacks (SOFC-CCS) [19]. CCS technology is currently considered to be stronger and better suited for integration into gas turbine processes, as possible differential pressures between the anode and cathode sides (e.g. under load changes) may be higher due to the cylindrical shape [5]. Targeted control of a low differential pressure means that new generations of planar SOFC stacks now can achieve service life cycles of more than 30,000 hours [14], [20], [22].

The BTU-CS [4] MAXEFF research program investigates the suitability of planar SOFC stacks for the hybrid pressure-charged hybrid application with respect to a cluster design (SOFC Stack Grape Technology / SOFC-SGT) [1]. A recent Mitsubishi-Hitachi Power Systems, LTD (MHPS) project demonstrates SOFC-CCS technology in a hybrid combination with gas turbines [5]. By contrast, the

IOP Conf. Series: Materials Science and Engineering **501** (2019) 012007 doi:10.1088/1757-899X/501/1/012007

TFC 1.0 strategy relies on the available market for planar SOFC stacks and follows the SOFC-SGT strategy. The tubular SOFC stacks of the MHPS technology show a high differential pressure resistance compared to planar systems. This article is based on a MGT-SOFC system architecture (type TFC 1.0) for the study with planar stack systems.

A research and development-oriented collaboration between industrial companies and research institutes will establish the most suitable SOFC technology for hybrid systems in the future. Figure 1 visualizes the 270kW TFC1.0 system. In this case, the SOFC stacks are connected to grapes and the already industrialized micro turbine is shown. [1]

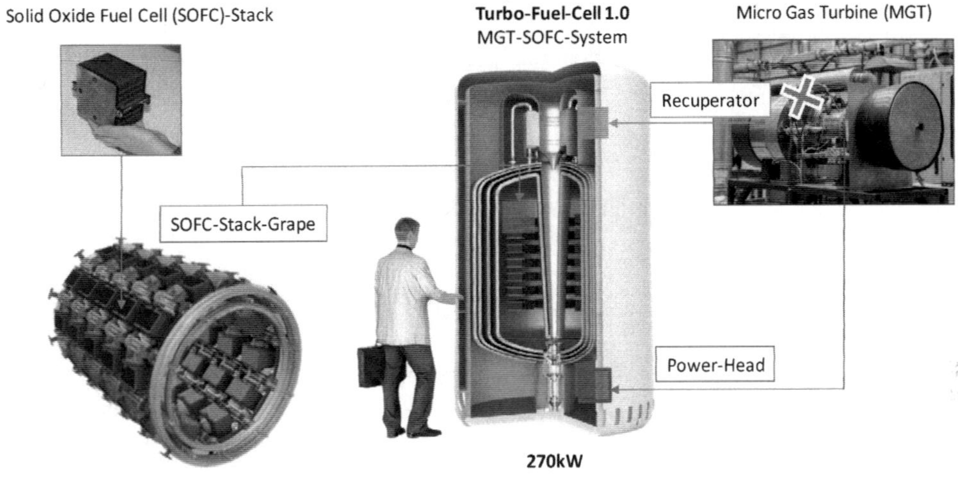

Figure 1. Visualization of a MGT-SOFC System with SOFC-SGT [1]

The System architecture examined in this article is based on the TFC 1.0 type [1]. The complexity of the high-temperature heat exchanger fixes the temperature difference between the SOFC inlet temperature and the recuperator outlet temperature. The complexity of the high temperature heat exchanger determines the temperature differential between the SOFC inlet temperature and the recuperator outlet temperature. For this reason, SOFC technology with reduced temperature and internal reforming is examined in this study.

2. Fundamentals and heat balancing in the cycle with the TFC-1.0 technology

A realization of the TFC-1.0 is planned with the industrialized SOFC components available today. In future, however, an internal reforming process (exothermal) can reduce the operating temperature of the SOFC cell, under the condition that sufficient electrolyte conductivity is possible. At present, electrolyte membranes of yttrium-stabilized zirconia (YSZ) are used [16]. Above 750°C, this material is a useful oxide ion conductor, which is the reason for the currently high SOFC operating temperature (700°C to 900°C). At this temperature range, the technology generates sufficient performance and achieves a very high efficiency. The fuel cell heat loss and the heat from the afterburner (burning the non-converted fuel) is the equivalent to the traditional combustion chamber in the MGT process [2]. Part of this heat energy is needed for the reforming process or pre-reforming process (with a complete implementation inside the SOFC-Stack). To complete the MGT-Process the heat quantity ΔQ_{Input} is necessary. It is determined using the SOFC efficiency chain ($\eta_{SOFC}^{rev} \cdot \eta_U \cdot \eta_I \cdot \eta_C$). The heat quantity is the not in electricity converted chemical energy described with the formation enthalpy of the fuel as follows [2]:

$$\Delta Q_{Input} = \Delta^R H \cdot (1 - \eta_{SOFC}^{rev} \cdot \eta_U \cdot \eta_I \cdot \eta_C) \qquad (1)$$

This technology is equipped with an innovative multi-layer pressure vessel [21]. The housing temperature of less than 80°C is possible due to the multilayer insulation, while the operating temperature of the SOFC in the first application, with a reforming process outside the SOFC, is 860°C.

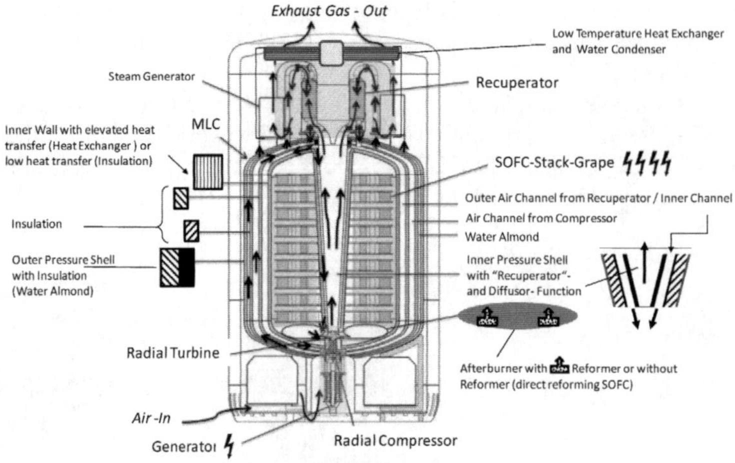

Figure 2. Visualization of a TFC-1.0 System in a vertical MGT Axis-Architecture and the main components of the process. The MLC technology allows minimal heat losses

Figure 2 illustrates the architecture of the actual system. In this design, the entire system is thermal-optimized, because all heat losses through the ducting of the MLC (multi-layer container) are low and returned into the process. In the latest design, the inner MLC-wall is used for the high-temperature heat exchange to obtain the required SOFC inlet temperature (heat balancing in the cycle).

For general understanding of the MGT-SOFC process, Figure 3 shows the basic structure of the fundamental equations.

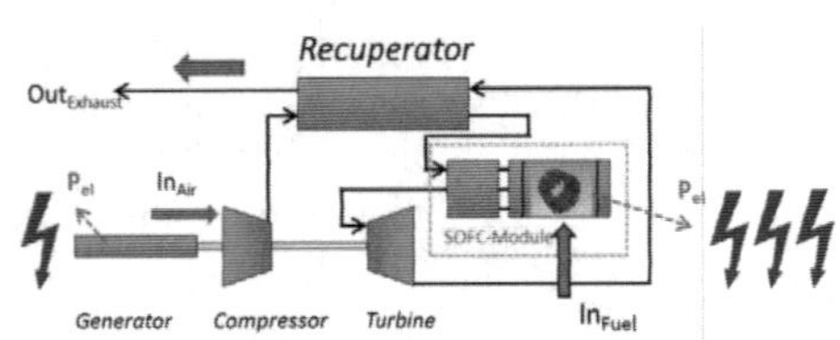

elementary charge, reaction entropy, faraday constant, reversible voltage

$$\eta_{SOFC}^{rev} = \frac{\Delta^R G}{\Delta^R H} = 1 - T_{SOFC}\cdot\frac{\Delta^R S}{\Delta^R H} = -\frac{n\cdot F\cdot U_{rev}}{\Delta^R H}$$

$$\eta_{MGT-SOFC}^{real} = \left(\eta_{MGT}^{real} + \eta_{SOFC}^{rev}\cdot\eta_U\cdot\eta_I\cdot\eta_C\cdot\left(1-\eta_{MGT}^{real}\right)\right)\cdot\eta_{Sys} \quad (2)$$

$\eta_U = \frac{U_{real}}{U_{rev}}$ electrochemical efficiency; $\eta_I = \frac{I_{real}}{I_{th}}$ current efficiency; $\eta_C = \frac{\dot{m}_{F,conv}}{\dot{m}_{F,in}}$ fuel conversion efficiency

Figure 3. Sketch of the MGT-SOFC process (left) and the fundamental equations (right) [2]

3. The X-value (for the quantification of heat balancing in the cycle) and the extended definition taking the SOFC inlet- and outlet temperature into account

To estimate the design effort of a MGT-SOFC-process, the heat ratio X was introduced in [2] as the ratio between the exchanged heat quantity ΔQ_{HeatEx} and the required amount of heat $\Delta Q_{Input} = \Delta^R Q + \Delta^I Q + \Delta^F Q - \Delta Q_{Reformer}$ to complete the MGT-cycle.

$$X = \left(\frac{\Delta Q_{HeatEx}}{\Delta Q_{Input}}\right) \quad (3)$$

To be on the safe side for the system design, the maximum amount of heat that the high-temperature heat exchanger supplies was determined from the temperature difference between the mean SOFC temperature and the recuperator outlet temperature (Figure 4 left).

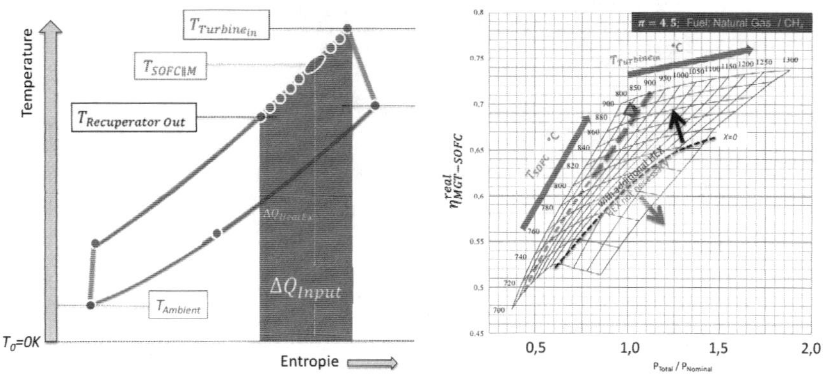

Figure 4. MGT-SOFC-cycle in a T-s diagram for explanation the X-Value (left) and TFC1.0 performance map with the X-value border line (right) [2]

Further, to clarify the heat balance on the SOFC stack, the definition of [1] was extended to the dimensionless temperature spread ratio across the cell. The dimensionless temperature spread of the SOFC cell is defined as follows:

$$\vartheta_{Spread} = \left(\frac{T_{SOFC\|Outlet} - T_{SOFC\|Inlet}}{T_{SOFC\|M}}\right) \tag{4}$$

With a fictitious heat capacity $\tilde{C}_p$, the relationship can be rewritten as follows:

$$T_{SOFC\|M} = \frac{1}{\vartheta_{Spread} \cdot \dot{m}_{overall} \tilde{C}_p} \cdot \left\{ \left[\left(\dot{m}_{LowSyn} \cdot \overbrace{\Delta H_{LowSyn}}^{**)} \right) + \overbrace{(\dot{m}_{Air-O2} \cdot \Delta H_{Air-O2})}^{cathode\ outlet \Rightarrow HEX} \right]_{SOFC-Outlet} - \left[\left(\dot{m}_{HighS} \cdot \underbrace{\Delta H_{HighS}}_{*)} \right) + \underbrace{(\dot{m}_{Air} \cdot \Delta H_{Air})}_{cathode} \right]_{SOFC-Inlet} - P_{el\_SOFC} - \dot{Q}_{heat\ loss} \right\} \tag{5}$$

*including formation enthalpy of the: **)High Syngas from the pre reformer or reformer ⇒ SOFC,*
**)LowSynGas ⇒ to the catalytic or conventional afterburner*

$$\dot{m}_{overall} = \underbrace{\dot{m}_{Air} + \dot{m}_{H_2O} + \dot{m}_{Fuel}}_{ref.outlet} = \dot{m}_{HighS} + \dot{m}_{Air} = \dot{m}_{LowSyn} + \dot{m}_{Air-O2} = (1-\beta) \cdot \dot{m}_{Air} + \dot{m}_{H_2O} \tag{6}$$

4. MGT-SOFC map extension by introducing a SOFC stack internal reforming

Internal reforming is a possible method for increasing the efficiency of a SOFC-System. No separate reforming apparatus is needed, a more even hydrogen formation and a higher methane conversion ratio are possible. But, for the reforming, catalyst deactivation reactions are possible and the combination of the electrochemical reaction and the reforming in one cell can reduce the flexibility in the operation of the fuel cell [9], [10], [17], [18]. With a combination of pre- and direct reforming systems, the requirement of a higher and even hydrogen production should be solved. However, the balancing of such a system needs more time and is a task of future development.

In [1], this mean temperature was given to allow a representation in a map (Figure 4 right). Here, the dimensionless power

$$\frac{P_{Total}}{P_{Nominal}} = \frac{(P_{MGT} + P_{el\_SOFC})_{Total}}{(P_{MGT} + P_{el\_SOFC})_{Nominal}} \approx \frac{\dot{m}_{Total\_T}\left(\eta^{real}_{MGT\_T} + \eta^{real}_{SOFC\_T} \cdot (1-\eta^{real}_{MGT\_T})\right) \cdot \eta_{Sys\_T}}{\underbrace{\dot{m}_{overall\_N}\left(\eta^{real}_{MGT\_N} + \eta^{real}_{SOFC\_N} \cdot (1-\eta^{real}_{MGT\_N})\right) \cdot \eta_{Sys\_N}}_{overall\ efficiency\ in\ the\ design\ point\ \approx 70\%}} \tag{7}$$

is plotted against the overall efficiency as a function of the mean SOFC temperature and the turbine inlet temperature.

The map shows that for an uncooled micro turbine (max. $T_{Turbine-in}$, the green limit curve) the x-value is in a range of 0,9 in the MGT-SOFC design point. The boundary line x = 0 is shown also. In the area below this curve, the MGT-SOFC process can perform without a high-temperature heat exchanger. Above this curve, a high temperature heat exchange is required [2], [19].

For the practical SOFC application of the planned TFC 1.0, a steam reforming process has been established [9], [15]. Therefore, this boundary condition was also selected in the work plan. It is mentioned that, in principle, the partial oxidation and the autothermic reforming can also be used for the generation of synthesis gas (HighSyn Gas).

If the SOFC stack allows a high temperature spread ϑ_{Spread} or if the average operating temperature $T_{SOFC\|M}$ is lower, less heat has to be exchanged at high temperature.

However, a high spread-ratio means higher thermal stresses in the stack, which is negative for the lifetime. A low SOFC operating temperature $T_{SOFC\|M}$ would be very good. However, here the problem, that the oxide ions conductivity decreases (border temperature 650°C), can occur. An inner reforming can help to lower the thermal load. Since less cathode air is required for internal stack cooling, system performance increases as more stacks are used (higher power density of the TFC1.0).

5. Heat balance through a coordinated distribution between the reforming process inside and outside the SOFC stack

From the field of possible MGT-SOFC processes (by taking into account the practical engineering conditions) the presented TFC architecture is favoured and further optimized. In addition to the realization of a very high efficiency, in particular the service life, the costs and a high power density are in focus.

A thermally adapted reforming process in the SOFC stack reduces the temperature increase on the anode side by heat removal. Decrease in the temperature distribution across the stack reduces the thermal stresses. As a result, the reformer is mutating into an adapted pre-reformer. This balancing process hast to be adjusted carefully, so the temperature cycle profile is exegetically well balanced and the operating temperature of the SOFC does not fall below the limit temperature.

Due to the internal reforming, the mass flow on the cathode side can be reduced because the required stack cooling is lowered. The resulting effect leads to a shift of the generated energy between the SOFC and MGT in favour of the SOFC. This means getting a higher relative system performance because of the lower excess air. With the same turbo-set the overall performance becomes better. Studies, calculated with the simulation system Ebsilon [12], also show that the system is experiencing a further increase in efficiency.

It has been determined by calculations that in a MGT-SOFC process, the amount of energy generated by the SOFC, with identical MGT, can be doubled. In addition, internal reforming reduces the energy required for the reformer. With the same MGT, the required reformer performance is almost identical and this with almost twice the output power of the entire process. In Figure 5 and Figure 6, both processes are presented as an example with an indication of the energy flows relative to the generated electrical energy (EP).

In the pictures, the dimensionless component power is shown in blue and the dimensionless cycle process temperatures are shown in green. The dimensionless power is built by the total power of the TFC and the dimensionless temperature is formed with the turbine inlet temperature. The following definitions apply:

$$\mathfrak{P} = \frac{|\pm P_{component}|}{P_{System}} = \frac{|\pm P_{component}|}{(P_{MGT} + P_{el\_SOFC})_{System}} \equiv Blue\ numbers \tag{8}$$

$$\vartheta_{i\_cycle} = \left(\frac{T_{i\_cycle} - T_0}{T_{Turbine\_in} - T_0}\right) \equiv Green\ numbers\ ,\ with\ T_0 = 273{,}16\ K \tag{9}$$

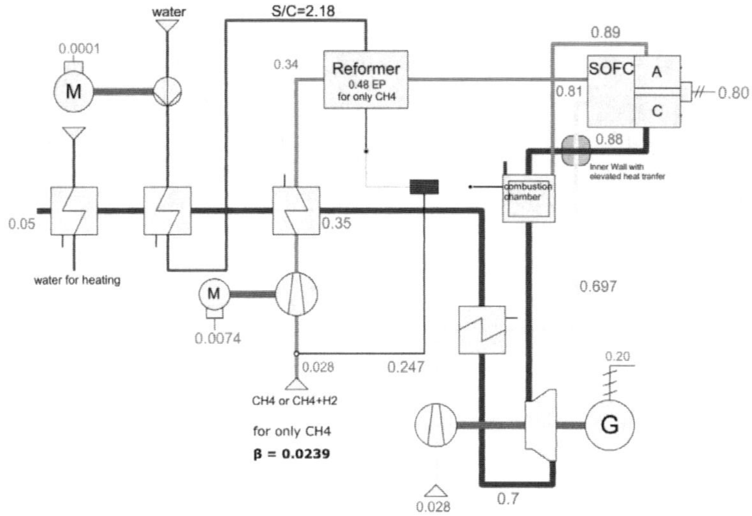

Figure 5. MGT-SOFC with external reforming

The bypass in Figure 6 is used to regulate the hydrogen concentration in the anode inlet for CH_4 / H_2 gas mixtures at the primary inlet, since the pre-reformer otherwise reformed too much of the CH_4.

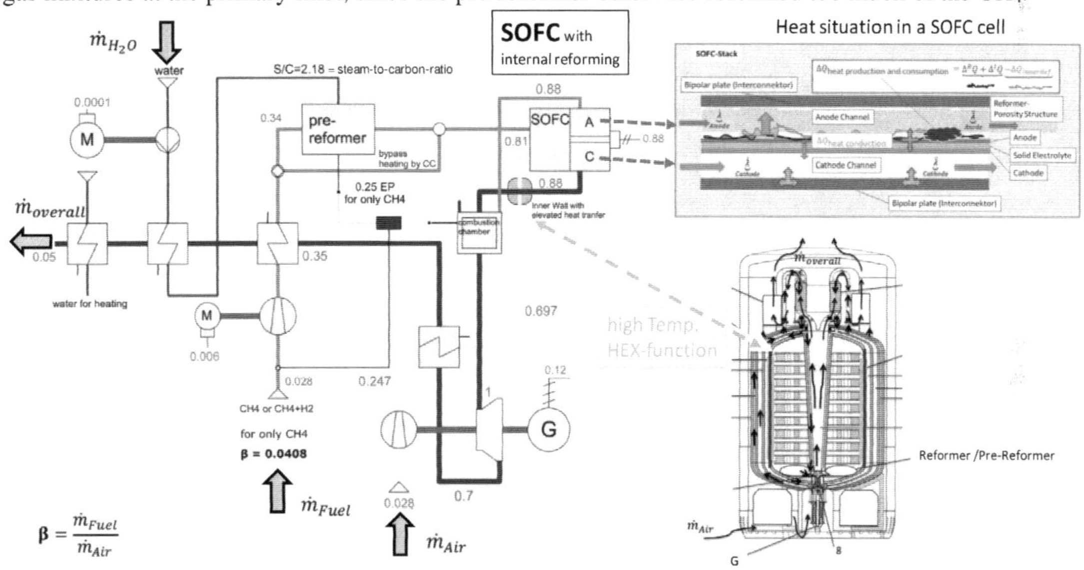

Figure 6. MGT-SOFC with internal reforming TIT(Turbine Inlet Temperature)

6. Heat situation and heat balancing in a SOFC stack with external and internal reforming

Heat Transfer problems can be described by dimensionless number models [11].

In SOFC-cells, oxygen ions flow through the membrane and affect the heat transfer processes in the interfaces of the cathode and the anode. On the cathode side, the heat transfer compared to a channel without membrane is larger *(*depending on the oxygen-ion flux*)*. At the anode, the reverse case occurred *(*the heat transfer decreases with increasing oxygen-ion flux*)*. This influence on the heat transfer can be described by the Graetz-Number. It is the reciprocal of the known Fourier-Number. It expresses the ratio of the convective heat transferred $\breve{Q}$ to the heat dissipated $\bar{\bar{Q}}$ and can be described by the Peclet-Number and the dimensionless channel length as follows [11]*:*

$$Gz = \frac{\check{Q}}{\bar{\bar{Q}}} = {}^{1}/_{Fo} = Pe \cdot \frac{1}{\tilde{x}} \tag{10}$$

with $\tilde{x} = \frac{x \cdot S_u}{4A} = \frac{x}{d_E}$ (S_u=scope, A = area, x= channel coordinate, d_E = equivalent diameter)

A description can be done by modifying the Pe-Number in such a way that there is a simplified modification via the Re-Number with a Pr-Number dependent only on the flow material properties as a function of x/l_c (l_c = channel length). The result is approximately the following relationship:

$$Pe(x) = \underbrace{\mathbb{R}e(x)}_{Correction}\ \mathrm{Pr}(x) \approx \mathbb{R}e\left(\frac{x}{l_c}\right) \cdot \left(\frac{4\kappa_E}{9\kappa_E - 5}\left(1 - \frac{x}{l_c}\right) + \frac{4\kappa_O}{9\kappa_O - 5} \cdot \frac{x}{l_c}\right) \tag{11}$$

For the Reynolds number, a correction term is used on the membrane wall side.

Here, the cumulative oxygen ion mass (proportional to the current flow) per channel region is introduced in relation to the channel mass flow and as a function of the running length.

$$\mathbb{R}e = Re\left(1 + \breve{C}\right) = \frac{u \cdot d_E}{v}\left(1 + \breve{C}\right) \tag{12}$$

$$\breve{C} = \frac{\sum \dot{m}_{o^-}}{\dot{m}_{Cannel} \cdot l_c} x \text{ with } \breve{C} = \breve{C}_{Anode}, \breve{C} = -\breve{C}_{Kathode} \text{ and } \underbrace{\breve{C} = 1}_{other\ channel\ sites}$$

It applies to Graetz number for the anode side, the cathode side and the other channel walls:

a) Membrane side (anode and cathode *)):

$$Gz = {}^{1}/_{Fo} = \frac{d_K}{x} Re\left(1 \overset{*)}{\pm} \frac{\sum \dot{m}_{o^-}}{\dot{m}_{Channel} \cdot l_c} x\right) \cdot \left(\frac{4\kappa_E}{9\kappa_E - 5}\left(1 - \frac{x}{l_c}\right) + \frac{4\kappa_O}{9\kappa_O - 5} \cdot \frac{x}{l_c}\right) \tag{13}$$

b) The other channel walls:

$$Gz = {}^{1}/_{Fo} = \frac{d_K}{x} Re \cdot \left(\frac{4\kappa_E}{9\kappa_E - 5}\left(1 - \frac{x}{l_c}\right) + \frac{4\kappa_O}{9\kappa_O - 5} \cdot \frac{x}{l_c}\right) \tag{14}$$

The heat transfer through the membrane and the interconnectors can be calculated as a simple heat conduction process [11], [13]. For the heat transfer calculations well known relationships for the laminar pipe flow can be used. In this case, a relationship of Churchill and Oze [8], experimentally and numerically validated by Berg [7] is recommended.

Because the Re-Numbers of the flow channels in planar SOFC-Stacks are very low (Re = 1 to 50; Pr = 0,7), the x*-value

$$x^* = \frac{x}{D} \frac{1}{Pe} \tag{15}$$

is already in the area of the formed flow, so that approximately the Nu-Number is 3,65 and the heat transfer coefficient α only depends on the thermal conductivity as a function of the gas composition (anode or cathode site) which is in the channel a function of x [11]. With known gas composition at the inlet and outlet it can be written for the anode (A) and cathode (C) channel:

$$\underbrace{\alpha(x)}_{A,C} \approx 3{,}65 \cdot \frac{1}{d_E} \cdot \underbrace{\lambda(x)}_{A,C} = 3{,}65 \cdot \frac{1}{d_E} \cdot \left[\underbrace{\dot{\lambda}}_{A,C}\left(1 - \frac{x}{l_c}\right) + \underbrace{\ddot{\lambda}}_{A,C} \frac{x}{l_c}\right] \tag{16}$$

with $\dot{\lambda}$ = *thermal conductivity at the entrance* and $\ddot{\lambda}$ = *thermal conductivity at the outlet*

Whether the buoyancy effect and the layer position of a SOFC stack must be taken into account can be estimated using the known Gr-number (Grashof-number) of the SOFC-channels [11]. Here, it was found that the buoyancy effect caused by the temperature difference on the cathode side and the anode side causes no effect, since the Gr-number is negligible. Forming the Gr number with the difference in density between the educts and the products on the anode side, one must estimate this influence, since the Gr-number has a slightly negligible value. There is no significant effect on the cathode side.

The estimate for the anode side is for the vertical SOFC layers alignment. When horizontally aligned, the effect is much smaller.

The estimation is made with a further correction of the Gz-number for the anode side. An equivalent Re-number with the correction term $\mathfrak{C}_{Gr}$ with the Gr-number is used.

$$Gz = {}^{1}/_{Fo} = \frac{d_K}{x} \cdot \sqrt{\underbrace{\left| Re^2 \underset{E,O}{\pm} \mathfrak{C} \right|}_{Re_{equivalent}}}\left(1 \overset{Anode}{\mp} \frac{\sum \dot{m}_{o^-}}{\dot{m}_{Channel} \cdot l_c} x\right) \cdot \left(\frac{4\kappa_E}{9\kappa_E - 5}\left(1 - \frac{x}{l_c}\right) + \frac{4\kappa_O}{9\kappa_O - 5} \cdot \frac{x}{l_c}\right) \tag{17}$$

with the correction term $\mathfrak{C}_{Gr} \approx 0{,}4 \cdot Gr$ and with
$E = +$, for the condition "*anode entrance up*" and $O = -$, for the condition "*anode outlet up*" ,
because the products on the membrane side have a higher density than the educts

For the heat transfer calculations, known relationships for laminar pipe flow can be used again (see above).

Because the equivalent Re-Numbers of the flow channels are in a range of ~180, the Nu value of 3,65 can be assumed after a length of about 2,4% (in the case of the examined own SOFC stacks, Figure 7) and the above α(x) relationship can be combined.

As proven, the determination of the temperature distribution and thus the stress distribution is essentially dependent on the unique heat transfer coefficient, the heat conduction processes and the amount of heat released in the cell. This amount of heat is released on the anode surface. In a first approximation, it is dependent on the oxygen-ion flux, since a uniform distribution of the educts is assumed. In a realistic simulation, however, the exact composition of the gases have to be known. As a first approximation, however, the simplified approach is sufficient. If a reformation process takes place within the cell, the amount of heat required for this purpose must be deducted. For the heat production and the consumption (with Reformer) applies:

$$\Delta^R H \cdot (1 - \eta_{SOFC}^{rev} \cdot \eta_U \cdot \eta_I) - \Delta Q_{innerRef.} = \Delta Q_{SOFC\_inside} = \underbrace{\Delta^R Q + \Delta^I Q}_{\substack{reversible \\ irreversible}} \underbrace{- \Delta Q_{innerRef.}}_{anode\ site} \quad (18)$$

In the case of SOFC-Stack internal reforming, the anode side can be provided with a permeable porous catalyst material or it can also be done for the reasons of stability in a SOFC cell without internal reforming or on the cathode side. This would result in heat transfer conditions, which differ from the conditions in a simple channel. The heat flows in the porous layer can be relatively easily interconnected with the heat transfer processes in the fluid. This is dependent on porosity-value ψ. The same applies to the cathode channel, if e.g. higher mechanical stabilities and better heat transfer by means of a porous layer would be required in future cell constructions.

Channel-side porosities are an obstruction. Assuming the channel without porosity, at the same gas mass flow, the Graetz-number will increase depending on the equivalent diameter of the remaining flow area of the porosity. Since the Reynolds number is very low in this case, the Nusselt number after a short channel length has the value 3,65 [11], [13]. In a simplified model, the heat transfer processes can be imaged via a parallel-serial heat conductor circuit with the porosity material. For the solid material and the fluid in a porous reformer structure, a total thermal conductivity can be formed to a first approximation. This depends on the porosity ψ and for such a structure can be written in general.

$$\lambda_{total} \approx \mathscr{B} \cdot \lambda_{serial} + (1 - \mathscr{B}) \cdot \lambda_{parallel} = \mathscr{B} \cdot \lambda_{serial} + (1 - \mathscr{B}) \cdot \lambda_{parallel} \quad (19)$$

For λ_{Solid}-value of the solid material and the $\lambda(x)$-value of the fluid are determined $\lambda_{s,p}$-values for the parallel and serial interconnection. Inserted into the relationship, the adaptation factor B gives the value of the porous fissile material of the reformer. For a sponge-like structure the adaption factor is about 0,54 [6]. With this relationship and with the above heat transfer relationship, with a modified equivalent diameter, the heat transfer problem in a porous layer (e.g. reformer) can be easily described. The modified equivalent diameter is determined by the porosity of the layer. For the dimensionless channel length thus applies:

$$\tilde{x} = \frac{x \cdot S_u}{4A} = \frac{x}{d_E} = \frac{x}{4} \cdot \frac{\overbrace{A_V}^{\substack{specific \\ surface}}}{\underbrace{\psi}_{porosity}} \quad (20)$$

For the determination of the specific surface, models of the literature can be taken. [6]

In the heat balancing procedure, after the mean temperature of the SOFC has been determined, a heat balancing is carried out within the stack in such a way that the temperature gradient is low. Thereafter, the adjustment is made with the heat balancing of the cycle. This involves the coordination of the pre-reformer/internal reformer distribution and the adjustment of the HEX-function of the inner MLC wall.

IOP Conf. Series: Materials Science and Engineering **501** (2019) 012007 doi:10.1088/1757-899X/501/1/012007

7. Outlook

This publication gives an insight view into a research and development area of cycle process fuel cell tuning and optimization. This area is currently in its infancy and is part of the TFC 1.0 development strategy.

As shown in chapter 5, the entire cycle is treated in a multidisciplinary way and the calculation programs are coordinated. At the moment, this design process is not yet completed. The SOFC stack calculation module is in development. First results of the thermal simulation part can already be shown. In the first approach, the SOFC cell is assembled with different material in thin layers which have different in mechanical and thermal properties. Therefore, the temperature distribution of the SOFC in cell level and also in stack level are essential for stack design and implementation in each system.

The prediction of temperature distribution of the SOFC relates to complicate mass and heat transport phenomena in the cell level, which can be done using FSI technique *(*Fluid–structure interaction*)*. However, these methods are very complicated and time consuming and focused only on the cell level. To avoid the mechanical error due to thermal deformation, the simulation must cover the SOFC stack level. Therefore, a simplified FEM model of the thermal analysis stack has been developed specifically for this publication as an example. Parallel to this, various test stacks under charged conditions are examined in the MAXEFF test facility *(*Fig. 7*)*. Thus, a wide model validation can take place. As a case example, a stack of multiple cells is replicated for a steady state thermal FEM analysis using ANSYS. The stack is placed in the middle of a chamber surrounded with air with the Temperature $T_{Chamber}$. The dimensionless temperatures are defined as follows.

$$\vartheta_i = \frac{T_i - T_{Chamber}}{T_{Chamber} - T_{SOFC||Inlet}} \qquad (21)$$

On the cell level, heat transfer and heat generation is assumed, modelled in a similar way to a test stack *(*Fig. 7*)*. Convective transfer in the channels of cathode and anode sides are modelled using the coefficients in previous chapters. Thermal-Fluid-Elements simulate the gas flow in the channel and they are coupled to the channel wall surface to provide the gas temperature for the convection calculation in the channel. Specific heat generation on the anode surfaces is defined using the simplification of chapter 6. All thermal contacts are initially considered as ideal and the gas manifolds are neglected for the sake of brevity.

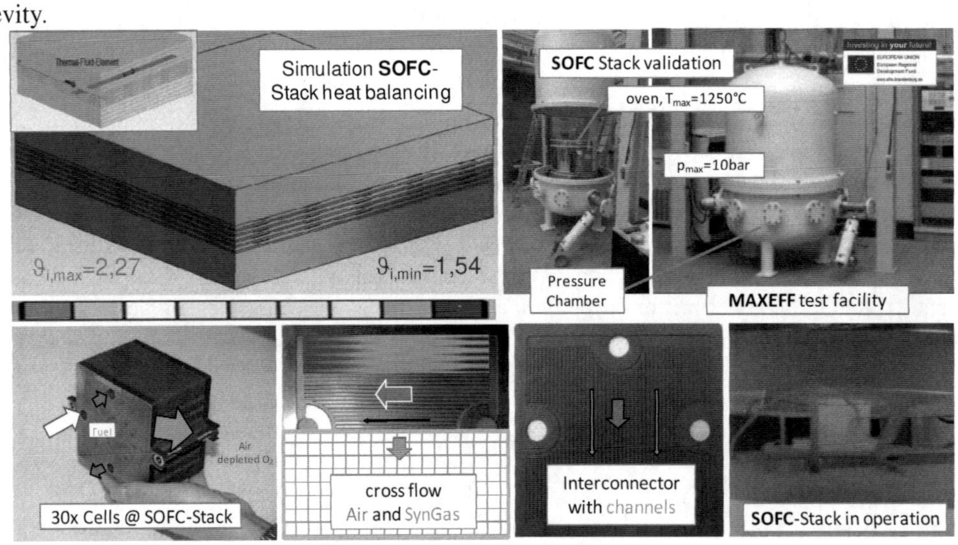

Figure 7. Test-facility at BTU-CS

The development of a detailed model of the fuel cell stack used is underway. Refinements in this simulation technique can be further improved with a feasible effort. Heat generation can be defined locally on the anode surface using external computation data. The thermal contact can be specified more precisely to enhance the simulation accuracy. Thermal radiation, which could have a significant impact, can also be implemented. By using a thermal fluid element model, the design of the flow channel orientation and flow parameters can be optimized to reduce stack deformation. This will lead to the next steps to precisely optimize the heat balance of a direct reforming fuel cell.

8. Conclusion

In this paper, the dependencies for the realization of a future MGT-SOFC system were shown and possibilities for further performance enhancement are presented. For this increase in efficiency, the heat balancing is an important aspect. For a preliminary design, a careful heat balancing of the cycle must be made. This has already be shown in [1], [2], [20]. In this case, the heat balancing between the reforming outside and inside the SOFC-Stack plays a role. It could be shown that with a good tuning a significant increase in performance with the same turbo set is possible. In addition, the overall efficiency of the hybrid process increases. Since a good matching is only possible with a further heat balancing of the SOFC stack, this process was also shown. The heat conditions in the stack were explained and first calculation results were shown. Furthermore, it has been proven that the Nu-numbers for the small Re numbers within the stack can be clearly defined. This provides further confidence in modelling to optimize the heat-balancing situation. The research and development activities are part of the realization strategy of the energy converter of the future (the Turbo-Fuel-Cell 1.0) [1].

References

[1] Berg H P, Himmelberg A, Lehmann M, Dückershoff R and Neumann M 2017 *The Turbo-Fuel-Cell 1.0 – family concept, Compact Micro Gas Turbine (MGT) – Solid Oxide Fuel Cell (SOFC) energy converters in the 100-500 kW electrical power range for the future* 8th TSME-International Conference on Mechanical Engineering (TSME-ICoME 2017) - IOP Conf. Series: Materials Science and Engineering 297 (2018) 012004

[2] Berg H P, Dückershoff R, Lehmann M and Prechavut N 2017 *MICRO TURBO-FUEL-CELL-TECHNOLOGY - Hybrid compact turbo machinery technology and thermodynamic aspects regarding design parameters of a high efficient MGT-SOFC-system* Proceedings of 12th European Conference on Turbomachinery Fluid dynamics & Thermodynamics ETC12, April 3-7, 2017; Stockholm, Sweden; Paper ID: ETC2017-266

[3] Prof. Dr. Berg und Kiessling GmbH (B+K GmbH), *https://www.bergundkiessling.com/*, Cottbus 2018 called 2018/09/29

[4] BTU Cottbus-Senftenberg, Lehrstuhl Verbrennungskraftmaschinen und Flugantriebe, Prof. Dr.-Ing. Heinz Peter Berg, *https://www.b-tu.de/fg-vfa/*, Cottbus 2018 called 2018/09/29

[5] Mitsubishi Hitachi Power Systems Global: *produkt information SOFC-MGT Hybrid Power Generation System, https://www.mhps.com/products/sofc/technology/*, YOKOHAMA, called 2018/09/30

[6] Dietrich B 2010 *Thermische Charakterisierung von keramischen Schwammstrukturen für verfahrenstechnische Apparate*, Dissertation Karlsruher Institut für Technologie Fakultät für Chemieingenieurwesen und Verfahrenstechnik, Karlsruhe 2010

[7] Berg H P 1991 *Experimentelle Bestimmung des örtlichen inneren Wärmeübergangs von Turbinenleit- und –laufschaufeln mit Hilfe der Analogie zwischen Wärme- und Stoffübergang*, Dissertation, Universität Darmstadt, 1991

[8] Churchill S W and Ozoe H 1973 *Correlations for Laminar Forced Convection in Flow over an isothermal Flat Plate and in Developing and Fully Developed Flow in an Isothermal Tube* Journal of Heat Transfer, Bd. 95

[9] Forschungszentrum Jülich *Reformierung von Erdgas, http://www.fz-juelich.de/iek/iek-3/DE/Forschung/BGE/Brennstoffzellenseiten/Reformierung/Reformierung%20von%20Erdgas/_node.html*, called 2018/09/20

[10] Shekawat D, Spivey J J and Berry D A 2011 *FUEL CELLS: TECHNOLOGIES FOR FUEL PROCESSING*, chap. 4 by Rostrup-Nielsen J R and Hansen J B: *Steam Reforming for Fuel Cells*, Elsevier, Amsterdam 2011

IOP Conf. Series: Materials Science and Engineering **501** (2019) 012007 doi:10.1088/1757-899X/501/1/012007

[11] Herwig H 2000 *Wärmeübertragung von A-Z*; Springer-Verlag, Heidelberg
[12] Steag System Technologies 2018 *Ebsilon Professional, User manual* Essen
[13] VDI 2006 *VDI-Wärmeatlas*, Springer-Verlag, Heidelberg
[14] Heddrich M P 2012 *Thermodynamische Analyse von SOFC-Systemkonzepten und experimentelle Validierung* Dissertation Fakultät für Natur- und Materialwissenschaften der Technischen Universität Clausthal
[15] Boder M 2000 *Katalytisch modifizierte Materialien für die interne Reformierung von Kohlenwasserstoffen in der Hochtemperatur-Brennstoffzelle* Arbeitsbericht zum DFG-Forschungsvorhaben DI 696/3-1 Karl-Winnacker-Institut der DECHEMA e.V., Frankfurt Main
[16] Ningbo SOFCMAN Energy 2017 *http://www.sofcman.com/index.html* Ningbo City, Zhejiang Province, PR.China
[17] Timmermann H, Fouquet D, Weber A, Ivers-Tiffée E Hennings U and Reimert R *Internal Reforming of Methane at Ni/YSZ and Ni/CGO SOFC Cermet Anodes* FUEL CELLS 06, 2006, No. 3-4, 307–313
[18] Dicks A L 1996 *Hydrogen generation from natural gas for the fuel cell systems of tomorrow* Journal of Power Sources 61 (1996) I 1.3-124
[19] Berg H P and Krienke C 2015 *Mikrogasturbinen-SOFC-Prozesse mit hohem Wirkungsgradpotential*, "12. MAGDEBURGER MASCHINENBAU-TAGE"-Universität Magdeburg 30.09. - 01.10.2015
[20] Yang C 2012 *Mass and Heat transfer processes in Solid oxide fuel cell (SOFC)* Project Report 2012 MVK160 Heat and Mass Transport, Lund, Sweden
[21] Patent DE 10 2017 107 003.6 *Exergieverlust reduzierender, mehrhäutiger Druckbehälter für den Einsatz von druckbeaufschlagten Hochtemperaturbrennstoffzellen*, Cottbus 2017
[22] Kurzweil P 2013 *Brennstoffzellentechnik - Grundlagen, Komponenten, Systeme, Anwendungen.* 2., überarbeitete und aktualisierte Auflage. Wiesbaden : Springer-Verlag ISBN: 978-3-658-00084-4

IOP Conf. Series: Materials Science and Engineering **501** (2019) 012008 doi:10.1088/1757-899X/501/1/012008

Design and testing functional model compacting machine for produce new shape biofuels

Ľ Šooš[1], J Bábics[1], J Beniak[1], Peter Križan[1], Pavel Kovač[2], and Miloš Matúš[1]

[1] The Slovak University of Technology, Faculty of Mechanical Engineering, Námestie slobody 17, Bratislava, 812 31, Slovak Republic
[2] University of Novi Sad, Faculty of Technical Sciences, Trg D. Obradovica 6, 21101 Novi Sad, Serbia

* Corresponding Author: lubomir.soos@stuba.sk

Abstract Compaction technologies have been known for over 130 years. Among the best-known compaction technologies are included briquetting, pelleting and compacting. The end products of such technologies are briquettes, pellets and granules. Recent times have seen a worldwide "boom" in the use of these refinings, especially with regard to pellets.
One great advantage of pellets in comparison with classic fuels is that according to need the three decisive parameters '***shape, size and density***' of the resultant fuel can be changed during production. This leads to an optimization of costs for their production and storage and transportation requirements, but most of all, prerequisites in terms of optimal energy recovery.

One of the most discussed issues in the production of refined biofuels is the shape and size of the refined fuel. The quadrangle, n-angle or cylinder do not have optimal shapes from the point of view of automated transport in the fuel combustion process. From this standpoint the ball is the optimal shape, but it is not possible to create such pressings in a continuous process and present-day pressing and compacting machines lack a "fuel life" phase under pressure with simultaneous cooling. The endurance and cooling process has of course a great influence on the stability of the fuel produced.
The goal of the submitted paper is the development, design and testing of a new, patent-protected compacting machine construction which will allow the production of pressings biofuels of an almost circular shape. In contrast with readily available compacting – compression machines, such a compactor also provides cooling of the pressings under high pressure.

1. Defining the problems

If biofuel is to be a full-value substitute for fossil fuels, it must fulfill three basic criteria – environmental, energy and economic.

The beginning technologies for the adaption of organic waste for its later energy recovery, is its compaction under high pressure and temperature without the addition of a binder. By using these technologies, it is possible to gain refined fuel in an optimal shape and state suitable for transport, storing and later combustion. Among the best-known compaction technologies belong briquetting, pelleting and compacting, [1].

Briquetting technology is the oldest of the three. At a given time, there is usually created just one briquette. The briquettes are relatively large, (minimal characteristic dimension 30 mm) of a cylindrical, cuboid or n-angular shape. They have a smaller ratio of surface to volume and so burn more slowly and for a longer time. They are suitable for fireplaces and hand-supplied furnaces. For larger surface to volume ratios, briquettes with larger dimensions (characteristically dimension > 80 mm) are produced with openings, [1].

With progressing automation in the combustion process, it is necessary to harmonize the factors that affect this process. Modern 21st century fuels must, in addition to the mentioned factors, also provide comfort. This is the major reason why, in recent times, the pelleting technology has been the most dynamically expanding compaction technology. The main advantage of this technology is the possibility of a fully automatic process of fuel delivery during the combustion process, even for small family furnaces. The cylindrical shape of the fuel can present certain difficulties. For problem-free feeding from the supplier to the furnace, the minimal diameter of the conveyor should be four times longer than the length of the pellet. A second factor is the wearing of the machine equipment and its parts since the rate of matrix and roller wear is given by the surface to volume ratio of the fuel. Pellets have a high ratio and also the wear rate of the contact parts of the machine and openings in the matrix is high. Combustion rate of the fuel is another factor. Pellets are characterized as having a high surface to volume ratio. Consequently the pellet surface quickly oxygenates and therefore also burns quickly, [3].

Of interest too is the technology of compaction during which a material with the required fractions and dampness is compacted between two reverse-rolling smooth or grooved cylinders are pressed against one another fig. 1. The results of compaction are granules or compressed strips that can be used for further processing, particularly in the metallurgy and chemical industries. The almost round shape of briquettes is of course suitable from the standpoint of automatic feeding, but the technology is not suitable for pressing waste from wood, because under pressure it lacks the maintaining phase. In this phase cooling of the briquette must take place, and as a result the binder, lignin, moves from the plastic to the solid state, by which the briquette acquires the required solidity, [3].

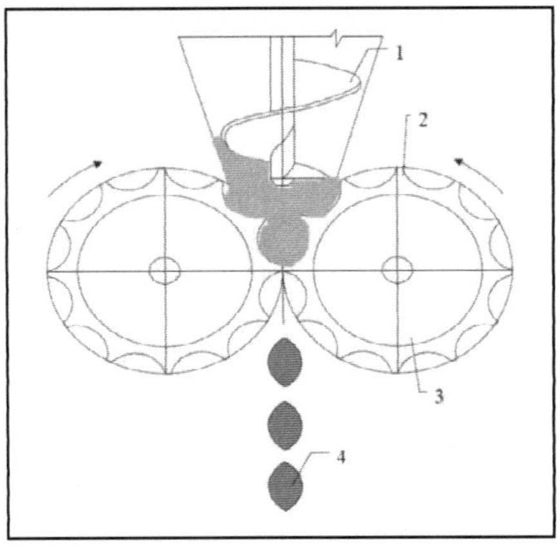

Fig.1 – Scheme of compacting machine.
1-feeding screw, 2- grooved bandages,
3- opposing cylinders, 4-granule

A very significant factor of modern fuel is a suitable shape for automatic feeding of the fuel to the furnace.

For the already mentioned problems with automation in the process of combustion, space has been created for producing a convenient shape of fuel from the standpoint of automatic feeding. It is currently desirable to maintain the suitable shape of a briquette with a low surface to volume ratio. Such a newly-designed must also have a maintenance phase for cooling the briquette under pressure, by which the need for an additional binder, a disadvantage to compacting, is eliminated.

2. Briquette shape

The concept of an ideal briquette arose from a comparison of the properties, advantages and disadvantages of the briquettes of individual technologies. It is an attempt to create a briquette that would contain the advantages of pelleting, briquetting and compacting while at the same time eliminating the disadvantages of the individual technologies. A new briquette shape would be produced on the principle of compacting, which means being pressed between two rollers. In contrast with compacting, the new technology would have to allow for the briquette to be for a certain time under pressure, by which cooling would occur, as is common in pelleting and briquetting. Therefore, there arose the need to design a shape and, following that, a compactor machine that would manage to make new-shaped briquettes.

Briquetting **Pelleting** **Compacting**

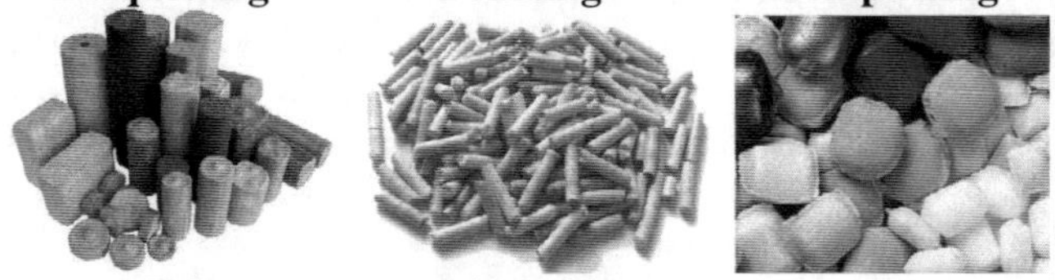

Briquettes were valuated by:

- briquette burning process
- automatization of combustion process
- machine wear
- briquette storage and transit

Fig.2. Scheme of idealized granule, [4]

On the basis of the input requirements and later theoretic analysis we created the almost-round briquette, which is designed as the intersection of two half cylinders each rotated through 90°and is suitable for automatic feeding. With such a briquette it is possible to regulate the ratio of the surface to its volume. The edges at the intersection of the two cylinders allow optimal lighting of the briquette, [4].

3. Kinematic

After the design of a suitable shape for the briquette, we continued with a combination of potential solutions and to the design of a kinematic compactor that would be capable of producing such briquettes. The result of this combination was a proposal for a completely new principle "Ring compaction machine". In 2010 we received a granted patent SK 287505 B6, [2] for the stated principle. The invention belongs to the area of compaction of bulk organic and inorganic materials.

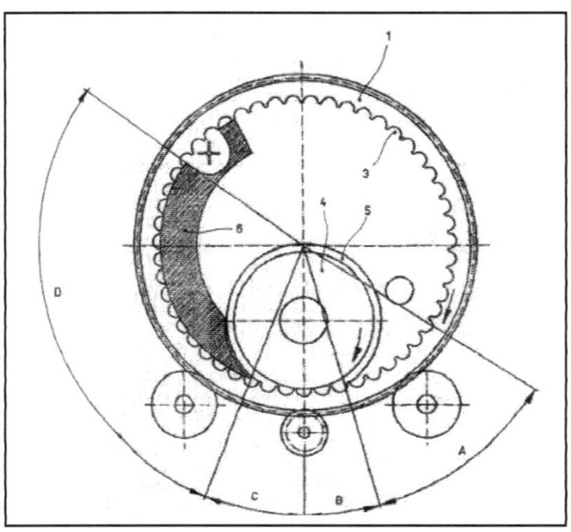

Fig. 3 - Main arrangement of the ring press with interior press disc

The essence of the new patented solution is large ring 1 which has in its interior circumference a great number of slots 3 in semi-cylindrical shape whose axes are parallel with the axis of rotation of the ring. The compression instrument is also disc 4, which on its outside circumference has a cylinder slot 5, which fits into the slots of ring 1 and is turned 90° from the slots of the ring. So both the ring and the compression disc have independent movement. The hold-off stage under the pressure at which the molds are cooled is ensured by the calibration struts 6. In stage A there takes place the filling of the press chambers that appear during rotation, in stage B compression, in stage C partial expansion and finally in stage D gradual cooling of the briquette under pressure, and its calibration. The effect of ´ cooling of the binder brings the binder (lignin) from the plastic to the firm state, by which the briquette achieves the required rigidity, [2].

4. Design and production of functional model

A further task arising out of the work on the overall project was designing the compacting machine itself. At the beginning, an analysis of kinematics of motions and the combination of possible structural solutions was carried out. In order to achieve the required briquette shape and kinematics of motions we designed a functional model of the compacting machine as shown in figure 4.

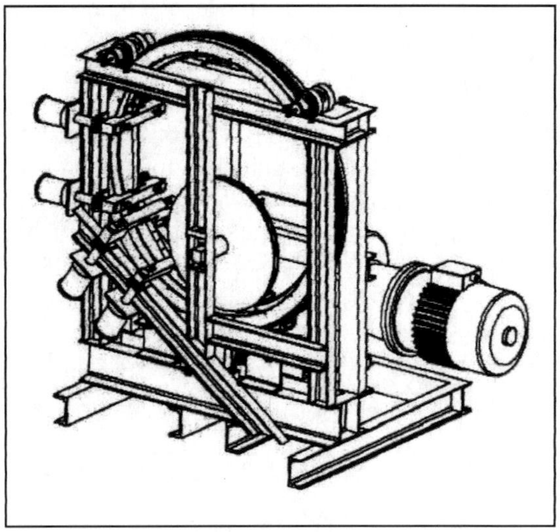

Fig. 4 - Main arrangement of the ring press with interior press disc

Design of force elements was based on a theoretical analysis of the forces calculated during the densification process. On the basis of such forces, calculations were made for the required input of the main drive, input of the drive of the compacting disc and the drive of the feeding screw.

Another very important task related to designing the functional model was testing of strength and shaft sag of the compacting disc, as shown in figure 5 and the overall solidity of the frame, figure 6. In figure 5 we can see the area of the maximum shaft sag indicated in red. This is the area directly under the compacting disc. The maximum shaft sag is 0.303 mm in the direction of the loading force. In figure 6 the largest deformation of the frame in a direction of pressing force can be clearly seen. The largest shift is displayed in the upper part of the frame, 0.566 mm.

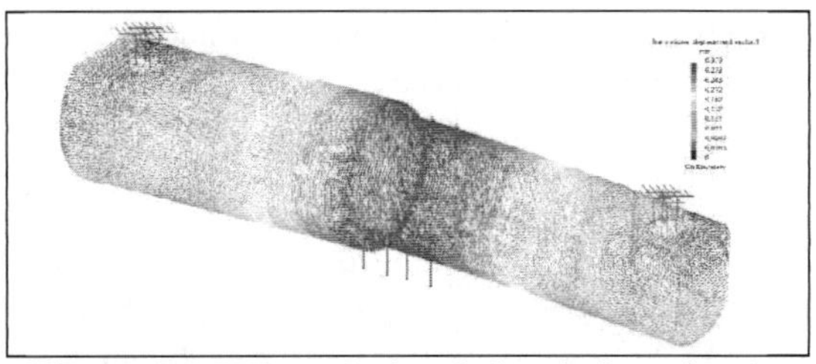

Figure 5 MKP analysis of the shaft under the compacting disc, [5].

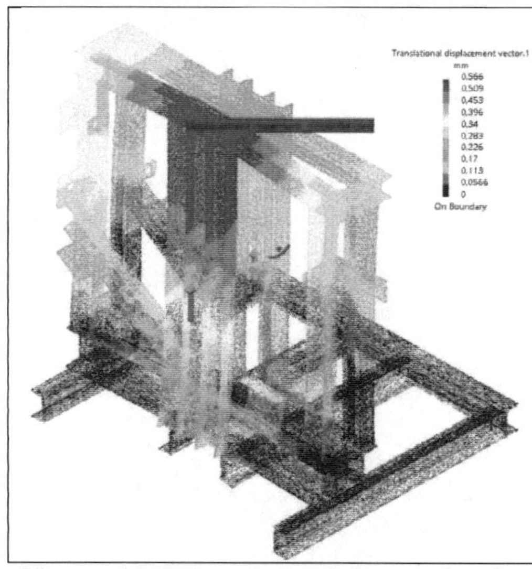

Figure 6 MKP analysis of frame of a ring compacting machine with disc, [5].

The final task in this part of the project was to prepare the manufacturing sketches and the final manufacture of the functional model, figure 7.

5. Testing of functional model

In this part of the task two basic types of testing were performed on the functional model. The objective of the first set of measurements was to determine the required density of a briquette on the functional model. The second very important task in this experiment was to determine the maximum pressing force for the set density of the briquette.

5.1 Setting of briquette density

The foundation of this testing was to determine the required volume of the briquette $V\ [m^3]$ and the assumed theoretical density of the briquette $\rho\ [kg.m^{-3}]$. With a calculated volume of the briquette of $V=1.531.10^{-6}\ m^3$ and a set theoretical value of density $\rho=1{,}000 kg.m^{-3}$ the weight of one briquette will be $m=1.531.10^{-3}\ kg$. For those values we subsequently pre-set the frequencies of revolution of the feeding screw while maintaining the revolutions of ring and compacting discs at a constant level. Subsequently,

IOP Conf. Series: Materials Science and Engineering **501** (2019) 012008 doi:10.1088/1757-899X/501/1/012008

we conducted 3 sets of measurements, while gradually raising the revolutions of the feeding screw. For each measurement, 10 random briquettes were selected and measured, as shown in Table 1.

5.2 Determination of pressing force

First of all, it was necessary to calibrate the strain gauge. We loaded the shaft under the compacting disc of the hydraulic compactor DHL 15, made a reading of the measured force from the IMADA dynamometer and subsequently, we allocated the respective tension to the value of the force, figure 9. We repeated the process of calibration four times and subsequently made an average of the values; all the measured values can be seen in the figure below.

Following the measurement, we moved the calibration curve through the average dependency, as can be seen in figure 10. We mounted the calibrated shaft onto the device and began the process of compacting sawdust. The courses of tension were recorded by means of an Omega ON-DAQXL device.

Figure 7 The actual structure of a ring compacting

n=30 min^{-1}	n=35 min^{-1}	n=40 min^{-1}

Figure 8 The quality of briquettes at various feeding screw revolutions.

Table 1 Weight of briquettes

Weighted briquette	Weight (g) n=30 min^{-1}	Weight (g) n=35 min^{-1}	Weight (g) n=40 min^{-1}
1	1.164	1.341	1.356
2	1.247	1.389	1.192
3	1.371	1.251	1.346
4	1.276	1.319	1.375
5	1.352	1.295	1.346
6	1.194	1.249	1.356
7	1.293	1.372	1.381
8	1.316	1.298	1.329
9	1.24	1.382	1.382
10	1.42	1.146	1.361
Average weight	**1.2873**	**1.3042**	**1.3424**

Each measurement provided the output in the form of a text file, including the respective values that were then processed in Excel, thus identifying the maximum measured value. In the third measurement, we achieved a maximum tension of 1.95 V at a briquette density of ρ = 876.51 *[kg.m^{-3}]*, figure 11.This value was inserted into the equation of the trend line of the calibration diagram. By inserting the values in the equation *y=60.241x-2.3084 [kN]* the final values of the maximum pressing force were achieved, Table 2.

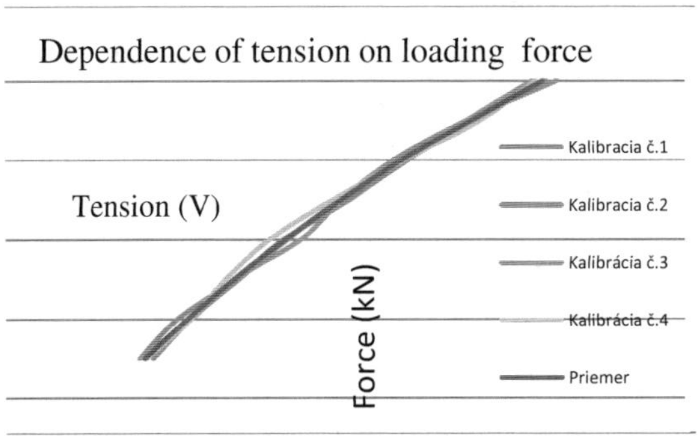

Figure 9 Dependence of loading force on strain gauge tension

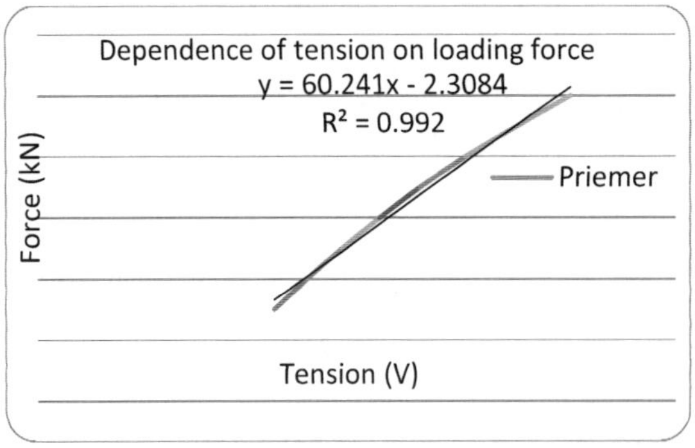

Figure 10 Average dependence of loading force

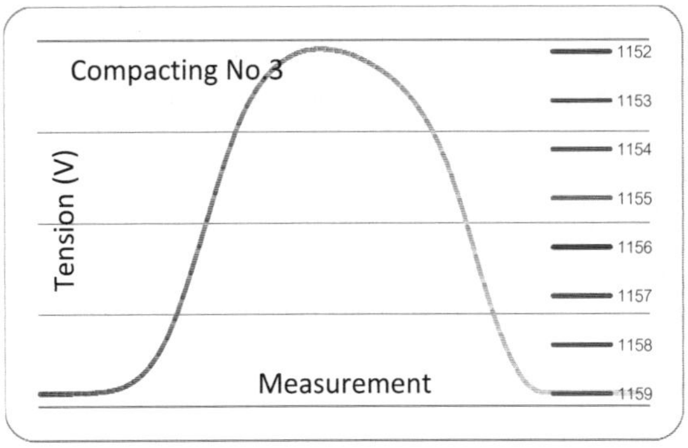

Figure 11 Dependences of tension during pressing, No. 3

Table 2 Measured maximum pressuring force

Average density *ρ* *[kg.m^{-3}]*	840.27	851.57	876.51
Measured tension *[V]*	1.809	1.948	1.951
Maximum pressing force *F [kN]*	108.9	115.05	115.22

6. Discussion and Conclusion

The goal of the presented paper is the design and function testing of a new progressive conception of a compaction machine for the production of a new optimised shape for refined biofuels. In their design for a new briquette shape and the design of a machine for their manufacture, the authors started out from a wide-ranging analysis of the advantages and disadvantages of known compacting technologies. For the new shape we created a principally new kinematic compacting press, designed and made a functioning model for testing the principle and attempted the production of new briquette shapes. There is still a lot of work to be done to optimise and perfect the construction. It will be necessary to perform a number of verification calculations and measurements, synchronize the rotation speed of the filling screw with the speed of the ring and press disc in order to get the optimized amount of raw material into the press chamber, and therefore the optimal briquette density. Work still has to be done on the easy removal of the briquettes from the ring as well as on the device's performance when managing several rings next to each other. The first experimental and function tests have already shown the justification of the new construction and the correctness of the path taken.

Acknowledgements

The research presented in this paper is an outcome of project No. APVV-16-0476 "Research and development of the progressive design of the high speed rotor mounting in spinning machines" funded by the Slovak Research and Development Agency.

References

[1] Ľ. Šooš, S. Korgo, The Effect of Wood Humidity on Briquet Quality, *5th Conference on Environment and Mineral Processing (part. 2)*, pp. 577-583, 2002

[2] Ľ. Šooš, Kontinuálny spôsob lisovania biomasy do optimálnych výliskov a zhutňovací stroj, Patent number: SK 287505. – Date of acquisition: 14. 10. 2010

[3] Šooš, Ľ. - Matuš, M. - Beniak, J.- Križan, P.: Development of the compaction machine for the production of new shapes of pressed biofuels. In: 8th TMSE ICOME 2017, 297 (1/2018) doi:10.1088/1757-899X/297/1/012008.

[4] Blaško, P.: Progressive construction of an annular pellet machine, Diplom project, STU Bratislava, 2018, 58 p.

IOP Conf. Series: Materials Science and Engineering **501** (2019) 012009 doi:10.1088/1757-899X/501/1/012009

Investigation of motorcyclist and pillion passenger injuries using numerical simulations

H. Sudyoddee[1], M. Behr[3], M. Llari[3], S. Koetniyom[1,2] and J. Carmai[1,2*]
[1] The Sirindhorn International Thai-German Graduate school of Engineering (TGGS)
[2] Automotive Safety and Assessment Engineering Research Centre
King Mongkut's University of Technology north Bangkok, 1518 Pracharat 1 Rd., Bangkok, 10800, Thailand
[3] IFSTTAR, LBA UMR T24, Aix Marseille Universite', Marseille, France

[*] Corresponding Author: julaluk.c@tggs.kmutnb.ac.th, +6625552000 ext. 2913

Abstract. Road traffic fatalities involving powered-two wheeler (PTW) are the highest portion in Thailand. Few studies are available concerning the head impact loading and head trauma in case of real PTW accidents. This work, therefore, aims to improve the understanding on kinematics and head trauma of PTWs rider and pillion passengers. By selecting a real-world PTW accident provided by the Central Institute of Forensics Science database, reconstruction of the accident was performed using multibody dynamics simulations. Parametric studies which focused on vehicle speed and position of rider and passenger at impact was conducted to obtain the best correlations. Then, head impact velocity, impact direction and head impact area have been extracted and implemented in the finite element head model to simulate the head trauma by impacting directly the head model on the vehicle structure. The motorcyclist's head and the pillion passenger's head impact conditions were retrieved and used as initial condition for the finite element head impact simulation. Head and brain injuries were analyzed.

1. Introduction

PTWs fatality rate in Thailand remains very high among the other type of vehicles. It has been reported that 73% of traffic deaths involved powered-two wheelers [1]. Head injury is the major cause of death [1]. The mechanisms and characteristics of injuries sustained by the PTW casualties are research interest to support improvement of countermeasures and protective system. Many studies have focused on a single rider [2] but in real accidents pillion passengers are also among the casualties. Limited studies are available concerning the head impact loading and head trauma for real PTW accidents. The objective of this study is to improve understanding of kinematics and head trauma of PTW rider and pillion passenger by using a combination of two computer simulation tools. Computer simulations of PTW-to-car collision provide an effective approach for studying injury mechanisms of motorcyclist and pillion passenger [2-4]. There are various literatures which have used multibody simulations to reconstruct the accidents [2-5] and study pedestrian kinematics [5]. Multibody dynamics simulation is not computer intensive and therefore a good tool for trend or parametric studies. However, the rigid body models cannot give details of the injuries suffered by human body. Finite element simulation can provide insight into the injuries of human body but it is time consuming. A combined approach of multibody dynamics and finite element simulations can be an effective way to study both rider and pillion kinematics as well as head injuries.

2. Methodology

Figure 1 illustrates the methodology used in this study. Combined approaches of multibody simulation and finite element simulation were employed to investigate the kinematics and head injury of both motorcyclist and pillion passenger involved in a real world PTW accident. A multi-body system generally consists of rigid and flexible bodies joined together by kinematic joints or force elements. It

IOP Conf. Series: Materials Science and Engineering **501** (2019) 012009 doi:10.1088/1757-899X/501/1/012009

is sufficiently flexible to construct a multi-body model of a human body or vehicle with various kinematic joints and discrete bodies of particular size and shape. In this study multibody simulation was used for accident reconstruction and parametric studies. Three parameters were specified including car speed, PTW speed and impact angle. With parametric studies, an optimal impact configuration for the selected PTW accident case was obtained. Head impact velocity was extracted and adopted as boundary condition for finite element human head impact simulation. The human head consist of a deformable skull and brain which can facilitate the analysis of head traumatic injury of the rider and pillion passenger.

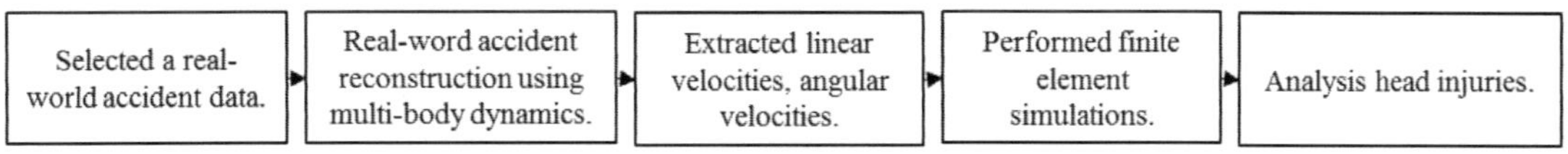

Figure 1. Research methodology diagram

3. A real-world accident reconstruction and parametric studies

3.1. *The real-world accident data*

A real-world accident data provided by the Central Institute of Forensics Science database was selected for the reconstruction. The accident involved both rider and pillion passenger. The report [6] presents evidence where a passenger car model Toyota Corolla Altis collided with a PTW (Honda series Nova 125 RS Sonic). The final resting point of the PTW and the position of the passenger car are shown on the map in figure 2. The PTW did not decelerate before collision. There were two PTW victims. The motorcyclist was 31 years old, 1.85 m tall. The medical document reported that rider was seriously injured around the face, lacerated in the region of right ear lope, mouth, neck, right abdomen, and left hand. In addition, his femurs were broken on both sides as well as a fracture of the right wrist. He died at the scene with his body tangled with the PTW. Adult pillion passenger had serious injuries and was taken to hospital. Figure 3 shows the impact location on the car. The left door deformed due to the PTW front wheel impact as illustrated in figure 3(a). The top corner of the windshield cracked due to the head impact as shown in figure 3(b). According to figure 3 (c), the PTW front wheel deformed significantly.

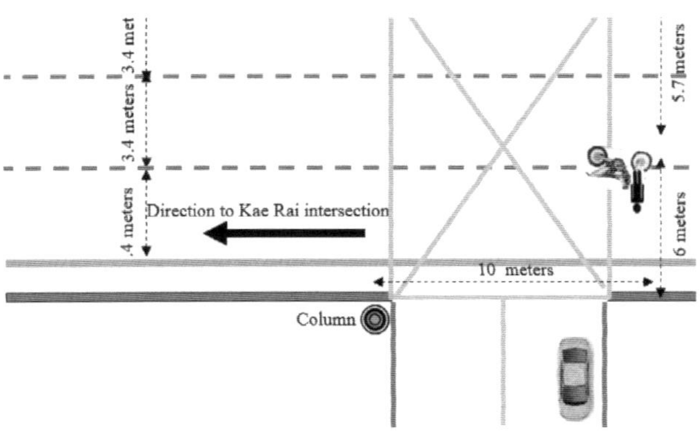

Figure 2. The map shows that the rest point of the PTW and the parking of the passenger car.

(a) (b) (c)

Figure 3. Illustrates (a) deformed door (b) windshield crack (c) deformed front wheel and handle bar

3.2. *Multi-body model for PTW accident reconstruction*

The accident scenario can be reconstructed in 3D space by establishing the model of the car, PTW and human bodies together with their movement and the load of the rider and passenger. MADYMO (mathematical dynamic model) is a multi-body dynamics software that was employed for the reconstruction. It can simulate the dynamics response and study the behaviour of interconnected rigid and flexible bodies. It is a common reconstruction tool found in literature [2-5]. The multi-body model of the car and PTW in accident cases were set up according to their real shape and mass in MADYMO. The rider and pillion passenger models were adopted from a previous investigation [4].

3.2.1. Rider and a pillion passenger multibody dynamics models

A Hybrid II 50th-percentile-male anthropomorphic test device (ATD) represented the motorcyclist consisting of 74 ellipsoids as show in figure 4. The rider was validated [4] and was modified for appropriate use in this study. The height was 1.85 m and the estimated weight was 70 kg. The adult pillion passenger was a Hybrid III 5th-percentile-female ATD that was made of 74 ellipsoids as shown in figure 4. The weight of 70 kg and height 1.76 m were estimated due to lack of information on the pillion passenger.

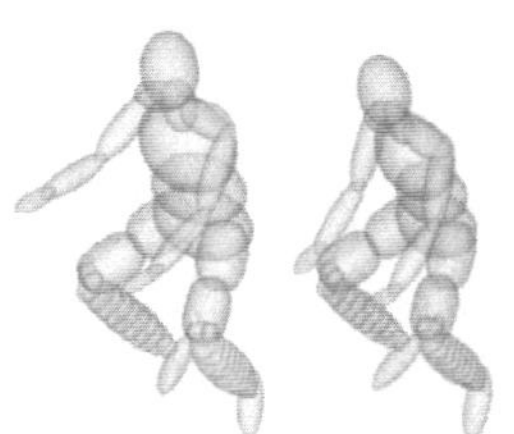

(a) *(b)*

Figure 4. The multi- body system of the motorcyclist (a) and the adult pillion passenger (b).

3.2.2. Car and a PTW multibody model

The significate parameters for setting up the vehicle model are the external dimensions such as width, height, and length as shown in figure 5 and specified in table 1. A car model consisted of 39 ellipsoids. It was a simplified model of the whole car. The PTW was modified from a 1985 Kawasaki Ninja which was developed by [4]. The characteristics were obtained from literature [7]. The external appearance and specification of PTW model are shown in figure 5 and in table 1. PTW consist of 19 ellipsoids.

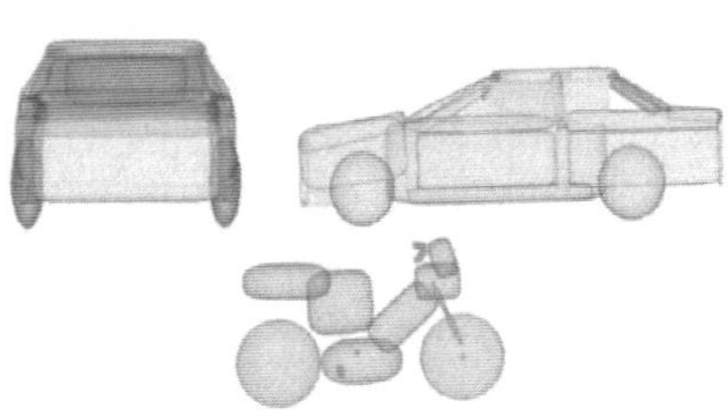

Figure 5. Multi-body dynamics of Toyota Corolla Altis (upper) and Honda Nova 125 RS Sonic (lower).

Table 1. Specifications of Toyota Corolla Altis and Honda Nova 125 RS Sonic.

Titles	Details of Toyota Corolla Altis	Details of Honda Nova 125 RS Sonic
Overall length	4,540 mm	1,900 mm
Overall width	1,760 mm	660 mm
Overall height	1,470 mm	960 mm
Wheelbase	2,600 mm	1,250 mm
Weight	1,310 kg	100 kg
Front tire	195/65R15	60/100-17M/C33P
Rear tire	195/65R15	80/90-17M/C44P

3.2.3. Impact simulation set-up

PTW model was directed towards the car model. Rider and passenger were positioned as shown in figure 6. The impact point of rider' s head was at the corner of the windshield as evident from the accident scene. A red marker sensor was created at that point for head tracking. The impact point of passenger's head was at the car roof, marked with a pink sensor as in figure 6. The important parameters for simulating the accident are the speed, direction, impact area, impact angle between the passenger car and powered-two wheeler. Final position of the vehicle and the influence of deceleration of the vehicle are also considered. Based on the variation of these parameters, a number of simulations were performed to obtain the optimum configuration, which gave the best correlation with the accident scene.

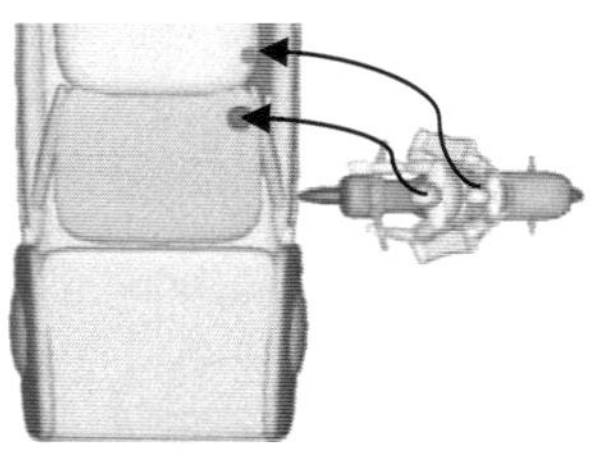

Figure 6. PTW to car impact set up with rider and passenger on board.

3.2.4. The parametric studies

The parametric studies consisted of 1,500 different conditions. The 10 different speed of the passenger car were 3, 3.5, 4, 4.5, 5, 5.5, 5.55, 6, 6.5, and 7 m/s. PTW speed were 13, 13.5, 14, 14.5, 15, 15.2, 15.5, 16, 16.5, and 17 m/s. The 15 different impact angle between the passenger car and PTW were 20°, 30° 40°, 50°, 60°,70°, 80°, 90°, 100°, 110°, 120°, 130°, 140°, 150° and 160° respectively as shown in figure 7. MADYMO software version 7.5 was used.

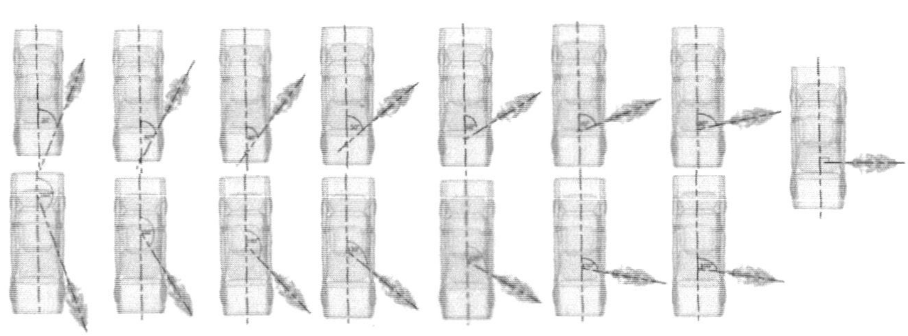

Figure 7. Impact angle between the passenger car and the PTW.

Table 2 illustrates example simulation results from parametric studies for car velocity of 5.5 km/h. For each simulation the head contact location of rider and passenger were identified. If both rider's and passenger' s head hit the pre-defined targets i.e. marked in green in table 2, they were taken into consideration for further kinematic comparison with the accident data. The optimum configuration was for the car speed of 5.5 m/s (20km/h), PTW speed of 15.28 m/s (55km/h) and impact angle of 90°. Figure 8 shows kinematics of rider and passenger during the impact in this optimum configuration.

Table 2. The parametric studies when the speed of the passenger car was 5.55(m/s), variety of impact angle and the speed of PTW.

Pre-crash velocity of passenger car(m/s)	Impact angles (degree)	Pre-crash velocity of motorcycle(m/s)									
		13	13.5	14	14.5	15	15.28	15.5	16	16.5	17
5.55	20	M	M	M	M	M	M	M	M	M	M
5.55	30	M	M	M	M	M	M	M	M	M	M
5.55	40	M	M	M	M	M	M	M	M	M	M
5.55	50	M	M	M	M	M	R	R	R	R	R
5.55	60	M	M	M	M	M	R	R	R	R	R
5.55	70	M	M	M	M	M	R	R	R	R	R
5.55	80	M	M	M	M	M	R	R	R	R	R
5.55	90	M	M	M	M	P	RP	RP	R	R	M
5.55	100	M	M	M	M	M	M	R	R	M	M
5.55	110	M	M	M	M	M	M	M	M	M	M
5.55	120	M	M	M	M	M	M	M	M	M	M
5.55	130	M	M	M	M	M	M	M	M	M	M
5.55	140	M	M	M	M	M	M	M	M	M	M
5.55	150	M	M	M	M	M	M	M	M	M	M
5.55	160	M	M	M	M	M	M	M	M	M	M

P represents passenger's head contacts the pink mark.
R represents rider's head contacts the red mark.
RP both heads contact their sensors.
M both miss the points.

The first contact occurred at 0.03s between the front wheel of the PTW and the left front door of the passenger car. This is supported from the accident photo that shows damage to the front wheel and suspension of PTW (figure 3(c)). Also, the left door of the vehicle is damaged (as shown in figure 3 (b)). The second collision occurred at 0.05 s, with the rider's abdomen impacting against the right handle bar of the PTW. The right knee also hit the left door of the vehicle structure. The rider's chin contacted the windshield and the chest against A pillar at time 0.08 s. These matched with the crack mark on the windshield and the accident injury report. At 0.13s of simulation, the pillion passenger's head contacted the car roof structure. At 1.05s, the rider was in between the PTW and the car, which was also the final position in the picture reported from the accident scene.

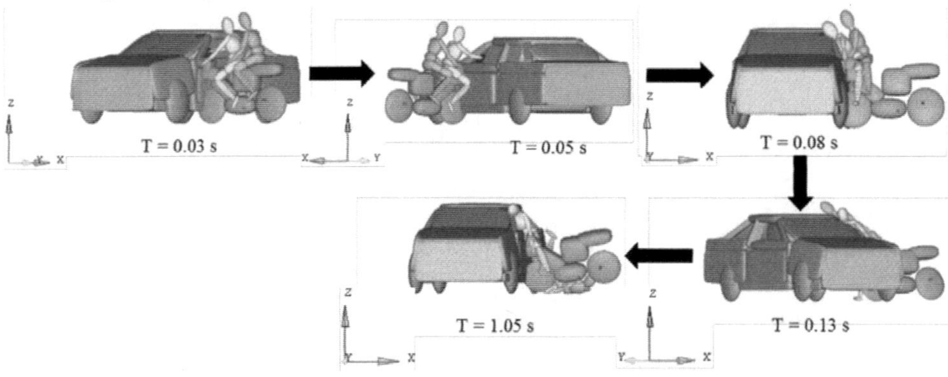

Figure 8. The diagram illustrates kinematics of the rider and the passenger during the crash.

For analysis on head injury, the linear velocities in x, y and z directions together with the angular velocities just before impact were extracted from the multibody dynamic simulation. They are summarised in table 3.

Table 3. The results of the velocity.

	Motorcyclist	Pillion passenger
Impact time	0.071 s	0.118 s
Linear velocity in x axis	-9.33288 m/s	-4.02261 m/s
Linear velocity in y axis	5.52556 m/s	5.93356 m/s
Linear velocity in z axis	-4.77071 m/s	-0.631232 m/s
Angular velocity in x axis	-13.0804 rad/s	-9.38231 rad/s
Angular velocity in y axis	-21.8953 rad/s	-7.74171 rad/s
Angular velocity in z axis	3.81019 rad/s	3.84717 rad/s

4. Finite element model of human head for impact

A deformable finite element human head was taken from the Total Human Body Model for Safety developed validated by [8]. The head consists of skull and brain. The material model for skull was elastic-plastic with damage and for the brain was viscoelastic with material properties according to [9]. Each head was placed at the position just before the impact as shown in figure 9. Head velocities illustrated in table 3 was applied to both the rider and passenger head models.

(a) (b)

Figure 9. The illustrations of (a) set-up rider's head (b) set-up pillion passenger's head.

4.1 Head injury analysis

The Head Injury Criteria (HIC) value for the rider was 1980, higher than the acceptable threshold of 1000 for AIS 3+ injuries. This implied high risk of skull fracture. However, for the pillion passenger HIC was only 56.84, suggesting a low risk of skull fracture. Figure 10 (a, b) shows the skull stress of the rider. Maximum stress was 219 MPa at the left temporal. This is above the criteria of 80 MPa for skull fracture [5]. High stress also occurred around the chin area that matched with the injury report of the accident. Figure 10 (c) shows distribution of strain on the brain. Maximum strain of 1.09 occurred on the left white matter cerebrum. This value is much higher than the criteria of 0.2[10]. It implies contusion of the brain. Figure 11 shows the stress on the passenger's skull. The maximum stress occurred on the left mandible which did not exceed the fracture limit. The passenger was at low risk of skull fracture. The maximum brain strain was 0.3 which is just above the brain injury criteria. Moderate risk of brain injury is possible.

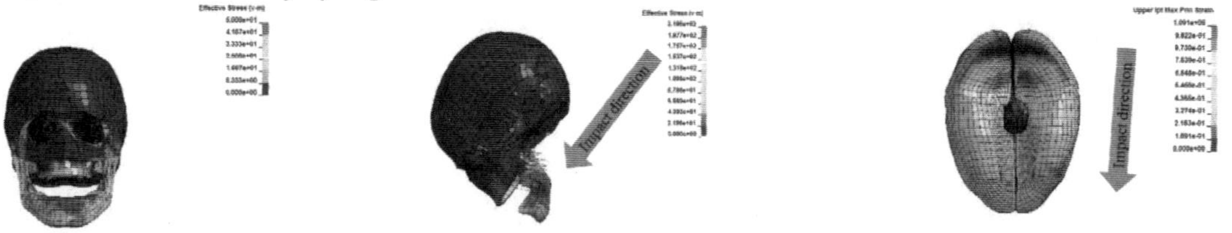

(a) (b) (c)

Figure 10. Distributions of (a, b) stress on skull and face (c) strain on the rider's brain.

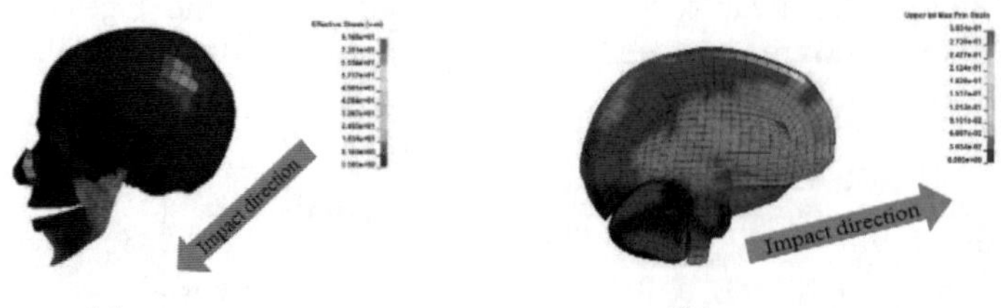

Figure 11. Distribution of (a) stress on the passenger's skull (b) strain on the passenger's brain.

5. Conclusion

A real-world road motorcycle accident was reconstructed using multibody simulation. Parametric studies were performed to obtain the impact configuration which correlated best. The optimal configuration was at the car impact speed of 5.5 m/s, PTW speed of 15.28 m/s and impact angle of 90°. The impact direction, head impact location and velocity were extracted and used as initial conditions for FE simulation. Finite element injury simulation revealed high risk of skull fracture and severe brain injury for the rider while low risk of skull fracture and moderate brain injury for the pillion passenger. The combination of multibody simulation and finite element simulation gives good understanding of kinematics and head injury experienced by the rider and pillion passenger.

Acknowledgments
The authors would like to thank CIFS for an in-depth motorcycle accident data and LBA, IFSTTAR for supporting MADYMO software.

References

[1] World health organization *World health statistics 2015.*
[2] Serre T, Masson C, Perrin C 2012 *Accident analysis and prevention* 49223-228.
[3] Guo L, Jin XL, Zhang XY, Shen J, Chen YJ, and Chen JG 2008 Study of injuries combining computer simulation in motorcycle-car collision accidents, Forensic International 17790-96.
[4] Llar M, Godio-Raboutet Y, and Arnoux PJ 2016 Analyse of a powered two-wheeler (PTW) passenger kinematics and impact conditions, *ICILSM*
[5] Montoya D, Thollon L, Llari M, Perrin C, and Behr M 2017 Head injury criteria in child pedestrian Accidents *Int. J. Crashworthiness* **23** 497-506
[6] Central Institute of Forensics Science database *Internal case report* 2016.
[7] Frank T, Smith J, Hansen D, and Werner S 2009 Motorcycle Rider Trajectory in Pitch-Over Brake Applications and impacts, *SAE Int. J. Passeng. Cars - Mech. Syst.* 31-42
[8] Iwamoto M, Nakahira Y, Tamura A, Kimpara H, Watanabe I and Miki K, 2007, Development of Advanced human models in THUMS, *Proc. The 6th European LS-DYNA Conference*, 2007.
[9] Galford JE and McElhaney JH. 1970 A viscoelastic study of scalp, brain and dura. *Journal of Biomechanics* **3** 211-222.
[10] Feist F and Klug C 2016, Numerical study on the influence of the upper body and neck on head kinematics, *Proc. Int. Conf. on IRCOBI (Malaga:IRCOBI)* 162-184.

IOP Conf. Series: Materials Science and Engineering **501** (2019) 012010 doi:10.1088/1757-899X/501/1/012010

Design of Attitude Control System for Stratospheric Balloon Gondolas by Sliding Mode Control

T Nakano*

Department of Mechanical Engineering, National Institute of Technology: Oita College, 1666, Maki, Oita, 870-0152, Japan

* Corresponding Author: t-nakano@oita-ct.ac.jp

Abstract. This paper reports consideration of efficient design method for the azimuthal-attitude control of stratospheric balloon gondolas utilizing only a motorized pivot. To maintain sufficient robustness against variation of the model parameters, applying of sliding mode controller is tried and proposed control system is validated by numerical simulation studies.

1. Introduction

Balloon experiment is known as a powerful method for various scientific observations and engineering experiments; a large gas-balloon is used to transport a payload (gondola) equipped with observational instruments or experimental devices to stratospheric altitude. In one example, balloon-borne telescopes have been developed and used for many astronomical and planetary observations. The balloon workshop was held at NASA Glenn Research Center in 2012, and this workshop concluded that balloon platform has potential to solve many science questions in the planetary sciences field[1].

Attitude control is one of the key technologies on balloon experiments requiring to point the observational instruments on the gondola in the target direction. The dynamics of the gondola suspended from the balloon envelope via a rope has one-degree of freedom around vertical axis. Control technologies for rotating and stabilizing the gondola in azimuthal direction have been studied through the past experiments. Attitude-control methods applied in practical use can be classified into two ways: a method using a reaction wheel for actuating the gondola and a motorized pivot at top of the gondola for desaturating the reaction wheel[2,3]; other one using only a motorized pivot to twist the suspension rope and rotate the gondola[4,5]. Attitude control with only a motorized pivot has an advantage on simplifying and downsizing the gondola system. However, vibration of the gondola depending on the torsional spring characteristic of the rope affects the control performance. This characteristic unpredictably changes by atmospheric environment during the flight, so robustness for such parameter variation is essential to control-system design. Therefore, achieving high responsiveness and accuracy is difficult with this control method, and it has been applied on only some flight-experiment requiring relatively rough-accuracy attitude control.

This paper will study about a performance-improvement method of azimuthal-attitude control system with only a motorized pivot. By solving the above-mentioned problem about robust stability, this control system will be able to be applied for more balloon gondolas, and balloon experiments will be able to conduct with inexpensive cost. To substantialize enough robustness against variation of parameters, applicability of sliding mode control (SMC) will be studied. Firstly, dynamics model of the control system with a pivot will be built. After that, a control system based on SMC to realize the stability will be considered, and effectiveness of the proposed control method will be confirmed by numerical simulations.

2. Dynamics model of the control system

Figure 1 shows the dynamics model of the balloon gondola considering only rotational motion in azimuthal direction. θ_b is the angle of the bottom of the balloon-envelope, θ_p is the angle of the pivot shaft, and θ_g is the angle of the gondola. At stratospheric altitude, the balloon envelope rotates irregularly by impact of disturbance torque. This disturbance rotation is transmitted to the gondola via the suspension rope. Azimuthal-attitude control system of the gondola must have satisfactory performance to compensate the disturbance rotation. The pivot shaft at the top of the gondola is rotated by the control torque τ_m of the motor.

Motion of the gondola in the azimuthal direction is described as:

$$J_g\ddot{\theta}_g = \tau_m \, , \tag{1}$$

where, J_g is the moment of inertia of the gondola.

Dynamics of the pivot shaft is also given as following equation by approximating the suspension rope as a torsional spring system with a damper:

$$J_p\ddot{\theta}_p + c_r(\dot{\theta}_p - \dot{\theta}_b) + k_r(\theta_p - \theta_b) = -\tau_m \, , \tag{2}$$

where, k_r and c_r are the coefficient values of torsional spring and damper, respectively.

On the many actual balloon gondolas in some previous flight missions, the actuation systems of pivot were often equipped with an angular-velocity control system for its motor. In this research, control system utilizing the angular velocity of the motor as a control input was modeled with the assumption that the motorized pivot has the angular-velocity control system.

Equation (1) can be rewritten using the equation (2) as:

$$\ddot{\theta}_g = \frac{1}{J_g}\tau_m = -\frac{J_p}{J_g}\ddot{\theta}_p - \frac{c_r}{J_g}(\dot{\theta}_p - \dot{\theta}_b) - \frac{k_r}{J_g}(\theta_p - \theta_b) \tag{3}$$

Relational expression between θ_p, θ_g, and the angle of the pivot motor θ_m is given as:

$$\theta_p = \theta_g - \theta_m \tag{4}$$

Following expression can be obtained by inserting the θ_p expressed in the equation (4) into the equation (3):

$$\ddot{\theta}_g = -\frac{J_p}{J_g}(\ddot{\theta}_g - \ddot{\theta}_m) - \frac{c_r}{J_g}(\dot{\theta}_g - \dot{\theta}_m - \dot{\theta}_b) - \frac{k_r}{J_g}(\theta_g - \theta_m - \theta_b) \tag{5}$$

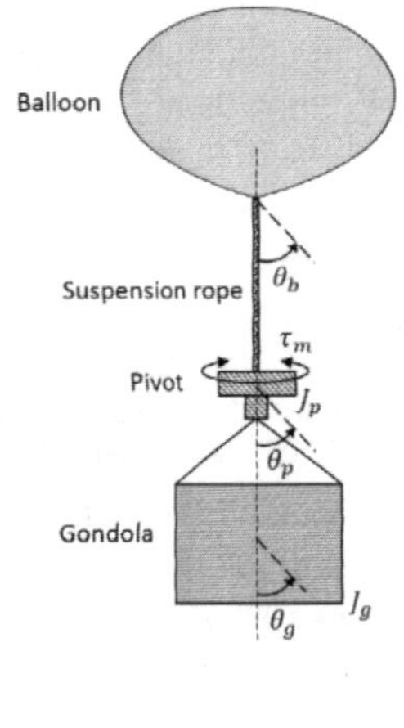

Figure 1. Dynamics model of the azimuthal-control system.

In general, the moment of inertia of the pivot J_p is extremely smaller than it of the gondola. By approximation as $J_p/J_g \fallingdotseq 0$, equation (5) can be represented as:

$$\ddot{\theta}_g = -\frac{c_r}{J_g}(\dot{\theta}_g - \dot{\theta}_m) - \frac{k_r}{J_g}(\theta_g - \theta_m) + \frac{c_r}{J_g}\dot{\theta}_b + \frac{k_r}{J_g}\theta_b \tag{6}$$

Through these processes, the control system using $\dot{\theta}_m$ as a control input can be defined. In this research, responsiveness of the motor driving system was assumed as:

$$\ddot{\theta}_m = -\alpha\dot{\theta}_m + \alpha\dot{\theta}_m^* , \tag{7}$$

where, $\dot{\theta}_m^*$ is a target value of $\dot{\theta}_m$, and α represents characteristic of the motor driving system.

From equation (6) and equation (7) are written in the state equation as:

$$\begin{aligned} \dot{\boldsymbol{x}} &= \boldsymbol{A}\boldsymbol{x} + \boldsymbol{B}u + \boldsymbol{D}\boldsymbol{d}\,, \\ \boldsymbol{y} &= \boldsymbol{C}\boldsymbol{x}, \end{aligned} \tag{8}$$

where,

$$\boldsymbol{x} = \begin{bmatrix}\theta_g & \dot{\theta}_g & \theta_m & \dot{\theta}_m\end{bmatrix}^T, \quad u = \dot{\theta}_m^*, \quad \boldsymbol{d} = [\theta_b \quad \dot{\theta}_b]^T \tag{9}$$

$$\boldsymbol{A} = \begin{bmatrix} 0 & 1 & 0 & 0 \\ -k_r/J_g & -c_r/J_g & k_r/J_g & c_r/J_g \\ 0 & 0 & 0 & 1 \\ 0 & 0 & 0 & -\alpha \end{bmatrix}, \quad \boldsymbol{B} = \begin{bmatrix} 0 \\ 0 \\ 0 \\ \alpha \end{bmatrix},$$

$$\boldsymbol{C} = [1 \quad 0 \quad 0 \quad 0], \quad \boldsymbol{D} = \begin{bmatrix} 0 & 0 \\ k_r/J_g & c_r/J_g \\ 0 & 0 \\ 0 & 0 \end{bmatrix}. \tag{10}$$

3. Define of Parameter Conditions and Control Requirement

In this chapter, parameter conditions of the dynamics model and limitation of the control actuator will be defined for design and verification of the control system.

3.1. *Nominal model*

Nominal parameter values of the dynamics model were defined as:

$$J_g = 100\ kgm^2, k_r = 1.0\ Nm/rad\,, c_r = 0.2\ Nm/(rad/s)\,, \alpha = 10.$$

These values were determined referring some previous studies.

3.2. *Parameter variation*

The pivot applies torque to suspension rope and twist it; it also induces torsional vibration in azimuthal direction. Therefore, characteristic variation of the suspension rope expressed as k_r and c_r extremely affects control stability and performance. Control system must be designed considering such parameter variation to keep enough stability and desirable responsiveness.

In this research, performance validation of the designed control system was performed in three cases, as follows:

- Case-1: $k_r = 1.0\ Nm/rad,\ c_r = 1.0\ Nm/(rad/s)$ ··· conditions of the ground test.
- Case-2: $k_r = 0.2\ Nm/rad,\ c_r = 0.2\ Nm/(rad/s)$ ··· conditions of the flight with a low k_r
- Case-3: $k_r = 2.4\ Nm/rad,\ c_r = 0.2\ Nm/(rad/s)$ ··· conditions of the flight with a high k_r

In the actual developments of balloon-gondola system, ground tests are often conducted to evaluate the performance of the control system; the balloon gondola is suspended from a crane via a short suspension rope. Case-1 represents conditions in the ground test. Damping coefficient c_r in the ground test is usually much larger than it in the flight is. Parameter values of Case-1 were set referring previous development of the gondola system[6]. Case-2 and Case-3 represent flight situations with parameter variation of k_r. The suspension rope tends to become rigid because of low-temperature environment at the flight altitude. On the other hand, torsional spring coefficient varies by total weight of suspended gondola. Robustness for variation of k_r is necessary to conduct azimuthal-attitude control successfully under various situations. Here, variational-range of k_r were defined as Case-2 and Case-3 based on the previous study of azimuthal-control system[4].

3.3. Limitation of the control device

Limitation value of the rotational speed of the motor was set as:

$$|\dot{\theta}_m| \leq 40\ deg/s$$

4. Study of the control system design utilizing an optimal-servo controller

Firstly, applying an optimal servo control on the azimuthal-attitude control system was considered. In this chapter, design process of the controller will be introduced briefly, and its validation by numerical simulation will be presented.

4.1. Control law

The optimal-servo controller to regulate the azimuthal angle θ_g according to the target value θ_g^* was designed. An augmented system with an integral element is defined as:

$$\begin{gathered} \boldsymbol{e} = \theta_g^* - \theta_g, \quad \boldsymbol{\omega} = \int_0^t e\, d\tau, \quad \widetilde{\boldsymbol{x}} = \boldsymbol{x} - \boldsymbol{x}_\infty, \\ \tilde{u} = u - u_\infty, \quad \widetilde{\omega} = \omega - \omega_\infty, \\ \begin{bmatrix} \dot{\widetilde{\boldsymbol{x}}} \\ \dot{\widetilde{\boldsymbol{\omega}}} \end{bmatrix} = \begin{bmatrix} \boldsymbol{A} & 0 \\ -\boldsymbol{C} & 0 \end{bmatrix} \begin{bmatrix} \widetilde{\boldsymbol{x}} \\ \widetilde{\boldsymbol{\omega}} \end{bmatrix} + \begin{bmatrix} \boldsymbol{B} \\ 0 \end{bmatrix} \tilde{u}, \quad \boldsymbol{e} = [-\boldsymbol{C} \quad 0] \begin{bmatrix} \widetilde{\boldsymbol{x}} \\ \widetilde{\boldsymbol{\omega}} \end{bmatrix}, \end{gathered} \tag{11}$$

where, $\boldsymbol{x}_\infty$, u_∞ and ω_∞ are steady-state value of $\boldsymbol{x}$, u, and ω, respectively. Servo controller can be constructed by design an optimal regulator using this augmented system.
A performance index J_a including design parameters Q_{11}, Q_{22}, R_e is represented as:

$$J_a = \int_0^\infty (e(t)^T Q_{11} e(t) + \widetilde{\omega}(t)^T Q_{22} \widetilde{\omega}(t) + \tilde{u}(t)^T R_e \tilde{u}(t)) dt \tag{12}$$

Optimal servo controller can be designed by solving $\tilde{u}$ which minimize the performance index J_a. The concreate derivation process is introduced in the reference[7].

4.2. Numerical simulation

Numerical-simulation studies were performed to validate the designed control system in four cases; Nominal case and Case-1~3 shown in previous chapter. In these studies, target values of azimuthal-attitude control were set as 30 degrees with reference to the initial position, and rotational status of the balloon envelop $\dot{\theta}_b$ and θ_b were set as zero. Numerical-simulation environment was built as a continuous-time system. Design parameters Q_{11}, Q_{22} and R_e must be tuned properly through trial and error process. Here, these parameters were tuned as $Q_{11} = 50$, $Q_{22} = 5$, and $R_e = 5$ in order to achieve higher responsiveness as much as possible with avoiding output saturation of the motor.

Firstly, controller was designed using parameter values of the ground test environment shown as Case-1. Figure 2 shows the simulation results (Simulation-1). Left one and right one indicates the history of the gondola azimuthal angle, and rotational speed of the motor, respectively. In this simulation, attitude control of Case-2 became unstable. At the flight altitude, control becomes easily unstable because of low damping coefficient of gondola dynamics caused by low density of the atmosphere. Therefore, it is possible that controllers optimized for ground-test condition induce undesirable response in the flight. Next, controller was designed applying the model parameters of nominal model. Figure 3 shows the simulation result (Simulation-2). The controller could not stabilize attitude control of Case-2. As this result indicates, a low torsional spring coefficient (it means that low rigidity of the suspension

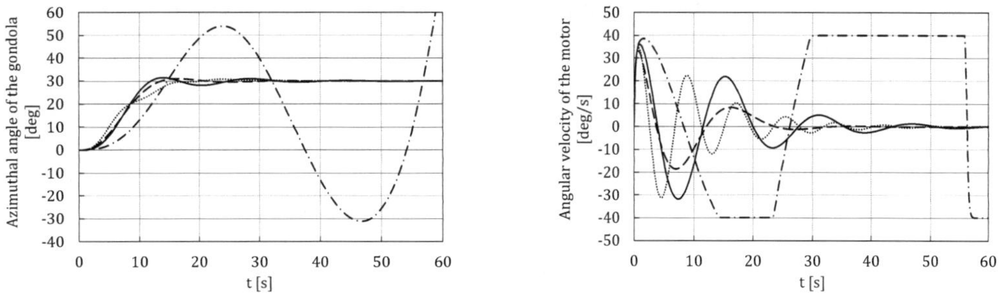

Figure 2. Results of the Simulation-1 (Nominal case: solid line, Case-1: dashed line, Case-2: chain line and Case-3: dotted line).

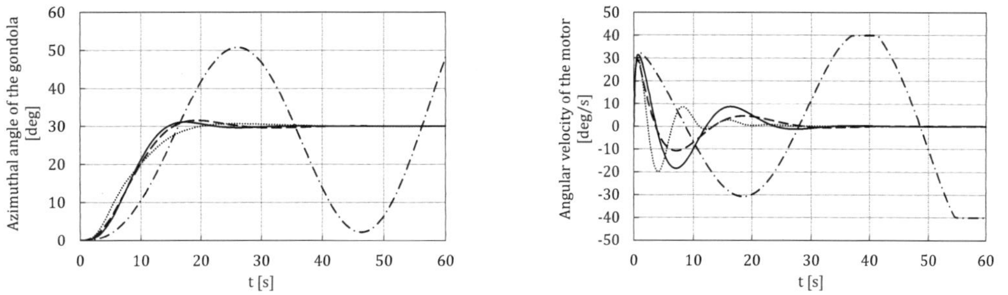

Figure 3. Results of the Simulation-2 (Nominal case: solid line, Case-1: dashed line, Case-2: chain line, and Case-3: dotted line).

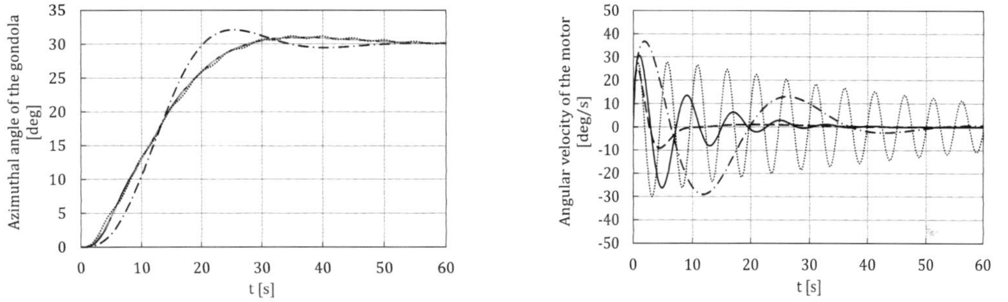

Figure 4. Results of the Simulation-3 (Nominal case: solid line, Case-1: dashed line, Case-2: chain line, and Case-3: dotted line).

rope) is expected to cause instability of the control. Furthermore, a controller was designed with applying the model parameters of Case-2. Figure 4 shows the simulation result (Simulation-3). Azimuthal angle of the gondola appears to converge to the target value in all cases. However, output of the motor had undesirable oscillation, especially in Case-3. These results indicate that control systems adjusted to a low torsional spring coefficient have possibility to become unstable by high torsional spring coefficient. A series of simulation studies confirmed that there is difficulty to achieve sufficient robustness against parameters variation with only the optimal-servo control.

5. Study of applying a sliding-mode controller

This chapter will describe a validation about design of azimuthal-attitude control system with applying a sliding-mode control (SMC)[8]; it is one of non-liner control methods which has a potential to achieve enough robustness against errors of model parameters of the control system.

5.1. Control law

IOP Conf. Series: Materials Science and Engineering **501** (2019) 012010 doi:10.1088/1757-899X/501/1/012010

In this research, SMC for the augmented system shown as in equation (11) was considered. SMC system includes two control inputs; a liner control input for obtaining desirable control characteristics on a switching hyperplane, and a non-liner control input for keeping the system state on the switching surface. An equivalent control input u_{eq} in sliding mode state can be represented with a switching hyperplane matrix $\boldsymbol{S}$ as:

$$\sigma = \boldsymbol{Sx}\,, \tag{13}$$
$$u_{eq} = -(\boldsymbol{SB})^{-1}(\boldsymbol{SAx} + \boldsymbol{SQr})\,,$$

where, σ is a switching function. Here, in order to design $\boldsymbol{S}$, a deriving method giving an arbitrary stability margin was applied. The matrix $\boldsymbol{S}$ can be determined by solving a Riccati equation shown as:

$$\begin{aligned} &PA' + A'^{T}P - PBB^{T}P + Q = 0, \quad S = B^{T}P, \\ &A' = A + \varepsilon I, \qquad \varepsilon \geqq 0, \end{aligned} \tag{14}$$

where, Q was set as $Q = I$. ε is a design parameter indicating an arbitrary stability margin. Control input u of SMC can be expressed with the liner input u_l and non-liner input u_{nl} as:

$$u = u_l + u_{nl} = -(\boldsymbol{SB})^{-1}(\boldsymbol{SAx} + \boldsymbol{SQr}) - k(\boldsymbol{SB})^{-1}\frac{\sigma}{|\sigma|}, \tag{15}$$

where, u_l is equal to u_{eq}, and u_{nl} is a non-liner control input. If a Lyapunov function V achieving $\sigma \rightarrow 0$ is defined as:

$$V = \frac{1}{2}\sigma^2\,, \tag{16}$$

the system can reach the sliding mode by using the control gain k which is satisfied with sufficient condition of $\dot{V} < 0$. In this research, a smoothing function was applied on the non-liner input for reduction of chattering of the control input as:

$$u_{nl} = k(\boldsymbol{SB})^{-1}\frac{\sigma}{|\sigma| + \eta}, \tag{17}$$

where, η is a parameter to adjust smoothing effect.

5.2. Numerical simulation

SMC for the azimuthal-attitude control system was designed according to the process described in previous section. Parameter values of Case-3 were used as condition of the control system design. The stability margin ε were chosen as 0.06 through trial and error to realize same followability of response as the designed control system in previous chapter. Gain value k of the non-liner control was set to 50 to satisfy reaching condition of sliding mode against change of the model parameter. Smoothing parameter η was set to 0.1.

Numerical-simulation study was performed to evaluate control characteristics of the designed control system. Target value of the control was set as 30 degrees, and rotational status of the balloon envelop $\dot{\theta}_b$ and θ_b were set as zero in this simulation. Figure 5 presents the results of the simulation (Simulation-4) indicating the gondola azimuthal angle, rotational speed of the motor, and control inputs of SMC. As Figure 5 shows, azimuthal angle of the gondola converged successfully to the target value in all cases. The rotational speed of the motor also maintained enough stability in all cases. As the graph of u_{nl} in Figure 5 shows, it was found that non-liner control functioned properly to compensate influences of the change of model parameters.

From these results, it was confirmed that SMC method is effective for azimuthal-attitude control of the balloon-gondola which has variation of the parameter, and sufficient robustness can be realized certainly.

6. Conclusion

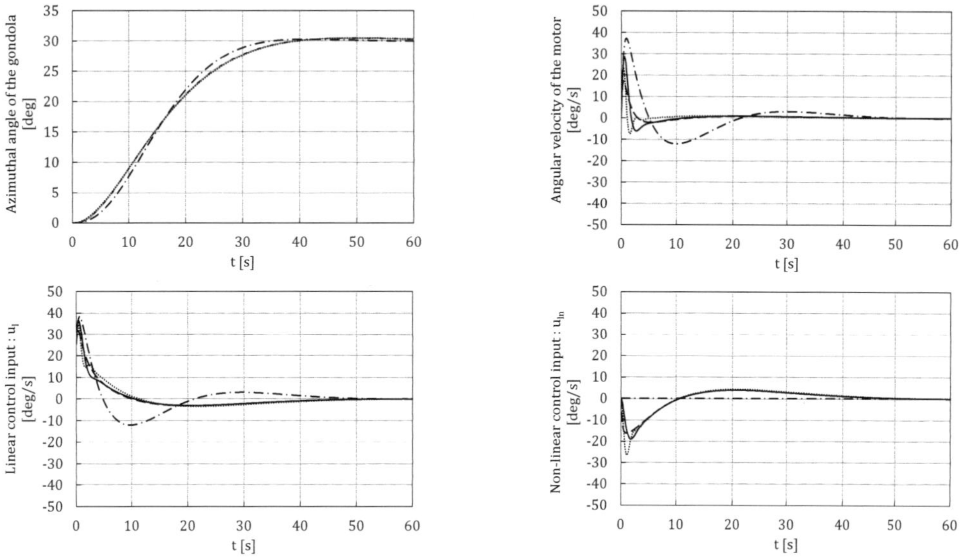

Figure 5. Results of the simulation-4 (Nominal case: solid line, Case-1: dashed line, Case-2: chain line, and Case-3: dotted line).

This paper reported consideration about the design of the azimuthal-attitude control system for stratospheric balloons using only a motorized pivot.

Firstly, the overview of the azimuthal-attitude control system was presented, and the derivation process of the numerical model was described. In this research, numerical model of the control system using angular velocity of the motor as a control input was constructed. Parameter conditions and control requirements for design of the control were defined based on the actual balloon-gondola systems. Characteristics variation of the suspension rope is predicted to influence the control performance seriously. Therefore, some simulation cases with different characteristic values of the suspension rope were defined considering the ground-test situation and flight environment.

Next, azimuthal-attitude controllers were designed and verified. Optimal-servo control was applied on the control system, and its control performance was confirmed through some numerical-simulation studies. This study revealed that it is difficult to maintain robustness against change of the parameters with only an optimal-servo controller. After that, applying SMC was considered. Designed control system with SMC was also validated by numerical-simulation study. It was confirmed that SMC realize enough robustness for variation of the characteristics of the suspension rope. From this research, the azimuthal-control system with SMC could be proposed, and the basic information about its availability was obtained.

References

[1] Kremic T, Hibbitts K, Young E, Landis R, Noll K and Baines K 2013 *Proc. IEEE. Aerosp. Conf.* 1-8.

[2] Chingcuanco A, Lubin P, Meinhold P and Tomizuka M 1990 *Dyn. Sys. Meas.Control.* **112**(4) 703-10.

[3] Lecinski A, Card G, Knölker M and Hardy B 2017 *J. Astro. Instru.* **6**(2) 1-15.

[4] Bando N, Fuke H, Shoji Y, Doetinchem P, Hailey C, Sakai S and Hashimoto T 2015 *Aerosp. Tech. Japan., JSASS* **14** 59-65.

[5] Matko D, Yajima N and Hinada M 1996 *Rep.ISAS.* **665** 1-53.

[6] Shoji Y, Taguchi M, Nakano T, Maeda A, Imai M, Gouda Y, Watanabe M, Takahashi Y, Sakamoto Y and Yoshida K 2016 *Trans. JSASS.* **14**(ists30) 95-102.

[7] Ikeda M and Suda N 1988 *Trans. SICE.* **24**(1) 40-46.

IOP Conf. Series: Materials Science and Engineering **501** (2019) 012010 doi:10.1088/1757-899X/501/1/012010

[8] Nonami K and Den K 1994 *Sliding Mode Control* (Tokyo: Corona Publishing Co., Ltd.)

IOP Conf. Series: Materials Science and Engineering **501** (2019) 012011 doi:10.1088/1757-899X/501/1/012011

Conceptual Design of Tactical Solar Power UAV

W Jiajan, J Kampoon*, J Klongtrujrok, and I Yuthayanon

Department of Aeronautical Engineering Navaminda Kasatriyadhiraj Royal Air Force Academy, 171/1 Phaholyothin Rd, Saimai District, 10220, Bangkok, Thailand

*Corresponding author: janewit_kamp@rtaf.mi.th, +662-534-8057, +662-534-8057

Abstract. The purpose of this research is to study and design a concept of the tactical solar power UAV. The mission requirements derived from military applications: Intelligence, Surveillance, and Reconnaissance)ISR(which including a longer operation time)8 hours(, 350-1000 meters service ceiling, and less than 5 kg for take-off weight. This paper, a conceptual design of solar power UAV mainly based on Noth's methodology. A simple program coding created for sizing solar power UAV by using Scilab. In addition, the vortex lattice method)XFLR5 program(also employed to evaluate the basic aerodynamic characteristics for three different low-Reynolds number airfoils. The wing incident angle was then designed. From the conceptual design results, the coding indicated that the solar power UAV is the conventional type of aircraft, which is provided 4 meters in wingspan)Aspect ratio 13(, 3.88 kg for take-off weight, and 69.13% for solar area to wing area ratio.

1. Introduction

The Royal Thai Air Force (RTAF) is responsible for the preparation of its capabilities to defend the kingdom using the air operations and the operations related to those capabilities. RTAF aims at being one of the best air forces in ASEAN by 2020. In addition, the policy of RTAF commander-in-chief year 2016-2018, particularly in the intelligence aspect focused on the enhancements of intelligence, surveillance and reconnaissance (ISR) mission. It is obvious that UAV is the useful platform for the ISR military operations. Currently, there are some problems about RTAF's UAV. Firstly, most RTAF's tactical UAV have limited endurance because of using power solely from battery. Secondly, the high frequency of take-off and landing affect the life cycle period of the avionics and payload that result in high risk to damage of those payload. Lastly, to participate in the global warming concern, which help reducing the amount of CO_2 emission to the atmosphere.

To date there are several approaches to have UAV stay flying theirs mission in the sky as long as possible like aerial refueling [1-3], energy gaining from environment by using efficient maneuvering [4-5], air launch [6] and alternative energy such as solar cell [7-11], fuel cell [12] etc.. Each approach has their own strength and weakness depend on mission requirements and figure of merit. However, for tactical UAV the exploitation of solar power is the most efficient approach.

Therefore, this paper proposes the conceptual design of the solar power UAV for tactical mission in order to enhance the endurance. The mission requirements derived from military applications specific to ISR mission which including a longer operation time, i.e. greater than 8 hours, 350-1000 meters service ceiling, and less than 5 kg for take-off weight.

The rest of this paper organized as follow, the first section explains the mission requirement and conceptual design methodology. The next section deals with airfoil selection. After that, initial sizing of wing and tail is described. Finally, conclusions are drawn for further studies.

2. Solar Power UAV

The configuration of the solar power UAV is indifference to battery powered UAV in terms of aerodynamics efficiency, which is dependent on the mission requirements and the design criterion. The only different is the propulsion system, i.e. the photovoltaic cell or solar cell are placed on the upper surface of wing, fuselage or empennage to harvest the energy from sun ray and convert to electrical energy during the day. The obtained electrical energy supply to the motor-propeller, the onboard electronic and the excess energy charge to battery. The first solar power UAV, Sunrise I with 3.2 m wingspan, soared in the sky on 1974 [13]. Since then, there are several successes in design and development of the solar power UAV [14-18].

2.1. *Mission Requirements*

The mission requirement is the important input that drives the conceptual design phase in order to come up with the fundamental 3-D drawing of the UAV. This paper focuses on ISR, which generally refers to continuous monitoring target by using imagery obtained from camera payload [18]. The mission profile of the tactical solar power UAV illustrated in figure 1. This UAV is hand launched to take-off, then climb to cruise altitude, then loitering over target area performing ISR, then cruising back to base, after that descending and net landing, respectively. It is obvious that the tactical solar power UAV spent most of the time in cruising and loitering flight phase, which is steady and level flight.

The list of performance requirements are as the followings,

1. Take-off weight less than or equal to 5 kg
2. Camera payload weight 0.17 kg
3. Wingspan greater than or equal to 4 m
4. Operating altitude 300 – 1,000 m
5. Endurance greater than or equal to 8 hours
6. Cruise speed range between 8-9 m/s
7. Stall speed 6 m/s

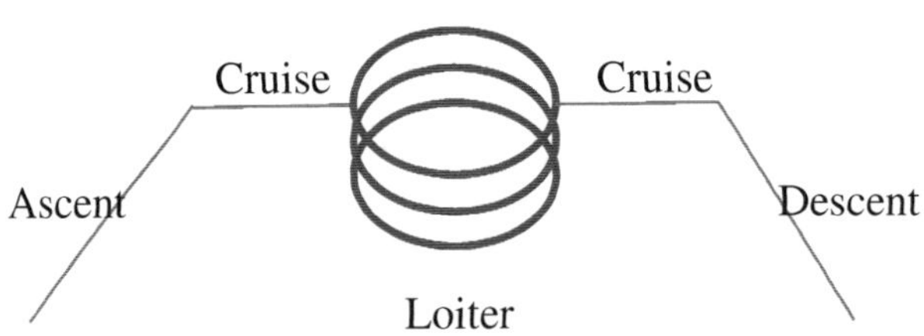

Figure 1 The Mission Profile of the tactical Solar Power UAV

2.2. *Conceptual Design of Tactical Solar Power UAV*

The principles of the UAV design are similar to the principles developed for the design of manned aircraft. The design process is divided into three phases, i.e. the conceptual design, the preliminary design and the detail design. Each phase has different inputs, calculations and outputs in systematic method [19]. The size of UAV varies according to the required mission and theirs performance as well. However, the conceptual design of the solar power UAV has many aspects different from traditional conceptual design methods [20-21]. There are two approaches in the conceptual design method [17], which are the iterative approach and the analytical approach. This paper used the analytical approach because of the ease of implementation and the proof of successful flight of Sky-Sailor. This method relies on the analytical equations describing the characteristics of each component on board UAV, i.e. Maximum Power Point Tracker (MPPT), motor-propeller, solar cell etc. and 2 balance equations that is to say, mass balance and energy balance as shown in figure 2. The total take-off weight of the UAV is the sum of the mass of each component on board, e.g. payload, structure, propulsive unit, solar cell, MPPT, battery. Once the total take-off weight is obtained, the aerodynamic lift and drag can be estimated. These aerodynamic forces then converted to power required for steady and level flight. This amount of mechanical energy is then balance by the electrical energy required for daily operation time, which come from solar panel.

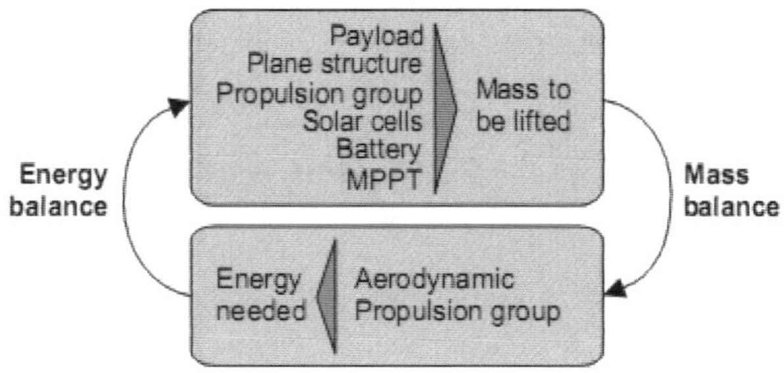

Figure 2 The Mass balance and Energy balance [17]

As mention in section 2.1, the tactical solar power UAV mostly fly in steady and level flight, such that lift force need to balance the total weight of the UAV in order to fly at constant altitude and drag force need to balance the thrust from electric motor and propeller in order to fly at constant airspeed. In addition, the electrical energy consumption needs to balance by the solar energy obtained from the solar panel. Therefore, the amount of solar energy must be greater than the electrical energy required to fly steady and level flight, moreover safely complete the mission. Figure 3 show the schematic diagram of the analytical conceptual design methodology in details. This diagram illustrated the mathematical calculation as a loop to solve for realistic solar power UAV's configuration in terms of wing aspect ratio (AR), wingspan (b) ,and mass (m), given a complete list of all the parameters shown in the table A1 - A2 in the appendix.

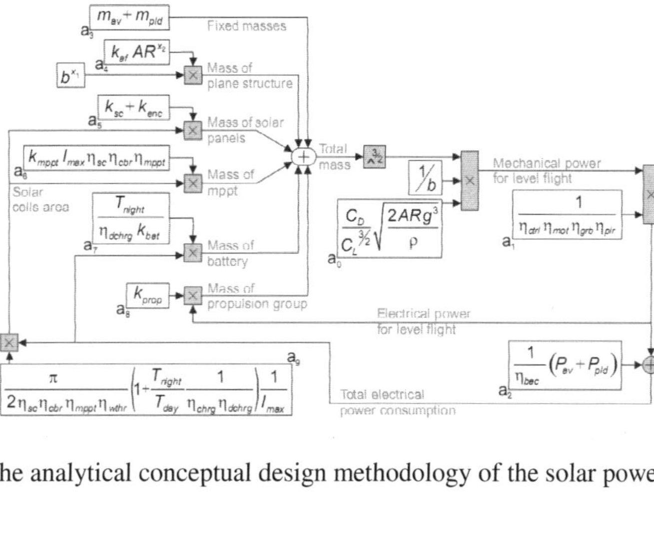

Figure 3 The analytical conceptual design methodology of the solar power UAV [17]

IOP Conf. Series: Materials Science and Engineering **501** (2019) 012011 doi:10.1088/1757-899X/501/1/012011

The conceptual design methodology explained in section 2.1 originally coded in MATLAB. This paper, the design algorithm is converted and written in Scilab, which is comparable to MATLAB and freeware under GNU license.

2.3. *AERO Sun Surfer Design*

The AERO Sun Surfer, tactical solar power UAV, was designed base on the mission and performance requirements in section 2.1 and the methodology in section 2.2. This prototype aircraft is required to fly at least 8 hours in scattered cloud sky condition. The payload for ISR mission is EO/IR camera, which weighs 170 g and consumes power 1.5 W.

The possible configurations of the UAV represent by, i.e. wingspan, aspect ratio, speed, power and the percentage of wing area required to install the solar cell. The results show in figure 4-5. Figure 4 (a) shows the possible wingspan versus total mass of the solar power UAV. Figure 4 (b) shows the magnifying version of figure 4 (a) for more details, notice that this graph focuses on the 4 m wingspan because UAV of this size is suitable for hand launch take-off. The corresponding total mass of 4 m wingspan is 3.88 kg for aspect ratio 13. Notice also that, the increasing aspect ratio results in the higher total mass of the UAV.

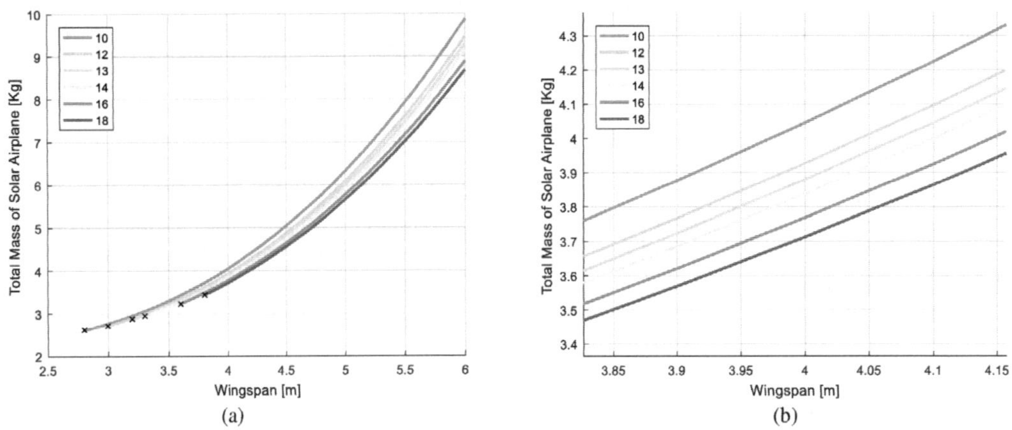

Figure 4 The possible solar power UAV's wingspan versus total mass with variation of aspect ratio

The results in figure 5 show various important design variables versus wingspan and corresponding to aspect ratio. Figure 5 (a) shows UAV's speeds vary with wingspan for a given aspect ratio and the 'x' mark denotes the minimum UAV's speed. Figure 5 (b) shows wing areas vary with wingspan for a given aspect ratio. Figure 5 (c) shows the mechanical power generated by propeller, varying with wingspan for a given aspect ratio and figure 5 (d) shows the percentage of wing area need to install solar cell, varying with wingspan for a given aspect ratio.

From the various candidates, the considerations and figure of merits were trade-off, i.e. mobility, manufacturing cost. Finally, the AERO Sun Surfer conceptual prototype has the total mass 3.88 kg. The wingspan equal to 4 m with aspect ratio equal to 13 with cruise speed equal to 8.144 m/s, wing area equal to 1.231 m^2, required 69.13% of wing area to install solar cells in order to fly steady and level flight, and the corresponding power from propeller is 14.14 W.

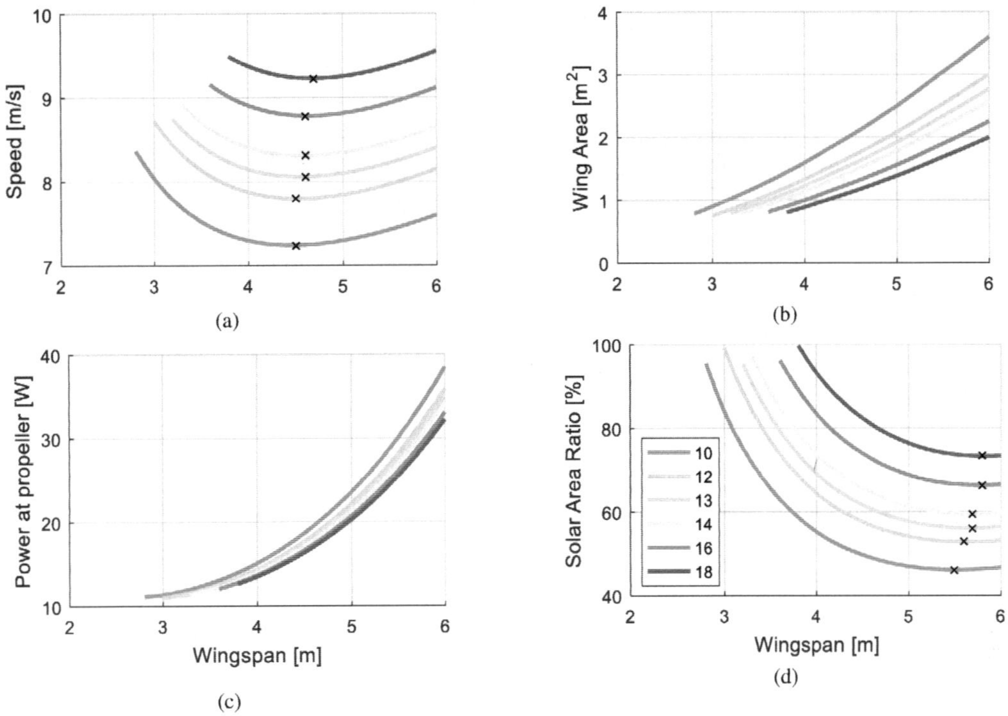

Figure 5 The possible solar power UAV's speed, wing area, propeller power, and area required for solar cell versus wingspan for a given aspect ratio

3. Airfoil Selection

Recall that the mission requirement for the tactical solar power UAV is to fly ISR mission for at least 8 hours. The conceptual design configuration obtained from section 2.3 is then passing to the preliminary design phase. To make this UAV soar to the sky, the wing is the main contributor component, therefore this section explains some detail of airfoil selection for the tactical solar power UAV. Recall also that this UAV spend time mostly in cruising and loitering, this conceptual configuration design to cruise at slow speed. This is the main reason of selecting 3 low Reynolds number airfoil, i.e. S1223, SG6043, and SD7032 to perform aerodynamic characteristics analysis with XFLR5 program. The basic aerodynamic required compose of i) lift coefficient, ii) drag coefficient, and iii) pitching moment coefficient versus angle of attack. The airfoil sections show in figure 6, where figure 6 (a) - (c) are the cross section contour of S1223, SG6043, and SD7032 respectively. The analyses consider the low speed flow where the Reynolds number equal to 250,000.

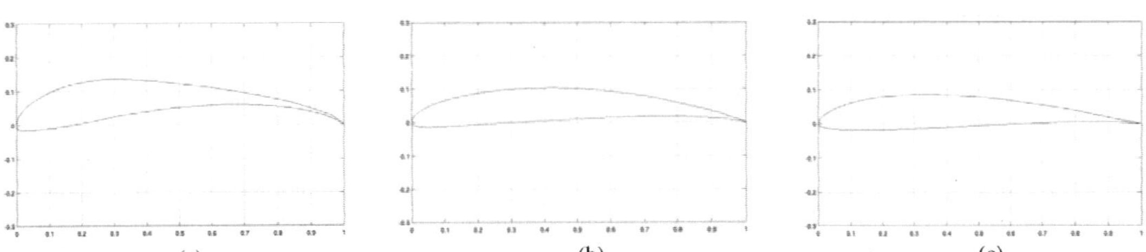

Figure 6 The low Reynolds number airfoil section contour of S1223, SG6043, and SD7032

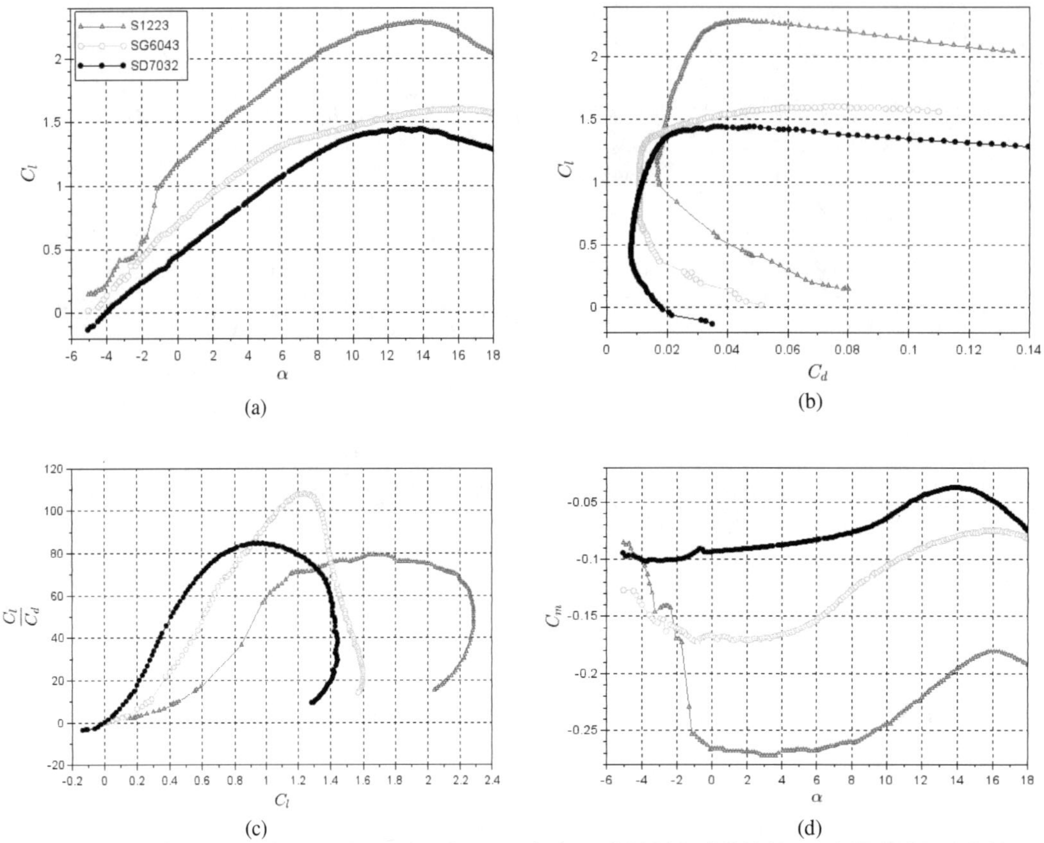

Figure 7 The basic aerodynamic characteristics of S1223, SG6043, and SD7032 airfoil

The results show in figure 7 (a) - (d). Figure 7 (a) shows the comparison of lift coefficient vary with respect to the angle of attack from -5 degree to 18 degree among these threes. It is obvious that S1223 airfoil gives the maximum lift coefficient at angle of attack 14 degree, which is equal to 2.25. SG6043 airfoil is the runner up and gives the maximum lift coefficient at angle of attack 16 degree, which is equal to 1.6. SD7032 airfoil is the third, which gives the maximum lift coefficient equal to 1.4 at the angle of attack 14 degree. Figure 7 (b) shows the comparison of drag polar or the plot showing between lift and drag coefficient of these airfoils. Notice the fact that, despite the highest lift coefficient of S1123 airfoil nonetheless drag coefficient is also the highest, which is the lowest aerodynamic efficiency among the threes. SD7032 and SG6043 have lift coefficient range from 0.6 – 1.0 and 0.7 – 1.3 in cruising condition, however SD7032 has lower drag coefficient range from 0.01 – 0.13 compare to 0.013 – 0.014 of SG6043 corresponding to the cruising condition. Figure 7 (c) shows the lift-to-drag ratio or aerodynamic efficiency of airfoil shape. It reveals that at the design point, where lift coefficient equal to 0.8, SD7032 airfoil gives the highest lift-to-drag ratio, 82 for this case and 78 for SG6043 airfoil. Figure 7 (d) shows the comparison of pitching moment coefficient vary with respect to the angle of attack of three airfoils. It can be noticed that SD7032 has the lowest negative pitching moment during linear operating spectrum among the threes, which means this airfoil has the least nose down moments. The results show that SG6043 airfoil is the suitable airfoil to be selected for utilizing as wing section of the tactical solar power UAV because of having wider range of lift coefficient, while drag coefficient is almost constant corresponding to this range of lift.

IOP Conf. Series: Materials Science and Engineering **501** (2019) 012011 doi:10.1088/1757-899X/501/1/012011

4. Initial Wing and Tail Sizing

In order to design the wing for solar power UAV, the important consideration is the shape and area for installing solar cell on the upper surface of the wing. There are many possible wing planform configurations, however the straight wing planform seem to be the simplest configuration for calculation and installing solar cell. Therefore, the straight wing with aspect ratio equal to 13 and wingspan of 4 m is considered as starting candidate for initial wing sizing. This section explains the initial wing and tail sizing by using XFLR5.

Given wingspan, aspect ratio, and wing area therefore the chord can be calculated. All these design parameters need to be input in the XFLR5 to estimate the basic aerodynamic characteristics and wing incidence angle at design point. Recalls that the initial lift coefficient in conceptual design phase is assume to be 0.8. Having define the vortex lattice method for Reynolds number 250,000 and varying angle of attack from -5 to 15 degree, the results showed that at angle of attack equal to 2 degree gives lift coefficient equal to 0.808 and drag coefficient equal to 0.017. Therefore, the wing incidence angle is set to 2 degree. Figure 8 shows the physical of the wing and the results from XFLR5.

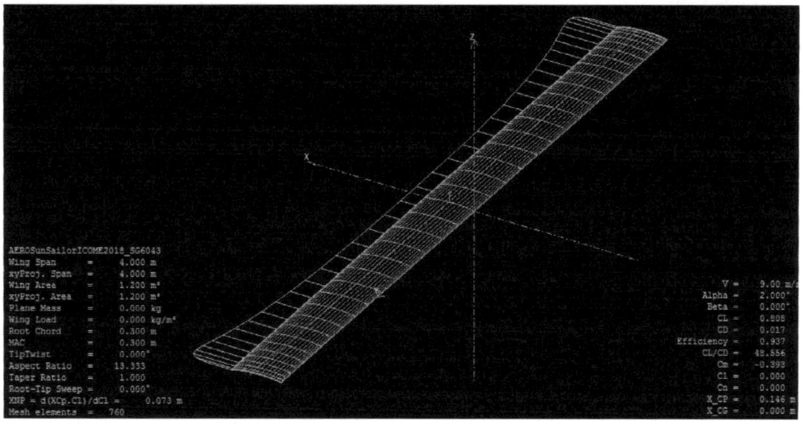

Figure 8 The basic aerodynamic characteristics of the straight wing with SG6043 airfoil

Generally, the conventional aircraft configuration is unstable without having empennage or tail. The tail is the most important component to make the whole aircraft stable or flyable. However, there are several tail configurations, i.e. T-tail, Y-tail, V-tail, H-tail etc., to be selected for the UAV. The consideration is based on the mission and performance requirements. This paper chose T-tail configuration because of two reasons, the one is this type of empennage avoiding the effect of downwash of airflow downstream from the wing. The other is the horizontal stabilizer, which is placed on top of the vertical stabilizer, expose the whole area to the sun, and hence give some extra area for solar cell installation. However, the higher weight is the penalty of this type of empennage. This point should be taken into consideration in the design review phase. As mention above, the empennage is the most important component to yield the stability of UAV. However, it will be given more details analysis in the preliminary design phase paper, particularly, in stability and control analysis of this UAV. Therefore, the first layout of the conceptual design phase of the tactical solar power UAV is shown in Figure 9. This prototype named AERO Sun Surfer. It was designed to have the conventional configuration, where the wing is the straight wing and is the top wing with 4 m span and 0.3 m chord length. The whole wing area is 1.231 m^2, which 69.13% of this area is occupied by solar cell. The solar cell efficiency is 22%, which is a flexible mono crystalline type. This type of solar cell is not expensive; therefore, it is suitable for academic research project. It has T-tail shape empennage. Table 1 shows basic data of the AERO Sun Surfer UAV.

IOP Conf. Series: Materials Science and Engineering **501** (2019) 012011 doi:10.1088/1757-899X/501/1/012011

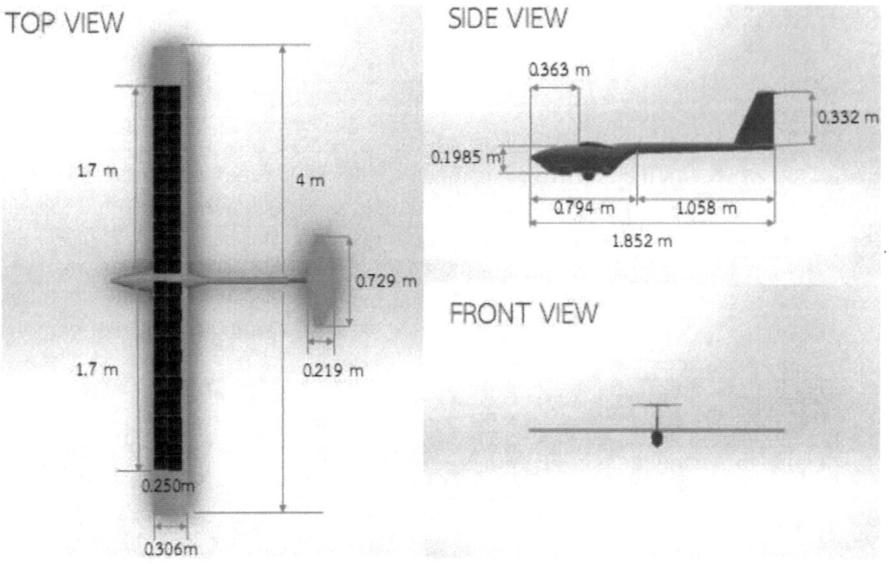

Figure 9 The conceptual layout of the AERO Sun Surfer UAV 3-D drawings

Table 1 The Basic data of AERO Sun Surfer UAV.

Main Geometric Data	AERO Sun Surfer
Wingspan	4.00 (m)
Wing Chord	0.30 (m)
Wing Area	1.23 (m^2)
Solar Area	0.85 (m^2)

Conclusions

This work presented the conceptual design of AERO Sun Surfer, the tactical solar power UAV, which has longer endurance to complete ISR mission and to enhance the performance of RTAF's existing tactical UAV. The conceptual design utilized Noth's methodology. This methodology has been proven for several successful flight of solar power UAV range from small to large scale UAV. However, the further step is to move all efforts to preliminary design phase where each part of UAV will be closely look into more details.

Acknowledges

This work was supported by the Aeronautical Engineering Department, Navaminda Kasatriyadhiraj Royal Air Force Academy.

Appendix

Table A1 The Variable of AERO Sun Surfer UAV.

Variable	Value		Description
AR	13		Aspect ratio
b	4	[m]	Wing span
m	3.88	[kg]	Total mass

Table A2 The Parameter of AERO Sun Surfer UAV (constant)

Parameter	Value		Description
C_L	0.8		Airfoil lift coefficient
C_{Dafl}	0.0013		Airfoil drag coefficient
C_{Dpar}	0.006		Parasitic drag coefficient
e	0.9		Oswald's efficiency factor
I_{max}	650		Maximum irradiance
k_{bat}	190.3600	[J/kg]	Energy density of lithium-ion
k_{sc}	0.32	]kg/m²[	Mass density of solar cells
k_{enc}	0.26	]kg/m²[	Mass density of encapsulation
k_{mppt}	0.00042	]kg/W[	Mass to power ratio of MPPT
k_{prop}	0.008	]kg/W[	Mass to power ratio of propeller group
k_{af}	0.44/9.81	]kg/m³[	Structural mass constant
m_{av}	0.85	]kg[	Mass of autopilot system
η_{bec}	0.65		Efficiency of step-down converter
η_{sc}	0.222		Efficiency of solar cells
η_{cbr}	0.90		Efficiency of the curved solar panels
η_{chrg}	0.95		Efficiency of battery charge
η_{ctrl}	0.95		Efficiency of motor controller
η_{dchrg}	0.95		Efficiency of battery discharge
η_{grb}	0.97		Efficiency of gearbox
η_{mot}	0.85		Efficiency of motor
η_{mppt}	0.97		Efficiency of MPPT
η_{plr}	0.85		Efficiency of propeller
Pav	1.5	]W[	Power of autopilot system
x1	3.1		Airframe mass wingspan exponent
x2	- 0.25		Airframe mass aspect ratio exponent
m_{pld}	0.17	]kg[	Payload mass
η_{wthr}	0.7		Irradiance margin factor
P_{pld}	9	]W[	Payload power consumption
$T_{reserve}$	2*3600	[s]	Reserve time
T_{day}	8*3600	[s]	Day duration (Operation time)
Rho	1.1655	[kg/m³]	Air density (500 m)

References

[1] Katz J 2017 *J. Aircraft* **54** 2311-2316.

[2] Kampoon J and Dogan A 2010 Guidance of Receiver Aircraft to Rendezvous with Tanker in The presence of Wind, *AIAA Guidance, Navigation, and Control Conference*, August, Toronto,Ontario,Canada.

[3] Blake W, Dickes E and Ginras D 2004 UAV Aerial Refueling – Wind Tunnel Results and Comparison with Analytical Predictions, *AIAA Atmospheric Flight Mechanics Conference and Exhibit*, August , Providence, Rhode Island.

[4] Bonnin V and Toomer C 2013 Energy-Harvesting Mechanisms for UAV Flight by Dynamic Soaring, *AIAA Atmospheric Flight Mechanics Conference*, August, Boston, Massachusetts.

[5] Makovkin D and Langelaan J 2014 Optimal Persistent Surveillance using Coordinated Soaring, AIAA Guidance, *Navigation, and Control Conference*, January, National Harbor, Maryland.

[6] Berend N, Bourgaie M, Defoort S, Hermetz J and Tallec C 2006 Innovative Air-Launch System using A Multirole UAV, 57th *International Astronautical Congress*, October, Valencia, Spain.

[7] Hosseini S and Mesbahi M 2016 *J. Guidance Control and Dynamics* **39** 1980 – 1993.

[8] Guo Z, Chen X, Hou Z and Guo J 2011 Development of a Solar Electric Powered UAV for Long Endurance Flight, *11th AIAA Aviation Technology*, Integration, and Operations (ATIO) Conference, September, Verginia Beach, Virginia.

[9] Maleki M 2011 Conceptual Design Method for Solar Powered Aircrafts, *49th AIAA Aerospace Sciences Meeting including the New Horizons Forum and Aerospace Exposition*, January, Orlando, Florida.

[10] Stern T, Steele K, Yang L and Kruse M 2018 High Efficiency Wing Integrated Solar Arrays for UAV Applications, AIAA Propulsion and Energy Forum, *Joint Propulsion Conference*, Cincinnati, Ohio.

[11] Hartney JC 2012 Conceptual Design of a Model Solar-Power Unmanned Aerial Vehicle, *50th AIAA Aerospace Science Meeting Including the New Horizons Forum and Aerospace Exposition*, January, Nashville, Tennessee.

[12] Ofoma U and Wu C 2004 Design of a Fuel Cell Powered UAV for Environmental Research, *AIAA 3rd Unmanned Unlimited Technical Conference, Workshop and Exhibit*, September, Chicago, Illinois

[13] Rosales J, Derra M, Guierrez L, Kreutz K, Rodiquez A, Sagarnaga L and Gross A 2017 Low-Cost High-Endurance Solar-Powered Unmanned Aerial Vehicle, *35th AIAA Applied Aerodynamics Conference*, June, Denver, Colorado.

[14] Malaver AJR., Gonzalez LF, Motta N and Villa TF 2015 Design and Flight Testing of an Integrated Solar Powered UAV and WSN for Remote Gas Sensing. In *IEEE Aerospace Conference.*

[15] Morton S, D'Sa R, and Papanikolopoulos N 2015 Solar Powered UAV: Design and Experiments. In *IEEE International Conference on Intelligent Robots and Systems (IROS).*

[16] Oettershagen P 2017 High-Fidelity Solar Power Income Modeling for Solar-Electric UAVs: Development and Flight Test Based Verification. ETH Zurich.

[17] Noth A 2008 Design of Solar Powered Airplanes for Continuous Flight (Doctoral dissertation, ETH Zurich).

[18] Jay Gundlach. *Designing Unmanned Aircraft Systems A Comprehensive Approach.* American Institute of Aeronautics and Astronautics,Inc.,Virginia,VA, 2014.

[19] Sadraey, M., A systems Engineering Approach to Unmanned Aerial Vehicle Design, 10th AIAA Aviation Technology, Integration, and Operations (ATIO) Conference., September (2010), Forth Worth, Texas.

[20] Daniel P. Raymer. *Aircraft Design: A Conceptual Approach.* American Institute of Aeronautics and Astronautics,Inc.,Virginia,VA, 2006.

[21] Jan Roskam. *Aircraft Design* DAR Corporation, Lawrence, Kansas, 2000

IOP Conf. Series: Materials Science and Engineering **501** (2019) 012012 doi:10.1088/1757-899X/501/1/012012

Automated Spark Ignition Timing Controller (ASITC) For Internal Combustion Engine Control Unit

M S Ajmir [1], AB Elmi [1], A Mohammad Nazir [2] and J Muhammad Nabil Asyraf [3]

[1]School of Aerospace Engineering, Engineering Campus, Universiti Sains Malaysia 14300 Nibong Tebal, Penang,Malaysia

[2]School of Electrical and Electronic Engineering,Engineering Campus,Universiti Sains Malaysia,14300 Nibong Tebal, Penang, Malaysia

[3]School of Mechanical Engineering, Engineering Campus, Universiti Sains Malaysia,14300 Nibong Tebal, Penang, Malaysia

* Corresponding Author: meelmi@usm.my

Abstract. Engine controller is a component that determines the engine running smoothly. With low cost, the Automated Spark Ignition Timing Controller (ASITC) can be design for the suitability of the study. The ASITC is to control spark ignition. For this purpose, Arduino microcontroller is used. It is equipped with potentiometer and 1.8 in TFT display. Through ignition timing, a misfire condition and knocking is actuate by advance spark. Instrustar DAQ is used to record data from spark ignition, oxygen sensor and knocking sensor. From the data obtained, signal processing will be done to get the desired feature through the filtering process. This paper is to study the ASITC to be made for specific features as needed.

1. Introduction

In a few decades, number of vehicle models and production grow rapidly. The pollution that cause by industrial revolution contribute to environmental is inevitable for the future generation. Research from Mansha et al.(2012) [1] mention about secondary pollution from exhaust emission can produce more hazardous pollution. It has observed that engine condition making the environment worst reported by Gequn (2013)[2] and Kakaee (2013) [3].

Inside the engine system, spark plugs are the main part that ignite in the combustion chamber. Libin et al. (2016) [4], Mansha et al.(2012) [1] and Jenson et al.(2015) [5] suggested spark plugs can be manipulated to overcome the imbalance combustion. Moreover, when the condition of incompletes condition due to different air fuel mixture, octane number of fuel and ion current supplied. These scenario as variables are the contribution to pollutants, but it can be solve by developing the multiple variable controller to actuate the right timing to the spark plugs as mentioned by Laurain et al (2015) [6].

Reference Naik(2004) [7], engine controller would able to identify misfire condition based on OBD II legislation. The use of pressure sensor is expensive Roos(2017) [8]. Therefore, by using knocking sensor is an alternative to get the misfire condition, Ajmir et. al (2018) [9]. This microcontroller would able to custom the function of ECU in the automobile system. It may varies from spark plug timing, air fuel ratio, engine cooling, and others condition to be controlled and monitored by the microcontroller.

This paper will describes the design process and initiative study of spark plug timing controller by identifying the source to actuate the sequence during the process. The controller is for users to control spark timing either to actuate misfire and knocking condition or normal engine running condition. It is design for single cylinder engine.

2. Methodology

This section, we begin the study by focusing the related sensor. Crank position sensor (CPS) from the engine is use to actuate spark ignition timing. The frequency is use to design spark plug controller for the engine. Zaim et al. (2016)[10]is manipulating fuel injector to identify the para meters of misfire condition. Thru the experimental procedure, the design of spark plug controller is applied. While signals from oxygen sensor are recorded to be analyzed. Meanwhile in this project, an accelerometer is use to identify the misfire condition comparable to oxygen sensor signals. In order to execute the misfire condition, the accelerometer is mount to the nearest position of the engine piston.

Figure 1, show the project flowchart. A controller is design to initiate spark plug timing. It is to identify the ideal angel of spark plug timing based from the CPS signals. In other hand, the controller is use to actuate the misfire condition for further analysis.

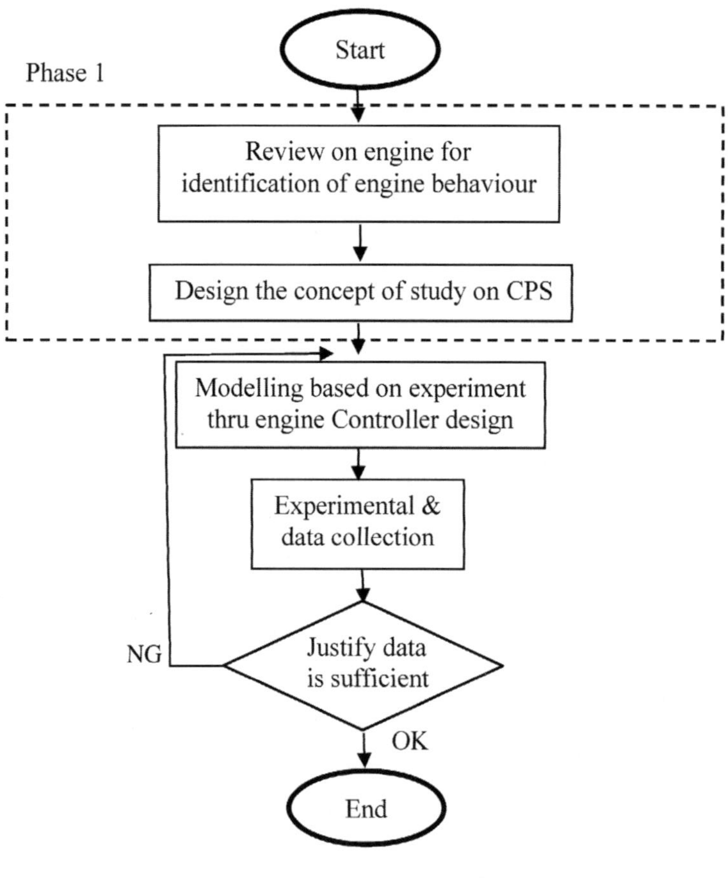

Figure 1: Project Flowchart

2.1. Experiment Setup

A Yamaha LC135, water-cooled 135cc single cylinder engine at ISI Laboratory, School of Aerospace Engineering, and University Science Malaysia cylinder is mount on an engine bench. The bench is design to suit the usage laboratory experimental work. Signals of CPS are record to identify frequency to spark plug timing activation. The data will be use to design the engine controller. Figure 2 shows the equipment for the experiment.

Figure 2: Experiment setup.

A switch will be controlling the artificial misfire condition. The Arduino controller will be controlling the spark timing. Therefore, the switch for artificial misfire condition is located with the same controller. The accelerometer mounted on the engine will record the signal during the artificial misfire condition is activated.

Figure 3 show the knock sensor is mount to the engine block, red and black wire is attach to the oscilloscope to record signals. There are few wire wrapped with a black tape. It contains charging circuit and CPS wire circuit. CPS wire circuit is attach to oscilloscope to identify the signals. The signals are identify to design the Arduino controller for the engine.

Figure 3: Knock sensor attached to the engine and CPS wire for signals identification.

IOP Conf. Series: Materials Science and Engineering **501** (2019) 012012 doi:10.1088/1757-899X/501/1/012012

3. Results and Discussion

Figure 4(c) show signals from CPS is record. The signals indicate an acceleration from 1000 RPM to 6000 RPM. The fluctuations show it produced alternate current, constantly within 5V. Different signals recorded for engine idling condition. Figure 4(a) show, where idling is the condition of the engine run at low speed constantly. Furthermore, figure 4(b) shows signal at high speed, about 6000 RPM. CPS signals show constant signals voltage but different width of peak-to-peak. It show the speed of the engine, whether it is on idling condition at low speed, acceleration from low speed to high speed or constantly in high-speed condition.

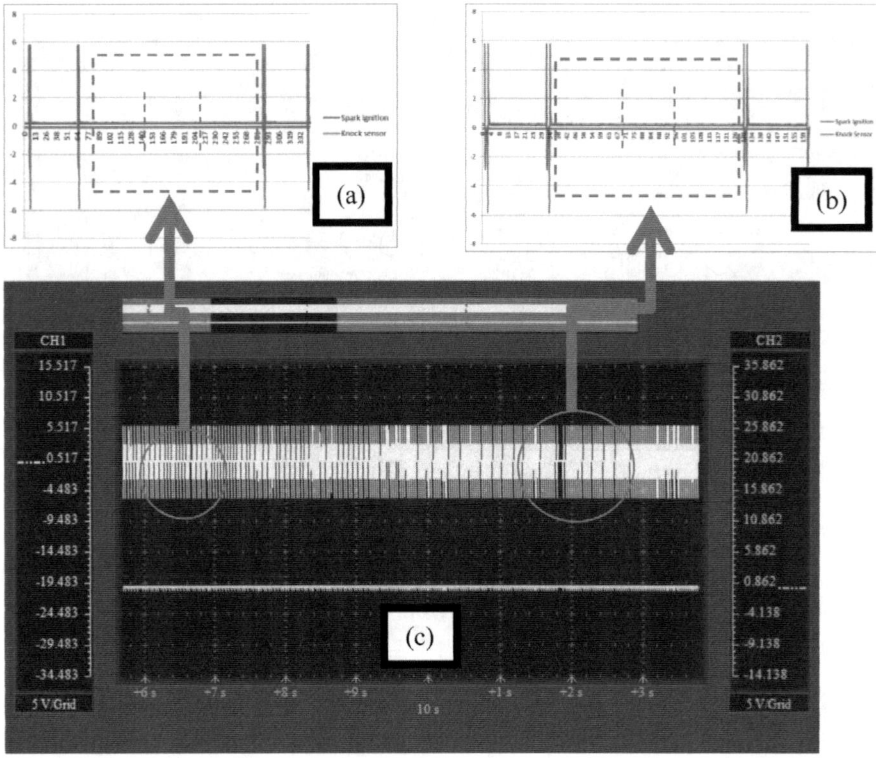

Figure 4: (a) low speed; (b) high speed; (c) signals range 1000-6000 RPM

From the signals recorded, it is possible to design spark plus timing controller based on the signals from the CPS. The CPS produced a constant voltage of 5V but different fluctuation between peak-to-peak conditions. The signals show the speed of the engine, where it is possible to actuate the spark plug timing, based on the peak-to-peak distance.

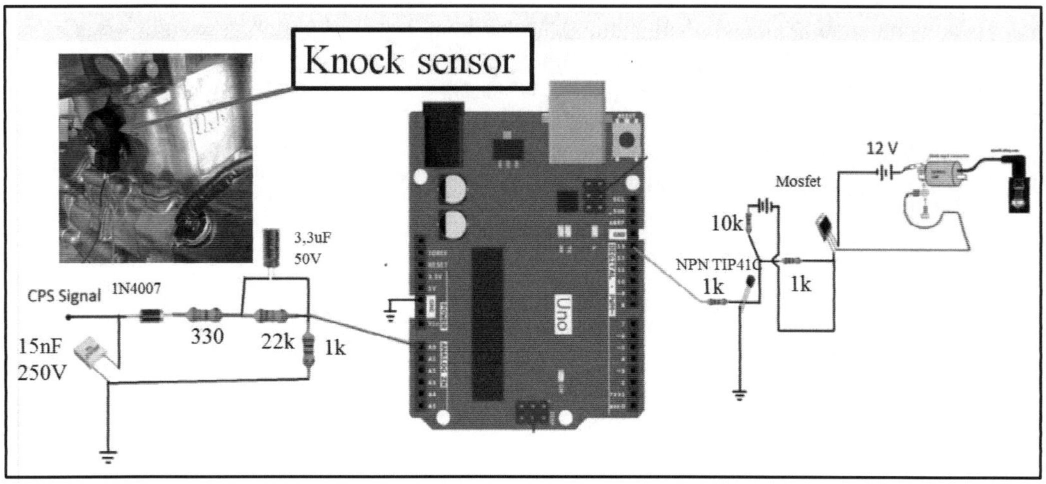

Figure 5: Arduino controller circuit design.

Figure 5, show the circuit design to actuate spark plug timing using Arduino controller. The CPS signals are used the actuate the spark plug timing. In other hand, to actuate misfire condition, a Arduino programmed based is used. The programmed is set to switch off the spark plug for a few times every 5-10 second. The condition are made to meet the misfire condition based to OBD II legislation refer to Bahram et al (2013)[11], Chu et al.(2005) [12].

4. Conclusion

Based on the analysed CPS signals, the peak-to-peak signal is the main trigger to spark plug timing using Arduino controller. Signal within 5V is detect to trigger the spark. Our signal manage to detect the peak-to-peak signal to initiate good timing based on the air fuel ratio and engine speed. Through model based, misfire condition can be design through the control unit. Over testing for hours, accuracy of 75% detection model based misfire condition. The knock sensor would able to detect misfire condition. In the same monitoring condition of knocking condition. For further analysis, a program can be add to comply the OBD II legislation to notify the misfire and knocking condition.

Acknowledgments

The authors would like to thanks the Ministry of Higher Education, Malaysia for the financial support of this research through Hadiah Latihan Persekutuan (HLP) and Innovative System and Instrumentation (ISI) Research Team, School of Aerospace Engineering, Engineering Campus, Universiti Sains Malaysia for support of this work.

References

[1] M. Mansha, E. M. Shahid, and A. H. Qureshi, 2012. *Pak. J. Engg. & Appl. Sci.* **11**, no. x, pp. 114–128,
[2] G. Shu, J. Pan, and H. Wei, 2013 *Appl. Therm. Eng.* **51** no. 1–2, pp. 1297–1306,
[3] J. Zareei and A. H. Kakaee 2013 *Eur. Transp. Res. Rev.* **5**, no. 2, pp. 109–116,
[4] L. Jia, J. D. Naber, and J. R. Blough, 2016. *J. Combust.* vol. 2016,
[5] J. Abraham and A. R. Bhende, 2015 *Int. J. Recent Innov. Trends Comput. Commun.* **3**, no. 2, pp. 129–132,
[6] T. Laurain, J. Lauber, and R. M. Palhares 2015 IEEE 10th Conference on Industrial Electronics and Applications (ICIEA)
[7] S. Naik, 2004 *Int. J. Adapt. Control Signal Process* **18**, no. 2, pp. 181–198,
[8] A. Roos, 2017. *Faculteit Economie Universiteit Gent*

[9] M. S. Ajmir. J. M. Nabil. A. A.B Elmi, A. Mohammad Nazir, 2018. *proceeding iCompEx'18.*
[10] M. Z. M. Pauzi, E. A. Bakar, and M. F. Ismail, 2016. *IOP Conf. Ser. Mater. Sci. Eng.* **114**, p. 012140,
[11] B. Bahri, A. A. Aziz, M. Shahbakhti, and M. F. M. Said, 2013 *Latest Trends Circuits, Control Signal Process* pp. 178–183,
[12] Y. Wang and F. Chu, 2005 *Mech. Syst. Signal Process* **19**, no. 4, pp. 900–912,

IOP Conf. Series: Materials Science and Engineering **501** (2019) 012013 doi:10.1088/1757-899X/501/1/012013

Improvement of windshield laminated glass model for finite element simulation of head-to-windshield impacts

J. Prasongngen[1], I. P. A. Putra[2], S. Koetniyom[1, 2], and J. Carmai[1, 2*]

[1]The Sirindhorn International Thai-German Graduate school of Engineering (TGGS)
[2]Automotive Safety and Assessment Engineering Research Centre
King Mongkut's University of Technology north Bangkok, 1518 Pracharat 1 Rd., Bangkok, 10800, Thailand

* Corresponding Author: julaluk.c@tggs.kmutnb.ac.th

Abstract. The automotive windshield, with which pedestrians come into frequent contact during car accidents, has been identified as one of the main sources for severe pedestrian head injuries. Finite element simulation of head-to-windshield is widely used for the design and evaluation of car safety performance. Most of the windshield models used in the simulations for pedestrian head injury studies are simplified model. However, accurate modelling of windshield mechanical behaviour is necessary for good prediction of head injury during car-to-pedestrian collision. The purpose of this paper is to improve the model of automobile windshield for simulation of a collision between a pedestrian and a car. The laminated glass windshields of various thickness were tested according to UNECE R43. The deformation and crack pattern were recorded for validation of the finite element model. The finite element models of two layers and three layers laminate glass were developed. Element deletion method and share node techniques were employed for crack propagation pattern. Appropriate mesh size and shaped were obtained. The simulation results were compared with the experimental results of ball drop tests. The simulation results of the laminated glass model with two layers, 1mm-triangle mesh shows good agreement with the experimental results. This laminated glass model was then implemented to the windshield of the car. Simulation of headform impact tests according to the EEVC protocol on pedestrian protection were performed. The results in terms of head deceleration, crack pattern and windshield energy absorption were compared with the test results. Good agreement can be seen.

1. Introduction

Almost half of all fatality in the world's road accident involves motorcyclists, cyclists and pedestrians [1]. The accident of vehicle-pedestrian collisions has been identified as one of the main reasons for severe pedestrian head injuries causing life-long disability or death [2].

Windshield protection and performance under head impact during a collision are legal requirements for the manufacturers. The safety glass for automobile windshields must follow UNECE R43 [3]. It requires a ball drop test and headform impact at various heights, to evaluate penetration, failure resistance and break pattern. The European New Car Assessment Program (Euro NCAP) [4] require headform impact tests to verify that the windshield assembly is less dangerous for pedestrians.

However, the automobile windshield models used in the simulations are all laminated glass components which are complex composite materials consisting of glass, polymer foils and adhesives (Polyvinyl butyral or PVB). Numerical models for windshield are not predictive with respect to

IOP Conf. Series: Materials Science and Engineering **501** (2019) 012013 doi:10.1088/1757-899X/501/1/012013

analysing their deformation and failure under impact loading conditions. The windshield model must be able to predict the initial failure of the laminated glass and the crack propagation. For the improvement of the predictability for windshield, the behaviour of glass was studied using finite element modelling. The glass is modelled as a linear elastic material. Once the principle strain reaches a critical value, the glass fails and the interlayer is carried loads. The PVB-interlayer is modelled as a hyper-elastic material [5-7].

The object of the study is to improve predictive capabilities of windshield model by including crack propagation. This is to enhance accuracy of pedestrian-to-car collision simulation. This paper first addresses laminated glass impact tests based on UNECE R43 [3]. The finite element model of laminated glass was developed and compared with experiment tests for validation. The laminated glass model was then employed for headform-to-windshield impact test according to EEVC test protocol [8].

2. Laminated glass impact tests

The laminated glass specimens of 300 mm × 300 mm with two different thicknesses were prepared for rigid ball drop tests as shown in figure 1. The specimen was placed between a steel frame and a rubber frame. The steel box was used as a base structure. The ball/impactor is made of steel with a diameter of 82 mm and has a mass of 2.26 kg in free motion as shown in figure 1. High speed camera was set up to record crack patterns. The test conditions and specimen thickness are illustrated in table 1.

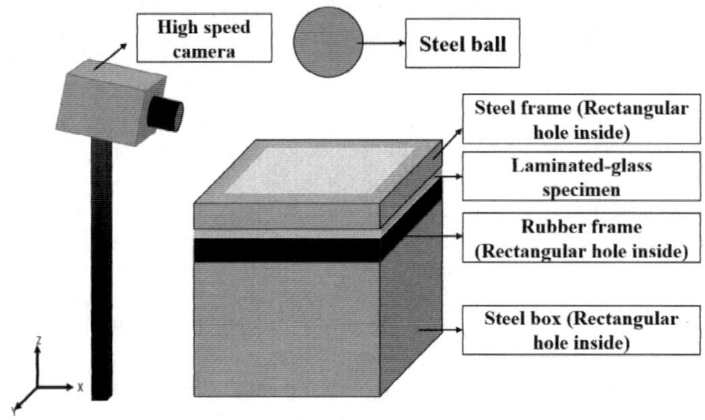

Figure 1. Ball-laminated glass experiment setup.

Table 1. Impact test conditions.

		Thickness of layers (mm)		
Test number	Drop height (m)	Total	Glass	PVB
1	2	6.18	2.90	0.38
2	4	6.40	2.82	0.76
3	2	8.14	3.88	0.38

The glass maximum deflections were measured together with the impact positions. The results of the experiment of the laminated glass-ball impact are shown in the table 2. Test number 1 and 3 were tested at the same drop height but with different specimen thickness. The thicker specimen in test no. 1 reveals larger deflection compared with the one on test no. 3. The ball penetrated the specimens for both tests. Different maximum deflections can be seen for the same thickness specimens but with different drop height as shown in test no. 1 and 2. The ball did not penetrate the specimen in test no. 3 due to the lower drop height. The crack patterns are also shown in table 2. The experimental results will be used to validate the finite element model of laminated glass developed in the next section.

Table 2. Experiment result of the ball impact test.

	Experiment		
Type	No. 1	No.2	No.3
Final crack		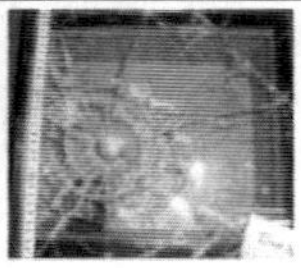	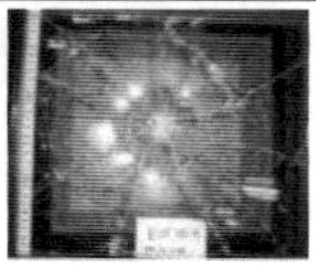
Final condition	Penetrated	Penetrated	Not penetrated
Impact position x, y (mm)	127, 176	93, 125	174, 165
Maximum deflection (mm)	34.30	39.70	22.66

3. Development of laminated glass impact model

Two and three layers laminated glass models were created as shown in figure 2. The two-layer model consisted of Glass-PVB whereas three-layer model consisted of Glass-PVB-Glass. The glass behaviour was modelled using *MAT_MODIFIED_ PIECEWISE_LINEAR_PLASTICITY and PVB behaviour was modelled using *MAT_PIECEWISE_LINEAR_PLASTICITY in LSDYNA [9] with material properties taken from a typical car windshield model [10] as summarized in table 3. Adhesion between glass layer and PVB later was modelled using the shared node technique [11]. This technique assumes that the glass layer and the PVB layer are perfectly bonded to each other and there are no relative displacements between glass and PVB. The crack characteristic was modelled using the Element Deletion Method (EDM) which is also known as the erosion contact method. The basic idea of this method is to remove the mass elements of the global mass matrix to represent the failure of these elements. This method has been implemented in the LS-Dyna software.

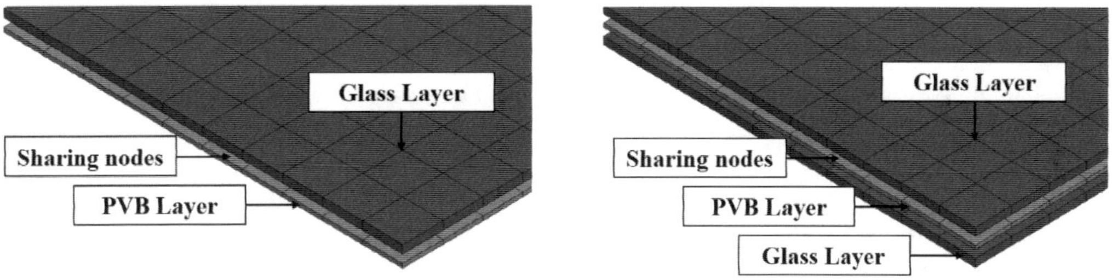

Figure 2. Two- and three-layers shell elements of laminated glass.

Table 3. Material properties [10].

Part	Young Modulus (GPa)	Density (kg/m^3)	Poisson's ratio	Plastic Strain at Failure
Glass	70.0	2,500	0.22	0.001
PVB	2.5	1,000	0.35	2.5

The laminated glass impact test was set up as shown in figure 3. The ball with 2.26 kg mass was model as rigid. Initial impact velocity corresponding to each drop height was assigned to the ball. All degree of freedom at 4 edges of specimen are fixed. The contact between ball and laminated glass was designd with a friction coefficient of 0.

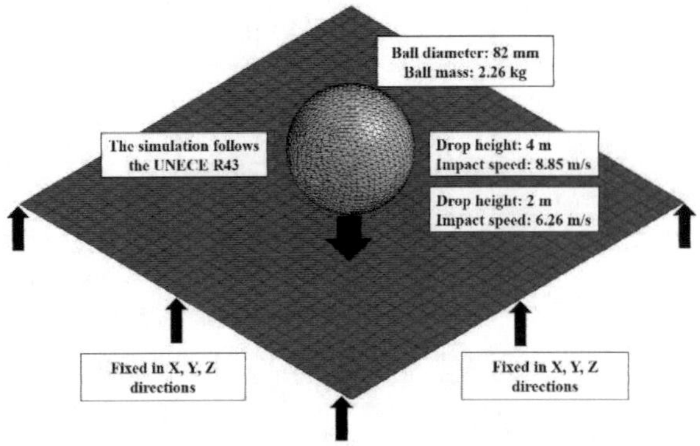

Figure 3. Simulation setup of ball laminated glass impact.

4. Mesh sensitivity study and model validation

The finite element simulations of test no. 2 were conducted with triangular and rectangular mesh. Each shape varied in sizes of 1 mm, 2 mm, 3 mm, 4 mm and 5 mm. In addition, all set up condition were performed for both two-layer and three-layer laminated glass models. The baseline windshield laminate glass model was also simulated for comparison. Total of 22 simulations were performed.

The crack pattern results of different mesh shape and size for two-layer and three-layer model are shown in figure 4 and 5 respectively. The crack characteristic was measured using the circular patterns as shown in figure 4 and 5. The circle diameters, maximum deflection, energy absorption obtained from various mesh sizes and shapes for two-layer and three-layer models are illustrated in table 4.

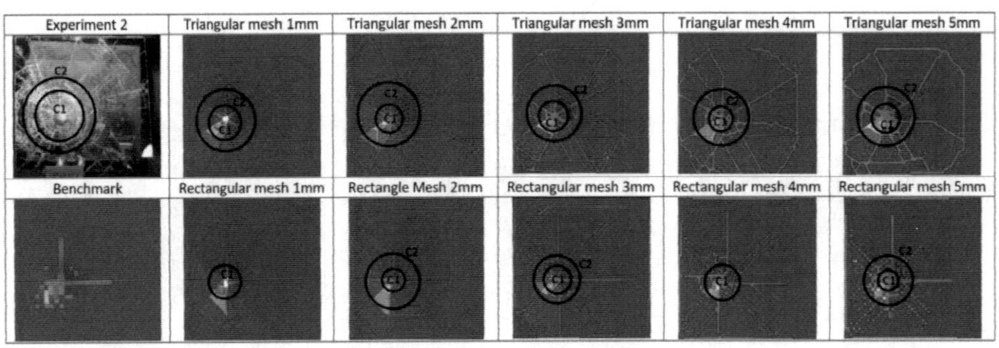

Figure 4. Comparison of crack pattern results between triangular and rectangular shape for two-layer laminated glass model.

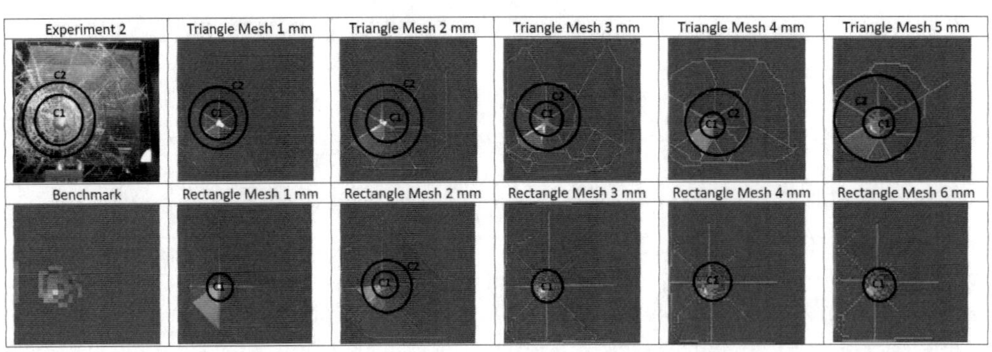

Figure 5. Comparison of crack pattern results between triangular and rectangular shape for three-layer laminated glass model.

From the simulation results in table 4, it was found that the triangle mesh gives better comparison with the experiment than the rectangle mesh for both two-layer and three-layer models. Smaller mesh size gives closer deflection and final crack patterns to the experimental results. It is also found that the maximum deflection and crack patterns obtained from the two-layer model compared better with the experimental results for all mesh sizes and shapes. The triangular shape of size 1 mm had the highest validation of 97.45%. Circular pattern of C1 was 96.20%, C2 was 98.21%. The predictive capability of the laminated glass model has been improved when comparing with results from the baseline laminate glass model without crack propagation. 21.5% improved in the maximum deflection. The energy absorption is higher than that obtained from the baseline model.

However, the results from the triangular shape of 1 mm were closest to that of the experiment. Reduction in mesh size resulted in an increase of the number of elements and therefore long simulation time. Triangular shape of 1 mm simulation time was 350 minutes whereas the mesh size of 5 mm took only 5 minutes. This project will select the triangular shape of 5 mm for improving the automobile windshield. This can save duration of simulation by 70 times with an accuracy of 93.97%. The difference in maximum deflection between 1 mm and 5mm triangular shape is only 3.5 %.

Table 4. FE simulation results of mesh sensitivity.

	Maximum deflection (mm)		Energy absorption (J)		Circular Diameter of 2L (mm)		Circular Diameter of 3L (mm)	
Type and size of mesh	2L[a]	3L[b]	2L	3L	C1[c]	C2[d]	C1	C2
Experiment 2	-	39.70	-	-	-	-	71.00	123.00
Benchmark	30.37	25.02	137.21	119.05	-	-	-	-
Triangle Mesh 1 mm	38.69	36.71	170.87	162.18	73.70	125.20	76.35	148.00
Triangle Mesh 2 mm	38.08	34.58	172.81	162.47	64.08	143.46	74.71	148.10
Triangle Mesh 3 mm	37.70	33.91	168.37	158.62	73.60	119.50	48.72	141.70
Triangle Mesh 4 mm	37.60	32.22	164.89	154.24	59.45	117.60	65.00	136.10
Triangle Mesh 5 mm	37.31	32.86	160.71	156.36	69.60	124.40	54.40	155.60
Rectangle Mesh 1 mm	36.56	33.51	159.35	143.44	66.34	-	71.08	-
Rectangle Mesh 2 mm	35.15	31.17	148.54	134.66	66.00	126.58	50.65	109.47
Rectangle Mesh 3 mm	35.17	31.22	143.30	134.81	71.10	-	64.47	-
Rectangle Mesh 4 mm	35.78	30.36	142.58	132.22	89.52	-	73.00	-
Rectangle Mesh 5 mm	36.69	30.76	140.20	131.95	58.50	120.55	72.75	-

[a]2L is 2 layers
[b]3L is 3 layers
[c]C1 for the first circular pattern
[d]C2 for the second circular pattern

The laminate glass model with 5 mm triangle mesh was also validated with experiment test no. 1 and no. 3. Comparisons of maximum deflection and crack patterns are shown in table 5. Reasonable agreement can be seen.

Table 5. FE simulation two layer of triangular 5 mm result

	Final crack		Maximum deflection (mm)	
Experiment	Test	Simulation	Test	Simulation
No. 1	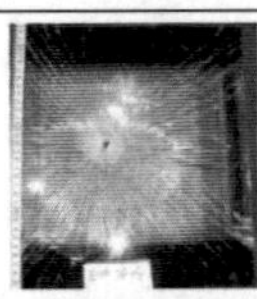	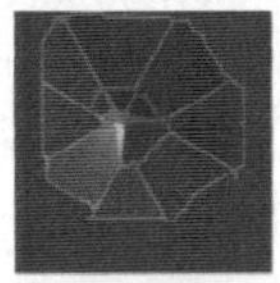	34.30	32.69
No. 3		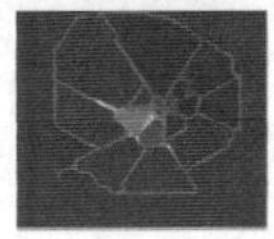	22.66	28.03

5. Development of headform-to-windshield impact test

5.1. Headform-to-windshield impact model set-up
The two-layer laminated glass model with triangle mesh of 5 mm validated in the previous section was adopted to model the car windshield. The headform of 4.8 kg mass was modelled according to EEVC test protocol [5]. An accelerometer was attached at the C.G. of the headform to measure the linear acceleration. Figure 6 shows the setup of headform-to-windshield impact simulation. All edges of windshield were fixed in all direction. Headform initial velocity was 11.1 m/s. It impacted at the middle of the windshield.

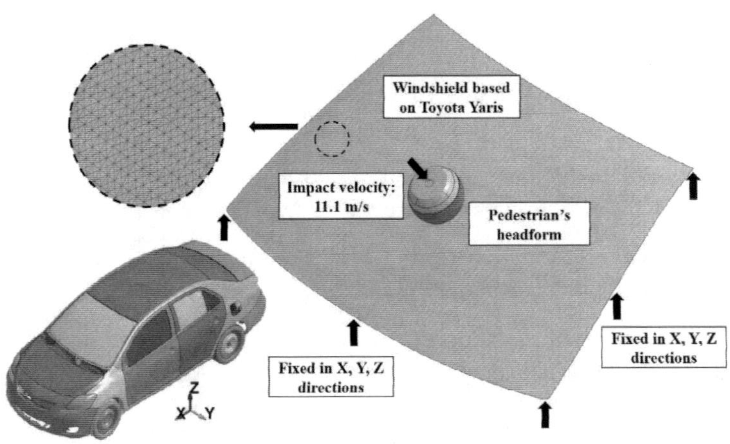

Figure 6. Windshield-headform impact simulation setup.

5.2. Validation results
The comparison of final crack pattern between the test [12] and the simulation are shown in figure 7. The result in term of headform's linear acceleration between the simulation and the test is presented in graph shown in figure 8. Overall, the results of the simulation showed good agreement with the result from the test. When the headform impacted the windshield in the centre position, the maximum value of headform's linear acceleration from the simulation was lower than the test results. The value in the test was 1,190 m/s^2 and the simulation was 1,100 m/s^2 with a validation of 92.44%. The overall picture's trend also showed good agreement with the test result.

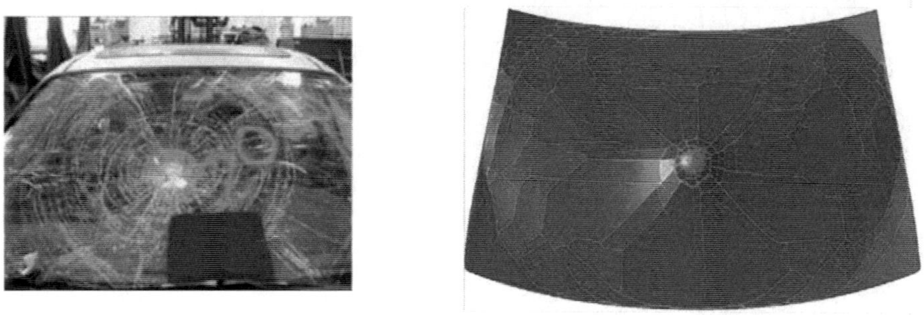

(a) (b)

Figure 7. Final crack pattern of windshield between (a) test and (b) simulation.

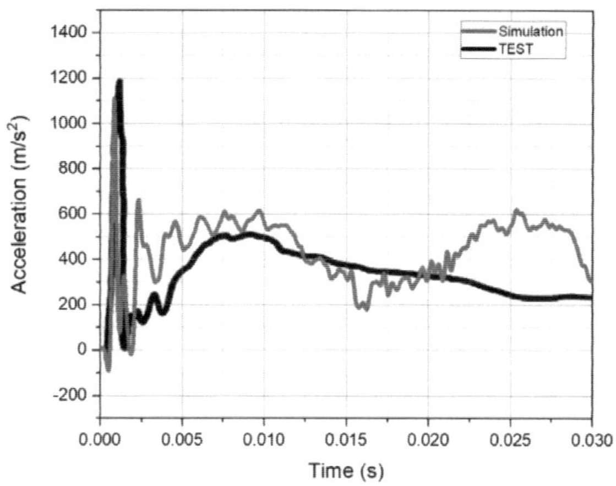

Figure 8. Comparison of headform's linear acceleration between the test and simulation.

6. Conclusion

Laminated glass models were set up in combinations of glass and PVB with different mesh shapes, mesh sizes and number of layers based on the LS-DYNA code. Adhesion between glass and PVB was modelled using shared node technique. Element deletion method was employed to include crack behaviour. Mesh sensitivity analysis was conducted. The results indicated that the triangular shape, size 1 mm model is the most accurate in representing a laminated glass. The laminated glass impact simulation shows good agreement with experimental result. Laminated glass model was employed for headform impact test. The simulation result was compared with experimental results. Good agreement of the head acceleration and crack pattern is revealed. Based on these results, vehicle designers should consider the design of the windshield for improvement in pedestrian protection.

Acknowledgments
Authors would like to thank V.M.C. Safety Glass Thailand Co., Ltd., for facilitating the glass-impact experiment.

References

[1] World Health Organization 2015 Global status report on road safety p 1-9
[2] Otte D 1999 Severity and mechanism of head impacts in car-to-pedestrian accidents In: *Proceedings of the Int.l conf. on the biomechanics of impacts* p 329-41

[3] United Nation Economic Commission for Europe 2012 Regulation 43: Uniform provisions concerning the approval of safety glazing materials and their installation on vehicles p 42-52
[4] European New Car Assessment Programme 2013 Pedestrian Testing Protocol p 1-53
[5] P A Bois D, Kolling S and Fassnacht W 2003 *Computational Materials Sci.* **28** 675-83
[6] Timmela M, Kollingb S, Osterriederc P, P A Boisd D 2007 *Int. J. Impact Engin.* **34** 1465-78
[7] Peng Y, Yang J, Deck C and Willinger R 2013 *Int. J. Impact Engin.* **57** 27-35
[8] European Experimental Vehicles Committee (EEVC) 1994 Proposals for methods to evaluate pedestrian protection for passenger cars p 2-40
[9] Livermore software Technology Corporation 2007 LS-DYNA KEYWORD USER'S MANUAL Version. 971 p 1393-2206
[10] Reichert, R., Mohan, P., Marzougui, D., Kan, C. et al., "Validation of a Toyota Camry Finite Element Model for Multiple Impact Configurations," SAE Technical Paper 2016-01-1534, 2016
[11] Chen S, Zang M, Wang D, Zheng Z, Zhao C 2016 Finite element modelling of impact damage in polyvinyl butyral laminated glass *Composite Structures* p 1-11
[12] Rooij van L, Bhalla K, Meissner M, Ivarsson J, Crandall J, Longhitano D, Yukou Takahashi, Dokko Y, Kikuchi Y 2003 Honda R&D Co., Ltd., Japan 5-6

IOP Conf. Series: Materials Science and Engineering **501** (2019) 012014 doi:10.1088/1757-899X/501/1/012014

Design of a CFRP composite monocoque: simulation approach

Tatthep Kanketr*, Ekkarin Phongphinnittana and Baramee Patamaprohm

Department of Mechanical and Aerospace Engineering

*Corresponding Author: tatthep.k@gmail.com, Telephone Number 0-2555-2000 ext. 8315, Fax Number 0-2586-9541

Abstract. This article deals with the design of CFRP composite monocoque chassis for Formula Student race car. The design objective is to maximize a specific torsional stiffness of the monocoque and also satisfy the safety requirement assigned by Formula SAE. The sandwich structure has been used regarding to its high flexural rigidity per weight. The thickness and stacking sequences of composite plies have been optimized for each particular zone of the monocoque chassis using the FEM simulation.

1. Introduction

Formula SAE is a design competition for engineering students to have an opportunity to apply their knowledge to design a downscale formula race car, so called a formula student. Among various components, chassis is the main one that affects the behavior of a race car and provides a safety for drivers. The chassis needs to have a sufficient stiffness and satisfy the safety standard qualified by Formula SAE. Currently, most of chassis in this competition are a tubular steel space frame which has limitations of weight. The replacement of space frame by a lightweight structure will improve a performance of a race car [1]. A monocoque made from carbon fiber-reinforced polymer (CFRP) composite seems to be a good candidate thanks to its very high strength-to-weight ratio. Although the use of CFRP composite structure for monocoque chassis in race car or high-performance car is very common, the detail on thickness and stacking of composite is always confidential. Moreover, this material is very costly and highly depends on its fabrication process. Thus, the simulation-based design is preferable to optimize the material parameters. A hybrid monocoque has been chosen in this study to economize the budget of the production process. This type of monocoque refers to a composite cockpit with conventional steel tube sub-frame (Figure 1). One of the key properties of racing cars is a torsional stiffness of chassis. High torsional stiffness will result in a good handling performance especially in cornering [2]. This property has been studied for different tube cross-section of space frame [3] and depends on a class of vehicle as shown in Table 1. In order to provide a high torsional stiffness, the composite sandwich structure has been used regarding to its high flexural rigidity per weight [4]. This structure consists of composite skin and core materials. The foam core material has been chosen since it has a good energy absorption and relatively low price comparing to a high-performance honeycomb core. This study aims to optimize the thickness and stacking sequences of composite plies for each particular zone of the monocoque chassis. The FEM simulation has been used in cooperation with analytical method in order to maximize its specific torsional stiffness and simultaneously satisfy the safety requirement of the Formula SAE. The results from this study will be used in the future work for a crash simulation of this monocoque.

Table 1. different chassis torsional stiffness of vehicle class

Vehicle	Torsional stiffness (N · m / degree)
FSAE	1000 – 5000
Passenger	5000 – 20000
Sports	15000 – 40000
Formula One	10000 – 100000

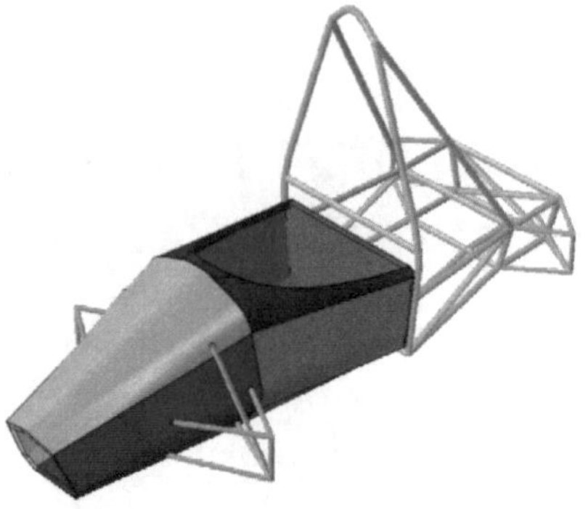

Figure 1. hybrid monocoque

2. Materials

In the study, the sandwich structure has been used thank to its high flexural rigidity per weight. The increased distance between two rigid skins by lightweight core can notably improve the structural stiffness while slightly adds an additional weight to the structure. The sandwich structure consists of skin and core material as shown in Figure 2 and 3. The CFRP composite is designated for a skin material while the closed-cell rigid foam is used as a core material. The mechanical properties of these two materials will be implemented in the FEM simulation.

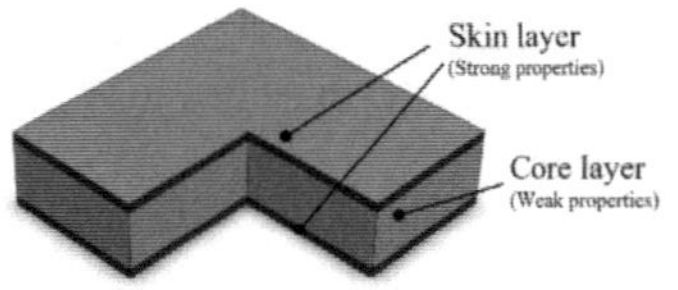

Figure 2. Components of sandwich structure

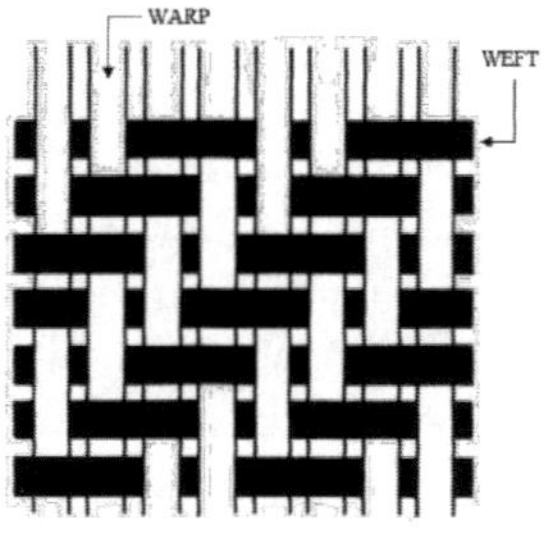

Figure 3. 2x2 twill weave fabric

2.1 Composite

The CFRP composite for skin material is fabricate from a carbon fiber woven fabric with 3k, 2x2 twill weave pattern and thickness of 0.25mm as a reinforcement phase and epoxy resin as a matrix phase. The 3k means 3,000 filaments per tow and the number of fibers in warp and weft direction is equal (2x2) [2]. The hand lay-up process was used to fabricate specimens for mechanical properties characterization. The curing took about 3-4 hours at room temperatures. The mechanical properties were characterized from the previous study [5] and summarize in Table 2.

Table 2. Properties of Carbon composite

Materials Properties	E_1 (GPa)	E_2 (GPa)	E_3 (GPa)	ν_{12}	ν_{23}	ν_{12}	G_{12} (GPa)	G_{23} (GPa)	G_{31} (GPa)	S11 (MPa)	S22 (MPa)	S33 (MPa)
Carbon	58.04	58.04	3.5	0.04	0.38	0.38	2.70	1.25	1.25	577.04	577.04	58.87

2.2. Foam

For a core material, PVC closed-cell rigid foam from DIAB has been chosen. The advantage of closed-cell rigid foam is energy absorption by its plastic deformation. It simultaneously enhances the crash performance of the monocoque chassis. The thickness of foam core is fixed to 20 mm. Its mechanical behavior and material properties have been characterized from the previous study and summarize in Table 3 and Figure 4.

Table 3. Properties of PVC foam

Materials Properties	E (MPa)	ν $(=\nu_{plastic})$	S_{yield} (MPa)
Rigid Foam	34.15	0	0.88

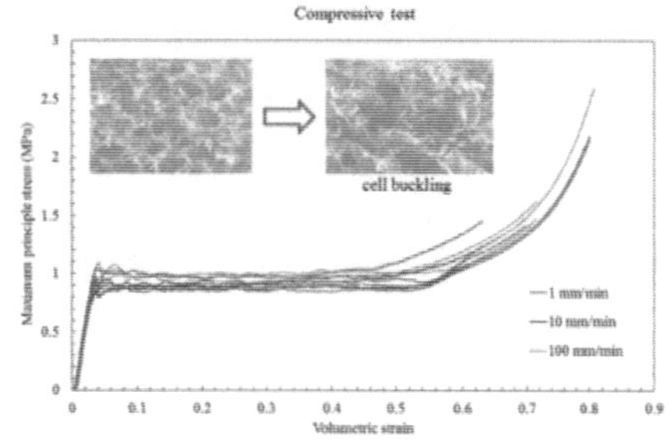

Figure 4. PVC foam behavior under compression tests

3. FEM Simulations

According to the FSAE Structural Equivalency Spreadsheet (SES) [6], there are the minimum requirements of mechanical properties including flexural rigidity (EI), area (A), yield tensile strength (YTS), ultimate tensile strength (UTS), max load at mid-span to give UTS for 1m long tube, max deflection at mid-span to give UTS for 1m long tube, and energy absorbed up to UTS, known as the baselines, for alternative material to be used in chassis design [5]. All properties baselines are equivalent to space frame steel tube properties for safety issues. Each material has different baseline requirements. For composite materials, the minimum bending or flexural modulus of the T3.30 Laminate test in SES is required.

The simulation approach is preceded in this study. Two stages of FEM simulations have been carried out using Abaqus/CAE. The simulation of three points bending test on sandwich structures with different design of composite stacking needs to be initially performed in order to ensure that each design is satisfied the flexural modulus baseline. Then, the potential designs will be used in the second simulation stage which is the composite monocoque simulation. The objective of this simulation is to investigate structural torsional stiffness of monocoque when using the given composite architectures. For Formula SAE, the monocoque structure can be divided into 7 sections as shown in the Figure 5. and the Table 4. Each section also requires different baseline.

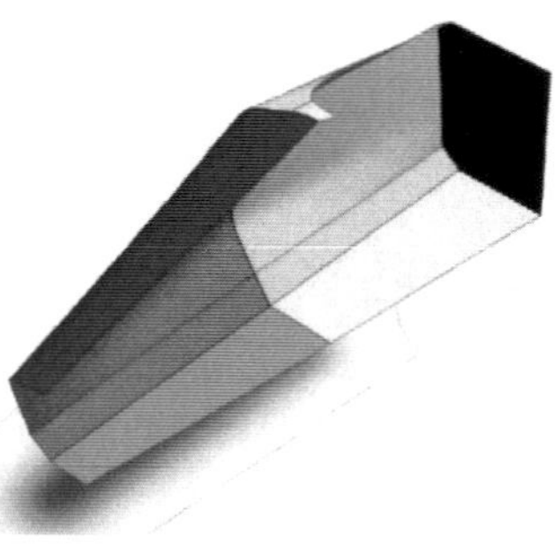

Figure 5. Monocoque different sections

Table 4. Overview of monocoque section

Colour	Section Name
	Front Hoop Bracing (FHB)
	Front Bulkhead Support Structure (FBSS)
	Front Floor (FF)
	Top Cockpit (TC)
	Side Impact Structure Side (SISS)
	Side Impact Structure Floor (SISF)
	Back Support (BS)

IOP Conf. Series: Materials Science and Engineering **501** (2019) 012014 doi:10.1088/1757-899X/501/1/012014

3.1 Sandwich structure bending simulation

The numerical sandwich structure specimen was created according to the FSAE structure Equivalency Spreadsheet (SES), Laminate test result T3.30. The width of sandwich panel is recommended to 275 mm and the span length between supports is set to 400 mm. The foam core thickness is fixed to 20 mm. The sandwich panel contains 3 layers of composite on each side with the thickness of 1 mm per layer. The 3D model was created in Abaqus with C3D8R element type. The global mesh size was set to 3 mm which leads to 202,400 elements. All support pins and loading pin of bending test were defined as a rigid body. Only sandwich panel specimen was allowed to deform. The outer composite layers were assigned a different fiber orientation defined as SSP (Stacking Sequence Pattern) for each simulation. In this study, 4 SSPs have been investigated and summarized in Table 5. The two support pins were fixed and a 7500N force was applied at the loading pin as shown in Figure 6. The sandwich pattern design from this simulation stage was guaranteed to meet the SES sheet requirement.

Table 5. Stacking sequence of sandwich structure

Code	Stacking sequence
SSP1	[0 / 0 / 0 / Core / 0 / 0 / 0]
SSP2	[0 / 45 / 0 / Core / 0 / -45 / 0]
SSP3	[0 / 30 / 0 / Core / 0 / -30 / 0]
SSP4	[0 / 60 / 0 / Core / 0 / -60 / 0]

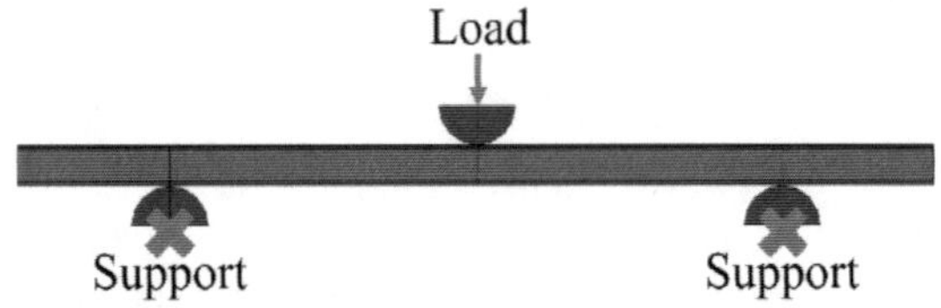

Figure 6. 3-points bending model and boundary conditions

3.2 Monocoque torsional stiffness simulation

The hybrid model was used to simulate the torsional stiffness of the monocoque chassis. This model uses shell elements for the composite structure part and 2-node linear beam section element for the steel tubes both of sub-frame and suspension components as shown in Figure 2. The composite sandwich properties including core thickness, skin layer thickness and stacking sequence of composite layers were assigned to shell elements while steel properties and cross-sectional geometries of space frame tubes were given to 2-node linear beam section elements in Abaqus. The boundary conditions were set as a typical torsional stiffness test; the rear A-beams were fixed at both sides and the force was applied to two font A-beams as shown in Figure 6. The displacement under loading was obtained from the simulation to calculate the angle of twist using eq.1. Finally, the torsional rigidity was determined by eq.2 using the angle of twist and the applied force.

$$\mathrm{TR} = \frac{Fl}{1000 \cdot \theta} \quad (1)$$

$$\theta = \frac{2|\delta|}{l} \times \frac{180}{\pi} \quad (2)$$

where

TR =Torsional Rigidity $(N \cdot m / \deg ree)$

d = displacement in vertical axis (mm)

l = lenght between left A-beam to right A-beam (mm)

F = testing load (N)

θ = angle twist

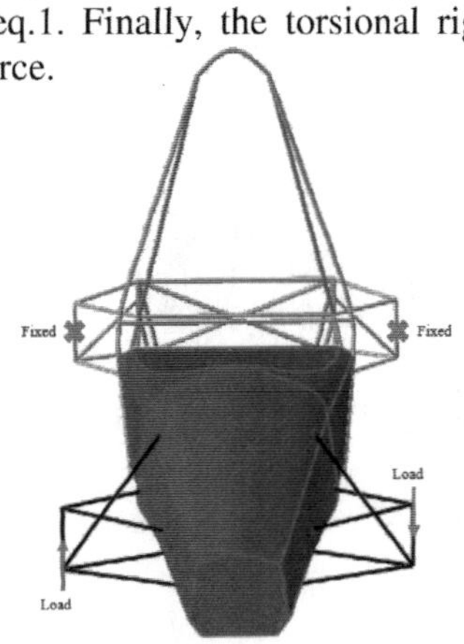

Figure 7. Boundary condition of torsional stiffness simulation

4. Simulation results

4.1 Sandwich structure bending result

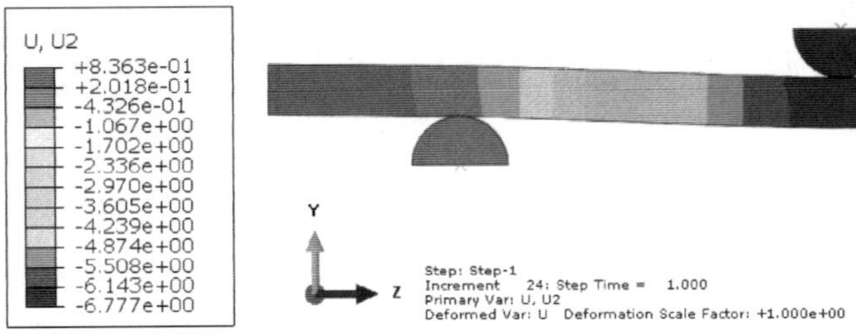

Figure 8. 3-point bending contour of a vertical deflection (Y-axis)

The vertical deflection of 3-point bending simulation is shown in Figure 7 as the contour displacement in Y-axis. The bending of flexural modulus of the 4 SSPs is defined as a slope of applied force-displacement curves (Figure 4). The results show that all SSPs pass the minimum requirement from SES laminate testing at 190 N/mm.

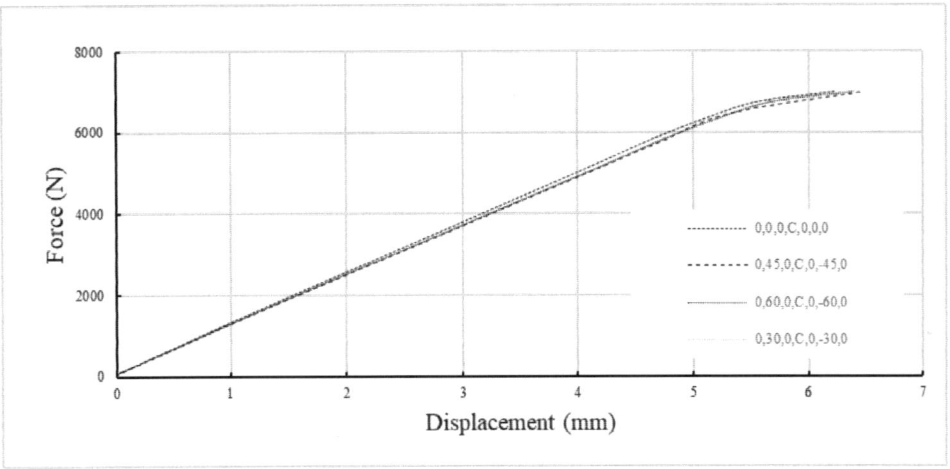

Figure 9. Simulation results of 3-points bending test

In addition to the minimum flexural modulus of sandwich structure used in monocoque chassis (the first stage simulation), 4 out of 7 sections of monocoque which are greatly affect the torsional stiffness; Front Hoop Bracing (FHB), Front Bulkhead Support Structure (FBSS), Side Impact Structure Side (SISS) and Side Impact Structure Floor (SISF) have to pass their minimum requirements assigned by FSAE SES. Each particular zone has different baseline which relatively compare the sandwich structure to a number of steel tube as shown in Table 6. We noted that the SISS and SISF were considered together known as Side Impact Structure (SIS). The required mechanical properties; EI and YTS for each zone have been calculated and compare to its baselines in term of percentages (Table 4). The linearity limit at 6200 N in Figure 4 has been inputted to FSAE Structural Equivalency Spreadsheet for YTS calculation and compare to YTS baseline (steel). The percentages of EI and YTS for the three mandatory zones need to be over 100% so that they can be used in monocoque model to simulate the torsional stiffness.

IOP Conf. Series: Materials Science and Engineering **501** (2019) 012014 doi:10.1088/1757-899X/501/1/012014

Table 6. The result of structure equivalent in percent

Zone	Number of tubes	EI	Yield tensile strength
FHB	1	100.3%	161.1%
FBSS	3	101.3%	154.0%
SIS	3	115.8%	NA

4.2 Monocoque torsional stiffness result

The simulation result of vertical displacement for angle of twist calculation and local in-plan shear stress ($S_{12,local}$) for Model 1 is demonstrated in Figure 6. Table 5 is the summary of all models angle of twist and torsional stiffness along with their composite designs.

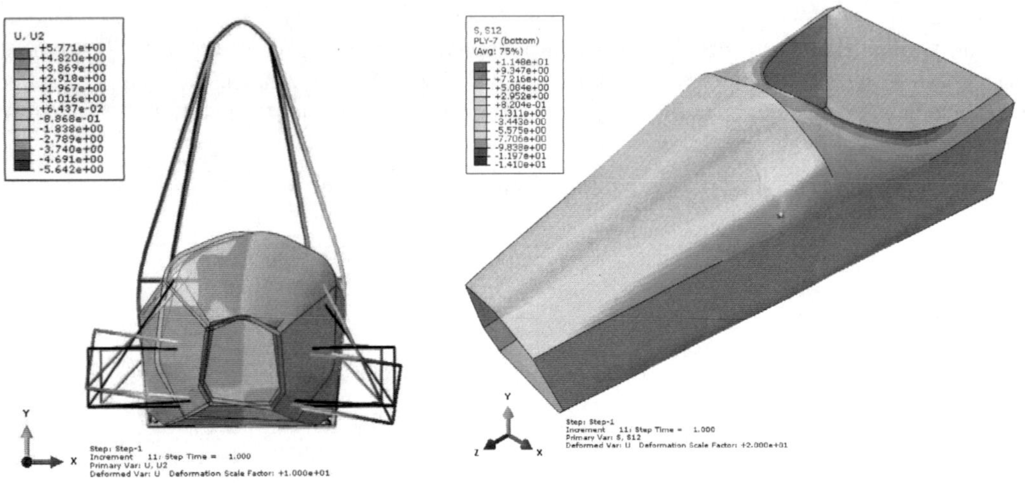

Figure 10. Contour plot of a Y-deflection and shear stress

Table 7. Summary of torsional stiffness in different patterns.

	FHB	FBSS	FF	TC	SISS	SISF	BS	Degree (θ)	Torsional stiffness ($N \cdot m/degree$)
Model 1	SSP1	SSP1	SSP1	SSP1	SSP1	SSP1	SSP1	0.6505	5336.63
Model 2	SSP1	SSP1	SSP2	SSP2	SSP1	SSP1	SSP1	0.5710	6079.57
Model 3	SSP2	SSP1	SSP1	SSP2	SSP2	SSP1	SSP1	0.5204	6671.41
Model 4	SSP3	SSP1	SSP1	SSP3	SSP3	SSP1	SSP1	0.5538	6267.43
Model 5	SSP4	SSP1	SSP1	SSP4	SSP4	SSP1	SSP1	0.5596	6204.73
Model 6	SSP2	SSP2	SSP2	SSP2	SSP2	SSP2	SSP2	0.4900	7085.46

IOP Conf. Series: Materials Science and Engineering **501** (2019) 012014 doi:10.1088/1757-899X/501/1/012014

5. Discussion

The results from the 3-points bending simulation show that the orientations of fiber which interprets by composite stacking do not affect the flexural modulus as long as they contain the 0° woven ply (1.2% max. difference). This can be explained by the nature of composite that preferable supports a given load in fiber direction so that the 0° woven ply will support almost all bending load. On the other hand, the fiber orientation has highly effect on the torsional stiffness as the results of monocoque simulation. The models 1-7 show the modification of torsional stiffness depending on their SSPs. The maximum torsional stiffness was obtained from Model 6 at 7085 Nm where the SSP2 [0/45/0/Core/0/-45/0] was applied to all monocoque section. This shows a good agreement with the theory of composite mentioning that the ±45° composite stacking sequence is preferable to withstand the torsional load. Comparing to the torsional stiffness of metallic space frame at 1900-2000 N·m, the composite monocoque is far better in this concern and will definitely provide a better handling for the race car. However, in the real-world application, the monocoque does not have only torsional load but also bending load or even impact load in case of crash. The use of Model 6 may not be optimized for all loading. Hence the criteria to design a composite monocoque should also have considered together the other structure properties, especially the energy absorbed capacity which relates directly to the safety issue of drivers.

6. Conclusion

This study is to design the composite monocoque using simulation approach to maximize its torsional stiffness. The conclusion can be made as follows:

- The flexural modulus from all proposed composite stacking patterns is relatively the same since all designs contain a 0° woven ply which is a dominated ply to support bending stress
- Fiber orientations which represent by composite stacking have an effect of torsional stiffness of monocoque structure
- FEM simulation simplifies the difficulty to optimize a composite stacking that satisfy the safety standard and maximize the torsional stiffness at the same time
- The future work will focus on the effect of core material thickness which also relate to the crash energy absorption of the monocoque

References

[1] Takahashi N, Kageyama Y, and Kawamura N 2011 Research of Multi-Axial Carbon Fiber Prepreg Material for Vehicle Body (No. 2011-01-0216). SAE Technical Paper.

[2] Sampo E 2010 Chassis torsional stiffness: analysis of the influence on vehicle dynamics. In SAE 2010World Congress&Exhibition.

[3] Liu CH, Li G, Ma YH and Yang XG 2018 Torsional Stiffness Comparison of Different Tube Cross-Sections of a Formula SAE Car Space Frame. In *MATEC Web of Conferences* 153 04002 EDP Sciences.

[4] Gay D and Hoa SV 2007 Composite materials: design and applications. CRC press. vol 4-chapter 4, 69-81.

[5] Patamaprohm B, et al, "Improving the impact resistance of CFRP composite structures", The 32nd *Conference of Mechanical Engineering Network of Thailand*, 3-6 July 2018, Mukdahan,Thailand

[6] SAE International [Internet]. Pennsylvania: 2018 FSAE structural Equivalency Spreadsheet V1.3 Available from: https://www.fsaeonline.com/cdsweb/gen/DocumentResources.aspx

IOP Conf. Series: Materials Science and Engineering **501** (2019) 012015 doi:10.1088/1757-899X/501/1/012015

Development and evaluation of advanced seat and helmet for safer motorcycle users

S Wannapor*, S Koetniyom and J Carmai

The Sirindhorn International Thai-German Graduate School of Engineering, King Mongkut's University of Technology North Bangkok, Bangkok, Thailand

* Corresponding Author: Sthaphorn.w-mae2015@tggs-bangkok.org

Abstract. The purpose of this research is to develop the prototypes of advanced motorcycle seat and protective equipment of compulsory helmet wearing through experimental evaluation. In this prototype, suitable sensors for detecting pillion, strapping chin and wearing helmet of motorcycle users are integrated with the existing motorcycle and helmet. These sensors were developed and investigated through the experimental testing procedures. For the evaluation of pillion seat sensor, the hip of dummy hybrid III 50th percentile male was used with the discrete values of mass on the sliding machine at each position of pillion seat. To conduct the experimental evaluation of strapping chin and wearing helmet conditions, the EN960 european standard headform were used at each position. Consequently, the functional detective range of the developed prototypes are identified and specified at pillion seat, chin strap and helmet wearing conditions. Additionally, these results are planned to develop an advanced system for supporting motorcycle users and increasing awareness of injury reduction.

1. Introduction

Motorcycle is very popular mode of transport in Southeast Asia [1, 2] because the motorcycle has agility, convenience and effectiveness in city traffic conditions such as the inaccessible public transportation. However, motorcycle is the least safe form when compared with other transportations because of its lack of protection around the rider with greater chances of severe injuries in case of crash. In fact, 34% of total road traffic deaths was caused by motorcycle accidents in Southeast Asian region which was one of the highest when compared to any other region of the world. Thailand uses a great number of motorcycles and confronts with high rate of motorcycle user fatalities and injuries. It is well accepted that the road traffic accident occurs due to at least one of the three elements: human error, road and environment defects or vehicle defects [3].

In the past, the technology and innovation were continuously developed to have sustainable growth in Japan motorcycle industries [4]. However, the safety innovation is one of the key factors that can protect from the road accident such as motorcycle AFS (Adaptive Front-lighting System) [5], smart helmet [6-11]. For the smart helmet, various sensors are used to detect the wearing condition of rider. Such system can enforce only the rider to wear the helmet. Therefore, the system should be compulsory for both rider and pillion to wear the helmet for protection of their severe injuries or death. In another user enforcement in passenger car, warning technologies were developed for enforcement of the driver and passenger such as seat belt reminder (SBR).

Even though the safety technologies and innovation were developed to protect road users from accident, such evaluation should be conducted for their limitation and characteristics to inform the related users and organization. For example, the control variation of antilock brake system (ABS) for motorcycles were evaluated in term of benefit and effectiveness from 350 real-world accidents using various parameters such as collision speed, initial speed distance of falling to collision point and braking distance etc. [12]. In addition, 13 motorcycle models with and without ABS technology during 2003–2008 were compared with fatal crashes per 10,000 registered vehicle years in order to understand the effectiveness of such technology and innovation [13].

Therefore, the purpose of this research is to develop and evaluate the prototypes of advance motorcycle seat and protective equipment for enforcing helmet wearing through the experiment.

2. Development of advanced seat and helmet

In this research, the prototypes of advanced motorcycle seat and helmet was developed. It consists of two main parts, i.e. pillion and helmet parts. The ergonomics of human and operating conditions of motorcycle users were considered in this development. The suitable sensors were attached with helmet and seat in order to detect the pillion, strapping chin and wearing a helmet. Moreover, the selected sensors should be integrated and combined in the existing motorcycle and helmet.

2.1. Detection of the pillion

These consists of flex sensor and press pad mechanism. The attachment of flex sensor 2.2" inches length with resistance 70 Ohm as shown in figure 1(a) is used to detect the pillion presence by the flex sensor bending due to mass gravity. Flex sensor is attached by superglue and double-sided of tape. Press pad mechanism is attached above the flex sensor as shown in figure 1(b, c) and used to distribute the mass of pillion at pillion seat. Press pad mechanism is designed by considering the width of human sit bones to ensure that flex sensor is bent while the pillion sitting on the seat [14].

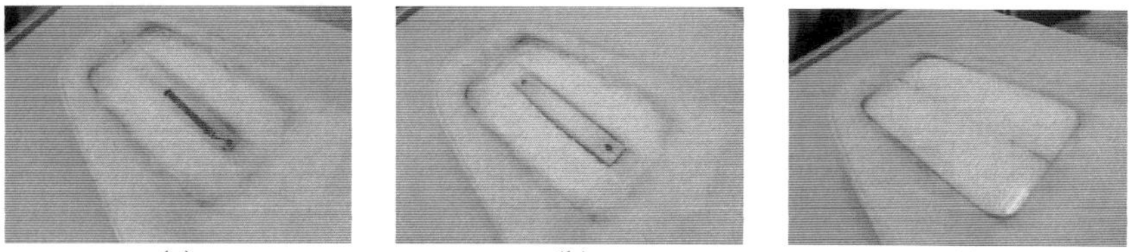

(a) (b) (c)

Figure 1. Attachment of flex sensor and press pad mechanism.

2.2. Detection of the strapping chin and wearing helmet

These consists of flex sensor and infrared sensor. The same flex sensor 2.2 inches length in the detection of the pillion are used and attached by the super glue and double-sided tape at the left side of chin strap. Flex sensor is used to detect the strapping chin curve of motorcycle users. To detect the wearing helmet condition by motorcycle users, the infrared sensor was attached inside the helmet at the position of ear by superglue and double-side tape. The detection of infrared sensor is 2 to 30 cm of distance and 35 degree of detection angle. The attachment of flex and infrared sensor are shown in figure 2.

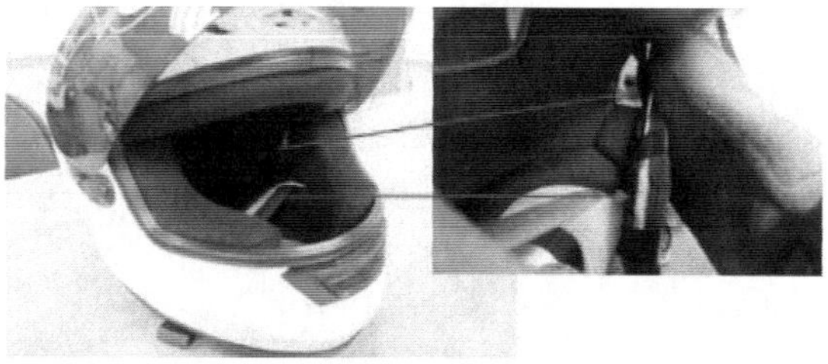

Figure 2. Attachment of flex and infrared sensor.

3. Experimental evaluation

The experimental evaluation were performed in the laboratory by using the hip of dummy hybrid III 50th percentile male and EN960 standard headform. The complete hip of dummy hybrid III and EN960 european standard headform were accomplished by CNC milling machine, which are made from high-density foam. The dummy hybrid III and EN960 european standard headform are the most widely used crash test dummy in the world for the evaluation of automotive safety restraint systems in the frontal crash test. They are also used to evaluate the protective helmet under repeatability of testing conditions.

3.1. Pillion seat

The experiment setup as shown in figure 3, are composed of hip of dummy hybrid III, mass, sliding machine and tape ruler. The hip of dummy hybrid III was used for evaluating the pillion seat sensor. The sliding machine is pressed on a hip of dummy hybrid III via H-point. The discrete value of mass is usually applied on the pressed sliding machine to represent the mass of motorcycle users. The tape ruler is attached on the floor to indicate the positions of pillion seat.

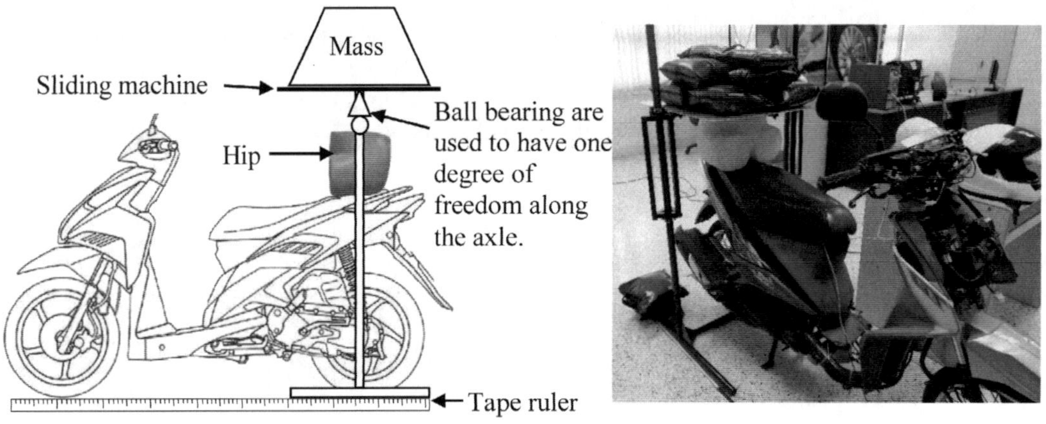

Figure 3. Experiment setup of pillion seat sensor.

In the experimental procedure, the veracity of the experimental results relies on a precision of mass and position of pillion seat. Therefore, the mass and position of pillion seat should are measured to achieve their precision values under calibration. The positions of pillion seat were divided as shown in figure 4, by measuring in the X direction from the end of pillion seat as the reference point (R). The input parameters for the experimental cases are given in table 1.

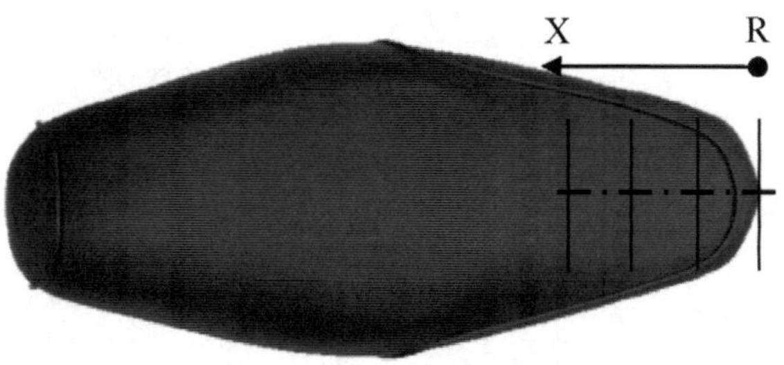

Figure 4. The positions of seat evaluation.

IOP Conf. Series: Materials Science and Engineering **501** (2019) 012015 doi:10.1088/1757-899X/501/1/012015

Table 1. Experimental cases of pillion seat sensor with input parameters.

Test cases	Position (cm)	Mass (kg)
1	$X_1 = 10$	30.8
2		40.8
3		50.8
4		60.8
5		70.8
6		80.8
7	$X_2 = 20$	30.8
8		40.8
9		50.8
10		60.8
11		70.8
12		80.8
13	$X_3 = 30$	30.8
14		40.8
15		50.8
16		60.8
17		70.8
18		80.8

3.2. Strapping chin and wearing helmet

The EN960 european standard headform was used for evaluating the strapping chin and wearing helmet sensors. The plane of EN960 headform as shown in figure 5, are composed of BB' plane, AA' plane, reference plane and basic plane. The BB' plane (2.54 cm from reference plane) is the additional plane in order to analyze the problem of interest. These planes are deemed to correspond to the level of the lower edge of the headband of helmet.

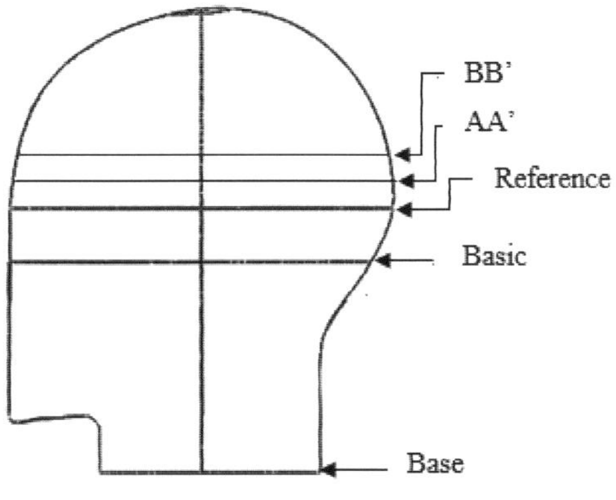

Figure 5. EN960 european standard headform plane.

In the experimental procedure, a helmet is worn with headform as shown in figure 6, by placing the helmet leading edge as the same level with BB' plane and AA' plane. Meanwhile, there are three strapping chin characteristics i.e. fit, slack and release conditions. The round gauge [15] with cylinder diameter of 2.54 cm was used to represent the gap measurement between the chin of headform and chin strap with slack condition. In the experimental preparation, the size of headform should have the same fit condition with the size of helmet. To conduct the experimental work, the input parameters for the experimental cases are given in table 2.

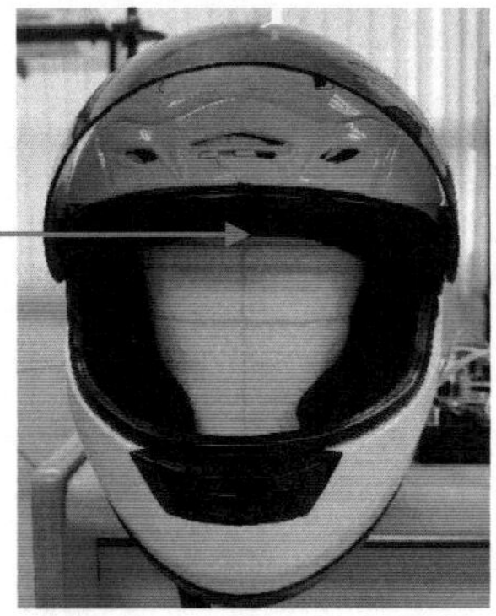

Figure 6. Experiment setup of strapping chin and wearing helmet sensor.

Table 2. Experimental cases of strapping chin and wearing helmet sensors with input parameters.

Test cases	Position of wearing helmet	Chin strap condition
1	AA' plane	Fit
2		Slack
3		Release
4	BB' plane	Fit
5		Slack
6		Release

4. Data recorder

The data recorder was used to record the output of resistance change of the sensors. The sampling rate at 0.5 Hz is used for monitoring and recording the data.

5. Experimental results

To identify and specify the pillion seat, the chin strap and wearing helmet conditions, the resistance change of the sensors was principally considered.

5.1 Pillion seat

The resistance change of pillion seat sensor at $X_1 = 10$, $X_2 = 20$ and $X_3 = 30$ cm positions are shown in figure 7. Before applying the mass on the seat, the resistance of the sensor starts from 35 Ohm. Position X_1, in the graph can be seem that the resistance was not changed with 30.8 kg and unequal changed with 40.8, 50.8, 60.8, 70.8 and 80.8 kg. In the position X_2 the result indicated that the resistance was greatly changed equally with 30.8 to 80.8 kg. However, the resistance was changed from 30.8 to 80.8 kg in the position X_3, which was lower than that in the position X_2.

5.2 Strapping chin and wearing helmet

The resistance of strapping chin and wearing helmet sensor are shown in figure 8. Before strapping chin and wearing a helmet, the resistance of the sensors start from 64 Ohm and 211 Ohm respectively. For the strapping chin and wearing a helmet at AA' plane and BB' plane conditions, it seems that the resistance values of fit and slack conditions were changed from the original to low value. Whereas the resistance of releasing chin strap condition was higher than that of fit and slack conditions. The different resistance value is the indicator of strapping or unstrapping chin conditions. In term of wearing a helmet, the resistance was changed from the original to low values.

IOP Conf. Series: Materials Science and Engineering **501** (2019) 012015 doi:10.1088/1757-899X/501/1/012015

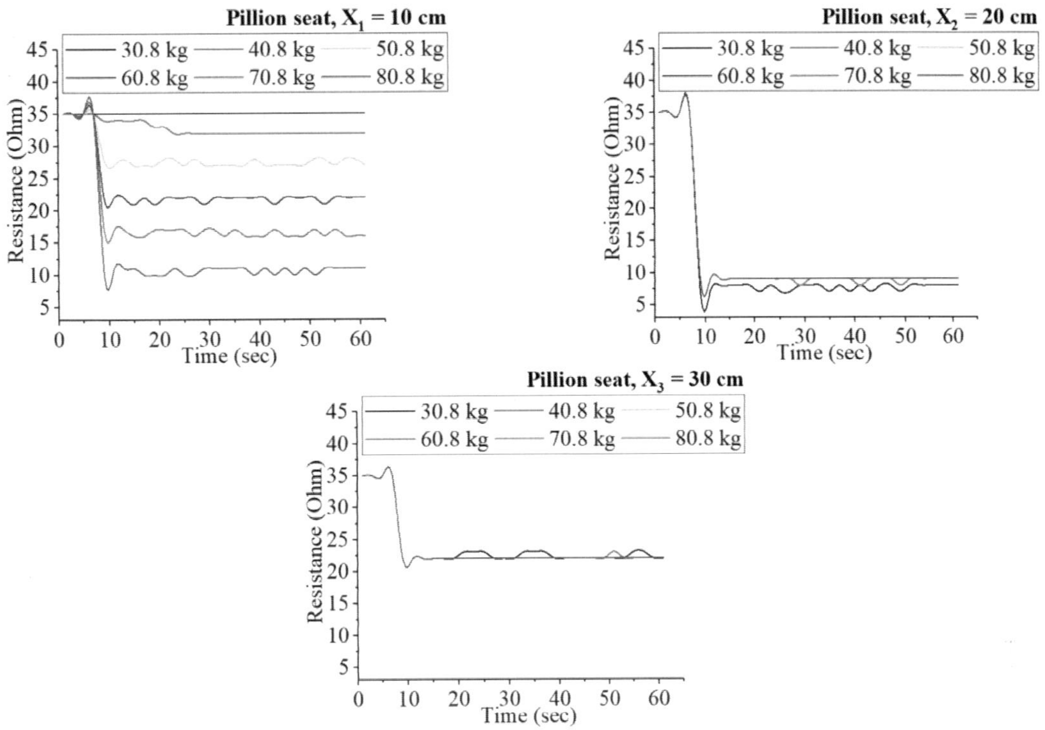

Figure 7. The resistance change of pillion seat sensor.

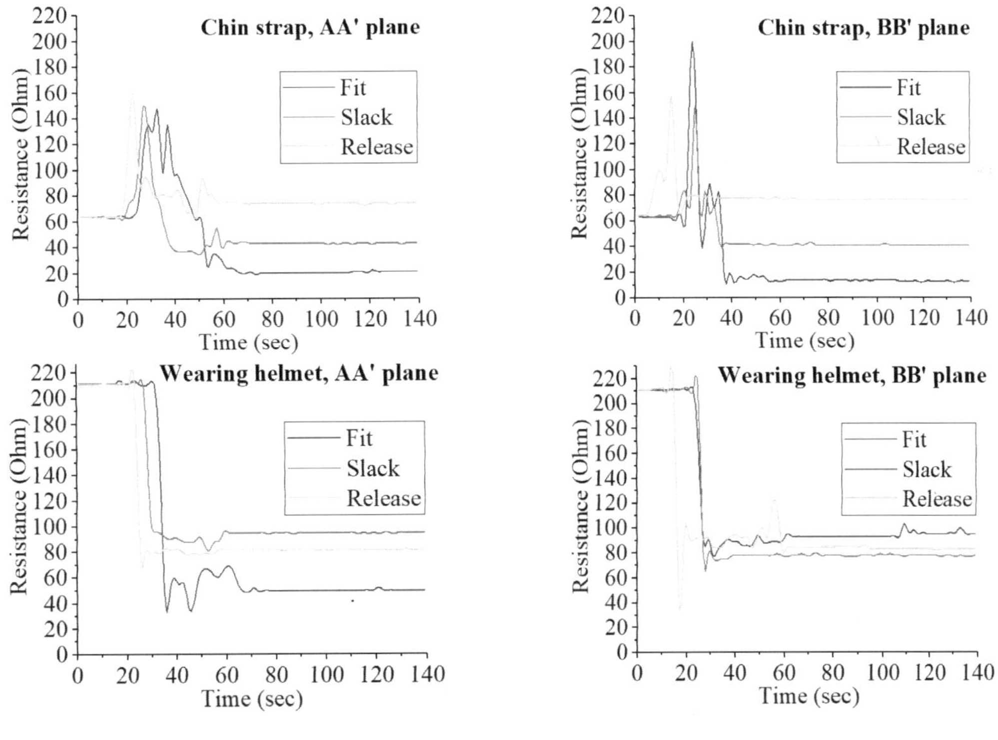

Figure 8. The resistance change of strapping chin and wearing helmet sensor.

IOP Conf. Series: Materials Science and Engineering **501** (2019) 012015 doi:10.1088/1757-899X/501/1/012015

6. Discussion

The evaluation of advanced seat and helmet in this research was performed by the static experiment. Actually, the daily motion of sitting position in motorcycle is variable. The motorcycle users might adjust the position to sit on the seat and change the position of wearing helmet. However, the system could be conducted in the dynamic experiment in the future.

7. Conclusion

In this research paper, an experiment is performed to evaluate the functional detection range of pillion seat, chin strap and wearing helmet sensor by using the hip of dummy hybrid III 50th percentile and EN960 headform standard.

The parameters to evaluate the pillion seat are sitting position and mass. From the results, the resistance of pillion seat sensor at the end range of seat are change unequally. It is the chance that the pillion seat sensor cannot indicate the pillion is presence.

For strapping the chin, the evaluation parameter is the style wearing condition. From the result, the resistance is not constant due to the adjustment of helmet with the headform. After adjustment, it can indicate the use of chin strap by the fit and slack conditions. Also the sensor can indicate the unstrapping chin condition.

The wearing position does not affect the wearing helmet condition. However, it can indicate a wearing of helmet and unhelmet with all rotated conditions.

Acknowledgments

The authors would like to thank The Sirindhorn International Thai-German Graduate School of Engineering for laboratory and tools.

References

[1] World Health Organization, "Motorcycle-related injuries in children in the South-East Asia Region" WHO 2014, pp.1–17, 2014.
[2] World Health Organization, "Global Status Report on Road Safety 2015: Supporting a Decade of Action," WHO 2015 Geneva Switzerland, pp.8–9, 2015.
[3] Laksanakit C, Taneerananoon P and Wichettapong K 2016 *Engineering Journal.* **20** 200-210
[4] Yamamura E and Sonobe T 2005 *Journal of Evolutionary Economics.* **15** 166-186
[5] Nakano S, Shimada T, Nakaya A, Dohmoto M and Nagata K 2006 *SAE Technical paper* 7
[6] Jerlin Jose S. T, Rahul A. S and Sajin S 2017 *i-manager's journal on Wireless Communication Network.* **6** 28
[7] Sivasubramaniyan K, Santhosh Murali S, Shalini K, Shanmuga Priya S, and Vijay P 2018 *IJIRTSE.* **4** 22-26
[8] Muthiah M, Aswin Natesh V, and Sathiendran R K 2015 *Proc. Int. Conf. on Smart Sensors and System*
[9] Magno M, D' Aloia A, Polonelli T, Spadaro L and Benini L 2016 *Int. Conf. on Sensor Systems and software* vol 205 pp 55-67
[10] Dhulavvagol P.M., Shet R, Nashipudi P, Meti A.S., and Ganiger R 2018 *Int. Conf. on Cognitive Computing and Information Processing* vol 801 pp 346-357
[11] Thedailystar. (2018, May). Smart helmets cloud safe lives [Online]. Available: https://www.thedailystar.net/backpage/smart-helmets-could-save-lives-1496911.
[12] Roll G, Haffmann O and Konig J 2009 Esv conference smarter-usa.org
[13] Eric R. Teoh 2011 *journal traffic injury prevention.* **12** 169-173
[14] Wu X, Rakheja S and Boileau P.-E. 1999 *International Journal of industrial Ergonomics.* **24** 545-557
[15] Klug C, Feist F and Tomasch E 2015 *IRCOBI Conf. Proc. on the Biomechanics of injury* pp 136-155

Influence of Metal Oxide Ashes on Soot Oxidation Kinetics and Nanostructure using Electron Microscopy and Thermogravimetric Analysis

P Koko[1,*], P Karin[1], E Saenkhumvong[2], C Charoenphonphanich[2],

W Phairote[2], N Chollacoop[3] and K Hanamura[4]

[1]Automotive Engineering Program, International College, King Mongkut's Institute of Technology Ladkrabang, Bangkok, 10520, Thailand
[2]Faculty of Engineering, King Mongkut's Institute of Technology Ladkrabang, Bangkok 10520, Thailand
[3]National Metal and Materials Technology Center, National Science and Technology Development Agency, Pathumthani 12120, Thailand
[4]Department of Mechanical Engineering, School of Engineering, Tokyo Institute of Technology, Tokyo 152-8550, Japan

* Corresponding Author: E-mail: phyozin.mech@gmail.com

Abstract. According to increasingly stringent regulations on particulate emission from automotive vehicles, diesel engine must be equipped with Diesel Particulate Filter (DPF) to trap the Particulate Matter (PM) which are very harmful to human health. Diesel particulate matters are composed primarily of unburned hydrocarbon (soot) and metal oxide ashes as solid fraction. DPF can trap PM with higher filtration efficiency and the process which can burn the soot into carbon dioxide is called regeneration process. Although regeneration process can burn the soot effectively, incombustible ashes will be remained inside the DPF channel causing engine back pressure. These metal oxide ashes are mainly derived from lubricant additives, engine wear and trace metals from diesel fuel. In this article, different nanostructures of diesel soot and metal oxide ash derived by diesel blending lube oil condition were briefly compared using Transmission Electron Microscopy (TEM) image analysis. Electron Dispersive X-ray Spectroscopy (EDS) analysis was introduced to investigate the chemical composition of particulate matters. Thermogravimetric Analysis (TGA) was also conducted to compare the oxidation kinetics of pure diesel soot and the influence of metal oxide ash on soot oxidation kinetics. Contamination of metal oxide ashes promoted soot oxidation rate due to the presence of metallic additives from lube oil acting as a catalyst on soot oxidation kinetics.

1. Introduction

Particulate Matter (PM) emitted from diesel engines are very harmful to human health and must be removed due to stringent emission regulations. PM are mainly composed of solid fraction (soot and ash) and soluble organic fraction such as (sulfates and nitrates) organic compounds. Diesel Particulate

Filter (DPF) is one of the most effective aftertreatment system which can perform higher PM trapping efficiency. There are honeycomb rectangular channels where the exhaust gas flows through its porous walls. The trapped PMs must be burned out during vehicle running and a chemical oxidation process which can oxidize the trapped soot into carbon dioxide is called regeneration process. Although regeneration process can burn the soot effectively, the unburned metal oxide ashes will be remained along the inlet channels causing engine back pressure. These metal oxide ashes are originated mainly from lubricant additives and a very few amounts from engine wear and fuel trace metals [1 - 4].

Conventional lubricant additives such as anti-oxidants, rust and corrosion inhibitors, viscosity index modifiers, anti-wear (AW) agents, detergents and dispersants are well known additives which all are responsible to get better performance of engine lubricating oil. Among them, detergents help to prevent the high temperature deposits on the surface of metal walls, and it assists to neutralize acids that form in the lube oil. They are typically composed of calcium and magnesium as chemical metallic compounds. During engine combustion, lube oil can also participate, and these metallic compounds leave as an ash deposit when the oil is burned [5, 6].

P. Karin et al., mentioned that nanostructure and oxidation kinetics of biodiesel PM compared to Diesel PM. A primary soot particle has two distinct parts: an inner core and an outer shell. The size of each crystallite and agglomeration might be strongly related to Brownian force of gas molecule, drag and shear forces of fluid dynamics, electrostatics forces of charge elements. Nanostructure of soot is composed of curve line crystallites while the ash nanostructure shows straight line structures [7, 8].

During engine combustion, the lubricants can also participate, and the contamination of lube oil can vary the morphology and oxidation kinetics of particulate matters. Some literatures have already investigated not only physical characterization like morphology and particle size distribution but also chemical composition focusing on lube oil related particulate matters [9 - 11]. Y. Wang et al., found that the anti-wear lubricant oil additives changed the nanostructure of emitted particles and led to particles with a more disordered nanostructure. Moreover, oil-related particles have larger aggregate size than diesel particles [12].

In the previous literatures, oxidation kinetics of carbon black and engine's PMs were briefly compared. Engine's PMs are easier to oxidize than carbon black nanoparticles because of containing unburned hydrocarbon. Oxidation kinetics of PM is dependent upon both physical (shape of reactant substance) and chemical composition (oxygen, unburned hydrocarbon and others). The calculated apparent energy of diesel engine's PMs were in the range of 117-130 kJ/mol. H. Jung et al., also investigated that the influences of metals derived from lube oil on oxidation kinetics of soot particles by dosing 2% lube oil and determined apparent activation energy. By using Arrhenius equations, frequency factor of oil dosed PM were about twice as large as without dosing [13 - 15].

This research mainly compares nanostructure of diesel soot and metal oxide ash which are mainly derived from lubricant additives. The designed lube oil consisting excess amount of Calcium (Ca) additives was blended 10% directly into diesel fuel to obtain more ash formation among PM composition. Oxidation kinetics of diesel PM and diesel blending lube oil PM were also investigated to compare the influence of metal oxide ash on soot oxidation kinetics.

2. Experimental

A direct injection compression engine (Kubota – RT140) was used to generate the particulate matter. Engine specification are described in Table 1. Eddy Current Engine Dynamometer was used to control the desired engine speed and load conditions. Lab-View program controlled the engine dynamometer. A soot collector which contains metal netting inside was used to collect the PM powders.

Table 1 Engine specification

Items	Details
Engine Type	1-cylinder, Direct Injection, CI engine
Bore x Stroke	(97 x 96) mm
Displacement	709 cm^3
Compression ratio	18:1
Power	9.2 kW @2400 RPM
Injection Timing	19°CA bTDC
Injection Pressure	22 MPa

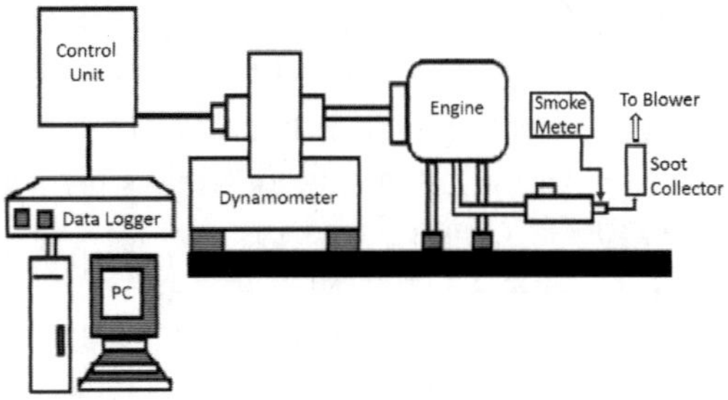

Figure 1. Schematic diagram of engine setup

Table 2 Properties of test fuel

Fuel Properties	Unit	Result
Chemical Formula	-	$C_{14.2}H_{28}$
Auto Ignition Temperature	°C	288
Calorific Value	kJ/kg	46180
Heat of Vaporization	kJ/kg	250
Viscosity	mm^2/s	3.0
Stoichiometric A/F ratio	-	12.3
Density	kg/m^3	844.8

Table 3 Properties of blending lubricant

Test	Method	Unit	Result
Density @15°C	ASTM D4052-15	g/mL	0.8960
Density @30°C	ASTM D4052-15	g/mL	0.8866
Viscosity @40°C	ASTM D445-15a	mm^2/s	154.8
Viscosity @100°C	ASTM D445-15a	mm^2/s	15.05
Magnesium	ASTM D6481	%wt	0.1816
Zinc	ASTM D6481	%wt	0.0798
Phosphorous	ASTM D6481	%wt	0.0856

Schematic diagram of the engine setup is shown in figure 1. To conduct the experiment, commercial diesel containing ($C_{14.2}H_{28}$) was used as an ideal fuel and secondly, designed lube oil was blended 10% by mass directly into diesel fuel to obtain more ash formation among PM composition. This blending lubricant was supported by Bangchak Corporation and it contains excess amount of Calcium (Ca) additives since $CaSO_4$ is the most abundant sulfate compound found in diesel ashes inside used DPF. The fuel properties and blending lubricant oil properties are briefly described in Table 2 and 3. Density and viscosity of blending lubricants were tested under ASTM D4052-15 and ASTM D445-15a respectively. According to ASTM, viscosity is measured as kinematic viscosity and usually described in data sheets at 40°C and 100°C. For the engine lubrication system, conventional lubricating oil 15W-40 API CI-4/SL synthetic engine oil was used.

3. Results and Discussion

3.1 Morphology of ultrafine agglomerate particles

Transmission Electron Microscopy (TEM) image analysis was used to determine the ultrafine agglomerate primary particle nanostructures as shown in figure 2. Soot powders were collected under 2400 RPM engine full load condition since this engine condition is properable to generate the soot efficiently according to previous smoke intensity results. TEM image results showed that the agglomerate structure of ultrafine particles from diesel blending lube oil condition were not significant different compared to neat diesel engine's ultrafine particle. The primary particles from both conditions were grouped each other with similar agglomerate structures and results have not shown much differences.

3.2 Nanostructures of soot and metal oxide ash

Figure 3 compare TEM images of different nanostructures of soot (pure carbon) obtained from the neat diesel condition and metal oxide ash which is derived from diesel blending lube oil condition. The nanostructure of diesel soot primary particle is a spherical shape composed of curve line crystallites as described in figure 3 (a).

Different from soot, in figure 3 (b), metal oxide ash nanostructure is not a spherical shape composed of parallel straight line hatch patterns which shows similar nanostructures among metals. Due to the fact that blending lubricant oil contains metallic additives and these additives cannot be burned during engine combustion and finally it might be changed into metal oxide ashes. Therefore, particulate matters from diesel blending lubricant condition shows distinct straight lined patterns of ash collaborating with unburned hydrocarbon (soot).

3.3 Elemental composition of particulate matters

To investigate the chemical elemental composition of particulate matter, electron dispersive x-ray spectroscopy collaborating with SEM was introduced. X-ray spectrum detect the sample and distinguish the elements according to their particular atomic number. Regarding to the results of SEM-EDS as described in figure 4, particulate matters from neat diesel conditon shows only carbon (soot) and oxygen as elemental composition. However, particulate matters from diesel blending lube oil condition contain not only carbon and oxygen but also unburned metallic elements such as Calcium (Ca), Phosphorous (P), Sulfur (S), Silicon (Si) and Zinc (Zn) respectively. Among metallic additives, elemental composition amount of Ca expressed highest weight percentage due to the fact that the blending lubricant was designed with excess Ca additives. Therefore, EDS result pointed out that metallic additives from engine lubricating oil cannot be burned by engine combustion and these unburned metallic additives might be transformed into metal oxide ash which cannot be burned by regeneration process inside diesel particulate filter. On the other hand, these incombustible ashes might promote soot oxidation kinetics since ash are mainly originated from metallic additives of lube oil.

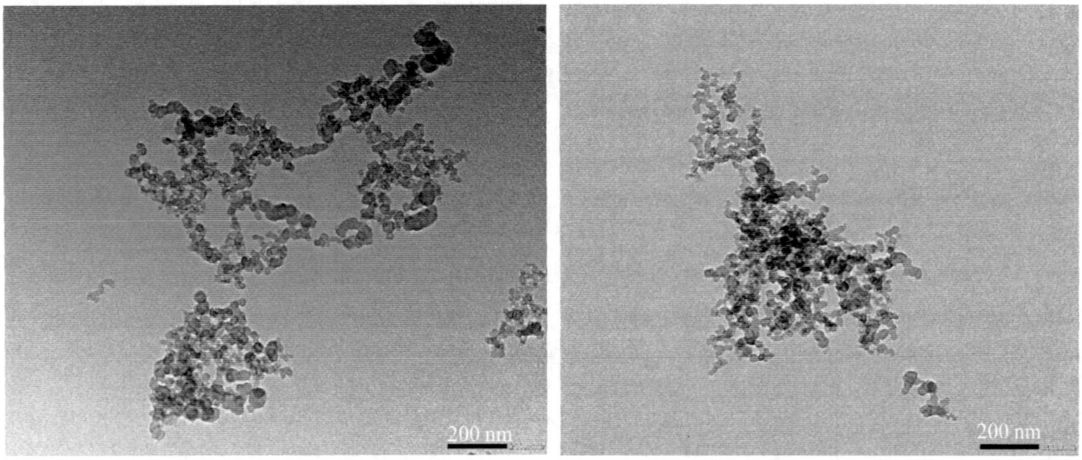

)a(PMs from neat diesel)b(PMs from diesel blending lube oil

Figure 2. TEM images of ultrafine agglomerate particles from small engine (a) diesel and (b) diesel blending lubricating oil condition under 2400 RPM engine full load condition

IOP Conf. Series: Materials Science and Engineering **501** (2019) 012016 doi:10.1088/1757-899X/501/1/012016

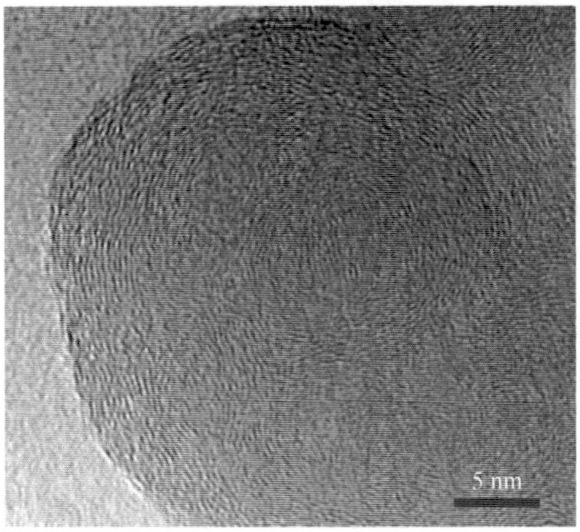

(a) Nanostructure of diesel soot

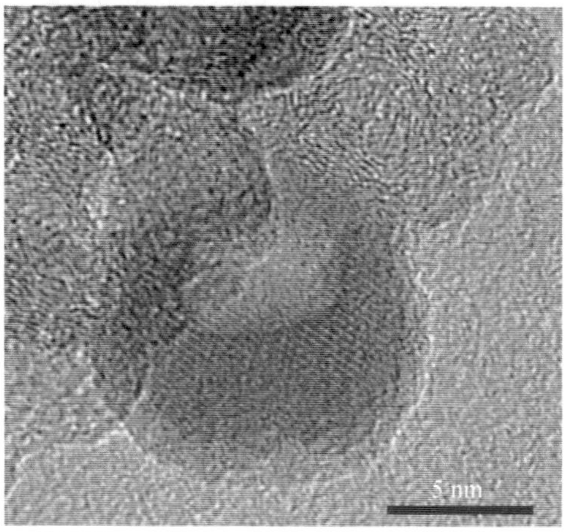

(b) Nanostructure of metal oxide ash

Figure 3. TEM images of nanostructure of (a) diesel soot and (b) metal oxide ash

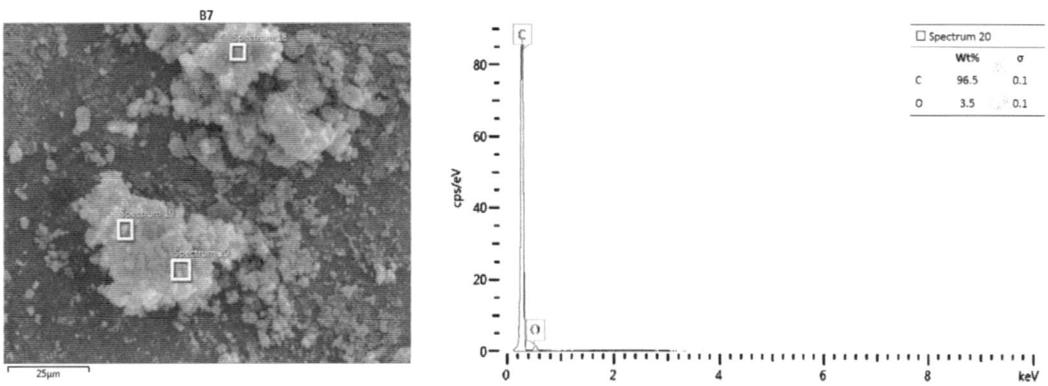

)a(Chemical composition of diesel PM

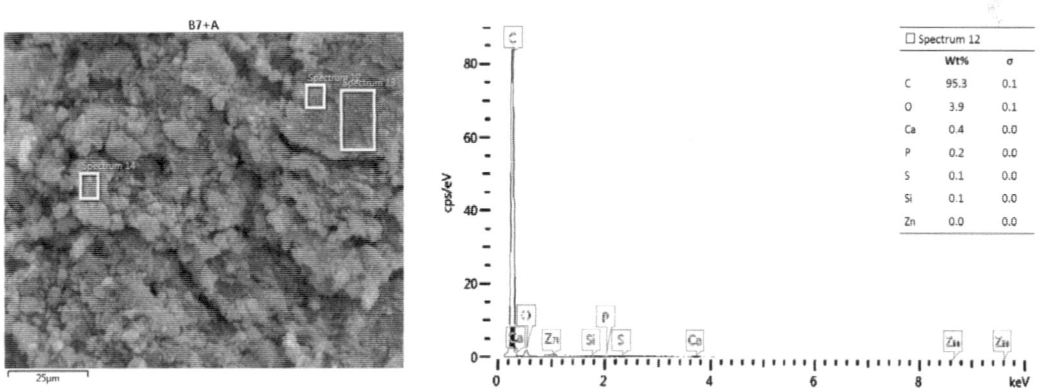

)b(Chemical composition of diesel blending lube oil PM

Figure 4. Elemental analysis of (a) Diesel PM and (b) Diesel blending lube oil PM by using electron dispersive x-ray spectroscopy (SEM-EDS) showing unburned metallic lubricant additives derived from engine lubricating oil

IOP Conf. Series: Materials Science and Engineering **501** (2019) 012016 doi:10.1088/1757-899X/501/1/012016

3.4 Oxidation kinetics of particulate matters

Thermogravimetric analysis (TGA) was conducted to investigate the influence of contamination of metal oxide ash on soot oxidation kinetics. Mass conversion results of diesel particles and diesel blending lube oil particles were compared under different operating temperatures (600 °C and 625 °C). Mass conversion rates can be calculated to describe the faster oxidation kinetics of diesel blending lube oil PMs. Isothermal method was used to maintain the operating temperature. Nitrogen was introduced before air is injected and the sample was oxidized until 90 minutes. Particles were treated with increasing temperature about (20 – 30) minutes by nitrogen atmosphere in order to occur vaporization process. After vaporization, air (oxygen and nitrogen) was start injected and the system was maintained under isothermal to initiate PM oxidation process. Chemical oxidation reaction of PM mass conversion results can be expressed by Eq.1 and the amount of mass conversion over time can be

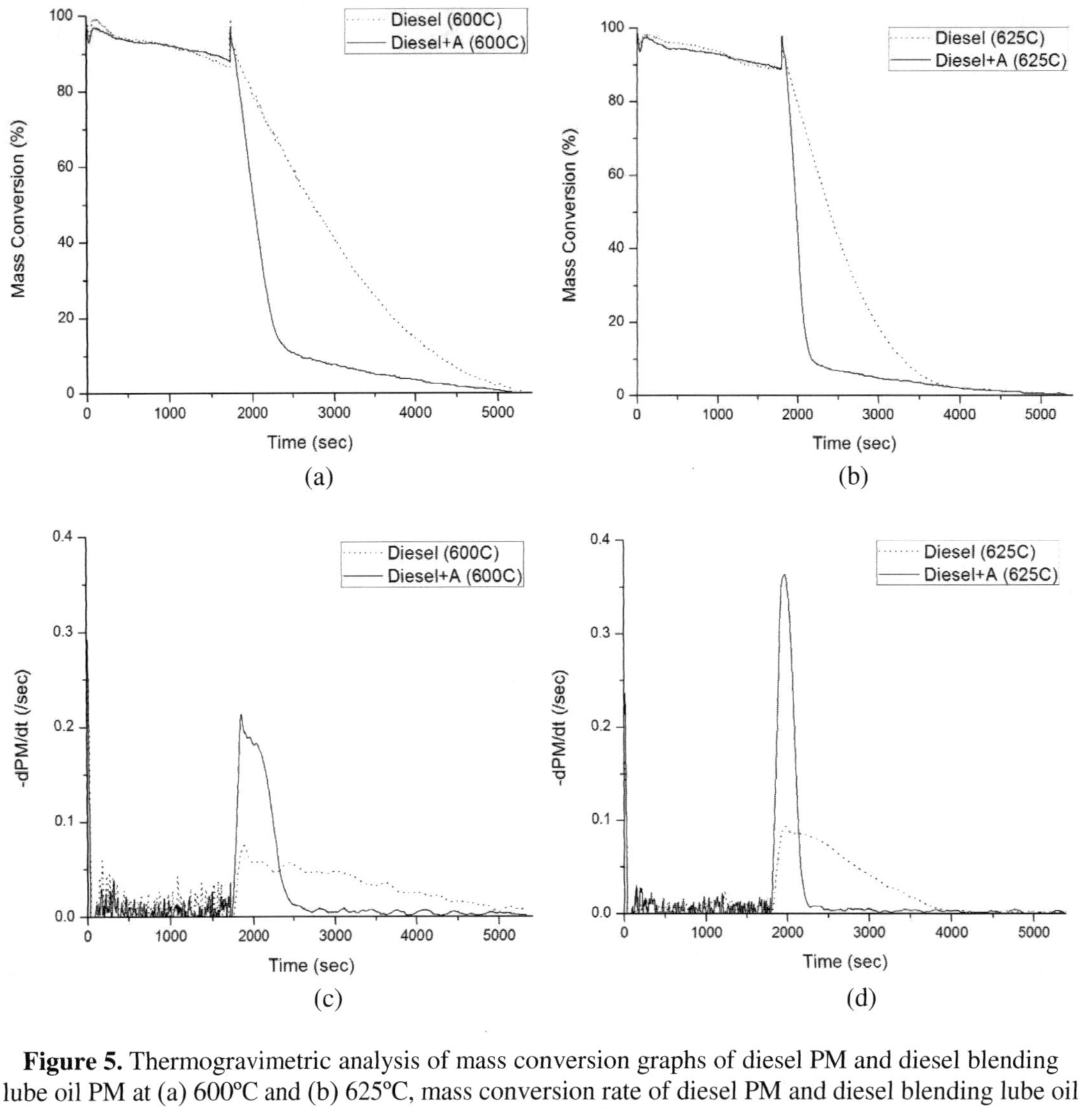

Figure 5. Thermogravimetric analysis of mass conversion graphs of diesel PM and diesel blending lube oil PM at (a) 600°C and (b) 625°C, mass conversion rate of diesel PM and diesel blending lube oil PM at (c) 600°C and (d) 625°C

$$C + O_2 \rightarrow CO_2 \quad (1)$$

$$-\frac{d\,[C]}{dt} = k[C]^n[O_2]^m \quad (2)$$

calculated as mass conversion rate by using Eq.2, where C is represented as PM mass, t is time, n and m are reaction order of PM and oxygen during oxidation process.

Regarding to thermogravimetric analysis results as described in figure 5 (a and b), particles from diesel blending lube oil condition (represented as Diesel+A) were easier to oxidize than neat diesel's particles. The unburned lube oil metallic additives might be participated as an oxidative catalyst resulting faster oxidation kinetics under both operating temperatures 600°C and 625°C. As shown in figure 5 (c and d), mass conversion rates were also calculated to mention sharply for the higher mass conversion rate of diesel blending lube oil PMs.

4. Conclusion

Morphology of ultrafine agglomerate particles, different primary particle nanostructures of lube oil derived metal oxide ashes compared to diesel soot and influence of incombustible metal oxide ashes on soot oxidation kinetics were briefly discussed. The agglomerate structure of ultrafine particles from diesel blending lube oil condition has not shown significant difference compared to agglomerated ultrafine particles from neat diesel condition. Nanostructure of primary soot particle is a spherical shape composed of curve line crystallites while the metal oxide ash has not shown a spherical shape composed of parallel straight-line hatch patterns. Nanostructure of metal oxide ash is like the nature of common metals' nanostructures since these metal oxide ashes are also derived from unburned metallic additives of engine lubricating oil.

According to the chemical characterization results, PM from neat diesel condition composed mainly of carbon and oxygen as elemental composition while PM from diesel blending lube oil condition contained not only carbon and oxygen but also unburned metallic additives from engine lubricating oil. Particulate matters from diesel blending lube oil condition were easier to oxidize than PMs from neat diesel condition. Metallic additives from engine lubricating oil cannot be burned by engine combustion and it might be changed into metal oxide ash. These metal oxide ashes promote oxidation kinetics resulting faster mass conversion rate. Moreover, although accumulated metal oxide ashes tend to reduce effective filtration length of diesel particulate filters, on the other hand, these incombustible ashes might assist to promote the soot oxidation kinetics during regeneration process of diesel particulate filter.

5. Acknowledgements

The author would like to acknowledge to International College, King Mongkut's Institute of Technology Ladkrabang (KMITL) for research funding and Bangchak Corporation (BCP) for supporting designed lubricating oil to accomplish this research.

6. References

[1] Peter Eastwood 2008 Particle Emission from Vehicles, John Wiley & Sons, Ltd, England.

[2] Myung C L and Park S 2012 *International Journal of Automotive Technology* **13(1)** 9-22.

[3] Mohankumar S and Senthilkumar P 2017 *Renewable and Sustainable Energy Reviews* **80** 1227-1238.

[4] Myung C L, Ko A and Park S 2014 *International Journal of Automotive Technology* **15**(2) 2203-2218.

[5] Leslie R Rudnick 2008 Taylor & Franscis Group, London, New York.

[6] Tornehed P and Olofsson U 2011 *Proceedings of the Institution of Mechanical Engineers, Journal of Automotive Engineering* **225**(D) 1055-1066.

[7] Karin P, Oki H, Hanamura K and Charoenphonphanich C 2012 *Journal of Research and Applications in Mechanical Engineering* **1**(2) 3-8.

[8] Liati A, Dimopoulos Eggenchwiler P, Müller G, Schreiber D and Aguirre M 2012 *Atmospheric Environment* **49** 391-402.

[9] Vouitsis E 2007 *SAE Technical Paper* 2007-24-0110.

[10] Karin P, Supanamok C and Hanamura K 2016 *Journal of Research and Applications in Mechanical Engineering* **4**(2) 126-134.

[11] Tan P 2017 *Journal of Environmental Sciences* 354-362.

[12] Wang Y, Liang X, Shu G, Dong L, Sun X, Yu H and Wang Y 2015 *Tribology International* **92** 379-386.

[13] Karin P 2015 *International Journal of Automotive Technology* **16**(2) 211-219.

[14] Karin P 2017 *International Journal of Automotive Technology* **18** (1) 31-40.

[15] Jung H 2003 *SAE Technical Paper* 2003-01-3179.

IOP Conf. Series: Materials Science and Engineering **501** (2019) 012017 doi:10.1088/1757-899X/501/1/012017

Design and Development of Truck Rear Underrun Protection Device

S. Pooudom[1], S. Chanthanumataporn[1,2*], S. Koetniyom[1,2], and J. Carmai[1,2]

[1] The Sirindhorn International Thai-German Graduate school of Engineering (TGGS)
[2] Automotive Safety and Assessment Engineering Research Centre
King Mongkut's University of Technology north Bangkok, 1518 Pracharat 1 Rd., Bangkok, 10800, Thailand

* Corresponding Author: saharat.c@tggs.kmutnb.ac.th

Abstract. Underride truck crashes occur when a passenger car rides under the rear of a large truck. These kinds of accidents are especially dangerous because of the size difference between trucks and passenger cars. Therefore, it is necessary to install the Rear Underrun Protection Device (RUPD) at the rear of trucks for safety enhancement. The aim of this research is to design an effective RUPD which is strong enough to prevent a passenger car falling under the rear of the truck and able to absorb large amount of crash energy to minimize accident severity. The novel RUPD design, equipped with energy absorption guard and proper support structure is proposed in this paper. The study on the angle and separation distance of support structures is performed to find the proper design for the support structure of RUPD. The internal energy and displacement of various design of support structures are investigated by performing Finite element crash simulation in LS-Dyna. The FMVSS 223 standard is utilized as reference for the standard design of RUPD. The result shows that the support structure angle of 20 degree yields optimal displacement and energy absorption. Moreover, the new proposed RUPD is able to avoid the A-pillar deformation and eliminates underride altogether.

1. Introduction

The difference in size and height between passenger cars and trucks can allow the underride of the car in rear end collision, in which the entire vehicle will go underneath the truck. The passenger compartment is crumpled and completely absorbs impact energy. This kind of accident results in severe passenger injury and death. It has been estimated that the energy-absorbing front, rear and side under-ride protection could reduce deaths in car to truck impacts by about 12% [1].

The UNECE R58 and FMVSS 223 standards were used as the reference for evaluating the capability of the RUPD to protect underride collisions between cars and trucks rear-end. Based on the UNECE R58 and FMVSS 223 standards the static test is performed on each point of the Rear Underrun Protection surfaces. The UNECE R58 [2] standard focuses on the deflection distance on the Rear Underrun Protection surfaces. Nevertheless, the FMVSS 223 standard concerns at both the deflection distance and the energy absorption on the Rear Underrun Protection surfaces. The UNECE R58 and FMVSS 223 standards have the same goal, in which the Rear Underrun Protection have to be strong enough to prevent the automobile sliding under the truck and enable the automobile's crumple zone to absorb collapse energy.

IOP Conf. Series: Materials Science and Engineering **501** (2019) 012017 doi:10.1088/1757-899X/501/1/012017

The main requirements of rear underrun protective systems are essentially two: avoiding the underride of the car and absorbing the greatest amount of energy so as to reduce the deceleration peaks and, consequently, the injuries suffered by the passengers.

In any designing or redesigning process of a new device, it is also necessary to follow the regulations governing its use. In FMVSS 223 [3], the regulation imposes that (as shown by figure 1.):

- The ground clearance of the horizontal member shall not exceed 560 mm when measured at each support to which the horizontal member is attached.
- Distance between RUPD and tail end must not exceed 305 mm.
- Length of RUPD must not exceed more than 100 mm on each side of truck.
- Cross-sectional guard must not less than 100 mm.

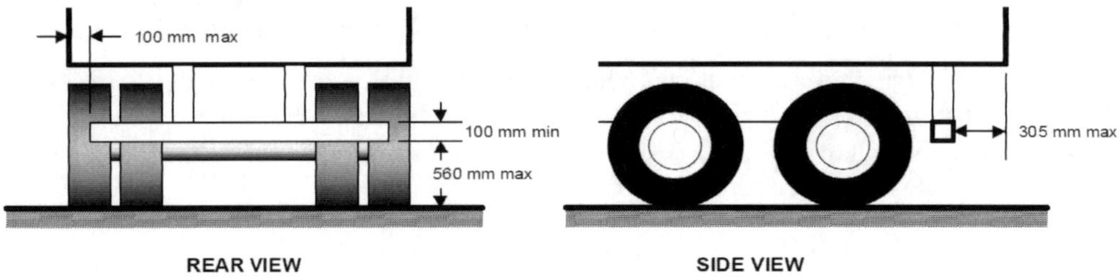

Figure 1. Dimensional limit of a RUPD [3].

The Underrun Protection device is not only a high-strength structure but also able to absorb collision energy in order to secure the car occupants. The Department of Transportation and the National Highway Traffic Safety Administration (NHTSA) has previously stated that the more rigid the Underrun Protection devices, the more likely they are to injure passengers. On the other hand, the more collapse of Underrun Protection, the more harmful it will be to passengers.

The objective of the study is to design and develop the safer RUPD, which is able to absorb collapse energy while at the same time it must be strong enough to protect rear underride truck crashes. In this paper, the effect of RUPD support structures is focused by investigating the angle and separation distance of support structures in order to define the proper design for the support structures of the safer RUPD.

2. Development of safer RUPD

The common standard rear underrun protection device [4] is consisted of two support structures and a crossbar as shown in figure 2. The support structures are used to rigidly connect the RUPD cross bar with the truck frame. In this paper, the support structures has been focused in designing for improve the strength and energy absorption ability by varying the support structures angle and the separation distance between support structures. The RUPD crossbar is generally designed to be rigid structure for underride protection. However, in this study, with the objective to enhance crashworthiness, the crossbar structure is designed with a six-stage compression of steel beams, so as to allow good energy absorption during crushing [5].

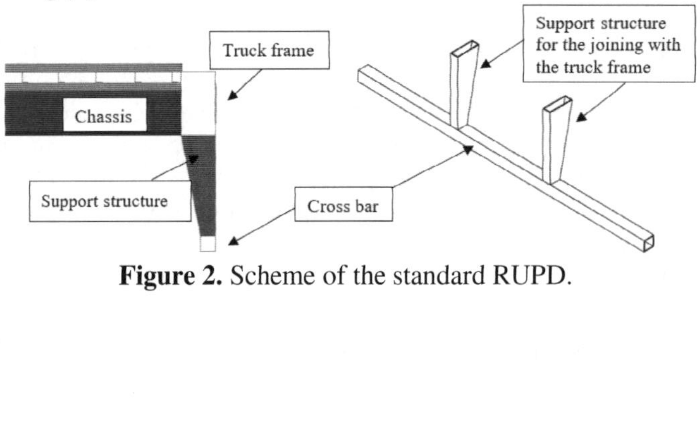

Figure 2. Scheme of the standard RUPD.

2.1. Support structure angle

In the process of support structure development, the variation of the support structure has been designed to study the effect of support structure angle and define the proper angle used for design the safer RUPD. The angle of support structure (θ) as shown in figure 3(a) were design to be 15, 20, 25, 30, 45, and 60 degree respectively. However the separation distance between support structures is 748 mm. corresponding to the width between chassis rails as shown in figure 3(b). Based on the standard RUPD, each support structure is attached under the chassis rail since the installation is quite convenient and secure.

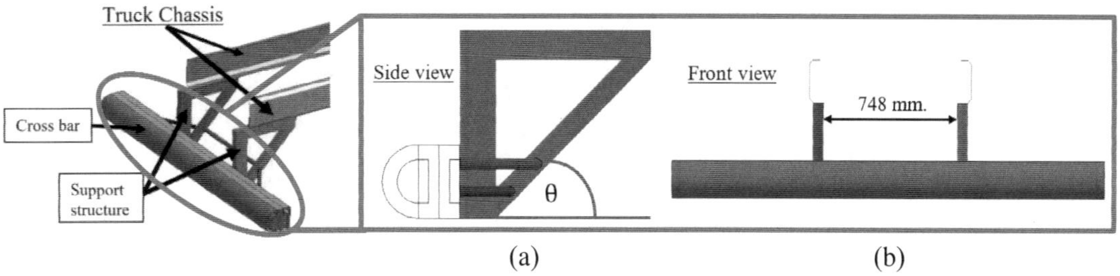

Figure 3. (a) Support structure angle, (b) separation distance between support structures.

2.2. Separation distance between support structures

The effect of separation distance between support structures is investigated with the hope to enhance the crashworthiness from the offset collisions. The distance between support structures has been increased from the standard distance of 748 mm. (30% of crossbar length) to the wide one of 1232 mm (50% of crossbar length). Moreover the support structures are installed under the truck's bed instead of the chassis rail. The upper part of the support structures has been redesigned to install the support structure under the truck's bed as shown in figure 4.

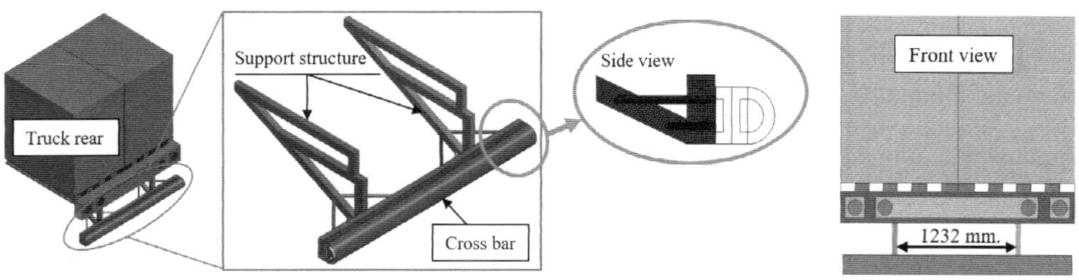

Figure 4. Scheme of proposed RUPD.

3. Finite Element model of car

Vehicle Finite element model of 2010 Toyota Yaris passenger sedan adopted from the National Crash Analysis Center (NCAC) [6] was utilized in this research as the car model. The detail model of this vehicle can be seen in the figure 5. The utilized vehicle model has been validated against the full scale crash test performed by NCAP [6]. As shown in the figure 6, the validation focuses on the comparison of the NCAP test high speed camera picture and simulation picture, in which the crush depth in the front of the car and the acceleration at different locations of the car are compared. The simulation results are found to be consistent and in good agreement with the NCAP crash test data.

Figure 5. The detail FE model of Toyota Yaris passenger sedan 2010.

Figure 6. Comparison between NCAP test and simulation results [6].

4. Result and discussion

4.1. Effect of support structure angle

The effect of support structure angle was investigated in the condition of 50% offset crash as shown in figure 7(a). The car travels forward with speed of 56 km/hr. and the ground clearance of the RUPD is 37 cm. as shown in figure 7(b).

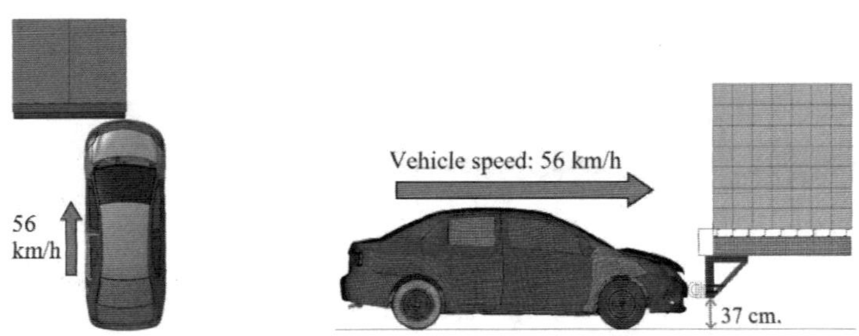

(a) (b)

Figure 7. (a) 50% offset, (b) test speed and ground clearance.

Table 1. Material property for proposed RUPD model.

Material	Young modulus	Density	Yield stress	Poisson's ratio
Mild steel	210 GPa	7.85 g/cm²	200 MPa	0.3

The internal energy and displacement of the proposed RUPD with various support angles are presented in figure 8 and 9 respectively. In the first stage of investigation, the support structure angle is defined to be 15, 30, 45, and 60 degrees as shown in figure 8(a) and figure 9(a) to approximately observe how much energy was absorbed by the device and how much bending displacement in each support structure occur during car crashes. The results in figure 8(a) and figure 9(a) shows that the support structure at an angle of 15 degrees has the highest capability of energy absorption with the least bending displacement.

In the second stage, the support structure angle of 15, 20, 25, and 30 degrees has been defined for deeper investigation. The obtained results as presented in figure 8(b) and figure 9(b) reveals that the energy absorbed by the support structure of angle 20 degrees is almost identical to that of 15 degrees. Also, the displacement of the support structure of 20 degrees is quite identical to that of 15 degrees as well as lesser than that of 25 and 30 degree. Therefore, the 20 degree support structure has been defined for the safer RUPD.

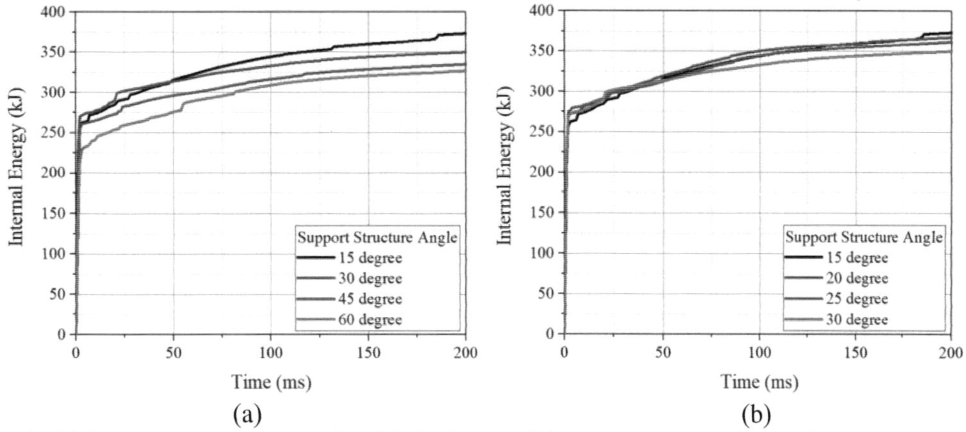

Figure 8. (a) Internal energy for 15, 30, 45, 60 degree, (b) Internal energy for 15, 20, 25, 30 degree.

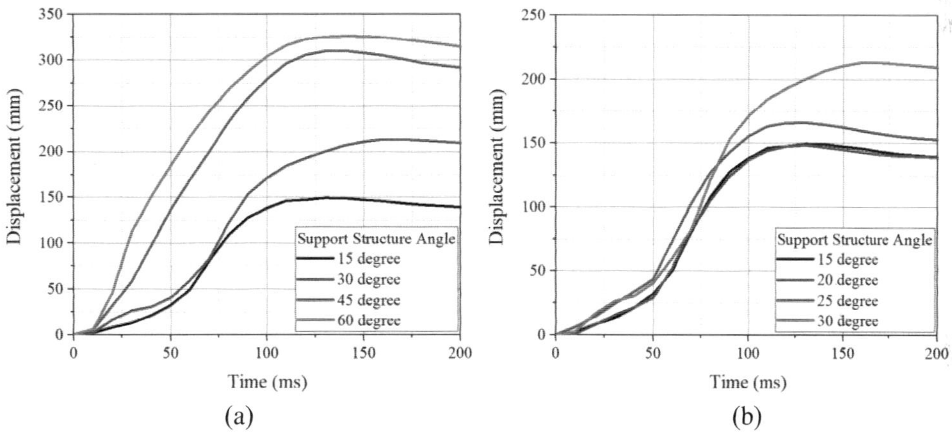

Figure 9. (a) Displacement for 15, 30, 45, 60 degree, (b) Displacement for 15, 20, 25, 30 degree.

4.2. Effect of separation distance between support structures

The effect of separation distance between support structures has been investigated by comparing two condition of separation distances, the standard distance of 748 mm. (30% of crossbar length) and the wide distance of 1232 mm (50% of crossbar length) as shown in figure 10.

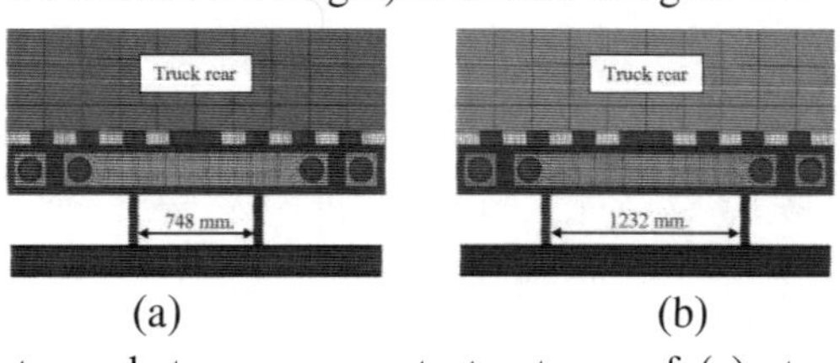

Figure 10. Separation distance between support structures of (a) standard separation, (b) wide separation.

IOP Conf. Series: Materials Science and Engineering **501** (2019) 012017 doi:10.1088/1757-899X/501/1/012017

The crash simulation is set up for three conditions, full width, 50% overlap, and 30% overlap as shown in figure 11. The vehicle speed was set equal to 56 km/h and the ground clearance for all conditions is 37 cm. as shown in figure 12.

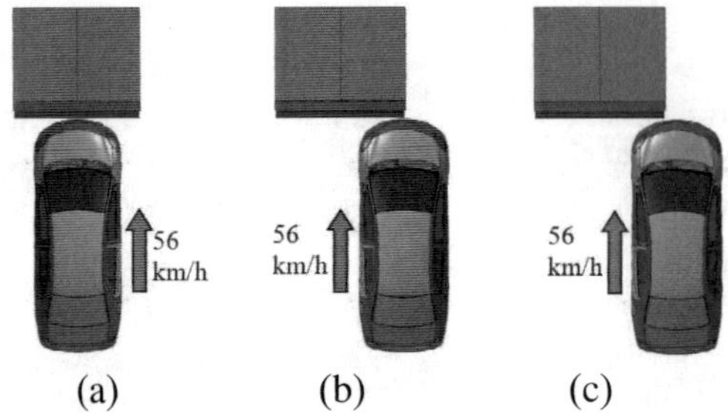

Figure 11. (a) Full width, (b) 50% overlap, (c) 30% overlap.

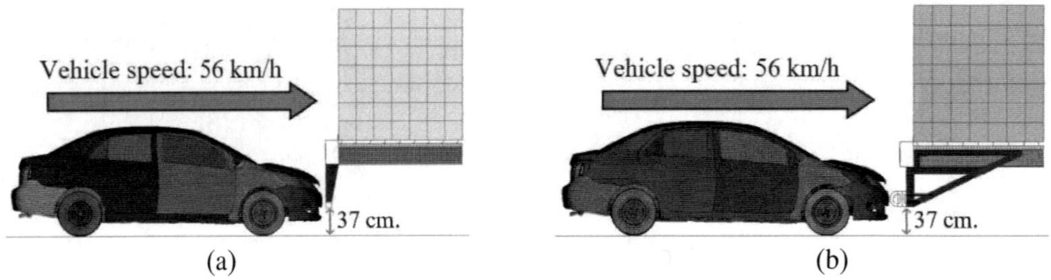

Figure 12. The crash simulation set up in side view of (a) standard RUPD and (b) proposed RUPD.

The results of six crash tests are concluded in table 2 and the maximum crash intrusion scenario of the test vehicle were shown in figure 13-15. The results indicate that there is no underride which occurs in the condition of full width crash. Moreover, the longitudinal A-pillar deformation is equal to zero for both standard and wide separation RUPD.

For the condition of 50% overlap crash, zero longitudinal A-pillar deformation is measured. None of the underride happens for both standard and wide separation RUPD. As shown in figure 14, although the deformation of the standard separation RUPD is greater than that of wide separation RUPD, the longitudinal A-pillar deformation is equal to zero for both cases.

Table 2. Toyota Yaris sedan front into the standard separation and wide separation RUPD crash simulation results.

Conditions	RUPD	Guard performance	Underride	Max. longitudinal A-pillar deformation (cm)
Full width	Standard separation	Good	None	0
	Wide separation	Good	None	0
50% overlap	Standard separation	End bent forward	None	0
	Wide separation	End bent forward	None	0
30% overlap	Standard separation	End bent forward	Severe	24
	Wide separation	End bent forward	None	0

For the condition of 30% overlap crash, the underride present in the standard separation RUPD of which the longitudinal A-pillar deformation is equal to 24 cm. On the other hand, none of underride happens for wide separation RUPD and there is no longitudinal A-pillar deformation. As shown in figure 15, the standard separation RUPD unfortunately allows the car underride the truck because the car does not crush the support structure but crush only the cross bar which is not strong enough to save the car from underriding. The wide separation RUPD can absorb more energy than the standard separation RUPD which leads to no deformation in passenger cell and no underride happens after crash. Nevertheless, the crossbar of both RUPD still bends forward due to offset crash.

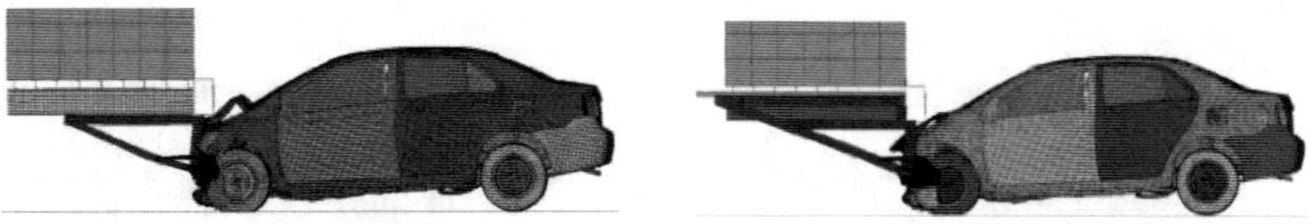

(a) (b)

Figure 13. Full width tests with (a) standard separation RUPD and (b) wide separation RUPD.

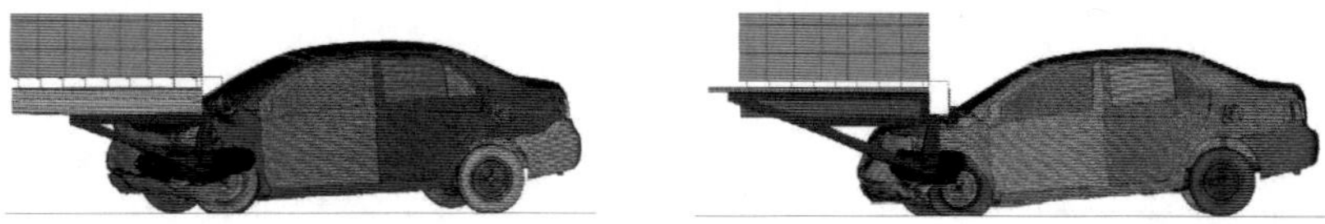

(a) (b)

Figure 14. 50 percent overlap tests with (a) standard separation RUPD and (b) wide separation RUPD.

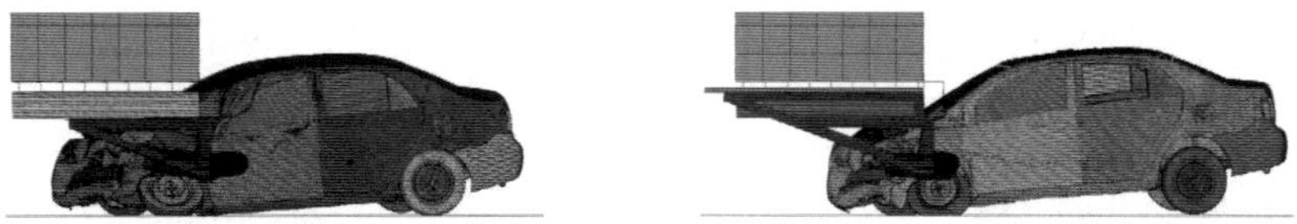

(a) (b)

Figure 15. 30 percent overlap tests with (a) standard separation RUPD and (b) wide separation RUPD.

After the investigation of the separation distance between support structures, the proposed RUPD with the wide separation design was further investigated by comparing the crash result with those of standard RUPD. The results of six crash tests were shown in table 2 and the maximum crash intrusion scenario of the 2010 Toyota Yaris passenger sedan was shown in figure 16-18. The results indicate that there is no underride which occurs in the condition of full width crash. Moreover, the longitudinal A-pillar deformation is equal to zero for both standard and proposed RUPD.

For the condition of 50% overlap crash, severe underride is observed for the standard RUPD while the longitudinal A-pillar deformation is equal to 57 cm. No underride happens for proposed RUPD and no longitudinal A-pillar deformation is measured as shown in figure 17.

Table 3. Toyota Yaris sedan front into the standard and proposed RUPD crash simulation results.

Conditions	RUPD	Guard performance	Underride	Max. longitudinal A-pillar deformation (cm)
Full width	Standard RUPD	End bent forward	None	0
	Proposed RUPD	Good	None	0
50% overlap	Standard RUPD	End bent forward	Severe	57
	Proposed RUPD	End bent forward	None	0
30% overlap	Standard RUPD	End bent forward	Catastrophic	101
	Proposed RUPD	End bent forward	None	0

For the condition of 30% overlap crash, the catastrophic underride can be seen for the standard RUPD while the longitudinal A-pillar deformation is equal to 101 cm. as shown in figure 18(a). On the other hand, no underride happens for the proposed RUPD and the longitudinal A-pillar has no deformation as shown in figure 18(b). Nevertheless, the cross bar of both guards still bends forward due to offset crash.

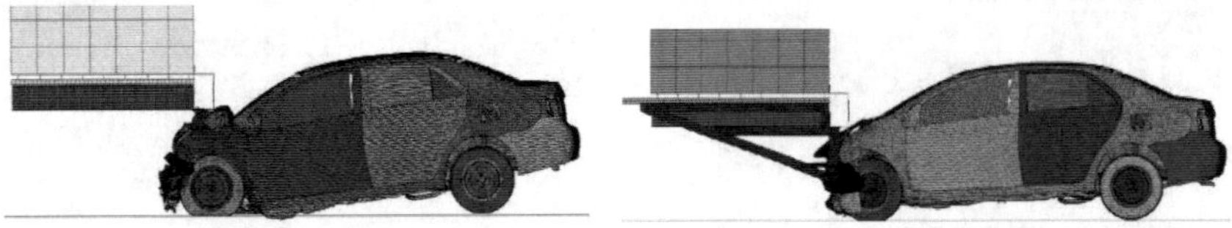

(a) (b)

Figure 16. Full-width tests with (a) standard RUPD and (b) the proposed RUPD.

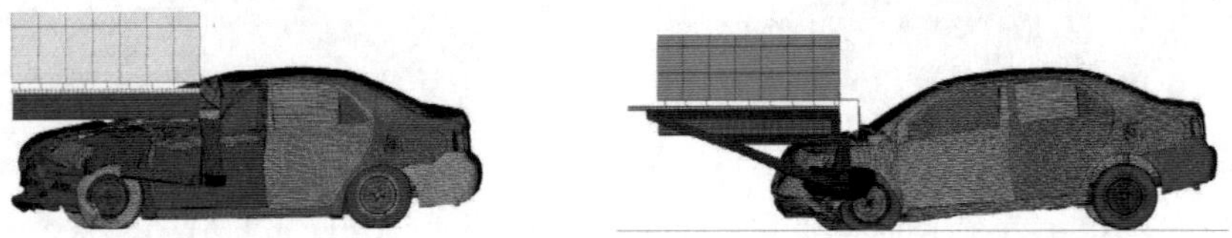

(a) (b)

Figure 17. 50 percent overlap tests with (a) standard RUPD and (b) the proposed RUPD.

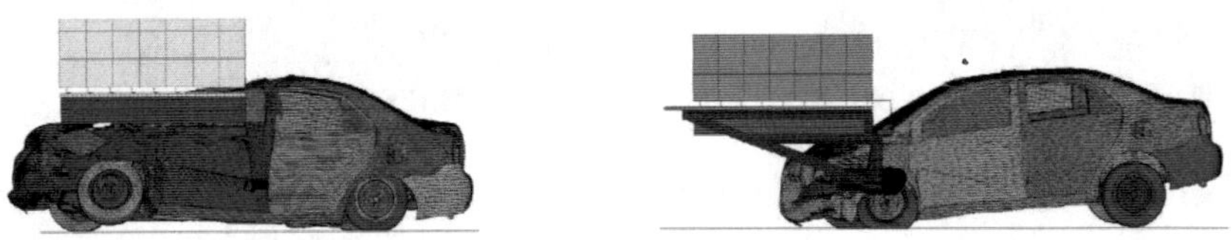

(a) (b)

Figure 18. 30 percent overlap tests with (a) standard RUPD and (b) the proposed RUPD.

5. Conclusion

The proposed RUPD model were set up with different support structure angle using LS-DYNA. The crash simulations for studying the effect of support structure angle and separation distance were conducted. The results indicated that the support structure angle of 20 degrees is the optimal one for the proposed RUPD and the wide separation support structure (50% of cross bar length) can protect the car from underride for full width, 50%, and 30% overlap crash. Thus the optimal condition was employed for designing the proposed RUPD. In addition, the crash test results of the standard RUPD were compared with those of proposed RUPD. It is observed that the proposed RUPD shows better performance than the standard one. Based on these results, RUPD designers can consider these factors for designing the support structure of RUPD to improve crashworthiness in car to truck rear end collisions.

References

[1] Z. F. Albahash, M.N.M. Ansari 2016 A review on rear under-ride protection devices for trucks *International Journal of Crashworthiness*

[2] United Nations Economic Commission for Europe, ECE Regulation No. 58: "Uniform provisions concerning the approval of rear underrun protective devices (RUPD)", unece.org/trans/main/wp29/wp29regs41-60.html

[3] National Highway Traffic Safety Administration 2009 Title 49 Code of Federal Regulations (CFR) Part 571, rear impact protection Washington, DC: Office of the Federal Register

[4] M. L. Brumbelow, Insurance Institute for Highway Safety, United States 2011 Crash test performance of large truck rear underride guards *NHTSA DOT HS.* 11-0074
[5] A. Chege, Kshitij, A Kale, M. R. B. Agrewale, K. C. Vora. 2017 Design and development of impact energy absorbing bumper *J.Sci. & Eng.research* **8** 2229-5518
[6] D. Marzougui, R. R. Samaha, C Cui, S. Kan, The National Crash Analysis Center, The George Washington University 2012 Extended Validation of the Finite Element Model for the 2010 Toyota Yaris Passenger Sedan

IOP Conf. Series: Materials Science and Engineering **501** (2019) 012018 doi:10.1088/1757-899X/501/1/012018

Development of Motorcycle helmet for pre-school Children Using Metal Foam

N. Prasartthong[1], S. Koetniyom[1,2], and J. Carmai[1,2*]

[1]The Sirindhorn International Thai-German Graduate school of Engineering (TGGS)
[2] Automotive Safety and Assessment Engineering Research Centre King Mongkut's University of Technology north Bangkok, 1518 Pracharat 1 Rd., Bangkok, 10800, Thailand

* Corresponding Author: julaluk.c@tggs.kmutnb.ac.th

Abstract. Thailand is second in the road fatality rankings published by World Health Organization in 2015. Almost 80% of traffic deaths involves motorcycles and many of the passengers are children. However, there are still no appropriate head protection equipment for small child pillions. The child helmet should also be lightweight and have high energy absorption. This paper aims to develop a child helmet by using metal foam. The impact tests of the smallest size commercial motorcycle helmet were performed in TIS 369-2557 standard. The finite element model of an existing child helmet was developed and employed to simulate the helmet impact tests using both rigid dummy head and deformable child head. The helmet model was validated with experimental tests to obtain the baseline model. Good agreements can be observed. In addition, the deformable child head can give detail information of brain injury experience by the child during impact. The existing helmet model was redesigned with aluminium foam and was compared with the baseline model after impact tests. Improvement in terms of helmet weight and head injury mitigation can be seen.

1. Introduction

The use rate of motorcycles in Thailand and Southeast Asia are very high compared to other type of vehicles. In addition, it is the type of vehicles that causes the highest risk of severe injuries and fatalities. Thai Ministry of Public Health 2015 report on injury surveillance data from 33 injury surveillance sentinel sites showed transport accidents are leading causes of severe injury and death of children under 15 years old (39.17% of severe injuries and 61.46% of deaths) [1]. Save the Children Thailand has reported that there are 1.3 million child pillion passengers on motorcycles [2]. However, there are still no appropriate head protection equipment for small child pillions, nevertheless most of helmets sold for children are reduced-size adult helmets. Inappropriate head protection equipment for small child pillions bring the injuries to head and neck that could cause severe injury and disability among child pillions passengers. Most of existing helmets sold in the market are reduced-size adult helmets which are not appropriate for small children (4-7 years old). Children have different anthropometry, injury mechanism from adult [3]. Their cervical spine is weaker than adults and therefore the child helmet should be lightweight and at the same time must have high energy absorption to mitigate the head acceleration. In order to reduce helmet weight, common practice is to use lower density foam liner or to reduce the thickness of the foam liner. This can lead to lower head protection performance. One of the advanced material which has been used as energy absorber in automotive industry is metal foam. It is light weight and has high strength to weight ratio as well as good energy absorbing capability [4,5]. The automotive industry has applied metal foam to many parts of car such as the crush box [5]. Pinnoji et. al [6], did research that focused on the weight of motorcycle helmet and came out with metal foam for shell part in the helmet. The results showed metal foam gave a better performance when compared to the normal shell. Caserta et. al [7], did a study in enhancement of motorcycle helmet with aluminium honeycomb as a material for helmet liner and it provided a better result in term of protection. To reduce the helmet weight and maintain the head protection performance according to the standards, the paper, therefore, aims to develop a child helmet by using metal foam. The effectiveness of the proposed helmet will be compared with the existing small helmet for children.

2. Helmet Impact

2.1. Helmet impact test set up

Helmet impact test of a commercial child helmet was performed for finite element model validation. The test conditions are based on TISI 369-2557 [8] which adopted impact test condition for half helmet from DOT FMVSS 218 [9]. Four locations, including crown, front, rear and side, of a half helmet must have impact tests. The helmet impact testing machine and all equipment needed are shown in figure 1. The DOT headform size small and an accelerometer to measure the peak acceleration were required. The helmet was fit to the headform. Two types of anvil including flat anvil and hemisphere anvil were employed. The headform together with the helmet were raised to 1.83 m and 1.37 m for flat and hemi anvil respectively. They were dropped to impact the target below. With this height, the impact speed will be approximately 6 m/s.

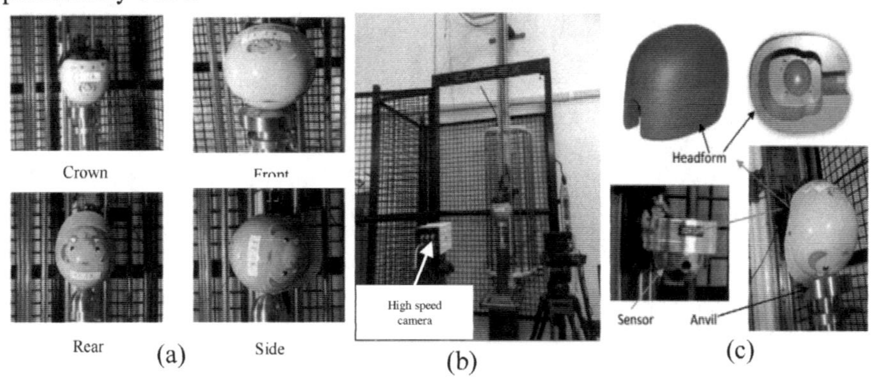

Figure 1. Shows child helmet impact test at four locations (a)crown, front, rear, side. (b) helmet impact testing machine and (c) equipment for helmet impact test.

2.2. Impact test results

Linear acceleration-time graph of each test is shown in figure 2. The peak acceleration and Head injury criteria (HIC) [10] of all four impact configurations are summarized in table 1. DOT FMVSS 218 certainly adopted this maximum force as a guide to set peak linear acceleration at 400g. The peak acceleration obtained from the front impact was higher than the maximum allowable value of 400g. This helmet failed in the impact test. Improvement of performance in the frontal area is strongly required.

Table 1. Summary of the helmet impact test results.

Impact Location	Drop height (m)	Impact speed (m/s)	Anvil type	Peak Acc(g)	HIC	Pass or fail
Crown	1.832	5.8320	Flat	250.4	1473	Pass
Front	1.83	5.8122	Flat	435	2206	Fail
Side	1.372	5.0596	Hemi	212.8	961	Pass
Rear	1.370	5.0829	Hemi	220.1	859	Pass

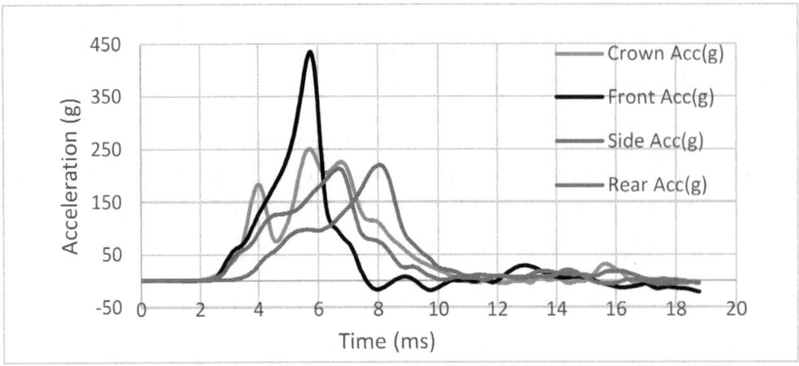

Figure 2. Shows acceleration-time graph of helmet impact testing experiments

3. Development of finite element model for helmet impact testing and validation

3.1. Finite element model set up

The finite element model was developed for the same type and brand of the small helmet used in the impact test. The current helmet was 3D scanned to obtain the correct profile, shape and dimension for creating a finite element model as shown in figure 3a. The model was created in LS-DYNA R7.1.3. The outer shell was made of ABS. Its behavior was described using Plastic kinematics material model [11] in LS-DYNA with a mass density of 1200 kg/m^3, Young's Modulus of 4000 MPa and Poisson's ratio of 0.37 [12]. The outer shell was modeled using shell element of 3.3 mm thickness. The foam liner was made of EPS. Its compression behavior was described using the low density foam material model in LS-DYNA with a mass density 90 kg/m^3, Young's Modulus of 25 MPa and Poisson's ratio of 0. The liner was modeled using tetrahedral solid elements with non-uniform thickness varied from 12 mm at the front to 20 mm in the vertex as shown in figure 3a. In order to simulate the interfaces between the headform-helmet and helmet-anvil, a surface-to-surface type of contact with a friction coefficient of 0.35 and 0.5 were used respectively [15].

Table 2. Material properties for outer shell and foam liner in helmet. [12]

Helmet component	Mass density (kg/m^3)	Young's modulus (MPa)	Poisson's ratio	Thickness (mm)
Outer shell	1200	4000	0.37	3.3
Foam liner	90	8.64	0	12-20

In addition, Figure 3c shows how to set up model for the both types of Anvil. Crown and front location were against flat anvil, side and rear were against the hemisphere. The impact velocity was employed as an initial condition. Therefore, the helmet was positioned as close as possible to the anvil and known impact speed from the test was assigned to the helmet for each case. The anvil was modelled as rigid elements and restricted from movement in all directions. Two types of heads were employed which were validated child dummy head in P6 series. [13] and the other is the validated head of child human body [14], as shown respectively in figure 3. P6 made from steel covered by the rubber, to represent the skin and the equivalent weight. For deformable head, which includes skull and the brain part as shown in figures 3, were modeled with actual mechanical behaviors that are able to provide injury of the brain while the P6 can only provide kinematics responses.

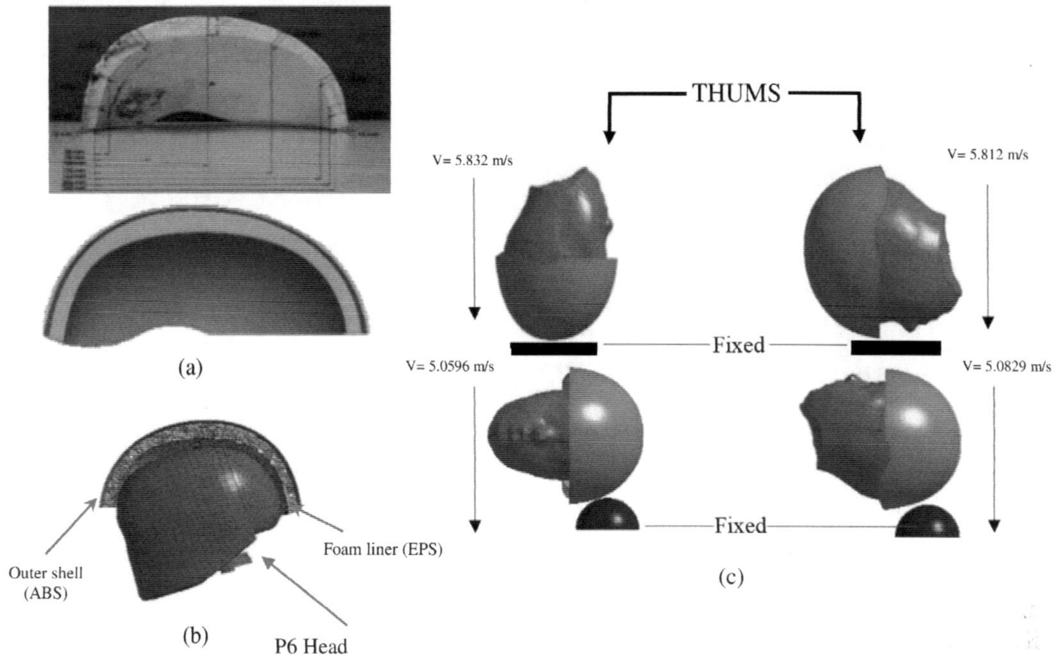

Figure 3. Shows (a) the cross section from the sagittal plane of the foam liner and CAD model (b) Finite element model for the two main parts of helmet (c) helmet impact models set up.

3.2. Model validation

The resultant acceleration was extracted from the accelerometer at the Center of Gravity (C.G.) of the rigid head and from the node at the C.G. of the deformable human head. Both models were validated by comparing the simulation results with the impact experimental results. Table 3 shows comparison of peak acceleration and HIC.

Table 3. Comparison of peak accelerations and HIC obtaining from the experiments and simulations.

Impact Location	Peak acceleration (g)			HIC		
	Experimental test	P6	THUMS	Experimental test	P6	THUMS
Crown	250.41	244.32	268.74	1473	2284	1339
Front	434.96	437.35	468.59	2206	5191	5297
Side	220.09	192.69	218.04	961	848.6	1246
Rear	212.77	223.63	198.89	859	1527	1233

The peak accelerations and HIC obtained from the experiments and the simulations with a rigid child head show small difference. Both values obtained from simulation with the rigid head are a bit lower than the experimental tests except front and rear location, the difference is less than 12.5%. Peak accelerations obtained from simulation with the deformable head shows slight discrepancy with the experiments. The highest difference in this value is at the crown location with a difference of 7.7%. Large difference of HIC value obtained at the front location is seen even though the peak linear acceleration shows little difference. This is because the impact time interval during the first peak obtained from both simulations are greater than that of the experiments. HIC is a function of both linear acceleration and contact time interval. However, the DOT FMVSS 218 standard prefer peak acceleration as passed criteria.

4. Modification of helmet model using metal foam

Implementation of metal foam as part of helmet has great ability in impact energy absorption. It also possesses low density with good shear strength and fracture strength. Since metal foam will be used in the design of child helmet, the compression test of the aluminium foam is required to obtain the material properties. The stress-strain curve obtained from compression tests are shown in figure 4. Table 4 summarizes mechanical properties of the tested aluminium foam. The aluminium foam mechanical behavior was model using crushable foam material model

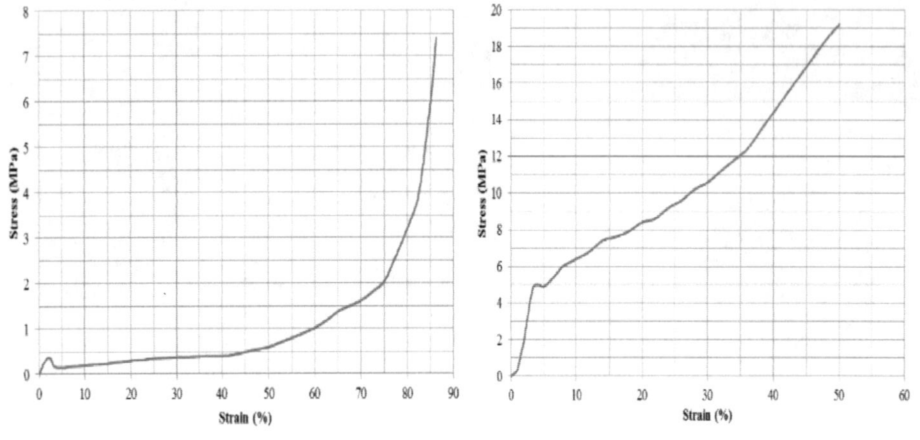

(a) (b)

Figure 4. Average stress in relation to strain for aluminium foam relative densities of (a) 0.051 and (b) 0.185.

Table 4. Mechanical properties of aluminium foam.

Basic Product Specifications		Compressive Mechanical Specifications		
Density	Relative density	Densification Strain	Plateau Stress	Initial Yield Stress
140 kg/m^3	0.051	73%	0.59 MPa	0.40 MPa
510 kg/m^3	0.185	43%	0.40 MPa	5.25 MPa

4.1. Design variation

In order to discover potential benefits of metal foam, the existing helmet model was redesigned into six types with two different densities of aluminium foam as shown in table 5. Two thickness of outer shell were studied. The same impact test conditions were performed with the P6 dummy head.

Table 5. Summaries the design variation and corresponding total mass.

Model	Outer shell mat.	Inner liner mat.	Total Mass (Kg)
Baseline	ABS	EPS	0.346
1	ABS	Al foam 140 kg/m^3	0.380
2	ABS	Al foam 510 kg/m^3	0.729
3	Al foam 140 kg/m^3, 3.3 mm thick	EPS	0.115
4	Al foam 510 kg/m^3, 3.3 mm thick	EPS	0.197
5	Al foam 140 kg/m^3, 4.3 mm thick	EPS	0.131
6	Al foam 510 kg/m^3, 4.3 mm thick	EPS	0.254

4.2. Simulation results

Table 6 shows comparison of peak acceleration and HIC for each simulation cases. Introducing metal foam to 5 helmet models led to reduction in the peak acceleration for all impact locations except the rear location. The peak accelerations obtained from all six model at the front location are less than the baseline model and below the maximum allowable value of 400g of TISI standard. Model no. 3 shows the acceleration of 303.58 g. For the rear location, the peak acceleration is more than the baseline model. However, they are all still lower than the allowable value. The HIC values also followed the same trend as the peak accelerations. Model no. 3 with aluminium foam outer shell of 140 kg/m^3 density and 3.3 mm thickness exhibit optimum performance when taking mass of the helmet into consideration. The mass can be reduced by 66.76%.

Table 6. Peak acceleration and HIC results from P6.

Model	Peak acceleration(G)				HIC			
	Impact Location				Impact Location			
	Crown	Front	Side	Rear	Crown	Front	Side	Rear
Baseline	244.32	437.35	192.69	223.63	2284	5191	848.6	1527
Model No.1	192.16	333.22	173.68	271.94	1188	3409	776.2	2086
Model No.2	290.44	341.53	195.63	287.66	2674	3568	1065	2268
Model No.3	203.25	303.58	200.96	266.35	1744	3056	1084	2166
Model No.4	207.78	311.32	208.61	260.92	1846	3270	1197	2096
Model No.5	197.94	370.12	186.49	328.53	1647	3697	1009	3317
Model No.6	227.99	311.35	181.71	323.19	1836	3266	999.7	3335

4.3. Injury analysis

The helmet model no. 3 was employed to simulate the impact test with deformable head to investigate the brain injury as shown in figure 5 and figure 6. The skull stress is above 80 MPa for the cases without helmet and with the baseline helmet, this indicate high risk of skull fracture [16]. However, the simulation with model No.3 shows large reduction in brain pressure that is much lower than 80 MPa. Very high strain occurred at the brain stem for the case without helmet. The maximum brain strain value of this case is 0.98 which is much larger than the 0.2 threshold [17] It implies severe brain contusion. The head with helmet shows reduction in brain strain especially with the redesigned helmet models where the maximum brain strain is lower than 0.2. Low risk of brain injury is observed for the case with the redesigned helmet model no. 3.

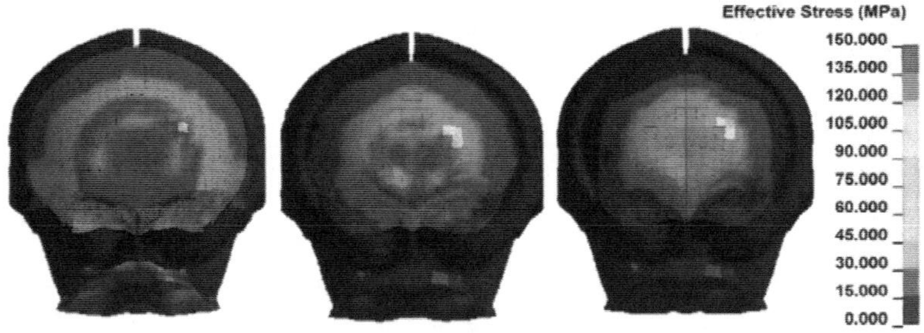

(a) (b) (c)

Figure 5. Shows the Von Mises stress distribution on the skull at the front impact location from (a) deformable head, (b) baseline model, (c) re-design model No.3

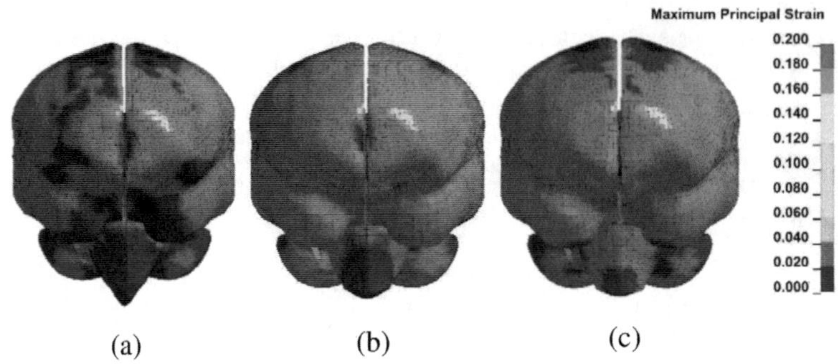

Figure 6. Shows strain distribution that occur in the brain at the front impact location form (a) deformable head, (b)baseline model (c) re-design model No.3.

5. Conclusion

Finite element model of the child helmet was developed and validated with the experimental results. The existing helmet failed the impact test at the front location. The modification of the helmet was done by introducing metal foam with two densities and two shell thickness for both outer shell and liner. Different designs of child helmet were evaluated by their performances and weight. Introducing aluminium foam to the helmet mitigated the peak acceleration experienced by child's head. The optimum design was with low density aluminium foam as outer shell of 3.3 mm thickness. It improved the head linear acceleration, HIC while reducing the mass. This modified child helmet with a mass reduction of 66.76 % passed impact tests for all locations. The skull stress is lower than the fracture limit whereas the brain strain is also lower than the brain injury criteria. The modified helmet can mitigate brain injury.

6. Acknowledgements

The authors would like to thank King Mongkut's University of Technology North Bangkok for the financial support of this study (KMUTNB-61-GOV-B-39).

References

[1] Damnakkaew K 2015 *Annual Epidemiological Surveilance Report* (Nonthaburi: Ministry of Public Health) p 201
[2] Save the children Thailand 2014 *The 7% Project* online document: http://www.7percent.org/about/child%20motorcycle%20helmet%project%20thailand.pdf
[3] Patlak M and Joy J 2002 *I Is Soccer Bad for Children's Heads?: Summary of the IOM Workshop on Neuropsychological Consequences of Head Impact in Youth Soccer* (Washington DC: National Academy Press) p 3
[4] Ashby M F Flecks N Evans A and Wadley H N G 2000 *Metal Foams: A design Guide* (London: Butterworth-Heinemann)
[5] Cymat Technologies Ltd *Aluminium foam technology applied to automotive design* (Toronto: CYMAT)
[6] Pinnoji P, Mahajan , Bourdet N, Deck C and Willinger R 2009 *Int. J. Impact Eng.* **37** 274–284
[7] Caserta G, Iannucci L and Galvanetto U 2011 *Compos. Struct.* **93** 2748–2759
[8] TIS 369-2557 2014 *Thai Industrial Standard Protective Helmets for Motorcycle Users* (Bangkok: TIS)
[9] U.S. Department of Transportation 2011 *National Highway Traffic Safety Administration Laboratory Test Procedure for FMVSS No. 218 Motorcycle Helmets* (Washington DC: NHTSA)
[10] Eppinger R, Sun E, Bandak F, Haffner M, Khaewpong N, Maltese M, Kuppa S, Nguyen T, Takhounts E, Tannous R, Zhang A and Saul R 1998 *Development of Improved Injury Criteria for*

IOP Conf. Series: Materials Science and Engineering **501** (2019) 012018 doi:10.1088/1757-899X/501/1/012018

the Assessment of Advanced Automotive Restraint Systems – II (Washington DC: NHTSA)
[11] Livermore Software Technology Corporation (LSTC) 2007 *LS-DYNA Keyword User's Manual volume 1 version 971* (California: LSTC)
[12] Fernandes F, Alves de Sousa R, Willinger R, Deck C 2013 *Proc. Int. Conf. on IRCOBI* (Gothenburg: IRCOBI) pp 431- 442
[13] European Union 2005 *Regulation No 44 of the Economic Commission for Europe of the United Nations (UN/ECE) — Uniform provisions concerning the approval of restraining devices for child occupants of power driven vehicles ('child restraint systems')*
[14] Iwamoto M, Nakahira Y, Tamura A, Kimpara H, Watanabe I and Miki K 2007 *Proc. On The 6th European LS-DYNA Conference* (Gothenburg: LS-DYNA Conference) pp 47-56
[15] Klug C, Feist F and Tomasch E 2015 *Proc. Int. Conf. on IRCOBI* (Lyon: IRCOBI) pp 136-155
[16] Montoya D, Thollon L, Llari M, Perrin C and Behr M 2017 *Int. J. Crashworthiness.* doi:10.1080/13588265.2017.1345602
[17] Feist F and Klug C 2016 *Proc. Int. Conf. on IRCOBI* (Malaga: IRCOBI) pp 162-184

Development of field test procedure to assess force performances of used tires

Jirawat Dockyeam

The Sirindhorn International Thai-German Graduate School of Engineering (TGGS) King Mongkut's University of Technology North Bangkok,Thailand

* Corresponding Author: jirawat.d-mae2015@tggs.kmutnb.ac.th

Abstract. Based on United Nation Economic Commission for Europe (UNECE) regulations, there are no existing standard procedures to identify the force performance of used tire which can be physically degraded by time. In vehicle safety, these tire forces in lateral and longitudinal directions are the key factors to response the driver decision during the brake application and maneuver conditions. Therefore, the objective of this research is to develop the used tire performances based on three vehicle testing standards such as brake test, steady-state circular test and severe lane-change test respectively. To assess the tire force performances, two tire categories which are used in those tests with the driving and pedal robot for the sedan vehicle are classified as new tires from 2018 and used tires from 2015, 2014, 2013. To identify tire force performances, various indicators for each test are determined. For example, the standard brake test from UNECE R13H is used to investigate tire friction force in the longitudinal direction. The severe lane-change manoeuvre and steady state circular tests are used to identify the lateral tire force from ISO 3888-2 and ISO 4138 respectively. As a result, it can be concluded that the tire force performances from new and used tires are different based on various indicators.

1. Introduction

The automobile industry is an important sector in developing countries like Thailand. According to the 11th National Economic And Social Development Plan (2012-2016) [1], the automobile industry development is of special interest, especially the parts supplier for domestic and international use. Furthermore, as per the 12th National Economic and Social Development Plan (2017-2021), the government will be establishing the National Automobile and Tire Test Centre. This center can support government policy regarding automobile and parts development [2].

According to United Nations Economic Commission for Europe (UNECE), UNECE R. 117 is set up as a standard for tire testing. This standard test is divided into rolling resistance and wet adhesion [3]. These tire testing standards do not consider the forces in the lateral and longitudinal directions [4]. However, the tire forces are related to vehicle testing standard UNECE r13-h for longitudinal vehicle direction, while the lateral vehicle direction was indicated by ISO 4138 or 3888. [5-7].

At present, tire research for vehicle users focuses mainly on tire parameters affecting the ride and comfort that involves noise, rolling resistance [8,9]. Rolling resistance study indicates in term rolling resistance coefficient while noise test focuses on noise level. However, there are some research that are related to tire degradion over time [10]. Peel strength and elongation rate determines the level of degradation.

IOP Conf. Series: Materials Science and Engineering **501** (2019) 012019 doi:10.1088/1757-899X/501/1/012019

However, there are no standard and research works for tire testing that involves a full vehicle model [11]. Thus, the combination of the brake, steady state circular and severe lane-change manoeuvre tests can be used to identify the tire performance based on the tested vehicle. These tests pattern correspond to vehicle dynamics that affects the tires forces in longitudinal and lateral direction. Therefore, the objective of this research is to develop the field test procedure and assess used tire that relates to degradation and tire aging. Furthermore, tire testing is an important thing to do for measuring the tire quality such as performance of the road adhesion to reduce the accidents as well.

2. Methodology

Three different procedures are developed to evaluate tire performance in terms of lateral and longitudinal forces based on vehicle field tests. To identify the tire performance in longitudinal direction, the braking test procedure are developed under UNECE R13-h. For Steady state circular test and severe lane-change maneuver test, the tire performance in lateral direction are involved. Three different tests procedure are described in details below:

2.1. Brake test

The criteria of brake testing of the United Nations Economic Commission (UNECE) is referred as number R13-H (type-0) and is described as the brake performance testing. The brakes have been tested with the vehicle transmission engaged and also disengaged. The vehicle test speed and pedal force are shown in table 1.

Table 1. Condition of vehicle used in brake test

Vehicle Condition	Test speed	Pedal force
Transmission engaged	100 km/h	80 – 120 N
Transmission disengaged		

2.2. Steady state circular test

This test is designed to measure the steady state lateral acceleration of the vehicle. The tests are to be conducted as per ISO 4138, on a closed skid pad around a number of marked circles of 8 m radius as shown in figure 1. In this test, the vehicle is required to begin at selected steer angle and run in left turn and accelerate around the prescribed circle. The vehicle stopped acceleration when the it started to deviate from the track. In the actual test, vehicle was selected steer angle at 16 degree for circle 8 m radius.

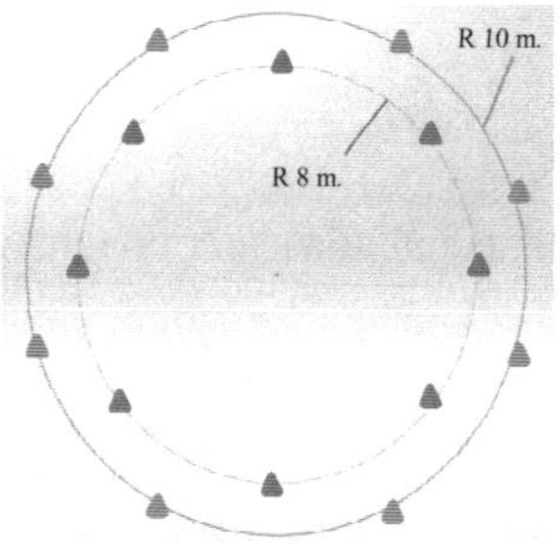

Figure 1. Dimension of test track for steady-state circular test .

2.3. Severe lane-change manoeuvre test

The severe lane-change manoeuvre is a dynamic procedure of fast transition from the initial straight trajectory to the parallel laterally shifted one, which is next followed by the turn to the trajectory identical with the initial one. The test track on which the severe lane-change maneuver tests were performed was determined according to ISO 3888 standard as shown in figure 2. The Initial speed can not deviate by more than 5 km/h from test speed. Starting from a yaw velocity equilibrium condition, a steering input is applied as rapidly as possible to a preselected value and maintained at that value for several seconds until the measured vehicle motion variables have reached steady state. No change in throttle position can be made, although speed may decrease. In this test, vehicle was running at 40 km/h and 50 km/h. All test runs were performed at least 3 times.

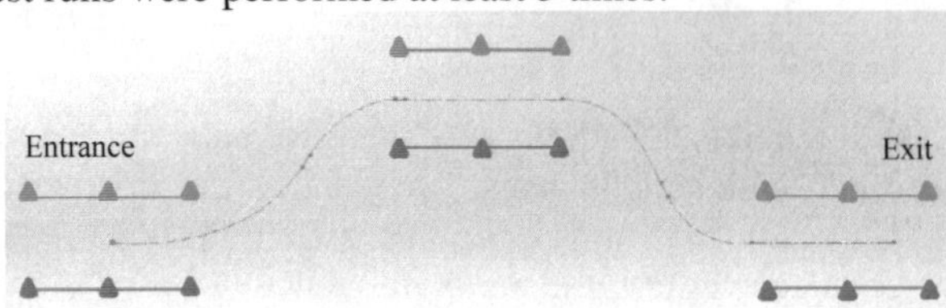

Figure 2. Dimension of test track for steady-state circular test .

3. Equipment Preparation

3.1. Test vehicle

The test vehicle used was a passenger car (M1), front engine, front wheel drive (FWD) as shown in Figure 3 .

Figure 3. A sedan type of passenger car and test tires, which was used for this testing .

3.2. Test tires

The tires used in the tests was divided into two categories, new tires from 2018 and used tires from 2015, 2014, 2013. New tires were used as benchmarks for performance testing. The size of the tires in this tests was 215/60/R15 .

3.3. Weighing Pad

To measure the weight distribution of test vehicle on each wheel & axle, the weighing pads was used. The distribution of weight was used to calculate Center of Gravity (C.G.) of the vehicle as shown in Figure 4 .

Figure 4. The vehicle weight measurement on weighing pad.

From the above instrument, it is observed that weight on front right and front left wheel was 366 kg and 345.5 kg respectively, and weight on rear right and rear left wheel was 235 kg and 211 kg respectively. Gross weight of vehicle is 1157.5 Kg.

3.4. Systems for Motion Measurement

Inertial Measurement Unit is used for motion measurement during the testing. This unit provides highly accurate measurements of velocity, pitch, roll, and yaw, using three yaw rate sensors and three accelerometers. IMU was calibrated by the manufacturer. It is located at ±20 cm away from the center of gravity (C.G.) as shown in Fig 5.

Table 2. Specification of instrumentation used in tests

Instrument	Quantity	Measurement range	Measurement accuracy
Gyroscope Kyowa G-SAT-A-900	Yaw rate	±15.71 rad/s (±900 deg/s)	±0.5%
Triaxial Accelerometer Kyowa AMA-A	Lataral Acceleration, Deceleration	±98.07 m/s^2 (±10 G)	±1%
GPS Garmin GPS-18x	Longitudinal Velocity	< 3 meters	0.2 km/h

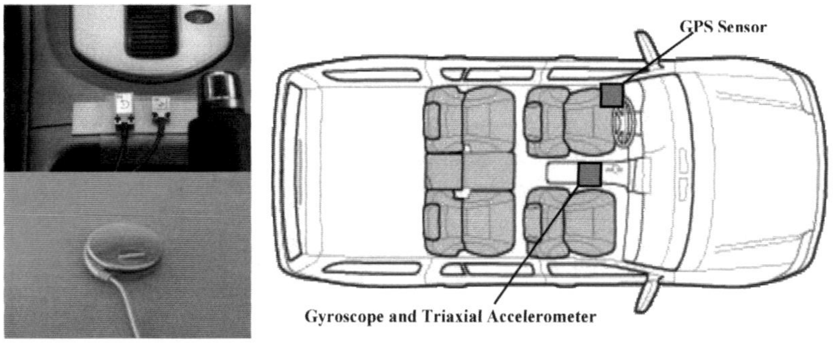

Figure 5. Instrument (left) and it's location in the tested car (right).

3.5. Driving Robots

The driving robots can be installed in almost any vehicle and can deliver repeatable and accurate control. This tests used steering robots and pedal robots to control the test vehicle.

3.5.1 Steering Robots.

The steering wheel robots were used to control the steering wheel of test vehicle in severe-lane change and steady state tests. For both test procedures, the steer angle are significant to validate the vehicle. This equipment can control steering wheel up to a maximum angle of 360 degree at a velocity of 5 m/s

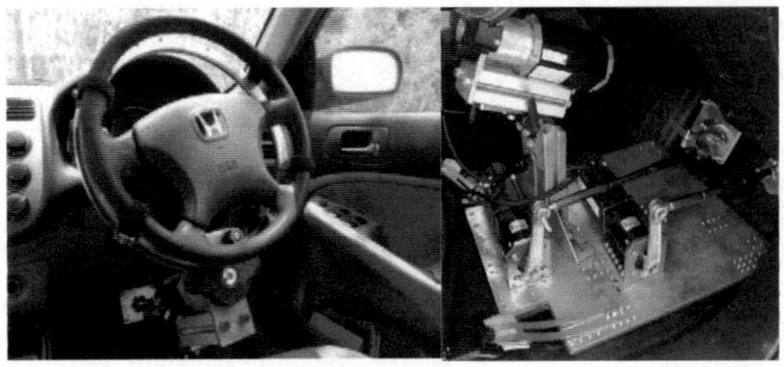

Figure 6. The steering wheel and pedal robots,which was selected for this testing.

3.5.2. Pedal Robots.

The pedal robots included throttle and brake actuator as shown in figure 6. The throttle actuator has been used to control speed for all tests procedures. Furthermore, brake test procedure requires pedal force, which was provided by a brake actuator. This equipment can control a range of pedal forces.

4. Results and Discussion

4.1. braking test

In all braking tests results, the transmission is disengaged and engaged conditions as shown in figure 7. The maximum average braking deceleration occurred for new tires at 7.50 m/s2 while transmission engaged . The lowest average braking deceleration for used tire (2013) was when transmission disengaged. In addition,the maximum value in engaged condition was higher than in disengaged condition. Therefore, brake deceleration relied on tires aging and transmission conditions. Furthermore, tires aging is a significant parameter for braking deceleration.

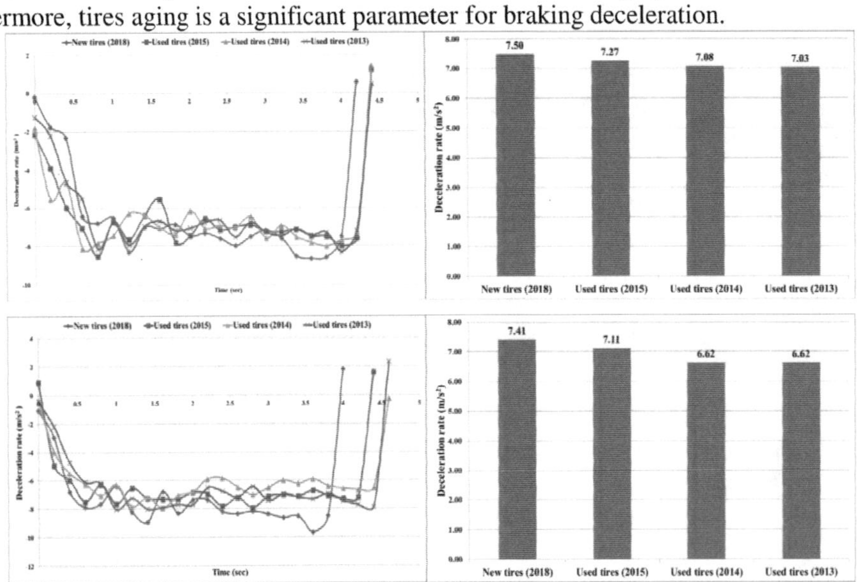

Figure 7. Time histories of braking deceleration at transmission engaged conditions (top) and time history of braking deceleration at transmission disengaged conditions (bottom).

4.2. steady state circular testing

From steady state circular testing, the condition of test with circular (8 m.) radius. new tires (2018) have the maximum lateral acceleration at 8.25 m/s^2, 33.49 km/h. The lowest lateral acceleration appeared with used tires (2013) during 6.8 m/s^2, 28.75 km/h . Therefore, lateral acceleration relied on vehicle speed and tire aging. Moreover, lateral acceleration decreases due to tire aging. Thus, it was a significant parameter for decreasing lateral acceleration under the same steer angle.

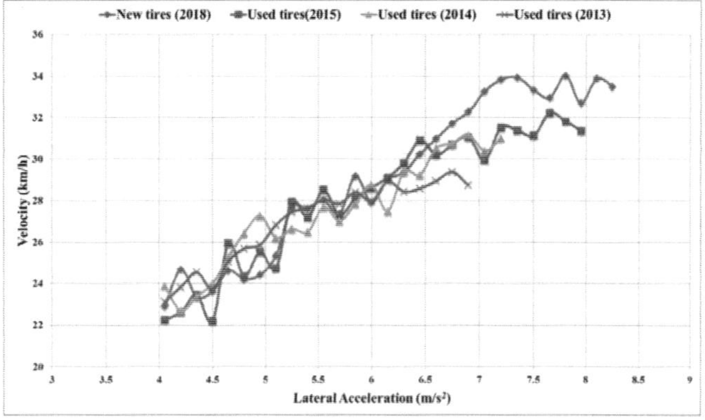

Figure 8. Vehicle speed for Lateral acceleration at constant radius.

4.3. Results of severe lane-change testing

Under the same condition of test with different velocity(40,50 km/h) and age of tires, it was discovered that the maximum yaw rate occurred on new tires (2018) at 50 km/h as shown in figure 9 (right). The lowest yaw rate appeared on used tires (2013) at 40 km/h. Therefore, yaw rate relied on vehicle speed and tire aging. Moreover, tire aging was significant parameters for decreasing yaw rate under the same velocity.

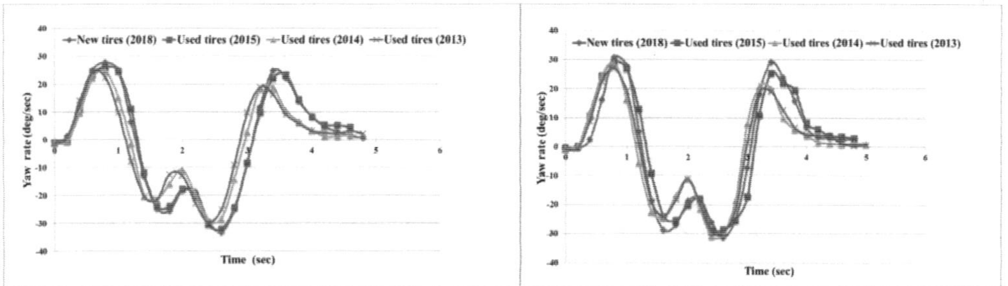

Figure 9. Time histories of maximum yaw rate at 40 km/h (left) and time history of maximum yaw rate at 50 km/h (right).

5. Conclusion

This study is present the procedure to measure the performance of tires. The tests performed on tires showed that different aging resulted in significantly different braking deceleration, lateral acceleration and yaw rate values. All the tests covered lateral force and longitudinal force.

In the all braking tests at 100km/h conditions with transmission engaged and disengaged, the results show approximately 4 % decreasing in deceleration rate, that affected from engine deceleration. In addition, braking deceleration from used tires when compare with new tires is reduced approximately 2-3% in every age group and effect significantly to vehicle braking performance.

For the steady-state circular tests in fixed steer angle conditions, the lateral acceleration of new tires was almost linear. In case of used tires the characteristics of lateral acceleration change was strongly non-linear. This non-linearity points show poorer stability in the range of lateral acceleration. In addition, lateral acceleration at terminal speed in used tires was reduced approximately 5-7 % in every age group. Therefore, the tests revealed that, compared with new tires, the majority of used tires showed a lower cornering performance of vehicle.

The results of tests prove that the differences in the manner of driving a car with tested tires are relatively significant, particularly in reference to maximum yaw rate in the path of severe lane-change manoeuvre. The test showed the maximum yaw rate for used tires were lower than new tires. In addition, maximum yaw rate from new tires to used tires was reduced approximately 2-9 % in every age group. Thus, the tests revealed the majority of used tires showed a lower handling performance of vehicle when compared with new tires.

The result of new tires and used tire had the same trend for both longitudinal and lateral tire force. Therefore, this research fulfills all of the conditions of the real road driving to develop the tire testing model.

References

[1] The Eleventh National Economic And Social Development Plan. (2012-2016). Title Strategy for Restructuring the Economy toward Quality Growth and Sustainability, The manufacturing sector plays a leading role in the Thai economy but overall productivity in this sector has not developed greatly. Thailand: Office of the National Economic and Social Development Board Office of the Prime Minister

[2] The Twelfth National Economic And Social Development Plan. (2017-2021). Title Flagship Projects, National Automobile and Tyre Test Center. Thailand: Office of the National Economic and Social Development Board Office of the Prime Minister

[3] United Nations Economic Commission for Europe Regulation No. 117: Uniform provisions concerning the approval of tyres with regard to rolling sound emissions and/or to adhesion on wet surfaces and/or to rolling resistance unece.org/trans/main/wp29/wp29regs101-120.html

[4] Stone R and Ball K J. 2004 Tire Force Generation *Brakes and Tires (Automotive Engineering Fundamentals)*, ed 3rd (USA: Palgrave Macmillan) chapter 9 pp 429-433

[5] United Nations Economic Commission for Europe Regulation No. 13-H: Uniform provisions concerning the approval of passenger cars withregard to braking unece.org/trans/main/ wp29/wp29regs0-20.html

[6] International Organization for Standardization No. 4138: Passenger cars Steady-state circular driving behavior iso.org/standard/54143.html

[7] International Organization for Standardization No. 3888-2: Passenger cars Test track for a severe lane-change manoeuvre iso.org/standard/57253.html

[8] Clar-Garcia D, Velasco-Sanchez E, Campillo-Davo N, Campello-Vicente H, Sanchez-Lozano M. 2016 *Appl. Acoustics* **110** 2332

[9] Duez B. 2016 *TransportationResearch Procedia.* **14** 1051 – 1060

[10] Baldwin M J, Bauer R D and Ellwood R K 2006 *Polymer Degradation and Stability* **92** 103-109

[11] United Nations Economic Commission for Europe Regulation No. 30: Uniform provisions concerning the approval of pneumatic tyres for motor vehicles and their trailer unece.org/trans/main /wp29/wp29regs21-40.html

IOP Conf. Series: Materials Science and Engineering **501** (2019) 012020 doi:10.1088/1757-899X/501/1/012020

Design Optimisation of a 6-RSS Parallel Manipulator via Surrogate Modelling

Kantapon Tanakitkorn*

Faculty of International Maritime Studies, Kasetsart University, Si-Racha, Chonburi, 20230, Thailand

* Corresponding Author: kantapon.ta@ku.th

Abstract. Parallel manipulators are robots with the capability to perform motion up to six degrees of freedom. This work presents a surrogate-based optimisation process for designing a 6-RSS parallel manipulator that is driven by six rotary actuators; the kinematic performance of the manipulator is defined by the volume of the total orientation workspace and the global conditioning index. The surrogate modelling technique was demonstrated by simulation to be highly beneficial in exploring the design landscape thoroughly even for a high-dimensional case with six design variables. Computational expense in the optimisation process was reduced by sequentially searching the surrogate models instead of the original performance indices. One platform geometry with the best compromise between the two contradicting indices was obtained in accordance with the weighted-sum function. Despite the fact that this work considers a small-scale 6-RSS parallel manipulator, the presented surrogate-based optimal design framework is generic and can be applied to different types of manipulators.

1. Introduction

Parallel manipulators (PMs) are robots with several closed-kinematic chains of actuators that are connected to the same rigid body. These robots have several advantages over their serial counterparts, such as better load-carrying capacity, higher rigidity, and higher positioning accuracy [1, 2]. This work consider a 6-RSS type manipulator whose base plate and mobile plate are connected by six identical link chains; each chain consists of a revolute actuator (R) and two spherical joints (SS) [3].

Due to the revolute actuation, the platform could create more nonlinear effects to the kinematic chain when compared to the conventional prismatic actuation ones. This problem was addressed by [4], and a kinematic model of a 6-RSS manipulator was analysed for improving quality of trajectory tracking control. A closed-form dynamic equations of 6-RSS manipulators was derived by [5] to take into account the effect of inertia as well as the external forces on the platform. In spite of the nonlinearity, the platforms with rotary actuators have been proven to be practically useful for highspeed manipulation [6].

Regarding the classical optimal design methodologies, different design criteria are considered as objective functions for establishing the best platform design. Performance of different optimisation algorithms were compared and discussed in [7]. The concept of Pareto efficiency may be applied when multiple design criteria are considered simultaneously [8]. A workspace analysis is the primary consideration in a platform design since PMs have relatively small and irregular-shaped workspaces when compared to serial manipulators. The workspace of a manipulator may be identified numerically by discretising the Cartesian space into a finite number of small grids; each grid is checked against a number of criteria, e.g. that grid is reachable and the joint limits are not violated, and the workspace is formed by connecting all the valid grids together [9]. In addition, a local conditioning index (LCI) [10] is typically used to gives a measure for how close a particular configuration of the platform is to the singular configuration where the platform loses its stiffness or the kinematic chains degenerate. The overall dexterity over a given workspace is referred to as the global conditioning index (GCI) [8]; this global index is computed numerically by averaging the LCI of the discrete grids over the workspace. For further discussion of various performance indices, please refer to [11].

Although simple in concept, the discretisation method is the most computationally expensive part of the optimisation process. This is the case especially when the total orientation workspace is considered because, for every discrete position, the platform must be evaluated at all orientations in a given range [12]. Such a high computational expense can be alleviated by using surrogate models (SMs) as substitutes for the true objective functions. The approximation models can be based on the simple regression models such as polynomial regression [13], but Kriging surrogates appear to be the most popular choices for design search and optimisation [14, 15]. The preferred search algorithm is then used on the SM to avoid evaluating those expensive objective functions, speeding up the optimisation process.

The main aim of this paper is to present a procedure for establishing the optimal platform design via the surrogate modelling approach. To this end, the relevant parameters for a 6-RSS platform are described in Section 2. Kinematic performance evaluation is defined in Section 3. The surrogate-based optimisation framework is explained in Section 4. This is followed by the results and discussion in Section 5. The conclusions are presented in Section 6.

2. Platform Geometry

The PM currently studied is illustrated in Figure 1. It is designed to consist of standard components, these are: 1) a base plate with a radius of r_b, 2) a mobile plate with a radius of r_m, 3) six identical RC servos mounted to the base plate with horn lengths of l_h, and 4) six identical connecting rods with lengths of l_r (each rod has magnetic ball joints on both ends).

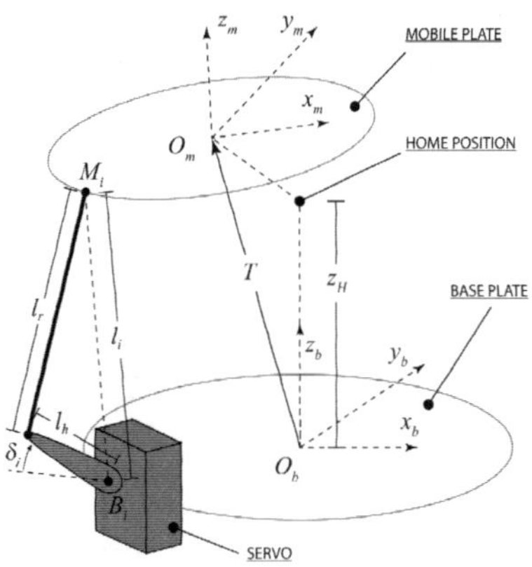

Figure 1. 6-RSS parallel manipulator.

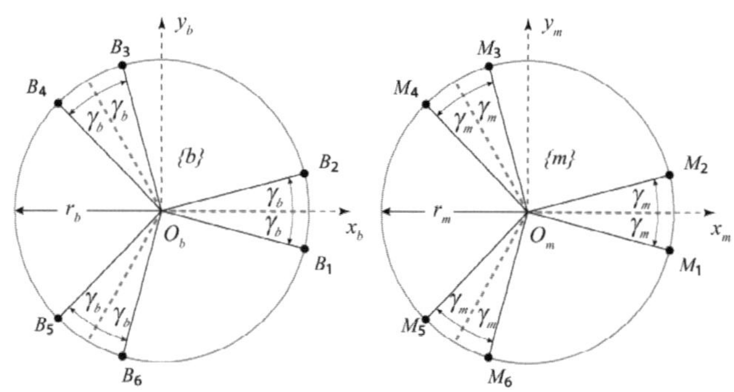

Figure 2. Reference frames that are fixed to the base plate (left) and the mobile plate (right).

Reference frames that are fixed to the centres of the base plate, O_b, and the mobile plate, O_m, are denoted by $\{b\}$ and $\{m\}$, respectively. The joint attachment points on the base plate, B_i, and on the mobile plate, M_i, are grouped into three pairs and are distributed evenly on their respective circles (see Figure 2). Parameters γ_b and γ_m denote the angle spacing between the joints for each pair on the base plate and the mobile plate, respectively.

Joint positions on the base plate expressed in $\{b\}$ for $i = 1,2, \dots ,6$ are:

$$ {}^bB_i = r_b\left[\cos(\Gamma_{b,i}) \quad \sin(\Gamma_{b,i}) \quad 0\right]^{\mathrm{T}}, \tag{1} $$

where

$$\Gamma_{b,i} = [i - 2 - \mathrm{mod}(i, 2)]\frac{\pi}{3} + (-1)^i \gamma_b. \quad (2)$$

Likewise, joint positions on the mobile plate expressed in $\{m\}$ for $i = 1,2,\ldots,6$ are:

$${}^mM_i = r_m\left[\cos(\Gamma_{m,i}) \quad \sin(\Gamma_{m,i}) \quad 0\right]^{\mathrm{T}}, \quad (3)$$

where

$$\Gamma_{m,i} = [i - 2 - \mathrm{mod}(i, 2)]\frac{\pi}{3} + (-1)^i \gamma_m. \quad (4)$$

The parameter γ_h refers to the angle of the servo horn that is shifted from the radial vector on the base plate; δ_i denotes the angle of the ith servo horn relative to the horizontal plane. For simplicity, a pivot of the ith servo is defined to be coincident with B_i. In this regard, the endpoint of the servo horn expressed in $\{b\}$ is:

$${}^bH_i = {}^bB_i + l_h[\cos(\beta_i)\cos(\delta_i) \quad \sin(\beta_i)\cos(\delta_i) \quad \sin(\delta_i)]^{\mathrm{T}}, \quad (5)$$

where β_i is the angle that the sweeping plane of the ith servo horn makes to the *x-axis* of $\{b\}$:

$$\beta_i = \Gamma_{b,i} - (-1)^i \gamma_h. \quad (6)$$

The mobile plate's configuration (position and orientation) in Cartesian space is converted to an angle for each servo horn in joint space with the following inverse kinematic model [16]:

$$\delta_i = \sin^{-1}\left(\frac{\mathcal{L}}{\sqrt{\mathcal{M}^2+\mathcal{N}^2}}\right) - \tan^{-1}\left(\frac{\mathcal{N}}{\mathcal{M}}\right), \quad (7)$$

where

$$\mathcal{L} = l_i^2 - l_h^2 + l_r^2, \quad \mathcal{M} = 2l_h(z_{m,i} - z_{b,i}), \quad \text{and} \quad (8)$$
$$\mathcal{N} = 2l_h\sin(\beta_i)(x_{m,i} - x_{b,i}) - 2l_h\cos(\beta_i)(y_{m,i} - y_{b,i}).$$

The parameters $x_{m,i}$ and $x_{b,i}$ denote the x components of ${}^bM_i$ and ${}^bB_i$, respectively, and so forth. Equation (7) is used to verify whether or not the mobile plate is able to attain a certain configuration.

3. Kinematic Performance Evaluation

3.1. Workspace

This paper considers the total orientation workspace, $\mathcal{W}$, that describes how much translational movement is allowed while a parallel robot is still able to perform any orientation in a prescribed set [12]. To identify $\mathcal{W}$, it is necessary to start by discretising the potential range of translational motion with a regular grid; this yields a finite number of positions to be evaluated. Then, each position is tested with the eight representative orientations which are all combination of ± 10 degrees of roll, pitch, and yaw.

A position is said to be a member of $\mathcal{W}$ if all of the eight representative orientations are attainable. The volume of $\mathcal{W}$ is computed by summing the volumes of all the tiny elements. For generality of the analysis, the resulting workspace volume is non-dimensionalised by dividing it by the volume of a sphere of radius r_b.

3.2. Global Conditioning Index

The global conditioning index (GCI) is commonly used for describing the overall performance over the entire workspace [8, 11]:

$$\mathrm{GCI} = \frac{\int_{\mathcal{W}} \kappa(J)^{-1} \mathrm{d}\mathcal{W}}{\int_{\mathcal{W}} \mathrm{d}\mathcal{W}} \quad (9)$$

where $\boldsymbol{J}$ is the Jacobian matrix [4] and $\kappa(\boldsymbol{J})$ is the condition number [8]. The total orientation workspace, $\mathcal{W}$, is determined as explained in Section 3.1. A higher GCI means that the platform can be controlled more precisely.

3.3. Mechanism Interference Detection

The feasibility of the design is limited by the interference between the manipulator's parts. To ensure sufficient clearance between the parts, the interference detection algorithm is implemented as follows. On each servo, two fixed-radius circles are assigned to both ends of the servo horn and one circle to the rear part of the servo body. The joints on the mobile plate are also assigned with the circles. The mechanism interference is detected if any of the circle pair intercept, or the servo horns cross. In which case both $\mathcal{W}$ and the GCI are set to zero in order to differentiate the infeasible designs from the feasible ones.

4. Surrogate-Based Design Optimisation

The conventional optimisation methods rely on the direct search of the true objective functions. This framework may not be practical when dealing with expensive objective functions (either computationally or financially or both). However, it is possible to approximate those expensive functions by suitable math functions that are relatively less expensive to evaluate; this approximation is known as surrogate models (SMs) or response surface models (RSMs). The optimisation then performs on the SM to avoid evaluating the expensive objective function.

The surrogate-based optimisation methods involve three stages. First, the true function is sampled at multiple points. This initial data set is used for constructing an SM in the second stage. The search and refine strategy is performed on the SM in the last stage, and it iterates until the solution converges. This procedure is summarised in Figure 3, and the three states are explained in the following subsections.

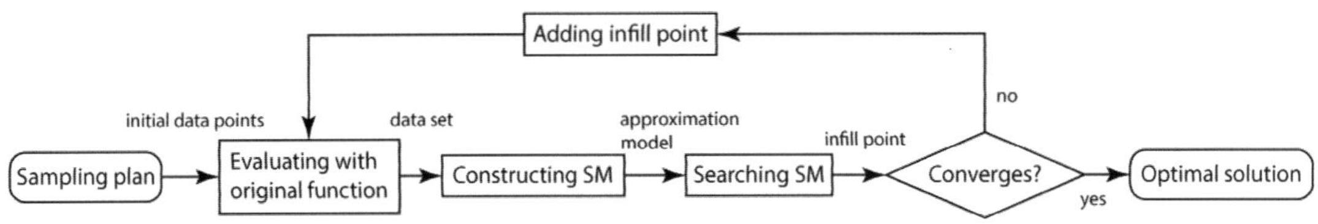

Figure 3. Three fundamental stages of surrogate-based design optimisation.

4.1. Sampling Plan

The surrogate modelling is a process of building an approximation of the original expensive functions. In this regard, like other model fitting tools, an initial set of sample data of an original function is crucial for constructing an SM. The most straightforward way to obtain a *space-filling* data set is to sample the design space evenly and uniformly with the full-factorial sampling plan [17]. Take for instance a 2-dimensional (2D) design space, such a sampling technique will produce a rectangular grid of points. This sampling plan, however, is not efficient because the points overlap when projecting on an axis. Some of these points may be considered repetitive and could be discarded to save computational cost.

Latin hypercube sampling (LHS) is a more efficient alternative that is commonly employed to sample the design space without repetitions. There may be a number of ways to arrange the points in the LHS manner, and in some case the points may bias towards one area. To ensure the data points thoroughly spread over the design space, the minimum distance between all point pairs should be maximised [13]. Since this optimal LHS is *space-filling* and does not generate repetitive points, it allows limited number of samples to gaining as much information of the expensive function as possible.

4.2. Constructing a Surrogate Model

Various types of SMs are presented in the literature. These models are relatively less expensive to evaluate but is sufficiently good in predicting the original function. Kriging is typically the model of choice because it is highly flexible and works well in most situations even with highly-nonlinear, multi-modal functions [18]. Another advantageous of the Kriging model is that it can provide uncertainty estimation of the approximation [13, 14]. This measure is crucial for searching and refining the SM (this point will be discussed in the following subsection).

It has been shown that the platform workspace volume can be improved with a sacrifice of GCI, and vice versa [8]. This means that the ultimate 6-RSS platform design does not exist. Instead, the process will give a range of nondominate designs in the form of a Pareto front [8, 9]; all such designs are valid and optimal but they have different weights for different indices. It is typical, in practice, to combine the two objectives using a weighted-sum function, i.e., $f = w_1 \times f_1 + w_2 \times f_2$, where $w_1 + w_2 = 1$. A Kriging surrogate is then constructed for representing the weighted-sum function that defines the compromise between the two contradict objectives.

4.3. Searching and Refining a Surrogate Model

Once an initial SM has been built, the optimum can be located using one of the existing optimisers such as genetic algorithm (GA). It is important to note that this point is only optimal for the approximation and may not necessarily be so for the true function. In this regard, the point must be considered as an *infill point* which is then evaluated by the original function and augmented into the data set. The SM is refined accordingly and the new optimum is located. By repeating this process, the predicted optimum will converge. The major concern for this pure exploitation infill criterion is that the SM search and refinement will be concentrated on the first optimal region predicted by the SM. As a consequence, the search is highly likely to be trapped in this region, and the global optimum is not guaranteed.

On the other hand, choosing the maximum mean squared error (MSE) as a next infill point will help exploring the uncertain regions of the SM where the global optimum might be hiding [13]. However, this criterion is pure exploration that only yields a more accurate SM. At some point this exploration criterion needs to be stopped and followed by the exploitation search scheme, but at which point can be a tough decision.

Kriging is the Gaussian process-based model that permits the calculation of the expected improvement (EI) [14]. This measure corresponds to the magnitude of improvement expected from refining the SM at any given point. The maximum EI criterion offer a great compromise between exploring uncertain regions and exploiting the current optimum [14, 18]. More importantly, since the points that have been sampled have EI $= 0$, they will not be sampled again; this implies that the design landscape will be sampled throughout and the global optimum will be found eventually. Using EI as an infill criterion also provides a consistent performance for most situations when compared to other advanced infill criteria [19]. Due to these reasons, the maximum EI infill criterion is considered in this paper.

5. Design Optimisation via Surrogate Modelling

The 6-RSS kinematic model was implemented in MatLab. The surrogate model MatLab toolbox employed in this work is available online at: https://optimizationcodes.wordpress.com. Each platform design was evaluated within a domain of size 0.25m × 0.25m × 0.10m. The domain was discretised with a regular grid. Every position within the domain was tested with the eight representative orientations. Then, the total orientation workspace and the GCI were computed numerically. Supported by Figure 4, a grid size of 2% of the base plate radius is sufficiently small; this setting offers reliable values for both the workspace volume and the GCI with an affordable computing time of just under 360 seconds (obtained with a ThinkPad T450, Intel Core i5-5300U, 8GB RAM) per one platform design.

The surrogate-based optimisation approach was applied for predicting performances of the 6-RSS manipulator over the variation of the six design parameters. The first two were the joint angles on the base plate (γ_b) and the mobile plate (γ_m). Other four more design parameters were as follows: home position (z_H), mobile plate radius (r_m), servo horn's length (l_h) and servo azimuth angle (γ_h). The LHS were used for generating 600 initial samples (100 samples per one design variable). These sample were evaluated in terms of $\mathcal{W}$ and the GCI. The interference detection step was also included in the evaluation to penalise the designs that were mechanically infeasible. For each data point, the two objectives were summed with weights of 0.3 and 0.7 for $\mathcal{W}$ and the GCI, respectively. A Kriging SM of the weighted-sum function were constructed accordingly. More infill points were added by sequentially searching and refining the SM with the GA based on the EI of the weighted-sum function.

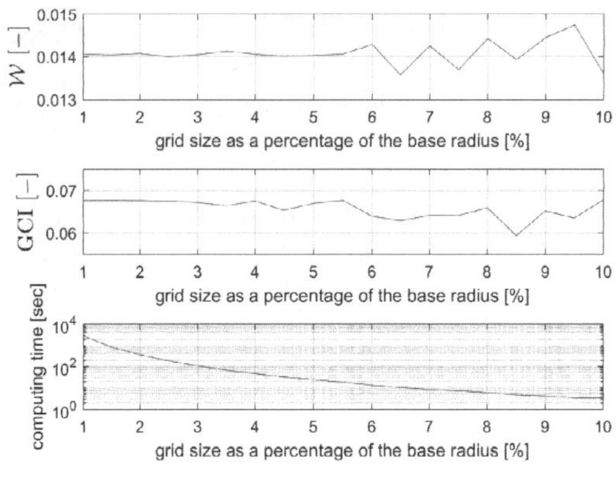

Figure 4. Grid convergence plot.

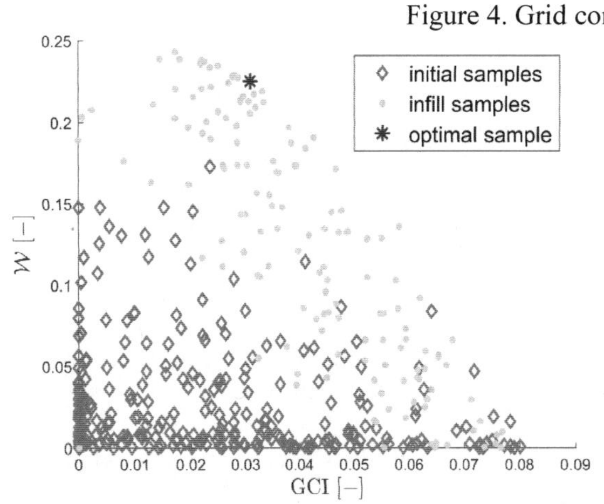

Figure 5. Pareto front on the objective space and the optimal design regarding the weighted-sum function.

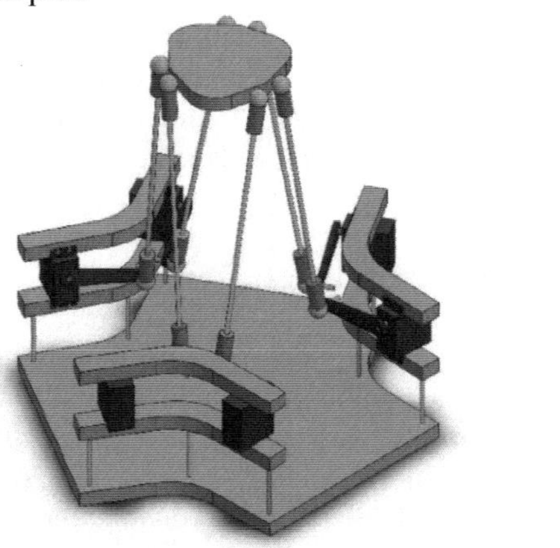

Figure 6. The optimal design of the 6-RSS PM: $l_h = 0.05$ m, $r_b = 0.1$ m, $r_m = 0.042$ m, $z_H = 0.18$ m, $\gamma_b = 25°$, $\gamma_m = 14°$, and $\gamma_h = 153°$.

The SM approach has shown to be highly beneficial in exploring the design landscape thoroughly, even for this case with six design variables. Regarding Figure 5, with just 150 infill points (green dots) the boundary of the objective space was pushed further from the initial data set (red diamonds) to the Pareto front where one objective could not be improved without sacrificing the other objective. On the contrary, directly searching on the true performance functions would not be effective because, for example, the GA search with 100 individuals for 20 generations would require up to 2000 samples to be evaluated with the computationally intensive functions. The presented SM-based approach yielded the optimal manipulator configuration as shown in Figure 6.

6. Conclusions

This paper presented a surrogate-based design method for a 6-RSS PM whose performance was characterised by the volume of the total orientation workspace and the GCI. The existing toolbox was used for generating a space filling sample points as well as for constructing surrogate models of the manipulator performances over a range of design parameter variation. The GA optimiser was then employed to sequentially refine the approximation models and search for the optimal design. Since the surrogate models were relatively less expensive to evaluate when compared to true platform performance evaluation, searching on the approximation models could significantly reduce computational cost in the optimisation process.

The SM approach has been demonstrated to be highly beneficial even for a complex case involving six design variables; it showed to explore the design landscape thoroughly and the Pareto front was identified with just 600 initial samples and 150 infill points. One platform design with the best compromise between the two contradict indices were derived regarding the weighted-sum function.

References

[1] Dasgupta B and Mruthyunjaya T 2000 *Mechanism and Machine Theory* **35** (1) 15–40.
[2] Taghirad H D 2013 *Parallel Robots: Mechanics and Control* (Boca Raton: CRC Press)
[3] Lau T and Zhang Y 2013 New design of 6-RSS parallel manipulator with large rotation workspace *Int. J. Mechanisms and Robot. Syst.* **1** (2/3) 221–239.
[4] Dumlu A and Erenturk K 2013 *Robotica* **32** (04) 643–657
[5] Li K and Wen R 2011 Closed-form dynamic equations of the 6-RSS parallel mechanism through the Newton-Euler approach *Proc. 3rd ICMTMA* **1** 712–715.
[6] Poppeova V, Uricek J, Bulej V and Sindler P 2011 *Technical Gazette* **18** (3) 435–445.
[7] Lou Y, Zhang Y, Huang R, Chen X and Li Z 2014 *IEEE Trans. Autom. Sci. Eng.* **11** (2) 574–584.
[8] Lara-Molina F A, Rosario J M and Dumur D 2010 *The Open Mechanical Eng. J.* **4** (1) 37–47.
[9] Altuzarra O, Pinto C, Sandru B and Hernandez A 2011 *J. Mechanical Design* **133** (4) 041007.
[10] Xie F, Liu X -J and Wang J 2011 Performance evaluation of redundant parallel manipulators assimilating motion/force transmissibility *Int. J. Adv. Robot. Syst.* **8** (5) 113–124.
[11] Patel S and Sobh T 2015 Manipulator performance measures - a comprehensive literature survey *J. Intell. Robot. Syst.* **77** (3-4) 547–570.
[12] Jiang Q and Gosselin C M 2009 Maximal singularity-free total orientation workspace of the GoughStewart platform *J. Mechanisms and Robot.* **1** (3) 034501.
[13] Forrester A I, Sbester A and Keane A J 2008 *Engineering Design via Surrogate Modelling* (Chichester: John Wiley & Sons, Ltd).
[14] Bhosekar A and Ierapetritou M 2018 *Comput. Chem. Eng.* **108** 250–267.
[15] Skinner S N and Zare-Behtash H 2018 *Appl. Soft Comput. J.* **62** 933–962.
[16] Szufnarowski F 2013 Stewart platform with fixed rotary actuators: a low cost design study *Advances in Medical Robotics*
[17] Queipo N V, Haftka R T, Shyy W, Goel T, Vaidyanathan R and Kevin Tucker P 2005 *Progress in Aerosp. Sci.* **41** (1) 1–28.
[18] Xiao S, Rotaru M and Sykulski J K 2012 Exploration versus exploitation using kriging surrogate modelling in electromagnetic design *COMPEL* **31** (5) 1541–1551.
[19] Parr J M, Keane A J, Forrester A I and Holden C M 2012 Infill sampling criteria for surrogate-based optimization with constraint handling *Engineering Optimization* **44** (10) 1147–1166.

IOP Conf. Series: Materials Science and Engineering **501** (2019) 012021 doi:10.1088/1757-899X/501/1/012021

Estimation of Yield Surface on Iron Powder Compaction for Finite Element Analysis using Drucker-Prager Cap Model

R Ogawa [1], T Yoshimura [1], N Yagi [1], K Mizuta [1], and Y Taniguchi [2,*]

[1] Student of National Institute of Technology, Nara College, Nara 639-1080, Japan
[2] Department of Mechanical Engineering, National Institute of Technology, Nara College, Nara 639-1080, Japan

* Corresponding Author: taniguchi@mech.nara-k.ac.jp

Abstract. Shear failure yield line in Drucker-Prager yield surface for conventional atomized iron powder has been estimated with critical stress state and uniaxial failure stress state by single shear test and uniaxial failure test for green compact. Lateral force measurement test was also conducted using newly devised instrumented die compaction system to identify compaction yield characteristics as Cap yield surface. By those estimations, materials input parameters for Drucker-Prager Cap model have been determined successfully. Finite element analysis of simple closed-die compaction has been conducted and good correspondence in variation of compaction pressure with densification has been recognized.

1. Introduction

Powder metallurgy, PM method is one suitable method to produce metal parts with high accuracy and high cost performance as it can process metal parts with certain mechanical properties in complex shape such as gear, sprocket, pulley, and synchronizer hub. In the metal powder compaction process using closed-die system, which applies to mass production for various PM parts, failure of green compact such as slip-crack often occurs especially for compaction of multi-step part. Since slip-crack usually caused by improper powder flow on the boundary of high density part and low density part during consolidation, it is important to optimize compression velocity in tools for each compaction step including ejection of solidified green compact [1]. Application of finite element numerical simulation method reduces time and cost in design and trials of the metal powder die compaction processes, as it is able to predict magnitude of density distribution and shear stress concentration occurrence. A number of yield criteria with constitutive equation such as Shima-Oyane model and Drucker-Prager model have been suggested to describe consolidation behaviour that treats powder as continuum materials [2-4]. Materials input parameters necessary in use of numerical simulations are usually estimated by several material testing, such as uniaxial closed-die compaction test, triaxial compaction test and direct shear test, which are commercially used in soil mechanics [5]. However, degrees of die compaction pressure for usual metal powders such as iron for sintered alloy are at least 10 times higher in comparison with soil mechanics and pharmaceutical powder compaction industry. Triaxial compaction method is most preferable to estimate yield surface and constitutive behaviour of granular materials as it can possible to measure both of elastic and plastic properties. Nevertheless, it is hard to develop for metal powders due to high strength in consolidation, so a few multiaxial compaction testing system with high rigidities exists. Practical way of determining key material parameters at low cost is using instrumented die compaction system with mathematical treatment to convert the applied stresses to a multiaxis stress yield properties [6].

Application of numerical simulation method including shear failure criterions is most effective to predict crack occurrence during compaction, unloading and ejection stage. Drucker-Prager Cap, *DPC* model provides both shear yield behaviour and volumetric deformation with hydrostatic stress for granular materials. When stress state is encountered for critical state corresponding to a high deviatoric stress but a moderate hydrostatic stress, consolidated powder fractures as weak mechanical links between particles cannot resist applied shear stress. Such stress states on compression-shear plane with several compact densities provide critical state line, *CSL*, which is division point in both consolidation and failure occurrence. Although single shear testing is most simple method to estimate shear failure characteristics of granular materials, it was not applicable for metal powder, because at high pressure region the direction of shear plane does not correspond to the shear direction [7]. We have developed improved single shear testing method for metal powder to measure stress state in shear failure plane on high density state [8]. As results, *CSL* of iron powder compact with corresponding density ratio has been successfully estimated as described in previous work [9].

In this study, material input parameters in *DPC* model is estimated to carry out numerical simulation of iron powder compaction process using commercially available finite element method (FEM) solver ANSYS. Firstly, *CSL* estimated in previous work is converted in the stress plane with first invariant stress I_1 and second invariant deviatoric stress J_2. Then shear failure lines in *DPC* model is estimated with *CSL* and uniaxial failure stress state of green compact. For the identification of compaction yield surface in *DPC* model, lateral force measurement was conducted using newly devised instrumented die compaction system. Thus the whole of yield surface is able to determine by combining both stress state, shear failure occurrence and compaction proceeding. Elastic properties are also identified by measuring unloading characteristics, so variations of volumetric plastic strain during compaction were calculated and compaction hardening characteristics has been identified. Finally, FEM simulation of cylindrical die compaction using commercial available software ANSYS has been carried out and validated.

2. Definition of yield surface in DPC model

2.1. Shear failure line

Yield criterion of *DPC* model in ANSYS is a function of stress invariants I_1, J_2 and J_3. Shear failure function Y_s is given by:

$$Y_s(I_1, \sigma_c) = \sigma_c - A exp(B I_1) - \alpha I_1 \tag{1}$$

Where σ_c is the cohesion related hardening parameter with consolidation. A, B and α are material parameter which defines shear yield stress respectively. Equation (1) means that shear failure envelope is simplified as linear relationship with σ_c and α for $A = B = 0$ as shown in figure 1. So, shear yield surface is expressed as conical shape having center axis of mean stress σ_m in three principal stress space. α means slope of shear failure line.

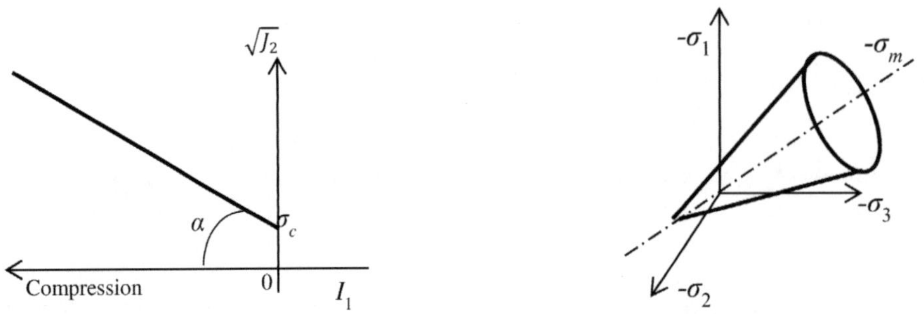

(a) Shear failure line simply defined by cohesion and slope

(b) Shape of shear yield surface drawn in principal stress space

Figure 1. Definition of shear failure criterion in DPC model

2.2. Compaction yield line

The Compaction cap function Y_c is formulated using the shear failure function as:

$$Y_c(I_1, K_0, \sigma_0) = 1 - H(K_0 - I_1)\left(\frac{I_1 - K_0}{R_c Y_s(K_0, \sigma_0)}\right)^2 \tag{2}$$

where: H means Heaviside function, R_c = Ratio of elliptical I_1 to J_2 axis, and K_0 = Key flag indicating the current transition point at which the compaction cap surface and shear failure surface intersect. σ_0 expresses subsequent yield surface which is defined as evolution of cohesion. Note that the Heaviside function roles as follows;

1. When I_1 is greater than K_0, the compaction cap takes no effect on yielding. It means that only shear failure function Y_s is effective.
2. When I_1 is less than K_0, the yielding may only happen in the compaction cap portion, which is shaped by both the shear failure function Y_s and cap function Y_c as expressed by equation (2).

2.3. Hardening functions

The cap hardening law is defined by describing the evolution of the parameter X_0, which is the current intersection point of the compaction cap and the I_1 axis. In other words, hardening is expressed by increasing volumetric plastic strain ε_v^p with consolidation proceeding that is defined as:

$$\varepsilon_v^p = W_1^c[exp\{D_1^c(X_0 - X_i)\} - 1] \tag{3}$$

where: X_i = Initial value of X_0 at which the cap function takes effect. W_1^c = Maximum possible plastic volumetric strain which is defined from densification curve of the powder material. D_1^c = Hardening parameter. So that the evolution of K_0 provides the isotropic hardening of compaction cap as state variable, which can be implicitly described using X_0 and σ_0 given below:

$$K_0 = X_0 + R_c Y_s(K_0, \sigma_0) \tag{4}$$

2.4. DPC model in $I_1 - \sqrt{J_2}$ plane

According to above yield functions with some assumptions in theory of plasticity, *DPC* yield function is described as follows;

$$F(I_1, J_1, J_2, J_3, K_0, \sigma_0) = J_2 - Y_c(I_1, K_0, \sigma_0)Y_t(I_1, \sigma_0)Y_s^2(I_1, \sigma_0) \tag{5}$$

where Y_t is expansion cap function defined similar to equation (2). The yield surface with subsequent yield locus can be drawn schematically in $I_1 - \sqrt{J_2}$ plane as shown in figure 2.

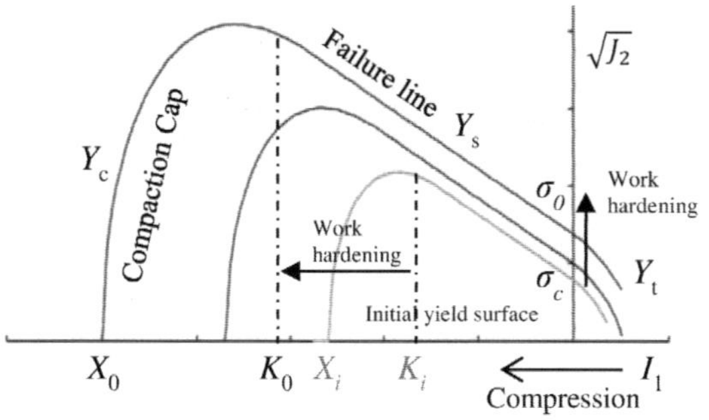

Figure 2. Yield locus of *DPC* model: Consists of three segments which are different yield functions, unite to express both of shear failure and consolidation characteristics in powder materials. Work hardening implemented by state variable K_0 with volumetric plastic strain ε_v^p and evolutions of cohesion σ_0 defines subsequent yield condition.

Thus, the materials input parameters are summarized as follows; R_c: Compaction cap parameter, X_i: Compaction cap yield pressure, σ_c: Cohesion yield parameter, α: Shear envelope linear coefficient, W_1^c: Limiting value of volumetric plastic strain, D_1^c: Hardening parameter respectively.

3. Experimental procedures

3.1. Estimation of input parameters

Three kind of material testing is suggested in this study, single shear test, uniaxial failure test, and uniaxial closed-die compaction test with lateral force monitoring using newly devised instrumented die compaction system. Uniaxial failure test of green compact provides a point of stress state in failure by uniaxial principal stress (σ_1, $\sigma_2 = \sigma_3 = 0$). The uniaxial failure stress state is obtained as $\sqrt{J_2} = \sigma_1/\sqrt{3}$ and $I_1 = \sigma_1$ at current compact density. Single shear test of green compact provides *CSL*. Since *CSL* has been identified in previous work [9], we are able to estimate K_0 directly. Therefore, slope of failure lines α and evolution of cohesion σ_0 with densification proceedings can be estimated by equation (1).

Lateral force measurement provides a point of stress state in densification by simple triaxial principal stresses (σ_1, $\sigma_2 = \sigma_3 = \sigma_L$, σ_L: Lateral pressure) during compaction. Therefore the stress state is given by:

$$I_1 = \sigma_1 + 2\sigma_L \tag{6}$$

$$\sqrt{J_2} = |\sigma_1 - \sigma_L|/\sqrt{3} \tag{7}$$

3.2. Experimental conditions

Conventional atomized iron powder (Kobe Steel, Ltd. 300M) with 1.0 mass % zinc stearate which is an additive as interparticle lubricant has been used. It is necessary to prepare green compact specimens with proper density for both of uniaxial failure test and single shear test. The specimens in its density ratio ρ, defined as the ratio of the compact density to material density, of 0.70, 0.75, 0.80 and 0.85 was used. The shape of the specimens is cylindrical in 10 mm diameter with 15 mm height for uniaxial failure test, and 20 mm diameter within 10 mm height for single shear test.

Principle of the newly devised instrumented die compaction system is shown in figure 3. The green compact which shape is square column with 8 mm × 8 mm sectional dimension can be compacted. Compaction load and lateral force during compaction including unloading stage were measured by load cells. Punch strokes were adjusted to obtain green compacts of 0.70, 0.75, 0.80 and 0.85 in density ratio after ejection respectively. Weight of powder was fixed in 3.5 g so that about 8 mm square cube shape is obtained at $\rho = 0.85$. Compression punch velocity is 0.5 mm/s and punch strokes is stopped for 5 sec before unloading with 0.1 mm/s in each conditions.

IOP Conf. Series: Materials Science and Engineering **501** (2019) 012021 doi:10.1088/1757-899X/501/1/012021

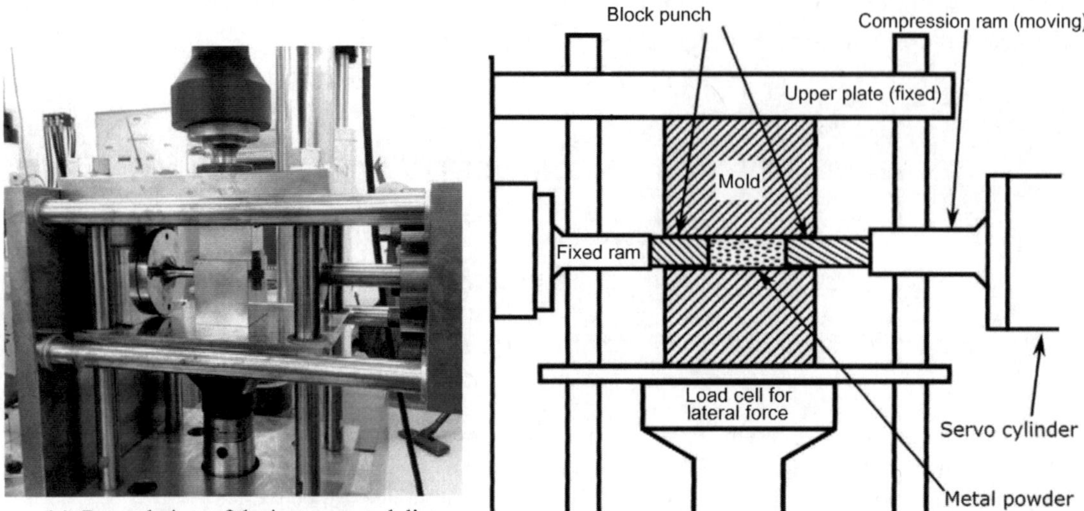

(a) General view of the instrumented die compaction system

(b) Schematic view of die set in the system

Figure 3. Principle of instrumented die compaction system

4. Results and discussions

4.1. Estimation of compaction cap surface by lateral force measurement

Figure 4 shows results of instrumented die compaction test. Lateral force/pressure clearly remains as residual components after unloading as shown in figure 4(a) and (b). As results variation of $\sqrt{J_2}$ which is non-negative value reverses its direction at the point of stress state $\sigma_1 = \sigma_2 = \sigma_3$ during unloading stage as shown in figure 4(d). Further unloading is accompanied by increasing shear stress component with elastic recovery. It is considered that plastic volumetric expansion occurs eventually when $\sqrt{J_2}$ changes beyond the secondary peak, then σ_1 becomes 0. These loading properties seems to be the same for each density ratio. Stress state corresponding to the compaction cap surface is indicated at the maximum compaction stress for each density ratio.

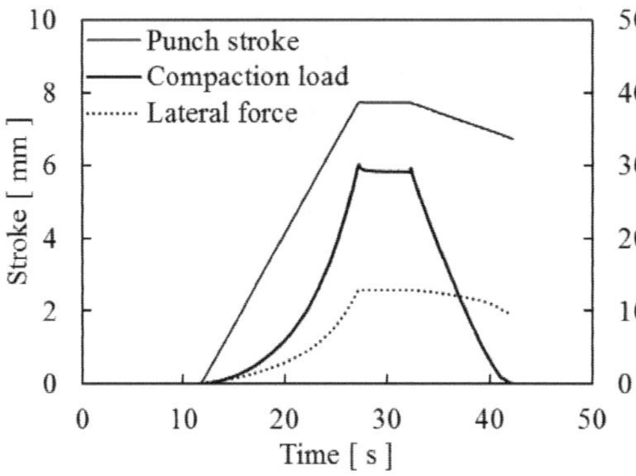

(a) Variations of punch stroke and forces on operation time (Density ratio ρ = 0.85)

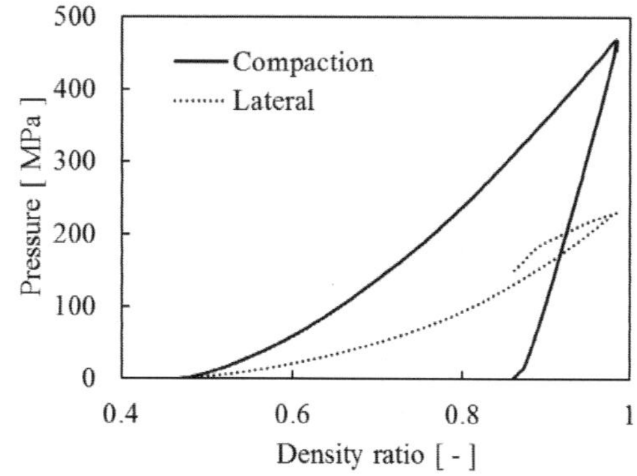

(b) Variations of compaction and lateral pressure with densification

(c) Photograph of green compact after ejection (Density ratio ρ = 0.85)

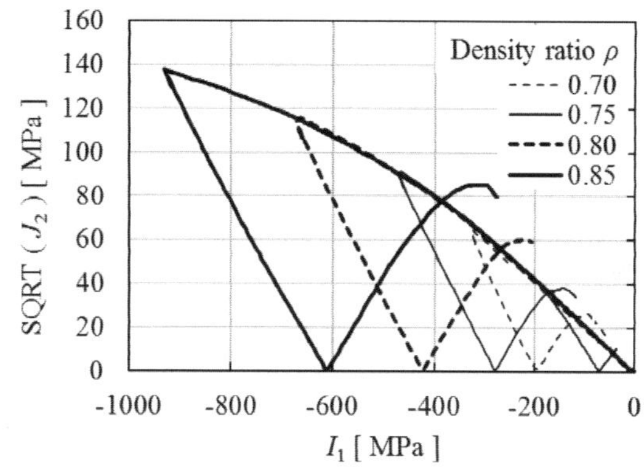

(d) Densification curve mentioned in $I_1 - \sqrt{J_2}$ plane

Figure 4. Results of instrumented die compaction test

4.2. Estimation of yield surface of DPC model

Concerning results of the estimation of shear failure yield line, the stress states which on the yield lines are plotted and yield surface has been drawn as shown in figure 5. Materials parameters, R_c = 3.8 and α = 0.2 with proper X_i and σ_c values which are suitable for each plots has been identified. It is considered that estimation of yield surface has been successfully achieved because the all plots are seems to be compatible with the yield functions although these plots were measured by different experiment respectively. This fact is well supported by unloading curve that ends on the *CSL*.

IOP Conf. Series: Materials Science and Engineering **501** (2019) 012021 doi:10.1088/1757-899X/501/1/012021

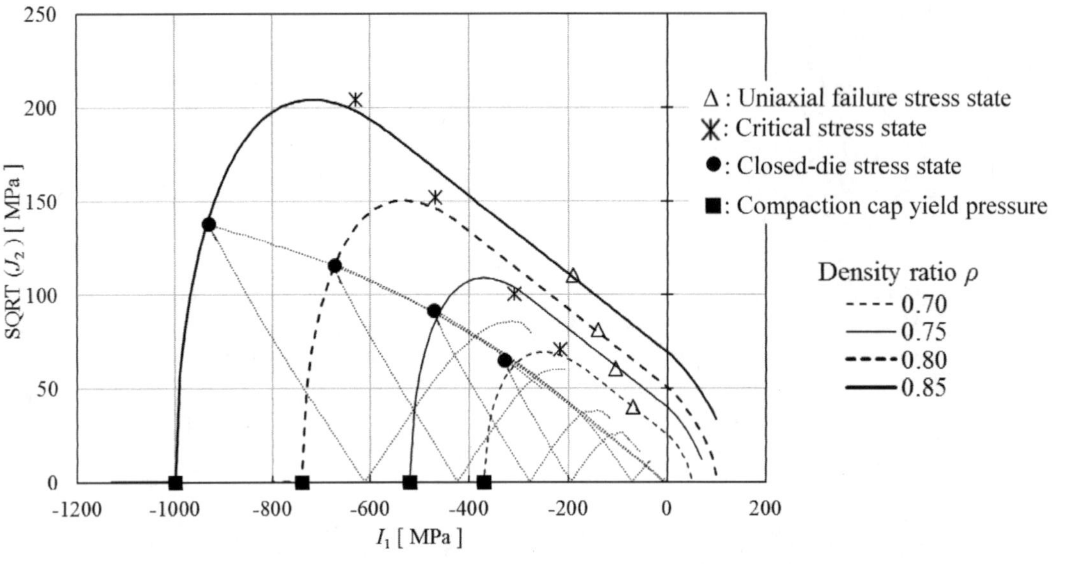

Figure 5. Estimated *DPC* surfaces for each green compact density

4.3. Cap hardening parameters

Cap hardening parameters has been identified by equation (3) as $W_1^c = 0.6$ and $D_1^c = 0.0045$. Figure 6 shows relationship between I_1 and ε_v^p with experimental result. Unfortunately, there is deviation because the influence of changes in elastic properties is not considered in this study.

4.4. FEM results and validation in densification curve

FEM analysis of cylindrical closed-die compaction has been carried out with the elastic modulus of 4245 MPa as constant value. Figure 7 shows equivalent plastic strain distributions inside of the green compact. Note that one-half of the shape has been shown as axi-symmetric analysis. It seems that the distribution tendencies are fairly simulated as strain concentrate at top corner is clearly recognized in contrast with lower strain observed at bottom corner due to single action compression.

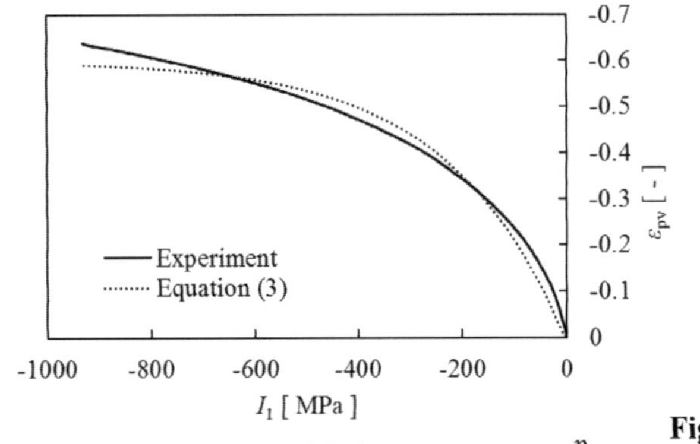

Figure 6. Relationship between I_1 and ε_v^p

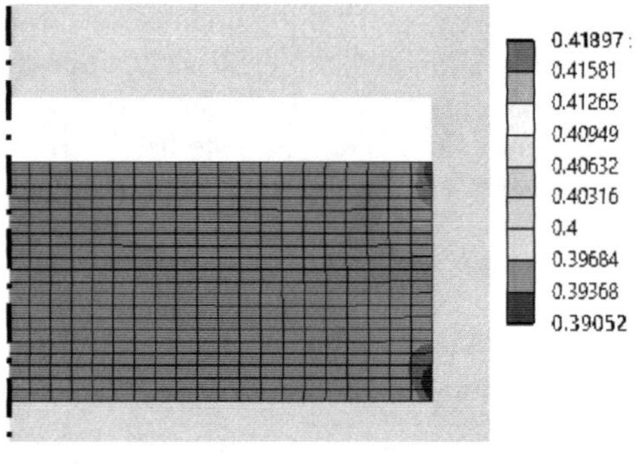

Figure 7. Equivalent plastic strain distributions by FEM analysis

Figure 8 shows variations of compaction pressure with mean density ratio increase. According to the deviation in ε_v^p, it is hard to simulate densification curve at high pressure region precisely. Unloading properties is quite different because of elastic modulus that changes with density in actual compaction.

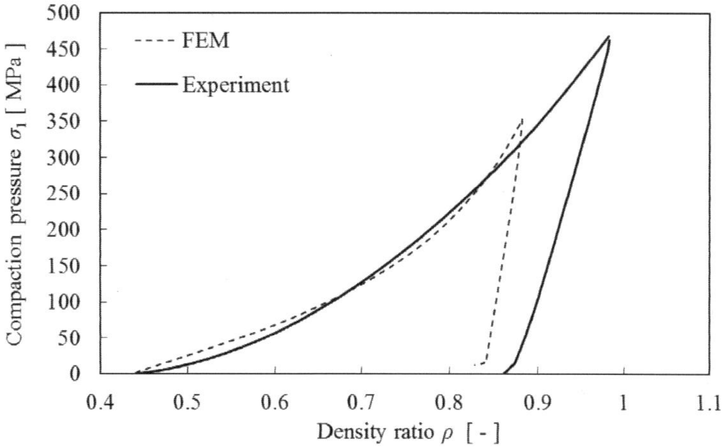

Figure 8. Variations of compaction pressure with density ratio increase

However, it is possible to predict compaction pressure and plastic strain distributions when density ratio is less than 0.88 as shown in figure 7 and figure 8. It seems that estimation of elastic properties becomes effective to optimize cap hardening parameters more precisely.

5. Conclusions

Materials input parameters of iron powder on shear failure yield and compaction cap yield functions for *DPC* model have been estimated from several material testing, uniaxial failure, single shear and lateral force measurement. It has been recognized that the all plots mentioned in $I_1 - \sqrt{J_2}$ plane are compatible with the yield functions so *DPC* yield surface has been successfully identified. Finite element analysis of simple closed-die compaction has been conducted and plastic strain distributions are fairly simulated. Unloading properties should be investigated to optimize cap hardening parameters more precisely.

Acknowledgments

A part of this work was financially supported by JSPS KAKENHI Grant No. 15K13851.

References

[1] Mosbah P, Shima S, Habraken A and Charlier R 1999 *J. Japan Society of Powder and Powder Metallurgy* **46(7)** 696-704
[2] Shima S and Oyane M 1976 *Int. J. Mech. Sci.* **18** 285-291
[3] Shima S and Saleh M 1993 *Mechanics of Materials* **16(1-2)** 73-81
[4] Drucker DC and Prager W 1952 *Quarterly of Appl. Math.* **10** 157-165
[5] Sinka I, Cocks A, Morrison C and Lightfoot A 2000 *Powder Metall.* **43(3)** 253-262
[6] Mosbah P, Bouvard D, Ouedraogo E and Stutz P 1997 *Powder Metall.* **40(4)** 269-277
[7] Shima S, Kotera K and Nishimura K 1999 *Adv. Tech. of Plasticity* **2** 1333-1338
[8] Taniguchi Y, Dohda K and Wang Z 2004 *Trans. Japan Society of Mech. Eng.* **70(696)** 2555-2560 (in Japanese)
[9] Taniguchi Y, Shiota K and Kodama K 2015 *Proc. 6th TSME Int. Conf. on Mech. Eng. (Phetchaburi)* AMM020

IOP Conf. Series: Materials Science and Engineering **501** (2019) 012022 doi:10.1088/1757-899X/501/1/012022

Fatigue failure of an idle gear shaft of a gearbox

Charnont Moolwan[1*] and Samroeng Netpu[2]

[1]Department of Industrial Engineering, Faculty of Engineering
[2] Program in Engineering Management, Graduate School, Kasem Bundit University, 1761 Pattanakarn Rd., Suanluang Bangkok 10250, Thailand

* Corresponding author. E-mail address: chanon.moo@kbu.ac.th

Abstract This paper reports the results of failure analysis of an idle gear shaft of a gearbox in a hot re-rolling steel mill in Thailand. The shaft failed prematurely after only about 6,000 hours of service which was very much lower than the expected working life of 40,000-50,000 hours. The results showed that the shaft failed by fatigue fracture. Beach marks on the fracture surface were clearly visible. Fatigue cracks were initiated at two corners of the keyway of the shaft. Relatively large final fracture area of the fracture surface indicated that the shaft was under a very high stress at the time of failure. The root cause of the fracture is poor design and poor machining of the keyway. Too small fillet radii at the bottom of the keyway gave rise to stress concentration resulting in excessive stresses at the keyway corners. These high stresses led to fatigue crack initiation at the keyway corners followed by crack growth, and final fracture.

1. Introduction

Shafts are extensively used in machines and numerous engineering components including gearboxes. Failures of shafts not only result in replacement cost, but also in process downtime. This could have a drastic effect on productivity and, more importantly, late delivery. In the case being investigated, for example, the downtime was four days and 2,500 metric tonnes of steels were lost before the failed shaft could be replaced.

A gear shaft is usually subject to a high torsion and a bending moment as well as cyclic stresses which, when combined, may cause fatigue in the shaft. These factors may be influenced by stress concentrations which could drastically decrease the fatigue life and lead to the fracture of gear shaft [1]. Fatigue fracture is one of the most common causes of shaft failure. Fatigue failures are insidious and are therefore important considerations in mechanical designs [2]. Despite preventive measures taken during the design stage, fatigue failure can still occur due to either defects introduced during fabrication and/or degradation of shafts during service [3]. The most common failure modes of gear shafts were found to be, in decreasing order of frequency, as follows: fatigue, impact fracture, wear, and stress rupture [4]. Stress concentrations resulting from various causes such as protrusions, undercuts, poor machining, inadequate fillet radii were found to play key roles in shaft fracture [5,6]. Protrusions and troughs on the surface of the relief groove on shaft due to poor machining gave rise to high stress concentration leading to crack initiation [7]. In general, fatigue strength of engineering components increases with decrease in surface roughness. Manufacturing errors such inclusions, porosities, quench cracks, retained austenite, soft ferrite are potential sites for fatigue cracks to initiate internally [8, 9, 10]. Fatigue damage often starts at the outer surface of component due to surface integrity problems resulting from manufacturing, and the presence of stress concentrations originating from surface topography [11]. The failed pinion shaft used in bowl mills was analyzing the cause for failure. Results revealed the initiation of cracks coming from the keyway corner. A small overload fracture zone was also observed in interior of the shaft suggesting low stress but high stress concentration torsional failure [12].

This paper aims at identifying the causes of failure of an idle gear shaft in a gearbox so that the reoccurrence of similar failure can be minimized or avoided in the future.

2. Background

The failed shaft was used in a gearbox in a continuous hot rolling steel re-bars mill in Thailand. The steel mill produces re-bars 12 mm to 20 mm diameter with a capacity of 35 metric tonnes per hour. The mill was designed for rolling steel billets with cross-sectional area of 120x120 mm to 130x130 mm square and six meters long. The gearbox was installed between pass number two and three, and was driven directly by a 900 kW AC electric motor. The rotational speed of the failed shaft was 20.8 rpm. The shaft failed prematurely by sudden fracture after approximately 6,000 hours in service. Normally, a gearbox has an expected working life of around 40,000-50,000 hours [13].

Details of the gearbox assembly showing the failed shaft and the location of the fracture are as shown in Figure 1a, and key dimensions of the failed shaft are shown in Figure 1b.

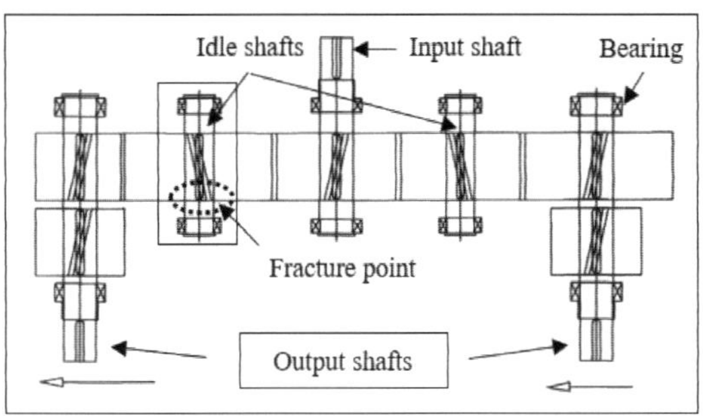

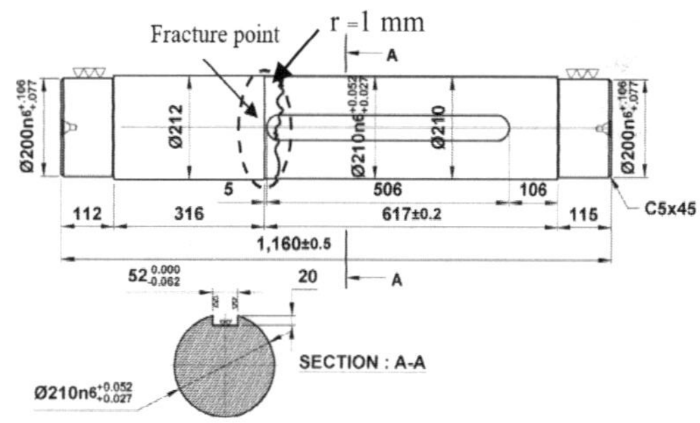

Fig. 1. (a) The gearbox layout showing failed shaft and fracture position (b) Key dimensions of the failed shaft

3. Investigation procedure

The failed shaft was first inspected visually and macroscopically. Relevant dimensions were measured and details of operating conditions noted. Fracture surface were examined visually as well as by using optical microscopy and scanning electron microscopy (SEM) techniques.

Hardness measurement at various distances from the crack origin of the keyway corner to the interior were carried out on a polished specimen using a Mitutoyo hardness tester model ARK 600 (steel ball diameter 1/16") with a load of 981 N. Chemical composition of the failed shaft material was analysed using a spectrolab spectrophotometer (Model: M8, Type: LAVWA 18A). Microstructure of the failed shaft material was examined under an optical microscope (LECO: IA32-Image analysis system).

Forces acting on the shaft and stress analysis were performed using bearinX software of Schaeffler Group. Applied stress was determined based on the data from actual operating conditions and relevant dimensions. The geometry of the model in the analysis reflected the actual dimensions of the shaft. The bending moment calculated using bending moment and shear calculator online civil engineer software.

IOP Conf. Series: Materials Science and Engineering **501** (2019) 012022 doi:10.1088/1757-899X/501/1/012022

4. Results

4.1 Visual examinations

4.1.1 Fracture surface examination

The fracture surface of the failed shaft has two distinct portions both of which are rather flat as shown in Figure 2. Beach marks are clearly visible in one zone of the fracture surface. Such fracture surface appearance indicates that the shaft most likely failed by fatigue fracture. The final fracture surface area is approximately 80% of total fracture surface indicating that the stress in the shaft, hence the loads, must have been very high. Fracture surface appearance, as shown in Figure 2b, also indicates the crack most likely initiated at two corners of the keyway of the shaft. Ratchet marks in the fracture surface suggest further that the fracture is fatigue fracture.

The overall fracture surface was rather flat indicating that the final fracture was brittle by nature and that the shaft was subjected to very high triaxial stress before failure.

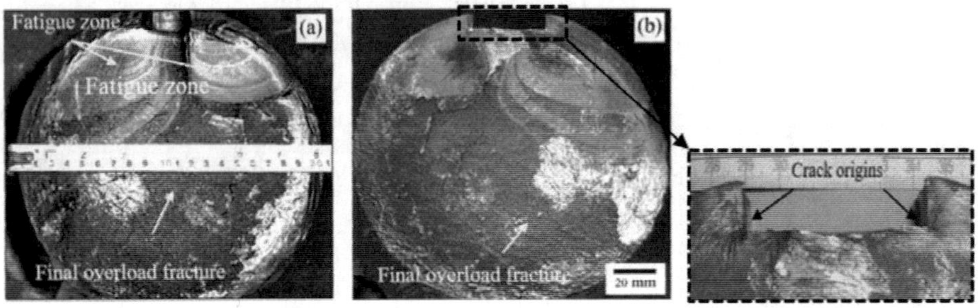

Fig. 2. The failed shaft the fracture surfaces

4.1.2 Keyway examination

The keyway of the failed shaft was examined in more detail to identify characteristics and features of the keyway that could lead to crack initiation. It was observed that the keyway surface had distinctive rough machining marks as shown in Figure 3a. The keyway corners are quite sharp as shown in Figure 3b. Rough surface and sharp corners of the keyway are possible causes for the failure of the shaft as such features could lead to stress concentration and crack initiation.

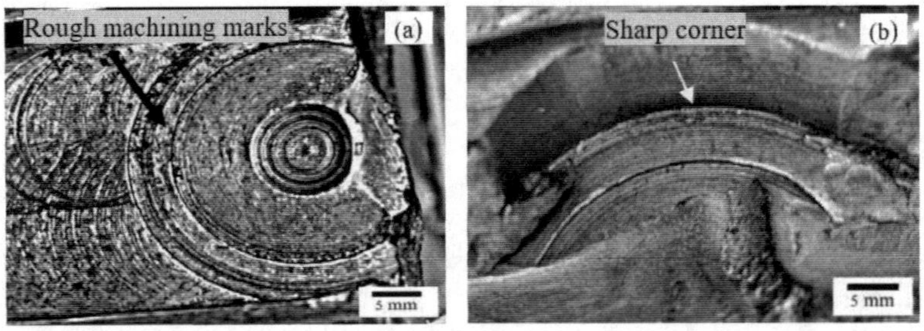

Fig. 3. Keyway (a) machining marks on surface (b) sharp corners

4.2 Keyway corners examinations

A sample from the material in vicinity of the keyway corner was cut, polished and examined to look for any abnormalities. The fillet radius of the keyway edge was found to be 0.228 mm as shown in Figure 4a which is much smaller than the recommended value of between 0.7-1.0 mm [14]. Two cracks at a keyway corners were found as shown in Figure 4b. This demonstrates clearly that the crack did indeed start at the keyway corners.

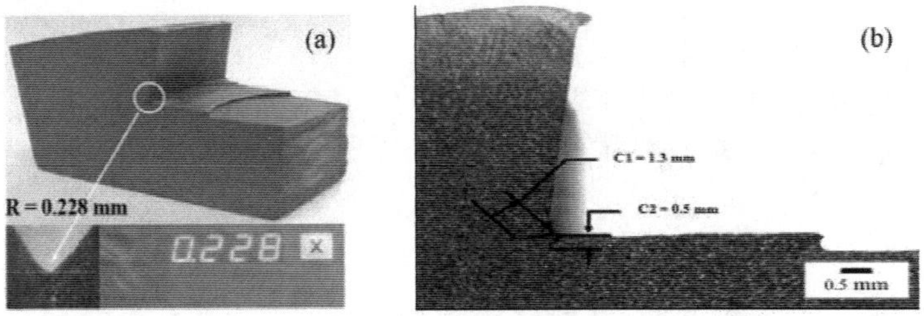

Fig. 4. Section the keyway showing (a) fillet radius (b) cracks at a keyway corner

4.3 *Hardness measurements*

The hardness values measured at various distances from the crack region of the corner of the keyway are as shown in Figure 5. The hardness values were found to be between 92.7-94.9 HRB (195-210 HB) which were within specified limits of commercial machinery steels [15]. The hardness values indicated shaft material was in annealed condition to improve machinability. The result indicated that correct material was selected.

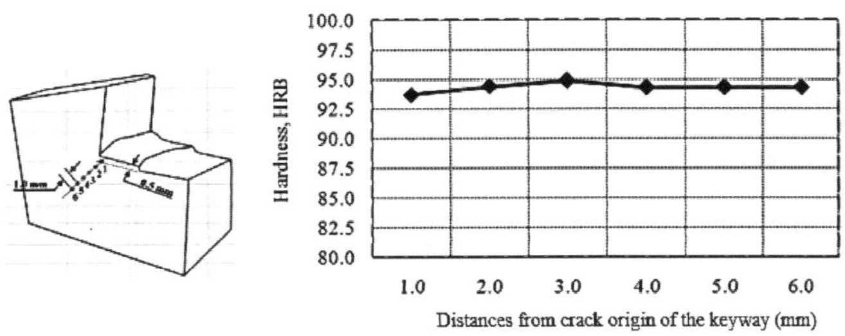

Fig. 5. Hardness values of the failed shaft

4.4 *Chemical composition analysis*

The average values of the chemical compositions of shaft are shown in Table 1. The compositions are that of low alloy Cr-Mo steel to JIS- SCM440 standard [16]. The SCM440 belongs to a class of high strength structural steels and is commonly and widely used in making shafts [17]. The result shows that proper material was used for making the shaft.

Table 1 Chemical compositions of the failed shaft and JIS-SCM440 (%wt)

Material	C	Si	Mn	Ni	Cr	Mo
Failed shaft	0.426	0.276	0.845	0.014	1.092	0.208
SCM440	0.38-0.43	0.15-0.35	0.60-0.85	0.25 (Max)	0.85-1.25	0.15-0.30

4.5 *Microstructural examination*

Examination of the polished and etched example from the shaft material in the vicinity of a keyway corner revealed that there was a crack which extended from the keyway corner through the sample as shown in Figure 6a. This indicated that the crack started at the keyway corner then propagated towards the center of the shaft.

The microstructure of the failed shaft was a mixture of pearlite and ferrite as shown in Figure 6b. The structure of the shaft is typical structure for low alloy medium carbon steel in an annealed condition [18].

Fig. 6. The microstructure (a) crack growth direction and crack origin (b) (high magnitude)

4.6 Force analysis

The transmitted torque (T) was calculated using equation 1 [19].

$$T = \left(\frac{9{,}550 P_o}{N} \right) \qquad (1)$$

where T is transmitted torque (Nm). P_o is power of electric motor (kW) and N is rotational speed of shaft in rpm.

In the present case the power of electric motor is 900 kW and the rotational speed of input shaft is 25.2 rpm. The transmitted torque carried by each shaft is 50% of the value in equation 1 as there are two output shafts in the gearbox. The forces on the shafts were calculated using bearinX software of Schaeffer (FAG) Company. The calculated values of F_{1x}, F_{2y} and F_{1z} are -79.75 kN, 165.30 kN, 446.50 kN, respectively, as shown in Figure 7.

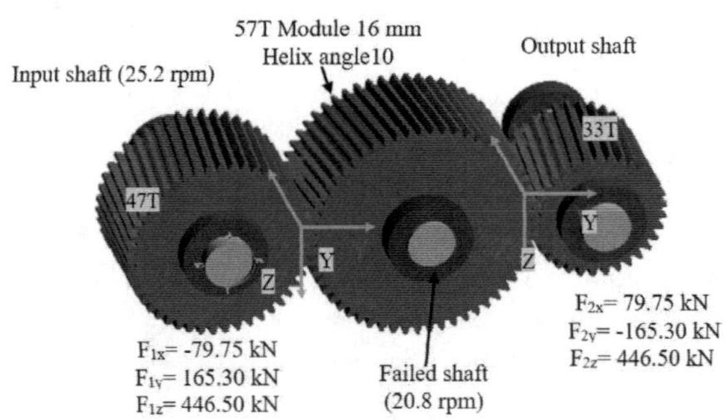

Fig. 7. Force analysis on the gear

4.7 Stress analysis

The shaft under consideration is being subjected to bending force. So as the shaft rotates there is a fluctuation of stress, the point which is subjected to bending moment value is found and shown in Figure 8. Bending stress (σ_b) at the keyway corner of failed shaft is found and shows in Table 2.

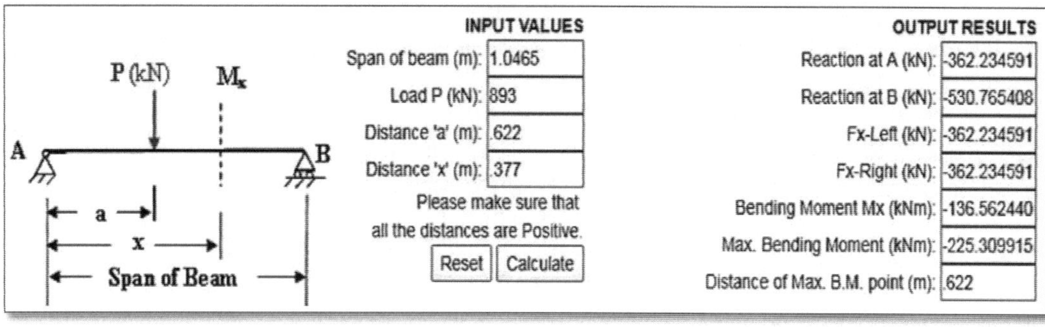

Fig. 8. Bending moment of the shaft

Table 2 Calculation values of bending stress

Description	Values	Unit
Geometric moment of inertia (I) = $\pi(d^4/64)$	9.541724×10^{-5}	m^4
c=dimeter of shaft (d)/2	0.105	m
Bending moment (M) at keyway corner	136.5	kNm
Bending stress (σ_b)=Mc/I	150.2	MPa

The maximum stress (σ_{max}) was calculated using equation 2 [20] and the stress concentration (k_t) was selected using chart of theoretical stress concentration factor [20]. Fatigue stress concentration factor (k_f) can be calculated using equation 3 [20].

$$\sigma_{\max} = k_f \times \sigma_b \tag{2}$$

$$k_f = 1 + q(k_t - 1) \tag{3}$$

$$q = 1/(1 + \frac{\sqrt{a}}{\sqrt{r}}) \tag{4}$$

where σ_{max} is maximum stress (MPa), σ_b is stress (MPa), k_t is stress concentration factor, r is fillet radius of the keyway corner, d is diameter of the failed shaft (mm), k_f is fatigue stress concentration and q is the notch sensitivity it can be defined from the Kunn-Hardarth formula in terms of Neuber's constant (a) and the notch radius (r). In this case, the measured notch radius at corner of keyway is 0.228 mm. The stepped change of diameters (D), diameter (d) of failed shaft is 212 and 210 mm respectively. Approximate fillet radius of stepped shaft is 1.00 mm as shown in Figure 1b. Stress concentration factor at change of diameter, k_t=3 (approximate) (stepped diameter with fillet r/d=1/210).

$$\sqrt{a} = \frac{104}{Sut} = 104/655) = 0.15877$$

$$\sqrt{r} = \sqrt{0.228} = 0.47740$$

$$q = 1/(1 + \left(\frac{0.15877}{0.47740}\right)) = 0.7504$$

The calculated value k_f was 2. 501. The calculated maximum stress included fatigue stress concentration factor (σ_{max}) is 375.6 MPa.

4.8 Material fatigue strength

Fatigue strength or endurance limit for steels can be approximated from tensile strength data using equation 5 [20].

$$S'_e = 0.50 S_{ut} \tag{5}$$

where S'_e is endurance limit of material, S_{ut} is tensile strength in MPa. From the chemical composition analysis and hardness measurement of shaft material, it was clear that shaft material was SCM440 steel in an annealed condition. The tensile strength of shaft material is assumed to be around 655 MPa [21]. The endurance limit of the steel is therefore approximately 327.5 MPa. The endurance limit of the shaft is expected to be lower than the above figure due to geometrical and other factors. The value can be calculated using equation 6 [19].

$$S_e = K_a K_b K_c K_d K_e K_f S'_e \tag{6}$$

where S_e is endurance limit of the shaft under investigation (MPa), K_a is surface factor, K_b is size factor, K_c is reliability factor, K_d is temperature factor, K_e is modifying factor for stress concentration, and K_f is miscellaneous effects factor. The values of various factors are as follows; $K_a = 0.75$, $K_b = 0.652$, K_c, K_d, K_e, $K_f = 1$ [20]. The endurance limit of this particular shaft, calculated using the above data, is approximately 160 MPa. The maximum stress value is high, exceed the fatigue strength of the material. It is no wonder that cracks initiated at the keyway corners.

5. Discussion

It can be seen in Figure 2 that the final fracture area is relatively large covering about 80% of the fracture surface. The large final fracture area indicated that the failed shaft was heavily stressed (workpiece of steel bar too low temperature) at the time of final fracture. The fatigue cracks initiated at both corners of the keyway and ratchet marks between two cracks can be readily observed. The results indicated the stresses at the corners of the keyway were very high which was confirmed by the calculation in the previous section. The result is agreement with that of other investigators [22].

A detail drawing of fillet radii of the keyway in drawing as shown in Figure 1 did not specify the value of fillet radii of the keyway. The shaft manufacturer cannot be blamed for machining the keyway with sharp corners. Inadequate fillet radii of the keyway resulted from poor design in detail design phase.

The keyway surface contained distinctive very rough machining marks as shown in Figure 3. This is the result of poor milling of the keyway. Poor design and poor milling, resulting in low values of fillet radii and rough surface, are potential stress raisers that could lead to high stress concentration being created at the keyway corners which eventually lead to the initiation of the fatigue crack.

6. Conclusions

1. The idle gear shaft of a gearbox under this investigation failed by fatigue fracture.
2. The fatigue cracks initiated at both corners of the keyway and propagated inwards until final fracture occurred.
3. Excessive stress combined with poor machining and inadequate fillet radii at the keyway corner, which gave rise to high stress concentration, are the main causes of the fracture.
4. It is recommended that extreme care must be taken in detail design phase of critical component such as detail specification of fillet radii of keyway corners and generous fillet radii should be provided. Machining must also be done with great care avoiding sharp corners and too rough surfaces.

Acknowledgement

The author thanks Dr. Pongsak Chaengkham, advisor of BKK steel mill Co., Ltd for his help in providing information about the history of the failed shaft and allowing the publication of this information.

References

[1] Yu ZW, Xu XL and Ding HX 2006 *Eng Fail Anal* **13** 1351-1357.
[2] TRC Engineering Services Technical Team, Investigating Material and Component Failure, July, 2004, http://www.tcreng.com/download/tcr-whitepaper-failure-analysis.pdf
[3] Berndt F and Bennekom A 2001 *Eng Fail Anal* **8** 135-144.
[4] ASM metals handbook, Failure analysis and prevention, vol. 11, 1996.
[5] V. B. Bhandari, Introduction to Machine Design, Tata McGraw-Hill, New Delhi, 2002.
[6] B. J. Hamrock, S. R. Schmin, and B. O. Jacobson, Fundamentals of Machine Elements, second ed., McGraw-Hill, 2005.
[7] S. Netpu, P. Srichandr, Failure Analysis of a Reducer Gearbox Shaft, *Proc. Int.Conf.on Mater. Metall. Tech.* Surabaya Indonesia 2009 (pp. FA10-13).
[8] Alban LE 1996 Failure of gears failure analysis and prevention. ASM handbook, vol. 11. OH: ASM International. p.587-601.
[9] Martin R and Karen S 1992 Fatigue failure of carburized steel gear from a helicopter transmission. In: Handbook of case histories in failure analysis. OH: ASM International. p. 228-30.
[10] Becker WT, Shpley JR, editors. Failure analysis and prevention, ASM handbook, vol.11.OH: ASM International; 2007. p. 719-20.
[11] Zahavi E. Torbilo V. Fatigue design-life expectacy of machine parts, Florida: CRC press; 1996
[12] Pavan AHV 2013 *Eng Fail Anal* **1** 103-109.
[13] S.C. Rulmenti, URB ball and roller bearings catalogue, 1993.
[14] Handbook of engineering fundamentals, John Wiley & Son, New York, 1975.
[15] Properties of Alloy Steels AISI 4140 http://www.efunda.com/materials/alloys/alloy_steels/show_alloy.cfm?ID=aisi_4140&show_prop= hard&Page_Title=Alloy%20Steel%20aisi%204140.
[16] Gloria Material Technology Corp., Alloy Steel (SCM440), http://www.gmtc.com.tw/pdf/02/e/Alloy%20steels%20SCM440.pdf.
[17] BANGKOK SPECIAL STEEL CO., LTD. http:/www.bssteel.co.th/product_en.html.
[18] W. F. Smith, Structure and Properties of Engineering Alloys, McGraw-Hill College, New York, 1992.
[19] Loewenthal SH Shafts, Couplings, Keys, ETC., MECHANICAL DESIGN HANDBOOK, McGraw-Hill, New York.
[20] J. E. Shigley, C. R. Mischke, and R. G. Budynas, Mechanical Engineering Design, seventh ed., McGraw-Hill, New York, 2004.
[21] R. L. Norton, Machine Design: An Integrated Approach, third ed., Pearson Prentice Hall, New Jersey, 2006.
[22] Bhattacharyya S and Banerjee A 2007 *Eng Fail Anal.* **4** 411-419.

IOP Conf. Series: Materials Science and Engineering **501** (2019) 012023 doi:10.1088/1757-899X/501/1/012023

Study of adhesive bonded joint failure in composite-metal joining

Kerati Suwanpakpraek[1,2], Baramee Patamaprohm[1,2],and Arisara Chaikittiratana[1,2*]

[1] Department of Mechanical and Aerospace Engineering, Faculty of Engineering,King Mongkut's University of Technology North Bangkok, 1518 Pracharat 1 Road, Wongsawang, Bangsue, Bangkok 10800, Thailand
[2] KMUTNB – TU Chemnitz Collaborative Center for Lightweight Structures Technologies (LiST), King Mongkut's University of Technology North Bangkok, 1518 Pracharat 1 Road, Wongsawang, Bangsue, Bangkok 10800, Thailand

[*] Corresponding Author: k.suwanpakpraek@gmail.com,
Telephone Number 0-2555-2000 ext 8315, Fax Number 0-2586-9541

Abstract. Currently, composites play an important role in the automotive industry where the weight reduction is a key of success due to its high strength to weight ratio. However, the replacement of metals with composites cannot be entirely archived, the joining between composite-metal is required. Bonding with adhesive is one of the most important methods in composite-metal joining. However, the accurate failure prediction of adhesive layers is still lacking due to the complicated state of stress at free-edges of specimens. The aim of this work is to investigate the failure behavior and analysis of acrylic adhesive which is particularly designed to provide superior bonding between composites and metals. Since an adhesive is generally a polymer, its failure depends on the hydro-static pressure. The chosen criterion needs to take into account this dependency. In this study, the Drucker-Prager failure criterion has been used. The parameters were characterized by tensile, compression and double lap shear tests. Then the FEM simulation was performed to predict a failure of single lap shear tests as a validation of model.

Keywords: adhesive bonding, Drucker-Prager failure criterion, composite-metal bonding, FEM simulation

1. Introduction

The environmental issue caused by carbon dioxide emission from fossil fuel usage effects directly to the modern automotive design. Consequently the automotive industry needs to find an appropriate solution and one of the best alternatives is the lightweight structure technology. The performance study tests claim that the reduction of automobile weight by 10% increase the fuel efficiency by 5-7 %. Composites are attractive material candidates to be applied to this technology since these materials present a very high strength to weight ratio, especially CFRP composites. However, the composites cannot be used for all parts of a car and they are needed to be assembled with the other parts which are typically metals. Thus, in order to successfully integrate the lightweight composite structural part to the automobiles, the composite-metal joining technology is essentially required. In general the joining technologies for fiber reinforced thermosetting polymer composite (e.g. glass fiber/epoxy resin composite) and metal consist of adhesive bonding, mechanical fastening and hybrid using together these two technics [1]. In this study, we aim to focus on the adhesive bonding between glass fiber/epoxy composite and steel. The advantages of adhesive bonding include a uniform load distribution across the joint section and no fiber damage during the preparation phase since none of drilling or forming is pre-required on the composite. The behavior and failure of adhesive bonding under static loading have been investigated. A "cohesive" failure referred to a failure on an adhesive layer has been expected to ensure an appropriate adhesive selection and surface preparation. For the analysis of adhesive failure, the Drucker-Prager criterion has been chosen since this criterion includes the effect of hydrostatic pressure to its failure surface. The tensile test, compression test and double lap shear test have been carried out for the characterization of model parameters. The single lap shear test coupling with FEM simulations have been served as a validation of model.

2. Materials and Behavior

2.1. Adhesive

The adhesive chosen for this study is Maxlok MX/T18 acrylic adhesive manufactured by LORD. This adhesive has been recommended by the manufacturer for composite-metal bonding. Its cure requires 24 hours at room temperature. A full cure of this adhesive is justified by an equable color visibility [2]. As a polymer based material, the mechanical behaviour of adhesive can complicate as elastoviscoplastic [3]. In addition, the failure of adhesive also depends on strain rate and hydro-static pressure [18]. Since the scope of this work focus only on the non-linear behavior and the effect of hydrostatic pressure, the effect of strain rate is not taken into account therefore the strain rate in each experiment need to be identical in order to isolate the viscosity of behaviour and simplify the characterisation phase. The non-linear behavior of adhesive is assumed to be a contribution of plasticity and the Drucker-Prager yield criterion is used since this model takes into account the effect of hydro-static pressure. The strains of plastic deformation are completely described by yield function, hardening rule, and flow rule [6]. Figure 1 shows the yield function in liner form plotting in Von Mises equivalent stress and hydrostatic pressure plan ($\sigma_v - p$ plane) [7]. The parameters of model are identified by tensile test, compression test and double lap shear test. The yield point of each test can be plotted as shown in Figure 1. The Drucker-Prager yield criterion has the following form:

$$f(\sigma_v, p) = \sigma_v - p\tan\beta - d = 0 \qquad (1)$$

where σ_v, p and d are Von Mises equivalent stress, hydrostatic pressure and cohesion of the material

The Von Mises equivalent stress can be calculated following this equation

$$\sigma_v = \sqrt{3J_2} = \sqrt{\left((\sigma_{11}-\sigma_{22})^2 + (\sigma_{22}-\sigma_{33})^2 + (\sigma_{33}-\sigma_{11})^2 + 6\left(\sigma_{12}^2+\sigma_{13}^2+\sigma_{23}^2\right)\right)/2} \qquad (2)$$

and the hydrostatic pressure is

$$p = -\frac{1}{3}I_1 = -\frac{1}{3}(\sigma_{11}+\sigma_{22}+\sigma_{33}) \qquad (3)$$

where σ_{11}, σ_{22} and σ_{33} are the normal stresses. σ_{12}, σ_{13} and σ_{23} are the shear stresses. I_1 and J_2 are the first invariant and the second deviatoric stress of the stress tensor respectively

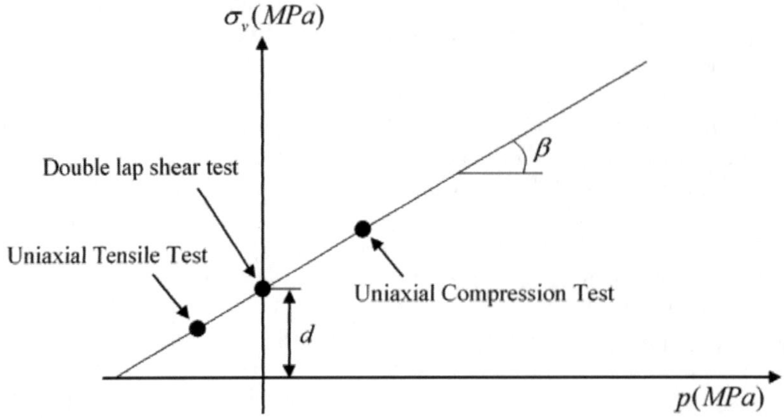

Figure 1. The Drucker-Prager yield and failure surface

The Drucker-Prager model is then applied for failure criterion with other set of parameters identified at the ultimate state of tensile test, compression test and double lap shear test. This decision is based on the assumption that these two surfaces are related to each other, the yield surface expend symmetrically to complete with failure surface (Figure 7) [17].

2.2 Adherends

Steel and composite adherends were used in this study. JIS SS400 carbon steel (non-alloy) is used as a steel adherend. As the steel, the mechanical behavior is isotropic. Another type of adherend is glass fiber-reinforced composite. The mechanical behaviour of this composite is orthotropic elastic and obeys the relationship written in the form of matrix notations as follows.

$$\{\varepsilon\} = [S]\{\sigma\} \tag{4}$$

$$\begin{bmatrix} \varepsilon_{11} \\ \varepsilon_{22} \\ \varepsilon_{33} \\ \gamma_{23} \\ \gamma_{13} \\ \gamma_{12} \end{bmatrix} = \begin{bmatrix} \frac{1}{E_1} & -\frac{\nu_{21}}{E_2} & -\frac{\nu_{31}}{E_3} & 0 & 0 & 0 \\ -\frac{\nu_{12}}{E_1} & \frac{1}{E_2} & -\frac{\nu_{32}}{E_3} & 0 & 0 & 0 \\ -\frac{\nu_{13}}{E_1} & -\frac{\nu_{23}}{E_2} & \frac{1}{E_3} & 0 & 0 & 0 \\ 0 & 0 & 0 & \frac{1}{G_{23}} & 0 & 0 \\ 0 & 0 & 0 & 0 & \frac{1}{G_{13}} & 0 \\ 0 & 0 & 0 & 0 & 0 & \frac{1}{G_{12}} \end{bmatrix} \begin{bmatrix} \sigma_{11} \\ \sigma_{22} \\ \sigma_{33} \\ \sigma_{23} \\ \sigma_{13} \\ \sigma_{12} \end{bmatrix}$$

where $\{\varepsilon\}, \{\sigma\}$ are the column matrix notation of stress and strain tensor and $[S]$ is the compliance matrix

The behavior of steel adherend is identified by tensile testing while a series of tensile tests in different fiber orientation ([0,90], [±45]) and a short-beam 3point bending test are used to identify the behavior of glass fiber-reinforced composites adherend.

3. Experimental investigation

3.1. Material characterization tests

An instron 5567 universal testing machine was used for all tests with a load cell of 30 kN capacity (accuracy ± 0.5%) and extensometer with gage length of 25 mm (accuracy ± 0.5%) for strain following propose [11]. All tests proceed under environment laboratory conditions.

3.1.1. Adhesive test

The adhesive requires the three difference tests in order to identify the parameters of Drucker-Prager yield criterion as stated previously. The tensile specimens were casted by silicone mold with the geometries according to ASTM D638[4] (Figure 2a). The compression specimens were casted by PVC mold with cylindrical geometries of 12.7 mm in diameter and a height of 25.4 mm according to ASTM D695-96 [5] (Figure 2b). The double lap shear specimens were produced according to ASTM D3528 [9] (Figure 2c). The adhesive was mixed with mixing gun and applied on both contact surfaces of double lap shear adherends. The specimens were then clamped by a paper clip to maintain the position and overlap distance of adherends. The adhesive thickness was controlled by 0.3 mm wires [10]. The specimens were cured 24 hours at room temperature conditions (28 ± 3° C and 75 ± 10% relative humidity).

3.1.2. JIS SS400 carbon steel tensile test

JIS SS400 structural carbon steel is a carbon steel (non-alloy). The tensile specimens were fabricated from a steel sheet with a thickness of 1.5 mm according to ASTM E8 [12] using a laser cutting machine. The specimens were loaded up to failure to characterize both of elastic and plastic behaviors.

3.1.3. Glass fiber-reinforced composite tests

The composite adherend in this study is made from E-Glass glass fiber woven fabric (WR600) and epoxy resin. The composite tensile specimens were fabricated with respect to the standard ASTM D3039 [13] using vacuum infiltration process. The composite specimens contains 8 woven fabric layers with stacking of $[0,90]_8$ and $[+45,-45]_8$. The tensile test on $[0,90]_8$ specimen characterizes the elastic modulus in longitudinal (E_{11}) and transverse (E_{22}) direction while a shear modulus in direction 12 (G_{12}) is identified from the $[+45,-45]_8$ specimen. The short-beam specimens for 3point bending test were fabricated according to ASTM D2344 [14] with the same process as tensile specimens. The number of layers was 12 and the stacking was $[0,90]_{12}$. This test characterizes a shear modulus in direction 13 (G_{13}) and 23 (G_{23}). All composite specimens are cured at room temperature (28 ± 3°C and 75 ± 10% relative humidity). for 24 hours.

3.2. Criterion validation test: Single lap shear test

A single lap shear test was used for validation of the Drucker-Prager yield criterion. The specimens were carried out according to ASTM D1002 [15]) Figure 2d(. For composite adherend, the number of layers were 10 layers with $[0,90]_{10}$ stacking. A metallic adherends were cut with a hydraulic press cutting machine. The two adherends were assembled for 20 mm overlap distance with an adhesive. Adhesive thickness was controlled by wires [10]. Paper clips were used to keep the adherends in place. The specimens were then cured for 24 hours at room temperature (28 ± 3° C and 75 ± 10% relative humidity).

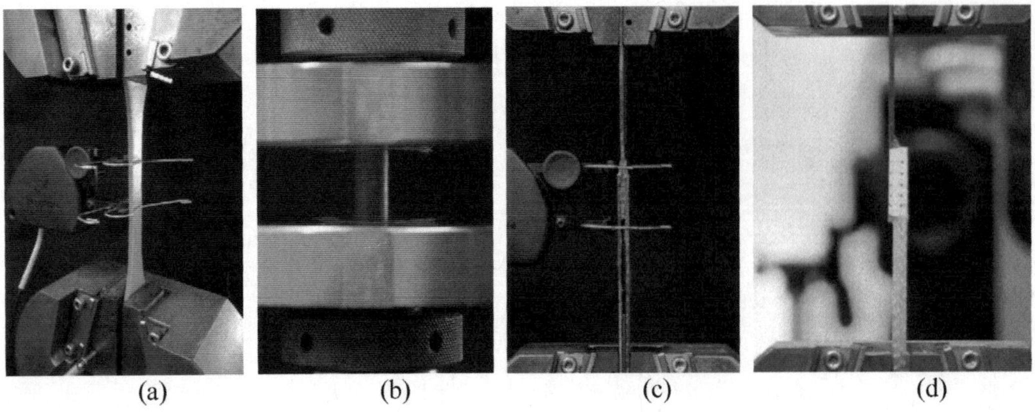

Figure 2. Different testing setups :(a) Tensile test, (b) Compression test, (c) Double lap shear test and (d) Single lap shear test

4. Static tests results

4.1. Adhesive test

The results of tensile tests are shown in Figure 3 and the identified parameters are summarized in Table 1. The linear slope of the experimental curve is a Young's modulus (E). A negative ratio of transverse and axial strain is a Poisson's ratio (ν). The tensile stress-plastic strain relation is also used for a hardening model. In order to implement into FEM simulation, a true stress-true plastic strain plot is required (Figure 4.). The eqs. 5-7 is used to calculate these true stress and true plastic strain. The result of compression tests are presented in Figure 5 and Table 1. Since the strain exceeds 5%, the stress-strain curves are used to describe its mechanical properties: The elastic modulus (E_1) is the linear slope between $\varepsilon = 0$ to 0.05. The post-yielded modulus (E_2) is identified by the tangential linear slope as show in Figure 5. The yield point is characterized by the intersection of slope E_1 and E_2. The inflection point is defined as a divergence of E_2 [16]. This point is considered to be a failure point of compression test. For double lap shear tests, the state of pure shear was assumed at the adhesive layer. However the deformation measured from the extensometer (Figure 6) is not an adhesive deformation alone but a combination of adhesive and adhered deformation. Hence, only maximum shear stress was preliminary achieved from the experiments (Figure 6 and Table 1). A cohesive failure has been observed as expected (Figure.8). This confirms a good adhesive selection. The Drucker-Prager yield and failure criteria can be finally identified from these three tests (Figure 7 and Table 2).

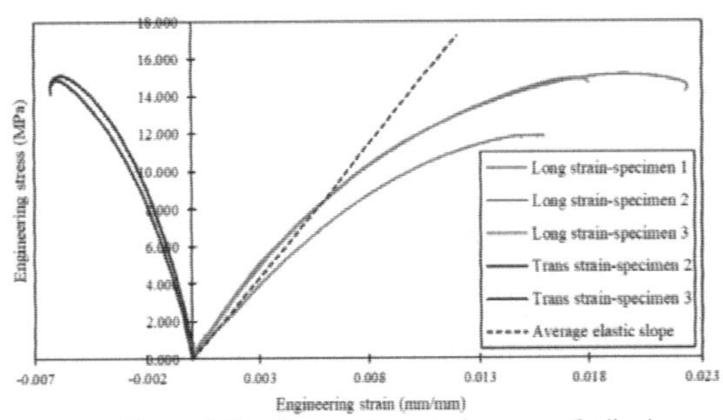

Figure 3. Tensile test: stress-strain curves of adhesive

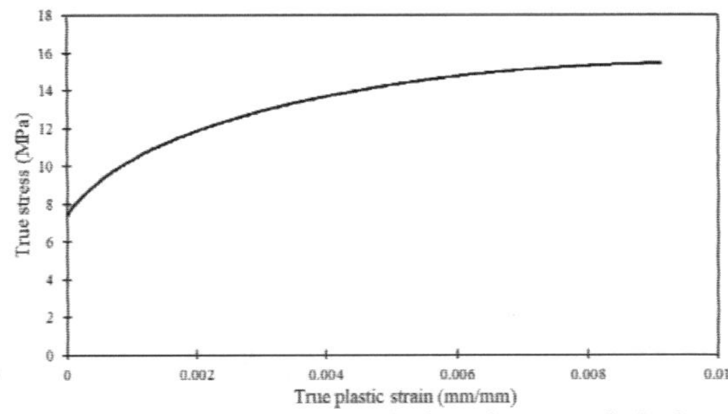

Figure 4. Tensile test: true stress-plastic strain curves of adhesive

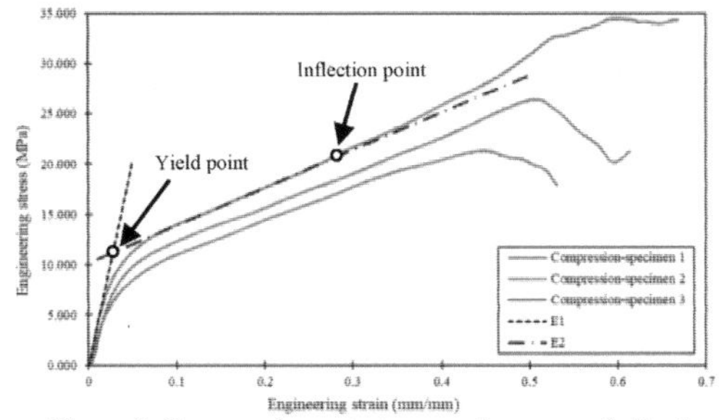

Figure 5. Compression test: stress-strain curves of adhesive

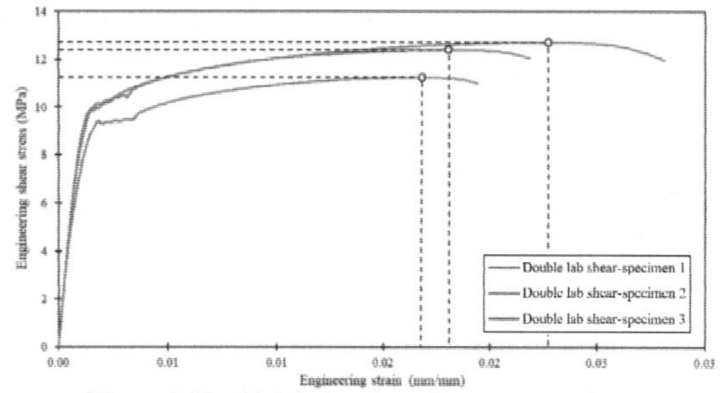

Figure 6. Double lab shear test: shear stress-strain curves

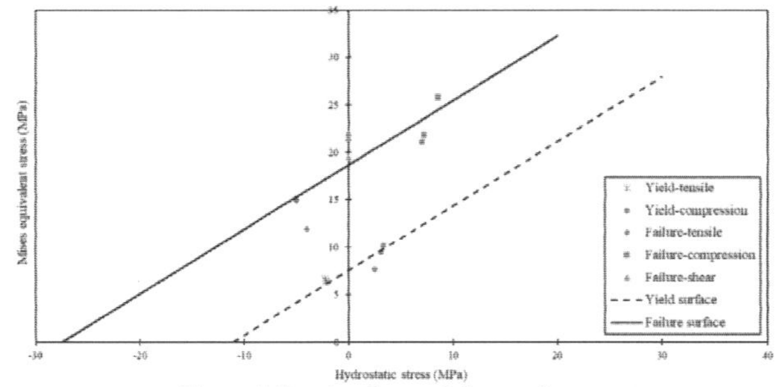

Figure 7. Drucker-Prager failure surface

Figure 8. A cohesive failure of double lap shear test

Table 1. Mechanical properties of adhesive.

Specimen	E-modulus E1 (MPa)	E-modulus E2 (MPa)	Yield strain ε_y (-)	Yield stress σ_y (MPa)	Failure strain ε_{fail} (-)	Failure stress σ_{fail} (MPa)	Poisson's ratio ν (-)
Tensile	1424.266	-	0.451	6.476	1.760	13.987	0.324
Compression	348.432	44.139	0.041	9.061	0.403	22.888	-
DB-lap shear	-	-	-	-	0.00023	12.128	-

(*) 3 specimens per test

Table 2. Summary of identified parameters for the adhesive.

$F = t - p\tan\beta - d$	β (degree)	d (MPa)
Yield surface	34.2	7.544
Failure surface	34.2	18.622

$$\sigma_{true} = F/S = Fl/S_0 l_0 = F(\Delta l + l_0)/(S_0 l_0) = F/S_0(1+\Delta l/l_0) = \sigma(1+\varepsilon) \tag{5}$$

$$\varepsilon_{true} = \int_{l_0}^{l} \delta l/l = \ln(l/l_0) = \ln((\Delta l + l_0)/l_0) = \ln(1+\Delta l/l_0) = \ln(1+\varepsilon) \tag{6}$$

$$\varepsilon^p = \varepsilon - \varepsilon^e = \varepsilon - E\sigma \tag{7}$$

4.2. JIS SS400 carbon steel test results

The Young's modulus (E) was identified from the tensile test and equals to 188.21 GPa (Figure 9, Engineering stress- strain curves). The Poisson's ratio (ν) was assumed as a typical value of metal and equals to 0.3. For the hardening characterization, a true stress- true plastic strain plot is required (Figure 10) using the eqs. 5-7. The curve from Figure 10 will be implemented directly in FEM simulation.

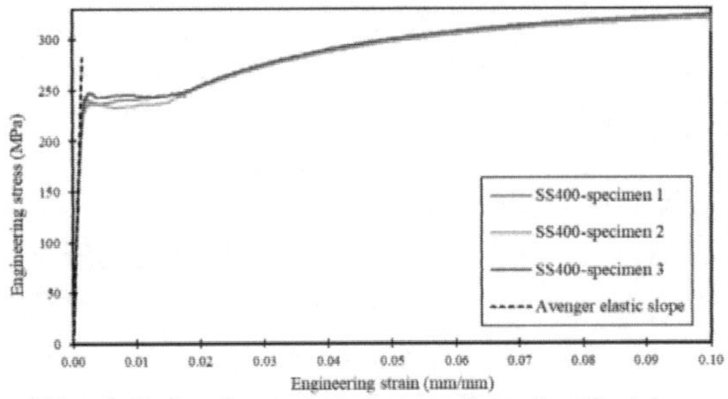

Figure 9. Engineering stress-strain curves in tension of metal

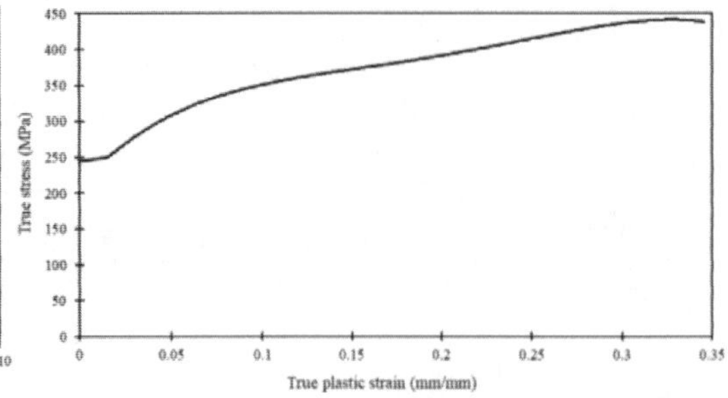

Figure 10. True stress-plastic strain curves in tension of metal

4.3. Glass fiber-reinforced composites test results

The results of tensile test are present in Figure 11 and 12. The linear slope of specimens [0, 90] in Figure 11 defines as elastic modulus in longitudinal (E_{11}) while the transversal modulus (E_{22}) is assumed to be equal to E_{11} due to the cross-ply stacking. The linear slope of specimens [+45,-45] in Figure 12 represents $E_{45°}$ and can relate to the in-plan shear modulus (G_{12}) using eq.8 [20]. The results of short beam 3 points bending test are shown in Figure 13. The linear slope in Figure 13 refers to $\Delta P/\Delta\delta$ in eq.9 [20] and can be used to calculate the out of plan shear modulus (G_{13}) and (G_{23}) which assumed to be equal due to the cross-ply configuration. All identified parameters are summarised in Table 3.

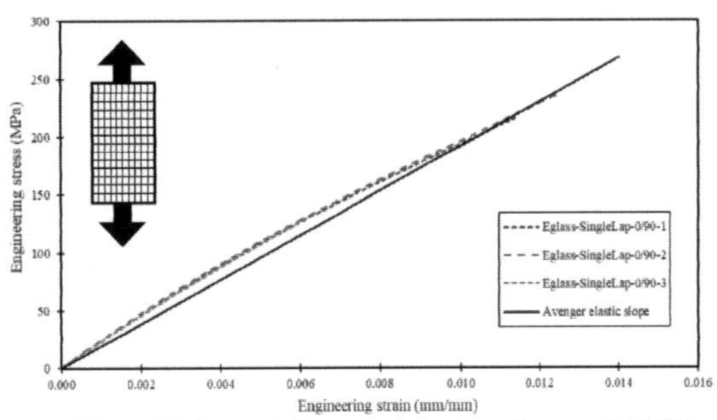

Figure 11 .Stress-strain curves in tension of composite[0,90]

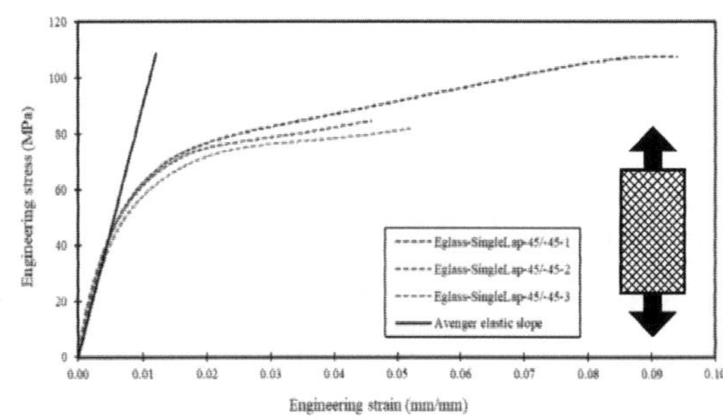

Figure 12. Stress-strain curves in tension of composite[45,-45]

$$\frac{1}{G_{12}} = \frac{4}{E_{45^\circ}} - \frac{1}{E_1} - \frac{1}{E_2} + 2\frac{\nu_{12}}{E_2} \tag{8}$$

$$\frac{1}{E_1} + \frac{6h_0^2}{5L^2G_{13}} = 4\frac{\Delta\delta}{\Delta P} \times \frac{b_0h_0^3}{L^3} \tag{9}$$

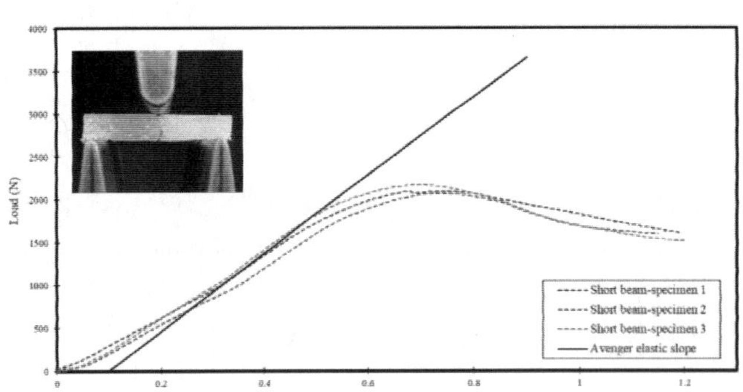

Figure 13. Load-displacement in short-beam 3 point bending test

Table 3. Summary of identified parameters for the composite.

Composite	$E_1 = E_2$ (GPa)	E_3 (GPa)[19]	G_{12} (GPa)	$G_{13} = G_{23}$ (GPa)	ν_{12} [19]	$\nu_{13} = \nu_{23}$ [19]
Glass/epoxy	19.16	3.5	2.93	0.83	0.04	0.38

5. FEM simulation

5.1 .Structural analysis

The single lap shear test simulation was performed to validate the Drucker-Prager model. Dimensions and boundary conditions of single lap shear specimens are shown in Figure 14. A 3D 8-noded solid element (C3D8) was used. A model has totally 165888 elements and 186120 nodes with the refinement zone at the edge of adhesive layer. The mesh sizes of adhesive layer are progressively increased from 7.5x10^{-4} to 0.037 mm^3. ABAQUS software was used to simulate the numerical test and taken into account the material nonlinearity. This test showed non-linear deformations due to the bending effects during the applied load.

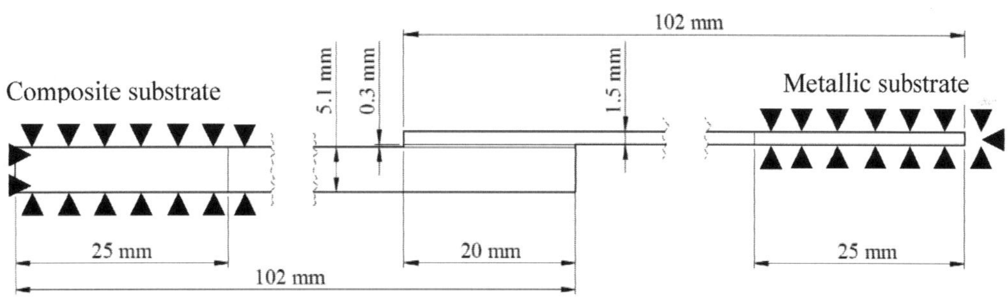

Figure 14. Single lap joint

5.2 .Validation of model

Figure 15 shows a deformation of single lap shear specimen by simulation. The red-circle zone of steel adherend proceeds a plastic deformation. The comparison of simulation and experiment is demonstrated in Figure 16. A good agreement was found up to 8 MPa after that the simulation appears to be slightly higher. The different can be explain by the failure of testing specimens which have been fail at the composite adherend (delamination) rather than adhesive layer as expected.(Figure 17) This means that the selected adhesive is stronger than the strength of composite intra-plies. It is confirmed a good adhesive selection.

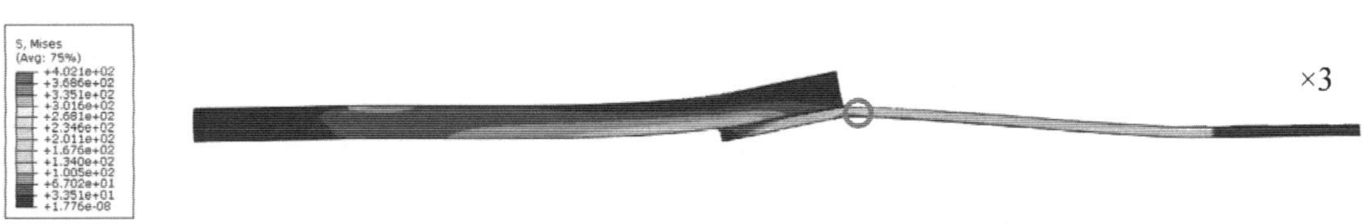

Figure 15. Finite element model of the single lap joint

IOP Conf. Series: Materials Science and Engineering **501** (2019) 012023 doi:10.1088/1757-899X/501/1/012023

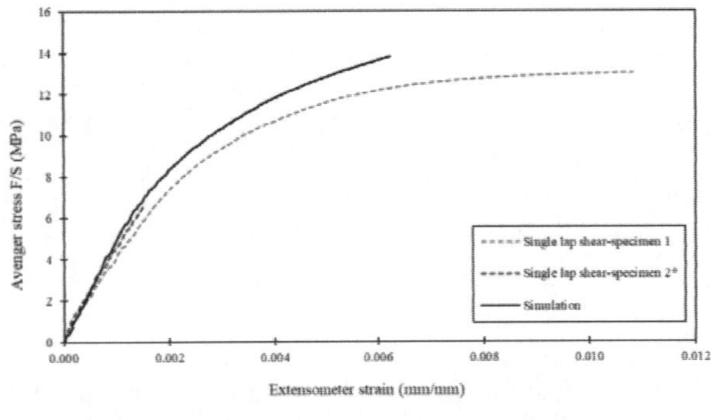

*Error in strain measurement at 7 MPa, the maximum stress of this test is 15.33 MPa

Figure 16. Validation of behaviour model by simulation :single lap shear test

Figure 17. The failure of single lap shear test

6. Conclusions

In this study, the adhesive behavior was found to depend on hydro-static pressure. Drucker-Prager yield criterion with isotropic hardening was used to define the plastic behavior of the adhesive. This criterion was also adapted experimentally as a failure criterion to estimate failure loads of adhesive. The numerical simulation has shown a good agreement of using Drucker-Prager for a yield criterion. For a failure prediction, the Drucker-Prager failure criterion has accurately predicted failure load (0.22% error) when considering all adhesive elements satisfied a failure criterion as a failure of structure (using post-processing method without any degradation rule on elements). However, for a high local stress concentration field, the characteristic length or even characteristic volume [21,22] were often used to predict a failure of structure rather than all elements failure. In that case, the single lap shear test can be used to identify a characteristic length and some other test needs to be proceeded in order to validate the identified length. The linear Drucker-Prager yield criterion model proposed in this study significantly differs from the higher order model e.g. hyperbolic Drucker-Prager model [17,18] at tri-axial state of stress (negative hydrostatic stress) which is a weakest state of stress for adhesive (Figure 18). However, all tests proceeded in this study cannot identify that tri-axial failure, a sophisticated Arcan-Mines test is required [17] in order to justify a suitable failure criterion.

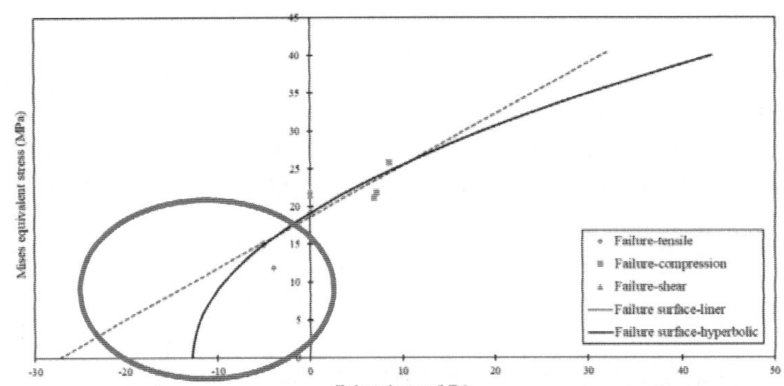

Figure 18. Linear and hyperbolic comparison the Drucker-Prager failure surface

References

[1] Jahn J, Weeber M, Boehner J, and Steinhilper R; 2016; Assessment strategies for composite-metal joining technologies–a review; Procedia CIRP; 50; 689-694.

[2] Lord Corporation Company; LORD TECHNICAL DATA]Internet[; 2018]cited 2018 Aug [31; Available from :https://www.lord.com/sites/default/files/Documents/TechnicalDataSheet/DS4048_Maxlok MX-Series.pdf.

[3] Mirkhalaf S M, Pires F A and Simoes R; 2016; An elasto-viscoplastic constitutive model for polymers at finite strains: Formulation and computational aspects; Computers & Structures; 166; 60-74.

[4] American Society for Testing and Materials; 2010; Standard test method for tensile properties of plastics; ASTM International.

[5] American Society for Testing and Materials (Filadelfia); 2002; Standard Test Method for Compressive Properties of Rigid Plastics; ASTM International.

[6] Allen P A; 2002; Hydrostatic stress effects in low cycle fatigue; A Dissertation Presented for the Doctor of Philosophy Degree; Tennessee Technology University.

[7] ABAQUS; 2016; ABAQUS Documentation; Version 2016; Dassault Systèmes; Rhode Island.

[8] Gay D and Hoa S V; 2007; Composite materials: design and applications; CRC press; p 292.

[9] American Society for Testing and Materials; 2008; Standard test method for strength properties of double lap shear adhesive joints by tension loading; ASTM International.

[10] Da Silva L F, Dillard D A, Blackman B and Adams R D ;2012; Testing adhesive joints: best practices; John Wiley & Sons.

[11] Instron Corporation; Instron Series 5500 Load Frames Including Series 5540, 5560, 5580]Internet[; 2005]cited 2018 Aug [31; Available from :https://docplayer.net/35203019-Instron-series-5500-load-frames-including-series-5540-5560-5580.html.

[12] ASTM American Society for Testing and Materials; 2009; Standard test methods for tension testing of metallic materials; ASTM international.

[13] American Society for Testing and Materials; 2006; Standard test method for tensile properties of polymer matrix composite materials; ASTM international.

[14] American Society for Testing and Materials; 2006; Standard test method for short-beam strength of polymer matrix composite materials and their laminates; ASTM International.

[15] ASTM International; 2010; Standard test method for apparent shear strength of single-lap-joint adhesively bonded metal specimens by tension loading (metal-to-metal); ASTM international.

[16] Angelidi M, Vassilopoulos A P and Keller T ; 2017; Ductility, recovery and strain rate dependency of an acrylic structural adhesive; Construction and Building Materials; 140; 184-193.

[17] Patamaprohm B, Phongphinittana E, Guinault C, Gantchenko V, Renard J, Jeandrau J P, Peyrac C and Lefebvre F; 2016; Study of adhesive joints under static and fatigue loading; Revue des composites et des matériaux avancés; 26 ; 63-85.

[18] Özer H and Öz Ö; 2017; The use of the exponential Drucker-Prager material model for defining the failure loads of the mono and bi-adhesive joints; International Journal of Adhesion and Adhesives; 76; 17-29.

[19] Patamaprohm B, et al; Improving the impact resistance of CFRP composite structures; The 32nd Conference of Mechanical Engineering Network of Thailand; 3-6 July 2018; Mukdahan; Thailand.

[20] Berthelot J M; 2012; Composite materials: mechanical behavior and structural analysis; Springer Science & Business Media; pp 195.449 –

[21] Whitney J M and Nuismer R J; 1974; Stress fracture criteria for laminated composites containing stress concentrations; Journal of composite materials; 8(3); 253-265.

[22] Hochard C, Lahellec N and Bordreuil C; 2007; A ply scale non-local fibre rupture criterion for CFRP woven ply laminated structures; Composite Structures; 80(3); 321-326.

Mechanical property and fracture mechanism of glass fiber reinforced polymer and carbon fiber reinforced polymer

T. Cuaiman[1,*], P. Karin[1], N. Ohtake[2], H. Akasaka[2], P. Larpsuriyakul[3], N. Prakymoramas[3], J. Rojsatean[3], D. Thanomjitr[3] and S. Kaewket[3]

[1]Automotive Engineering Program, International College, King Mongkut's Institute of Technology Ladkrabang, Bangkok 10520, Thailand
[2]Department of Mechanical Engineering, School of Engineering, Tokyo Institute of Technology, Tokyo, 152-8550, Japan
[3]National Metal and Materials Technology Center, National Science and Technology Development Agency, Pathumthani 12120, Thailand

* Corresponding Author: E-mail: tialcuaiman@gmail.com

Abstract. In the automotive field, most of the manufacturers are looking to replace the material steel, aluminium with lightweight material like carbon fiber or glass fiber composite. The purpose are due to their relatively high strength, higher chemical resistance, flexible usage temperature and higher stiffness than steel. In this study, mechanical properties of carbon fiber and glass fiber reinforced polymer were investigate. It was find that the tensile strength of carbon fiber composite is approximately 11% higher than that of glass fiber, almost twice in Young's modulus than that of glass fiber. Carbon fiber is two times higher than glass fiber in both flexural stress and young modulus of flexural three points bending test. Image analysis of fracture and damage were detect by field emission scanning electron microscopy (FESEM) in microstructure scale to observe the fracture mechanism. Observed different failure mode in fiber and resin. Chemical composition of composite and fibers were investigated by using electron dispersive x-ray (EDX) spectroscopy that gave out in 88 w.t % of carbon and 12 w.t % in carbon fiber twill (CFT) composite. On the other hand, glass fiber woven (GFW) composite contained 72.7 w.t % of C, 20.7 w.t % of O_2, and the rest contained Si, Ca, Al, Mg and Cl.

Keywords: Carbon fiber, Glass Fiber, Fracture Mechanism, Mechanical Properties

1. Introduction

Nowadays, most of the automotive manufacturer deeply invested to make research on carbon fiber and glass fiber due to their lightweight properties. The purpose of lightweight vehicles is for future automotive fields [1]. Vanishing of CO_2 footprint and preventing the particles matters (PM) emission from vehicles in our environment are also one of the most important point. Carbon fiber and glass fiber are the most popular materials not only in the automotive field but also in the field of marine, aircraft, military vehicles, sports, rockets and etc [2]. The main properties are lightweight, high stiffness, high temperature resistance and high strength compare to the same weight [3,4]. In composites, there are two main functions those are fiber and matrix. Fiber is the main product in the composite and matrix is one of the material that can create bonding strength between fibers [5]. Polymer matrix is the most widely used in the composite industry. There are two main matrix resins such as thermosetting and thermoplastic resins; those are the most useful resin in the composite field for laid up process laminar. Thermosetting resin is liquid form and it is irreversible resin [6]. After curing thermoset resin, the resin cannot turn back to the original liquid form and the resin can cured by chemical reaction at room temperature. Among several resin types in thermoset, epoxy is the most useful matrix because the matrix can lead composite to the best material of high bonding strength [7]. The composite need higher strength to resist the force and epoxy have to transfer stresses to fibers. Most of the composite starts to crack from matrix de-bonding, fibers delamination and pull out fibers from the matrix failure because matrix is not wet enough to bond one to another fiber. Then matrix cannot distribute the load uniformly to the fibers [8]. Dany Arnoldo Hernandeza [9] observed the main fracture of the composite is come from the interface zone (area between resin and fiber) which is affected by the tensile stress concentration on resin. It makes a crack initiation at the resin area, and then the force and crack are distribute to the fibers. Therefore, fiber is pull out then leads to break fibers. Issa A. Hakim [10] studied the porosity of the composite according to different vacuum poor, moderate and high vacuum. As the result, the poor vacuum composite has the highest quantity of porous that can make lower strength with faster crack then the toughness of fibers are decreasing. Energy absorptions are drop compare to moderate and high vacuum. From the view of an image, it is quite important to investigate the porous inside fiber and the structure of each fiber that can affect mechanical properties and fracture of composite. P. Karin [11] observed particle matters (PMs), average agglomerated particle diameter sizes are in the range of 50-500 nm size and average nanoparticle diameter size are in the range of 20-50 nm respectively. He used (TEM) transmission electron microscopy to investigate the size of PM. P. Karin [12] used TEM to observe particle matters size from biodiesel engine and diesel engine with different speed RPM. He investigated the average of nanoparticles size in approximately 48, 34 and 32 nm for carbon black for biodiesel and diesel engine's PMs respectively. P. Karin [13] has investigated carbon black by using electron microscopy analysis. Therefore, electron microscopy is one of the most important tool to study fracture mechanism. However, in case of composite, micro and nanostructures are not understand well, it induces to investigate the properties and fracture mechanism of composite with different type of base material.

2. Materials and Methodology

2.1. Materials

In this research, two types of materials were used with epoxy resin matrix; carbon fiber (twill weave and plain weave) based on polyacrylonitride (PAN) [8, 9] and E-glass fiber (woven roving and fabrics). Epoxy YD 582 and hardener TH7253 are use as the matrix in this composite. Listed materials in table 1 are supply by concrete composite.

Table 1. Materials type

No.	Material	Prod name	Composites name in this paper	Type
1	Carbon fiber twill	C135T	CFT	
2	Carbon fiber plain	PC6951500	CFP	Carbon
3	Glass fiber	Woven roving N323	GFW	
4	Glass fabrics	Glass cloth EW200	GFF	Glass
5	Epoxy (part: A)	YD 582		
6	Hardener (part: B)	TH 7253		Resin

2.2. Sample preparation

Orientation for all of the materials are bidirectional 0/90 degree. The fabrication process are wet lay up process based on the weave fibers pattern from manufacture. All of the fibers are lay up in the same orientation. In epoxy mixing ratio, there are two parts, part A: epoxy and part B: hardener are mix for 3:1 ratio based on fiber ratio and equations (1-3). The thickness of composite were control based on the fabric thickness of each layer. The same method is use for all composites with wet lay-up process. All of laid up conditions can see in table 2.

$$Hardener\ weight\ (g) = \frac{Weight\ of\ Fibers\ (g)}{4} \tag{1}$$

$$Epoxy\ weight\ (g) = weight\ of\ hardener\ (g)\ x\ 3 \tag{2}$$

$$Epoxy\ and\ hardener\ ratio = 3:1 \tag{3}$$

Composite was cut by computer numerical control (CNC) machine Mazak FJV-20 with the situation of spindle speed 150 mm/min and feed speed 90-100mm/min. Samples are prepared according to ASTM 638 for tensile test and ASTM 790 for flexural test specimens. Fabricated composite information of composites.

Table 2. Material ply and lay up status

No.	Material	Fabric layers	Fabric thickness (mm)	Fibers w.t (g)	Composite size (mm)	Comp w.t (g)
1	CF twill	10	0.31	261.1	320x320x3.2	512
2	CF plain	9	0.3	343.29	240x280x2.9	429
3	GF woven	10	0.3	408.1	320x320x3.2	636.06
4	GF fabrics	16	0.2	343.62	320x320x3.2	574.167

2.3. Mechanical testing

The specimens were test on a tensile test machine INSTRON 8801 with load cell 95KN, test speed 5mm/min for all composite with the same testing function at temperature 24 degrees Celsius with 47% RH. Another mechanical flexural test (three-points bending) is perform at 24C temperature and 49% RH with INSTRON 55R4502 machine. Testing rate speed is 5.2mm/min with cell load 10KN.

3. Result and Discussion

3.1. Tensile test results with different materials CFT, CFP, GFW and GFF

Figure 1 shows the average value of each composite. CFT/CFP has the highest tensile strength than that of GFW/GFF. The tensile strain of glass fiber is two times longer than carbon fiber elongation. CFT is the peak tensile stress around 450MPa while CFP is around 420MPa at the same tensile strain 1.3%. Maximum strength of GFW and GFF are decrease particularly compare to CFT /CFP. The elongation at break point are 1.37% for CFT and 1.31% for CFP respectively, while GFW and GFF elongation are 2.85% and 2.91% respectively. Young modulus is the main important point in mechanical test and properties. Therefore, the highest young modulus is around 37 GPa from CFP and it is almost similar to CFT composite with the value of 36 GPa. The value of CFT and CFP are almost two times higher than that of GFW and GFF's young modulus.

Tested materials data are summarize and compare to each sample properties in table 3. The specimens are not the same thickness, but the thickness will not affect to young modulus of materials because material properties are not depends on the thickness of samples. Unlike steel or alumina, the samples are sensitive and break at several points. Therefore, strain gauge was not use in this experiment. The elongations are calculate from sample crosshead displacement because it was collect at maximum strength of sample.

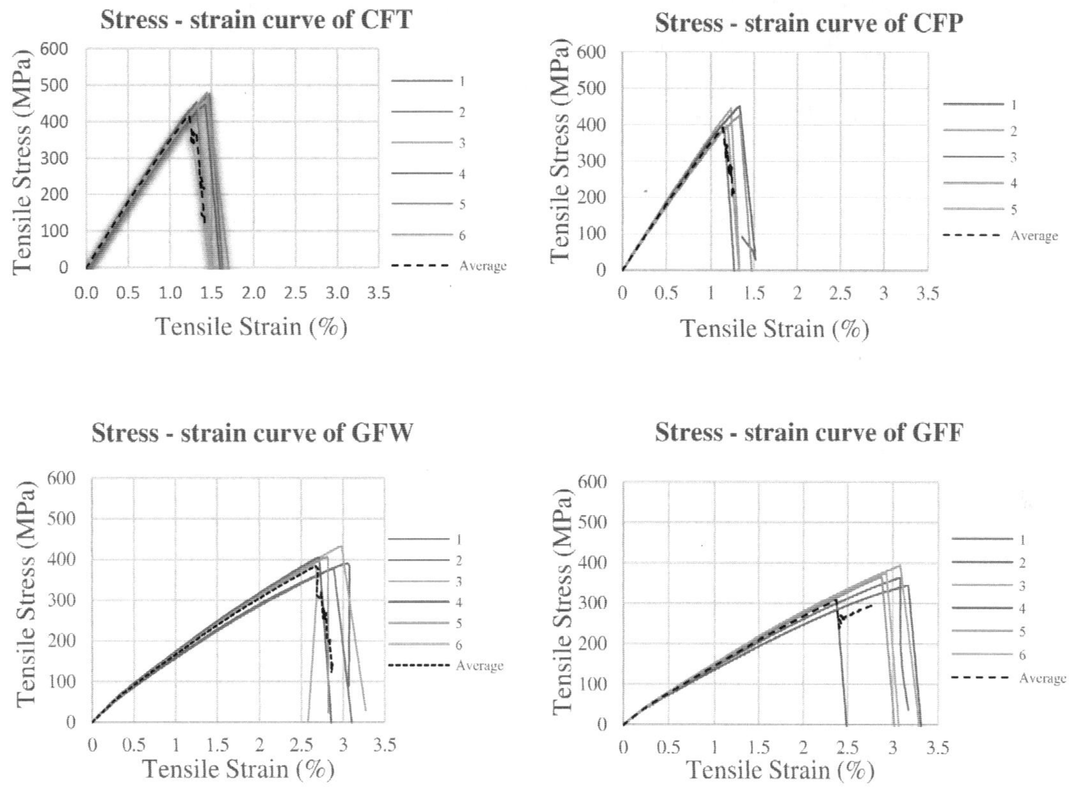

Figure 1. Stress-strain relationship of tensile test data of CFT, CFP, GFW and GFF

The specimens were disintegrate to be several pieces immediately after the peak points of tensile stress for CFT. It leads to brittle behaviour, but higher stiffness. Composite breakage of CFT and CFP are almost linear. Figure 1 also shows that stresses are immediate drop. For glass fibers, the nature of glass are flexible and higher elongation. The fracture behaviour of GFW and GFF are tougher comparable to CFT and CFP behaviour. After the maximum load are applied, the specimens break immediately in the weakest point of the sample where the fillet of specimens are located for all types of specimens. The failure mode of composite are as shown in figure 2.

Table 3. Comparison of composite material and their properties of tensile test

Samples	Thickness (mm)	Width (mm)	Grip distant (mm)	Maximum load (N)	Tensile strength (MPa)	Young modulus (MPa)	Elongation at break (%)
CFT	3.5	13.35	115	21,332.87	452.98	36,204	1.37
CFP	2.85	13.6	115	16,328.60	430.73	36,604	1.31
GFW	3.3	13.6	115	18,182.37	399.85	18,620	2.85
GFF	3.2	13.7	115	15,582.11	358.99	15,860	2.91

IOP Conf. Series: Materials Science and Engineering **501** (2019) 012024 doi:10.1088/1757-899X/501/1/012024

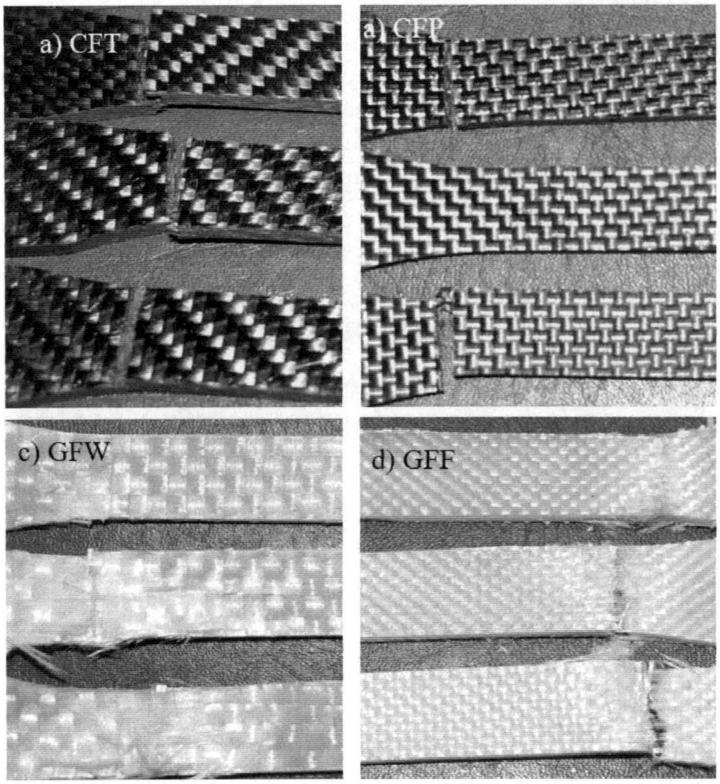

Figure 2. Fracture sample of tensile test a) CFT, b) CFP, c) GFW and d) GFF

3.2. Flexural Test

Three points bending flexural test condition are as mentioned; six pieces of each type are tested. The average flexural strength of all specimens are as shown in figure 3. Each type of samples are much different not only in flexure stress at maximum load but also in young modulus. The summary and properties of all flexural tested conditions are show in table 4. CFT and CFP have a big gap in maximum stress but almost the same flexure strain in elongation. GFW/GFF in this flexural test are drop gradually without breaking immediately. After the maximum load are applied, the fibers breaks unstably and tends to more elongation. In young modulus, CFT is the best one with the highest value of 47 GPa and the next follow by CFP with 42 GPa. Young modulus for GFW and GFF are lesser more than compare to CFT. On the other hand, glass fiber are very good in elongation that lead to the good toughness behaviour. The failure behaviour are also quite different in each type as shown in figure 4. Tension side of surface starts to crack first due to tension load and compression side are crack.

Table 4. Flexural strength and modulus of composites

Name	Maximum flexure load (N)	Flexure Stress at max load (MPa)	Modulus (automatic young's) MPa	Flexure strains at max flexure stress (%)	Thickness (mm)	Width (mm)
CFT	655.69	675.37	46,975	1.55	3.35	13.34
CFP	425.35	598.26	41,945	1.53	2.88	13.63
GFW	329.04	327.15	20,720	2.21	3.32	13.67
GFF	229.33	246.18	18,603	1.72	3.2	13.66

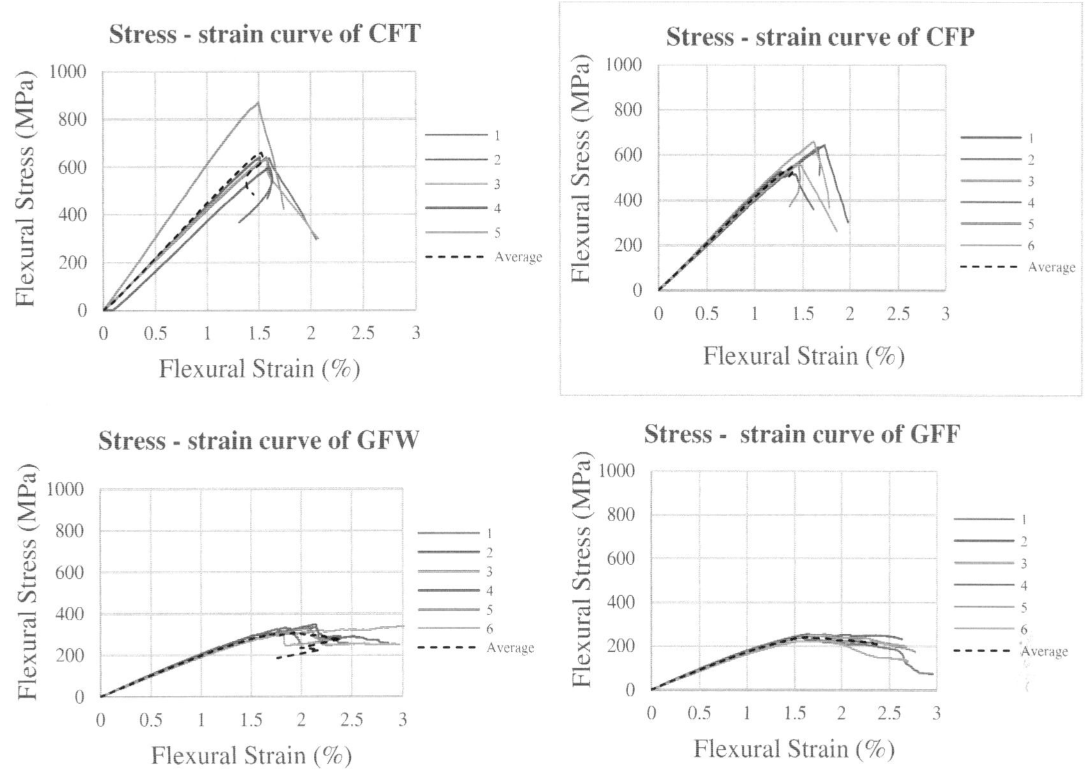

Figure 3. Flexural stress-strain curve for all composites

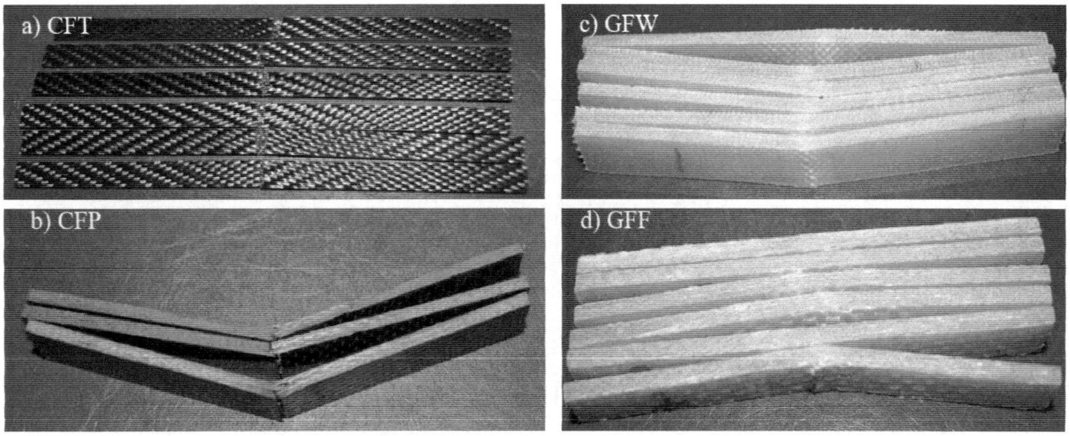

Figure 4. a) CFT, b) CFP, c) GFW and d) GFF specimens after the flexural test

3.3. FESEM image Analysis

Fracture mechanism of materials is very important role to improve the property or strength because it is extremely depend on the microstructure of interface. In this paper, the microstructure is describe based on tensile test specimens for all four types of composite. Firstly, the main fracture of all samples are come from the smallest and the weakest area of the specimens. The fracture mechanism of glass and carbon are quite different in crack area. Small area for CFT/CFF and quite a large area for

GFW/GFF with complex breakage. CFP fracture breakage shows 0 degree (the same direction with tension load) laid fiber are break almost linearly while 90 degrees across lay up fiber are break by each strand delamination in figure 5 a) and b). The main tension force of CFT and CFP are come from vertical direction laid up fibers, it lead to break resin and fiber. Longitudinal tension force pulls the fiber yarn to separate each filament that make each fiber delamination from bonding. As shown in figure 5 a), CFT is disintegrating into several pieces in the middle, but still hold together matrix and fiber on both edge sides. GFW/GFF is difficult to detect the breakage area because of breakage filament are complicated and breakage area of fibers is longer than that of carbon fibers. The main fracture behavior of GFW/GFF is delamination and pullout the fibers from the bonding matrix as shown in figure 5 c) and d). The matrix bond tightly, but the fibers are delaminate and pull out due to external from through fibers. Delamination of fiber will bring to stand alone each filament, therefore each filament are very easy to break. The middle filament are break first and the breakage are distribute to the both side of fibers, it is the main breakage of fiber in GFW and GFF. Additionally, voids can see in matrix area that comes from the fabricated process through air bubbles that didn't remove very well while lay up process are performed, it becomes voids in composite. Those voids are also one of the weak points to initiate crack in the matrix.

a) CFT composite

b) CFP composite

c) GFW

d) GFF

Figure 5. FESEM image a) CFW, b) CFP, c) GFW and d) GFF

3.4. EDS Analysis

FE-SEM machine with EDS system observed the element composition on the composite CFT and GFW with 60 CPS in figure 6. a) and b). CFT composite spectrum 1 proved that 88% w.t of

IOP Conf. Series: Materials Science and Engineering **501** (2019) 012024 doi:10.1088/1757-899X/501/1/012024

carbon, 11.7% of O_2 and 0.3% of calcium containing in composite. Therefore, especially carbon and oxygen are the most important element in this material to play at the main role of carbon composite. GFW has observed 72.7% w.t of C, 20.7% of O_2, 3.1% of Si, 2.2% of Ca, and a few of AL, Cl and Mg respectively in spectrum 1. Thus, element composition of epoxy can be assumed as C, O_2, Ca and Cl. Table 5 is the summary of EDS chemical composite in each GFT and GFW. On the other hand, breakage of CFT prove that it have brittle behaviour due to tension force to the fibers as shown in figure 5. Au containing percent is due to gold spray on sample for electron conductive purpose. Therefore, Au data was ignore in figure 6. A) and b). Al (Alumina) also EDS scanner may count from the stub, which use to stick the sample.

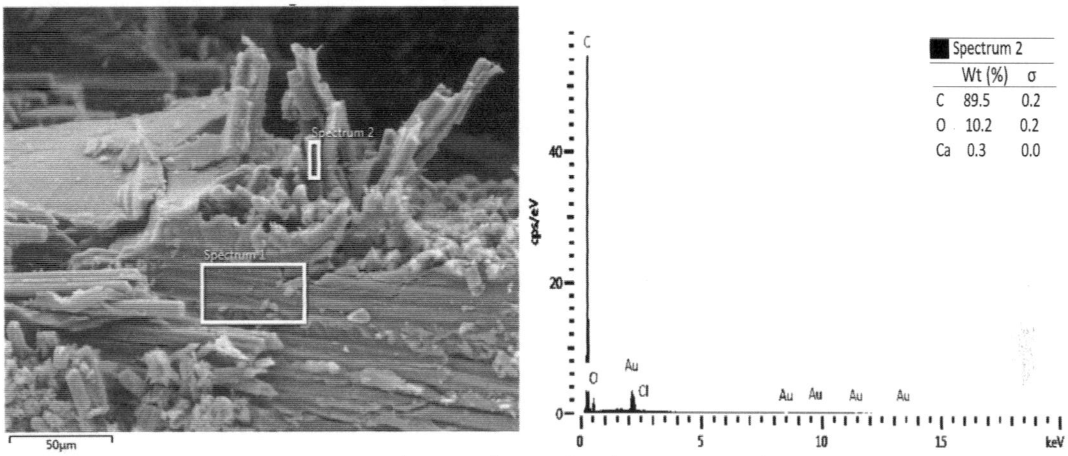

a) EDS element observatio of CFT composite

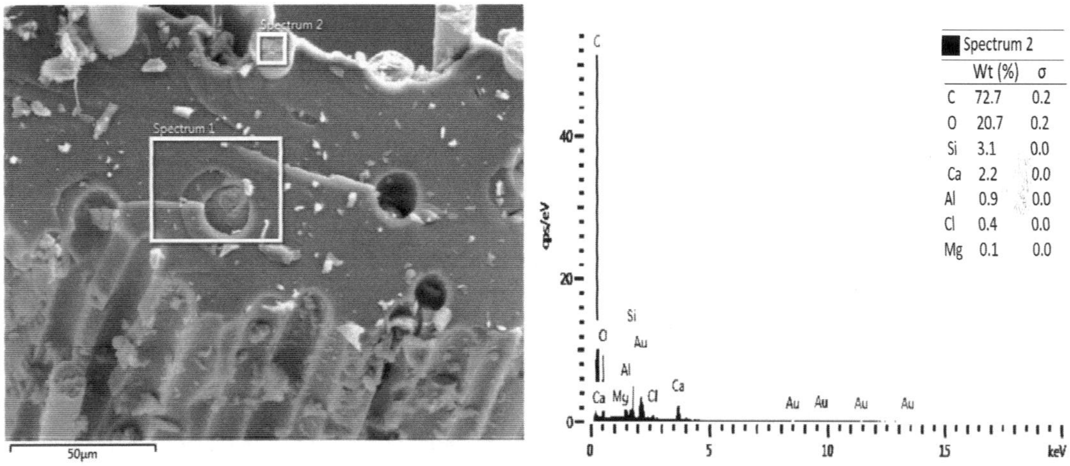

b) EDS element observation of GFW composite

Figure 6. Element composition of a) CFT and b) GFW composite by using electron dispersive x-ray spectroscopy (FESEM-EDS)

Table 5. EDS result of CFT/GFW composite and each filament

Element/ Composite or Fiber		Element (Weight %)						
		Carbon	Oxygen	Silicone	Ca	Al	Cl	Mg
CFT	Spect:1	87.9	11.7	0.3	-	-	-	-
GFW	Spect:1	72.7	20.7	3.2	2.2	0.9	0.4	0.1

4. Conclusion

The composite of carbon fiber and glass fiber reinforced polymer with an epoxy matrix were test in tensile testing machine, the results show carbon fiber composite (CFT/CFP) was the highest modulus and tensile strength. Approximately 11% of tensile strength are higher than that of GFW/GFF and almost double modulus are achieve in CFT and CFP with 36, 204MPa and 36,604MPa respectively.

Flexural test shows higher modulus than a tensile test with value 46,975 MPa and 41,945MPa for CFT and CFP respectively, meanwhile GFW and GFF are lower than CFT and CFP almost half.

Fracture surface shows, the main character of CFT/CFP are peel off fiber by each bundles and break therefore the breakage area are almost linear in each bundles. This experiments show that CFT/CFP have brittle behaviour. As a contrast, GFW and GFF are delaminate in each filament then each filament are crack after matrix transfer the load. Then the whole fibers are leads to failure. GFW/GFF has a lesser brittle deformation behaviour than CFT/CFP.

Acknowledgement

The author would like to thank and acknowledge to International College, KMITL for supporting research fund, National Science and Technology Development Agency (MTEC) for providing facilities and their big support, Somboon Advance Technology Co., Ltd. for providing Materials and Polyplastics for supporting equipment.

References

[1] Wulfsberg J, Herrmann A, Ziegmann G, Lonsdorfer G, Stöß N and Fette M 2014 *Procedia Eng.* **81** 1601-1607.
[2] Huang X 2009 *Materials (Basel)* **2** 2369-2403.
[3] Chawla KK 2016 Glass Fibers, *Ref. Modul. Mater. Sci. Mater Eng*, pp. 1-5.
[4] Shide S and Salve AV 2015 *Int. J. Eng. Res.* **4** 446–449.
[5] Patel A, Kravchenko O and Manas-Zloczower I 2018 P*olymers (Basel)* **10** 125.
[6] Enns JH 1993 Creep and mechanical properties of carbon fibre reinforced PEEK composite material.
[7] Acatay K. Akdeniz Kimya San. ve Tic. AS, Kemalpasa, Izmir, Turkey. 2017 Fiber Technol Fiber-Reinforced Compos, pp. 123.
[8] Sood M and Dwivedi G 2018 *Egypt J Pet.* **7** 75-783
[9] Hernandez DA, Soufen CA and Orlandi MO 2017 *Mater. Res.* **20** 951–961.
[10] Hakim IA, Donaldson SL, Meyendorf NG and Browning CE 2017 *Mater. Sci. Appl.* **8** 170.
[11] Karin P, Watanawongskorn P, Boonsakda J, Saenkhumvong E, Rungsritanapaisan S, Srivarocha S. 2017. Impact of Biodiesel on Small CI Engine Combustion Behavior and Particle Emission Characteristic, SAE Technical Paper 2017-32-0094.
[12] Karin. P, Boonsakda J, Siricholathum K, Saenkhumvong E, Charoenphonphanich C and Hanamura K 2017 *Int. J. Automot. Technol.* **18** 31-40.
[13] Karin. P, Supanamok C and Hanamura K 2016 *J. Res. Appl. Mech. Eng.* **4** 126-134.

IOP Conf. Series: Materials Science and Engineering **501** (2019) 012025 doi:10.1088/1757-899X/501/1/012025

Deflection of Composite Cantilever Beams with a Constant I-Cross Section

P Prombut* and C Anakpotchanakul

Department of Mechanical Engineering, Faculty of Engineering, Kasetsart University
50 Ngam Wong Wan Road, Chatuchak, Bangkok 10900, Thailand

* Corresponding Author: E-mail pongtorn.p@ku.ac.th

Abstract. Laminated composite beams are widely used as structural components in aerospace and wind energy industries. For airplane wings and wind turbine blades, the beams, commonly called spars, provide principal stiffness against bending deformation of the structures. The present work is focused on composite cantilever beams with a constant I-cross section. A unidirectional (UD) Glass/Epoxy lamina was chosen for composite material. The overall dimension of the beam is 50 x 55.2 x 2,500 mm. Various stacking sequences were assigned to the flanges and the web of the beam. A uniform distributed load was applied to the upper flange. The deflection results from First-Order Shear Deformation Theory (FSDT) and Finite Element Analysis (FEA) are in good agreement. The effective longitudinal modulus (E_x) of the flanges has strong influence on the bending stiffness and the beam deflection. Understanding important roles of materials and beam construction can help improving the design of I-beams. The validated FEA procedure can be extended to the analysis of realistic spars, which is based on an I-cross section plus curve, taper, and twist along the length.

1. Introduction

Composite materials have excellent specific stiffness and strength. Lightweight structures such as airplanes and wind turbine blades use composite beams, commonly called spars, to provide principal stiffness against bending deformation. The relationship between load and deformation of composite laminates is complex due to the orthotropic nature of the materials. The behavior at ply level requires a stiffness matrix derived from the classical lamination theory (CLT). The behavior at laminate level can be simplified by using effective in-plane and flexural modulus [1].

Theories of laminated composite beams can be divided into two classes. One is the class of classical beam theories (CBT), which are suitable for thin beams. The other is the class of shear deformation beam theories (SDBT), which take into account the shear deformation and rotary inertia. The latter are more accurate for thick beams [2]. Fundamental equations using a first-order shear deformation theory (FSDT) for rectangular beams were developed and verified with 3D finite element analyses (FEA) [2].

Composite beams with I-cross section are regarded as thin-walled, open-section, orthotropic beams. The I-beam is made from flat laminates that can have different properties. Effective modulus of the upper flange, lower flange, and web can be different. Therefore, equivalent tensile and equivalent bending stiffness of the cross-section must be determined [3]. Moreover, thick solid beams and thin-walled beams are subjected to a transverse shear strain that adds a shear deformation to the total beam deformation [4]. The first-order shear deformation theory (FSDT) takes this shear deformation into account with the assumption that the cross section of a beam remains plane but not perpendicular to the beam's neutral axis.

The present work is focused on composite cantilever beams with a constant I-cross section. The calculation of deflection using CBT and FSDT [3-5] was investigated. Simplified I-beams representing a wing's spar or a wind turbine blade's spar with various laminate stacking sequences were analyzed and compared with the results from FEA.

2. Composite laminate behavior

The relationship between external loads and deformation of composite laminates can be described by the classical lamination theory [1]. The stiffness matrix is commonly called ABD matrix.

$$\begin{Bmatrix} N_x \\ N_y \\ N_{xy} \\ M_x \\ M_y \\ M_{xy} \end{Bmatrix} = \begin{bmatrix} A_{11} & A_{12} & A_{16} & B_{11} & B_{12} & B_{16} \\ A_{12} & A_{22} & A_{26} & B_{12} & B_{22} & B_{26} \\ A_{16} & A_{26} & A_{66} & B_{16} & B_{26} & B_{66} \\ B_{11} & B_{12} & B_{16} & D_{11} & D_{12} & D_{16} \\ B_{12} & B_{22} & B_{26} & D_{12} & D_{22} & D_{26} \\ B_{16} & B_{26} & B_{66} & D_{16} & D_{26} & D_{66} \end{bmatrix} \begin{Bmatrix} \varepsilon_x^0 \\ \varepsilon_y^0 \\ \gamma_{xy}^0 \\ \kappa_x \\ \kappa_y \\ \kappa_{xy} \end{Bmatrix} \tag{1}$$

where [A]: Extensional stiffness matrix
[B]: Bending-extension coupling stiffness matrix
[D]: Bending stiffness matrix

Since the material properties are defined relative to fiber direction (coordinate 1-2), the effective modulus of a laminate relative to structural direction (coordinate x-y) have to be calculated from equation (2) to (4).

$$[a] = [A]^{-1},\ [d] = [D]^{-1} \tag{2}$$

$$E_x = \frac{1}{Ta_{11}},\ E_y = \frac{1}{Ta_{22}},\ G_{xy} = \frac{1}{Ta_{66}},\ and\ \nu_{xy} = -\frac{a_{12}}{a_{11}} \tag{3}$$

$$E_x^f = \frac{12}{T^3 d_{11}},\ E_y^f = \frac{12}{T^3 d_{22}},\ G_{xy}^f = \frac{12}{T^3 d_{66}},\ and\ \nu_{xy}^f = -\frac{d_{12}}{d_{11}} \tag{4}$$

where T is the laminate thickness, equations (3) are effective in-plane properties and equations (4) are effective flexural properties.

3. Deflection of composite cantilever beams

A cantilever beam with the cross section symmetrical with respect to the x-z plane is shown in figure 1. A uniform transverse load p_z (N/m) is applied in the symmetry plane.

The deflection of the beam is due to both bending and shear deformation. The maximum deflection at the free end of the beam can be determined using equation (5) [4].

$$w = w_b + w_s = \frac{p_z L^4}{8\overline{EI}} + \frac{p_z L^2}{2\bar{S}} \tag{5}$$

where L is the beam's length, $\overline{EI}$ is the equivalent bending stiffness and $\bar{S}$ is the shear stiffness.

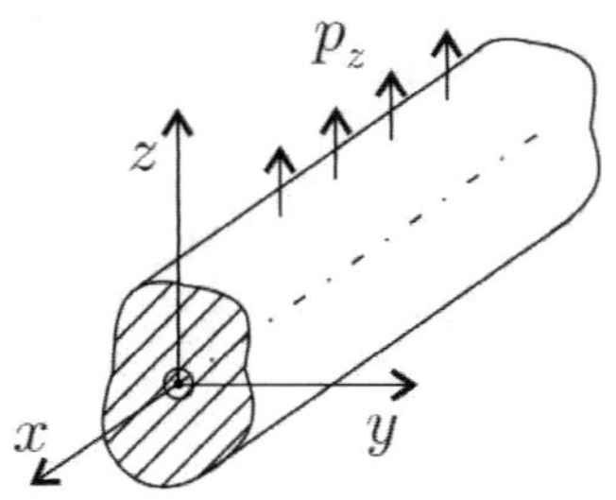

Figure 1. Orthotropic beam under transverse loading [4].

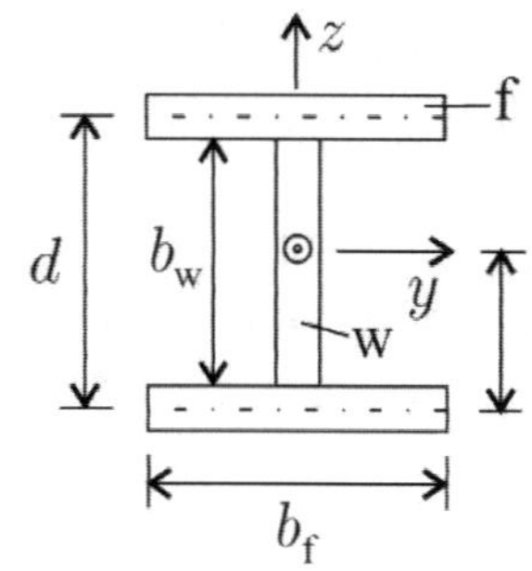

Figure 2. Cross section of composite I-beam [4].

The equivalent bending stiffness of a symmetrical thin-walled I-beam with each wall segment being a symmetric laminate can be determined using figure 2 and equation (6) [4].

$$\overline{EI}_{yy} = \frac{b_f}{(a_{11})_f}\frac{d^2}{2} + \frac{2b_f}{(d_{11})_f} + \frac{b_w^3}{12(a_{11})_w} \tag{6}$$

where the subscript *f* denotes the flange segments and the subscript *w* denotes the web segment.

The shear stiffness of the cross section is assumed to be from the web only. By the arrangement of the web segment, the transverse shear modulus G_{xz} in the beam coordinate is actually the in-plane shear modulus G_{xy} in the web laminate. The shear stiffness can be calculated by [6]:

$$\bar{S} = G_{xy} A_w \tag{7}$$

4. Composite I-beam analysis

The beam for deflection analysis has a constant I-cross section. One end is fixed and the other end is free, forming the cantilever beam configuration. The overall dimension of the beam is 50 x 55.2 x 2,500 mm as shown in figure 3. The flanges and the web have the cross section of 50 x 2.6 mm.

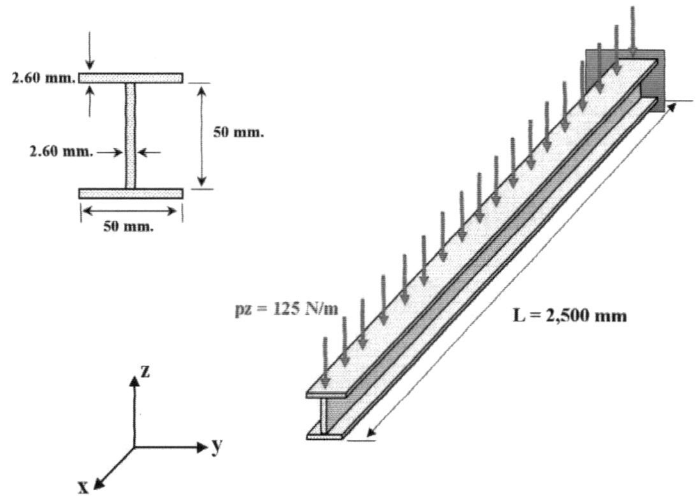

Figure 3. Cantilever beam under uniform transverse load.

IOP Conf. Series: Materials Science and Engineering **501** (2019) 012025 doi:10.1088/1757-899X/501/1/012025

A unidirectional (UD) Glass/Epoxy lamina was chosen for the material. The material properties are shown in table 1. Four stacking sequences were assigned to the flanges. They were selected to produce different level of deformation coupling [7]. Laminate F1 is for $[20/-70/-70/-70/20]_{2S}$, F2 for $[20_5]_{2S}$, F3 for $[0/90/0/90/0]_{2S}$, and F4 for $[90/45/-45/0/0]_{2S}$. The web has two stacking sequences: W1 is $[45/-45]_{5S}$ with high shear modulus and W2 is $[0/90]_{5S}$ with high extensional modulus. A uniform distributed load p_z = 125 N/m was applied to the upper flange.

Table 1. Properties of Glass/Epoxy UD lamina [5].

E_1 (GPa)	E_2 (GPa)	G_{12} (GPa)	ν_{12}	t (mm)
53.78	17.93	8.96	0.25	0.13

4.1. *Laminate and cross section properties,*

Effective modulus of the laminates used for the flanges and the web of the beams can be calculated using equations (3) and (4). Relevant results are shown in table 2. The values of the in-plane modulus and the flexural modulus of each laminate are approximately the same. For the laminate F4, E_x^f is noticeably less than E_x because the 0-degree layers are placed toward the middle of the laminate. The layers then become less efficient for bending resistance.

Table 2. Effective in-plane and flexural modulus.

Laminate	Layup	E_x (GPa)	E_x^f (GPa)	G_{xy} (GPa)	G_{xy}^f (GPa)
F1	$[20/-70/-70/-70/20]_{2S}$	28.42	28.92	10.92	10.94
F2	$[20/20/20/20/20]_{2S}$	38.72	38.72	9.99	9.99
F3	$[0/90/0/90/0]_{2S}$	39.64	40.07	8.96	8.96
F4	$[90/45/-45/0/0]_{2S}$	35.64	29.87	11.78	12.06
W1	$[45/-45]_{5S}$	24.98	24.93	16.02	15.93
W2	$[0/90]_{5S}$	36.05	38.75	8.96	8.96

The equivalent bending stiffness and shear stiffness are required for the calculation of the beam deflection. They can be determined using equations (6) and (7) respectively. Table 3 shows the equivalent stiffness of 8 I-beams obtained from a combination of the flanges and webs. The modulus E_x of the flange has strong influence on the bending stiffness $\overline{EI}_{yy}$. The E_x of laminate F2 is 36.24% greater than the E_x of laminate F1. This makes the bending stiffness of the beam F2W1 higher than the beam F1W1 by 31.97%. The modulus of the web has less effect on the bending stiffness. While E_x of laminate W2 is greater than W1 by 44.32%, the bending stiffness of the beam F1W2 is higher than F1W1 by only 5.14%. For shear deformation, the shear modulus G_{xy} of the web has direct effect on the shear stiffness $\bar{S}$ of the beam.

Table 3. Equivalent bending stiffness and shear stiffness of the cross section.

Beam	Flange	Web	$\overline{EI}_{yy}$ (N-m^2)	$\bar{S}$ (N)
F1W1	$[20/-70/-70/-70/20]_{2S}$	$[45/-45]_{5S}$	5,793	$2.083\text{x}10^6$
F2W1	$[20/20/20/20/20]_{2S}$	$[45/-45]_{5S}$	7,645	$2.083\text{x}10^6$
F3W1	$[0/90/0/90/0]_{2S}$	$[45/-45]_{5S}$	7,812	$2.083\text{x}10^6$
F4W1	$[90/45/-45/0/0]_{2S}$	$[45/-45]_{5S}$	7,090	$2.083\text{x}10^6$
F1W2	$[20/-70/-70/-70/20]_{2S}$	$[0/90]_{5S}$	6,091	$1.165\text{x}10^6$
F2W2	$[20/20/20/20/20]_{2S}$	$[0/90]_{5S}$	7,944	$1.165\text{x}10^6$
F3W2	$[0/90/0/90/0]_{2S}$	$[0/90]_{5S}$	8,112	$1.165\text{x}10^6$
F4W2	$[90/45/-45/0/0]_{2S}$	$[0/90]_{5S}$	7,389	$1.165\text{x}10^6$

4.2. *Finite element analysis (FEA)*

The beams were modelled with 2D shell element. After a mesh convergence study, the element size was chosen to be 2.5 x 2.5 mm. Each beam model has a fixed supports at one end and is free at the other end. The geometry, material, stacking sequences, and loading are described in the previous section. Figure 4 shows the beam's FEA model with coarse mesh to make the image clear.

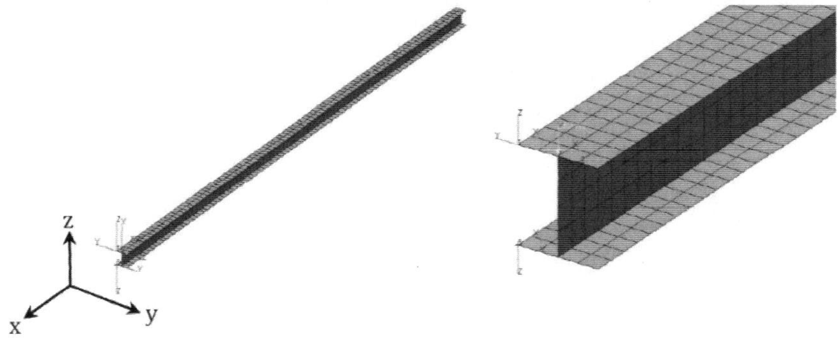

Figure 4. FEA Model of the composite cantilever I-beams.

4.3. *Maximum deflection at free end*

Deflection at the free end of the beams are presented in table 4. The CBT method does not take into account the shear deformation. Only the bending deformation (w_b) was calculated using equation (5). The FSDT method predicted the deflection using both bending and shear deformation from equation (5). The results from FEA are also included. Figure 5 shows a typical deformation result where the maximum deflection occurs at the free end of the beam. For the beam configuration in the present work, the shear deformation is negligible as the deflection values determined by CBT and FSDT are practically the same. The FEA method consistently predicts greater deflection compared to the FSDT. However, both method are still in good agreement with the maximum difference less than 11%.

Table 4. Deflection at free end of composite cantilever I-beams.

Beam	Flange	Web	Tip Deflection (mm)		
			CBT	FSDT	FEA
F1W1	$[20/-70/-70/-70/20]_{2S}$	$[45/-45]_{5S}$	105.36	105.55	116.00
F2W1	$[20/20/20/20/20]_{2S}$	$[45/-45]_{5S}$	79.84	80.03	88.80
F3W1	$[0/90/0/90/0]_{2S}$	$[45/-45]_{5S}$	78.13	78.32	86.80
F4W1	$[90/45/-45/0/0]_{2S}$	$[45/-45]_{5S}$	86.09	86.28	95.40
F1W2	$[20/-70/-70/-70/20]_{2S}$	$[0/90]_{5S}$	100.20	100.54	110.00
F2W2	$[20/20/20/20/20]_{2S}$	$[0/90]_{5S}$	76.83	77.16	85.50
F3W2	$[0/90/0/90/0]_{2S}$	$[0/90]_{5S}$	75.24	75.58	83.60
F4W2	$[90/45/-45/0/0]_{2S}$	$[0/90]_{5S}$	82.60	82.93	91.40

The modulus E_x of the flange has strong influence on the deflection. The E_x of laminate F2 is 36.24% greater than the E_x of laminate F1. As a result, FSDT predicts the deflection of beam F2W1 less than F1W1 by 24.18%. The modulus of the web has little effect on the deflection. While E_x of laminate W2 is greater than W1 by 44.32%, the deflection of the beam F1W2 is lower than F1W1 by only 4.75%.

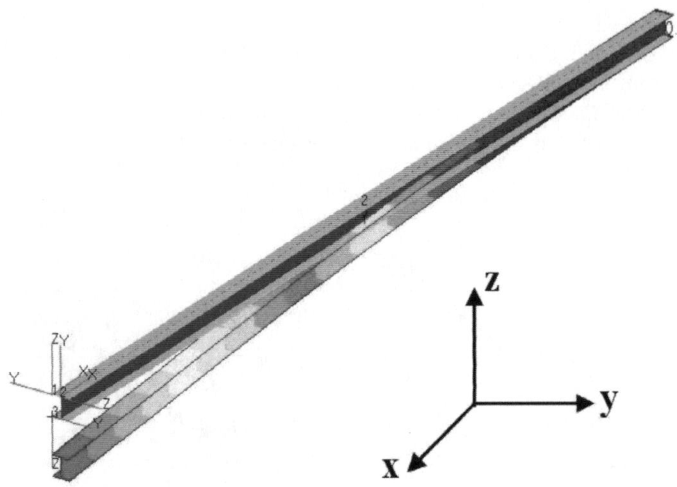

Figure 5. Deflection of composite cantilever I-beams from FEA.

5. Conclusion

Deflection of composite cantilever I-beams under a uniform transverse load was investigated. The deflection of the free end were determined using classical beam theory (CBT), first-order shear deformation theory (FSDT), and finite element analysis (FEA).

For the beam configuration in the present work, the shear deformation is negligible as the deflection values determined by CBT and FSDT are practically the same. The deflection results from FSDT and FEA are in good agreement. The FEA method consistently predicts greater deflection compared to the FSDT. However, the difference is less than 11%.

The effective longitudinal modulus (E_x) of the flanges has strong influence on the bending stiffness and the beam deflection. Understanding important roles of materials and beam construction can help improving the design of I-beams. The validated FEA procedure can be extended to the analysis of realistic spars, which is based on an I-cross section plus curve, taper, and twist along the length.

References

[1] Kaw A K 2006 *Mechanics of composites materials* (Florida: Taylor & Francis Group)

[2] Hajianmaleki M and Qatu M S 2001 Mechanics of composite beams *Advances in Composite Materials - Analysis of Natural and Man-Made Materials* ed P Tesinova (IntechOpen) chapter 22 pp 527-546

[3] Mohamed Ali J S, Hrairi A and Mohamad M 2015 *ARPN Journal of Engineering and Applied Sciences* **10** 10088-95

[4] Kollar L P and Springer G S 2003 *Mechanics of composite structures* (New York: Cambridge University Press)

[5] Vo T P and Lee J 2010 *Computers and Structure* **88** 347-56

[6] Damkilde L 2000 *Stress and stiffness analysis of beam-sections* (Lyngby: Technical University of Denmark)

[7] Sompong N and Prombut P 2017 *International Journal of GEOMATE* **12** 15-20

IOP Conf. Series: Materials Science and Engineering **501** (2019) 012026 doi:10.1088/1757-899X/501/1/012026

Static testing for composite wing of a two-seater seaplane

S Chinvorarat[1,*], B Watjatrakul[1], P Nimdum[1], T Sangpet[1], T Soontornpasatch[1], and P Vallikul[1]

[1]Department of Mechanical and Aerospace Engineering, King Mongkut's University of Technology North Bangkok, 1518 Pracharaj 1, Bangkok, 10800, Thailand.

*Corresponding author: sinchai.c@eng.kmutnb.ac.th

Abstract. The paper studied the strength and deformation characteristics of a composite wing of a two-seater seaplane to get the certificate of airworthiness which static testing is indispensable. The primary wing structure component includes upper and lower skins, leading edge, trailing edge, a root rib and main spar. The main purpose of static testing is to examine the bending strength of the wing. The testing results show good agreement with the FE analysis results, and the bending strength of wing is strong enough to support the limit loads (the maximum loads to the expected in service) without detrimental and permanent deformation, and without failure for at least 3s under the ultimate loads (limit loads multiplied by prescribed factors of safety), which meets the requirement of ASTM F2245-16c (Standard Specification for Design and Performance of a Light Sport Airplane).

1. Introduction

A seaplane is a powered fixed-wing aircraft capable of taking off and landing on water. Seaplanes that can also take off and land on airfields are in a subclass called amphibious aircraft. Seaplanes and amphibians are usually divided into two categories based on their technological characteristics: floatplanes and flying boats. A floatplane is a type of seaplane, with one or more slender pontoons mounted under the fuselage to provide buoyancy. By contrast, a flying boat uses its fuselage for buoyancy. For two-seater seaplane which maximum takeoff weight (MTOW) is less than 650 kilograms can be classified as a light sport airplane that meets certain regulations set by a Civil Aviation Authority of Thailand (CAAT).

To ensure flight safety and receive certificate of airworthiness for a light sport airplane, the static structural testing of the aircraft's wing must be performed and proved of safety. According to ASTM F2245-16c (Standard Specification for Design and Performance of a Light Sport Airplane), a structure should be designed to be able to withstand limit load (the maximum loads to be expected in service) without permanent damage or deformation upon unloading, and withstand ultimate load (limit loads multiplied by prescribed factors of safety) for at least 3 seconds without collapse. Therefore, the static test must verify the strength and stiffness of the wing structure under the design of load provision. Smith et al.[1] described the static testing of an ultralight airplane and replaced with masses to fulfill center-of-gravity requirements before testing. Wong et al. [2] presented the design of the test rig and the results of the static test of the Lockheed P-3 Orion Wing Leading Edge center section structure. Sullivan et al.[3] presented the results of the strength and stiffness characteristics of a carbon composite wing of an ultralight unmanned aerial vehicle. Sullivan et al.[4] described vibration testing of a full-scale carbon composite ultralight unmanned aerial vehicle. Kang et al. [5] performed static testing on the wing of a two-seater aircraft powered by Li-ion battery electric propulsion compliance with ASTM F2245-11.

This paper described the static testing and finite element (FE) analysis results of the seaplane wing. The static load is specified regarding the limit loads and the ultimate loads, which is the limit loads multiplied by prescribed factors of safety. The ultimate load safety factor is 1.5. The property of material

IOP Conf. Series: Materials Science and Engineering **501** (2019) 012026 doi:10.1088/1757-899X/501/1/012026

system, the result of finite element (FE) analysis, the static loading method and the testing results are described in the following section.

Figure 1. Seaplane (Amphibious)

2. Wing Structure

Composite materials have been widely utilized in airplane due to their high specific strength as compared to conventional isotropic materials. The two-seater seaplane main structure was made from all-composite which much lighter than traditional metal. The seaplane structural components were intentionally designed complying to ASTM F2245-16c which is the standard specification for design and performance of a light sport airplane. According to ASTM F2245-16c structure compliance section, the seaplane structure must be able to support limit loads (the maximum loads to be expected in service) without detrimental and permanent deformation, and ultimate loads without failure for at least 3 seconds. In order to verify the strength and deformation characteristics of the wing structural assemblies, the wing structural static testing was performed on a test rig following steps of applying limit loads prescribed in ASTM F2245-16c (positive limit loads not less than 4G and negative limit loads not greater than -2G).

The two-seater seaplane was designed with maximum takeoff weight (MTOW) less than 650 kilograms (1G), maximum empty airplane weight is 325 kilograms and each wing weights about 50 kilograms. The wing span of this plane is 9.73m, aspect ratio is 9.26, and the mean aerodynamic chord is 0.487m.

The wing structure is an integral type (for stress calculations the main spar is the only element responsible to resist the aerodynamic loads). Each wing has a mono I-beam spar, 10 ribs and a sandwich structure skin shell. A composite shell leading edge and a composite fairing for the aileron and flaps complete the wing structure. All the structural wing elements are made of composite and held together by adhesive. The wing loads are transferred to the fuselage by means of a wing strut attached to the spigot post and by the spar attached to the root rib.

As shown in Figure 2, the primary wing structure includes the main spar, ribs, the upper and lower skins, leading edge, trailing edge, and these structures are glued together by adhesive. However, the ailerons and flaps are removed in finite element (FE) analysis and static testing.

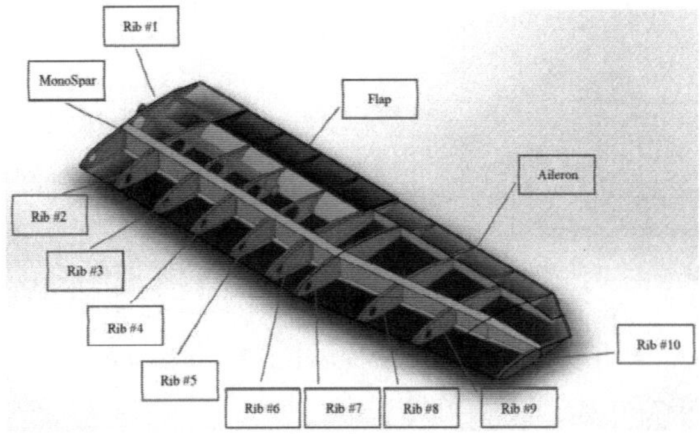

Figure 2. Wing Structure

The skins are made of a sandwich material. The spars and root rib are made of carbon composite. In order to provide the wing bending stiffness and form the carry-through structure, the root-rib/wing box interconnected by two steel screws bear the shear load and torsion.

The wing structure components are made of carbon fabric, glass fiber, foam, and parabeam 3D glass which the material properties are shown in Table 1. The skins are made of sandwich construction with a 2 mm foam core. The laminated ply pattern in the wing region is given in Table 2. Two M10 screws are used to mounted the root rib with the wing box, while another two M10 screws are used to mounted the strut with the wing and the fuselage.

Table 1. Material Properties of the Seaplane Wing

Material property	Carbon Fibre	Glass Fibre	ParaBEAM 3D Glass	Foam core [Isotropic]
Density [Kg/M3]	1.431	1.539	0.341	1.8
E11,[GPa]	36.59	11.58	0.255	0.085
E22,[GPa]	36.59	11.58	0.255	
G12,[GPa]	2.405	2.03	0.07	0.032
Possion	0.038	0.038	0.001	0.3
Max Stress [Mpa]	394.93	159.85	5.31	
Max Shear [Mpa]	49.26	37.95	2.24	

Table 2. Laminated Ply Pattern

ITEMS	NO. of plies	ORDERING	THICKNESS (mm)
SKIN	4	B-X-F-Y	2.9
SPAR - UPPER FLANK	24	A	7.2
SPAR - LOWER FLANK	12	A	3.6
SPAR - WEB	25	A-P-B	11.2
RIB #1	7	B-X-Y-B-X-Y-B	2.1
RIB #2	7	B-X-Y-B-X-Y-B	2.1
RIB #3	7	B-X-Y-B-X-Y-B	2.1
RIB #4	7	B-X-Y-B-X-Y-B	2.1
RIB #5	7	B-X-Y-B-X-Y-B	2.1
RIB #6	7	B-X-Y-B-X-Y-B	2.1
RIB #7	6	B-X-Y-X-Y-B	1.8
RIB #8	5	B-X-Y-X-B	1.5
RIB #9	7	B-X-Y-B-X-Y-B	2.1
RIB #10	4	B-X-Y-B	1.2
spigot post	38	B	11.4
Float	6	B-X-Y-X-Y-B	1.8
Carbon fabric 0-90, B : Carbon fabric +/-45 , X: Glass fibre 0-90, Y: Glass fibre +/-45, F: Foam, P : Parabeam			

3. Experimental

A test frame was used to serve as the support structure for static strength testing, as shown in Figure 3. It was made from 2 inches carbon steel square tube in "T" shape mounted firmly on the shop floor. A 400 x 150 mm "I" beam, 8 meters long, mounted with the shop column over the test frame with four chain hoists were used to create bending and pitching moment on the wing. The left wing of the seaplane with the wing strut was installed on the test frame as the normal configuration (zero angle of attack). Both positive/negative distribution load were simulated on the wing following ASTM F2245-16c recommendation. The -2G negative distribution load was applied on the wing by sand bags and the +4G positive load (limit load) up to +6G positive load (ultimate load) by the four chain hoists.

Figure 3. The seaplane wing mounted on the test stand with all equipment and instruments installed

3.1. Loading Method

There were four loading locations on the wing where each location was embraced by a wooden shroud hooked up with the hoist chain and the pulling force on each shroud was 30% location of mean cord. The distance of each loading location away from the fuselage centerline was shown in Fig.4. The wing loading and the total number of load cases were shown in Table 3. Starting with negative loads, the sand bags were placed on each assigned location until reaching -2G (limit load) following with positive loads by pulling each chain hoist through a load cell until reaching +4G positive load (limit load) and +6G positive load (ultimate load) respectively.

Table 3. Test load applied on the wing

Applied Load [Kgf]	-2G [-50%]	-1G [-25%]	1G [25%]	2G [50%]	3G [75%]	4G [100%] [Limit Load]	5G [125%]	6G [150%] [Ultimate Load]
	Negative Load		Positive Load					
Rib#2	-238.60	-108.60	109.25	217.06	303.09	391.83	488.25	581.71
Rib#4	-238.60	-108.60	105.65	214.12	296.17	393.71	485.83	580.53
Rib#6	-118.60	-53.60	47.73	93.77	127.16	163.80	205.86	243.86
Rib#9	-118.60	-53.60	47.48	99.97	127.57	165.92	203.29	239.97
Total	-714.40	-324.40	310.10	624.91	854.00	1115.25	1383.22	1646.07

3.2. Measurement Instrumentation

In order to measuring the strength of the wing, 4 strain gauges (rosettes type) were placed on difference locations as shown in Figure 4. The general-purpose strain gauges (KFG-10-120-d17-11L5M2S) were used to measure the strain having a nominal gauge length of 10 mm, resistance is 120±0.8 Ω and gauge factor is 2.07±1.0%. All strain gauges have been tested before bonding to desired location to complete the static testing. To obtain the wing deformation, 4 laser sensors (KEYENCE CMOS Multi-Function Analog Laser Sensor – IL series) were used to measure and record data simultaneously.

In order to collect the testing data, a computer data acquisition system (the NI PXIe-8840 PXI Express/CompactPCI Express embedded controller, PXIe-4330/4331 strain gauge module, and PXIe-4302 data acquisition module) was used to measure and record all strains and deformations.

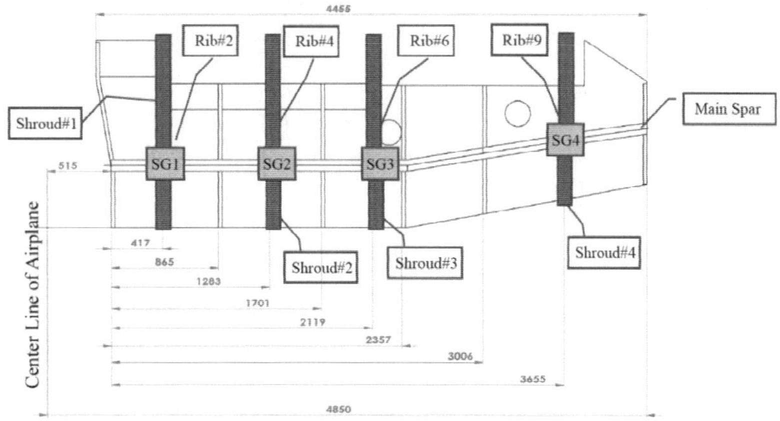

Figure 4. Location of four strain gauges installed on the wing spar

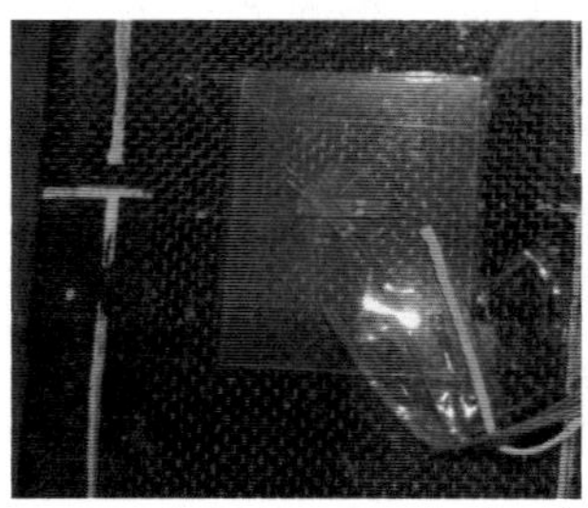

Figure 5. Strain gauges (rosettes type) glued to the wing skin

4. Finite Element Analysis

4.1. The model of finite element

In order to predict the wing structural deformation and strains under the critical loading conditions, ANSYS is used to establish finite element model. The model is established according to the actual wing structure. The continuum elements are used to the adhesive layer between structural elements (spar, root-rib/skin/spar, and skin/skin interconnections). All structural elements adopt Tri/Quad element, as shown in Fig.6. The FE model contains 8,963 elements and 9,963 nodes.

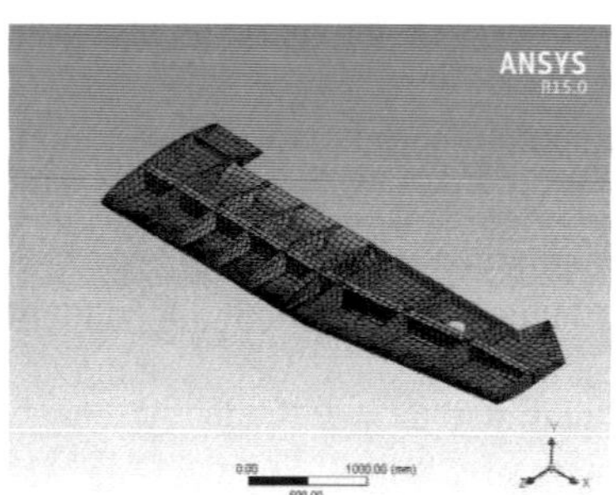

Figure 6. FE model of wing structure

4.2. Boundary Conditions and Loading

The boundary conditions agree with the static structural testing as identified with the side, up, and front directions defined as the X, Y and Z coordinates in Figure 6, respectively. The fixed nodes were assigned at the root of the main and minor spar, while the displacements of nodes are allowed (in Y direction) for the spigot (underneath the wing). The resultant force in the form of uniform distributed load was assigned at each load location, which was applied to the upper skin.

4.3. Static Analyses Method

To predict wing deformation as well as laminate strains under incremental loading to design ultimate load, geometric linear static FE analyses were performed. In order to identify the potential failure zones of the wing structure, the deformation and strain distributions were used.

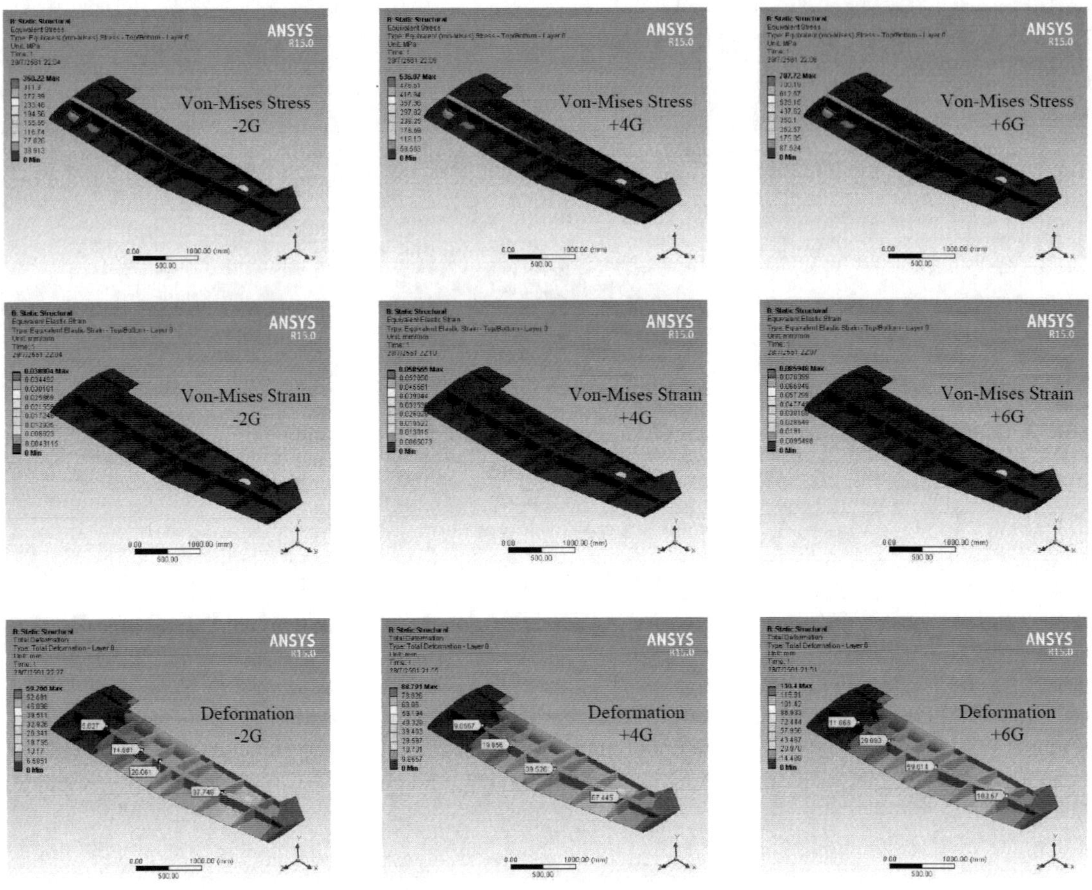

Figure 7. Predicted Stress-Strain-Deformation on various loading

5. Presentation of Results

Table 4 shows strain value under different load factor ranging from -2G to 6G (ultimate load). The strain value (Y- direction) occurs the most at the rib#2 closed to the root rib indicted that maximum stress occurs also in the neighborhood of main spar near the root rib. The finite element model shows that the maximum stress on 6G (ultimate load) condition is 707.72 MPa while on 4G (limit load) condition is 536.07 MPa. The comparison between the actual and predicted wing deformation on -2G, 4G, and 6G respectively shown on Figure 8 indicates small error at the tip of the wing. This concludes that the FE model is good enough to represent the wing structural testing.

Under the ultimate load, there are no other failures to appear on the wing, while the maximum deformations are 103.57 mm (measurement) and 104.66 mm (predicted). There is no detrimental or permanent deformation on the structure when unloading both positive and negative side. It is concluded that the bending strength of wing is strong enough to support ultimate loads without failure for at least 3s.

Table 4. Strain obtained from the measurement under different load factor

Load Factor	Rib#2,x (μm/m)	Rib#2,y (μm/m)	Rib#2,xy (μm/m)	Rib#4,x (μm/m)	Rib#4,y (μm/m)	Rib#4,xy (μm/m)	Rib#6,x (μm/m)	Rib#6,y (μm/m)	Rib#6,xy (μm/m)	Rib#9,x (μm/m)	Rib#9,y (μm/m)	Rib#9,xy (μm/m)
6G	63.70	-455.94	-260.49	201.00	-124.23	-0.07	-439.17	-45.34	-200.59	7.66	-122.96	-148.67
5G	52.96	-366.13	-206.67	166.71	-97.84	6.24	-369.00	-30.03	-159.97	9.12	-92.90	-116.38
4G	38.85	-279.63	-157.32	130.26	-76.73	2.17	-305.77	-17.84	-124.54	3.83	-66.46	-88.57
3G	29.73	-193.60	-106.46	101.24	-49.41	7.58	-230.78	-8.05	-85.80	7.16	-32.07	-52.89
2G	16.90	-116.89	-62.28	67.05	-30.92	5.85	-179.03	-3.23	-62.24	4.13	-16.22	-36.40
1G	9.32	-56.70	-26.78	35.20	-6.99	7.97	-89.62	-1.52	-33.06	3.42	-3.55	-13.89
-1G	2.14	-45.47	4.80	-17.86	-19.20	-41.77	91.51	0.09	48.06	4.36	10.30	13.29
-2G	12.60	-383.02	-19.94	-14.63	8.35	-95.56	206.43	14.04	77.98	8.15	43.20	15.96

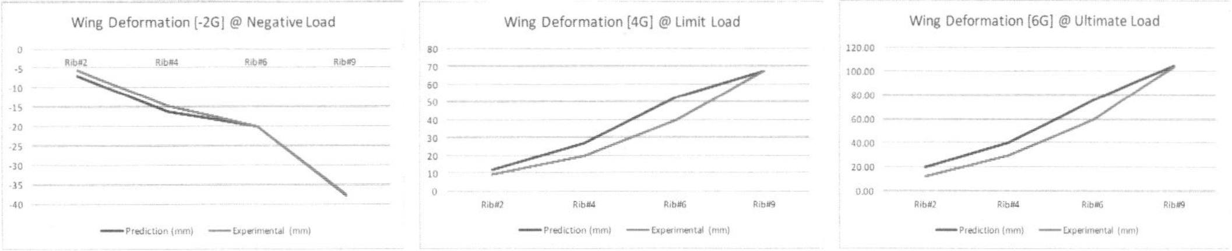

Figure 8. Comparison between Experimental and Predicted Wing Deformation

6. Conclusions

Static testing for composite wing of a two-seater seaplane was presented. A test frame was designed to serve as the support structure for the wing static testing. The deformation of wing and measured strain data were collected by computer. After unloading from the limit loads (the maximum loads to the expected in service), the ultimate loads (limit loads multiplied by prescribed factors of safety), and the negative load, there were no sign of detrimental or any form of failure happen to the tested wing. It has been concluded that the wing can support the limit loads and ultimate loads without appearing of permanent deformation and structural failure, therefore the wings can withstand ultimate loads condition without failure and comply with the requirement of the ASTM F2245-16c.

Acknowledgements

The authors would like to express our deeply appreciation to Thailand Research Fund (TRF) and Department of Mechanical and Aerospace Engineering, KMUTNB for their excellent and strongly support till the completion of this research.

References

[1] Smith H W 1988 *J. of Aircraft* **25**(1) 37–40
[2] Wong A and Luke G 1996 *Aeronautical and Maritime Research Lab. Rept*. DSTO-TR-0423, Melbourne, Australia
[3] Sullivan R and Simsiriwong J 2010 *J. of Aircraft* DOI:2514/6.2010-2807
[4] Sullivan R, Hwang Y, Rais-Rohani M and Lacy T 2012 *J. of Aircraft,* **46**(3) 814-20
[5] Kang Y, Liguo Z, Shude J, Yumei Y, and Wang J 2016 *Advanced Composites Letters*, Vol. **25**, Iss.**6**
[6] Fleuret C, Andreani A, Lain E, Grandidier J, L'heritie S, and Gorge A 2016 *Mechanics & Industry* **17**, 614
[7] ASTM F2245-16c, Standard Specification for Design and Performance of a Light Sport Airplane

IOP Conf. Series: Materials Science and Engineering **501** (2019) 012027 doi:10.1088/1757-899X/501/1/012027

Parametric Effects of Resistance Spot Welding between Li-ion Cylindrical Battery Cell and Nickel Conductor Strip

N. Phichai[1], P. Kaewtatip[1], V. Lailuck[2], S. Rompho[2], and M. Masomtob[2,*]

[1] Department of Mechanical Engineering, Faculty of Engineering, King Mongkut's University of Technology Thonburi, Bangkok, Thailand
[2] Materials for Energy Research Unit, National Metal and Materials Technology Center (MTEC), National Science and Technology Development Agency (NSTDA), Pathum Thani, Thailand

* Corresponding Author: manopm@mtec.or.th

Abstract. The resistance spot welding method is commonly used in the cylindrical battery packing because of production costs. However, its quality is lower than other welding methods. To improve the quality, three parameters which are slot geometries of nickel conductor strip, welding position and welding position sequence are studied in this paper using both simulations and experimental methods. According to different slot geometries, if the slot length is longer than the positive battery terminal diameter, the high welding current flows through the welding point. To weld different positions, it results in structural deformation of the positive battery terminal which directly affects the contact pressure. As different welding position sequences for three welding spots on the positive battery terminal, it is found that the sequence starting at the middle is better than which from the sides. In conclusion, these parameters can be used to improve the welding quality; nevertheless, there are still other parameters which will be studied in the future.

1. Introduction

Welding is commonly used in the battery packing industry for battery connection, especially resistance spot welding because of low production cost. However, the quality of resistance spot welding in battery packing is lower than that of other methods due to the high thermal and electrical conductivity properties of materials [1]. This causes low mechanical strength and high electrical contact resistance at the welded battery connection, which has significant impact on battery pack quality [2-3]. Nevertheless, proper welding parameters can be used to improve the welding quality.

Since cylindrical battery cells are widely used, this research focuses on parametric study of resistance spot welding between cylindrical cell and nickel conductor strip for welding quality improvement. There are three studied parameters being studied; the slot geometries of nickel conductor strip, welding positions and welding position sequence. Finally, best parameters setting is evaluated and presented.

2. Effects of resistance spot welding parameters on heat generation

Heat from electric current flow though connecting point is used to melt materials and form solid joint [4] as shown in Figure 1. The heat occurs in welding process based on Joule's raw [5] as Eq. (1) which requires proper parameters setting to obtain decent heat generation for good quality joint [3].

$$Q = I^2 \cdot R \cdot t \quad (1)$$

Where Q is heat input in generation (J), I is electric current in amperes (A), R is resistance in ohms (Ω)
t is time in seconds(s).

The parametric effect on heat generation can be separated to two categories; direct and indirect effect.

- Direct parametric effect; involving with heat generation directly though Joule effect which basically parameters are welding current and welding time.
- Indirect parametric effect; involving with heat generation though current flow concentration which relate to deformation and contact pressure in specific area.

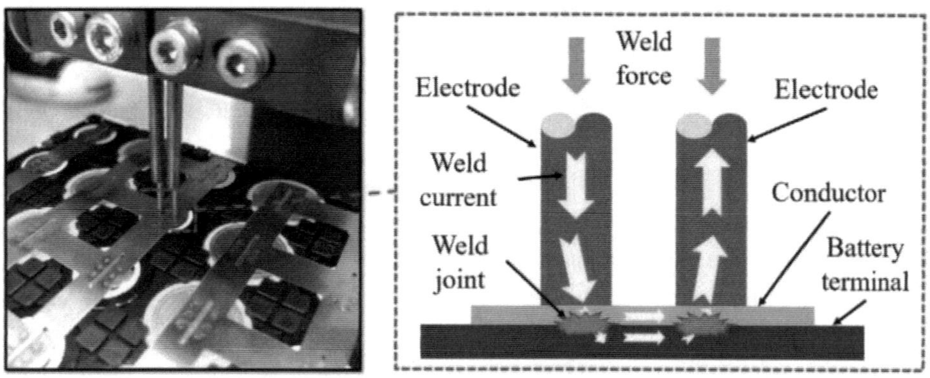

Figure 1. Schematic of resistance spot welding in cylindrical battery pack

3. Methods of parameters study

In this research, an experimental is conducted to define the effects of welding parameters on welded connection quality which are electrical contact resistance and ultimate shear force. Meanwhile, the simulation is used to observe deformation, temperature and current distribution during welding process that are difficult to observe with the experimental. The ratio between flow though and inflow workpiece current in simulation will be used to define the effects of parameters on welding performance.

3.1. Experimental setup

AP5KB AC spot welding machine is used for welding between 0.15 mm of slotted Nickel 200 strip and 0.3 mm of Nickel 200 strip with 1.5 mm of contact diameter electrodes to estimate results before conducting actual welding on real battery cells. The parameters setting on the spot welding machine are 2250 A of maximum welding current, two cycle welding time and 35 N of each electrodes pressing force. A current clamp meter and an oscilloscope are connected to electrode in order to measure current and see the voltage profile, which will be further used by the simulation to observe other behaviours during the welding process.

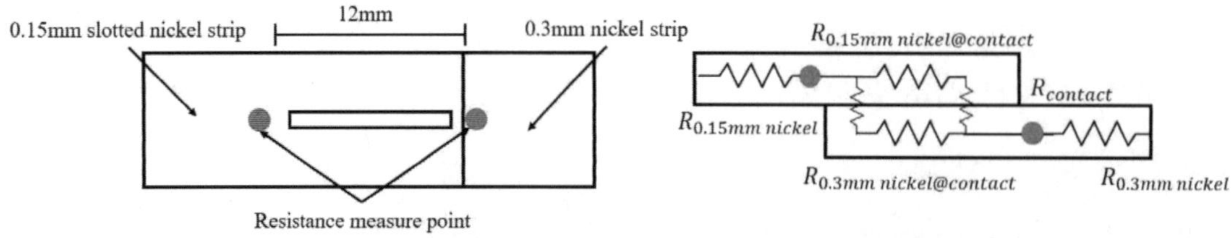

Figure 2. Schematic view and resistance diagram of welded specimen

To determine the contact resistance without taking into account the intrinsic resistance of the specimen materials, the measurement setup as depicted in Figure 2 is conducted with HIOKI BT3554 battery tester. The contact resistance can be calculated using Eq. (2).

$$R_{contact} = R_{measured} - R_{ideal} \tag{2}$$

where $R_{ideal} = \sum_{n=1}^{N}(R_{0.15mm\ @\ contact} \parallel R_{0.3mm\ @\ contact})$

$$= \left(\rho.\left(\frac{l}{A}\right)\right)_{0.15mm\ nickel} \parallel \left(\rho.\left(\frac{l}{A}\right)\right)_{0.3mm\ nickel}$$

$R_{contact}$ is electric contact resistance while $R_{measured}$ is measured resistance from HIOKI BT3554 battery tester and R_{ideal} is materials resistance from calculation where ρ is material resistivity, A is conductor cross section and l is conductor length.

3.2. Simulation setup

To simulate the actual welding process, the corresponding CAD models are imported together with the properties of the materials into the simulation with the following welding process analysis modules, namely *Solid Mechanic Module*, *Electric Current Module*, and *Heat Transfer Module*. The simulation process is divided into two steps in order to reduce computational time [3] as shown in Figure 2. The number of meshes in the given models is increased until the result of simulation converges.

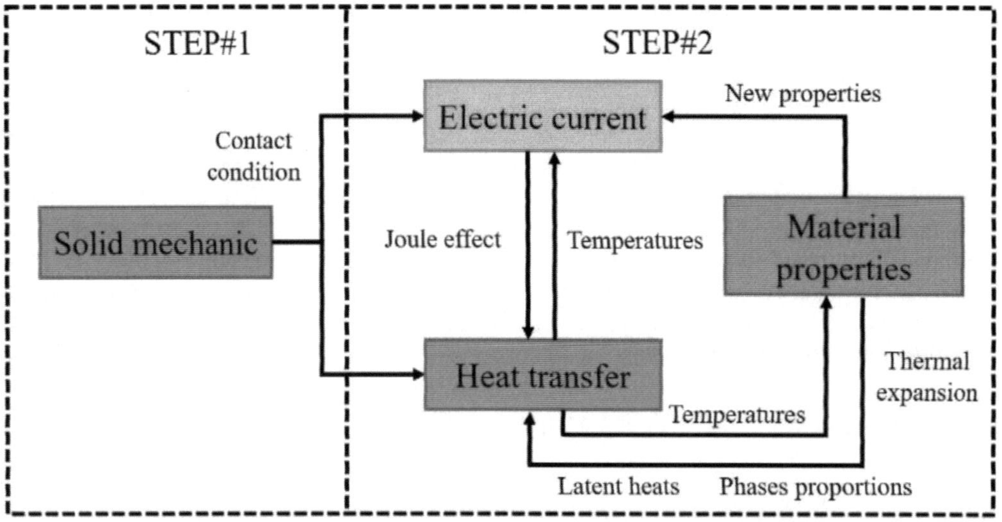

Figure 3. Coupling of resistance spot welding model in simulations

The contact conditions, which are surface roughness, contact pressure, Young's modulus and Poisson's ratio, are required by the electric current and the heat transfer module to calculate contact conductivity. Hence, the solid mechanic module in step#1 is used to obtain the contact condition between the materials and the deformation of the workpiece as a result of the electrode force. After that, the electric current module in step#2 will use the data from step#1 and the voltage profile from the experiment to calculate current distribution, which will be further used by the heat transfer module to calculate heat generation. However, surface roughness cannot be obtained from the solid mechanic module since it is a function of time spent by the actual welding process. Therefore, surface roughness needs to be adjusted until the current profile from the simulation converges to the actual welding process as shown by the example in Figure 4.

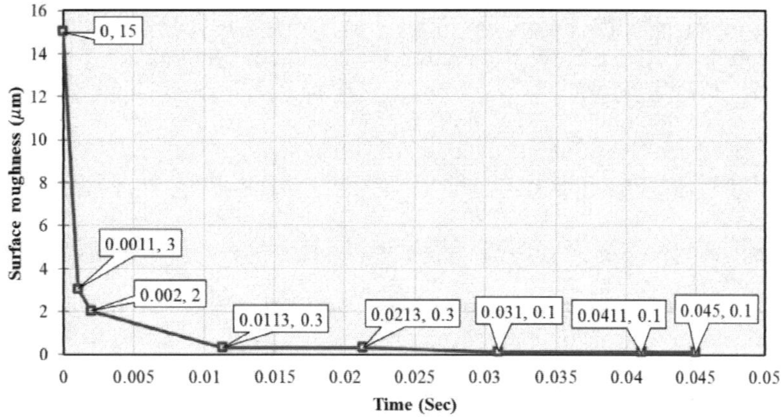

(a) Estimated surface roughness during process

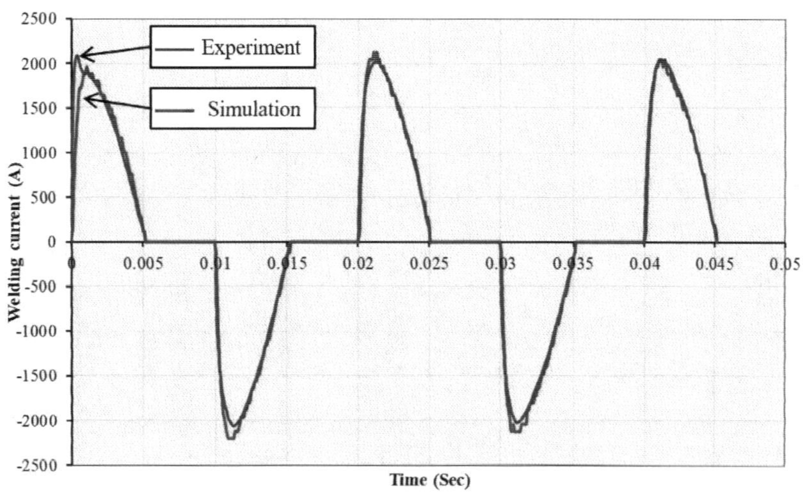

(b) Simulation current profile validation

Figure 4. Relation between surface roughness and current profile during welding in simulation

The contact conditions, which are surface roughness, contact pressure, Young's modulus and Poisson's ratio, are required by Electric Current Module and Heat Transfer Module to calculate electrical and thermal conductivity at the contact, respectively. Hence, Solid Mechanic Module in Step#1 is used to obtain the contact condition between the materials and the deformation of the workpiece as a result of the electrode force. After that, electric current module in Step#2 will use the data from Step#1 and the voltage profile obtained from the experiment to calculate current

IOP Conf. Series: Materials Science and Engineering **501** (2019) 012027 doi:10.1088/1757-899X/501/1/012027

distribution, which will be further used by Heat Transfer Module to calculate heat generation. However, surface roughness cannot be obtained from Solid Mechanic Module since it is a function of time spend by the actual welding process. Therefore, surface roughness needs to be adjusted until the current profile of the simulation is similar to the actual welding process as shown by the example in Figure 4.

4. Results and discussion

4.1. Effect of slot geometries in nickel conductor strip

The slot in a conductor strip can improve welding by restricting the current flow in such a way that it only flows along the path through the connecting strip and the battery terminal [3]. Figure 5 shows that increasing both length and width of the slot can increase current ratio, which means that better heat generation can be achieved, thereby yielding better welding performance and quality. However, the same dimension length of a slot on the current flow between the conductor strip and the battery terminal than the width of the slot. With increasing of slot geometries, the electrical conductivity is reduced because of less conductor strip area.

Because multiple spot welding is required for better connection, the length of slot geometry should be increased for multiple spots on terminal. In Figure 6 shows the amount of current flow is increased by length of slot especially when length of slot is longer the diameter of positive terminal area which is 9.7 mm in this case.

Normally, current flows along the path of least resistance. When the length of the slot in the conductor is greater than that of the battery terminal, most current is forced to flow along the path between the conductor strip and the battery terminal in the inner area, which has less resistance than the outer area. This results in larger current concentration in the welding area. Therefore, the length of the slot should be longer than the diameter of the terminal area to obtain best welding performance with minimum effect on the operation current.

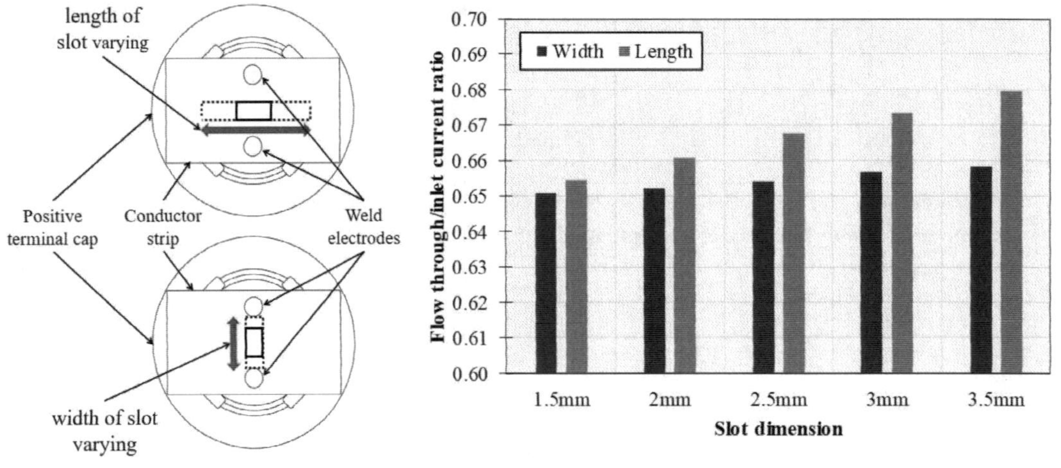

Figure 5. Slot geometric varying and comparing

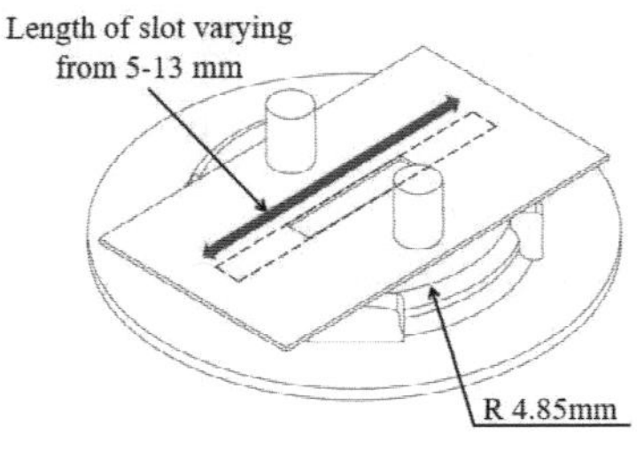

(a) Positive CAD model

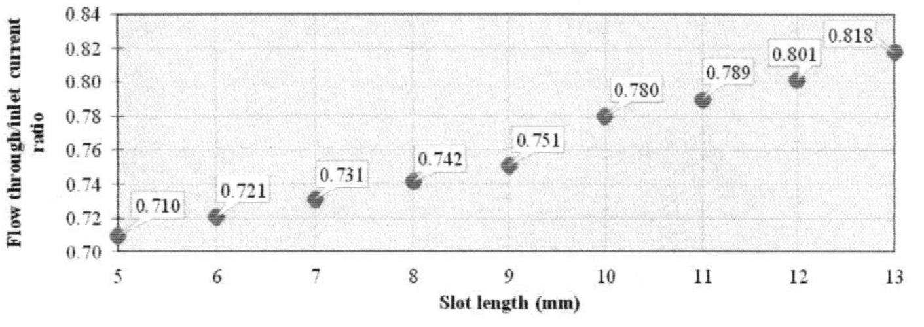

(b) Impact of slot length to welding performance comparing

Figure 6. Effect of slot length on positive terminal welding performance

4.2. Effect of welding position on battery terminal

The positive terminal experiences less deformation than the negative terminal when subject to an equal electrode compressive force thanks to its supporting columns [3]. However, the number of columns on the positive terminal depends on the design choice. Generally, there are two design choices, namely a design with an even number and one with an odd number of supporting columns. In this research, three supporting columns are used as an odd number of supporting columns, whereas four supporting columns are used as an even number of supporting columns. Figure *7* shows various deformation behaviours for different welding positions with different positive terminal designs where the distance between the two electrodes remains the same.

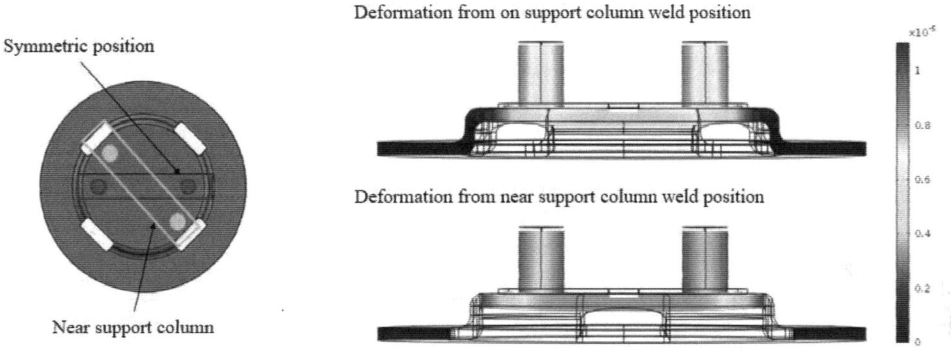

(a) Deformation of positive terminal with even number of support columns

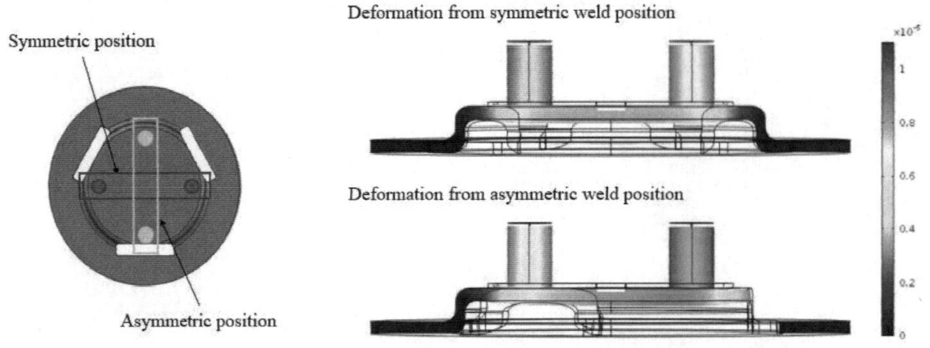

(b) Deformation of positive terminal with odd number of support columns

Figure 7. Deformation in metres from different electrode position on positive terminal

- In Figure 7 (a). shows the electrode welding position on the terminal with an even number of supporting columns. This design provides symmetric support and hence leads to uniform deformation by the electrode force. However, welding in the area near the supporting columns (yellow dots) results in more deformation than that in the area on the supporting columns (red dots).
- In Figure 7 (b). shows the electrode welding position on the terminal with an odd number of supporting columns. With this geometric design, it is possible to cause non-uniform deformation from the electrode force exerted by each electrode since the column provides asymmetrical support between the positions of the electrodes (yellow dots). Though, there is a position that provides symmetrical support located at the border of the supporting columns (red dots).

This means welding on terminal with an even number of supporting columns is easier than welding on terminal with the odd number of supporting columns thanks to better heat generation, which yields less deformation. Meanwhile, symmetric deformation on a terminal with an even number of supporting columns also provides symmetrical wear on the electrode from its symmetrical heat generation.

Moreover, welding on the positive terminal requires more attention than the negative terminal since it is possible to melt insulating ring that separates the positive and the negative side when welding at the edge of the terminal as shown in Figure 8. Therefore, welding at the caution positions can potentially cause an electrical short circuit and hence an explosion.

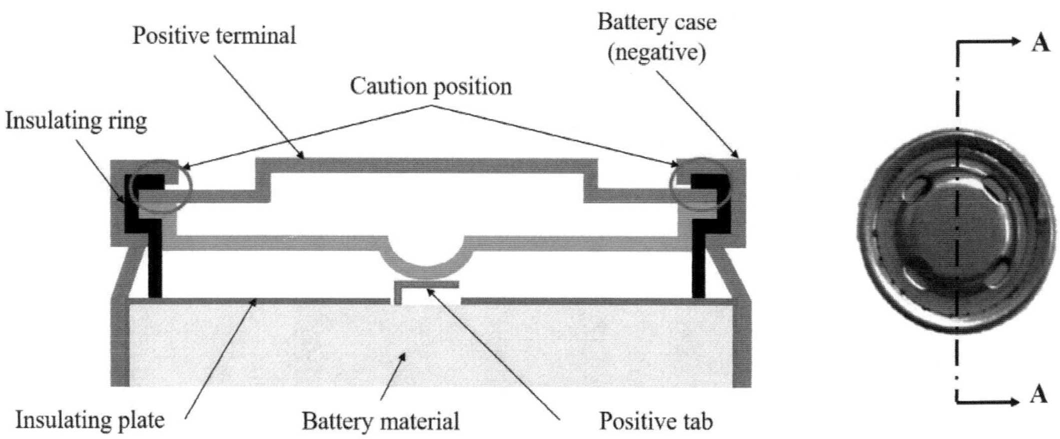

Figure 8. Cross section of cylindrical battery cell in horizontal

4.3. Effect of welding sequence on battery terminal

Based on the relation between the number of weld spots on a battery terminal and the electrical contact resistance, the design with four welding spots is a suitable choice [2]. However, only three welding spots can be performed on the positive terminal due to the limited positive terminal area. With this number of weld spots, there are three possible welding sequences as shown in Figure 9.

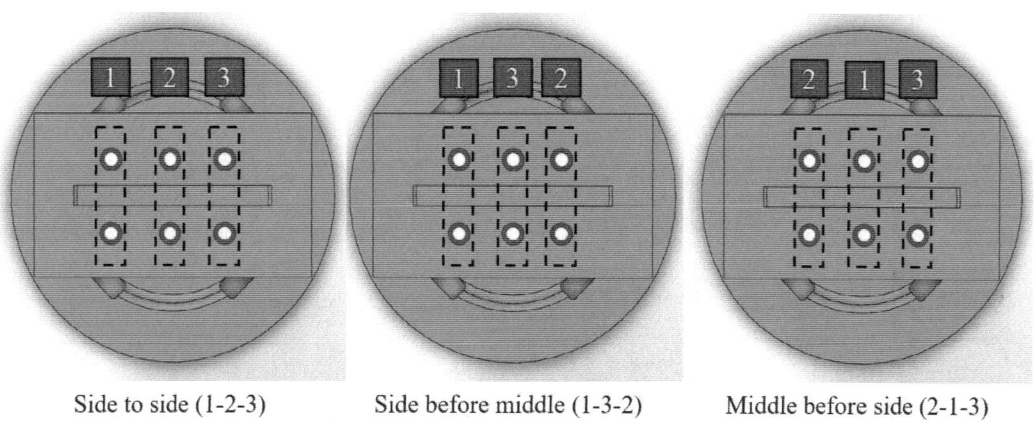

Figure 9. Three patterns for welding sequence on the positive terminal

The multiple spots on the battery terminal create a strong connection between the conductor strip and the battery terminal. During mechanical stress testing, the conductor strip itself is more likely to fail under mechanical stress before the weld joint. Therefore, electrical contact resistance is primarily considered in this case. Figure 10 (a) shows the results of (1) the middle-before-side sequence, which yields lowest electrical contact resistance, followed by (2) the side-before-middle sequence and (3) the side-to-side sequence sequences. The middle before side sequence can deliver the lower resistance from the first weld step; however, the electrical contact resistance reduction from further step is lesser than other sequences.

When looking at the current ratio from the simulation in Figure 10 (b), it shows the previous welding step compromises the welding current and the heat generation of the further step which corresponds to welding quality results in Figure 10 (a). The middle position is the most important position in any welding sequence. It can be compromised from both side weld joints if it is the last welded spot. Meanwhile, it can also compromise the weld current of the other positions if it is the first welded spot. The slot in the conductor also has an impact on this result since it affects the current flow concentration, especially in the middle of the slot area.

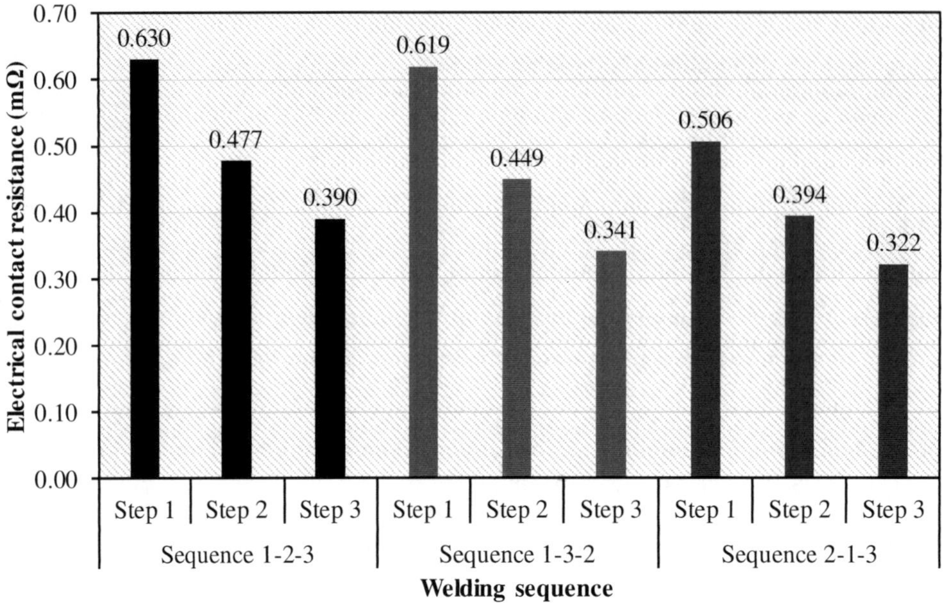

(a) Welding sequence effect on electrical contact resistance in each step from experiment

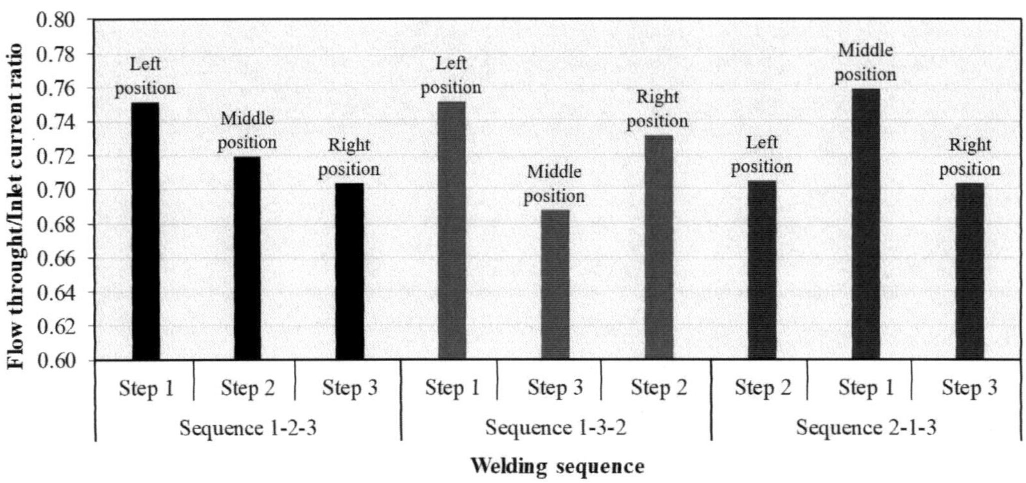

(b) Welding sequence step current ratio comparing from simulation

Figure 10. Effect of welding sequences on contact quality

5. Conclusion

In this research, simulation and experiments are conducted to investigate the effects of certain resistance spot welding parameters, namely welding position, welding position sequence, and geometry of the slot in a nickel conductor strip to improve battery pack quality. The results are compared to define the effects of these parameters. Furthermore, best parameters setting is evaluated and presented below.

- *Effects of slot geometries in a nickel conductor strip*

Increasing both width and length of the slot can improve welding performance. However, it also minimizes battery pack performance due to its smaller conductor area. Thus, the length of the slot should be longer than the diameter of the terminal area for the positive terminal welding while having enough width so that the slot can be visible, to yield optimal welding performance and allow for an easy welding operation with minimum negative effect. As for the slot in the negative terminal conductor, it should be located at the edge since it is the optimal welding position for the negative terminal.

- *Effects of welding position on a battery terminal*
  The welding on the positive the terminal with an even number of supporting columns results in uniform and less deformation than welding on the terminal with an odd number of supporting columns from its symmetric support. This behaviour results in good heat generation, which makes it convenient to weld on the terminal with an even number of supporting columns. Moreover, welding on the supporting columns results in less deformation because of its better support and welding at the edge of the positive terminal can cause an electrical short circuit and even a battery explosion.

- *Effects of welding sequence on a battery terminal*
  In multiple spot welding, the previous welding step compromises the welding current and the heat generation of the further step. To obtain optimal welding quality, the middle position, which is the worst compromised position, must be welded before the other positions to minimize such a compromise. Thus, the middle-before-side sequence should be performed on a battery terminal. The result may be different if the conductor strip contains no slot. Even though the obtained data can be used to improve the welding quality of resistance spot welding, there are still other parameters that can be studied for further research in future.

Even though the obtained data can be used to improve the welding quality of resistance spot welding, there are still other parameters that can be studied for further research in future.

Acknowledgement

This research is a part of researches supported by Thailand automotive institute (TAI) which the authors would like to gratitude including National Science and Technology Development Agency (NSTDA) for providing the necessary facilities. Both researcher and professor from National Metal and Materials Technology Centre (MTEC) and King Mongkut's University of Technology Thonburi (KMUTT) counsel useful knowledge in this work. Lastly, the author admiringly acknowledges for personal financial support from Thailand Graduate Institute of Science and Technology Scholarship (TAIST).

Reference

[1] Lee SS, Kim TH, Hu SJ, Cai WW and Abell JA 2010 *ASME 2010 Int. Manuf. Sci. Eng. Conf.* **1** 541–549.
[2] M. J. Brand, P. A. Schmidt, M. F. Zaeh, and A. Jossen 2015 *J. Energy Storage* **1** 7–14.
[3] M. Masomtob, R. Sukondhasingha, J. Becker, and D. U. Sauer 2017 Engin. J. **21** 457–473.
[4] Amada Miyachi America, "Fundamentals of Small Parts Resistance Welding," pp. 1–8, 2013.
[5] O. Andersson, "Process planning of resistance spot welding," 2013.

Analysis of Printing Pattern and Infiltration Percent over the Tensile Properties of PLA Printed Parts by a Fuse Deposition Modelling Printer

P Doungkom.[1, 2] **and K Jiamjiroch** [1*]

[1] Faculty of Engineering, Thammasat University, 99 Pahon Yo-Thin RD, Klong Nueng, Klong Luang, Pathumthani, 12120, THAILAND

[2] Faculty of Engineering and Technology, Pathuthani University, 140 Tiwanon Rd, Ban Klang, Amphoe Mueang, Pathumthani, 12000, THAILAND

* Corresponding Author: jkrit@engr.tu.ac.th

Abstract. Fuse Deposition Modelling (FDM) 3D printing is nowadays becoming more standard equipment in various fields of study as it can offer both low operating expenses and small investment cost; however, it has been accepted that the printed parts form this kind of printers is not too strong compare with the other Manufacturing technique. This study aims to analyze the effects of both Printing patterns and Infiltration percent on the tensile properties of the specimens obtained by an FDM printer. The specimens have been fabricated with three difference directions that are Horizontal (H), Crosswise (C), and Vertical (V) and within these groups, they are also vary the Infiltration percent at 20% 60% and 100%. After that they had been tested under a Universal tensile testing machine and then these obtained data had been analysed by an Analysis of Variance with 23 level. The result shows that both of Printing Patterns and Infiltration percent can be effect on Tensile strength of the specimens with significant different at 95% confidence level. Beginning with an increase of infiltration percent, it can lead to very strong effect on increasing of the tensile strength. In the case of print patterns, it can be concluded that the maximum stress is C, H and V, respectively. Because of the C pattern printing, tensile testing is a direction that is aligned with the printing of the layer. (For example, like pulling several ropes together)

1. Introduction

3D Printing, is one of additive manufacturing processes, being a prevalent production technology amongst modern-day designers as this technology has continuously progressed over the last ten years. Some of the advantages of 3D printing include the ability to mold complex work pieces that other technologies cannot do and it can offer a printed by only a single process [1]. As this technology is readily available for everyone, no skill workers can make a complex casting just one click. Even if there are many advantages of this technology, one of the main disadvantages of technologies is that it takes a long time to complete the printing parts. Fuse Deposition Modelling (FDM) is becoming famous as it offers both low investment cost and operating cost but its printed parts are weaker in comparison to more common plastic production processes, such as injection, blowing, extrusion, molding and CNC etc. [3] The heart of an FDM is an extruder, composted of feeding gears, heater and extruder. (see Figure 1) First, the feeding gears are going to convey a plastic wire into a nozzle. After that heater, attached closed to the end of extruder, has been energies. The plastic wire will be melt and then be extrude onto a platform layer by layer. The two sizes with the difference diameters of plastic wire, available on the market are 1.75 and 3.00 mm and the nozzle sizes may range from 0.20, 0.40, up to 1.00 mm. Plastic wires would be made from PLA and ABS, however, they could also Nylon, PETG, ASA, PC, ESD and etc.

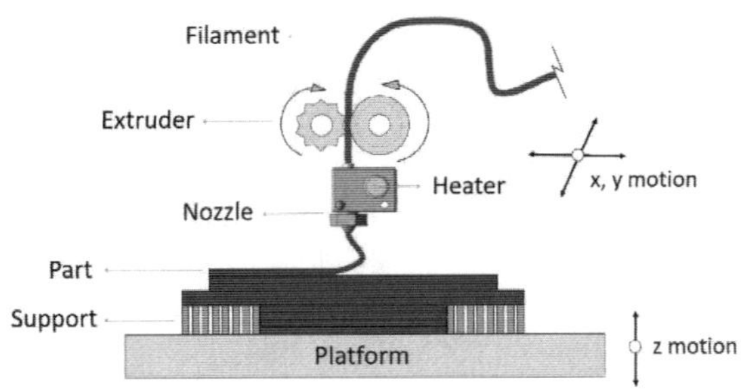

Figure 1 Fused Deposition model, FDM

Ashu Garg and Anirban Bhattacharya [4] investigated the failure behavior of the specimens resulting from tensile stress at the different layer thicknesses by the FDM 3D Printer. This research had been studied the layer thicknesses of the specimens to be 0.178 mm, 0.254 mm and 0.330 mm and they were performed a tensile strength by a Zwick/Roell Z050 tensile test machine. All of the specimens made form an ABS. The results showed that, the maximum elongation was approximately 4.6% on the specimens with layer thickness of 0.178 mm while the specimen of 0.330 mm offered the highest tensile strength at 34.5 MPa. John Ryan C. Dizon and Alejandro H. Espera, Jr. [5] studied the mechanical properties of polymers obtained from 3D printers from SLA, SLS [2], FDM and Polyjet.The types of polymers studied by the researcher are varied, such as ABS, Nylon, PPSF and PEI. The conditions of the specimen include the printing pattern and structural types of infill. The types of infill are 100 percent infill, honeycombs, drills and stripes. Tensile Testing by ASTM D638-14 for Plastics and ASTM D412-16 for Rubber and Thermoplastic Elastomers were tested. The results showed that the specimen with the honeycomb structure offered the highest tensile strength at break-point while crosswise has the highest tensile strength. Ravinder Sharma and Rupinder Singh [6] investigated the mechanical properties of PVC from FDM by parameter thickness, infill percentage and infill speed. The results showed that the highest yield stress was 3.47 MPa at the conditions of infill 100%; Layer thickness of 0.25 mm; and Deposition speed of 37 mm / min.

The work pieces, obtained from FDM, is well known that they are weakest compared with other traditional processes. Because 3D printing is not melting homogeneous, the printed layers may easier fracture. However, there are no any research has been purposed a mathematical modelling between a mechanical properties and both of percent of infill with difference printing pattern. This research focuses on the technique using a FDM 3D printing technique to investigate the mechanical properties of PLA filament materials. According to ASTM D638, the mechanical properties of specimens had been obtained by a universal tensile testing machine and then they were statistically analysed the by Analysis of Variance (Anova). The useful information from this study could allow any FDM users to understand the mechanical properties for designing any printed parts form this kind of 3D-printer.

2. Materials and Method

2.1 Plastic filament

The plastic filament used in this study is a Poly-Lactic acid; PLA, Brand AGRU. This filament is a thermoplastic made from natural materials with a low melting point, and is easy to degrade in the atmospheric condition. This PLA's properties have been listed in a Table 1.

Table 1. Mechanical properties of PLA Filament [7]

Parameter	Value	Unit
Diameter	1.75	mm
Diameter Tolerance	±0.05	mm
Melt Temperature	190-230	°C
Density	1.24	g/cm^3
Tensile Strength	60	N/mm^2
Color	Dark	

The 3D printer (FDM) used in the experiment developed in house with 300 × 500 × 300 mm printed part. This printer has a 0.4 mm single nozzle and Repetier-Host V 2.0.1 software is a control software in this experiment.

2.2 Specimen fabrication

In this experiment, the standard specimens were printed as referring to the American Society for Testing and Materials D638 (See Figure 2 Left). For all of the spacimens were printed as hexagonal honeycomb interior because it is very popular among designers. Moreover, this pattern has high stresses, good pressures and tensile strength compared to other types of infill pattern. [9] The print speed has set at 80 mm/s with 0.18 mm layer thickness and double shell lines. The infiltration percent were 20%, 60% and 100%. The printing patterns of the specimen follow these forms: Horizontal; H, Crosswise; C and Vertical; V (See Figure 3 Left and Right). In each of these forms, the specimens were fabricated on a different printer bases. Figure 3-right shows the front view of the specimen while it was printing. All of specimens were tested by a universal testing machine (INSTRON Model 5969) [8] with a 50 kN load cell. According to ASTM D638, the standard testing for mechanical properties was performed with pulling speed 5 mm/min at the 25 °C.

IOP Conf. Series: Materials Science and Engineering **501** (2019) 012028 doi:10.1088/1757-899X/501/1/012028

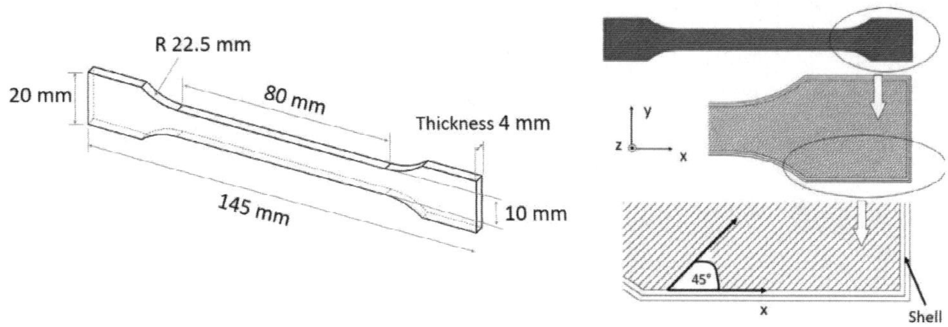

Figure 2 Left: Specimen Size (ASTM D638).
Right: Printing lines on the top and bottom shell (Horizontal; H)

In this study, there were nine conditions that are H20%, H 60%, H 100%, C 20%, C 60%, C 100%, V 20%, V 60%, and V 100%. Each condition was repeated by five replications. The Yield stress (MPa), percent elongation at the yield point (% Elongation at Yield), and elongation at break (% Elongation at Break) were their responses. In addition, printing time (Modeling Time) of each specimen might be recorded for father investigation.

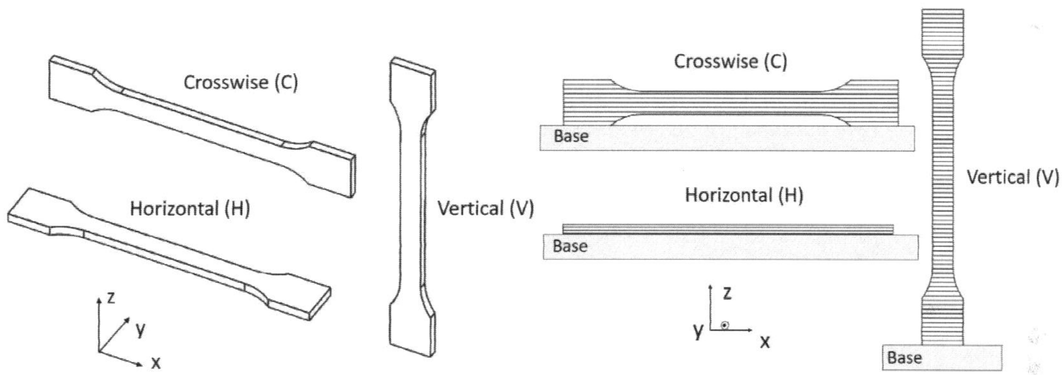

Figure 3 Left*:* Three types of printing pattern H, C, V. Right: Show each layer of printing pattern

3. Results and discussion

Table 2 shows the mechanical properties for the specimens. The maximum yield stress at 48.53 MPa printing condition at C100%s and minimum yield stress at 13.88 MPa printing conditions of C20%. At the same infill percentage, crosswise pattern (C) would offer the highest yield stress because the direction of force is in the same direction of printing direction. The highest and lowest percentages of Elongation at Yield were 2.90% and 1.34% at the C 100% and V 20% respectively. The highest and lowest recordings of Young's Modulus were 2398.57 and 1122.86 at the C 100% and H 20% respectively. Figure 4 shows the relationship between strain and stress; it can be seen that most fracture patterns are of a brittle nature. This type of fracture, when passed through a yield point, will have little or no stretching. However, the condition of the horizontal (H) printing pattern is the ductile fracture pattern. This fracture. This is very flexible or Elongation when pulled. The horizontal (H) printing patterns are shown in Figure 2 Right. Because of this pattern, the surface pattern is an alternate type of print path with an angle of reference (x) of 45°.

Table 2. Observed values of mechanical testing.

Exp. N	Printing Patterns	Infill (%)	Printing time (Minute)	Yield stress (MPa)	Elongation at Yield (%)	Young's Modulus (MPa)
	* Filament		-	45.89	3.47	2324.29
1	H	20	21	20.63	2.52	1122.86
2	H	60	37	24.06	2.58	1368.57
3	H	100	39	41.66	2.78	2171.43
4	C	20	45 (*Support)	27.77	2.59	1580.00
5	C	60	59 (*Support)	34.11	2.58	1981.43
6	C	100	61 (*Support)	48.53	2.90	2398.57
7	V	20	31	13.87	1.34	1441.43
8	V	60	42	19.49	1.52	1852.86
9	V	100	44	24.20	1.73	2020.00

* Filament is PLA Filament Cross-section circle area and Diameter 1.75 mm. (Pulling speed is 5 m/min)
*Support structure is used to support steep overhangs and Float of model. (Show in Fig 1).

The print time varies depending on the conditions as shown in Table 2. A low percentage of infill will take less time while the higher can lead to increase print time because the printer head needs longer time to travel with the higher density. The Crosswise (C) pattern took a longer printing time because it was necessary to print support parts, help in fabrication of the work pieces as same as a scuffle. And the less time is horizontal (H) pattern because it was no need to print any support.

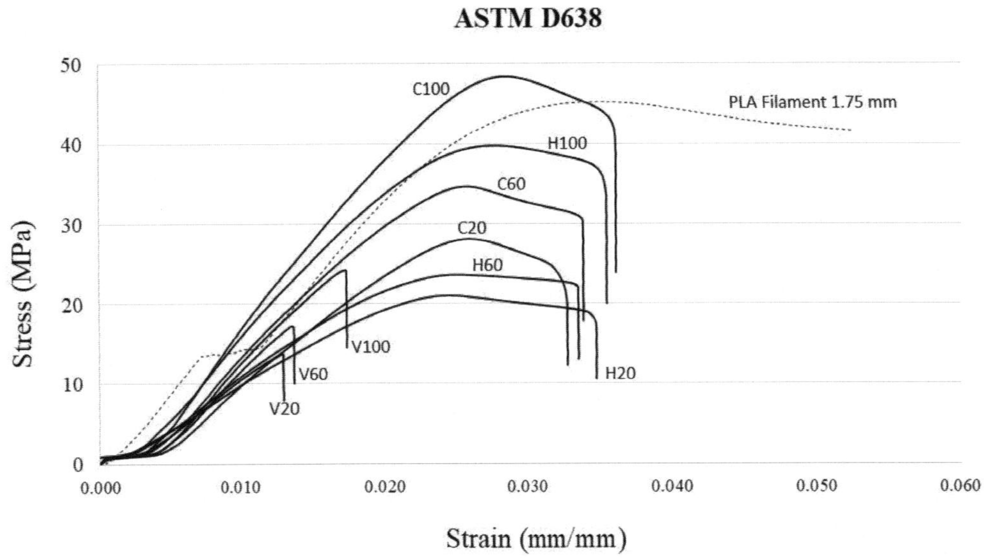

Figure 4 Relationship of Stress and Strain

Each printing pattern offered the different types of fracture. As shown in Figure 5, it was observed that fragment of the vertical (V) was perpendicular with the direction of force as same as the fragment happen on the non-elastic materials or brittle material. This can offer the less mechanical properties base on relatively low interfacial strength; therefore, it should avoid for loading a printed part in this direction. Facture pattern of Crosswise (C) and Horizontal (H) can lead to the same facture as happening in any elastic material. The most effective tensile stress was Crosswise (C), resulting in the fracture occurred unevenly because these specimens have the same load direction for each single filament.

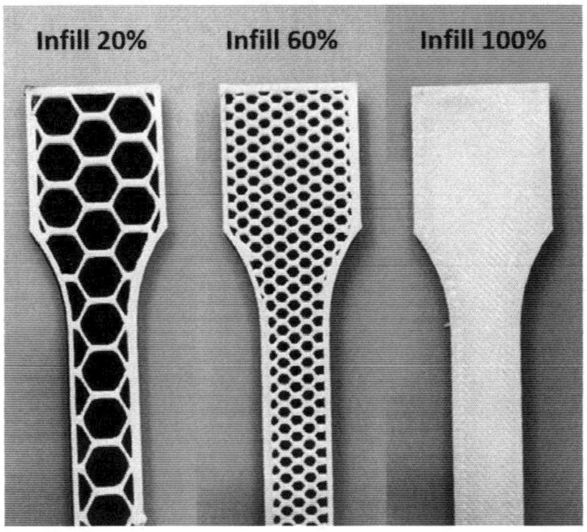

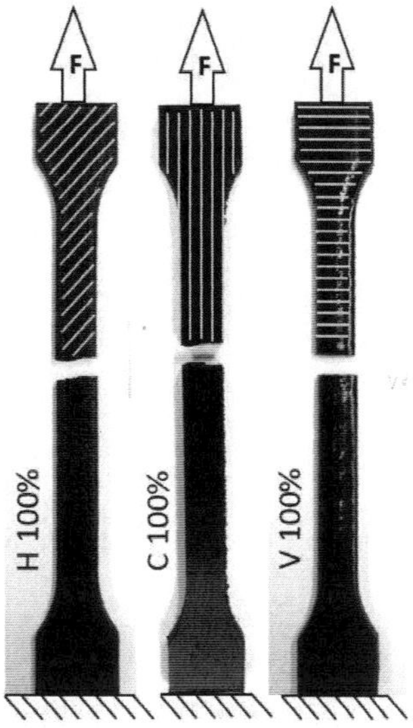

Figure 5 Left: Specimen of horizontal print; H at Infill% of a) 20% b) 60% and c) 100%
Right: Specimen of Infill 100% conditions after tensile test

3.1 Yield Stress

Table 3 shown an Analysis of Variance on Yield Stress (MPa). It can see that all of these factors including their relations are statically significant on Yield Stress (MPa). It can see that figure 6 shows the effect of both printing patterns at the differences of infill over Yield Stress. All of the samples offers lower Yield Stress than single Filament test as these can be evidence that all of the filaments do not perfectly melt. This can result in low Yield Stress compared to conventional melting processes.

Table 3 Analysis of Variance for Yield Stress (MPa)

Factor	DF	Adj SS	Adj MS	F-Value	P-Value
Patterns	2	3268.17	1634.08	511.28	0.00
% Infill	2	3345.69	1672.85	523.41	0.00
Patterns * % Infill	4	394.64	98.66	30.87	2.2×10^{-13}
Error	54	172.59	3.20		
Total	62	7181.09			

The highest Yield Stress at the same of infill was C, H and V, and the highest stress values at the same printing pattern were 100%, 60% and 20%, respectively. At 100 percent of infill, it offered a maximum Yield Stress but its average was lower than Yield Stress of single fiber.

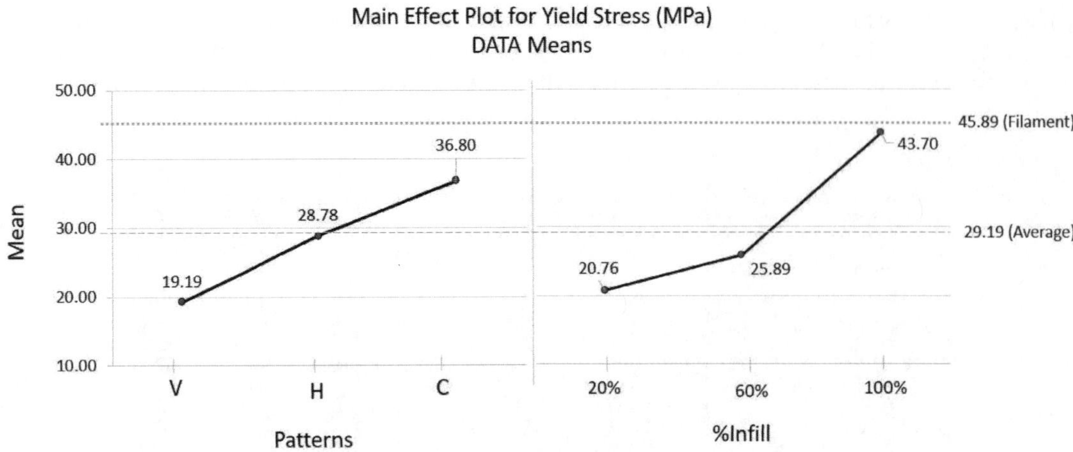

Figure 6 Data Mean for Yield Stress (MPa)

3.2 Young Modulus; E

Table 4 shows an Analysis of Variance on Young Modulus; E (MPa). Both factors had a significant effect on Young Modulus; E (MPa). Figure 7 shows the Means Young Modulus of each condition. The maximum Young Modulus was C, V and H, respectively, these were similar affect in a composite material at the low Share Modulus. The Maximum Young Modulus values are unsurprisingly 100%, 60% and 20%, respectively.

Table 4 Analysis of Variance for Young Modulus; E (MPa)

Factor	DF	Adj SS	Adj MS	F-Value	P-Value
Patterns	2	1963022.22	981511.11	103.62	0.00
% Infill	2	7020422.22	3510211.11	370.57	0.00
Patterns * % Infill	4	776702.59	194175.40	20.50	2.6×10^{-10}
Error	54	511514.29	9472.49		
Total	62	10271660.32			

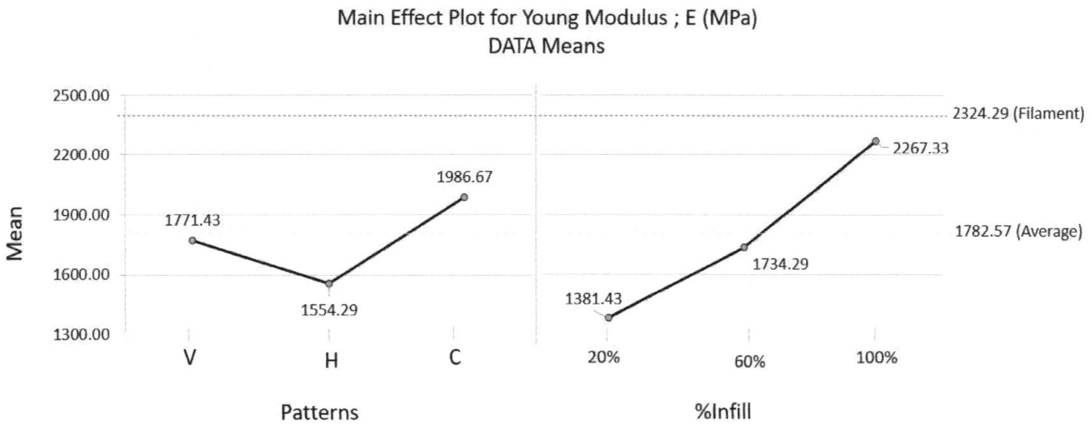

Figure 7 Data Mean for Young Modulus; E (MPa)

Figure 8 shows the effect of Yield Stress over percent of infiltration at the different printing patterns. It can see that all of test results can be model base on logarithmical model. Surprisingly, the maximum of Yield stress of Crosswise (C) was above the average of single filament test by 1 percent. However, this is not a statically significant at 95 percent of confidence level. Based on these results, it can lead to a drop of Yield stress by 20 percent of their maximum on both of Crosswise (C) and Horizontal (H) specimens and 10 percent for vertical (V) specimen.

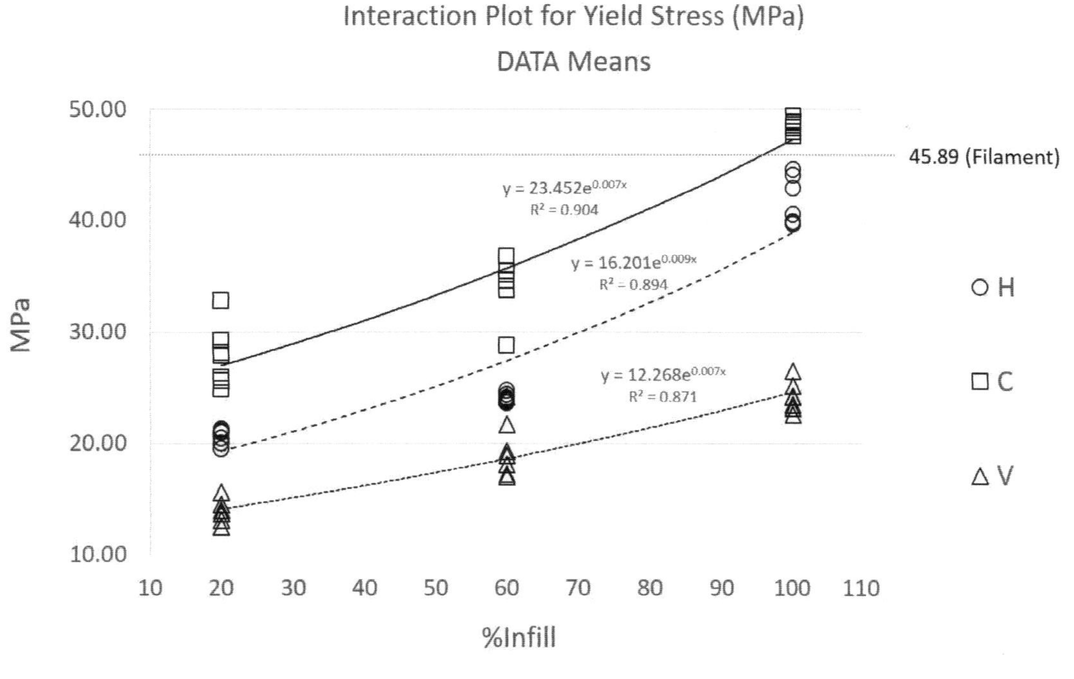

Figure 8 Interaction Plot for Yield Stress (MPa)

4. Conclusions

The results from this experiment show the mechanical properties of PLA specimens after performing a tensile testing. The experiments were clearly showed that at the C100 printing conditions, the maximum Yield stress is at 48.53 MPa and the V20 printing condition had the lowest Yield stress at 13.87 MPa. At the C100 printing conditions, the maximum Young's Modulus value was recorded at 2398.57 MPa and the H20 printing condition had the lowest Young's Modulus value at 1123.86 MPa. From this, it can be concluded that the type of printing pattern is one of the most parameter in FDM 3d printing, the highest yield stress would be happening in Crosswise (C), Horizontal (H) and Vertical (V), respectively. In addition, it could be separated the print pattern into two groups that are Crosswise (C) with Horizontal (H) and Vertical (V). In case of Crosswise (C) with Horizontal (H) at infill between 20% and 100% may offer increasing in Yield strength by 39 percent while Vertical (V) offers less yield strength as this pattern have a fracture like brittle material. The Infiltration percentages, that produced the highest yield stress is 100%, 60%, and 20%, respectively.

Acknowledgement

Funding from "Faculty engineering, Thammasat University" and also "a Research Funding Unit of Thammasat University" are gratefully acknowledged.

References

[1] Doungkom. P., Jiamjiroch K. ,Analysis of temperature distribution for a plastic injection nozzles in Fused deposition model; FDM, The 2017 *Technology Innovation Management and Engineering Science international conference* (TIMES-iCON2017), 21 November 2017, Mahidol University, Thailand, P.245-250.

[2] Sooksawat S., Punnawat P, Development of close loop control systems (PID) with SLA printer, Thesis of Mechanical Engineering Thammasat University 2016.

[3] Doungkom. P, Designs and Developments of powder Based 3D Printer, Master's thesis of Mechanical Engineering Thammasat University 2015.

[4] Ashu .G and Anirban B 2017 An insight to the failure of FDM parts under tensile loading: finite element analysis and experimental study, *Int. J. Mechanical Sci.* **120** 225–236.

[5] John Ryan C. Dizon and Alejandro H 2018 *Additive Manufacturing* **20** 44–67

[6] Ravinder S and Rupinder S 2018 Composites Part B **132** 237-243

[7] AGRU Kunststofftechnik GmbH, Semi-Finished Products. The Plastics Experts, Austria (2018). URL:https://www.agru.at/de/anwendungen/halbzeuge/agru-pla-filamente-fuer-3d-druck/,on 01/04/2018.

[8] Tensile Testing Machine Manual, 5960 Series Dual Column Table Frames, Operator's guide M10-16247-EN Revision E.

[9] Matter Unlocking Material Properties (3D Matter), URL:http://my3dmatter.com/influence-infill-layer-height-pattern/ ,on 03/04/2018.

IOP Conf. Series: Materials Science and Engineering **501** (2019) 012029 doi:10.1088/1757-899X/501/1/012029

Study on mat-surface fabrication utilizing fixed-abrasive tool (Application to brittle material)

K Yanagihara[1] and R Kita[1*]

[1] National Institute of Technology, Ariake College, 836855 Japan

* Corresponding Author: kiyoshi@ariake-nct.ac.jp

Abstract Mat-surface is a surface with microscopic irregularities whose dimensions are close to the wavelengths of visible light (400-700 nanometers).
The new method that utilizes fixed-abrasive tool to fabricate mat-surface without pre-masking and large scale back up facility, is proposed in our last report. The report clarifies that, mat-surface must be produced with fixed-abrasive process, If the relative motion between tool and work in grinding process can be realized as that in blasting,
In the reports, however, materials of workpieces which are ductile, such as steel and aluminium alloy, were selected. These materials are relatively easy to be indented. Therefore, in order to develop a new application of the proposed method, the method is applied to brittle material such as glass, it is investigated whether mat-surface can be obtained.

1. Introduction

In surface finishing process, surface without any reflex of light is often required. The surface is called "Mat surface" [1]. It is defined as a surface with microscopic irregularities whose dimensions are close to the wavelengths of visible light (400-700 nanometers).

In common industrial techniques, mat-surface is produced by using chemical etching, electrical or laser abrasion, loose abrasive lapping, or shot blasting. These conventional processes, however, demand pre-masking of workpiece and relatively large scale back up facility.

Therefore, in order to fabricate mat-surface without pre-masking and large scale back up facility, New method which is utilizing fixed-abrasive was proposed in our previous study [2-8].

Figure 1 shows the developed tool. It has two small diamond wheels. The wheels are fixed to the tool body thorough axles. Figure 2 illustrates how to use the tool and how the tool generates mat surface. In use of the developed tool, it is not pushed onto the work surface with high pressure like knurling but just touched on the surface. Although it seems like knurling process, our studies was clarified so far that pressure of the tool in the process is lighter than that in knurling [2].

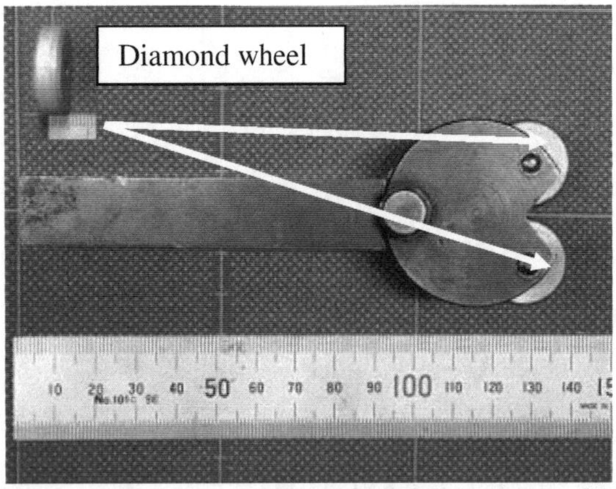

Figure 1. Driven-type mat-surface-fabrication tool [2].

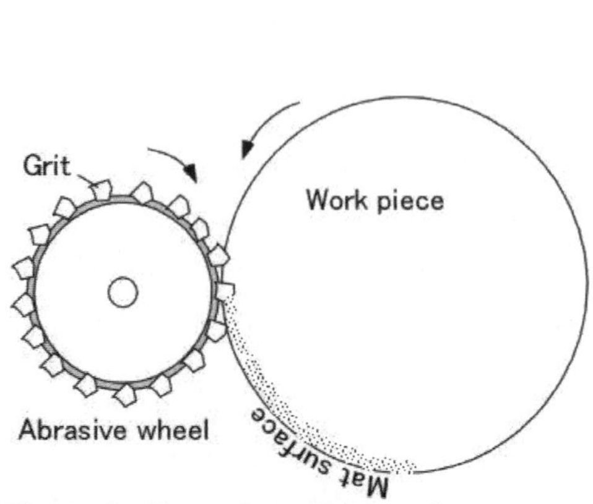

Figure 2. Illustration of Mat-surface generation with developed tool

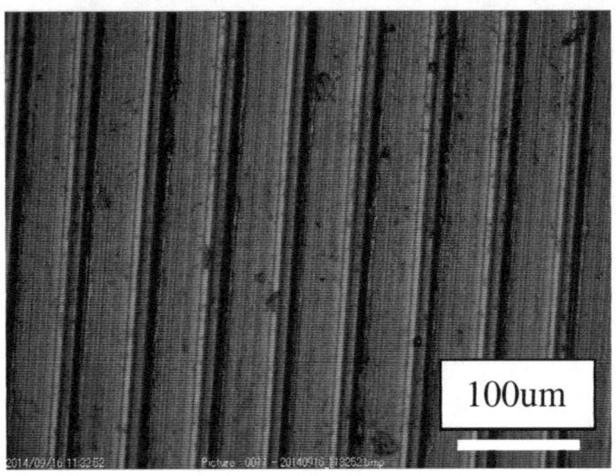

(a) Before

(b) After

Figure 3. Surface observation of workpiece with microscope [2].

Figure 3 (a), (b) shows the photographs of the surface of aluminum workpiece before, after the work processed with mat-surface-fabrication tool, respectively. In figure 3 (a): the work surface before experiment, some regular marks which are generated by cutter of lath are observed. On the other hand, in Figure 3 (b): after processed with the developed tool, the cutter marks are disappeared, and mat-surface is generated. Also, the topography of the obtained mat-surface is measured then. Random dimples are clearly observed over processed surface. These results proved that the idea to produce mat-surface by applying fixed-abrasive is right, and that the developed tool feasible [2].

On the other hand, however, ductile materials such as steel and aluminum alloy, were used in the reports. These materials are relatively easy to be indented by means of plastic deformation. Meanwhile the developed tool has never been applied to brittle material yet. Thus, the purpose of this study is to investigate whether the developed tool can produce mat-surface on surface of brittle material such as glass.

2. Experiment

2.1. Difference in mat-surface fabrication about glass and metal

Soda lime glass is selected for material of workpiece in this experiment. Table 1 is comparison of mechanical property between soda lime glass and aluminium alloy #2017 which was used in our former reports. Aluminium alloy which has rich ductility is easy to be indented. After processing it with mat-surface-fabrication tool, many random dimples that the most range of the peak and the volley

IOP Conf. Series: Materials Science and Engineering **501** (2019) 012029 doi:10.1088/1757-899X/501/1/012029

is from 10 to 15 um, are produced on the surface, then the surface demonstrates mat. Meanwhile, hardness of glass is about five times harder than that of aluminium although Young's modulus is almost the same, as shown in table 1. Surface of the grass is harder to be indented than that of aluminium. Cracks and Tips on glass surface are easier to be generated than that on aluminium surface because of the brittleness of glass. Therefore, the important key to obtain mat-surface on glass surface is to generate small tips without any generation of large cracks. This indicates that tool should be stroked on workpiece repeatedly as many as possible while relatively light weight is loading to tool.

Table 1. Mechanical properties of workpiece.

	Soda lime glass	Aluminum alloy 2017
Young's modulus (MPa)	7.16×10^4	6.9×10^4
Vickers hardness (N/mm^2)	5374	1029.4

2.2. Experimental set-up

Plate is selected as a workpiece in processing experiment of glass, although shaft is used in experiment of metal processing. Because plate shape is general on market and the surface accuracy and roughness is constant in quality. Because of the shape of workpiece, experiment employs numerical controlled milling machine: Yamazaki YZ-400MD.

Figure 4 shows the experimental set-ups. The tool of Figure 1 is installed on z-axis (vertical) of milling machine, and a glass plate is fixed with jig on vice. The tool is stroked x-axis direction (right and left in the figure) as feed. Also, the tool is free to move along z-axis direction, as shown in figure 5. Processing pressure can be changed by means of replacing weight on the tool from 6 to 12 N, as shown in the figure.

Experimental conditions are listed in Table 2. Three kinds of processing parameter, weight, feed rate, number of strokes are selected and varied to be investigated.

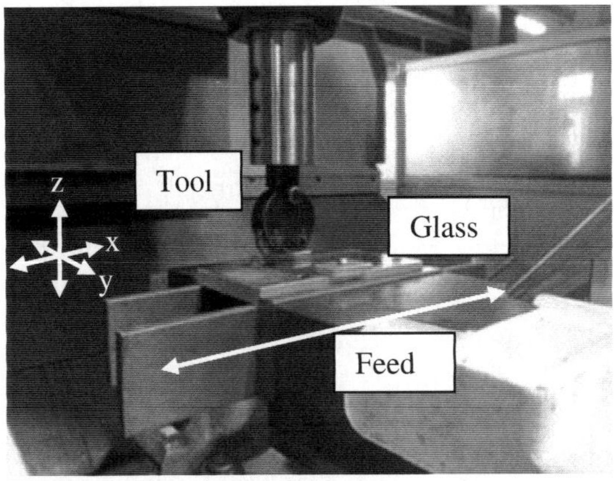

Figure 4. Experimental set-up.

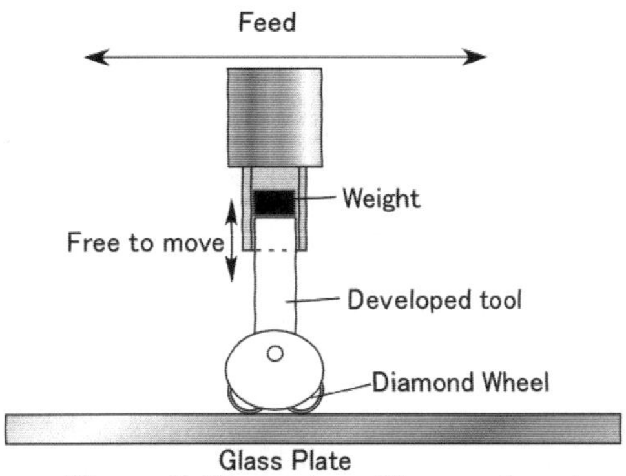

Figure 5. Illustration of how tool works.

Table 2. Specifications of diamond wheel and experimental conditions.

Tool specifications	Abrasive type	Diamond
	Grain size	(44um) #440
	Wheel diameter	20
Work material	Soda lime glass	
Processing parameter	Weight [N]	6, 9, 12
	Feed rate [m/min]	1, 2, 3, 4, 6
	Number of Strokes	1, 5, 10, 15, 20, 50

3. Results and discussion

3.1. Surface observation

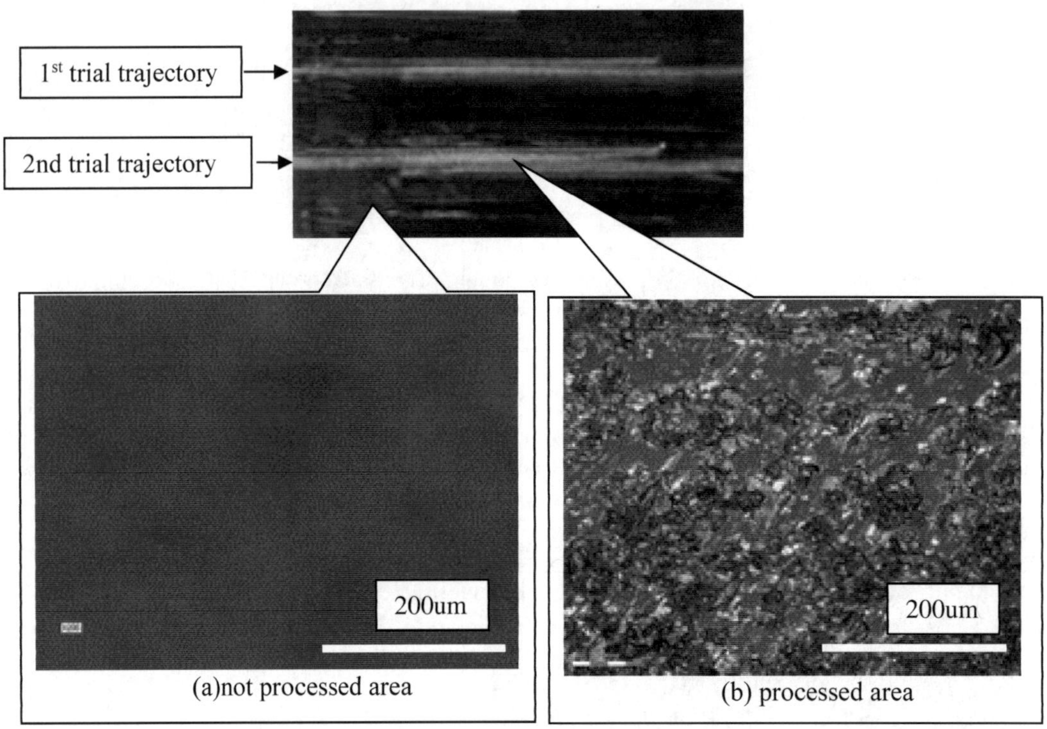

Figure 6. Obtained surface.

Figure 6 is the obtained surface after milling machine stroked a glass plate with the developed tool. To distinguish process or not process area, backside of glass plate is painted with black painting. Two lines that is shown as white line in the figure 6 can be observed. Magnified photo (a), (b) is the area of not-processed, processed area, respectively. In photo of (b), the surface seems to be covered with random dimples which is generated by means of not indentation but chipping.

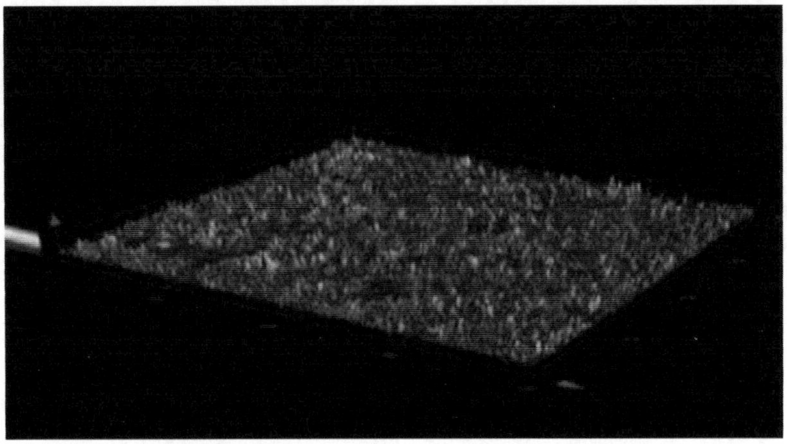

Figure 7. Topography of obtained surface.

The topography of the obtained mat-surface is also measured with Zygo New View 7100, as shown in Figure 7. Random surface with no direction that is produced by chipping are clearly observed over the processed surface. This proves that the fixed abrasive tool can produce mat-surface not only ductile material such as metal, but also brittle material such as glass.

3.2 investigation into optimum number of strokes of tool move

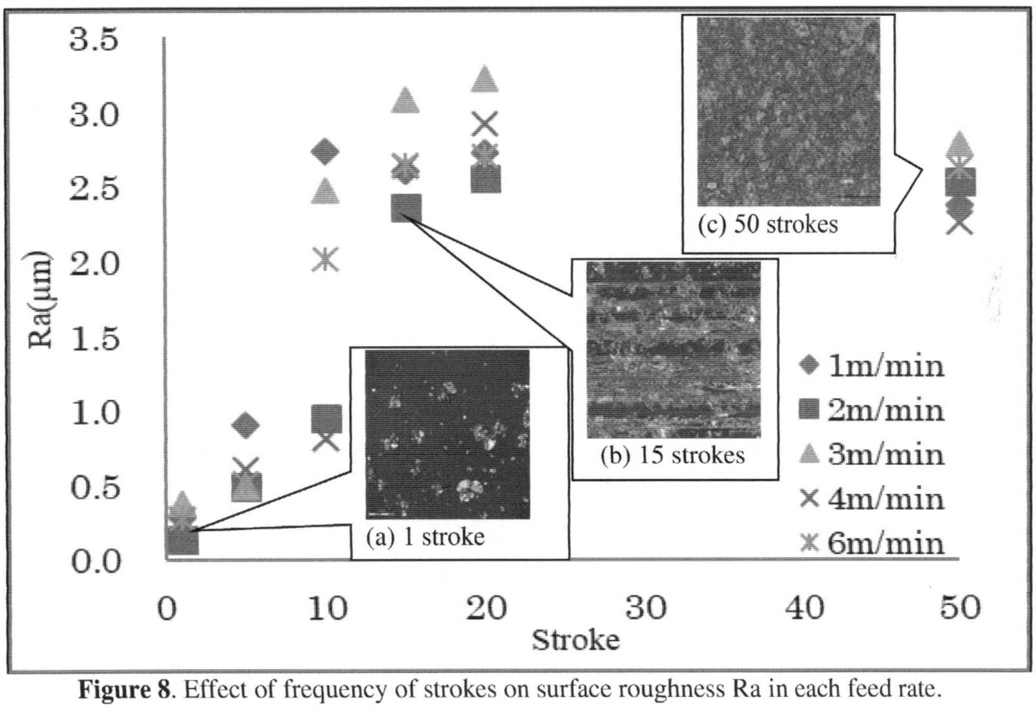

Figure 8. Effect of frequency of strokes on surface roughness Ra in each feed rate. Weight loaded on tool :12N

Figure 8 shows how frequency of strokes effects on surface roughness Ra of workpiece under the condition of weight:12N and different five feed-rates. In this figure, photographs of typical obtained surface in each stroke condition is also mapped. In the area from 1 to 5 strokes, chipping area is not generated so much on surface in figure 9 (a), thus Ra is small. In the area from 10 to 20 strokes, chipping area is not still sufficiently yet to cover the glass surface evenly, and some lines such as scratches is also seen in photo (b). Thus, distribution of Ra is broad. This concludes that the difference of stability in each process is large. Meanwhile, in area of 50 strokes, photo (c) demonstrates mat-surface, distribution of Ra is narrow. The result shows effect of number of strokes is heavier than that of feed rate, at least over 20 strokes must be desired to obtain stable mat-surface.

3.3 investigation into optimum weight of on tool

Figure 9 shows that effect of weight on tool on surface roughness Ra after tool is stroked 20 times on a specimen. Ra is in proportion to weight in each feed rate condition, since contact pressure of grit must be in proportion to weight and larger chipping must be also generated. Therefore, it results that heavier weight on tool as possible is desirable within range not to brake glass.

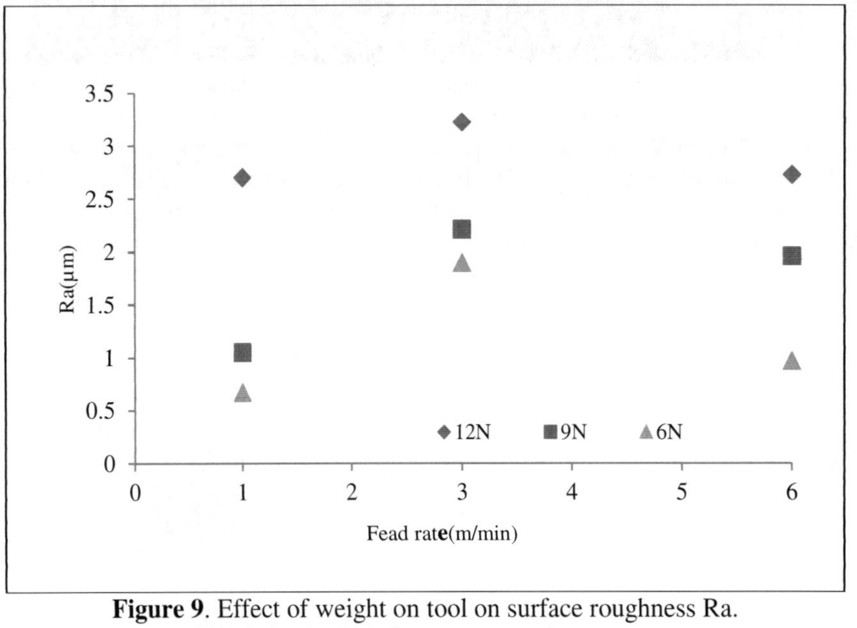

Figure 9. Effect of weight on tool on surface roughness Ra.
Tool is stroked 20 times on a specimen

4. Conclusion

In order to investigate whether the proposed mat-surface-fabrication tool can produce mat-surface on surface of brittle material, experiment that uses glass as workpiece is executed in this study. In fabrication experiment of mat-surface, successful result is obtained. The important keys to use the mat-surface-fabrication tool are obtained. The obtained results are following:

(1) our developed mat-surface-fabrication method and tool are effective to fabricate mat-surface of not only ductile material but also brittle material, since mat-surface on glass is obtained

(2) fabricated mat-surface on glass is consist of chipping. The mechanism is different from the mechanism of mat-surface fabrication on metal that is consist of indentation.

(3) effect of frequency of strokes of tool is larger than that of feed rate. At least over 20 times strokes must be desired to obtain stable mat-surface.

(4) it is desired that the weight loaded on tool is as heavy as possible within such the range that the tool cannot brake glass.

Acknowledgments

IOP Conf. Series: Materials Science and Engineering **501** (2019) 012029 doi:10.1088/1757-899X/501/1/012029

Technical Supports of the Technical Support Center for Education and Research of N.I.T., Ariake college is gratefully acknowledged.
A part of this work was supported by research grant of AMADA FOUNDATION.

References

[1] Matsuo K 1967 *Industrial Art News* **34 pp** 51-53 (in Japanese)
[2] Yanagihara K and Kita R 2017 *Proc. TSME ICOME-2017* p 0295
[3] Kita R and Yanagihara K 2014 *Proc. Std. M.M.G.S JSME Kyushu branch* (in Japanese)
[4] Kita R and Yanagihara K 2015 *Proc. of 2015 Ann. Mtg. of JSME* S1340103 (in Japanese)
[6] Kita R and Yanagihara K 2017 *Proc. of Std. M.M.G.S JSME Kyushu branch* (in Japanese)
[7] JP PAT.P 2015-021663
[8] JP PAT.P 2016-147369

IOP Conf. Series: Materials Science and Engineering **501** (2019) 012030 doi:10.1088/1757-899X/501/1/012030

Research on Shape and Dimensional Accuracy of FDM Produced Parts

J Beniak[1*], PKrižan[1], L' Šooš[1], and M Matúš [1]

[1] Slovak University of Technology in Bratislava, Faculty of Mechanical Engineering, Nam. Slobody 17, 812 31 Bratislava, Slovak Republic

Abstract. Production of technical parts for Mechanical Engineering field is very important. There are many conditions which the produced parts have to satisfy. Very important for functional parts are accuracy of final parts. This accuracy could be evaluated in different ways. For example it can be evaluated shape accuracy or dimensional accuracy. This paper is focused to Fused Deposition Modelling (FDM) technology which is used for Additive Manufacturing process. The parts in additive manufacturing are produced directly from virtual model and digital file which could be obtained by direct modelling with CAD software of by 3D scanning. Published paper deals realized research where the full factor experiment was prepared and measured shape and dimensional accuracy of produced specimens. There are designed two types of specimens with different shapes and dimensions. Digital models are processed with different different 3D printer settings. Measured values are evaluated by statistical methods.

* Corresponding Author: e-mail: juraj.beniak@stuba.sk; tel. +421 905 593 953

1. Introduction

For technical practice is very important question of production precision of the machine parts in view of their intended use. The use of additive production of additive manufacturing systems increase from year to year in the forefront also in the production of parts and components for final use. That is why in this area it is necessary to know the technology so much that it is possible to influence the preparation and process of production of components through additive manufacturing systems in order to ensure the required accuracy of the final part [1]. In the case of mechanical engineering practice there have to be taken into the consideration dimensional accuracy of produced parts, what is the most common, but it have to be counted with shape and positional tolerances. It is necessary to make sure that the produced shapes have defined positions between important entities.

There are presented in this this paper the experiments focused to dimensional and shape and positional precision measurement [8]. There are many ways how could be this precision measured. In this paper is presented the measuring process with use of optical measuring instrument combined with special software which is directly evaluate specified tolerance. It is new way how to make the measurement process more effective and make it easier.

This tolerance evaluation is also connected with the specified factors which are selected as possible factors which could influence accuracy of parts produced by additive production techniques. In this case is the production process focused to FDM (Fused Deposition Modeling) technology which work with semi melted thermoplastic material which is deposited layer by layer to build required 3D part.

2. Material and Devices

The main aim of this research is to specify what are the shape and positional tolerances and also dimensional accuracy produced specimens by FDM technology. First of all there were specified the factors which could be possible for influencing of produced parts. There are many factors which could work with tolerance influence during the 3D printing [2, 3]. After experiences and literature research we specified two main used factors.

Temperature necessary for plastic material melting and layer thickness. Each producer of filaments state just range of suitable temperatures, not the exact temperature. This is because different 3D printers work with different software and have different technical design, so there could be necessary different temperature setting, but also settings of many others parameters. The temperature affect the consistence or the flow of semi melted thermoplastic what will affect the diffluence of deposited fibres.

Layer thickness is affecting the dimensional accuracy, so we try to figure out if there is also influence to shape and positional tolerances [4]. Each 3D printer allows to set layer thickness and it is easy to change and affect the final quality of produced parts.

Each of this two factors have two levels. The specified values are set to marginal possible values. For PLA plastic, which have been selected as very easy used production material, the producer state the printing temperature from 180°C to 220°C. The layer thickness is set to 0,1 and 0,2mm.

Using of this selected factors and their specified levels, there have been prepared full factor experiment, so combination of each levels each factors together. Respecting Design of Experiment (DOE) there is prepared full factor experiment (Table 1).

From each of this combination are produced 5 specimens to ensure data samples for measurement and proper statistical evaluation. Each specimen or each measurement have been repeated 5 times on each specimen. So we collected 25 values for each measured parameter, what is really significant set of values.

Table 1. Design of Experiment (DOE)

Specimen (Experiment)	Layer Thickness (mm)	Printing Temperature (°C)
1	0,1	220
2	0,2	220
3	0,1	180
4	0,2	180

The specimens are presented on Figure 1. There are two types of designed models, just to measure the both rounded and angular shapes, to test the possibility of selected 3D printer. As it is visible, the objects have also the holes with appropriate shape. On the Figure 1 are also presented the dimensions of each model. The printed height of models is the same with value 40mm.

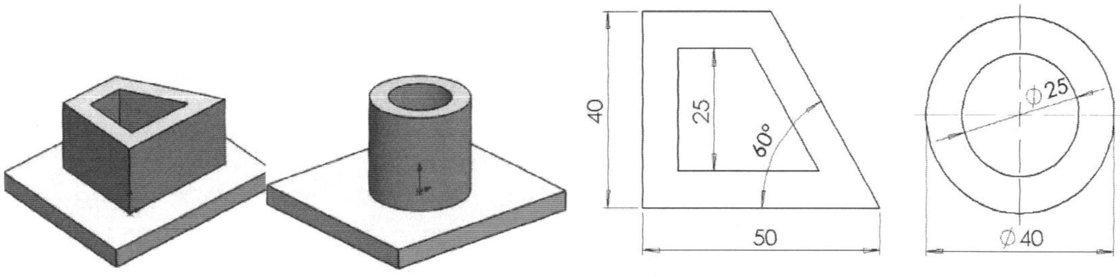

Figure 1. Designed 3D models selected for tolerance measurement.

Preparation of specimens are realized on FDM 3D printer Profi3DMaker (Figure 2 left) with following conditions:

- material of specimens – PLA
- printing speed – 50 mm.s^{-1}
- nozzle diameter – 0,3 mm
- number of bottom layers – 3
- density of bottom layers – 100%
- number of top layers – 3
- density of top layers – 100%
- number of wall contours – 2
- infill density – 5%
- infill structure – rectangular grid
- temperature of built platform - 50°C

Produced parts are subsequently measured. As is mentioned above, for measurement is selected sophisticated optical measuring device SmartScope MVP 200 (Figure 2 right).

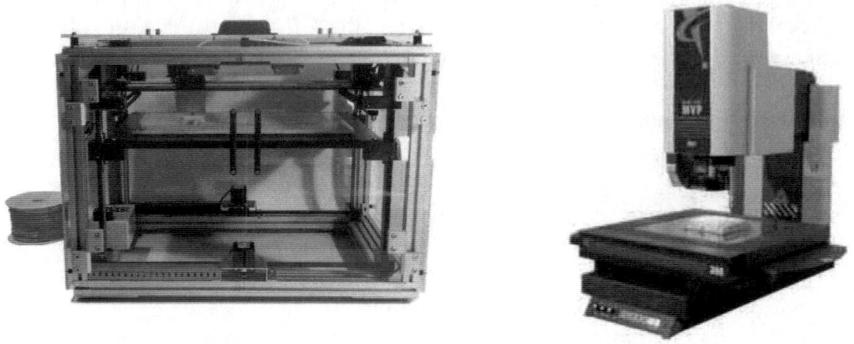

Figure 2. Profi3Dmaker 3D printer (left) [6] , Smartscope MVP 200 measuring instrument (right) [7],

Device MVP 200 work with measuring software Measure-X. This software automatically evaluate the final tolerance, it is just necessary to specify the appropriate lines or surfaces for correct measurement. The following Figure 3 and Figure 4 present labeling of measured tolerances. The dimension (for example "a" means the width of tolerance zone in which are the measured points.

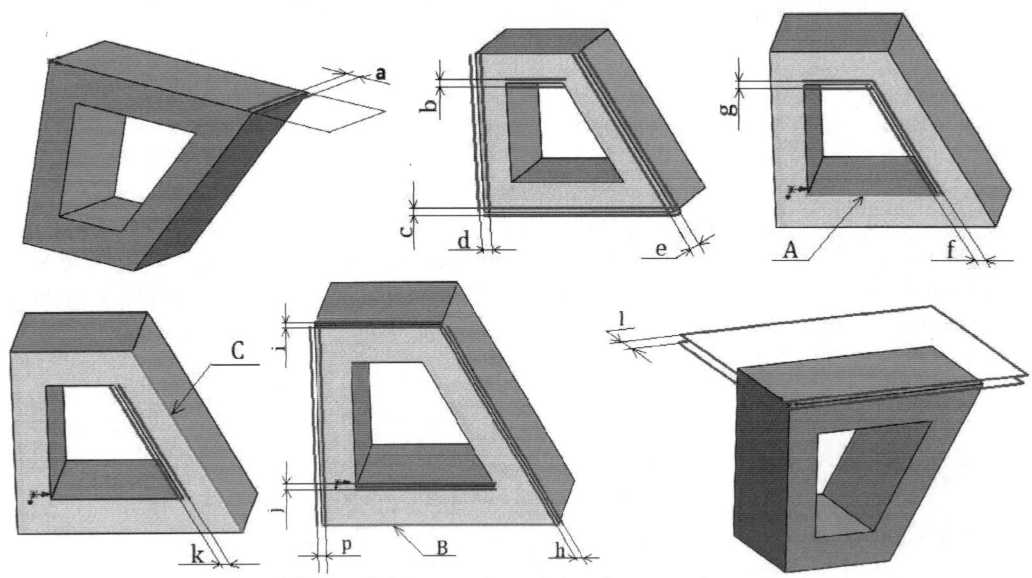

Figure 3. Measured precision for trapezium.

IOP Conf. Series: Materials Science and Engineering **501** (2019) 012030 doi:10.1088/1757-899X/501/1/012030

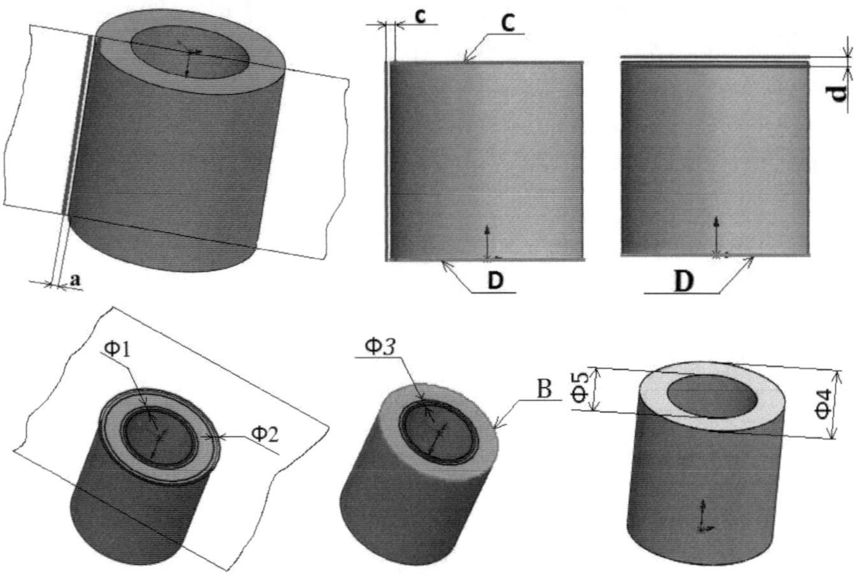

Figure 4. Measured precision for cylinder.

3. Measurement and evaluation of dimensional precision

The whole measurement is divided to measurement of dimensional precision and shape precision. In the case of trapezium there are measured dimension "*m*" and dimension "*n*" Measured values are presented in Table 2. Average values are calculated based on the equation (1). For statistical evaluation are also necessary at least standard deviation calculated by equation (2). Difference between theoretical designed value and average measured value are presented also on the Figure 5.

Table 2. Average values of trapezium dimensions

Exp.	Dimension *m* (mm)	Difference *m* (mm)	SD (mm)	Dimension *n* (mm)	Difference *n* (mm)	SD (mm)
1	49,86	-0,14	0,0749	39,87	-0,27	0,01573
2	49,955	-0,045	0,3991	39,96	-0,039	0,06515
3	49,8	-0,2	0,0618	39,85	-0,35	0,03554
4	49,96	-0,04	0,03977	39,83	-0,021	0,02344

The following is the formal mathematical formula for the arithmetic mean (a fancy name for the average).

$$\bar{X} = \frac{1}{n} * \sum_{i=1}^{n} x_i \qquad (1)$$

$\bar{X}$ = average (or arithmetic mean)
n = the number of terms (e.g., the number of items or numbers being averaged)
x_i = the value of each individual item in the list of numbers being averaged

Standard deviation values are calculated based on following equation:

$$SD = \sqrt{\frac{\sum(x_i - \bar{X})^2}{n-1}} \qquad (2)$$

Measured values are also evaluated by ANOVA statistical method [5]. The results from ANOVA analysis are as follow:

Dimension "*m*"

Thickness	F(1,96) = 6,75	p<0,010847	SS=0,02 MSe=0,01
Temperature	F(1,96) = 119	p<0,000001	SS=0,38 MSe=0,01
Thickness*Temperature	F(1,96) = 6,48	p<0,012475	SS=0,02 MSe=0,01

Dimension "*n*"

Thickness	F(1,96) = 24,1	p<0,000004	SS=0,04 MSe=0,01
Temperature	F(1,96) = 85,2	p<0,000001	SS=0,13 MSe=0,01
Thickness*Temperature	F(1,96) = 51,4	p<0,000001	SS=0,08 MSe=0,01

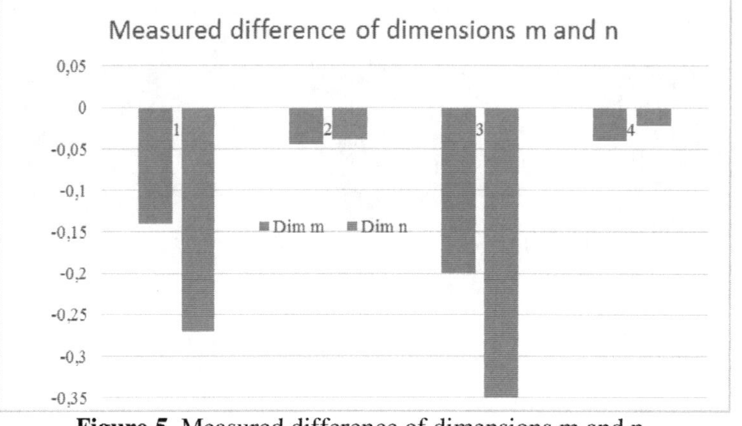

Figure 5. Measured difference of dimensions m and n.

Measured values from cylindrical specimens are presented in Table 3. Difference between theoretical designed value and average measured value are presented also on the Figure 6.

Table 3. Average values of cylinder dimensions

Exp.	Diameter ϕ4 (mm)	Difference ϕ4 (mm)	SD (mm)	Diameter ϕ5 (mm)	Difference ϕ5 (mm)	SD (mm)
1	40,27	+0,27	0,1675	24,44	-0,44	0,1363
2	40,25	+0,25	0,1336	24,5	-0,5	0,1841
3	39,85	-0,15	0,0433	24,82	-0,18	0,0209
4	39,84	-0,16	0,0265	24,87	-0,13	0,0263

Measured values are also evaluated by ANOVA statistical method. The results from ANOVA analysis are as follow:

Dimension "ϕ4"

Thickness	F(1,96) = 0,020;	p<0,888327;	SS=0,01; MSe=0,01
Temperature	F(1,96) = 364	p<0,000001;	SS=4,41; MSe=0,01
Thickness*Temperature	F(1,96) = 0,286;	p<0,593873;	SS=0,01; MSe=0,01

Dimension "ϕ5"

Thickness	F(1,96) = 0,002	p<0,965637;	SS=0,01; MSe=0,01
Temperature	F(1,96) = 264	p<0,000001;	SS=3,54; MSe=0,01
Thickness*Temperature	F(1,96) = 5,69	p<0,018994;	SS=0,08; MSe=0,01

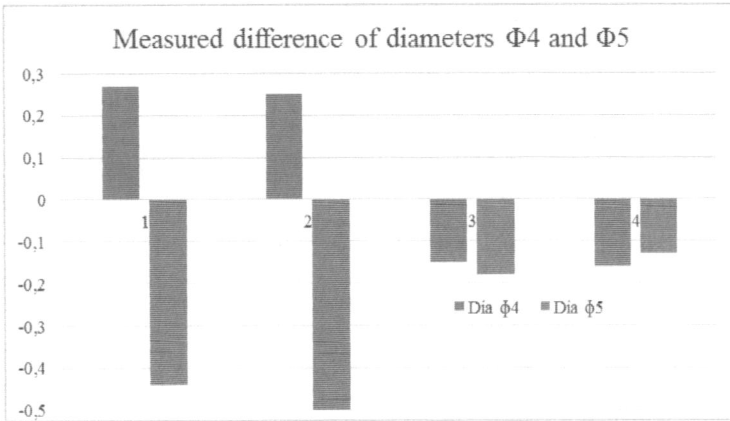

Figure 6. Measured difference of dimensions ϕ4 and ϕ5..

From this results we can see that the temperature as one of the factors is significant for dimensional precision. This is clear from the ANOVA analysis.

4. Measurement and evaluation of shape and positional tolerances

Based on the specified shape and positional tolerances from Figures 3 and 4, there was measured this tolerances. In the Table 4 are average values of tolerances specified on trapezium specimens. Average values of tolerances specified on trapezium specimens are in the Table 5.

Table 4. Average values of tolerances for trapezium specimens

(Exp)	Straight a (mm)	Straight c (mm)	Paralell i(B) (mm)	Paralell j(B) (mm)	Inclin f(A) (mm)	Inclin h(B) (mm)	Perpend p(B) (mm)	Flatness l (mm)
1	0,042	0,022	0,047	0,037	0,071	0,12	0,062	0,021
2	0,062	0,039	0,051	0,063	0,076	0,12	0,061	0,038
3	0,041	0,022	0,023	0,037	0,029	0,077	0,028	0,025
4	0,062	0,027	0,029	0,036	0,054	0,11	0,045	0,049

Table 5. Average values of tolerances for cylindrical specimens

(Exp)	Straight a (mm)	Paralell b(A) (mm)	Paralell d(A) (mm)	Perpend c(D) (mm)	Perpend c(C) (mm)	Round ϕ1 (mm)	Round ϕ2 (mm)	Concentric ϕ3(B) (mm)
1	0,065	0,11	0,19	0,13	0,11	0,29	0,21	0,15
2	0,084	0,096	0,19	0,21	0,14	0,15	0,24	0,14
3	0,05	0,1	0,22	0,14	0,13	0,16	0,074	0,11
4	0,084	0,13	0,2	0,2	0,14	0,092	0,094	0,058

From this measured values stated in Table 4 and Table 5 we can see that statistically the worst results are reached from the combination of factors temperature and layer thickness for experiment number 2 (0,2mm layer thickness and 220°C printing temperature) for trapezium specimens. We can also see that the main change of tolerance is caused by change of temperature. It is possible to explain that the

IOP Conf. Series: Materials Science and Engineering **501** (2019) 012030 doi:10.1088/1757-899X/501/1/012030

deposited material is diffluent more to the sides and this cause bigger inaccuracy. This statement is also based on ANOVA statistical analysis, which had similar results as in case of dimensional accuracy measurement.

5. Conclusion

Based on the presented average vales from tolerance and dimension measurement and based on the ANOVA analysis it is possible to state that the printing temperature have the significant influence for shape and dimensional tolerance of FDM produced parts. This is because with higher temperature the melted material is more liquid with easy flow. So the material flow to the side caused by higher liquid and gravity, so the dimensions and tolerances are more inaccurate compare to specimens produced by lower temperature. This statement is based also on the optical comparison of specimens. The specimens with high printing temperature have worst surface quality what also lead to the worst tolerances.

Acknowledgments

The paper is a part of the research done within the project KEGA 051STU-4/2018 – Innovative methods for education process improving in the field of additive manufacturing and computer-aided systems.

References

[1] Chua C K, Leong K F, Lim C S, Rapid Prototyping, Principles and Applications, *Nanyang Technological University*, Singapore, World Scientific Co. Pte. Ltd, 2003, ISBN 981-238-117-1.

[2] DUC PHAM OBE FREngAND STEFAN DIMOV, Rapid Prototyping, A time compression tool, *Manufacturing Engineering Centre, Cardiff University.*

[3] Beniak J, Križan P, Matúš M, Svátek M, *Ecological PLA plastic used for FDM Rapid Prototyping Technology*. In *SGEM 2015. 15th International Multidisciplinary Scientific Geoconference* : conference proceedings. Albena, Bulharsko, 18.-24. 6. 2015. 1. vyd. Sofia : STEF92 Technology, 2015, S. 117-123. ISSN 1314-2704. ISBN 978-619-7105-34-6.

[4] Anoop K. Sood, Raj K. Ohdar, Siba S. Mahapatra, Experimental investigation and empirical modelling of FDM process for compressive strength improvement, In *Journal of Advanced Research*, Volume 3, Issue 1, 2012, Pages 81-90, ISSN 2090-1232, https://doi.org/10.1016/j.jare.2011.05.001.

[5] Mendonsa C, Naveen KV, Upadhyaya P, Shenoy V D, Influence of FDM Process Parameters on Build Time Using Taguchi and ANOVA Approach, *International Journal of Science and Research (IJSR)*, ISSN (Online): 2319-7064

[6] Company information. Smartscope MVP 200. Available at: www.smartscope.com

[7] Printing Industry. Profi3DMaker. Available at: https://3dprintingindustry.com

[8] Svetlík J, Demeč P., Methods of Identifying the Workspace of Modular Serial Kinematic Structures. In: *Applied Mechanics and Materials* : CECOL 2012: 3rd Central European Conference on Logistics : Trnava, Slovakia, 28-30 November 2012. Vol. 309 (2013), p. 75-79. - ISBN 978-303785636-9 - ISSN 1662-9336

An Experimental Study on Human Characteristics under Acceleration of Vehicle

Taro Miyao[1] and Shoichiro Takehara[2]

[1] Dept. of Science and Technology, Sophia University of Graduate, Tokyo, 7-1 Kioi-cho, Chiyoda-ku, 102-8554, Japan
E-mail: t-miyao-p8m@eagle.sophia.ac.jp, 03-3238-3864
[2] Dept. of Science and Technology, Sophia University, Tokyo, 7-1 Kioi-cho, Chiyoda-ku, 102-8554, Japan
E-mail: stakeha@sophia.ac.jp, 03-3238-3863

Abstract. Recently, environmental consideration, energy-saving performance and low fuel consumption become important, therefore personal mobility is actively developed. When vehicle become small, it is expected that human body behavior has a big influence on running performance. The influence is related to stability of vehicle and safety of its driver. Simulation is used as a method to investigate the interaction between vehicle and its driver, and the body behavior can be reproduced in advance. However, since it is necessary to consider the validity for the use of simulation, experiments to capture body behavior are indispensable. This research is aimed to obtain knowledge about dynamics of driver riding inside the vehicle. Therefore, subjects are placed on a simple truck simulating a vehicle. The truck is applied lateral acceleration and acceleration in the direction of travel, and the motion of subjects was recorded with the motion capture camera. From the measured data, human characteristics are examined by comparing the body behavior of subjects for each condition.

1. Introduction

Recently, personal mobility is actively developed because environmental consideration, energy-saving performance and low fuel consumption become important. Such vehicles are expected to be friendly to person and the environment, possible to transport comfortable and effective short/middle distance. In vehicles of various shapes, it is assumed that the occupant's posture is various. In addition, it is necessary for driver's physical characteristics to be considered for design and development, because it is assumed that human motion has influence on the motion at a light weight vehicle. In order to evaluate human characteristics and motion of vehicle occupants, experiments by actual vehicle running, vibration experiments of the seat with a vibration, and the like have been carried out. [1,2,3,4,5] These experiments are concerned with behavior under steady conditions, these are not concerned with behavior under unsteady conditions. Then, it is important to obtain data on body behavior when lateral acceleration and acceleration in the traveling direction are applied. The lateral acceleration is assumed at the time of lane change or turning of the vehicle, and the acceleration in the traveling direction is assumed when the vehicle starts or stops. The purpose of this research is to gain findings of the body motion of the vehicle occupant. Therefore, experiments on body behavior were conducted under unsteady conditions.

IOP Conf. Series: Materials Science and Engineering **501** (2019) 012031 doi:10.1088/1757-899X/501/1/012031

2. Methods

Subjects were 11 persons aged from 22 to 24 years old. The sheet used in the experiment was inflexibility in order to eliminate the influence on the body behavior by flexibility such as an automobile seat. Figure 1 shows the schematic diagram of experiment setup. Ten motion capture cameras were used for measurement. The vehicle was given three types of acceleration by changing the elongation of the rubber band. Comparison of body behavior due to acceleration difference was conducted. Experiments were carried out at 1.2[m/s^2] which is the general deceleration of the automobile and 0.8[m/s^2] and 1.4[m/s^2] which are the deceleration around it. [(6)] However, even with the same rubber band elongation, there was a difference in acceleration depending on the weight of the subject. In this study, the weight was set to 75 kg using a weight in order to eliminate the difference in acceleration. Experiments were carried out by instructing subjects to open eyes or close eyes, to inform the timing of departure or not to notify. Instructions for open eyes and closed eyes are to investigate the influence on the body behavior by the subject's visual information. Whether to notify the timing of departure or not is for comparing the body behavior at the time of a sudden start of the vehicle and the body behavior at the time of recognizing the timing of starting the vehicle. For the analysis of experimental results, five data without data loss were selected. In addition, this research is approved by the Ethics Committee of Sophia University. Each subject was explained about informed consent before experiments.

Table 1. Combination of experimental conditions

Experimental condition	Eyes	Departure signal
1	Opened	✓
2	Closed	
3	Opened	
4	Closed	✓

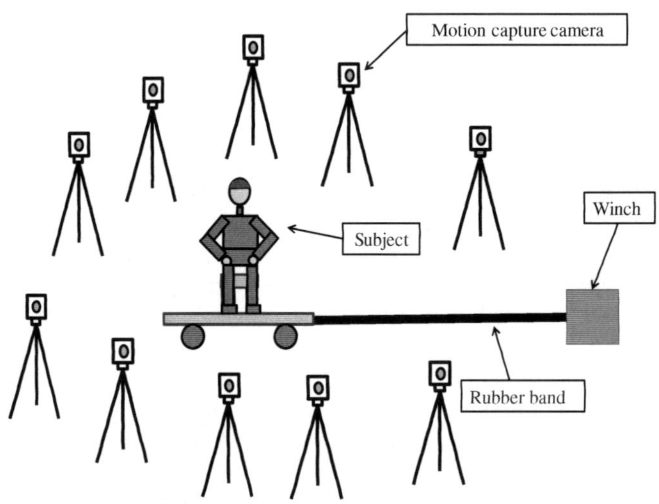

Figure 1. Schematic diagram of experiment setup

3. Result and discussion

3.1 *Difference of displacement by acceleration, body part and subjects*

Figure2 shows results of lateral experiments in condition1 by subject1. Figure2(a) shows displacement of six markers in acceleration of 1.4 [m/s^2]. Figure2(b) shows the displacement of the head top in acceleration of 0.8 [m/s^2], 1.2[m/s^2], 1.4[m/s^2]. The displacement of the head top is larger than the displacement of other portions (Figure2(a)). In addition, the displacement of head top increases as acceleration increases (Figure2(b)). Based on the above results, the experimental results were analyzed by displacement of the head top in acceleration of 1.4[m/s^2]. Figure3 shows the displacement of the head top of two subjects in condition1. The difference of head top at acceleration 1.2[m/s^2] and 1.4[m/s^2] is smaller as compared with the difference at acceleration 0.8[m/s^2] and 1.2[m/s^2] (Figure3(a)). This is thought to be because at low acceleration, the influence by acceleration is smaller than the influence by notify the timing of departure and vision, but beyond a certain acceleration, the influence by acceleration became larger than that by notifying the timing of departure and visual feedback. On the other hand, the difference of head top at acceleration 0.8[m/s^2] and 1.2[m/s^2] is smaller as compared with the displacement difference at acceleration 1.2[m/s^2] and 1.4[m/s^2] (Figure3(b)). Thus, as the acceleration increases, the displacement of the head top increases, but it can be confirmed that the influence of the acceleration on the displacement of the head top varies depending on subjects. Figure4 shows the displacement of the head top of the subject1 in the traveling direction in condition1 and condition2. It can be found that the head top displacement of the acceleration 1.4[m/s^2] is larger in condition1 than in condition2. This result shows that the subject showed different behavior if the experimental conditions were different. Therefore, findings consistent with experiences such as head displacement increased as acceleration increased was obtained. However, it was confirmed that the influence was affected by individuals and conditions.

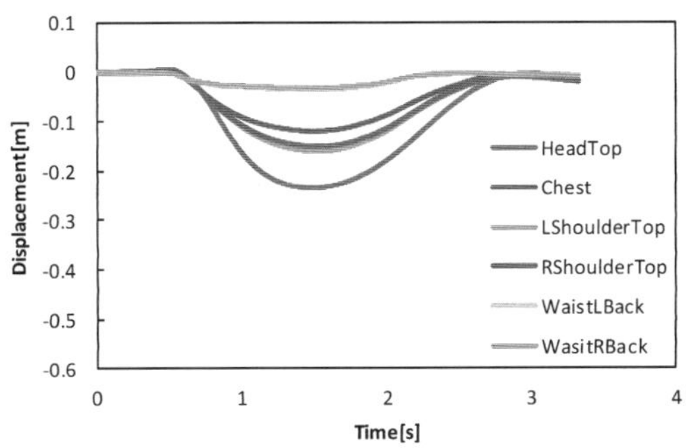

(a) Displacement of each marker

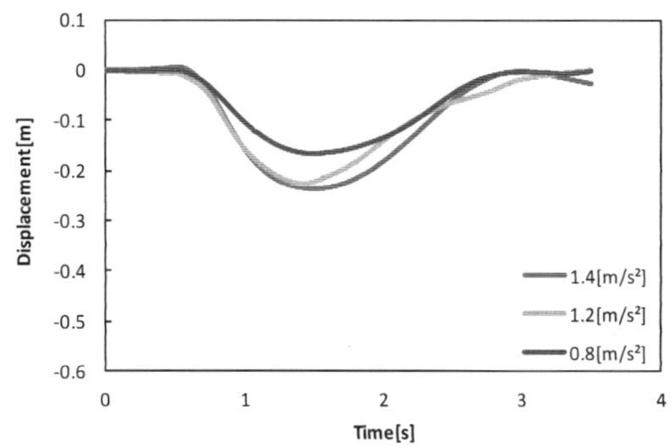

(b) Displacement of head top

Figure 2. Time history of displacement (Subject1)

IOP Conf. Series: Materials Science and Engineering **501** (2019) 012031 doi:10.1088/1757-899X/501/1/012031

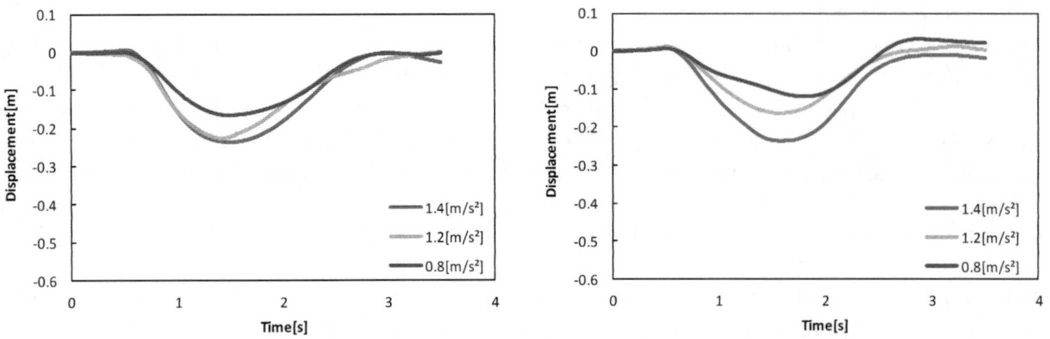

(a) Subject1 (b)Subject2

Figure 3. Time history of head displacement for each acceleration (Condition1)

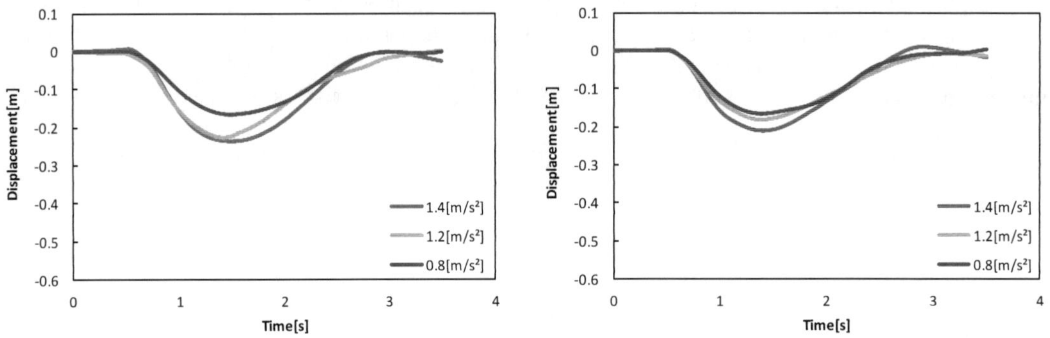

(a) Condition1 (b) Condition2

Figure 4. Time history of head displacement for each acceleration (Subject1)

3.2 *Relationship between acceleration and head displacement*

The relationship between the input acceleration of the two subjects and the displacement of the head is discussed in more detail. In figure5 to 8, the vertical axis shows the maximum displacement of the head, and the horizontal axis shows the maximum acceleration in the same experiment. In addition, the approximate straight line obtained by the least squares method is also shown. First, condition1 and condition4 are compared (Figure5). It can be found both subject1 and subject2 show that the inclination of the approximate straight line is smaller when eyes closed than eyes opened. Since it is impossible to obtain visual information when eyes closed, it is considered to be to control the head by stiffening the body in preparation for movement.

Second, condition2 and condition3 are compared (Figure6). It can be found that the slopes of approximate straight lines of subject1 and subject2 at the time of eyes opened and at the time of eyes closed are almost equal. This result shows that head displacement is considered to be proportional to acceleration under this condition.

Third, condition1 and condition3 are compared (Figure7). Subject1 is different from subject2, and it can be found that the inclination of the approximate straight line with notified the timing of departure is smaller than that without notified the timing of departure. This may possibly reduce the influence by stiffening the body before if there is notified the timing of departure. In addition, it can be found that displacement is bigger than that with notified the timing of departure without notified the timing of departure. This is thought to be because the reaction was delayed.

Finally, condition2 and condition4 are compared (Figure8). In both subjects, it can be found that the slope of the approximate straight line without notified the timing of departure is larger than the approximate straight line with notified the timing of departure, and the displacement is also large. It seems that this was caused by the same reason as before. Therefore, even for the same subject, variations in the inclination of the approximate straight line could be found by conditions. Even in the same condition, it was also found that the inclination of the approximate straight line is different for each subject. These results suggest that recognition of movement and visual information are influenced feedback to body behaviors. Furthermore, it is considered that the body behavior varies from subject to subject because this feedback varies depending on the personality owned by subjects themselves.

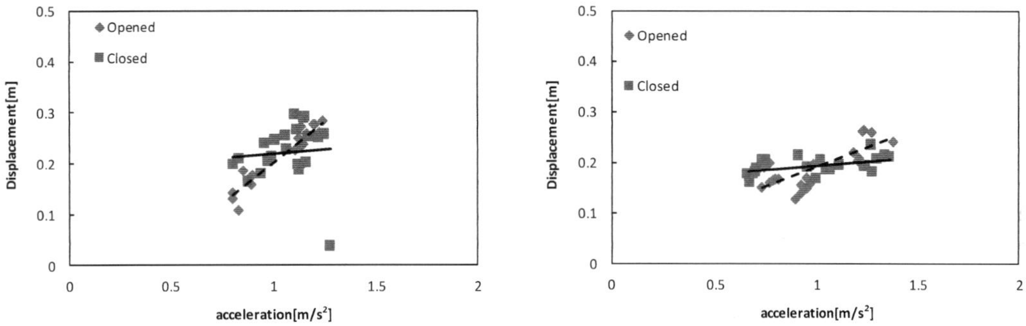

(a) Subject1 (b)Subject2

Figure 5. Comparison between condition1 and condition4

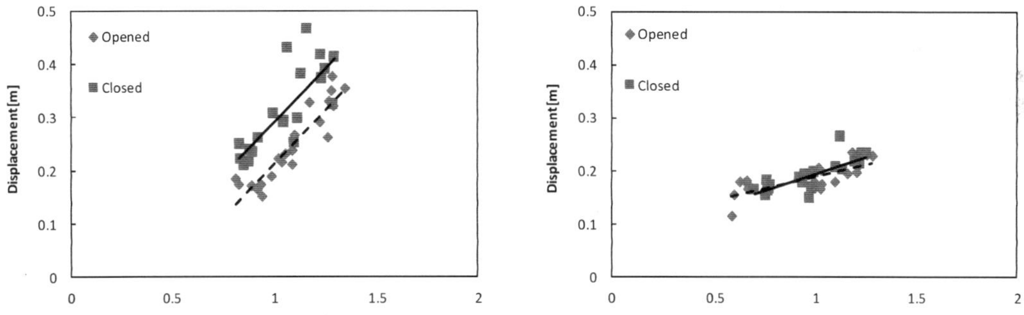

(a) Subject1 (b)Subject2

Figure 6. Comparison between condition2 and condition3

IOP Conf. Series: Materials Science and Engineering **501** (2019) 012031 doi:10.1088/1757-899X/501/1/012031

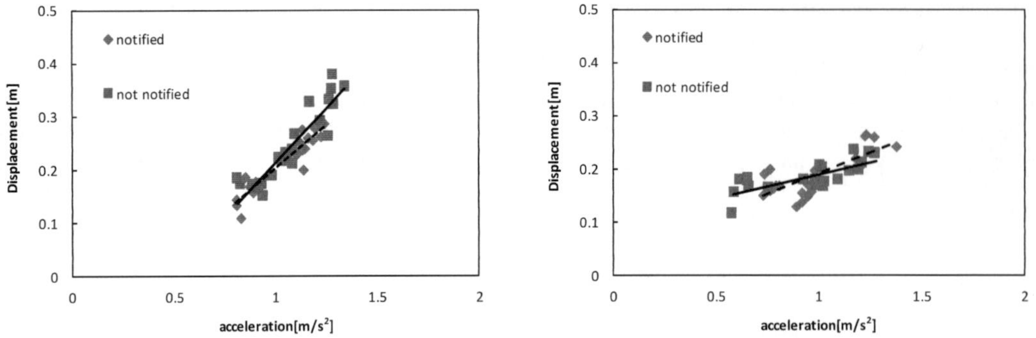

(a) Subject1 (b)Subject2

Figure 7. Comparison between condition1 and condition3

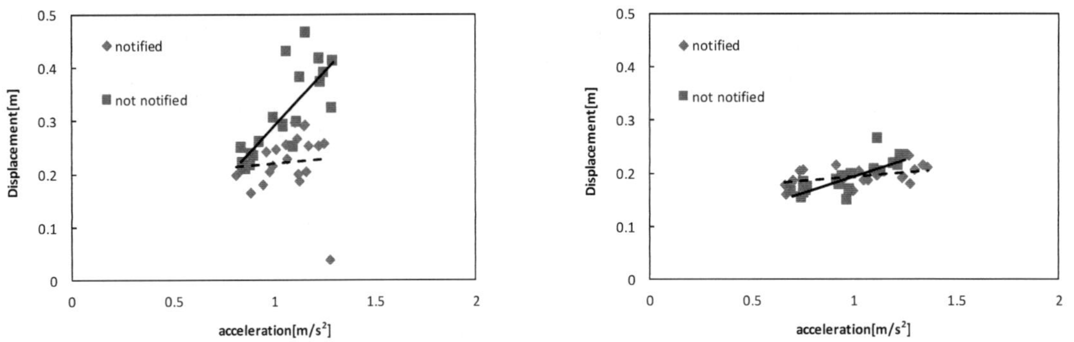

(a) Subject1 (b)Subject2

Figure 8. Comparison between condition2 and condition4

4. Conclusion

In this study, lateral acceleration is applied to the vehicle and the body behavior of the vehicle occupant was recorded. At the time of experiments, subjects were instructed whether to notify the timing of departure and whether to open eyes or close eyes. Characteristics of the body behavior were examined using these results. It is considered that the body is controlled in advance in preparation for movement with notified the timing of departure. On the other hand, without notified the timing of departure, it can be thought that the response was delayed because it could not prepare for the movement. In addition, results that the displacement increased proportionally with the acceleration was obtained. However, differences were found displacement and slope of approximate straight line depending on experimental conditions and subjects. Therefore, it is suggested that the individuality possessed by each person may influence feedback to the body behavior from external factors such as recognition to move and visual information.

References

[1] Akasaki K, Arakawa T, Oshinoya Y, Hasegawa S *2010 Journal of Japan Society of Mechanical Engineers Kanto Branch* **No.16** pp.157-158

[2] Aomura S, Tateno T, Abe M, Kawazoe K 2005 *J. Japan Soc. Mech. Eng. Kanto Branch* **11** 167-168

[3] Matsuda K, Yoshimura T, Tamaoki G 2009 *Dynamics & Design Conference* **9** 420-1-42

[4] Koizumi T, Tsujiuchi N, Abe H, Ninomiya J, 2007 *Dynamics&Design Conference* **7** 149-5
[5] Hayashi Y, Hase K, Takehara S, Torigaki T, Hirao A, Kudo Y and Yamamoto Y 2015 *Transactions of the Japan Society of Mechanical Engineers* **81**
[6] Kikuchi H, Okada A, Mizuno H, Kinuta Y, Nakamura T, Hagiwara G and Makimura K *J.Japan Soc. Civil Eng.* D**3** **68** 1196

IOP Conf. Series: Materials Science and Engineering **501** (2019) 012032 doi:10.1088/1757-899X/501/1/012032

Numerical Simulations of Skin Formation: Convergence of Moisture Transport and Stratum Corneum Thickness

Kohei Shobuda[1], and Katsuya Nagayama[1,*]

[1] Department of Mechanical Information Science and Technology, Kyushu Institute of Technology,

680-4 Kawazu, Iizuka, Fukuoka, 820-8502, Japan

* Corresponding Author: nagayama@mse.kyutech.ac.jp,Tel&Fax 81-948-29-7778

Abstract. As the epidermis adapts to different environments, desquamation of keratinocytes is affected by the water content in the epidermis. The thickness of the outermost layer of skin is determined by the interaction between exfoliation and moisture evaporation. The balance between these processes is the key to healthy skin, but has not been studied. The present study constructs a qualitative numerical model of stratum corneum thickness based on the interaction between exfoliation and evaporation processes.

1. Introduction

The dermatological mechanisms behind the smoothness of human skin are of great interest to the health and beauty industry. However, because most experiments that target human subjects are generally precluded by ethical constraints, the internal developmental mechanism behind skin roughness is understood only superficially from the data collected by noninvasive measurements. This study reports a numerical study of the formation of epidermis that uncovers insight about how skin develops.

The stratum corneum is the outermost layer of the human skin, and is also where new cells keratinize. This 'horny layer' is kept fresh by the sloughing of old keratin and replenishment with new cells. This peeling–regeneration cycle, called *turnover*, is known to be related to the moisture content in the stratum corneum [1]. For example, if the water content of the stratum corneum reduces the desmosomes that hold keratinocytes together degrade, and the stratum corneum becomes thicker than usual. The thickening of the outer layer suppresses the evaporation of moisture from the skin thereby enhancing water retention. In contrast, a high-humidity environment promotes desquamation. Therefore, epidermis turnover adjusts dynamically to different environments. Dry climates exert a remarkable aging effect on skin. Aging reportedly results from the decreased activity of the enzyme that degrades desmosomes in addition to the surrounding climate [1]. Many diseases are the result of maladaptation to the environment because epidermal turnover has failed.

Several computational models have been proposed for developing understanding of the mechanisms behind the development of human skin [2–5]. This article proposes a particle model that can handle complex biological phenomena, including cell interactions such as cell division, motion, deformation, and transition [6–8]. The method is suitable for simulating skin formation.

IOP Conf. Series: Materials Science and Engineering **501** (2019) 012032 doi:10.1088/1757-899X/501/1/012032

Assuming a healthy epidermis, the next sections analyze the process by which human skin converges to a constant thickness through the competing processes of water volatilization and exfoliation.

2. Methods

2.1. *Analysis flow*

Figure 1 is a flowchart of the present analysis. The simulation study period was 200 days. Transepidermal water loss assumed to be constant in our earlier study [9] was set as a variable in the present model to investigate epidermal turnover. The models in this study therefore use different values for the "water calculation", "desquamation calculation", and "update parameters" steps.

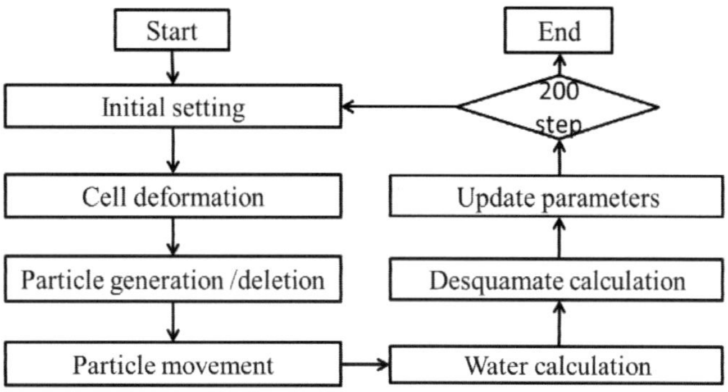

Figure 1. Analytic flow of the present study

2.2 *Initial setting*

To initialize the model, dermal layer and basal layer particles were placed in the calculation area ($300 \times 300 \times 300$ μ^3 grid), and keratinocytes were divided from the basal layer. The division frequency of the keratinocytes was set as a constant.

Epidermal cells are classified into four groups, depending on the progress of keratinization. Figure 2 shows the shapes of various particle types. Stem cells located closest to the body are called basal cells from which the spinal cells differentiate. Cells then differentiate further into granule cells, and eventually become outer layer horny cells. In the numerical simulation we differentiate the particles by thickness. In each step of the analysis, the thickness of each particle is decreased, and the number of particles that reach a certain thickness are updated.

Each particle moves to a position that balances the spring forces between particles [10]. New particles differentiated from the basal layer shift the neighboring particles, moving the whole mass upward. Periodic boundary conditions were assumed in our calculations.

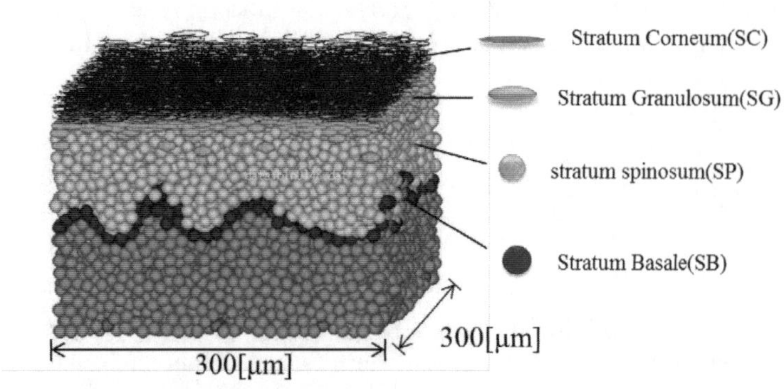

Figure 2. Particle model of keratinization

2.3 *Desquamation calculation*

Since the decomposition of desmosomes is proportional to the amount of skin moisture [1] desmosomes in the stratum corneum (SC) are decomposed by moisture. In the present model, each horny cell has its own desmosome content, which is updated after each water calculation step. The degree of reduction in desmosomes is determined by the moisture content of horny cells. Particles with desmosome content below the threshold are removed from the calculation area as extinct cells.

2.4 *Trans-epidermal water loss*

Skin roughness results from disturbances to the structure of the SC that reduce its water retention. Therefore, trans-epidermal water loss (TEWL) is a key indicator when evaluating the barrier function of the skin as well as the shape of the outer layer. Equations (1)~(3) describe moisture diffusion using Fick's law, a one-dimensional diffusion equation. Note that we defined C_0 as a constant assuming that the dermis provides a constant moisture supply.

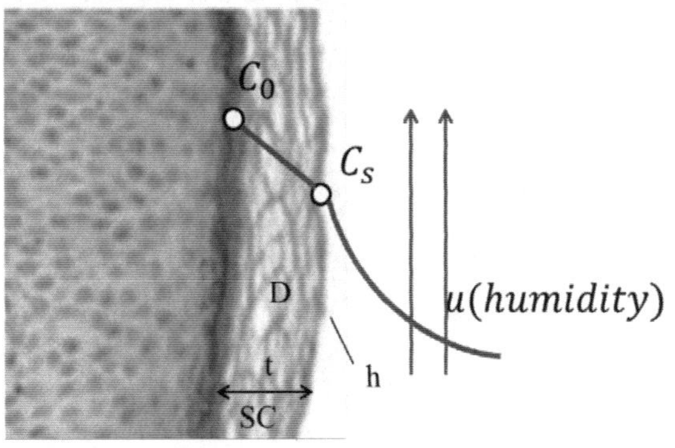

Figure 3
C_0 and C_s are water contents on the inside and outside of the SC, respectively. μ is humidity in the environment. D is the diffusion coefficient of water in the stratum corneum, h is a proportional constant, and t is the thickness of the SC.

Equation (1) determines the amount of water diffused from the SC to the surface. As the SC grows thicker less moisture evaporates. Assuming this phenomenon, we defined an inverse proportional relationship between TEWL and the SC thickness t. Equation (2) computes the amount of water that evaporates from the keratinous surface into the atmosphere. We introduced h as a parameter to describe the surface conditions such as the quality of the sebum barrier or the shape of the horny layer, which have some influence on evaporation rates. Equation (3) is derived by solving Equations (1) and

IOP Conf. Series: Materials Science and Engineering **501** (2019) 012032 doi:10.1088/1757-899X/501/1/012032

(2) for C_s and rearranging. This equation states the inverse proportionality between TEWL and SC thickness.

$$\mathrm{TEWL}_1 = -\mathrm{D}\frac{C_s - C_O}{t} \quad (1)$$

$$\mathrm{TEWL}_2 = \mathrm{h}C_s(1-\mu) \quad (2)$$

$$\mathrm{TEWL} = \frac{C_O}{\frac{t}{D}+\frac{1}{h(1-\mu)}} \quad (3)$$

3. Results and Discussion

In the simulation runs, the SC thickness increased and decreased in cycles, and eventually settled at a constant value. Figure 4(a) shows the placement of each kind of particle at different time periods in the analysis, and Figure 4(b) shows the water content of the particles at the same points.

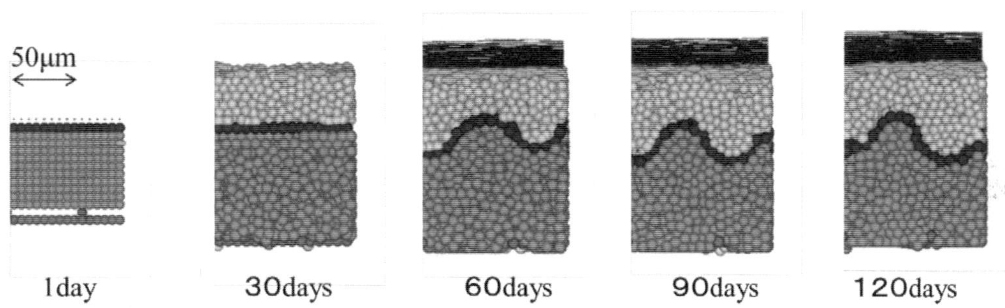

Figure 4(a). Placements of the various particle types at different times.

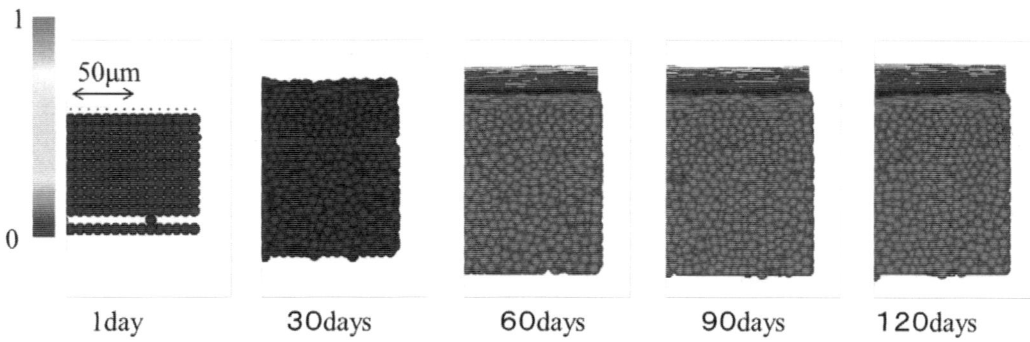

Figure 4(b). Water content of the particles at different times.

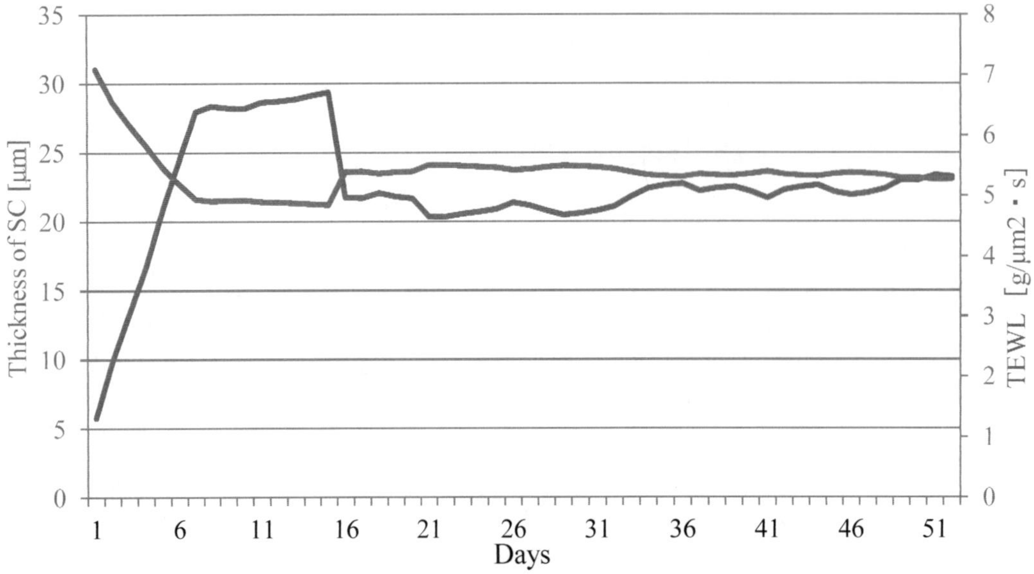

Figure 5. Convergence of turnover.

In Figure 5, the thickness of the SC and TEWL are plotted with day 1 as the step in which the first moisture calculation was performed. The thickness of SC gradually increases due to the increase of keratinous particles, and peeling began on day 7. SC thickness decreased more from days 14 to 16 than it did before and after because the horny cells were generated and peeled off before the moisture calculation step.

4. Conclusion

Healthy turnover is indispensable for the maintenance of healthy skin. During the turnover process, thickness of the SC is adjusted to the environment through the interaction between the water retention of the SC and the exfoliation rate, and the latter of which is determined by humidity in the environment. Many skin diseases follow the collapse of this mechanism due to keratin dysplasia and peeling. This article is built upon the model of moisture diffusion and exfoliation which we proposed in a previously published report. The model qualitatively reproduced the SC's water retention function by defining an inverse proportional relation between the SC thickness and TEWL. We modeled the relationship between peeling and water transport as linear. In our previous study, an upper limit was set on the number of SC cells, and any surplus was erased. However, this model yielded a constant thickness SC and became regardless of the environmental humidity. Therefore, we also altered the peeling model so the effects of moisture conditions in the SC on exfoliation rates could be studied. The revised model showed a convergence of SC thickness over time to a thickness determined by the model parameters. In the future, we plan to include a mechanism that describes excessive drying and peeling to consider the influence of inflammation and aging on the turnover mechanism.

5. References

[1] Junichi Kayma J 1999 *J. Soc. Cosmet. Chem. Japan* The Mechanism of Desquamation in the Stratum Corneum and Its Relevance to Skin Care ***33*(1)** [page numbers]

[2] Sütterlin, T., Huber, S., Dickhaus, H. and Grabe, N. (2009). Modeling multi-cellular behavior in epidermal tissue homeostasis via finite state machines in multi-agent systems. Bioinformatics, 25, 2057-2063.

[3] Sütterlin, T., Huber, S., Dickhaus, H. and Grabe, N. (2009). Modeling multi-cellular behavior in epidermal tissue homeostasis via finite state machines in multi-agent systems. Bioinformatics, 25, 2057-2063.

[4] Hirashima, T., Hosokawa, Y., Iino, T. and Nagayama, M. (2013). On Fundamental Cellular Processes for Emergence of Collective Epithelial Movement. Biology Open, 2, 600-666. doi: 10.1242/bio.20134523

[5] Adra, S., Sun, T., MacNeil, S., Holcombe, M. and Smallwood, R. (2010). Development of a Three Dimensional Multiscale Computational Model of the Human Epidermis. Open Journal PLoS ONE, 5, e8511.

[6] Nagayama, K., Nitta, J. and Miura, I. (2009). Numerical Analysis on Angiogenesis in Cancer Using a Particle Model. Theoretical and Applied Mechanics Japan, 58, 321-324.

[7] Hashiguchi, S. and Nagayama, K. (2011). Construction of a hair formation analysis method using the particle model, 2011 Bioengineering Conference, 23th Japan Society of Mechanical Engineers, 9D-01, Japan, 355-356.

[8] Nagayama K., Matsuoka S., Morisaki N. and Taguchi H. (2015). 3D Numerical Simulation of Hair Formation Process Using a Particle Model, Open Journal of Regenerative Medicine, February 6, 2015, http://file.scirp.org/Html/1-2390044_53849.htm

[9] Nagayama, K. andKurihara, T. (2018) Numerical Simulationof Skin Formation: The Relationship betweenTransepidermal Water Loss and Corneum Thickness. Journal of Applied Mathematics and Physics , 6, 1757-1762. https://doi.org/10.4236/jamp.2018.68150

[10] Kurihara, T., Nagayama, K. (2016) Particle Simulation on Human Epidermal Aging- Effect of Basal Layer and Cell Division Rate -, The 7th TSME International Conference on Mechanical Engineering, BME0001. http://hdl.handle.net/10228/00006245

IOP Conf. Series: Materials Science and Engineering **501** (2019) 012033 doi:10.1088/1757-899X/501/1/012033

Vibrational Analysis on Patellofemoral Joint Degradation of Swine's Knee

Nopdanai Ajavakom[1,*], Kakanand Srungboonmee[2], Kocharath Tanadumrongsak[1] and Visarut Limsowan[1]

[1]Department of Mechanical Engineering, Chulalongkorn University, Bangkok 10330, Thailand

[2]Center of Data Mining and Biomedical Informatics, Faculty of Medical Technology, Mahidol University, Nakorn Pathom 73170, Thailand

* Corresponding Author: nopdanai.a@chula.ac.th, Tel: (66) 02 218 6623, Fax: (66) 02 252 2889

Abstract. Knee's severe patellofemoral joint degeneration gives a lot of people difficulty to endure daily life because walking or any leg movement can cause serious pain. At present, physical examination, x-ray and MRI are often used to diagnose the condition; however, each of the methods has its own disadvantages. For instance, physical examination is highly dependent on the skill of the doctor who practiced the examination, which cannot avoid misdiagnosing the condition. X-ray can detect wounds and tears, but the accuracy is also not top-notch since the result of an x-ray is a 2-dimension picture. MRI is the most reliable method for diagnosing the condition; however, the cost is very high. We propose that other than x-ray and MRI, patellofemoral joint degeneration can be identified by analyzing vibrational signals obtained from an accelerometer attached to the patella while sitting and moving the leg up and down. At the beginning of the study, swine's knees are used to imitate human's knee. Swine's knees with various surface degradation levels are put on a machine that mimics the mechanism of knee motion. An accelerometer is mounted on the patella of the swine's knee while the machine is running to measure the friction and roughness induced vibration of the patella. The vibration results suggest that more surface degradation the higher vibration signal amplitude. Nevertheless, the method still needs a lot more improvement on the database and testing procedures, in order to make it accurate, dependable and affordable. Hence, with further analysis on patient's knees it is highly possible to determine whether patient's knee is degraded or not and at what degradation level.

1. Introduction

Knees are ones of the most important anatomies in our human body, they helps the walking, running, and jumping movements possible, while also supporting the entire body's weight. There are four main parts, which comprise a knee: femur, patella, tibia, and fibula. From long and endured usage of knees, they can become degraded, makes it painful to move, this can happens to anyone, however, athletics are more prone to this problem due to the overuse of the knees. Some develops Chondromalacia Patella (cartilage under the knee is softened) and some develops Osteoarthritis (wear and tear on the knee joints) just from walking. The degradation of knee is divided into 5 levels (or grades): 0 = healthy, 1 = minor spur growth/minor swell, 2 = greater spur growth/ knee density > 50%, 3 = obvious damage between

IOP Conf. Series: Materials Science and Engineering **501** (2019) 012033 doi:10.1088/1757-899X/501/1/012033

cartilages and bone/ knee density < 50%, 4 = cartilage is fully degraded [1-2]. Patient's knee degradation level can be confirmed by the use of MRI (Magnetic Resonance Imaging) [2-5], however, due to the expensiveness of the test, diagnosing the degradation of knees is very uncertain. Although, nowadays orthopedic doctors are able to detect the degradation by listening to the noise generated from the friction between the degraded knees [2-3], the accuracy of the test is only if the condition of the patient is beyond level 2 already, which at the time, might be too late to treat without undergoing a surgery. In addition, many people tend to wait until the problems presented and interrupted their daily life before checking for their health, this included their knees. Due to many reasons, such as, not considering that knees are working hard under the load applied by the users daily, or the diagnosis methods are costly, and if no problems presented, there is no need for them to check on them. If diagnosed before the knees entered pain-stage, pre-cautious and prevention for the further degradation due to lifestyle will be able to slow the inevitable degradation of the knees, thus create opportunity for patients to longer their knees' lifetime.

Knee degradation and especially the patellofemoral joint degradation is researched in clinical features of symptomatic patellofemoral joint osteoarthritis by Peat G et.al. [6], where the research includes statistics and data from real gathered samples of the plausible causes of knee degradation and the symptoms of such injuries. In addition, the study by Cannon A et.al. [7] studied and explained about how the patellofemoral resurfacing arthroplasty, which is the most common method to cure the injury.

Three main diagnosis methods are normally practiced nowadays. The first is a traditional physical examination, where a doctor, expertise with knees, will do a simple knee bending and stretching examination and observe the motion. This method is highly dependent on the doctor giving the examinations. Although it is the easiest and cheapest method, it is the most inaccurate method among the three. The second method is the x-ray, which is inaccurate due to its two dimensions result. Third is the MRI which is very accurate, but it is very expensive. Many researchers [8-10] suggest the methods of acquiring signals to diagnose the knee degradation whereas the signals are obtained from a large sample size. The patients, whom the hospital has already received the permission, are asked to sit on a rigid table in a relaxed position with the legs hanging on the air, then a miniature accelerometer is attached to the middle position of the patella. In another literature by Cai S et.al. [8], the method is also used, and they obtained data that have high accuracy of 88-93%.

Patellofemoral cartilage is the white and smooth articulating surface of the patella and femoral trochlea of the knee joint. Loading to the cartilage usually consists of both compression and shear. Compressive stress on the patella comes from contraction of quadriceps and patella tendon during knee flexion. Shear stress comes from the translational motion between the patella and femoral trochlea as the knee joint moves. Mechanical damage due to wear could alter the friction of the articular surface, causing vibration. In this study, friction-induced vibration from friction change of the articular surface is characterized in terms of progressive articular cartilage deterioration.

2. Swine Knee Joint: Experiment

We studied the vibration signals from the knee swine model where the patellofemoral joints of the pig are dissected and separated from the swine leg. The swine's knees are selected due to its similarity with human knees and their availability in many fresh meat markets. In our experiment, the swine's femur and patella are first cleaned by scraping off the meat on the bones. Four levels of femur's cartilage deterioration are then produced by means of mechanical rubbing from sandpaper so that its cartilage is thinning at the different levels until the bone is exposed. The femurs are rubbed 20, 40, 70, and 100 times with a constant force (figure 1a). As shown in figure 1b, a femur is tightly wrapped by elastic bands over our custom-made knee joint testing machine where the motion of knees is speed controlled. Then the patella is drilled at 5 locations, on the top for mounting the sensor using 4-mm drill, and 4 more on the sides of the patella for the screw mountings. After that, the side drills are screwed and hooked to the knee joint testing machine by dental elastic bands. Then the knee joint testing machine's motor driver is connected to the voltage power supply via wires and the computer via a USB cable. The vibration signals are then taken from each level of cartilage damage using the accelerometer attached on the patella, as the patella is translational moving at a constant speed. For each condition (no rub, 20, 40, 70 and 100 rubs), three sets of data are collected. Raw data from the measurement instrument is taken from the measurement instruments software in form of a text file and then converted into time-acceleration graphs. The filtering is then used, to minimize the residue occurred from noises In this research's filtering method is adapted from Cai S et.al. [11], where the finite impulse response (FIR) filter, a common filtering method, is used. The method used temporal statistics which are averaged to obtain the output data at various point of time. The paper suggests that cascade moving average filter is implemented. The method combines the two successive-placed moving average operators. The band power is calculated from the short time Fourier transform. Because VAG (vibroacoustic) signal obtained from the experiments are aperiodic and non-stationary, i.e. it contains varying frequencies that do not appear throughout the whole data. The power $P_{f_1 f_2} = \int_{f_2}^{f_1} S_x(f)df$ is used to calculated for the average power in signal $x(n)$, in range f_1 to f_2, where $S_x(f)$ is the power spectral density (PSD) and the unit of $P_{f_1 f_2}$ is Watt.

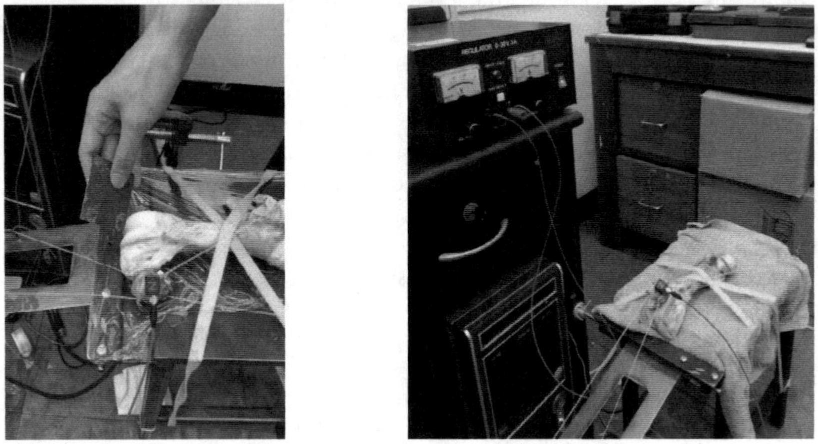

Figure 1. a) Using sandpaper to create wound b) swine knee subject setup.

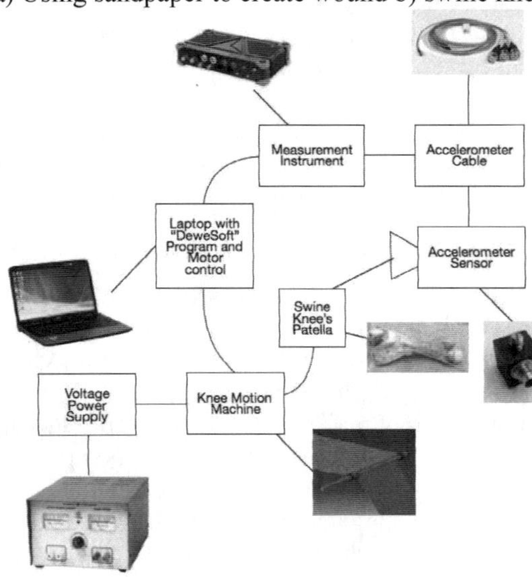

Figure 2. Experiment setup for swine knee data gathering.

3. Swine Knee Joint Testing: Results and Discussions

The shape of time-acceleration graph of no defect condition (figure 3a) is a distorted bell shape with spikes present along the curve. However, if we use the baseline wander filtered method, the graph (figure 3c) became a band with the present of spikes. The roughness of the bone and the vibration from the knee-motion machine could be the reason for the spikes shown on the curve. However, there are only a small amount of spikes on the curve which imply that the surface of patellofemoral compartments are quite smooth (figure 3b). Moreover, from the time-frequency graph (figure 3d), the intensity of the graph is not that high. The calculated band powers are, also, pretty low (see table 1). Therefore, we could say that the vibration produced from our knee-motion machine is quite low and the patellofemoral surface is quite smooth.

It can be observed that the amplitude and the number of spikes of the time-acceleration graphs of the swine knee after being rubbed by sandpaper 20, 40, 70, and 100 times (shown in figures 4a/c and 5a/c, for 40 and 100 rubs respectively) increase as the number of rubs increases. Hence, we can conclude that the smoothness of the time-acceleration graph depends on the condition of the swine knee. The amplitude and the spikes on the graph increase as the surface roughness increases as we rub more against the swine knee surface). For the time-frequency graph, we can see that the intensity of the graph seems to increase as the surface roughness increases so does the calculated band power.

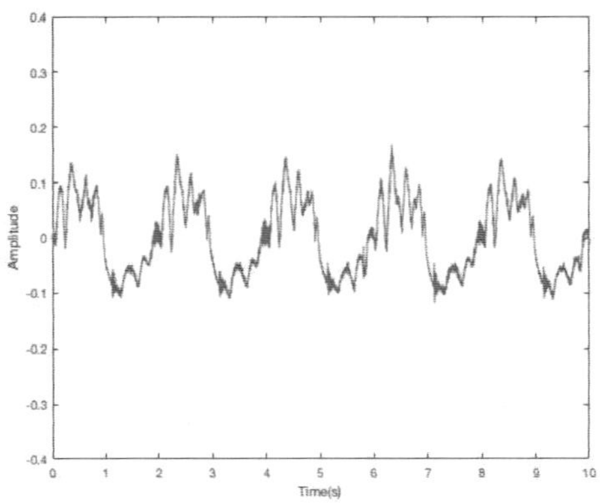

(a)

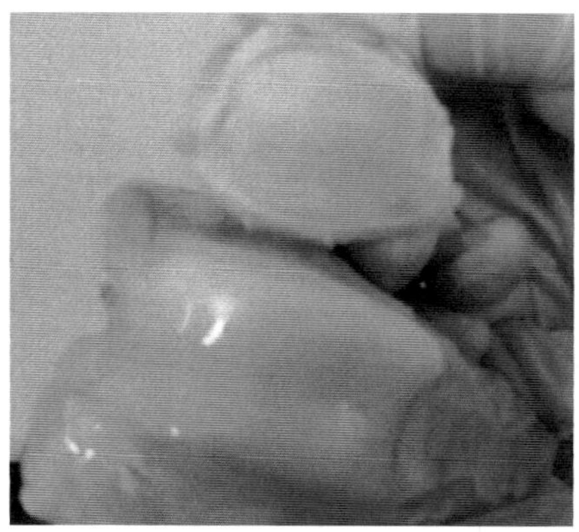

(b)

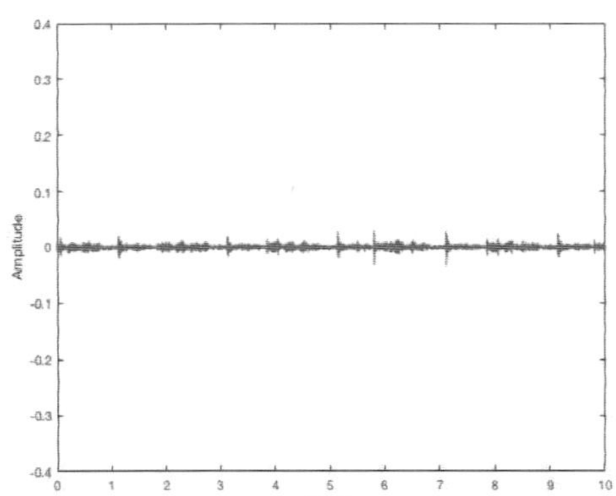

(c)

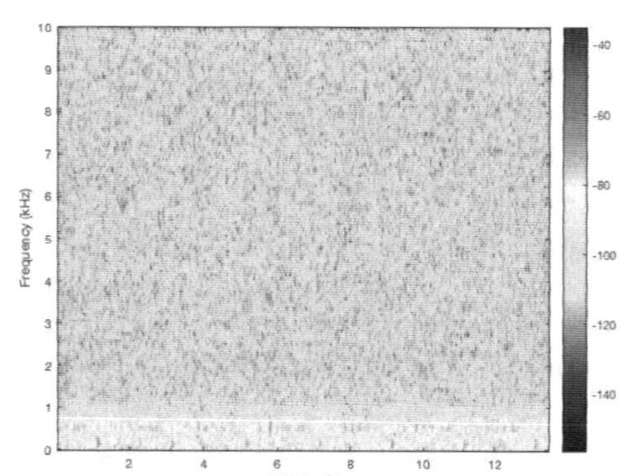

(d)

Figure 3. Swine knee with no defect: (a) Time [s]-acceleration [g] graph; (b) Femur surface; (c) Time [s]-acceleration [g] graph after baseline wander removed; (d) Time [s]-frequency [kHz] graph.

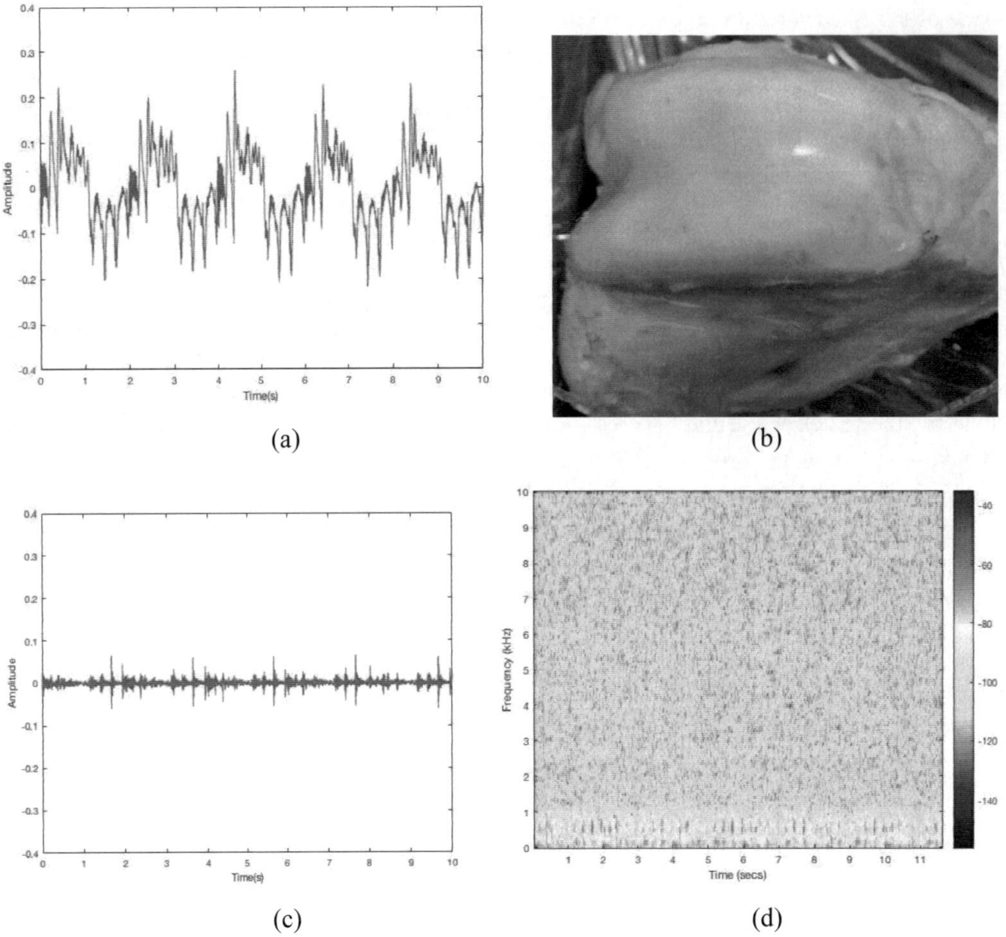

Figure 4. Swine knee rubbed by sandpaper 40 times: (a) Time [s]-acceleration [g] graph; (b) Femur surface; (c) Time [s]-acceleration [g] graph with baseline wander removed; (d) Time [s]-frequency [kHz] graph.

According to the calculated band powers shown in table 1, we can see that the trend of the calculated power spectral density of the signal vertically normal to the movement direction for each frequency range is proportional to the number of rubs as we create more cartilage damage. Especially, higher power spectral density is observed at the high frequency components (500-1,000 Hz) as the level of cartilage damage is progressing.

It is noted that there are many types of knee injury, such as tear, wear, effusion, disruption and etc., and each type of the injury generates different kinds of signal patterns, much more research and data collecting must be done in order for our findings to be applied and usable with real patients. Our further research would be done on real patients. Their patella's vibration signals will be collected, analyzed, and then verified with the MRI results. It is promising that, with some more studies, our proposed approach of using patella's vibration signals to determine the degradation levels of patient's knee would yield reliable and accurate outcomes.

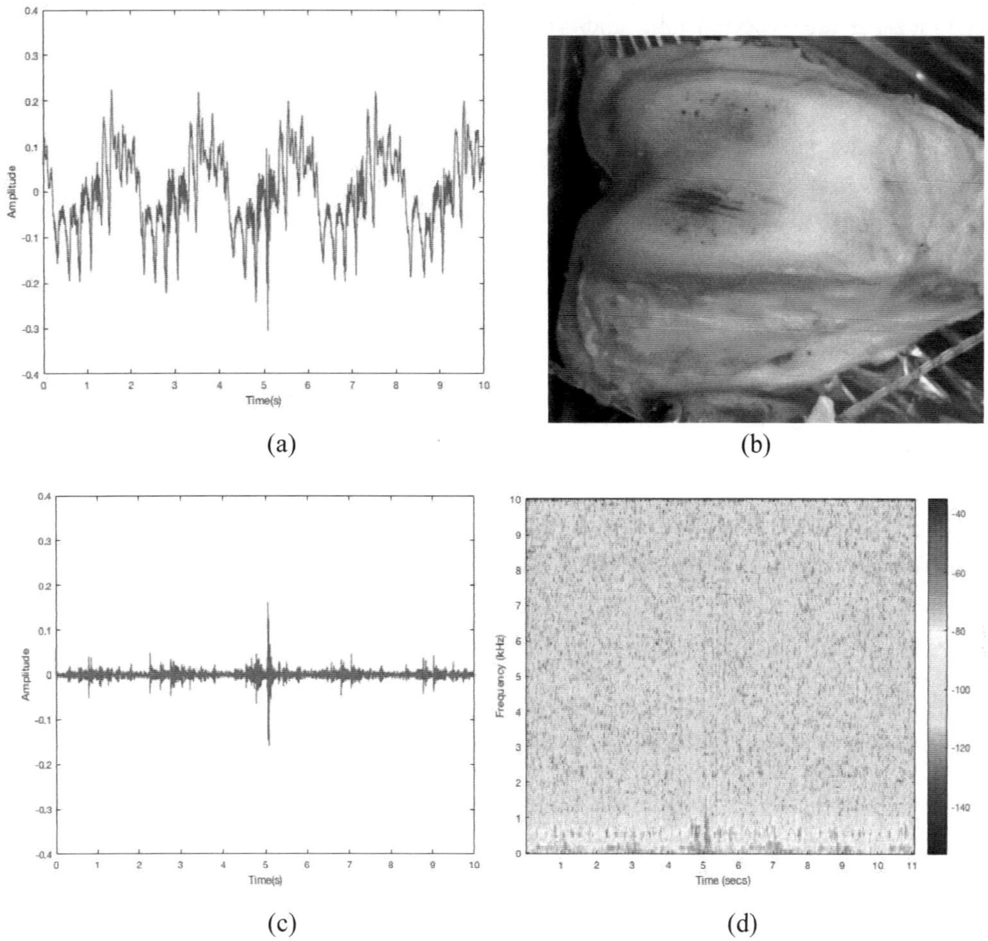

Figure 5. Swine knee after being rubbed by sandpaper 100 times: (a) Time [s]-acceleration [g] graph; (b) Femur surface (bone exposed); (c) Time [s]-acceleration [g] graph with baseline wander removed; (d) Time [s]-frequency [kHz] graph.

Table 1. Overall band power for sandpaper degraded swine knee experiment.

	Frequency range		
	P_1 (0 - 500 Hz)	P_2 (500 - 1000 Hz)	P_3 (200 - 500 Hz)
No defect	1.6814 W	0.8436 W	0.5537 W
20 times rubbed	2.5074 W	1.0652 W	0.3709 W
40 times rubbed	5.3856 W	2.6031 W	1.5299 W
70 times rubbed	6.0100 W	3.7656 W	1.7610 W
100 times rubbed	5.5685 W	3.8564 W	1.8173 W

4. Conclusion

In order to raise awareness for healthy lifestyle of one's knee before pain, a low-cost and effective method is needed. This study illustrated the potential use of the vibration signal for non-invasive detection of cartilage degeneration. An accelerometer is used to measure the vibration signals created by friction and surface roughness of the patella and femur of a swine knee. The signals are then analyzed to determine the severity of cartilage degradation. It can be seen that as the number of rubs on the cartilage increases the acceleration signals have more spikes as well as higher amplitudes. The vibration signal power in the frequency range of 0 to 1000 Hz also goes up as the cartilage becomes more damage. These results are quite promising with further research on human knees. That is we believe that the method of using accelerometer to measure vibration signals of the knee joint is worth the investment for some more studies to successfully implement on real patients.

References

[1] Wu B 2018 Stages of osteoarthritis of the knee *https://www.medicalnewstoday.com/kc/stages-osteoarthritis-knee-310579*
[2] Wu Y 2015 Knee Joint Vibroarthrographic Signal Processing and Analysis Springer Briefs in Bioengineering Springer-Verlag
[3] WebMD 2018 What is an MRI? *https://www.webmd.com/a-to-z-guides/what-is-an-mri#1*
[4] McCauley T R, Kier R, Lynch K J and Jokl P 1992 *American Journal of Roentgenology* **158.1** 101-5
[5] Gagliardi J A, Chung E M, Chandnani V P, Kesling K L, Christensen K P, Null R N, Radvany M G and Hansen MF 1994 y *American Journal of Roentgenology* **163.3** 629-36
[6] Peat G, Duncan R C, Wood L RJ, Thomas E and Muller S 2012 *Arthritis Res. Ther.* **14(2)** R63
[7] Cannon A, Stolley M, Wolf B and Amendola A 2017 *Iowa Orthop. J.* **28** 42–8
[8] Cai S, Yang S, Zheng F, Lu M, Lu Y and Krishnan S 2017 *Comput. Math. Methods Med.* **2013** 904267
[9] Shieh C H, Tseng C D, Chang L Y, Lin W C, Wu L F, Wang H Y, Chao P J, Chiu C L and Lee T F 2016 *Research Notes* **9** 352
[10] Aranchayanont T and Songsiri J 2015 Classification on knee medical condition from vibration signals Electrical Engineering Project Department of Electrical Engineering Faculty of Engineering Chulalongkorn University
[11] Cai S, Wu Y, Xiang N, Zhong Z, He J, Shi L and Xu F 2012 Detrending knee joint vibration signals with a cascade moving average filter Conf. *Proc. IEEE Eng Med Biol Soc.* **2012** 4357-60

IOP Conf. Series: Materials Science and Engineering **501** (2019) 012034 doi:10.1088/1757-899X/501/1/012034

Aerosol Deposition Characteristics During Exhalation Through Nose Treatment: Relationship between Impaction Parameter and Deposition Ratio

S Fujino[1*], T Yamamoto[2], Y Kobayashi[3]

[1] Advanced Course for Advanced Fusion Development, National Institute of Technology Gifu College, 2236-2 Kamimakuwa, Motosu, 501-0495, Japan

[2] Dept. Mech. Eng., National Institute of Technology Gifu College, 2236-2 Kamimakuwa, Motosu, 501-0495, Japan

[3] Dept. Otolaryngology-Head and Neck Surgery, Kansai Medical University, 2-3-1 Shin-machi, Hirakata, 537-1191, Japan

* Corresponding Author: 2017y32@gifu-nct.ac.jp, +81-58-320-1336, +81-58-320-1349

Abstract. Aerosol medicine exhalation through the nose treatment (ETN) has conducted as an effective treatment for Eosinophilic Chronic Rhinosinusitis (ECRS). In ETN, the patient inhales aerosol of inhaled corticosteroid (ICS) medicine from mouth using portable inhaler. Then a part of the aerosol still floas and remains in upper airway. When the patient exhales inhaled air through the nose, the aerosol is effectively transported on the walls of middle meatus and olfactory fissure. The mechanism of how ETN improves ECRS with asthma is still controversial even though ETN gets a lot of attention as a treatment method for ECRS with asthma. In this study, computational fluid dynamics (CFD) analysis of aerosol transport phenomena were performed based on patient CT data in order to evaluate the therapeutic effect of ETN numerically. Three cases were selected in this study, one is a patient who had ECRS with asthma and a history of endoscopic sinus surgery and the others are healthy cases. 3D anatomically accurate patient-specific models were reconstructed from the data obtained using multidetector CT scanner with medical imaging software package. The entire series was loaded into the software, and then the nasal-pharynx airway was identified in each of the axial images based on predefined threshold of -300 Housfield units relative to the surrounding tissue. The nasal-pharynx airway models were exported into a CFD meshing software package to generate discrete volume cells. This study used both a Euler-Lagrange particle transport model for aerosol transport and a model for complex intranasal turbulent flow. This study analyzed both inhaled state and exhaled state, and compared the aerosol deposition characteristics under conditions by changing the particle diameter and flow rate of aerosol. As a result of CFD analysis, amount of aerosol deposition depended on particle diameter and flow rate, and tendency of aerosol deposition was consistent in all cases analyzed in this study.
Furthermore, in the inhaled state, aerosol mainly deposited in nasal meatus areas where not contribute to treatment of ECRS. Exhaled state was more effective treatment for ECRS compared to the inhaled state. The results show advantages of ETN as well as CFD analysis will contribute further development of ENT.

1. Introduction

Aerosol medicine exhalation through the nose treatment (ETN) has conducted as an effective treatment for Eosinophilic Chronic Rhinosinusitis (ECRS) [1,2]. In ETN, the patient inhales aerosol of inhaled corticosteroid (ICS) medicine from mouth using portable inhaler. Then a part of the aerosol still floats and remains in upper airway. When the patient exhales inhaled air through the nose, the aerosol is effectively transported on the walls of middle meatus and olfactory fissure [3,4]. The mechanism of how

IOP Conf. Series: Materials Science and Engineering **501** (2019) 012034 doi:10.1088/1757-899X/501/1/012034

ETN improves ECRS with asthma is still controversial even though ETN gets a lot of attention as a treatment method for ECRS with asthma. In this study, computational fluid dynamics (CFD) analysis of aerosol transport phenomena were performed based on patient CT data in order to evaluate the therapeutic effect of ETN numerically.

2. Case Data

Three cases were selected in this study, one is a patient who had ECRS with asthma and a history of endoscopic sinus surgery (case 1) and the others are healthy cases (case 2, 3). Figure 1 shows the CT images of the patient (case 1) which were scanning using a multi-detector CT (Discovery ST, GE). A 3D anatomically accurate patient-specific model was reconstructed from the data obtained using multi-detector CT scanner with medical imaging software package, Mimics (Materialise Co.) [5]. The entire series was loaded into the software, and then the nasal-pharynx airway was identified in each of the axial images based on predefined threshold of -300 Housfield units relative to the surrounding tissue. Figure 2 is 3D nasal model. In the 3D model, NP is the nasopharynx, LETH 0, 1, 2 and RETH 0, 1, 2 are the ethmoid sinus region. LOFT is a region created for both suppressing the backflow phenomenon at the outlet boundary, and consequently stabilizing the convergence of the analysis.

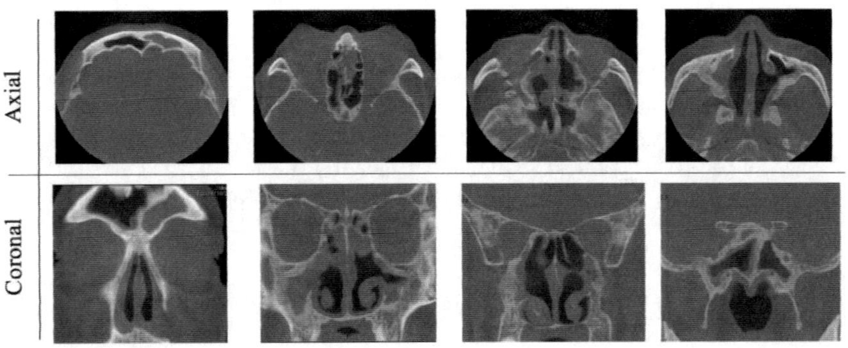

Figure 1. CT images of a patient who suffers Eosinophilic Chronic Rhinosinusitis.

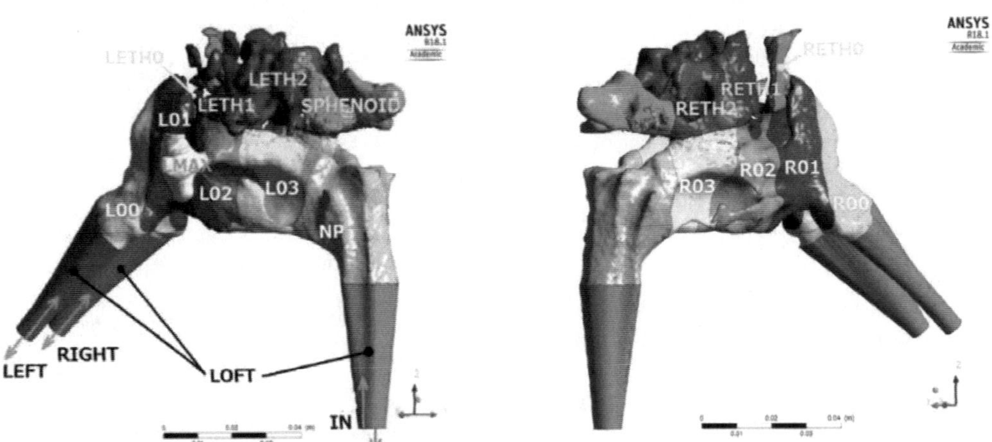

(a)Left side view (b)Right side view

Figure 2. 3D nasal model used in the study.

3. Computational Fluid Dynamics Analysis

The nasal-pharynx airway model was exported into CFD meshing software package, ICEM-CFD (ANSYS Co.) to generate discrete volume cells [6]. This study used both a Euler-Lagrange particle transport model for aerosol transport and a model for complex intranasal turbulent flow. This study analyzed both inhaled state and exhaled state, and compared the aerosol deposition characteristics under conditions by changing the particle diameter and flow rate of aerosol.

The condition of flow rate of through nose set at 10 l/min, 35 l/min, 50 l/min and 70 l/min, respectively. And the condition of particle diameter set at 1μm, 3μm and 10μm, respectively.

4. Evaluation of deposition characteristics of particles

As an index for evaluating the results of this analysis, we define the particle Deposition Efficiency (DE) and the Impaction Parameter (IP). Particle deposition rate is defined as in equation (1).

$$\mathrm{DE} = \frac{C_{dep}}{C_{in}} \qquad (1)$$

C_{dep} is the total amount of particles deposited on the entire nasal cavity - pharyngeal region or on the surface of each region in the nasal cavity, and C_{in} denotes the total amount of inflowing particles. Collision parameter is defined as in equation (2) [7].

$$\mathrm{IP} = {d_a}^2 \cdot Q \qquad (2)$$

d_a is the aerodynamic diameter and Q is the volumetric flow rate.

5. Result and Discussion

Figure 3 shows the CFD results of aerosol particle trajectory in both exhaled and inhaled states for patient case 1. It shows that the particles collide with each other in the nasopharynx with both exhalation and inhalation, and the particles are meandering. Figure 4 shows the relationship between deposition efficiency and impaction parameter in the whole nose for each case. Furthermore, in each figure, the experimental value of deposition efficiency [8] was compared with the analysis value. The vertical axis shows the particle deposition rate and the horizontal axis shows the impaction parameter. In all conditions, the trends are consistent. From the results, it is considered that the deposition efficiency of the particle depends on the impaction parameter. In addition, it was found that there is a tendency similar to the experimental value of deposition efficiency in all cases. Figure 5 shows the relationship between deposition efficiency and impaction parameter in the nasopharynx, ethmoid and nasal cavities of case 1. At exhalation, as the impaction parameter increases, the inertial collision in the nasopharynx is promoted, whereas in inhalation, particles due to gentle change in direction from the nostrils to the middle nasal passage. In addition to the inertial movement of the flow, peeling of the flow accompanying expansion and contraction of the inside of the nasal cavity acts, and it is presumed that particles gradually deposit before reaching the nasopharynx. As a result, in the inhaled state, it was found that aerosol mainly accumulates in the nasal cavities areas which does not contribute to treatment of ECRS. Therefore, exhaled states are considered more effective as a more effective treatment for ECRS compared to inhaled states.

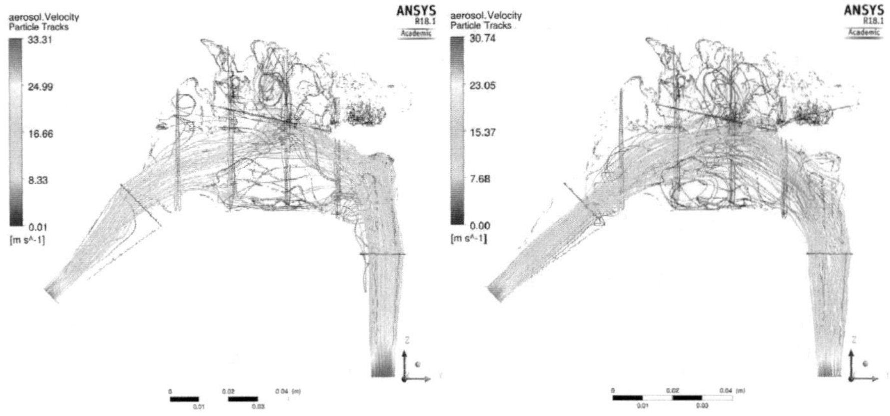

(a) Exhaled states (b) Inhaled states

Figure 3. Particle trajectories of analysis nasal model (case 1).

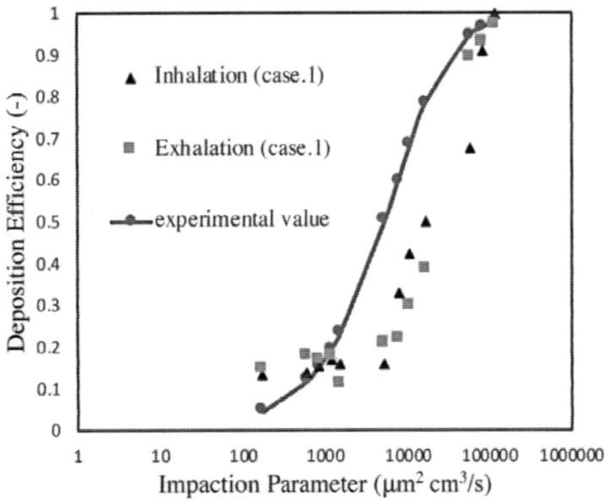

(a) Case 1

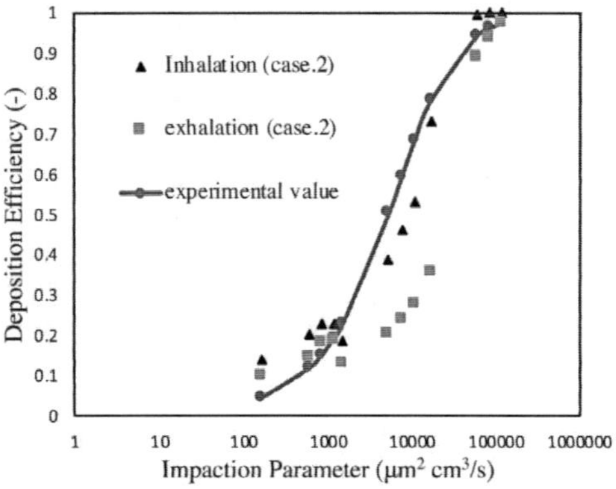

(b) Case 2

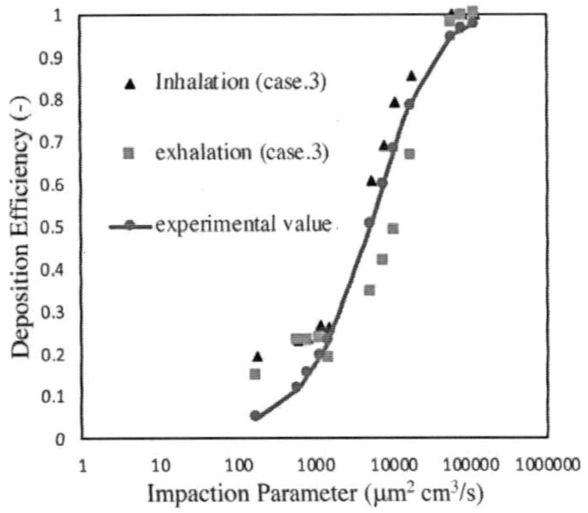

(c) Case 3

Figure 4. Relationship between deposition efficiency and impaction parameter (each cases).

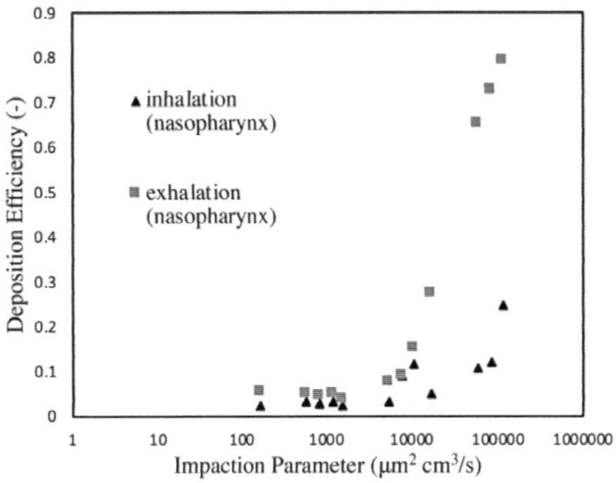

(a) Nasopharynx

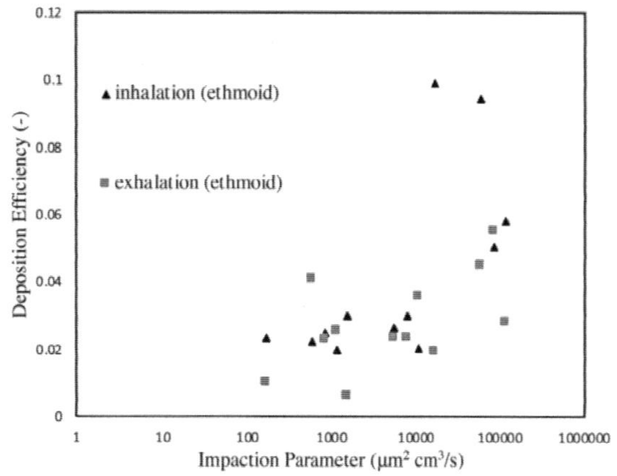

(b) Ethmoid

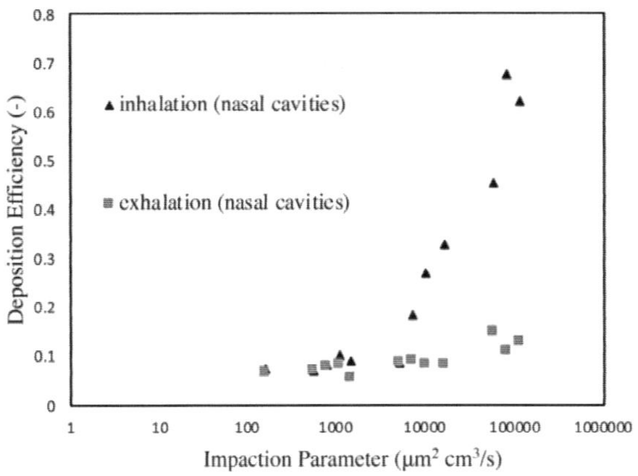

(c) nasal cavities

Figure 5. Relationship between deposition efficiency and impaction parameter (Nasopharynx, Ethmoid and nasal cavities).

6. Conclusion

In this study, computational fluid dynamics (CFD) analysis of aerosol transport phenomena were performed based on patient CT data in order to evaluate the therapeutic effect of ETN numerically. As a result, amount of aerosol deposition depended on particle diameter and flow rate, and tendency of aerosol deposition was consistent in all cases analyzed in this study. Furthermore, in the inhaled state, aerosol mainly deposited in nasal cavities areas where not contribute to treatment of ECRS. Exhaled state was more effective treatment for ECRS compared to the inhaled state.

Acknowledgement

This work was supported by JSPS KAKENHI Grant Number16K06098.

References

[1]Kohno T, Yamamoto Y and Kurokawa M 2006 Effect of inhaled corticosteroid on allergic rhinitis

[2]Asthema 19(4) pp 72-76.

[3]Kobayashi Y, Asako M, Kanda A, Tomoda K and Yasuba H 2014 A novel therapeutic use of HFA-BDP metered-dose inhaler for asthmatic patients with rhinosinusitis: *Case series. Int. J. Clin. Pharm. Th.* 52 pp 914-919.

[4]Kobayashi Y, Asako M, Ooka H, Kanda A, Tomoda K and Yasuba H 2015 Residual exhaled nitric oxide elevation in asthmatics in associated with eosinophilic chronic rhinosinusitis Asthma 52(10) pp 1060-1064.

[5]Hamada S, Matsumoto H, Kobayashi Y, Asako M and Yasuba H 2016 Nasal exhalation of inhaled beclomethasone hydrofluoroalkane-134a to treat chronic rhinosinusitis. J. Allergy Clin Immunology In practice 4(4) pp 751-753.

[6]Kuroda K, Yamamoto T and Hirose T 2015 CFD analysis for medicinal aerosol transport characteristics in nasal cavity flow *JSME Bioengineering Conference 2015* Niigata Japan.

[7]Materialize Inc. 2014 Mimics Users Guide.

[8]James T K, Bahman A, Julia S, Kimbell A W and Brian A W 2004 Particle Deposition in Human Nasal Airway Replicas Manufactured by Different Methods. Part I: Inertial Regime Particles *Aerosol Science and Technology* 38 pp 1063-1071.

[9]L Zhang and C P Yu 1993 Empirical Equations for Nasal Deposition ofInhaled Particles in Small Laboratory Animals and Humans *Aerosol Science and Technology* 19:1 pp 51-56.

IOP Conf. Series: Materials Science and Engineering **501** (2019) 012035 doi:10.1088/1757-899X/501/1/012035

The Analysis of Strain Distribution on Corrected Varus Knee under Walking Load Condition

Yanin Ruangrittisak[1], Kitti Aroonjarattham[2], Promtpong Anuchitchanchai[2], Chompunut Somtua[1] and Panya Aroonjarattham[1,*]

[1]Department of Mechanical Engineering, Faculty of Engineering, Mahidol University, Nakornpathom, Thailand 73170
[2]Department of Orthopaedics, Faculty of Medicine, Burapha University, Chonburi, Thailand 20131

* Corresponding Author: panya.aro@mahidol.ac.th

Abstract. The knee is one of the important joint of human body to carry the body load. The deformities of knee joint affect the human activities. To treat the deformity knee, close-wedge High Tibial Osteotomy (HTO) was one of the popular techniques to adjust the strain distribution on the deformity knee close to the normal by shifting the mechanical axis toward the lateral side. This research aims to evaluate the strain distribution on the varus knee corrected by zero degree, three degrees valgus and three degreesvarus close-wedge HTO under walking load. The result was shown the lowest strain distribution occur on the three degreesvarus model following by zero degree and three degree valgus model respectively. The surgeon should be correct the varus knee by three degreesvarus close-wedge HTO to get the strain distribution on the bone closet to the normal knee.

1. Introduction

Knee is one of the important joint to carry the body weight and activities load of human life. The mechanical axisof normal knee jointis an axis to transfer load from the body to foot is determined by drawing a vertical line from the center of femoral head to themiddle of femoral epicondyle and passes through the ankle joint [1]. The center of varusknee was shifted to the lateral side of mechanical axis and the hip-knee-ankle (HKA) was negative as shown in figure 1 but it becomes zero in case of neutral alignment [2].

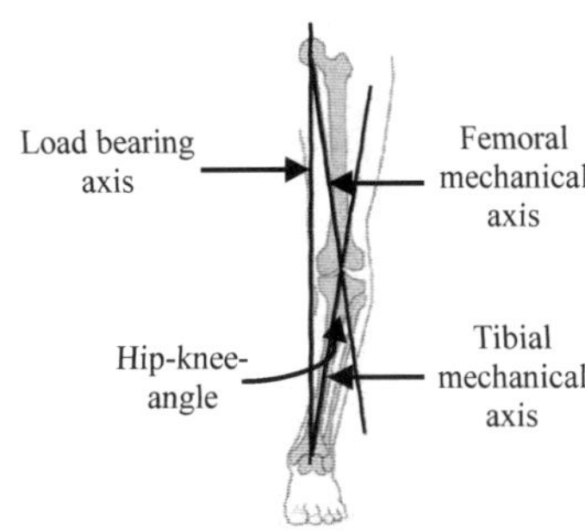

Figure 1. Load bearing axis and Mechanical axis of varus knee[3].

The correction of mechanical axis in varus knee at proximal tibia with a high tibial osteotomy (HTO) to adjust the strain distribution on the deformity knee close to normal had a good long-term result [4, 5, 6, 7]. From previous study, the varus knee corrected by close wedge HTO had a lower strain distribution than varus knee but higher than varus knee inserted total knee prosthesis under daily activities [8] but the angle of correction affect the strain distribution on the bone model [9]. This research aims to evaluate the strain distribution on the lower extremity after corrected the varus knee by close wedge HTO with zero degree, three degrees valgus and three degreesvarusof mechanical axis by finite element analysis to find the suitable angle of correction for support the surgeon's decision in surgical process.

2. Material and methods

2.1. Three-dimensional model

*2.1.1. Varus knee model*3D model of varus knee was scanned by computerized tomographic (CT) scanner and was reconstructed by ITK-SNAP program as shown in figure 2.

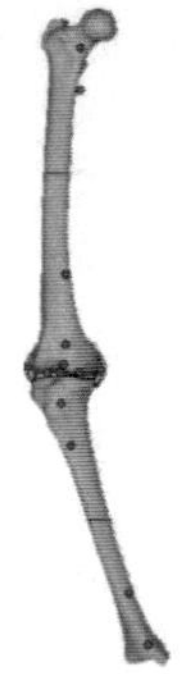

Figure 2.3D model of varus knee.

*2.1.2. Tibial condylar plate and fixation screws*To stabilize the bone after corrected by close wedge HTO technique, the tibial condylar plate was used to fix the cut proximal tibia. 3D tibial plate was constructed from CT data same as the bone and the screw fixation was created by SolidWorks software. The plate and screwswas shown in figure 3.

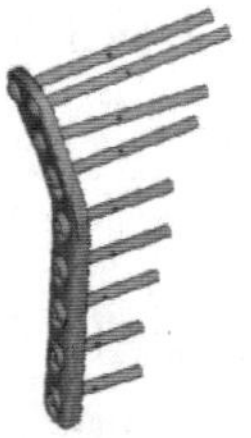

Figure 3.3D models of tibial condylar plate and screw fixations.

*2.1.3. Ligament and meniscus model*The knee joint was consisted of four ligaments as anterior cruciate ligament(ACL), posterior cruciate ligament (PCL), medial collateral ligament (MCL) and lateral collateral ligament (LCL)to stabilize the joint and meniscus to absorb the load transferred. The meniscus and ligamentswere created based on actual anatomy position by SolidWorks software as shown in figure 4.

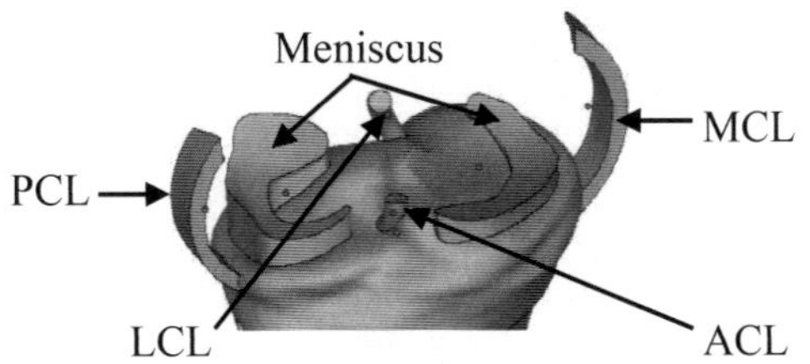

Figure 4.3D models of four ligaments and meniscus.

2.2. Virtual simulation

This research used the virtual simulation method to remove the partial of proximal tibia, correct the mechanical axis and insert the tibialplate and screw fixations to fix the tibia fragment. The varus knee model was corrected with three angle as zero degree, three degrees valgus and three degreesvarusby vary the degree of cutting proximal tibia. 3D models of corrected varus knee by close-wedge HTO were shown in figure 5.

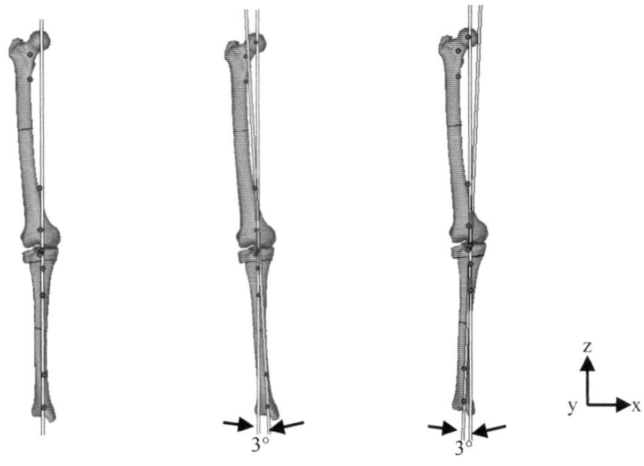

Figure 5. 3D models of corrected varus knee by close-wedge HTO: (a)Zero degree, (b)Three degrees valgus and (c)Three degrees varus.

2.3. Material properties

Material properties of cortical bone, cancellous bone, four ligaments, meniscus, screw fixation and tibial condylar plate were assumed homogeneous, isotropic and linear elastic [10, 11, 12]. The elastic modulus and Poisson's ratio of all material was shown in table 1.

Table 1. Material properties of all model[13].

Materials	Elastic modulus (MPa)	Poisson's ratio
Cortical bone	14,000	0.30
Cancellous bone	600	0.20
ACL, PCL and LCL	345	0.40
MCL	332.2	0.40
Meniscus	12	0.45
Stainless steel	200,000	0.30

2.4. Loading and boundary condition

The position of muscular forces for walking load for daily activities was act on the proximal femur. The body weight act on the top of femur and the model was fixed at the end of distal tibia as shown in figure 6. The magnitude of muscular force and body weight was separated in x-,y- and z-axis under walking condition were shown in table 2.

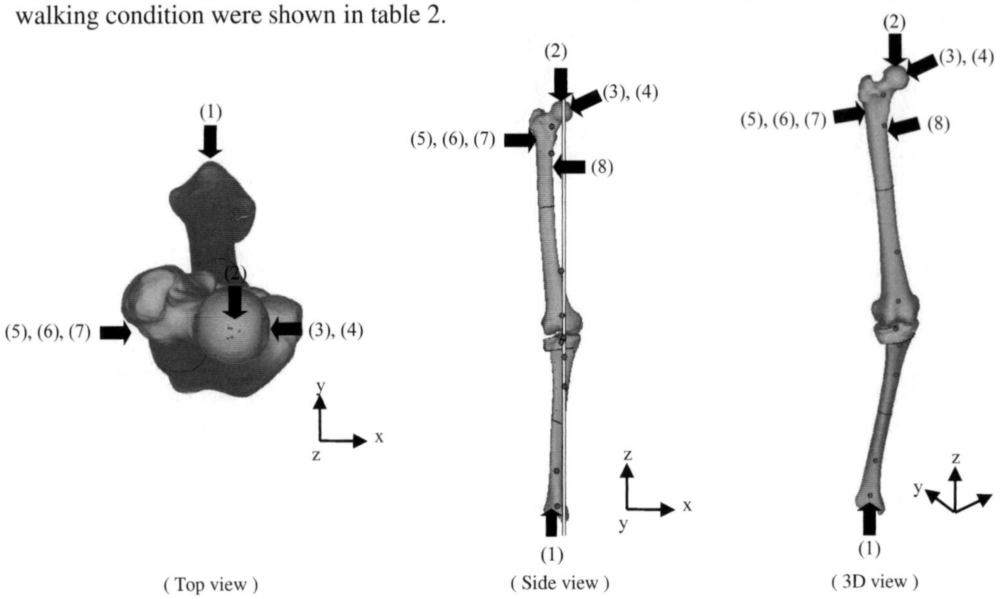

Figure 6. The position of muscular forces and body weight act on the proximal femur.

Table.2 The muscular force act on proximal femur under walking condition [14].

Position	Force	F_x(N)	F_y(N)	F_z(N)
1	Fix displacement	0.0	0.0	0.0
2	Body weight	0.0	0.0	-836.0
3	Hip contact	-54.0	-32.0	-229.2
4	Intersegmental resultant	-8.1	-12.8	-78.2
5	Abductor	58.0	4.3	86.5
6	Tensor fascia latae, proximal part	7.2	11.6	13.2
7	Tensor fascia latae, distal part	-0.5	-0.7	-19,0
8	Vastuslateralis	-0.9	18.5	-92.9

2.5. Finite element model

Four-node tetrahedral element was used in this study to create the mesh model as shown in figure 7 to analyze the equivalent of total stain distribution on bone model and von Mises stress distribution on tibial condylar plate and screw fixation. The femur model had a total of 10,463 nodes and 54,110 elements. The tibia model had a total of 28,092 nodes and 270,748 elements. The fixations had a total of 6,013 nodes and 25,512 elements.

(a) (b) (c)

Figure 7.3D mesh model: (a) Femoral bone,(b) Tibia bone and (c) Plate and screw fixation.

3. Result and discussion

All results was analysed by finite element analysis to evaluate the equivalent of total strain distribution on the varus knee corrected by close wedge HTO with three different degree as zero degree, three degrees valgus and three degreesvarus under walking condition. The maximumequivalent of total strain on three models was shown in table 3 and the strain distribution on femur and tibia were shown in figure 8 and 9 respectively.

Table.3 The maximum equivalent of total strain under walking condition (με).

Model	The maximum equivalent of total strain (με)
Zero degree	3,595.86
Three degrees valgus	3,644.01
Three degreesvarus	3,545.41

The maximum equivalent of total strain on the varus corrected by close wedge HTO with zero degree had 3,595.86microstrain and three degreesvarus had less than by 1.40% and three valgus had greater than by 1.34%. Themaximum equivalent of total strain of all models did not exceed 25,000 microstrain that safe for the bone damage after correct the varus knee [15, 16]. The surgeon should be correct varus knee by close wedge HTO with three degrees varus for the lowest maximum equivalent of total strain.

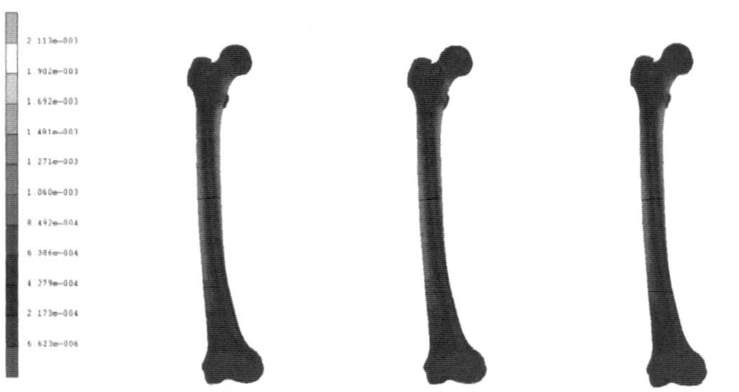

(a) (b) (c)

Figure 8.The equivalent of total strain distribution on femur under walking condition: (a) Zero degree, (b) Three degrees valgus and (c) Three degreesvarus.

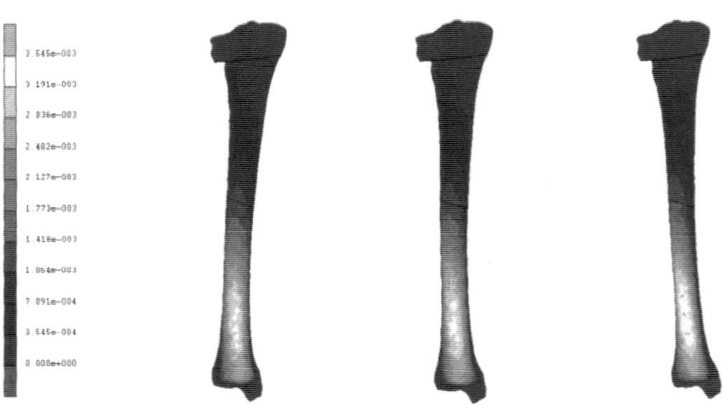

(a) (b) (c)

Figure 9.The equivalent of total strain distribution on tibia under walking condition: (a) Zero degree, (b) Three degrees valgus and (c) Three degreesvarus.

The maximum von Miss stress on tibial condylar plate and nine screw fixations for three different models was shown in table 5and the stress distribution on tibial condylar plate and set of screw fixation were shown in figure 10 and 11 respectively.

Table.4 The maximum von Miss stress on tibia condylar plate and nine screw fixations.

Model	Maximum von Mises stress (MPa)									
	Tibial condylar plate	Screw 1	Screw 2	Screw 3	Screw 4	Screw 5	Screw 6	Screw 7	Screw 8	Screw 9
Zero degree	421.49	20.96	50.00	106.00	66.35	15.24	16.87	17.65	27.16	195.44
Three degrees valgus	118.37	25.34	73.84	117.94	45.56	28.53	17.90	38.30	93.40	18.69
Three degreesvarus	129.61	23.98	25.48	103.97	81.98	14.97	14.38	22.67	25.07	105.55

(a) (b) (c)

Figure 10. The von Mises stress distribution on tibial condylar plate under walking condition: (a) Zero degree, (b) Three degrees varus and (c) Three degrees valgus.

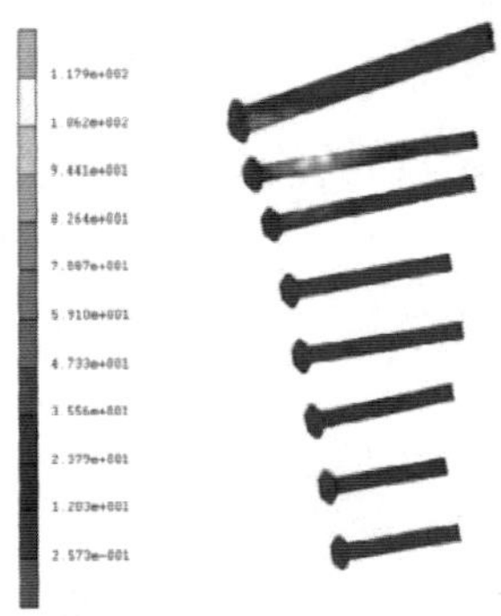

Figure 11. The maximum von Mises stress on the screw fixation under walking condition (Mpa) on model three degree valgus.

The highest maximum von Mises stress occurred on tibial condylar plate of zero degree model which was used to fix the top of proximal tibia. The maximum von Mises stressof all plate and screw modelsdid not reach the yield strength of stainless steel (750 MPa) [17] that safe for implant failure.

4. Conclusion

The prediction of strain distributed on lower extremity by finite element analysis was a popular tool for surgeon to plan the surgical treatment. The current investigation could lead to the acquisition of useful knowledge in order to establish of efficacious method to prevent the bone fracture after corrected the knee by close-wedge HTO[18]. The angle of correction did not affect the strain distribution on the bone model that showed less difference of maximum equivalent of total strain on three model lower extremities. The surgeon should be correct the varus knee by three degreesvarus close-wedge HTO to get the strain distribution on the bone closet to the normal knee.

Acknowledgments
The authors wish to thank the Faculty of Medicine, Burapha University and Biomechanics Analysis and OrthopedicDevice Design Laboratory (B-AODD LAB), Department of Mechanical Engineering, Faculty of Engineering, Mahidol University for their support with facilities.

References

[1] Lee D C and Byun S J 2012 High tibial osteotomy *Knee SurgRelat Res***24(2)** 61–69.
[2] Cooke D T, Harrison L, Khan B, Scudamore A, and Chaudhary A M 2002 *RheuInt* **22(4)** 160-164.
[3] Dynamic Chiropractic. Knee Osteoarthritis: Biomechanical Risks [Available on: http://www.dynamicchiropractic.com/mpacms/dc/article.php?id=55312] [Accessed on 30/1/2016].
[4] Coventry M B, Ilstrup DM and Wallrichs SL 1993 *J Bone Joint Surg Am* **75(2)** 196-201.
[5] Insall J, Joseph DM and Msika C 1984 *J Bone Joint Surg. Am* **66(7)** 1040-1048.
[6] Nagel A, Insall J N and Scuderi GR 1996 *J Bone Joint Surg. Am* **78(9)** 1353-1358.
[7] SprengerTR and DoerzbacherJF 2003 *J Bone Joint Surg Am* **85-A(3)** 469-474.
[8] Ruangrittisak Y, Anuchitchanchai P, Aroonjarattham K, Somtua C and Aroonjarattham P 2016 The strain distribution on varus knee corrected by close wedge high tibialosteotomy technique versus total knee replacement *Proc. 7th TSME Int. Conf. on MechEng*207.
[9] Aroonjarattham P, Aroonjarattham K and Suvanjumrat C 2014 *J Kasetsart (Nat. Sci.)* **48(2)** 263-282.
[10] Sitthiseripratip K, Van Oosterwyck H, Vander Sloten J, Mahaisavariya B, Bohez ELJ, Suwanprateeb J, Van Audekercke R and Oris P 2003 *Med Eng&Phy* **25(2)** 99-106.
[11] Serala B, Garc JM, Cegon J, Doblare M and Serala F 2004 *Injury* **35(2)** 130-135.
[12] Mahaisavariya B, Sitthiseripratip K and Suwanprateeb J 2006 *Injury* **37(8)** 778-785.
[13] Peraz A, Mahar A, Negus C, Newton P and Impelluso T 2008 *Med Eng&Phys* **30(6)** 755-760.
[14] Heller M O, Bergmann G, Kassi JP, Cales L, Haas NP and Duda G N 2005 *J Biomech* **38(5)** 1155-1563.
[15] Frost HM 1994 *The Angle Orthodontist* **64(3)** 175-188.
[16] Frost HM 2004 *The Angle Orthodontist* **74(1)** 3-15.
[17] Luo C A, Hua S Y, Lin S C, Chen C M and Tseng C S 2013 *Musculoskeletal Disorder* **14** 1-8.
[18] Bessho M, Ohnishi I, Matsuyama J, Matsumoto T, Imai K and Nakamura K 2007 *J. Biomech.* **40(8)** 1745-1553.

Biomechanical Study of Femoral Bone after Proximal Bone Tumor Removal

Panya Aroonjarattham[1], Peeranat Kijthaworn[1], Promchai Jamda[1], Tanakorn Supamongkol[1], Samart Suwanboriboon[2,*], Kitti Aroonjarattham[2] andChompunut Somtua[1]

[1] Department of Mechanical Engineering, Faculty of Engineering, Mahidol University, Nakornpathom, Thailand 73170
[2] Department of Orthopaedics, Faculty of Medicine, Burapha University, Chonburi Thailand 20131

* Corresponding Author: samartsuwanboriboon@gmail.com

Abstract. Proximal femur is connected between pelvis and the lower extremity to carry the body weight that transfer from the upper body to the lower part. The patients, had the proximal femur bone tumor had a problem in daily activities. The most treatment for bone tumor has two methods as radiation and surgery method. The surgeon removed the bone tumor, replaced with the other proximal femur and fixed by dynamic hip screw and four screws fixation. This study aim to analyse the postoperative results after surgery with finite element analysis, were divided into six cases as the same size proximal femur, 15% and 30% small proximal femur, 15% and 30% large proximal femur and the same size proximal femur constructed by polymethyl methacrylate (PMMA). The result were shown the maximum strain on the small size higher than another cases under walking and stair-climbing load and the same size proximal femur constructed by PMMA had the lowest maximum equivalent of total strain. The surgeon should be replacing the proximal femur with the same or the larger size of femur or the same size constructed by PMMA for a good clinical result.

1. Introduction

The patient with a giant cell tumor at proximal femur had a problem in daily activities[1, 2]. The most popular treatment was radiation and surgery depended on the surgeon's decision. This study was focus on the surgery method. After the surgeon removed the bone tumors, it will replace with another proximal femur to restore the function of hip joint. The problems of size and inclination angle [3] of femur affect the strain concentration that made the bone function loss and bonefracture.

This study aims to evaluate the strain distribution on the femoral bone after remove bone tumor at the proximal part and replace with another femur, which was fixing the same inclination angle. The replaced femur varied six sizes as original proximal femur, smaller two sizes and larger two sizes than the original bone and equal the original but construct from biomaterial namely polymethylmethacrylate(PMMA). The results are a guideline for the surgeon's decision to plan the surgery process.

2. Materials and Methods

IOP Conf. Series: Materials Science and Engineering **501** (2019) 012036 doi:10.1088/1757-899X/501/1/012036

2.1 Bone modeling

The femur and tibia bones were created from the computerized tomography (CT) data and reconstructed with ITK-SNAP program. The three-dimensional model was shown in figure 1.

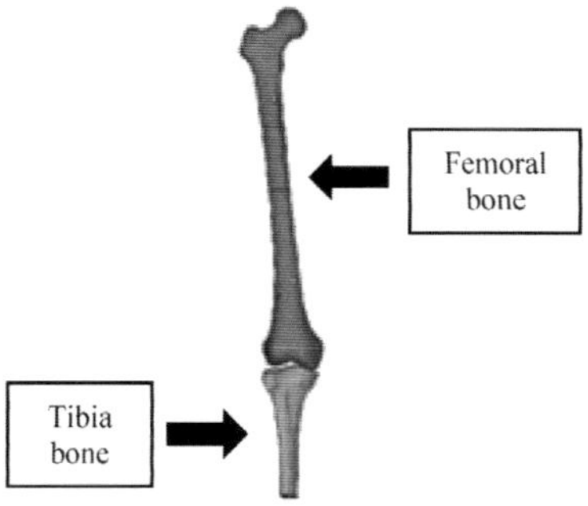

Figure 1.3D bone model.

At the knee joint, three-dimensional models of meniscus and four ligaments to connect between distal femur and proximal tibia as anterior cruciate ligament (ACL), posterior cruciate ligament (PCL), medial collateral ligament (MCL) and lateral collateral ligament (LCL)was drawn with SolidWorks software with actual shape and anatomical position. The model of meniscus and four ligaments were shown in figure 2 and 3 respectively.

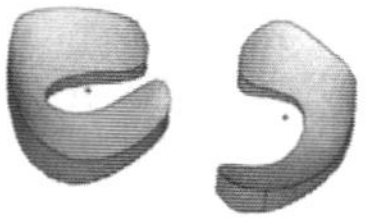

Figure 2.3D model of meniscus.

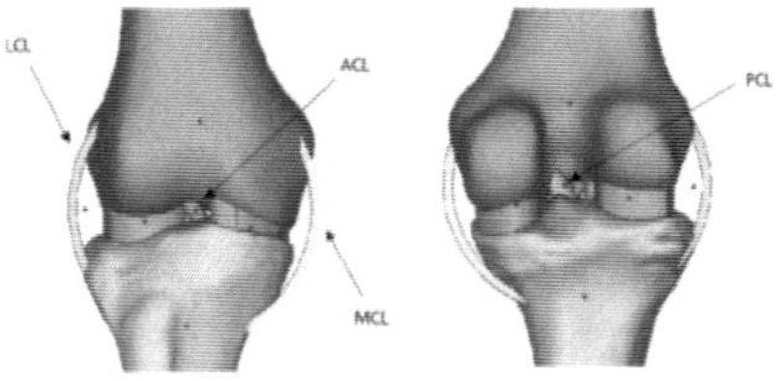

Figure 3.3D model of four ligaments.

2.2 Implant model

The implant used to fix the proximal femur after bone tumor removal was a dynamic hip screw included lag screw, side plate with barrel and cortical screws. The model of dynamic hip screw was created with SolidWorks software as shown in figure 4.

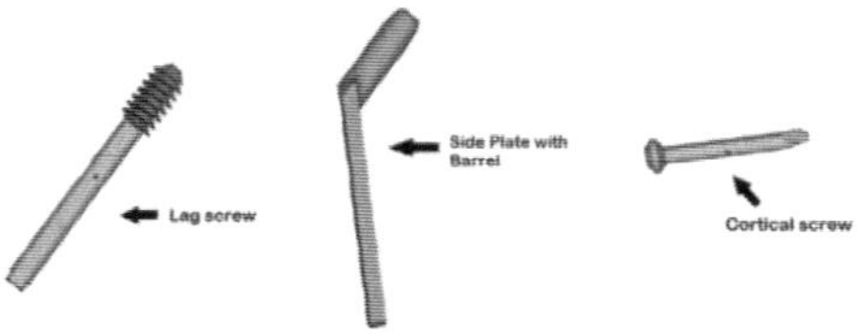

Figure 4.3D model of dynamic hip screw.

2.3 Material properties

All models in this study were assumed homogeneous, isotropic and linear elastic. Young's modulus and Poisson's ratio of materials were shown in table 1.

Table 1. Material properties of bone-implant model [4,5,6].

Material	Young's modulus (MPa)	Poisson's ratio
Cortical bone	14,000	0.30
Cancellous bone	600	0.20
Meniscus	12	0.45
ACL	345	0.40
PCL	345	0.40
MCL	345	0.40
LCL	332.2	0.40
Stainless steel	200,000	0.30

2.4 Boundary condition

The body weight act on the top of femoral head and the muscular force act on the proximal part under two conditions as walking and stair-climbing condition [7]. The position of body weight and muscular force was shown in figure 5 and 6 and the amplitude was shown in the table 2.

Figure 5. Fix displacement of boundary condition act on the proximal tibia.

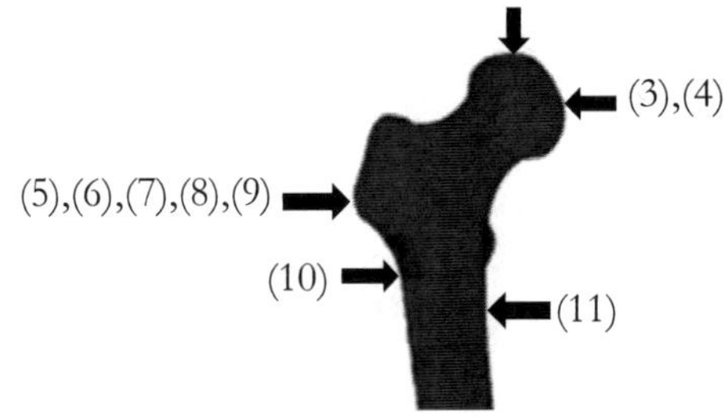

Figure 6. The position of body weight and muscular force act on proximal femur.

Table 2. The amplitude of force under walking and stair-climbing condition [7].

Position	Force	Walking			Stair-climbing		
		Fx (N)	FY (N)	Fz (N)	Fx (N)	FY(N)	Fz (N)
1	Fix displacement	0	0	0	0	0	0
2	Body weight	0	0	-836.0	0	0	-847.0
3	Hip contact	-54.0	-32.8	-229.2	-59.3	-60.6	-236.3
4	Intersegmental resultant	-8.1	-12.8	-78.2	-13.0	-28.0	-70.1
5	Abductor	58.0	4.3	86.5	70.1	28.8	84.9

6	Ilio-tibialtract,proximal part	-	-	-	10.5	3.0	12.8
7	Ilio-tibialtract,distal part	-	-	-	-0.5	-0.8	-16.8
8	Tensor fascia latae,proximal part	7.2	11.6	13.2	3.1	4.9	2.9
9	Tensor fascia latae,distal part	-0.5	-0.7	-19.0	-0.2	-0.3	-6.5
10	Vastuslateralis	-0.9	18.5	92.9	-2.2	22.4	-135.1
11	Vastusmedialis	-	-	-	-8.8	39.6	-267.1

2.5 Cases analysis

The replaced proximal femur varied five sizes: the smaller size was replaced proximal femur with 70 and 85 percentage, the larger size was replaced proximal femur with 115 and 130 percentage and the equal size constructed from PMMA as shown in figure 7. All cases analysis were shown in table 3.

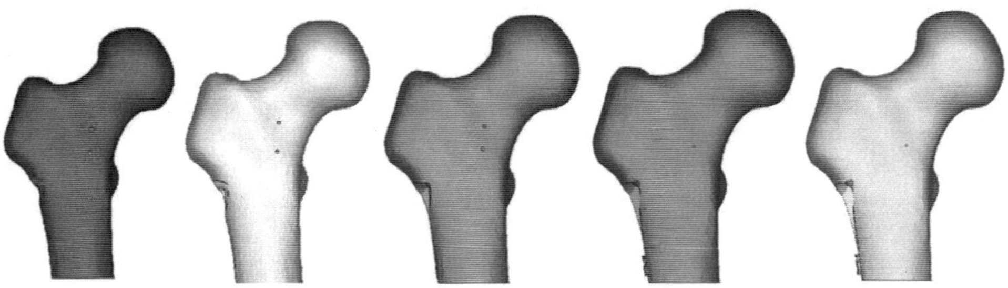

(a) (b) (c) (d) (e)

Figure 7. The model of proximal femur: (a) 70%, (b) 85%, (c) 100%, (d) 115% and (e) 130%.

Table 3. Cases analysis.

Cases	Model	Condition
1	70% Proximal femur	Walking
2	85% Proximal femur	Walking
3	100% Proximal femur	Walking
4	115% Proximal femur	Walking
5	130% Proximal femur	Walking
6	PMMA Proximal femur	Walking
7	70% Proximal femur	Stair-climbing
8	85% Proximal femur	Stair-climbing
9	100% Proximal femur	Stair-climbing
10	115% Proximal femur	Stair-climbing
11	130% Proximal femur	Stair-climbing
12	PMMA Proximal femur	Stair-climbing

2.6 Mesh generation

MSC software package was usedto generatethe mesh model of femur, tibia, meniscus and all ligaments with four-node tetrahedral element as shown in figure 8.

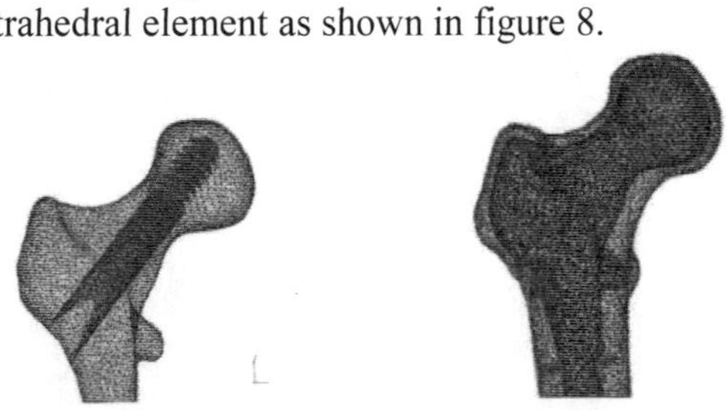

(a) (b)

Figure 8. Mesh model of: (a) Cancellous bone of femur and (b) Cortical bone of femur.

Total number of nodes and elements of 10cases analysis under walking and stair-climbing was shown in table 4.

Table 4. Total number of nodes and elements.

Cases	Nodes	Elements
1 and 7	84,412	363,917
2 and 8	88,591	384,262
3 and 9	90,502	393,322
4 and 10	101,052	443,140
5 and 11	110,479	489,758
6 and 12	96,060	427,306

3. Results and Discussion

3.1 The maximum equivalent of total strain on femur under walking and stair-climbing condition

The maximum equivalent of total strain on femur of all casesunder walking and stair-climbing condition was shown in the table 5.

Table 5.Maximum equivalent of total strain under daily activities.

Cases	Maximum equivalent of total strain (με)	
	Walking	Stair-climbing
70% Proximal femur	1,977.58	2,983.08
85% Proximal femur	1,762.85	2,350.92
100% Proximal femur	1,567.76	2,178.84
115% Proximal femur	1,273.99	1,312.73
130% Proximal femur	1,519.68	1,477.85
PMMA Proximal femur	1,134.97	1,256.61

The stain distribution on femoral bone of all cases under walking load was shown in figure 9. The strain concentration occurred at the region contact between the original bone and replaced femur because the changing sizes affect the strain distribution.The closest maximum equivalent of total strain of 100% proximal femur is the model of 130% proximal femur that the best option to change after removed the bone tumor. If considering the suitable size, the PMMA proximal femur is the same size of the original bone that was proper to replace for restore the hip joint function.

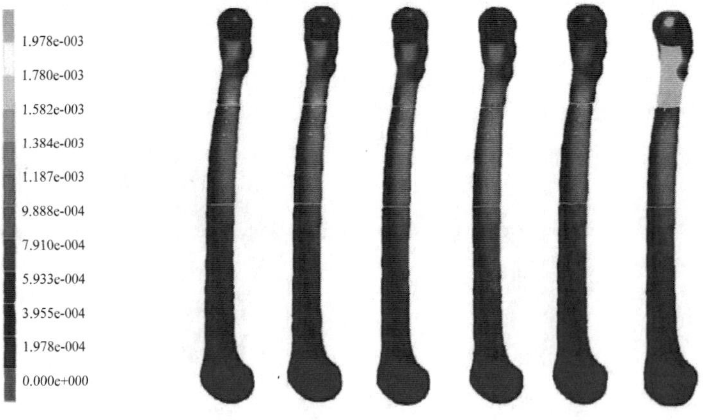

(a) (b) (c) (d) (e) (f)

Figure 9. The strain distribution on femoral bone under walking load: (a) 70% Proximal femur, (b) 85% Proximal femur, (c) 100% Proximal femur, (d)115% Proximal femur, (e) 130% Proximal femur and (f) PMMA Proximal femur.

The stain distribution on femoral bone of all cases under stair-climbing load was shown in figure 10.The strain concentration occurred at the region contact between the original bone and replaced femur same as under walking load. The maximum equivalent of total strain occurred on femoral bone under stair-climbinghigher than walking load because more muscular force act on proximal femur than under walking condition.

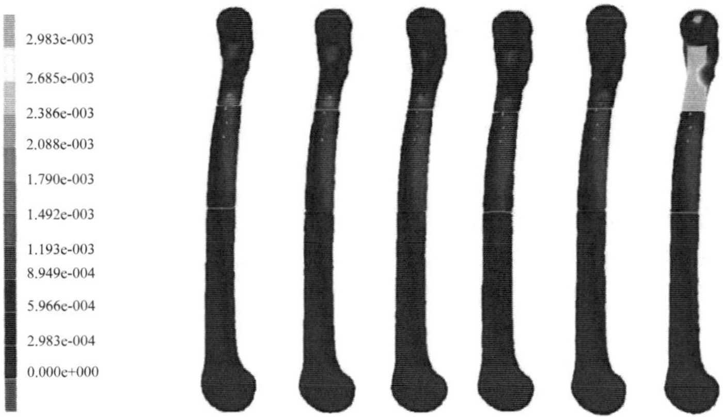

(a) (b) (c) (d) (e) (f)

Figure 10. The strain distribution on femoral bone under stair-climbing load: (a) 70% Proximal femur, (b) 85% Proximal femur, (c) 100% Proximal femur, (d)115% Proximal femur, (e) 130% Proximal femur and (f) PMMA Proximal femur.

4. Conclusion

The trend ofequivalent of total strain distributionhad higher value in smaller bone than larger bone that shownthe larger bones were suitable for treatment in bone tumor removal. All cases shown the maximum strain distribution lower than 25,000 microstrain (με) [7, 8] that safe for bone damage after replaced proximal femur. To reduce the chance of bone function loss for mismatchingsize, the PMMA proximal femur from reverse engineering and rapid prototype as shown in figure 11 was a best choice to replace for the proximal femur removal.

(a) (b)

Figure 11. The mirror proximal femur from reverse engineering and rapid prototype had shown the same size: (a) Original bone and (b) Mirror bone.

The mirror proximal femur was used to construct the silicone mold as shown in figure 12 for the surgeon to construct the PMMA proximal femur that had the same size and same inclination angle to protect the bone function.

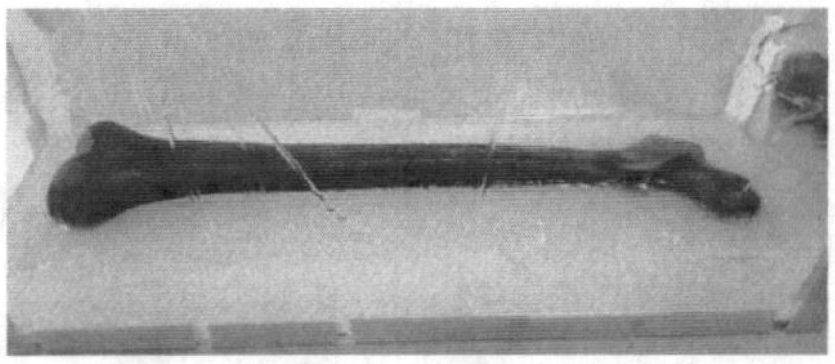

Figure 12. The silicone mold for construct the PMMA femur.

Acknowledgments
The author wish to thank the Faculty of Medicine, Burapha University and Biomechanics Analysis and Orthopaedic Device Design Laboratory (B-AODD Lab), Faculty of Engineering, Mahidol University for their kind support of the facilities.

References

[1]Silva P, do Amaral RA, de Oliveira LA, de Moraes FB and Chaibe ED2016*Rev Bras Ortop***51(6)** 739-743.
[2]Kirmani O, Muzaffar N, Ehsan M and Ahmad S 2013*Global Journal of Surgery***1(3)** 22-24.
[3]Mahaisavariya B, Sitthiseripratip K, Tongdee T, Bohez ELJ, Sloten JV and Oris P 2002*MedEng&Phy***24(9)** 617-622.
[4] Bendjaballah MZ, Shiraz-Adi A and Zukor DJ 1997*ClinBiomech***12** 139–148.
[5] Weiss JA and Gardine JC 2001*Crit Rev Biomed Eng***29** 1–70.
[6] Peraz A, Mahar A, Negus C, Newton P and Impelluso T 2008 *Med Eng&Phys***30(6)** 755–760.
[7] Frost HM 1994*The Angle Orthodontist***64(3)** 175-188.
[8] Frost HM 2004*The Angle Orthodontist***74(1)** 3-15.

IOP Conf. Series: Materials Science and Engineering **501** (2019) 012037 doi:10.1088/1757-899X/501/1/012037

Hemodynamic Characteristics in Aorta-Pulmonary Artery System

D Ichihashi[1], M Iinuma[1], T Yamamoto[2,*], M I Kori[3] and K B Osman[3]

[1]Advanced Course, National Institute of Technology Gifu College, 2236-2 Kamimakuwa, Motosu, 501-0495, Japan
[2]Dept. Mechanical Engineering, National Institute of Technology Gifu College, 2236-2 Kamimakuwa, Motosu, 501-0495, Japan
[3]IJN-UTM Cardiovascular Eng. Centre, 81310, UTM, Johor Bahru, Malaysia

* Corresponding Author: ytaka@gifu-nct.ac.jp

Abstract. The child patients of hypoplastic left heart syndrome undergoes major operations such as the Norwood procedure, the Bidirectional Glenn procedure and the Fontan procedure. Patent Ductus Arteriosus (PDA) stenting, which is instead of these procedures, is recently performed to prevent closure of ductus arteriosus and to alleviate the burden for the patient. This study clarified that hemodynamics in aorta-pulmonary artery system and mechanism of thrombus formation using Computational Fluid Dynamics (CFD) analysis and Particle Imaging Velocimetry (PIV) measurement. As a result of CFD analysis, low wall shear stress was observed at neighborhood of PDA stent. From results of CFD analysis and PIV measurement, the main blood flow was formed toward from aorta to pulmonary artery and there is no blood stagnation region in PDA.

1. Introduction

Hypoplastic left heart syndrome is one of the congenital heart diseases. The disease is characterized by hypoplasia of left atrium, mitral valve and aortic arch. In such the patient, blood flow balance is kept by patent ductus arteriosus (PDA) during unborn child. After birth, ductus arteriosus is naturally closed and then blood flow to lungs increase. This blood flow balance change causes various symptoms like lungs hyperemia, polypnea, oliguria and shock state. In order to treat the child patient some operations are required; i.e., the Norwood procedure and the Bidirectional Glenn procedure and [1, 2]. The Fontan procedure is finally performed to improve cardiopulmonary function. PDA stenting is recently performed to prevent from ductus arteriosus closing instead of the Norwood procedure and the Bidirectional Glenn procedure [3]. This surgery makes less burden on the patients compared with the conventional procedures. However, some clinical reports have reported that PDA stenting causes thrombosis around ductus arteriosus [4-6]. The detail mechanism why thrombus is formed has not been clarified until now. It is necessary to unravel blood flow balance of aorta-pulmonary artery system in the state of PDA. The purpose of this study is to elucidate hemodynamics of aorta-pulmonary artery system and thrombus formation mechanism using particle image velocimetry (PIV) measurement and computational fluid dynamics (CFD) analysis.

2. Method

2.1. *Numerical Simulation*

In order to estimate the influence of PDA stent on blood flow characteristics around PDA, CFD analysis is performed using CFX ver.18 (ANSYS Co.). The Analysed PDA model is shown in Fig. 1. As for boundary condition of CFD analysis, inlet flow rates at the aorta and pulmonary artery were set at 4.89 (L/min) and 2.11 (L/min), respectively. Outlet boundary conditions were given pressure Neumman boundary condition. Furthermore, wall boundary was given no slip condition. Shear stress transport (SST) model was used as a turbulent flow model. This analysis assumed blood as Newtonian fluid of which kinetic viscosity was set at 4.43×10^{-3} m²/s.

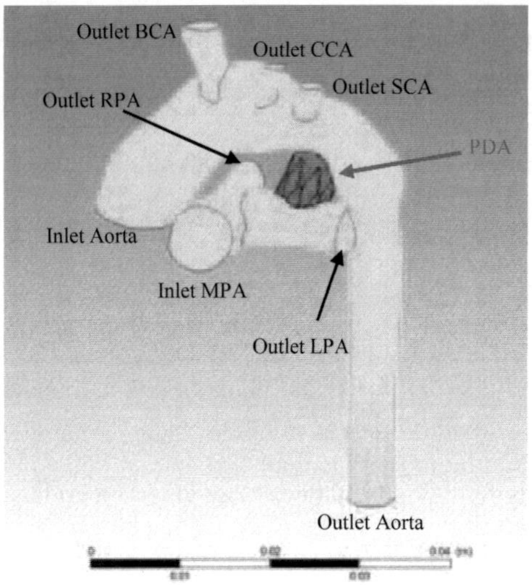

Figure 1. Analysis model of aorta-pulmonary artery system

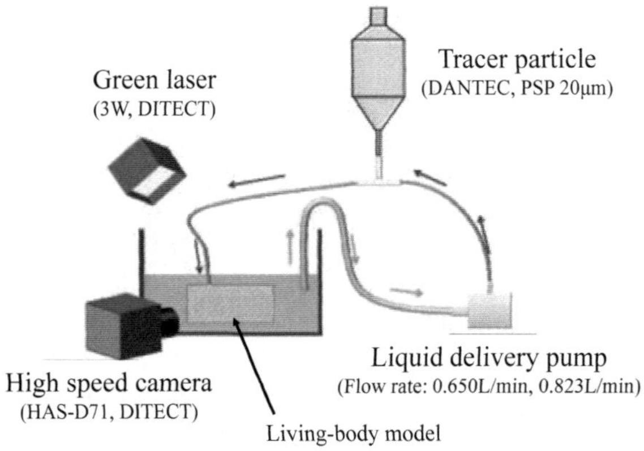

Figure 2. Experimental apparatus of PIV measurement.

2.2. *Particle Image Velocimetry*

PIV measurement has been developed as a calculation method for unsteady and instantaneous flow characteristics using image analysis techniques. Fluid motion is visualized by tracer particle in fluid flow, and then is numerically evaluated using digital image processing [7, 8]. The PIV apparatus in this study is shown in Fig. 2, and composes of a high-speed camera (DITECT D71, 800fps), a laser light source (DITECT 3W green laser), tracer particles (DANTEC PSP 20μm) and aqueous glycerine solution as working fluid. A living-body model of aorta-pulmonary artery system was made from silicon rubber and it is twice as large as the actual body scale (Fig. 3). Centrifugal pumps connected with both aorta and pulmonary artery inlets of the model. These flow rates are 0.650 L/min and 0.823 L/min, respectively.

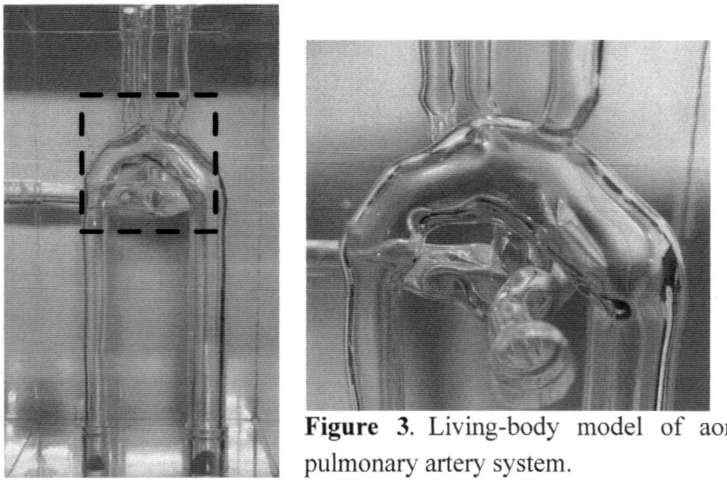

Figure 3. Living-body model of aorta-pulmonary artery system.

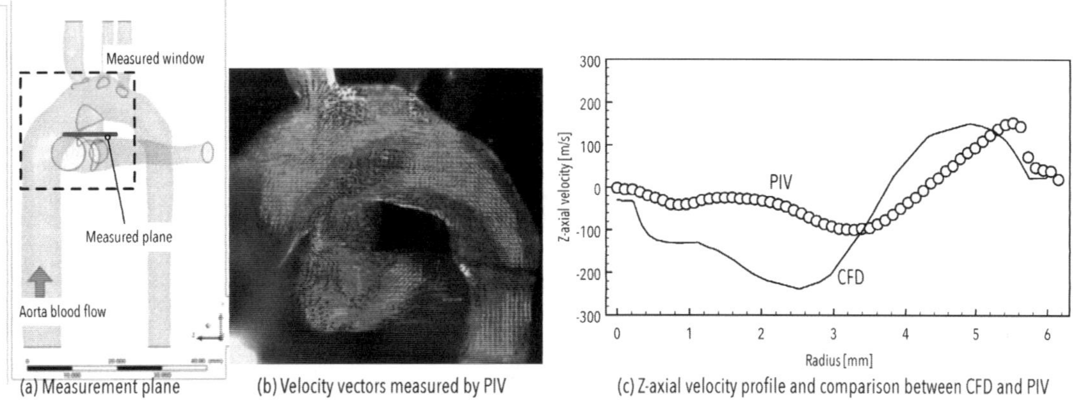

Figure 4. Measurement results of PIV experiment; velocity vectors and Z-axial velocity profile

3. Result and Discussion

Figure 4(a) and 4(b) show a 3-D data of original form of living-body model and velocity vector on a vertical section of PDA. As the result of the velocity vector, the main blood flow was formed toward from aorta to pulmonary artery. Figure 4(c) shows the measured Z-axial mean-square velocity profiles on a red line of Fig. 6(a), comparing with the result of CFD analysis. The results of CFD analysis and PIV measurement have not correspond quantitatively. However, the tendencies of the results show good agreement between them. This study assumed such the differences does not affect in following discussions significantly, using the results of PIV measurement and CFD analysis. Furtherer investigation and discussion are needed for the differences between PIV and CFD.

Fig. 4(b) and 4(c) reveal detail blood flow characteristics inside PDA; main blood flow was globally formed toward from aorta to pulmonary artery, at the same time, reverse direction flow was partially formed toward from pulmonary artery to aorta. In other words, aorta-pulmonary artery system is maintained blood flow balance by bilateral flow between aorta and pulmonary artery.

Figure 5 shows the streamlines in the aorta-pulmonary artery system calculated by means of CFD analysis. This result involves the influence of PDA stent installation. The laminar flow was observed in aorta and pulmonary artery, and complex and turbulence flows were observed around PDA. The main blood flow was formed toward from aorta to pulmonary artery as shown in the result of PIV measurement. Figure 6 shows wall shear stress (WSS) distribution in PDA. The areas alongside PDA stent showed partially lower WSS than other areas. This tendency implies that stagnation flow region is formed caused by the stent installation. The prior studies have reported that thrombus is easy to be formed at low WSS region [9-11]. In this study, CFD analysis has clarified that low WSS region is observed on the areas alongside the stent. These results in this study indicate that there is possibility of promoting thrombus formation due to PDA stent.

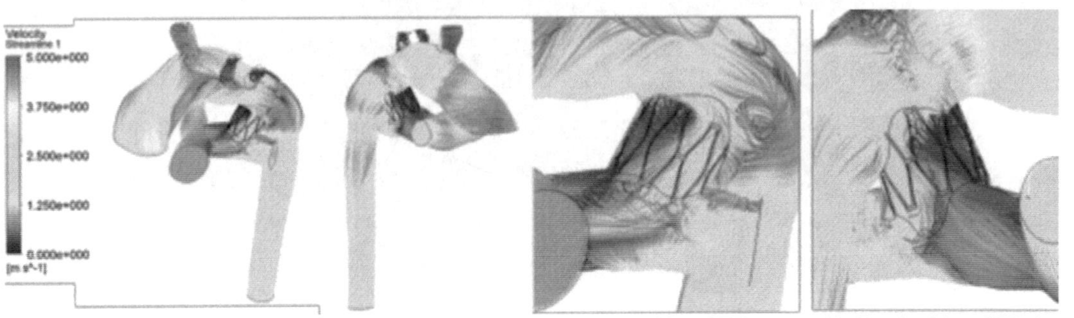

(a) Front side (b) Back side (c) Front side of PDA (d) Back side of PDA

Figure 5. Streamline of blood flow in PDA

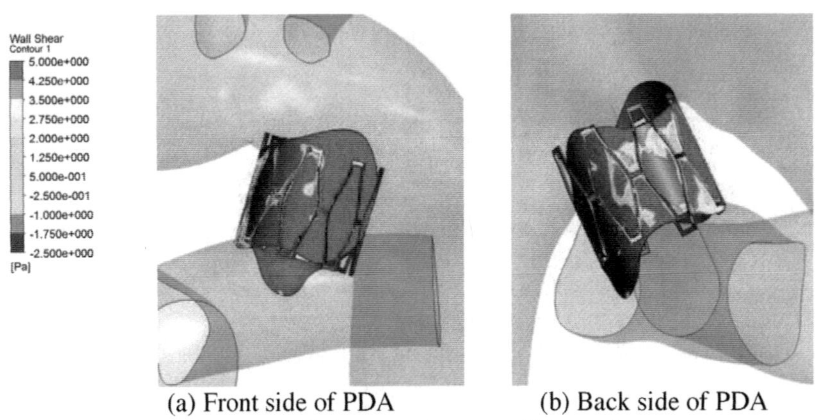

(a) Front side of PDA (b) Back side of PDA

Figure 6. Wall Shear Stress on PDA wall

4. Conclusion

This study conducted both CFD analysis and PIV measurement for blood flow characteristics in aorta-pulmonary artery system. As the results, this study found that (1) main blood flow was observed toward from aorta to pulmonary artery, at the same time, reverse direction flow was also observed toward from pulmonary artery to aorta, (2) low WSS were found in the area alongside PDA stent, meaning that low velocity (stagnation flow) region was formed there, and consequently (3) these results indicate that there is possibility of promoting thrombus formation due to PDA stent.

The results of CFD analysis and PIV measurements show agreement qualitatively and discrepancy quantitatively. Furtherer investigation and discussion is needed for the differences. In addition, this study simplified blood flow as continuous steady flow. In the next study, the author is going to elucidate pulsate flow characteristics in aorta-pulmonary artery system using Large Eddy Simulation (LES) model and stereo-PIV measurement system[] .

Acknowledgments
This work was supported by Koshiyama Science Foundation (2016) and Ogawa Science and Technology Foundation (2018).

References
[1] Sano S 2007, Recent development of congenital heart surgery in neonates and small infants, *Applied Journal of Japan Surgical Society* **108(6)** 339-343.
[2] Boucek M, Mashburn C and Kunz E 2005, Ductal Anatomy: A Determinant of Successful Stenting in Hypoplastic Left Heart Syndrome, *Applied Pediatric Cardiology* **26(2)** 200-205.
[3] Egan MJ, Trask AJ and Baker PB 2011, Histopathologic Evaluation of Patent Ductus Arteriosus Stents After Hybrid Stage I Palliation, *Applied Pediatric Cardiology*, **32(4)**, 413-417
[4] Bordones AD, Leroux M, Kheyfets VO and Wu YA 2018, Computational Fluid Dynamics Modeling of the Human Pulmonary Arteries with Experimental Validation, *Annals of Biomedical Engineering*, **46(9)**, 1309-1324
[5] Briceno JC Garcia-Torres AE, Silva AC 2004, A Modeling of the Hemodynamics of the Patent Ductus Arteriosus, *ASAIO J.* **50(2)**, 130
[6] Ceballos A, Morales RA, Divo E, Osorio R and Lauderdale F 2012, Computational Analysis of Hybrid Norwood Circulation with Distal Aortic Arch Obstruction and Reverse Blalok-Taussig Shunt, *The Annals Thoracic Surgery*, **94(5)**, 1540-1550
[7] Murai Y, Ido T, Ishikawa M and Yamamoto F 1998, Development of the post-processing Method for PIV Measurement Result Using Computational Fluid Dynamics Procedure, *Applied Transactions of the JSME*, **64(626)**, 109-116
[8] Busen M, Kaufmann TA, Neidlin M, Steinseifer U and Sonntag SJ 2015, In vitro flow investigations in the aortic arch during cardiopulmonary bypass with stereo-PIV, *Applied Journal of Biomechanics*, **48(10)**, 2005-2011
[9] Hashimoto S 1987, Effect of Velocity Gradient on Clot Growth and Erythrocyte Destruction, *J. Artificial Organs*, 16(**4**), 1641-1646
[10] Sukena R, Nakamura S, Naoi K, Ishima T and Obokata T 2000, Measurement of Pulsating Flow by Means of LDA Flow Rate Meter, *J. Mechanical Engineers JSME*, 1, 143-144
[11] Yoshida T, Okanaga H and Aoki K 2006, Flow Characteristics for Structure of Stenosis Tubes with Pulsating Flow, Tokai Univ. Report, **46(2)**, 67-72

IOP Conf. Series: Materials Science and Engineering **501** (2019) 012038 doi:10.1088/1757-899X/501/1/012038

Experimental Study of Microwave Ablation in Ex Vivo Tissues

P. Keangin*, P. Manop, T. Nonthakhamchan and M. Srisupanon

Department of Mechanical Engineering, Faculty of Engineering, Mahidol University, 25/25 Phutthamonthon 4 Road, Salaya, Nakhon Pathom, 73170, Thailand.

* Corresponding Author: E-mail: pornthip.kea@mahidol.ac.th

Abstract. Cancer is one of the leading causes of morbidity and mortality worldwide. Microwave ablation is a minimally invasive treatment procedure that uses heat from microwave energy to destroy cancer cells. The effectiveness of this technique is associated to the microwave power input and heating time of treatment during the process, as well as the type of cancer tissue. The microwave power absorbed of each cancer tissue type is different and effect on the choice of treatment conditions. This research aims to investigate the effects of tissue type and microwave power input on the temperature profile and the efficiency of cancer treatment during microwave ablation. An experiment in ex vivo different tissues of porcine during microwave ablation via microwave antenna is studied. The types of tissue studied include skin tissue, liver tissue and lung tissue. The microwave power input of 60 W, 80 W and 100 W are investigated. The heating time of 360 s is selected for the study. An infrared thermometer camera is used to measure the temperature profile of tissues. The results reveal that the temperature profile and the ablation size in the case of skin tissue are higher than the liver tissue and lung tissue, respectively. In addition, the temperature profile of all tissues increases with greater microwave power. This research provides the essential aspects for a fundamental understanding of heat transfer within porcine tissues in microwave ablation process and can be used as a guideline to improve the efficiency of cancer treatment.

1. Introduction

Nowadays, there are many people diagnosed with cancer and several factors contributing to the increase in cancer. In 2015, the death rate of cancer patients worldwide is as high as 8.8 million. For Thailand, cancer is the major cause of death among Thais for decades. The most common cancer is liver cancer, which has 6,596 cases of patients. Men have a higher incidence of liver cancer than women [1]. Smoking habits are the main cause of lung cancer. For one particular category of smoker, it is found that there is a 10 % chance of lung cancer [2]. There are many ways to treat cancer such as surgical resection [3], chemoembolization therapy [4] and radiation therapy [5] etc. All of the above methods have different advantages and limitations. Microwave ablation techniques have become an extensive choice for the treatment of cancer [6-7]. The key advantages of microwave ablation are consistently higher intratumoral temperatures, larger tumor ablation volumes and shorter treatment times include less susceptibility to the cooling heat-sink effects of air and blood flow [8]. There are few side effects in treatment and spent shortly time recuperating after treatment and the patient has a high opportunity of survival. The principle of microwave ablation is to pass the microwave energy to the tumor that needs to be removed by using a small microwave antenna insert into tissue that is cancerous. The new microwave ablation technique for cancer treatment is different from traditional treatments. The goal of microwave ablation treatment is to raise the temperature of the unwanted tissue (tumor) to therapeutic value of 50 °C to kill cancerous tissues within the body and affect metastases without the danger in nearby tissue [9]. Therefore, the topic of heat transfer as well as temperature distribution in human tissue for improved efficiency of microwave ablation treatment of cancer has been of interest for several years.

Research and academic papers on the process of destruction of cancer cells using microwave energy are widely. The research involved is divided into three groups are 1) Research and study on the destruction of cancer cells using microwave energy by tissue experiment 2) Research and study on the destruction of cancer cells using microwave energy by computer simulation and 3) Research and study on the destruction of cancer cells using microwave energy by tissue experiment and computer simulation. Research paper in the first group, such as Peralta H et al. [10] compared the surgery microwave ablation in ex vivo and in vivo with a similar microwave frequency. The experiment based on microwave power input of 50 W, 100 W and 150 W at heating time of 8 min. It was found that the size of the wounds from the microwave ablation in vivo surgery is larger than ex vivo surgery. In 2016, Liu W [11] reported a multi coaxial-slot antenna to form a figure-of-eight ablation area during microwave ablation. The microwave antenna worked at 2.45 GHz and also has a good impedance matching. In addition, the microwave antenna can solve the specific shape of the tumor and to realize a large ablation pattern, meanwhile has lesser damage to body (for the outer radius of the coaxial-slot antenna is less than 4 mm). Amabile C et al. [12] studied the microwave ablation of primary and secondary liver tumours with ex vivo, in vivo and clinical characterization. The experimental results found that the ex vivo data on bovine liver was more predictive of the actual clinical performance on liver malignancies than an in vivo porcine. The study of destroying liver cancer cells using microwave energy in the tissue to control the temperature distribution and monitor the impact on the tissues was proposed by Lopresto V et al. [13].

In the computational studies, the Pennes bioheat model is widely used for modeling the heat transfer in human tissues during microwave ablation due to simplification approach [14]. Wu X et al. [15] evaluated the effect of high frequency microwave antenna at 6 GHz and 18 GHz applied to liver cancer therapy against conventional microwave antenna at 915 MHz and 2450 MHz by a finite element model coupled electromagnetic field and bio-heat transfer equation. Tissue damaged region, temperature rise and distribution characteristics were analyzed. The results showed that the high frequency microwave antenna can cause less collateral damage, more concentrated ablation region and better material response than conventional microwave antenna. A thermoelastic deformation model of tissue contraction during thermal ablation was presented by Park C S et al. [16]. A finite difference method was considered to quantify the tissue contraction for a typical temperature distribution during thermal ablation.

Some previous studies on the destruction of cancer cells using microwave energy by tissue experiment and computer simulation were presented. Curto S et al. [17] developed a three-dimensional model of electromagnetic-thermal bioheat transfer with the finite element method to characterize power deposition and thermal ablation with asymmetrical insulated dipole antennas (single-antenna and dual-antenna synchronous arrays). Simulation results were validated against experiments in ex vivo tissue. In 2017, Deshazer G et al. [18] examined any differences of specific absorption rate (SAR) profile between predicting by simulation and computing from experimentally measured temperature profiles with an infrared (IR) camera. The SAR profile during the experimental microwave ablation changed, which were not present in simulation, suggesting inaccuracies in dielectric properties. In 2018, Ibitoye A Z et al. [19] studied the characteristics and features of four antennas (monopole, single slot, dual slot and sleeved antennas) for microwave ablation therapy to compare their efficiencies using analytical and experimental methods. The simulation was done to evaluate its reflection coefficient, power dissipation distribution, power dissipation density, SAR distribution and temperature distribution in tissue. The experiment was done on ex vivo bovine livers using 50 W lasting for 5 min and 10 min to determine ablation diameters, ablation lengths and aspect ratio. The sleeved antenna provided the best localization with the highest aspect ratio, the highest temperature distribution, large ablative diameter and the lowest backward heating.

Even though there are many researchers studied about microwave ablation in many years ago, a few studies on the effects of difference of tissue types and microwave power input in order to determine the efficiency of microwave ablation process by calculating ablation zone have done. In this study, the effects of tissue types and microwave power input on the temperature profile and the efficiency of cancer treatment during microwave ablation are presented. The variations in three tissue types: skin tissue, liver tissue and lung tissue of porcine are investigated. The microwave power input of 60 W, 80 W and 100 W for the duration of 360 s are studied. The obtained values will provide a basis for the development of microwave ablation process for tissues and organs.

2. Experimental Study

This section divided into two parts explains about the experiment consisting of experimental equipment and tissue samples and experimental setup.

2.1. Experimental equipment and tissue samples

Experiments are performed in ex vivo porcine skin tissue, liver tissue and lung tissue with a microwave applicator. The experimental equipment used in the experiment is as follows: 1) Microwave ablation system: is a microwave generator (Acculis, AngioDynamics, 60-140 W, 2.45 GHz) as shown in Figure 1. Microwave generator is connected to the microwave antenna. Microwave energy generated by microwave generator propagates in microwave antenna into the tissue from the slot to destroy cancerous tissue. 2) Microwave antenna: is a device for transporting microwave energy to a region where required to heat. The temperature of unwanted tissue (tumor) at a level above of 50 °C can effectively kill cancer cells [20-21]. The microwave antenna used in this experiment (Accu2i PMTA Applicator) is 19 cm length operating with internally cooled system for releasing internal heat as shown in Figure 2. 3) Fiber optic thermometer: is a temperature-measuring instrument. It has many advantages: capable of measuring the temperature in microwave heating process without any reaction and noise, capable of measuring the temperature in localised region and can show the difference between each measured points. In this experiments, fiber optic thermometer (FOT Lab Kit, LUXTRON) is 4-channel optic sensor which is 4-20 mA or 0-10 VDC output and able to measure the temperature in range of -100 °C to 330 °C and record data in PC software. 4) Infrared thermometer camera: is a device that captures infrared radiation (IR) emitting from an object to surroundings and forms a heat zone image. In the image, the hotter region is represented by dark color and the cooler region is represented by bright color. IR is a result of atom and molecular vibration and behaves like visible lights. The more molecules move will result in more temperature increases. Infrared thermometer camera will show a temperature profile of tissues after microwave ablation process. In addition, 5) Other devices include computer for recording data, thermometer knife and chopping board and temperature controlled ice bucket.

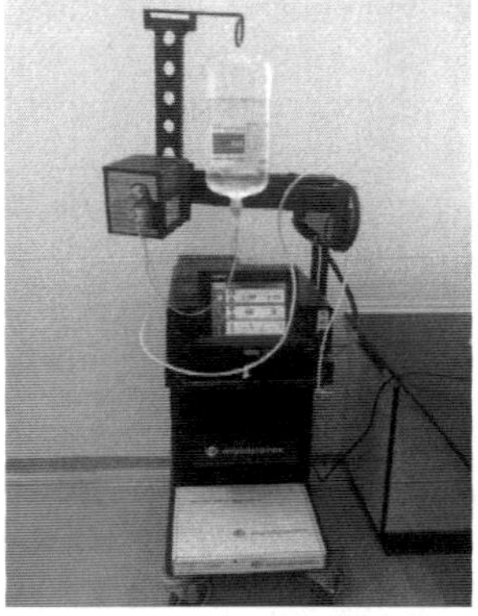

Figure 1. Microwave ablation system
(at Center of Scientific Equipment for Advanced Research, Thammasat University).

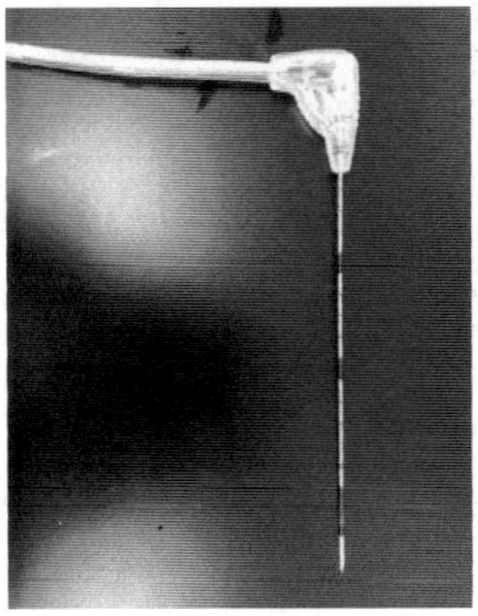

Figure 2. Microwave antenna.

In this research, the effects of three tissue types i.e. skin tissue, liver tissue and lung tissue on the temperature profile and the efficiency of cancer treatment during microwave ablation are studied. Each type of tissues has three samples for three microwave power inputs setting. Required each piece of tissue sample used is shaped as a rectangular prism with dimensions of 6 cm x 8 cm x 6 cm. Tissue samples are cooled to 10 °C in a temperature controlled ice bucket.

2.2. Experimental setup

Microwave ablation experiments are performed in ex vivo porcine tissue i.e. skin tissue, liver tissue and lung tissue. The experiment starts with the preparation of the tissue sample with size of 6 cm x 8 cm x 6 cm as shown in Figure 3. Then, immerse it in temperature controlled ice bucket that the initial temperature is equal to 10 °C. To validate the fiber optic, measure the temperature of water by the thermometer and the fiber optic cable to compared the temperature results as shown in Figure 4. For microwave ablation, a microwave antenna is inserted into a tissue sample to release microwave energy. Microwave antenna is inserted to a depth of 7.5 cm as shown in Figure 5. Microwave generator (Acculis, AngioDynamics) supplied microwave energy at the frequency of 2.45 GHz, which is amplified microwave power input of 60 W, 80 W and 100 W to the tissue sample. Connect the fiber optic cable to the tissue sample, set microwave antenna on the tissue sample and placed tissue sample in a bath. Then, turn on the microwave generator. For ex vivo experiments, microwave power input of 60 W, 80 W and 100 W are applied to each tissue for duration of 360 s as shown in Figure 5. The position of fiber optic cable is shown in Figure 6. Read and record the temperature from the fiber optic cable. After the microwave ablation process, take a temperature profile with infrared thermometer camera. Measure the ablation diameter and compute the ablation size. The experimental diagram is shown in Figure 7.

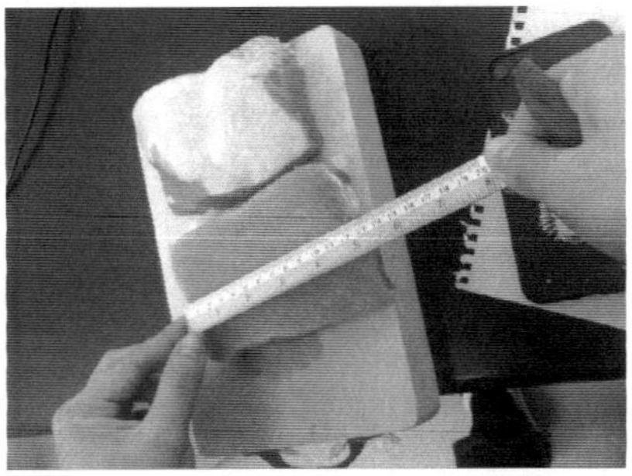

Figure 3. Prepare tissue sample.

IOP Conf. Series: Materials Science and Engineering **501** (2019) 012038 doi:10.1088/1757-899X/501/1/012038

Figure 4. Compare the temperature of water from the thermometer to the fiber optic cable.

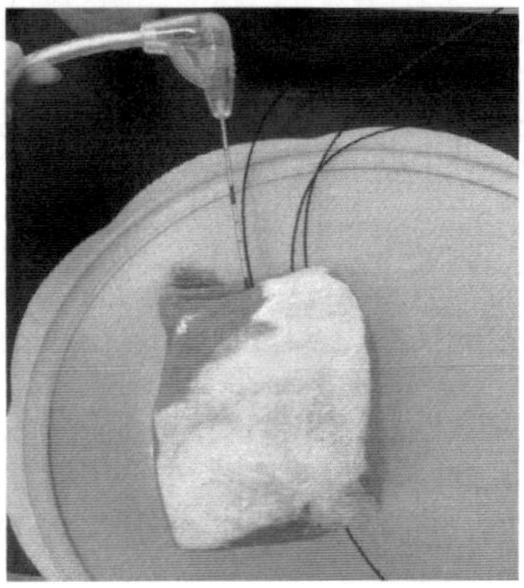

Figure 5. Tissue sample during experiment.

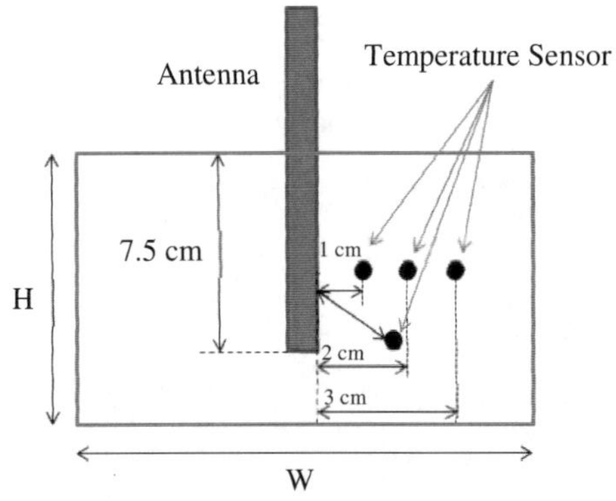

Figure 6. Position of antenna and fiber optic cables.

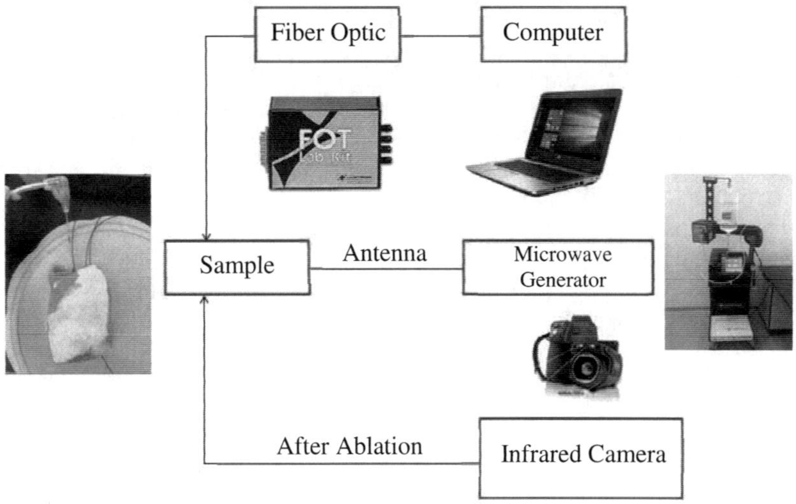

Figure 7. Experimental diagram.

In addition, the dielectric properties of skin tissue, liver tissue and lung tissue are measured using network analyzer (E5063A ENA Series) at frequency of 2.45 GHz in this study. The dielectric properties of skin tissue, liver tissue and lung tissue are exhibited in Table 1.

Table 1. The dielectric properties of skin tissue, liver tissue and lung tissue at frequency of 2.45 GHz.

Tissues	Dielectric Properties	
	Dielectric Constant, ε' (F/m)	Dielectric Loss Factor, ε'' (F/m)
Skin Tissue	54.20	17.97
Liver Tissue	48.90	14.44
Lung Tissue	30.01	7.10

3. Results and Discussion

In this study, the effects of the two parameters, namely, the types of tissue (skin tissue, liver tissue and lung tissue) and the microwave power input (60 W, 80 W and 100 W) have been investigated. A parametric study has been performed to estimate the effects of each of these variables and analyze their contributions in the temperature profile and the efficiency of cancer treatment during microwave ablation. Due to the actual ablation zone covers tissue areas where temperature exceeds 50 °C [20-21], the temperature of 50 °C as a threshold as tissue discoloration could be visually observed and measured for calculating the ablation size. In addition, the target ablation diameter of 2 cm and target ablation volume of 4.2 cm^3 are considered in this study [22].

The ablation zones in ex vivo skin tissue during microwave ablation with a single microwave antenna operating at 2.45 GHz at different microwave power inputs for duration of 360 s are shown in Figure 8. Figure 8(a)-8(c) show the ablation zones of the skin tissue at the microwave power inputs of 60 W, 80 W and 100 W, respectively. These figures illustrate the volumetric heating effect expected from microwave ablation, where a hot spot zone occurs at the position near the end tip and the slot of the microwave antenna. Microwave delivered from the antenna propagates inside the tissue and is transformed into the thermal energy by electromagnetic heating. The ablation zone curves form a nearly ellipsoidal shape around the slot and its highest value of temperature near the antenna slot. It then decreases with the distance from the antenna. The ablation zones within the skin tissue in the case of a higher microwave power provides a broader area of heat dissipation to the surrounding tissue near the slot of the antenna compared with result in the case of lower microwave power. Table 2 shows the dimensions of measured ablation diameter and ablation volume for different microwave power inputs in ex vivo skin tissue during microwave ablation.

The surface temperature profiles in ex vivo skin tissue during microwave ablation with a single microwave antenna operating at 2.45 GHz at different microwave power inputs for duration of 360 s are shown in Figure 9. Figure 9(a)-9(c) display the surface temperature profiles of the skin tissue at the microwave power inputs of 60 W, 80 W and 100 W, respectively. These figures show that the surface temperature profiles which correspond to the ablation zones quite well. The temperature profile within the skin tissue in the case of a higher microwave power provides a broader area of heat dissipation with higher temperature to the surrounding tissue near the slot of the antenna compared with result in the case of lower microwave power correspond to characteristic of the ablation zone.

The effects of microwave power input on the ablation zones in ex vivo liver tissue during microwave ablation with a single microwave antenna operating at 2.45 GHz for duration of 360 s are shown in Figure 10. Figure 10(a)-10(c) illustrate the ablation zones of the liver tissue at the microwave power inputs of 60 W, 80 W and 100 W, respectively. In the same way with the skin tissue, the ablation zone curves form nearly ellipsoidal shape around the slot. The highest value of temperature near the antenna slot and decreases by distance from the antenna. The ablation zones within the liver tissue in the case of a higher microwave power provides a broader area of heat dissipation to the surrounding tissue near the slot of the antenna compared with result in the case of lower microwave power. Table 2 lists the dimensions of measured ablation diameter and ablation volume for different microwave power inputs in ex vivo liver tissue during microwave ablation correspond to the skin tissue. It is however found that the experiment results of ablation zones and temperature pattern of skin tissue has a higher ablation volume of heat dissipation than the liver tissue. This is because the dielectric properties of skin tissue are higher than the liver tissue that becomes the reason of higher absorption of microwave energy as shown in Table 1.

Figure 11 shows the surface temperature profiles in ex vivo liver tissue during microwave ablation with a single microwave antenna operating at 2.45 GHz at different microwave power inputs for duration of 360 s. Figure 11(a)-11(c) show the surface temperature profiles at the microwave powers of (a) 60 W, (b) 80 W and (c) 100 W, respectively. These figures show that the surface temperature profiles which correspond to the ablation zones quite well. The temperature profile within the liver tissue in the case of a higher microwave power is higher than the case of lower microwave power that corresponds to in case of skin tissue. It is found that the temperature profiles in case of liver tissue have the same pattern with temperature profiles when compared with results for skin tissue. However, the temperature profiles for skin tissue provides a broader area of heat dissipation compared with results for liver tissue because the dielectric properties as shown in Table 1.

The last section discusses the effects of input microwave power on the ablation zones in ex vivo lung tissue during microwave ablation with a single microwave antenna operating at 2.45 GHz for duration of 360 s are shown in Figure 12. Figure 12(a)-12(c) illustrate the ablation zones of the lung tissue at the microwave power inputs of 60 W, 80 W and 100 W, respectively. Similar to the case of skin and liver tissues, the ablation zones of ex vivo lung tissue curves form nearly ellipsoidal shape around the slot. The highest value of temperature nears the antenna slot and decreases with the distance from the antenna. The higher microwave power generates the broader area of heat dissipation occurs. Table 2 illustrates the dimensions of measured ablation diameter and ablation volume for different microwave power inputs in ex vivo lung tissue during microwave ablation correspond to the skin and liver tissues. From Table 2, it can be seen that the ablation zones and temperature pattern of skin tissue has a higher ablation volume of heat dissipation than the liver tissue and lung tissues, respectively. This is due to the skin tissue has higher dielectric properties than the liver tissue and lung tissues, respectively which are the reason of highest absorption of microwave energy as shown in Table 1.

The surface temperature profiles in ex vivo lung tissue during microwave ablation with a single microwave antenna operating at 2.45 GHz at different microwave power inputs for duration of 360 s are

IOP Conf. Series: Materials Science and Engineering **501** (2019) 012038 doi:10.1088/1757-899X/501/1/012038

shown in Figure 13. Figure 13(a)-13(c) indicate the surface temperature profiles of the lung tissue at the microwave power inputs of 60 W, 80 W and 100 W, respectively. These figures show that the surface temperature profiles which correspond to the ablation zones quite well. The temperature profile within the lung tissue in the case of a higher microwave power provides a broader area of heat dissipation with higher temperature to the surrounding tissue near the slot of the antenna compared with result in the case of lower microwave power correspond to characteristic of the ablation zone. Once again, the findings revealed that the temperature profiles in case of lung tissue have the same pattern with temperature profiles in case of skin and liver tissues. When comparing temperature profiles in three tissues, it is observed that the temperature profiles for skin tissue provides a broadest area of heat dissipation compared with other tissues due to the dielectric properties as shown in Table 1. The comparison of the ablation zone size and ratio of ablation for each cases are illustrated in the Table 2.

The results in this Table 2 exhibit that the ablation volume in the case of skin tissue is higher than liver tissue and lung tissue, respectively, at the same microwave power input. In addition, the ablation volume of tissues increases with greater microwave power.

(a) (b) (c)

Figure 8. The ablation zones in ex vivo skin tissue operating at 2.45 GHz for duration of 360 s at different microwave power inputs of (a) 60 W, (b) 80 W and (c) 100 W, respectively.

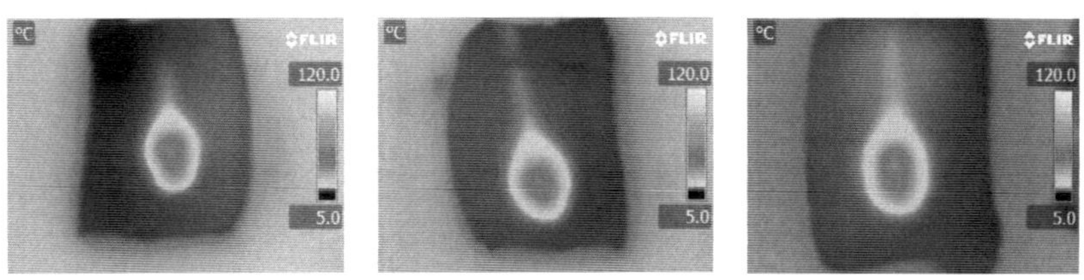

(a) (b) (c)

Figure 9. The surface temperature profiles in ex vivo skin tissue operating at 2.45 GHz for duration of 360 s at different microwave power inputs of (a) 60 W, (b) 80 W and (c) 100 W, respectively.

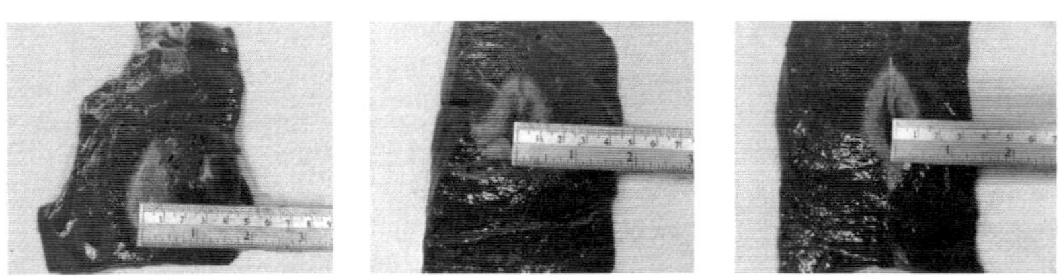

(a) (b) (c)

Figure 10. The ablation zones in ex vivo liver tissue operating at 2.45 GHz for duration of 360 s at different microwave power inputs of (a) 60 W, (b) 80 W and (c) 100 W, respectively.

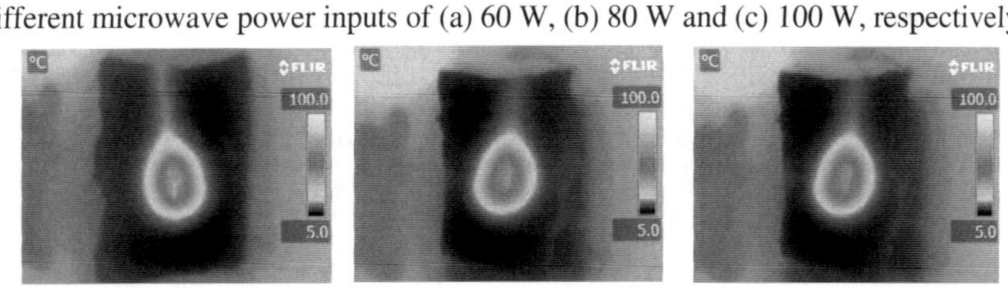

(a) (b) (c)

Figure 11. The surface temperature profiles in ex vivo liver tissue operating at 2.45 GHz for duration of 360 s at different microwave power inputs of (a) 60 W, (b) 80 W and (c) 100 W, respectively.

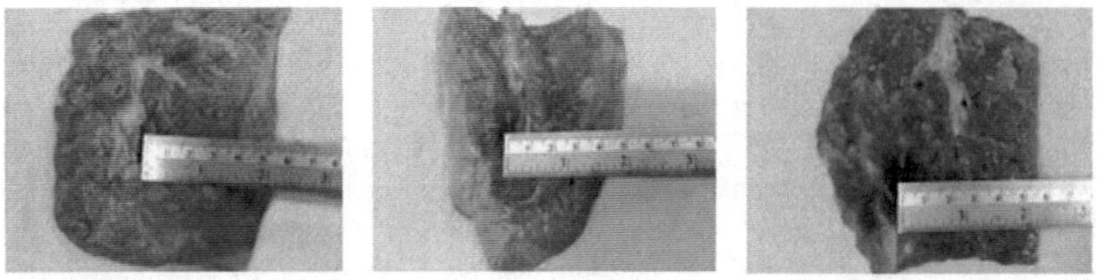

(a) (b) (c)

Figure 12. The ablation zones in ex vivo lung tissue operating at 2.45 GHz for duration of 360 s at different microwave power inputs of (a) 60 W, (b) 80 W and (c) 100 W, respectively.

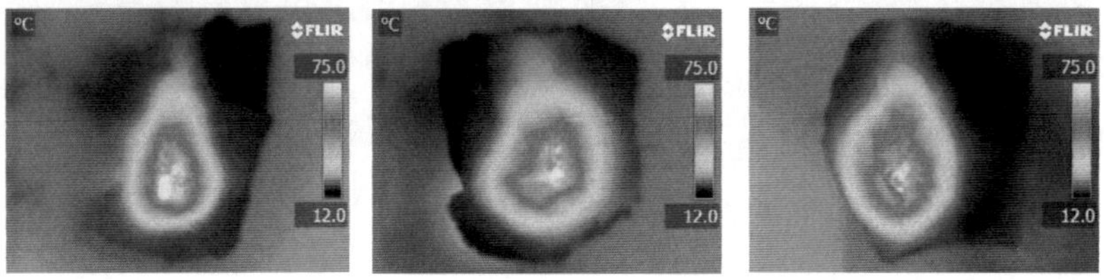

(a) (b) (c)

Figure 13. The surface temperature profiles in ex vivo lung tissue operating at 2.45 GHz for duration of 360 s at different microwave power inputs of (a) 60 W, (b) 80 W and (c) 100 W, respectively.

Table 2. Comparison of the ablation zone size and ratio of ablation for each cases.

Tissues	Power (W)	Time (s)	Target Ablation Diameter (cm)	Ablation Diameter (cm)	Target Ablation Volume (cm^3)	Ablation Volume (cm^3)	Ratio (Ablation Volume/ Target Ablation Volume) (-)
Skin Tissue	60	360	2.0	2.4	4.20	7.23	1.72
	80	360	2.0	2.6	4.20	9.20	2.19
	100	360	2.0	2.8	4.20	11.49	2.74
Liver Tissue	60	360	2.0	2.2	4.20	5.57	1.33
	80	360	2.0	2.4	4.20	7.23	1.72
	100	360	2.0	2.6	4.20	9.20	2.19
Lung Tissue	60	360	2.0	2.0	4.20	4.19	1.00
	80	360	2.0	2.4	4.20	7.23	1.72
	100	360	2.0	2.6	4.20	9.20	2.19

4. Conclusion

In this study, the experimental study of microwave ablation in ex vivo tissues is presented. The outcomes of this study clearly proved that the effects of the differences of types of tissue and microwave power inputs are well-grounded. The temperature profiles and the ablation volume in the case of the skin tissue are higher than the liver tissue and the lung tissue, respectively. Greater microwave power input not only results the broader ablation zone but also higher temperature distribution. This investigation provides the essential aspects for a fundamental understanding of heat transfer within tissues during microwave ablation include thermal treatment and can be applied to use in clinical tumour treatments to predict whether sufficient tissue volume has been ablated. Our future works will include investigation in in vivo and in vitro experiments.

Acknowledgments
The authors gratefully acknowledge the financial support for this work provided by the Thailand Research Fund (TRF) under the project No. MRG6180295 and Mahidol University.

References

[1] Bosch F X, Ribes J, Díaz M and Cléries R 2004 *Gastroenterology* **127**(5) S5–S16
[2] Peto R, Darby S, Deo H, Silcocks P, Whitley E and Doll R 2000 *BMJ* **321** 323–329
[3] Alejandro R G, Shouhao Z, Montserrat A R, Chan S, Steven W, Mouhammed H, Jose K Nancy P, Christopher W and Camilo J 2017 *Ann. Surg.* **268**(1) 172–178
[4] Yang Z, Chen G, Cui Y, Su T, Yu J, Xiao G, Han Y and Jin L 2018 *Cancer Biol. Ther.*1–7
[5] Fossati N, Karnes J, Boorjian S A, Moschini M, Morlacco A, Bossi A, Seisen T, Cozzarini C, Fiorino C, Chiorda B N, Gandaglia G, Dell'Oglio P, Joniau S, Tosco L, Shariat S, Goldner G, Hinkelbein W, Bartkowiak D and Briganti A 2017 *Eur. Urol.* **71**(6) 886–893
[6] Lucchina N, Tsetis D, Ierardi A M, Giorlando F, Macchi E, Kehagias E, Duka E., Fontana F, Livraghi L and Carrafiello G 2016 *Ann. Gastroenterol.* **29**(4) 460–465
[7] Prasanna T, Briggs G, Jain S and Yip D 2018 *Interact. Cardiov. Th.* **26**(3) 514–515
[8] Pfannenstiel A, Keast T, Kramer S, Wibowo H and Prakash P 2017 *Proc. SPIE 10066, Energy-based Treatment of Tissue and Assessment IX*, 1006601
[9] McGahan J P, Brock J M, Tesluk H, Gu W Z, Schneider P and Browing P D 1992 *J. Vasc. Interv. Radiol.* **3**(2) 291–297
[10] Peralta H, Pirani N and Clegg P 2006 *Radiology* 94-102
[11] Liu W and Lin X 2016 *Proc. IEEE International Workshop on Biomedical Circuit & System, (China)*
[12] Amabile C, Ahmed M, Solbiati L, Meloni M F, Solbiati M, Cassarino S, Tosoratti N, Nissenbaum Y, Ierace T and Goldberg S N 2017 *Int. J. Hyperthermia* **33**(1) 34–42
[13] Lopresto V, Pinto R, Farina L and Cavagnaro M 2017 *Int. J. Hyperthermia* **33** 83–100
[14] Pennes H H, 1948 J. Appl. Physiol. **1**(2) 93–122
[15] Wu X, Liu B and Xu B 2016 *Appl. Therm. Eng.* **107**(25) 501–507
[16] Park C S, Liu C, Hall K and Payne S J 2018 *Int. J. Hyperthermia* **34**(3) 221–228
[17] Curto S, Taj-Eldin M, Fairchild D and Prakash P 2015 *Med. Phys.* **42**(11) 6152–6161
[18] Deshaser G, Prakash P, Merck D and Haemmerich D 2017 *Int. J. Hyperthermia* **33**(1) 74–82
[19] Ibitoye A Z, Orotoye T, Nwoye E O and Aweda M A 2018 *Egypt. J. Basic Appl. Sci.* **5**(1) 24–30
[20] Sapareto S A and Dewey W C 1984 *Int. J. Radiat. Oncol., Biol., Phys.* **10**(6) 787–800
[21] Miller M W and Ziskin M C 1989 *Ultrasound. Med. Biol.* **15**(8) 707–722
[22] Rattanadecho P and Keangin P 2013 *Int. J. Heat Mass Transfer* **58**(1) 457–470

IOP Conf. Series: Materials Science and Engineering **501** (2019) 012039 doi:10.1088/1757-899X/501/1/012039

Verification of Solutions of Equations Governing Fluid Flows

T Watanabe[1,*]

[1]Graduate School of Informatics, Nagoya University, Nagoya, 464-8601, Japan

* Corresponding Author: watanabe@i.nagoya-u.ac.jp

Abstract. We have carried out the verifications of the solutions for the equations governing fluid flows. The equations solved are the similar equation of the boundary layer flow and the two-dimensional Navier-Stokes equations in the formulation of the stream function and the vorticity. The methods of the verification are the interval analysis and the Krawczyk operator based on the fixed point theorem. A modified Gaussian elimination method is introduced to solve the system of linear interval equations. The Krawczyk operator presents the results with a high accuracy. The modified elimination method gives the results with an accuracy of one or two order(s) of magnitude higher than those obtained by the original interval analysis.

1. Introduction

Numerical simulations have given useful results in many areas. Some of the results, however, predict feasible phenomena different from those found in the real world. In order to find and reduce these contradictions, the standards of the verification and validation [1] give guidelines to investigate the quality of the predictions. Numerical calculation suffers from numerical errors of the truncation error, loss of trailing digits, cancellation of significant digits and round-off error. When we numerically solve the differential equations in the real space, we also have a problem of the discretization error. These errors mainly result from the conversion of the real values to the floating-point representations defined by IEEE 754 [2]. In this paper, we pick up the round-off error in the numerical calculations. Two major methods, the interval analysis and the method based on the fixed-point theorem, help to estimate the round-off error [3]. These methods enclose a real number in an interval with bounds of floating-point values. The interval method ensures the intervals before and after the arithmetic operations and guarantees the accuracy. The method based on the fixed-point theorem accepts the constraints defining the phenomena and then it makes the intervals shorter and identify the narrowest interval that includes the exact value. We apply these methods to the two problems in fluid dynamics. One is the two-point boundary value problem of the self-similar boundary layer flow. The other is the two-dimensional lid-driven square cavity flow formulated by the stream function and the vorticity. The discretization method is the finite difference method. The solution method of the similar equation uses the Gaussian elimination method to resolve the system of linear equations presented by the Newton iteration method. In the elimination process, we modify the interval operation of the pivoting process and improve the results of the interval analysis. An iterative procedure of the Jacobi method is adopted to solve the Poisson equation with a diagonally dominant coefficient matrix in the problem of the cavity flow. The results obtained by these methods show that the method based on the fixed-point theorem gives a reasonable results with high accuracy.

2. Formulation

2.1. Interval analysis

In this subsection, we review the interval analysis. In this analysis, the interval X that includes the real value x has a representation of $[a,b]=\{x \mid a \le x \le b\}$ [4, 5]. Here, a and b denote the lower bound $\underline{x}$ and the upper bound $\overline{x}$, respectively. For simplicity, we express $[\underline{x},\overline{x}]$ as $[x]$.

In arithmetic between intervals X and Y,

$$X+Y=[\underline{x}+\underline{y},\overline{x}+\overline{y}], \tag{1}$$

$$X-Y=[\underline{x}-\overline{y},\overline{x}+\underline{y}], \tag{2}$$

$$X\times Y=[\min(\underline{x}\underline{y},\underline{x}\overline{y},\overline{x}\underline{y},\overline{x}\overline{y}),\max(\underline{x}\underline{y},\underline{x}\overline{y},\overline{x}\underline{y},\overline{x}\overline{y})], \tag{3}$$

$$X/Y=[\min(\underline{x}/\underline{y},\underline{x}/\overline{y},\overline{x}/\underline{y},\overline{x}/\overline{y}),\max(\underline{x}/\underline{y},\underline{x}/\overline{y},\overline{x}/\underline{y},\overline{x}/\overline{y})]. \tag{4}$$

When the division has an interval denominator including 0.0, it is assumed that the division returns an error. The midpoint of the interval $[x]$ is given by

$$mid([x])=mid([\underline{x},\overline{x}])=\frac{\underline{x}+\overline{x}}{2}. \tag{5}$$

In the manipulation of the intervals expressed in the floating-point representation, we need to convert the real values into floating-point values. In order to determine the lower and upper bounds of the result of the interval arithmetic, we first obtain the candidates of these bounds obtained from the arithmetic between two intervals. Then, we apply the round downward and round upward operations to these candidates and specify the superior of the lower bound and the inferior of the upper bound in the floating-point representation.

In the present study, we use PROFIL/BIAS [6] to perform the interval analysis, which is a numerical library including the operations among the interval scalars, interval vectors and interval matrices.

2.2. Self-similar equation of the boundary layer on a flat plate

The following equation derived by the Falkner-Skan transformation governs the self-similar boundary layer flow on a flat plate [7],

$$f'''+\frac{m+1}{2}ff''+m(1-(f')^2)=0, \tag{6}$$

with the two-point boundary conditions

$$\eta=0.0: \quad f=0.0, \quad f'=0.0, \tag{7}$$

$$\eta=\eta_e: \quad f'=1.0. \tag{8}$$

Here, η is the normalized normal coordinate from the flat plate and f is the normalized stream function. The prime denotes the derivative with respect to η. The streamwise velocity component is $U_e f'$, where U_e is the velocity component of the external stream at the edge of the boundary layer. With x that is the distance from the leading edge of the flat plate, the pressure gradient parameter m is $(x/U_e)(dU_e/dx)$, and it is assumed to be constant in the present study. The surface of the flat plate and the edge of the boundary layer are at $\eta=0$ and $\eta=\eta_e$, respectively.

We use the Keller's box method [8] to solve the differential equations. The third-order equation (6) has the following system of the first-order equations with new dependent variables of u and v,

$$f'=u, \tag{9}$$

$$u'=v, \tag{10}$$

$$v'=-\frac{m+1}{2}fv-m(1-u^2). \tag{11}$$

The boundary conditions are

$$\eta=0.0: \quad f=0.0, \quad u=0.0, \tag{12}$$

$$\eta = \eta_e : \quad u = 1.0 . \tag{13}$$

When we take $NJ + 1$ grid points from $\eta_0 = 0$ to $\eta_{NJ} = \eta_e$ on the η axis and derive the central difference equations of equations (9), (10) and (11), we obtain the $3 \times NJ$ system of nonlinear equations. Substituting $f_j^{n+1} = f_j^n + \Delta f_j$, $u_j^{n+1} = u_j^n + \Delta u_j$ and $v_j^{n+1} = v_j^n + \Delta v_j$ into these nonlinear equations and omitting the second-order terms of Δ, we obtain a system of linear equations with the boundary conditions of $\Delta f_0 = 0$, $\Delta u_0 = 0$ and $\Delta u_{NJ} = 0$, which uses the *n*-th approximate solutions of f_j^n, u_j^n and v_j^n and determines Δf_j, Δu_j and Δv_j with j from 0 to *NJ*. Updating the *n*-th approximate solutions with Δf_j, Δu_j and Δv_j gives the Newton method determining (*n*+1)-th approximate solutions. In the present study, the Gaussian elimination method is used to solve the system of linear equations.

2.3. *Lid-driven square cavity flow*

The lid-driven square cavity flow studied in the pioneering work by Ghia et al. [9] presents one of the benchmark problems. Here, we consider the cavity flow given by the following Navier-Stokes equations in the stream function ψ and vorticity ω formulation in the Cartesian coordinate $(x, y)^T$ with $0.0 \le x \le 1.0$ and $0.0 \le y \le 1.0$

$$\frac{\partial \omega}{\partial t} + \frac{\partial \psi}{\partial y}\frac{\partial \omega}{\partial x} - \frac{\partial \psi}{\partial x}\frac{\partial \omega}{\partial y} = \frac{1}{\mathrm{Re}}\left(\frac{\partial^2 \omega}{\partial x^2} + \frac{\partial^2 \omega}{\partial y^2}\right), \tag{14}$$

$$\frac{\partial^2 \psi}{\partial x^2} + \frac{\partial^2 \psi}{\partial y^2} = -\omega . \tag{15}$$

The boundary conditions are the unity velocity component in x coordinate direction on the driving lid at $y = 1.0$ and the no slip condition on the stationary walls at $x = 0.0$, $x = 1.0$ and $y = 0.0$.

The partial differential equations are discretized on the collocated square grid, and no singular problem at the corners between the moving and stationary walls [10] is considered. The Poisson equation (15) gives a system of linear equations whose coefficient matrix has a diagonal component of - 4.0 and other four non-diagonal components of 1.0. That is, at the grid points j and k in the x and y coodinates, respectively, the difference equation centered for $\psi_{j,k}$ can be expressed as

$$\psi_{j-1,k} + \psi_{j,k-1} - 4\psi_{j,k} + \psi_{j,k+1} + \psi_{j+1,k} = -h^2 \omega_{j,k} . \tag{16}$$

Here, h is the uniform grid spacing in x and y directions. While the direct method such as the Gaussian elimination method is applicable to solve the system of linear equations (16), the iterative method such as the Jacobi method is effective to resolve a large system in two or three-dimensional space. The diagonal dominancy of the coefficient matrix assures the convergence of the iterative process. Therefore, we assume that the *n*-th iteration solutions are similar to the (*n*+1)-th iteration solutions and

$$4\psi_{j,k}^n + 4\psi_{j,k}^{n+1} \cong 8\psi_{j,k}^{n+1} . \tag{17}$$

Then, we rewrite equation (16) as follows

$$\psi_{j-1,k}^{n+1} + \psi_{j,k-1}^{n+1} - 8\psi_{j,k}^{n+1} + \psi_{j,k+1}^{n+1} + \psi_{j+1,k}^{n+1} = -h^2 \omega_{j,k} - 4\psi_{j,k}^n . \tag{18}$$

This rewriting makes the coefficient matrix diagonally dominant and a favorable convergence is expected. Let denote the coefficient matrix, the constant vector and the unknown vector of equation (18) by A, b and x, respectively, and have an expression

$$Ax = b . \tag{19}$$

We decompose the coefficient matrix into the diagonal matrix D and the matrix A_{LR} with zeros on its diagonal

$$A = D + A_{LR}. \tag{20}$$

The n-th iteration solutions x^n give the (n+1)-th iteration solutions x^{n+1} by

$$x^{n+1} = D^{-1}(-A_{LR}\,x^n + b). \tag{21}$$

3. Solution Method

3.1. *Krawczyk operator based on the fixed-point theorem*

One of the methods to assure the accuracy of the numerical solution is the one that uses the interval analysis described in subsection 2.1. The other method regards the problem described by equations as a constraint satisfaction problem and uses the operator that makes the intervals of the solutions narrower. We express the governing equations with their boundary conditions as $f(x) = 0$, where each component of the vector function f represents one equation or one boundary condition. When approximate interval solution vector is assumed to be given by $[x]$, the iterative operator based on the Krawczyk method is given as follows [3, 5],

$$x_0 = mid([x]), \tag{22}$$

$$M = J_f^{-1}(x_0), \tag{23}$$

$$[J_k] = I - M\,J_f([x]), \tag{24}$$

$$[r] = x_0 - M\,f(x_0) + [J_k]([x] - x_0), \tag{25}$$

$$[x] = [x] \cap [r], \tag{26}$$

where J_f and J_f^{-1} are the Jacobian matrix of f and its approximated inverse matrix, respectively, and I is the identity matrix. The operation $\cap$ gives an intersection of two intervals. The interval $[x]$ obtained by equation (26) presents new iterative solutions.

3.2. *Modification of forward process in Gaussian elimination method*

The forward process of the Gaussian elimination method consists of the reduction of unknown variables. We now consider two equations

$$[a_{jp}][x_p] + \cdots + [a_{jq}][x_q] + \cdots = [b_j], \tag{27}$$

$$[a_{kp}][x_p] + \cdots + [a_{kq}][x_q] + \cdots = [b_k]. \tag{28}$$

In order to eliminate the unknown variable $[x_p]$, a new equation will be derived from, for example, equation (27)

$$\frac{[a_{kp}]}{[a_{jp}]}[a_{jp}][x_p] + \cdots + \frac{[a_{kp}]}{[a_{jp}]}[a_{jq}][x_q] + \cdots = \frac{[a_{kp}]}{[a_{jp}]}[b_j]. \tag{29}$$

This derivation is required only to make the coefficient of $[x_p]$ consistent between two equations and the application of the interval analysis that may introduces additional inaccuracy is not appropriate. Therefore, instead of equation (29), we use the equation obtained by the pointwise operation with midpoints of $a_{jp} = mid([a_{jp}])$ and $a_{kp} = mid([a_{kp}])$

$$\frac{a_{kp}}{a_{jp}}[a_{jp}][x_p] + \cdots + \frac{a_{kp}}{a_{jp}}[a_{jq}][x_q] + \cdots = \frac{a_{kp}}{a_{jp}}[b_j], \tag{30}$$

and modify the forward elimination process.

All the numerical simulations have been performed by the programs written in gcc and g++ on Linux with double precision calculations.

4. Result and Discussion

The floating-point calculation, the original interval analysis shown in subsection 2.1, the modified Gaussian elimination method (MGEM) and the Krawczyk operator are used to solve a system of linear equations for the self-similar equation in subsection 2.2. The results are shown in table 1, where the pressure gradient parameter m is 0.0 and - 0.05 and an interval of the form, for example, 0.33 [12, 24] means the interval of [0.3312, 0.3324]. The grid spacing is constant and it is 0.1. The intervals in the table show those of v_0 that corresponds the velocity gradient on the flat plate. The value of IT is the iteration numbers used to obtain a converged solution. Being compared with the result of the floating-point calculation, the original interval analysis has an accuracy of 2 or 3 digits. The result of MGEM gives the errors with one or three orders of magnitude smaller than those of the original interval method. This fact shows an approving effect of the current modification applied to the Gaussian elimination method. The Krawczyk operator offers a much higher accuracy when it is compared with the result of the original interval analysis or MGES. The decimal digits of the floating-point representation is 15.9 at most. Therefore, the result of the Krawczyk operator applied to the present similar equation shows an accuracy comparable with the resolution of the floating-point value.

Table 1. Solutions of the self-similar equation. $\eta_e = 8.0$.

	$m = 0.0$	$m = -0.05$
floating point calculation	0.3320414384213981, IT = 6	0.2135095226597685, IT = 6
interval analysis	0.33[12183451442928, 28570033968403], IT = 3	0.213[5073633921204, 622998266781], IT = 4
interval analysis with MGEM	0.33204[06663353578, 22105826957], IT = 5	0.2135[063433007140, 127020193930], IT = 5
Krawczyk operator	0.33204143842139[79, 84], IT = 4	0.213509522659768[3, 8], IT = 4

For the calculation of the lid-driven square cavity flow, we use the Krawczyk operator and the Jacobi method noted in subsection 2.3. The terms other than those in equation (15) is formulated by the central difference method with the original interval method. Figure 1 shows the contour of the midpoint of the stream function ψ obtained by the Krawczyk operator. The Reynolds number is 300 and the grid points in the x and y coodinates are 100. Figure 2 represents the profiles of the upper and lower bounds of the intervals of the stream function at x = 0.6. The result of the Krawczyk operator is depicted in figure 2 (a). On the end walls at y = 0.0 and 1.0 where the Dirichlet boundary condition of the velocity component is applied, the errors are near zero. At the center of the flow field,

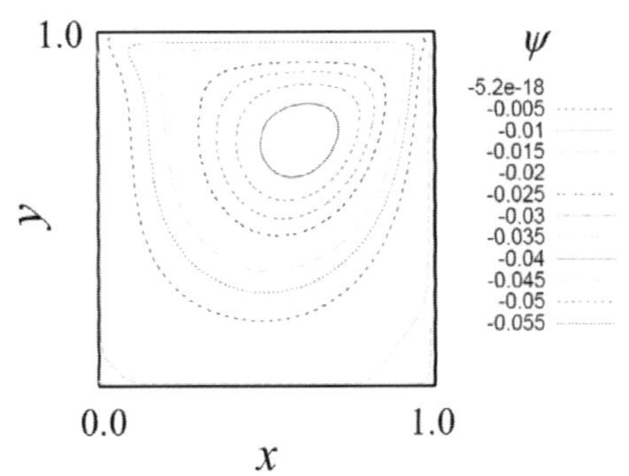

Figure 1. Solution of the lid-driven cavity flow obtained by using Krawczyk operator. The plots are the contours of the midpoints of ψ and the Reynolds number is 300.

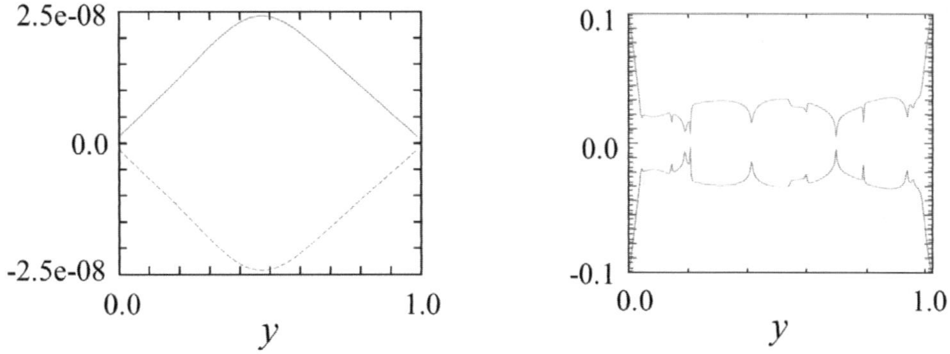

(a) Krawczyk method (b) Jacobi iteration method

Figure 2. Estimated interval of the solution of the lid-driven cavity flow. $y = 0.6$.

the interval is bounded by $\pm 2.5\times 10^{-8}$ and a favorable result is obtained. Figure 2 (b) shows the bounds obtained by the Jacobi method. While the solutions near the center of the flow are bounded within a certain range, the width between the bounds near the walls where the steep velocity gradient appears is large and the accuracy of the solution is low.

5. Conclusion

The interval analysis is applied two fluid dynamics problems. One problem is the ordinary differential equations governing the self-similar boundary layer flow. The other is the lid-driven cavity flow formulated with the stream function and the vorticity. The discretization method is the finite difference method. The Gaussian elimination method with the original interval analysis, the Gaussian elimination method with the modification of the pivoting process (MGEM), the Jacobi iteration method and the Krawczyk operator are used to solve the system of linear equations. For the solution of the self-similar boundary layer, MGEM gives the accuracy with one to three orders of magnitude higher than that of the original interval analysis. The accuracy obtained by the Krawczyk operation is almost comparable with the accuracy of the floating-point representation. For the solution of the lid-driven cavity flow, the Krawczyk operator presents the accuracy digits of half of the decimal digits of the floating-point representation. The result obtained by the interval Jacobi method does not offer any reasonable predictions of the flow near the stationary walls.

References

[1] Oberkampf W L and Roy C J 2010 *Verification and Validation in Scientific Computing*, (Cambridge: Cambridge University Press)
[2] Higham N J 2002 *Accuracy and Stability of Numerical Algorithms* 2nd ed (Philadelphia: SIAM)
[3] Jaulin L, Kieffer M, Didrit O and Walter É 2001 *Applied Interval Analysis* (London: Springer)
[4] Moore R E 1979 *Methods and Applications of Interval Analysis* (Philadelphia: SIAM)
[5] Neumaier A 1990 *Interval Methods for System of Equations* (Cambridge: Cambridge University Press)
[6] Knüppel O 1994 *Computing* 53 pp 277-87
[7] Cebeci T and Bradshaw P 1977 *Momentum Transfer in Boundary Layers* (Washington: Hemisphere)
[8] Cebeci T 2004 *Analysis of Turbulent Flows* 2nd ed (Oxford: Elsevier)
[9] Ghia U Ghia K N and Shin C T 1982 *J. Comput. Phys.* 48 pp 387-411
[10] Auteri F Parolini N and Quartapelle 2002 *J. Comput. Phys.* 183 pp 1-25

Effect of Geometric Design on Airflow Simulation Inside the Storage Room for Paddy

S Echaroj[1, *], N Pannucharoenwong[1] , P Vengsungnle[2], and M Wichangarm[3]

[1] Department of Mechanical Engineering, Faculty of Engineering, Thammasat University, 12120 Thailand.

[2] Department of Agricultural Machinery Engineering, Faculty of Engineering and Architecture, Rajamangala University of Technology Isan, 30000 Thailand.

[3] Department of Mechanical Engineering, Ubon Ratchathani University, 34190 Thailand

*Corresponding Author: snunkha@engr.tu.ac.th

Abstract. Stability of rice paddy organic content is essential for the production of high quality rice. For this reason it is important to design a proper storage room with sufficient ventilation and moisture removal capability. Airflow dynamics in terms of velocity profile were modelled through computational fluid dynamic (CFD) simulations in Fluent 6.3 program across the three different geometrical design of paddy storage room. The selection of turbulence model used in CFD is an important process due to the uncertainty of turbulence model in different operating scale. This research used a turbulence model and RNG k-ε model to calculate along the simulation of Discrete Ordinances (DO) type radiation and the heat flux was set to 800 W/m^2. The results of this study have shown that Model C with pyramid-like roof demonstrated the optimum moisture removal through convection mechanism with the lowest average velocity of 0.175 m/s and the average temperature of 365.88 K. The CFD technique could be suitable for the simulation of airflow, permitting decrease in time and cost in the evaluation of airflow inside storage facility for paddy rice.

1. Introduction

Rice or Oryza sativa L. is the principal food in most tropical continent including East, South, Southeast Asia and Latin America [1]. Countries including china, Japan, India, Pakistan, Vietnam and Thailand were found to supply 90% of rice consumed worldwide [2]. Due to significant economic expansion and population growth, the efficiency of rice production must incline by approximately 1% per year in order to guarantee sufficient food for everyone [3]. Thailand is the second largest exporter of rice with over 9.8 million tons in volume [4]. One of the main concern in rice production is during the storage period after cultivation of paddy rice. In Thailand, the moisture content of paddy rice immediately after harvest is 18 to 27% (wet basis), which is relatively high and likely to increase in conventional storage causing the quality of rice to decrease. The target moisture content of rice sold at mills need to be at least 14% (wet basis). In order to preserve the quality of paddy rice, the paddy content must be stabilized by appropriate storage construction. Inappropriate aeration system may resulted in problems associated with an increase in moisture content, insect and spoilage [5].

Many strategies have been proposed to preserve and at the same time remove moisture content of the paddy rice. For instance, Allaf et al. reported application of controlled pressure drop (DIC) process on post-harvest paddy rice. The DIC process included rapid heating of paddy rice at saturated pressure and then followed by abrupt reduction in pressure to roughly 4–5 kPa. Results showed that drying time of the paddy rice treated using DIC process was decreased six-folds compared with the untreated sample of paddy rice [6]. Pan et al. reported removal of moisture from medium grain rice through infrared radiation heating mechanism. A specific infrared power of 5,348 W/m^2 when applied was found decrease moisture content of rice samples by approximately 3.2 to 3.8 % at operation temperature of 63.5 °C [7]. The drying time of rice under infrared-assisted heating was only 58s, which was relatively small compared with hot air heating system [8].

For a natural drying process, heat and moisture transfer occurred mainly through force convection formed by rapid air flow inside the system. One of the main factors that affect air flow is porosity of the stored grain, which is related to the shape and size of the grain. A reduction in grain size was found to increase air flow resistance [9]. Currently, the main problem in storage facilities is the non-uniformity of air velocity which cause moisture and heat buildup inside the stored grain. Air velocity non-uniformity is the result of changes in porosity throughout the storage due to accumulation of small sized grain at the core of the facility and whole grain at the periphery [10]. Non-uniformity factor (NUF) can be calculated by equation 1 given the air velocity at the core and periphery of the storage room [11].

$$NUF = \frac{AF_p - AF_c}{AF_p + AF_c} \times 100 \qquad (1)$$

Where AF_p is air flow velocity at the periphery air and AF_c is the core or centre air flow velocity.

Simulation of non-uniform air velocity in natural drying process have been observed in many researches. One of the first air flow resistance model was developed in 1980 using the Ergun's equation [12]. Finite volume method was also developed and solved using computational fluid dynamics (CFD) to find the non-uniform distribution of air velocity. It was revealed that the placement of ring duct at the bottom of the storage facility resulted in a more uniform airflow distribution [13]. Baffles can be added to the ventilation system to create a more uniform air flow [14]. A RANS k-ε model for simulation of turbulent system was found to be accurate compared with experimental validation results [15]. Shahid et al. analyzed the performance of a passive approach to dry and cool simple battery pack. Results demonstrated that addition of a secondary inlet pletnum on top of battery packs was found to be the most efficient method to improve air flow velocity and the cooling rate [16,17].

The aim of this research is to investigate simulation of three different model of room for storage of paddy rice. A CFD analysis was employed to calculate air flow velocity and temperature inside the storage facility. A two equations turbulence system was employed in this research. Model A represented a doom-like shaped roof, while Model B and Model C contained gable-like and pyramid-like shape roof.

2. Methods and Materials

2.1. Numerical algorithm and governing equation

The governing equations were solve using a CFD software FLUENT through fundamental finite volume approximation. The system domain for air flow was set as steady-state, incompressible with a 3-D turbulent type. The governing equations are derived from two main mechanisms in operation including continuity (equation 2) and momentum (equation 3) .

$$\frac{\partial \rho}{\partial t} + \frac{\partial (\rho u_i)}{\partial x_i} = 0 \qquad (2)$$

$$\frac{\partial(\rho u_i)}{\partial t}+\frac{\partial(\rho u_i u_j)}{\partial x_i}=-\frac{\partial(P)}{\partial x_i}+\rho g_i+\frac{\partial\left[(\mu+\mu_t)\left(\frac{\partial u_i}{\partial x_j}+\frac{\partial u_j}{\partial x_i}\right)\right]}{\partial x_j}-S_i \tag{3}$$

$$P=p+\frac{2}{3}\rho k \tag{4}$$

$$S_i=-\left(D_{ij}\mu u+C_{ij}\frac{1}{2}\rho u^2\right) \tag{5}$$

Where ρ is the density (g/cm^3), t is time (s), u_i and u_j are the average superficial velocity components (m/s), k is the turbulent energy (m^2/s^2), g_i is gravity (m/s^2), μ is the viscosity of the medium (Pa s) and S_i is the source term equation or rate of strain tensor reflecting the normal viscous resistance (D_{ij}) and inertia resistance (C_{ij}) matrices for each material. A k-ε turbulent model (kinetic and turbulent dissipation) was used in the CFD analysis of air flow in the storage facility, as shown in equation 6 and 7. These 2 equation were employed to investigate the effect of turbulence on mean flow by solving the above governing equations entailing kinetic energy (k) and turbulent dissipation (ε) [18,19]. Other simulation have employed this turbulent model including agricultural engineering [20], air flow in greenhouses [17,21,22], and hydropower stations [14] .

$$\frac{\partial k}{\partial t}=\frac{\partial[\left(\mu+\frac{\mu_t}{\sigma_k}\right)\frac{\partial k}{\partial x_j}]}{\partial x_j}+\mu_t S^2-\rho\varepsilon \tag{6}$$

$$\frac{\partial \varepsilon}{\partial t}=\frac{\partial[\left(\mu+\frac{\mu_t}{\sigma_\varepsilon}\right)\frac{\partial \varepsilon}{\partial x_j}]}{\partial x_j}+\frac{\varepsilon}{k}(C_{1\varepsilon}\mu_t S^2-\rho\varepsilon C_{2\varepsilon}\varepsilon) \tag{7}$$

Where $C_{i\varepsilon}$, σ_k, and σ_ε are constants, S is the mean modulus, σ_k is a Prandtl number based turbulence constant and μ_t is the mean turbulent viscosity. These 2 equation turbulent technique were found to be better than standard method because it allow calculation and clear visualization of vortices

2.2. *Geometry and boundary conditions*

A three-dimensional type of rice paddy storage facility was employed for air flow and temperature simulation. The storage facilities are 3 meter wide, 8 meter long and 3 meter high. Three different model, drawn by ANSYS software, were used in the simulation as shown in Figure a, b and c. The roof use in this model was assumed to be a very excellent conduction of heat. Additionally, reflected heat radiation is indicated in the model as the hot layer at the bottom of the model. Air was assumed to enter from the left and right side of the storage facility.

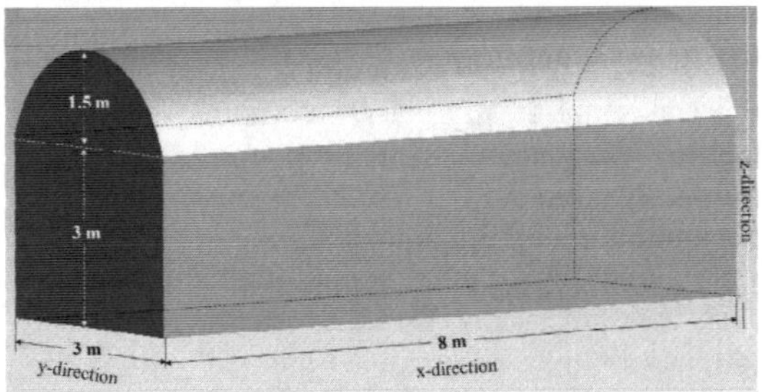

(a)

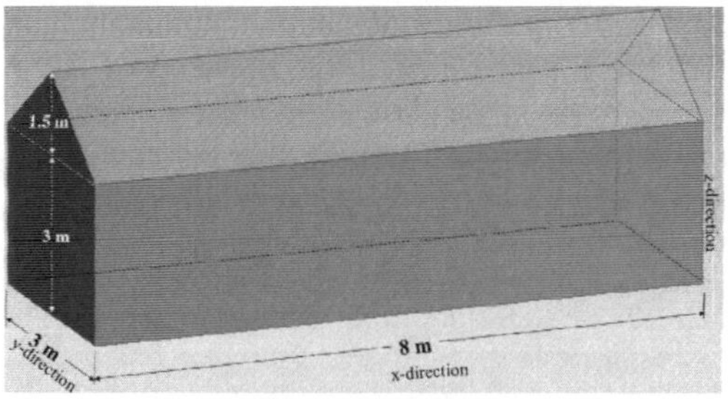

(b)

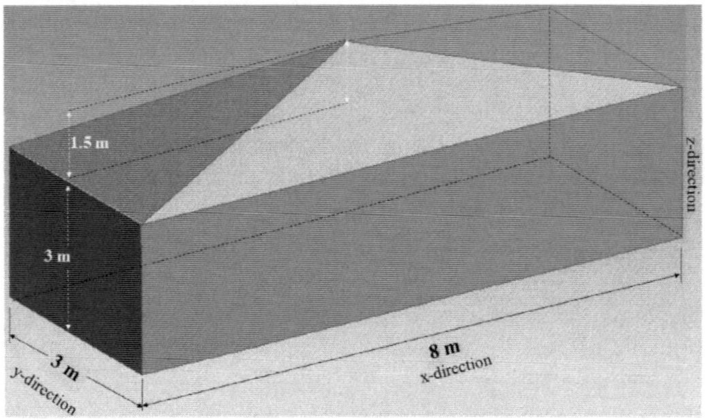

(c)

Figure 1. The geometry of the three different paddy rice storage facilities a) doom-like Model A, b) gable-like Model B and c) pyramid-like Model C.

2.3 Boundary conditions

The boundary condition at the entrance was calculated by using air volume which was assumed to be a single component model, incompressible and steady-state flow medium. The air pressure at the inlet and outlet was set at 101,325 Pa as shown in Table. Simulation was conducted based on density and heat flux at the walls were set at 800 W/m^2.

Table 1 Boundary conditions for CFD analysis

Boundary condition	Pattern
Inlet boundary condition	Pressure inlet 101,325 Pa
Outlet boundary condition	Pressure outlet 101,325 Pa
Solver	Density Base Type implicit
Time	Steady state
Near-wall treatment method	Standard wall function
Turbulence model	The RNG k-ε model
Radiation model	Discrete Ordinates (DO)
	Type Full Buoyancy Effect
Heat flux	Wall Heat flux 800 W/m^2

3. Results and discussion

The outcomes of CFD simulation from three difference model of paddy rice storage facility are shown in below. Figure 2 demonstrated the speed vector at different location inside the room. The colour and length of the vector is associated with the magnitude of speed vector. Air flow in the doom-shape (model A) room was relatively less intense compared with airflow in the other two models. A vague vortex was formed at the right side of Model A. Figure 2a demonstrated that airflow inside model B seemed to change direction more than Model A. This indicated better ventilation and airflow that is more intense, which correspond well with another research [23]. Model C, as shown in Figure 2c, seemed to have the most intense airflow with 7 small vortex and 2 larger vortex approximately 30 cm above the ground. These intense radial vortex airflow pattern created air curtain that effectively suppressed particle and increase ventilation inside the storage facility [24].

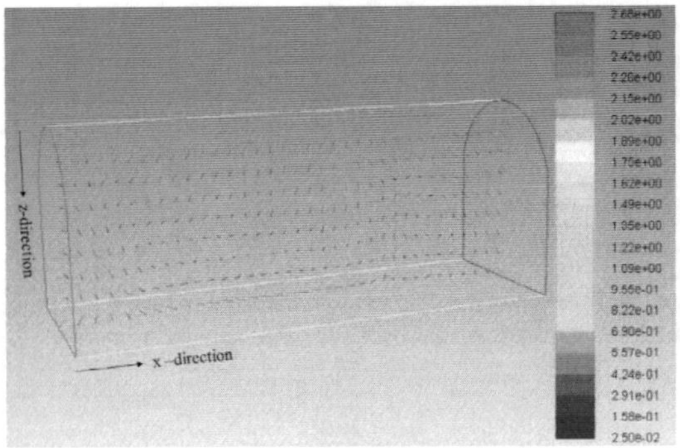

(a)

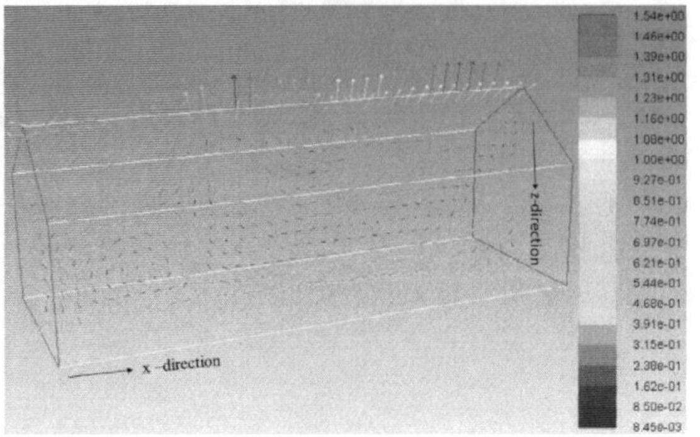

(b)

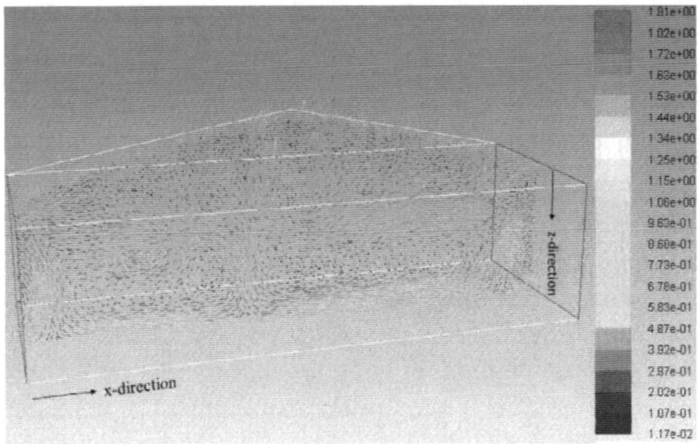

(c)

Figure 2. Speed vector for a) Model A, b) Model B and c) Model C.

Figure 3 illustrated the speed velocity (m/s) at all location considering the cross-sectional area inside the storage facility. These contour plots were found to correspond with the speed vector shown in Figure 2. Simulation of Model A revealed large area of relatively slow airflow (0.16 – 0.21 m/s) in the centre of the storage facility. Additionally, airflow along the floor in Model A was relatively slow as well indicating ineffective ventilation. The slow airflow area diminished as Model A was changed to Model B. Rapid airflow (0.37 - 0.48 m/s) emerges at the top left of Model B and drop down toward the floor along the x-direction. A significant decrease in rapid airflow area was illustrated in Model C, which have a pyramid-like shaped roof. Figure 4 compared airflow along the x-direction and 30 cm above the storage room floor of three different models. Model C demonstrated lowest air speed at all the location along the x-direction. The average air speed inside Model A and Model B were similar at approximately 0.31 m/s, while Model C gave half the average air speed compared with the other models. However, the air distribution inside Model C storage facility is clearly more uniform.

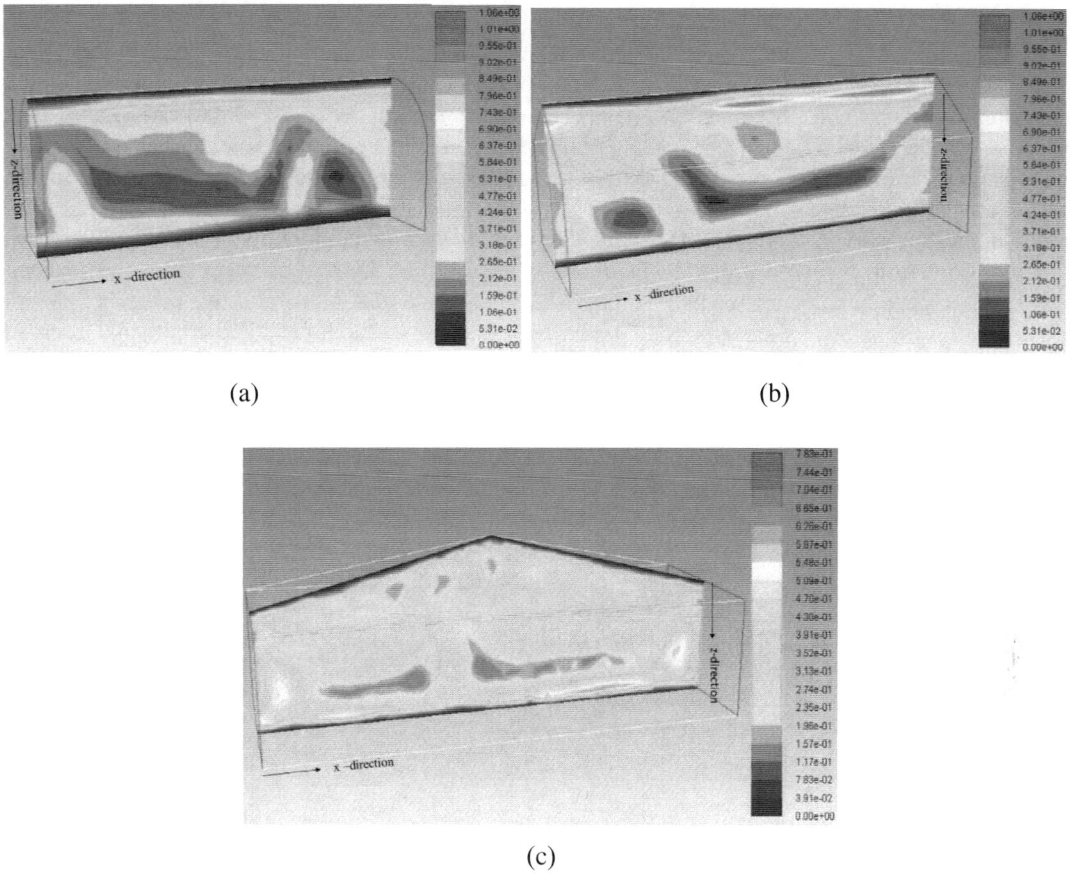

Figure 3. Cross sectional contour plot of air speed (m/s) inside a) Model A, b) Model B and c) Model C storage facility

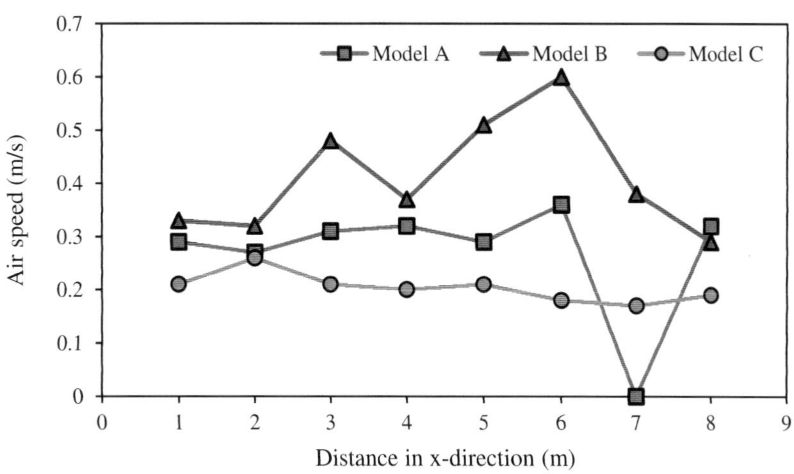

Figure 4. Air speed along the x-direction from the left-side of storage room and 30 cm above the floor.

Figure 5 demonstrated the air temperature profile of the three different models. In these models heat is transferred from the roof of the storage facility through convection and radiation. The temperature of the storage room's roof was 417 to 433 K which had been considered as an effective heat source for the simulation model. It was observed that the storage floor also emitted heat. The floor-leveled heat originated partially from reflected heat and also from radiation. For all models air temperature decreased gradually in form of uniform air temperature layer in the x-direction, which is the direction from the roof toward the floor as shown in Figure 5a - c. Simulated air temperature along the x-direction 30 cm above floor were shown in Figure 6. Model C was found to have higher air temperature at each location along the x-direction compared with the other models. This may be due to the low air velocity simulated in Model C. Lower air velocity area help maintain heat inside the storage room, while rapid air velocity initiate heat loss. The results corresponded well wither another research [25]. It can also be observed from Figure 6 that the temperature layer along the vertical x-axis from the roof changed more rapidly inside Model A and Model B compared with Model C. Air temperature and speed measurement have been conducted at different location inside the storage facility of Model A to validate the mathematical modelling concept use in the simulation. Validation revealed the collected data to be closed to simulated results with r-square close to 0.98.

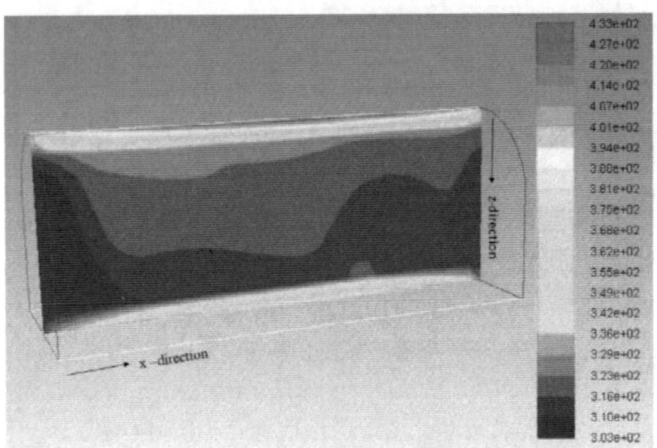

(a)

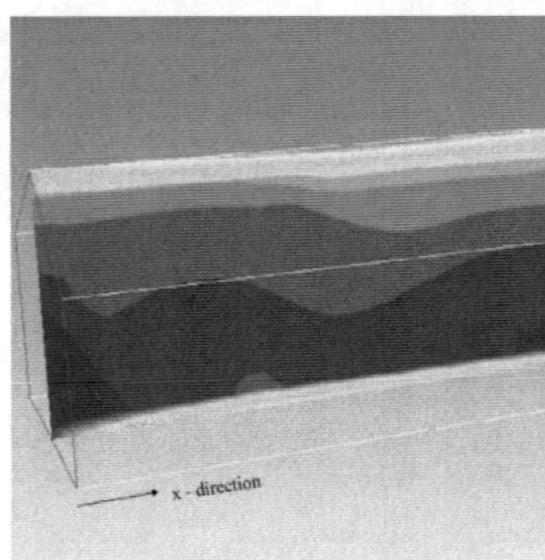

(b)

(c)

Figure 5. Cross sectional contour plot of air temperature inside a) Model A, b) Model B and c) Model C

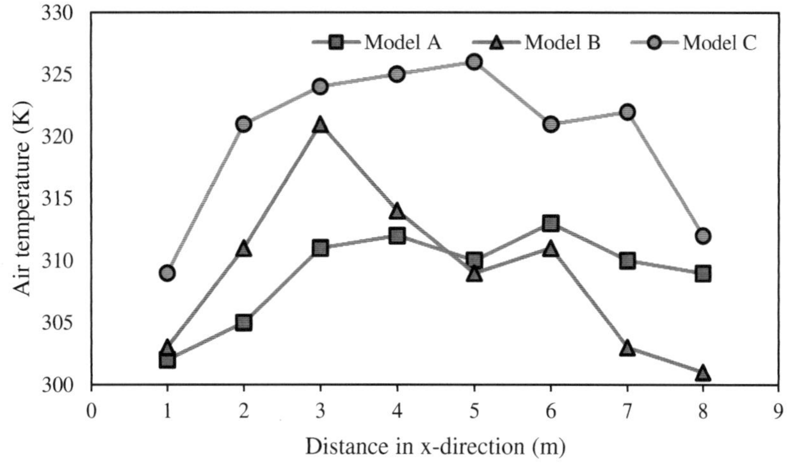

Figure 6. Air temperature along the x-direction from the left-side of storage room and 30 cm above the floor.

Conclusion

The effect of storage's roof geometry on the airflow pattern, speed and temperature were studied. Air speed vector demonstrated intense radial vortex inside Model C storage room with pyramid shape roof, indicating turbulent airflow behaviour. The air speed inside Model C was significantly lower than storage facility with dome and gable shape roof. However, airflow speed in Model C was found to be more uniform compared with the other two type of models. The low air speed in Model C resulted in an increase in convective heat transfer resistance which reduced heat loss and ultimately maintain higher air temperature compared with the average temperature inside storage facility with dome and gable shape roof. This research demonstrated the potential of a pyramid-like room to be used for storage and drying of rice paddy. Simulation using CFD analysis is an effective technique that can be employed to overcome time consuming and costly process that decision maker need to undergo in order to understand the effect of storage room facility.

Acknowledgments

The author would like to thank the Faculty of Engineering, Thammasat University for providing the necessary research funding for this project.

References

[1] Seck PA, Diagne A, Mohanty S, Wopereis MCS 2012 *Food Security* **4**(1) 7-24.
[2] Sumithra M, D SJ, Scott M, F MG 2014 *Annals of the New York Academy of Sci.* **1324**(1) 7-14.
[3] Normile D 2010 *Sci.* **321** 330-3.
[4] Sombunjitt S, Sriwongchai T, Kuleung C, Hongtrakul V 2017 *Agricul. Nat. Res.* **51**(5) 365-75.
[5] Bartosik R, Maier D 2006 *Transactions of the ASABE* **49**(4) 1095.
[6] Mounir S, Allaf K 2014 DIC-Assisted Hot Air Drying of Post-harvest Paddy Rice. In: Allaf T, Allaf K, editors. Instant Controlled Pressure Drop (DIC) in Food Processing: From Fundamental to Industrial Applications. New York, NY: Springer New York 45-55.
[7] Ragab K, Zhongli P, F TJ, S ESA, R HB, S EAM 2014 *J. Food Proc. and Preserv* **38**(1) 430-40.
[8] Ding C, Khir R, Pan Z, Wood DF, Venkitasamy C, Tu K 2018 *Food Chem.* **264** 149-56.
[9] S Chung D, Maghirang R, S Kim Y, S Kim M 2001 *Transactions of the ASAE* **44**(2) 331–336.
[10] Lawrence J, Maier DE 2011 *Biosys. Engin.* **110**(3) 321-9.
[11] Bartosik RE, Maier DE 2006 *Transactions of the ASABE*;49(4):1095-104.

[12] Lai FS 1980 *Transactions of the ASAE* **23**(3) 729.
[13] Garg D, E Maier D 2006 *9th International Working Conference on Stored Product Protection* 7-12.
[14] Ye W-B. 2017 *Appl. Ther. Engin.* **110** 573-83.
[15] Fuentes-Pérez JF, Silva AT, Tuhtan JA, García-Vega A, Carbonell-Baeza R, Musall M 2018 *Environ. Model. Soft.* **99** 156-69.
[16] Shahid S, Agelin-Chaab M 2018. *Ther. Sci. Engin. Prog.* **5** 351-63.
[17] Zhang Y, Kacira M, An L 2016 *Biosys. Engin.* **147** 193-205.
[18] Liu N, Wang W, Wang Y, Wang Z, Han J, Wu C 2017 *Appl. Ther. Engin.* **125** 1209-17.
[19] Jiru T, Bitsuamlak G. 2010 Int. J. Ventilation **9**(2) 131-47.
[20] Bustamante E, García-Diego F-J, Calvet S, Estellés F, Beltrán P, Hospitaler A 2013 Energies **6**(5) 2605.
[21] Bournet PE, Ould Khaoua SA, Boulard T 2007 *Biosys. Engin.* **98**(2) 224-34.
[22] Rohdin P, Moshfegh B. 2007 *Building and Environ.* **42**(11) 3872-82.
[23] Khatchatourian OA, Binelo MO. 2008 *Biosys. Engin.* **101**(2) 225-38.
[24] Nie W, Wei W, Cai P, Liu Z, Liu Q, Ma H 2018 *Advanced Powder Technol.* **29**(3) 835-47.
[25] Ferrua MJ, Singh RP. 2011 *Procedia Food Sci.* **1** 1239-46

Design and Flow Simulation of Concentrator Augmented Wind Turbine

S K Thangavelu*, CY Goh and CV Sia

Faculty of Engineering, Computing and Science, Swinburne University of Technology, 93300 Kuching, Sarawak, Malaysia

* Corresponding Author: sthangavelu@swinburne.edu.my

Abstract. In order for wind technology to compete with conventional sources of energy in terms of energy production costs, researchers are working on different ways to increase the energy density in wind and therefore augment wind turbine power output. Concentrator augmented wind turbines (CAWTs) have been thought to be one of the ways to augment wind turbine power output and hence improve the cost effectiveness of wind energy. This concept involves enclosing the wind turbine rotor with a duct that reduces in cross-sectional area downstream of the blade-plane and hence increasing mass flow rate through the turbine. No efforts have been done to commercialize this concept because little is known about the wind flow behaviour in the CAWT, its influence on turbine power output and the optimum concentrator design parameters. The main objective of this research is to propose a new design for CAWT and optimize the concentrator design parameters. Furthermore, the effect of wind velocity and air pressure on the proposed design concepts are also investigated. Four different CAWT design concepts were designed and simulated using Solidworks and Computational Fluid Dynamics (CFD) tool Ansys Fluent, respectively. The wind velocity and air pressure vs. the position of convergent nozzle were plotted. Results showed an outstanding improvement in wind speed, which is 16.1 m/s (5.4 times increment), over the 3 m/s inlet wind velocity in the design concept 3 (CAWT 3.0) with the nozzle angle 20°.

1. Introduction

Wind energy is second largest source of energy and it is the fastest growing renewable energy resource around the world. The main problem of using wind energy is the uncertainty of wind pattern. Low wind speeds result in less energy per unit volume of air passing through a turbine. This leads to higher electricity production cost from wind energy than from fossil fuels [1]. In order for wind technology to compete with conventional sources of energy in terms of energy production costs, researchers are working on different ways to increase the energy density in wind and therefore augment wind turbine power output. Various power augmentation systems have been studied to improve the performance of wind rotors by increasing the energy density of the wind. One of them is the ducted augmentation system [2]. Diffuser Augmented Wind Turbine (DAWT) and Concentrator Augmented Wind Turbine (CAWT) are able to provide an opportunity to extract power from unstable low wind speeds.

A DAWT would have a duct surrounding the wind turbine blades that increase in cross sectional area in the stream wise direction. The pressure behind the turbine will drop due to the wind turbine being enclosed by the diffuser, thus the wind velocity approaching the wind turbines will be increased [1-3]. CAWTs have been thought to be one of the ways to augment wind turbine power output and hence improve the cost effectiveness of wind energy. This concept involves enclosing the wind turbine rotor with a duct that reduces in cross-sectional area downstream of the blade-plane and hence increasing mass flow rate through the turbine. The concept of a CAWT has been considered to improve the efficiency of the wind turbines by increasing the wind speed upstream of the turbine. Since the power produced by a wind turbine is proportional to the cubic power of the upstream wind velocity approaching the rotor, doubling wind velocity results in eight-fold increase in wind turbine power output. Thus, if a concentrator (convergent nozzle) is inserted around or in front of the turbine, to capture air from a larger area and deliver it to the rotor through a smaller area, more power will be produced for a given turbine diameter and wind speed, thus exceeding Betz limit. The mass flow rate of air passes through a given cross section of the concentrator in a unit length of time is given by ρAV, where ρ is density (kg/m^3) of the air, A is the cross sectional area (m^2) and V (m/s) is the flow velocity. Applying the continuity, $\rho A_1 V_1 = \rho A_2 V_2$. Where the subscripts 1 and 2 refer to concentrator inlet and outlet, respectively. Since $A_1 > A_2$ [4].

Anzai et al. [5] analysed the effect of the concentrator and its optimum design. It shows that the outlet diameter of the concentrator is smaller than the rotor diameter; and the inlet diameter is much larger than the rotor diameter. A wind turbine concentrator was designed, manufactured, and tested under different moist air density and velocity conditions by Orosa et al. [6]. The experimental and simulated results show an increase in the velocity ratio whenever the air density is reduced. Results showed that this wind turbine concentrator showed ideal working conditions during the summer time [6]. Ghajar & Badr [7] presented a small experimental set-up that comprised building, testing and evaluating a collector-and diffuser-augmented wind turbine (CDAWT). Studies were conducted on a bare, shrouded, collector augmented, diffuser-augmented and finally collector-and diffuser-augmented wind turbine. CDAWT increased the augmentation ratio by 56% over that of the DAWT.

Rus [8] presented the experimental study on the increase of the efficiency of vertical axis wind turbines by equipping them with wind concentrators. For the improvement of the power coefficient of the wind turbine, a concentrator (curtain) was used in order to cancel the negative moments that affect the rotational movement of the rotor and to increase the speed of the airflow at the entry into the rotor. Shonhiwa and Makaka [4] recently suggested from their review that CAWTs are a promising way of increasing the power output in low wind speed areas. It is important to solve the problem of receiving wind from one direction and develop power output models that take into account the fact that wind turbines operate in unsteady flow conditions most of the time. This would be useful in designing CAWTs systems. There is still a need to understand various flow features that may be present in the system such as turbulence, eddys, veer and wake effects and their influence on power output so that this can be incorporated into the power output model. Before making efforts to turn the CAWT concept into a commercial competitive product, more research work has to be done to come up with optimum concentrator designs. More experiments need to be done to optimise the concentrator wall length and the incident angle since they have much influence on concentrator frictional losses. To reduce time and cost on experimental work, numerical modelling can be used and then later validated with experimental results.

Most of the studies of the ducted rotor concern the effect of the diffuser while little research has been done concerning the concentrator (nozzle). No efforts have been done to commercialize this concept because little is known about the wind flow behaviour in the CAWT, its influence on turbine power output and the optimum concentrator design parameters. The main objective of this research is to propose a new design for CAWT and optimize the concentrator design parameters. Furthermore, the effect of wind velocity and air pressure on the proposed design concepts are also investigated.

2. Methodology

The new design of CAWT is designed by using Solidwork software with different parameters. The CAWT design concept 1 (CAWT 1.0) is shown in Figure 1. The CAWT 1.0 is a combination of two different sized concentrators (double nozzle). The total length of the concentrator (L) is fixed as 1000 mm and the inlet diameter (D_1) is fixed as around 340 mm. The angle for second concentrator is fixed as 12°. The outlet diameter (D_2) is 260 mm and the concentrator angle (θ) will be tested for different angles, which are ~7°, ~10°, and ~12°.

CAWT design concept 2 (CAWT 2.0) is shown in Figure 2. The CAWT 2.0 is the combination of streamline shaped nozzle and small streamline shaped splitter inside the nozzle. It is believed that the wind velocity will be increased due to the geometry of the nozzle and presence of the splitter. The total length of the nozzle (L) is fixed as 1000 mm while the inlet diameter (D_1) is fixed as 400 mm and the angle θ will be tested for different angles, which are ~6°, ~9°, and ~12°.

CAWT design concept 3 (CAWT 3.0) is shown in Figure 3. The CAWT 3.0 is the combination of 2 different sizes of cone shaped nozzles. It is believed that the wind velocity will be increased when concave shape of nozzle is used. The total length of the nozzle (L) is fixed as 1000 mm and the inlet diameter (D_1) is fixed as 800 mm. The angle of outer nozzle is fixed as 15°and the angle θ will be tested for different angle which are ~10°, ~15°, and ~20°.

CAWT design concept 4 (CAWT 4.0) is shown in Figure 4. The CAWT 4.0 is the combination of 2 different shapes of nozzles. It is believed that the flange shape on the front nozzle would create right amount of wind behind the front nozzle, thus increasing the wind velocity. In addition, the presence of the second nozzle would create higher increase in the wind velocity. The total length of the nozzle (L) is fixed as 1000 mm while the inlet diameter (D_1) is fixed as 800 mm and the angle θ will be tested for different angles, which are ~12°, ~18°, and ~24°.

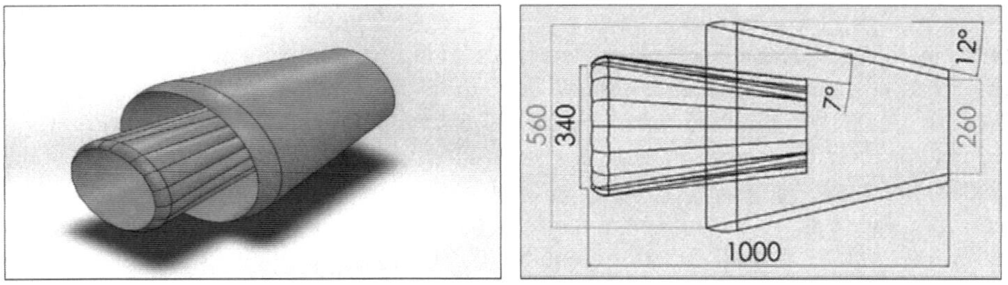

Figure 1. CAWT Design Concept 1 (CAWT 1.0)

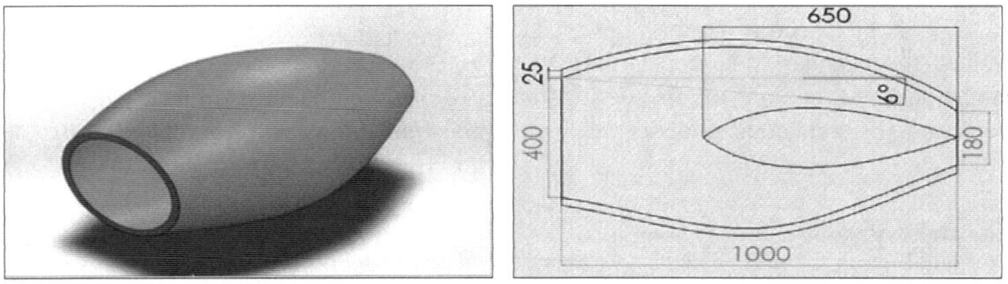

Figure 2. CAWT Design Concept 2 (CAWT 2.0)

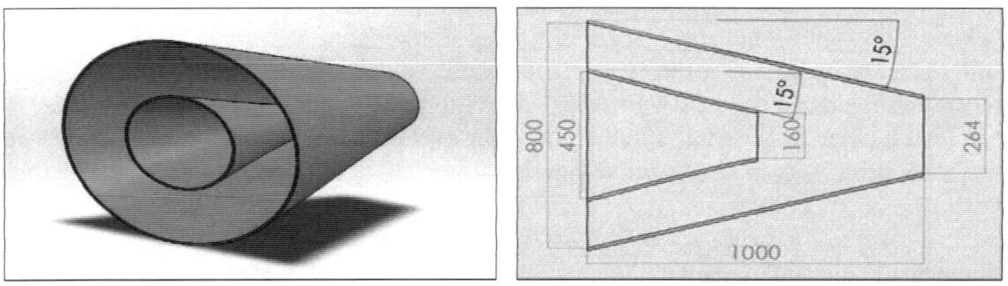

Figure 3. CAWT Design Concept 3 (CAWT 3.0)

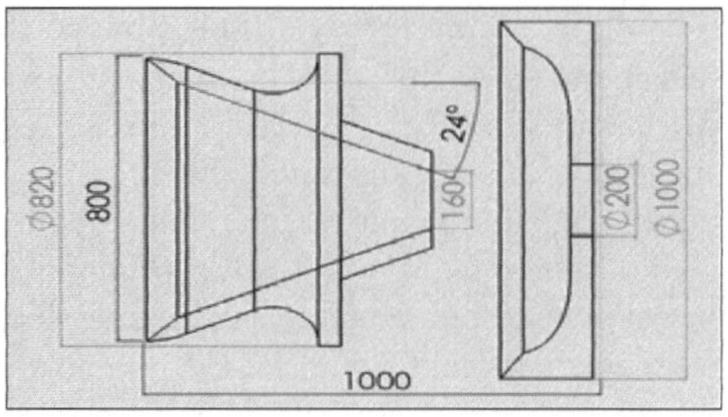

Figure 4. CAWT Design Concept 4 (CAWT 4.0)

Ansys Fluent is a CFD simulation tool, which is used for flow simulation. The air velocity and pressure can be investigated easily by using this tool. Boundary condition have been taken for simulation are: The initial wind velocity is set as 3 m/s; The inlet pressure is assumed as atmosphere pressure (0 kPa); Realizable k-epsilon model is used for the simulation and Single Reference Frames (SRF) model is used; and the airflow is assumed as incompressible steady state flow. Density based flow solver is used.

The turbulent flows can be described as the fluctuating velocity fields of the fluid. These fluctuations will be combined with transported quantities, for instance, energy, momentum and species concentration. Ansys Fluent provides three types of k-epsilon models which are standard k-epsilon model, Renormalization-group (RNG) k-epsilon model and Realizable k-epsilon model. Realizable k-epsilon model is strongly commended for use in this research as the flow features include strong vortices, rotation and streamline curvature. Hence, Realizable k-epsilon model is chosen for all the simulations in this research project, in order to obtain a more accurate data and results for concentrator wind turbine simulation.

Wall functions are a set of formulas that in effect of connecting the variables at near wall cells and quantities on the wall. The wall function contains the laws of the wall for mean temperature and velocity and principles for the near wall turbulent quantities. There are few cell types in the meshing process i.e. tetrahedron, hexahedron, prism, and pyramid. In this study, hexahedron mesh is used for the meshing because it provides higher quality solutions with less cells compared to other types of mesh. In addition, hexahedron meshes show reduced numerical diffusion when it is aligned with the flow. Besides, the boundary adaption is used to refine the cells in every edges of the concentrator. Flow simulations are done in order to investigate and identify the pattern of the air flow surrounding the concentrator. Different concentrator angle θ for each of the different concept CAWT are tested and simulated.

3. Results and discussion

In the CAWT 1.0, the wind velocity at the middle of the concentrator with various nozzle angles θ (12°, 16°, 21°) was simulated and recorded. Table 1 shows the various velocity increments from the inlet to outlet of the nozzle for each angle θ. The graphs of velocity and pressure versus the location in the nozzle are shown in Figure 5. It was found that the largest increment of wind velocity with 3 m/s of inlet velocity occurred for the angle of 21°. Based on the Figure 5 (a), it can be seen that the pattern of three different graphs are the same. The highest wind velocity for three different angles nozzle is achieved between the distances of 0.4 m to 0.5 m. Within the same range, the air pressure has also reached the lowest point, as shown in Figure 5 (b). However, this concept is not the optimum design as the increment of velocity is low. Further design improvements are needed in order to obtain a better result.

Table 1 Wind velocity increment in CAWT 1.0

Concentrator (nozzle) angle θ	**~12°**	**~16°**	**~21°**
Highest wind velocity achieved (m/s)	4.385882	4.33528	5.593
% Increment (initial velocity 3 m/s)	146.2%	144.5%	186.43%

Table 2 Wind velocity increment in CAWT 2.0

Concentrator (nozzle) angle θ	**~6°**	**~9°**	**~12°**
Highest wind velocity achieved (m/s)	4.5449	5.15721	5.67145
% Increment (initial velocity 3 m/s)	151.5%	182%	189%

Table 3 Wind velocity increment in CAWT 3.0

Concentrator (nozzle) angle θ	**~10°**	**~15°**	**~20°**
Highest wind velocity achieved (m/s)	12.3087	8.11366	16.1
% Increment (initial velocity 3 m/s)	410.3%	270.5%	537%

Table 4 Wind velocity increment in CAWT 4.0

Concentrator (nozzle) angle θ	**~12°**	**~18°**	**~24°**
Highest wind velocity achieved (m/s)	6.5684	7.1407	8.16496
% Increment (initial velocity 3 m/s)	219%	238%	272%

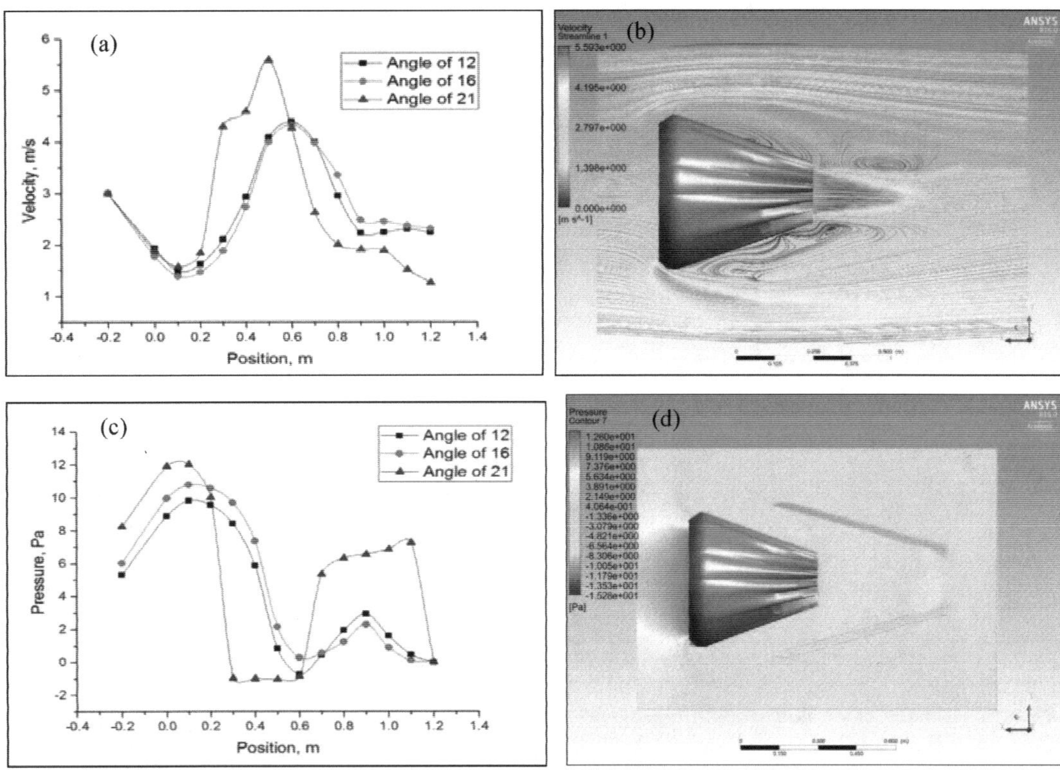

Figure 5. CAWT 1.0 (a) velocity (b) velocity simulation (c) air pressure (d) pressure simulation

Table 2 and Figure 6 show the comparison of velocity for three different angle of CAWT 2.0. The wind velocity at the outlet of the nozzle with various nozzle angles θ (6°, 9°, 12°) was simulated and recorded. The highest velocity of 5.67 m/s was achieved for the angle of 12°. Figure 6 shows the graphs for the three angles having the same trend, which is decreasing at the inlet and increasing at the outlet until it reaches the highest value. The wind velocity was increased at the outlet of the nozzle. Thus, this concept is not recommended due to low increment in velocity. Improvement on dimension and design is required in order to achieve a better performance.

Table 3 and Figure 7 shows velocity and pressure simulation of CAWT 3.0. The wind velocity at the outlet of the nozzle with various nozzle angles θ (10°, 15°, 20°) was simulated and recorded. Table 3 & Figure 7 show a remarkable result with the highest velocity of 16.1 m/s and the increment of velocity is 537% with 3 m/s of inlet velocity. Figure 7 shows all three graphs having the same pattern, which achieved the highest velocity at the output of the nozzle. According to the continuity equation, a decrease in area would lead to an increase in velocity. The cone shaped nozzle focuses the air from the inlet towards the outlet. The smaller diameter of outlet causes the wind velocity increments, thus more energy from the wind can be extracted. Figure 7 shows that the highest velocity and lowest air pressure occurred at the outlet. It can be concluded that CAWT 3.0 is an optimum design for a nozzle augmented wind turbine.

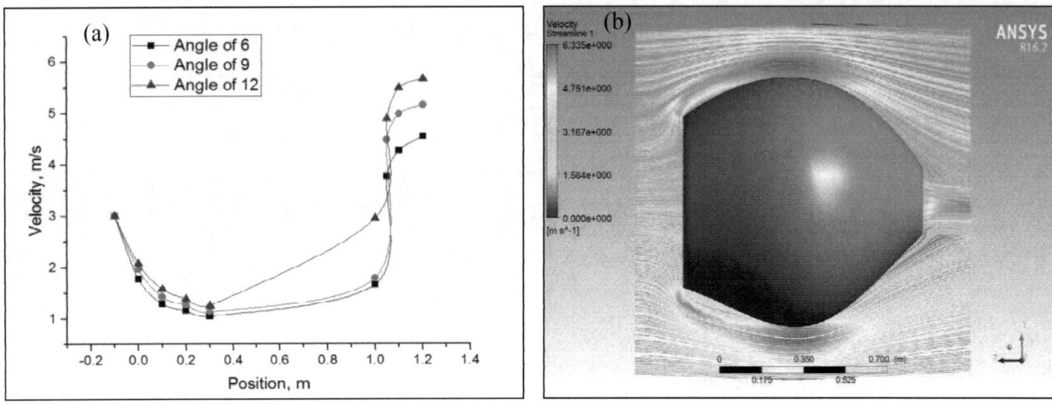

Figure 6. CAWT 2.0 (a) velocity (b) velocity simulation

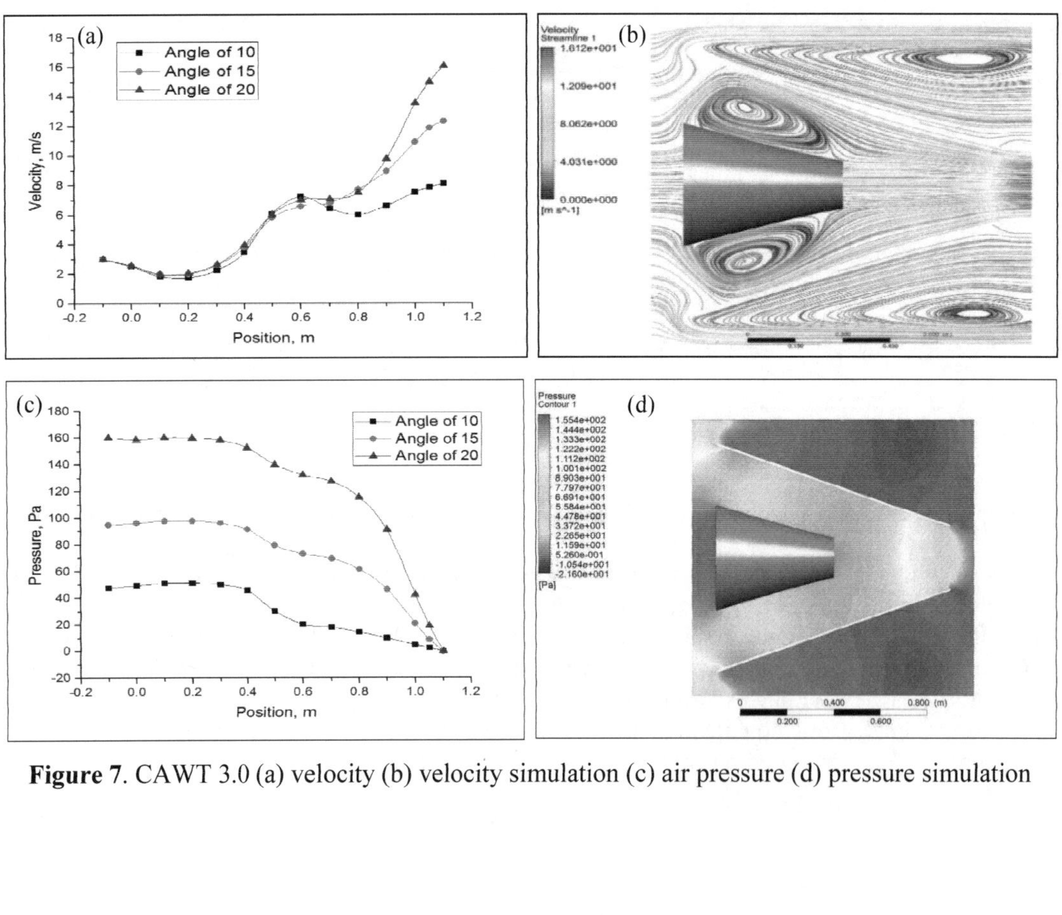

Figure 7. CAWT 3.0 (a) velocity (b) velocity simulation (c) air pressure (d) pressure simulation

Table 4 and Figure 8 show the comparison of velocity for three different angle of CAWT 4.0. The wind velocity at the outlet of the nozzle with various nozzle angles θ (12°, 18°, 24°) was simulated and recorded. Table 4 shows the comparison of highest velocity achieved with the three nozzle angles. The highest velocity of 8.16 m/s is achieved with the angle of 24°. The highest velocity is achieved at the same location where vortices are formed. It can be concluded that the vortices do influence the increment of velocity. The highest velocity occurred within the range of 0.6 to 1.0 m. However, the increment of velocity is lesser than CAWT 3.0, thus this concept is not recommended. With all the data that have been collected and obtained, it can be concurred that there are lots of room for improvement. The formation of vortices will contribute towards the increment of velocity and drop in air pressure. The formation of vortices creates a vacuum space in order to draw more air towards itself, thus wind velocity is increased. However, with wrong angle, the vortices would be formed inside the nozzle causing reduction in wind velocity at the outlet. It can be seen that the formation of vortices can be done by two different designs, either mixing the slow and fast flow or using the flange. Both ways have shown that the vortices can be formed with a right angle [7-9].

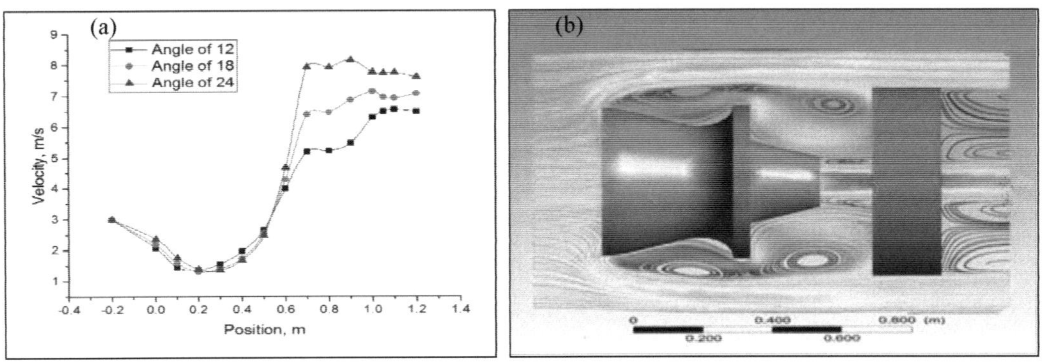

Figure 8. CAWT 4.0 (a) velocity (b) velocity simulation

Conclusion

This research analysed the various designs of nozzles for wind turbine by using Solidworks and Ansys Fluent. The results from the Solidworks and Ansys Fluent are collected and discussed. The CAWT 3.0 with angle θ of 20° is the ideal design with an outstanding result of 16.1 m/s air velocity with 537% of increment (initial 3 m/s). The behavior of the flow is based on Bernoulli's principle and continuity equation.

References

[1] Kannan T S, Mutasher S A and Lau, Y H K 2013 *Journal of Engineering Science and Technology*, **8** (4) 372
[2] Matsushima T, Takagi S and Muroyama S 2006 *Renewable Energy* **31** (9) 1343
[3] Ohya Y, Karasudani T, Sakurai A, Abe K I, and Inoue M 2008 *Wind Engineering and Industrial Aerodynamics*, **96** (5), 524
[4] Shonhiwa C and Makaka G 2016 *Renewable and Sustainable Energy Reviews* **59** 1415
[5] Anzai A, NemotoY and Ushiyama I 2004 *Wind Engineering* **28** 605
[6] Orosa J A, García Bustelo E J, Oliveira A C 2012 *Energy Sources, Part A: Recovery, Utilization, and Environmental Effects* **34** (13) 1222
[7] Ghajar R F and Badr E A 2008 *International Journal of Mechanical Engineering Education* **36** 58
[8] Rus L F 2012 *Journal of Sustainable Energy* **3** (1)
[9] Shikha S, Bhatti TS and Kothari D P 2005 *International Journal of Energy Technology and Policy* **3** 394

Alternative Constraint Handling Technique for Four-Bar Linkage Motion Generation

S Sleesongsom[1] and S Bureerat[2]

[1]Department of Aeronautical Engineering and Commercial Pilot, International Academy of Aviation Industry, King Mongkut's Institute of Technology Ladkrabang, Bangkok, 10520, Thailand
[2]Sustainable and Infrastructure Research and Development Center, Department of Mechanical Engineering, Faculty of Engineering, KhonKaen University, KhonKaen City, 40002, Thailand

*Corresponding author: Suwin Sleesongsom: suwin.se@kmitl.ac.th

Abstract. This paper proposes an extension of a new concept for four-bar path generation and a new penalty constraint handling technique from our previous work to a motion generation problem. The proposed technique could neglect the timing constraint for problems without prescribed timing, which is found to be an efficient technique for a four-bar path generation problem. Later, the remaining constraints are solved with a new constraint handling technique that also gives high performance. The technique is one kind of penalty function techniques. The present work proposes to combine the previous two methods for solving the four-bar motion generation problems. The comparative study, optimisation of motion generation problems with a new and a traditional penalty techniqueare solved by using self-adaptive population size teaching-learning based optimization (SAP-TLBO), is presented. In this study, three new motion generation test problems are used to test the proposed technique. The results show that the new technique gives acceptable results and can be applied with the motion generation problems without prescribed timing.

1. Introduction

Some applications of a four-bar linkage are window wipers, door closing mechanisms, and rock crushers among others that can be easily found in our daily life. Approximately, there are more than one thousand applications found in manufacturing [1]. This is a reason why this mechanism has been studied continuously until the present time [2-7]. The dimensional synthesis is one kind of mechanism design, which can be subdivided into three types i.e. path generation, motion generation and function generation. In this research, we adapt the previous techniques [1, 8] used for a path generation problem to study the motion generation [9], but the design problem is adopted from our previous work [1]. The path generation or path synthesis problem is a technique used to find the optimum dimensions of linkage where a point on a coupler link moves along the desired path [1,8], while motion generation or motion synthesis has a set of desired points and desired angles as a goal [9]. The design problem is set in the form of an optimisation problem with a penalty constraint handling technique [9]. Later an efficient technique to avoid the constraint of a crank angle sequence is proposed [1]. This technique has been proved to enhance the performance of the path generation synthesis. Thus, it makes sense to combine with the new constraint handling technique [8] and the new combination can enhance the performance of the motion generation problem. The proposed optimiser in this study is an efficient technique for solving path generation problem [1]. Then, in this research the performance of the proposed technique is studied. The present work is different from the very recently work by the same authors [10] in that they proposed the new constraint handling technique without using a penalty technique. The work only applied with the path synthesis problem, which is an efficiency technique.

The rest of this paper starts from Section 2 that details the position analysis of a four-bar linkage in brief, objective function, and the constraint handling technique. The design problems and optimiser are presented in Section 3. The design results are detailed in Section 4. The conclusions and discussion of the study are summarized in Section 5.

2. Methodology

2.1 Position analysis

A model of a four-bar linkage in this study is composed of four simple links connected with 4 simple joints. A variety of linkage types occur when assigning anyone link to be a frame. This linkage has one degree of freedom, which can operate with one actuator. The kinematic diagram of this linkage is shown in Fig. 1. The position analysis of the four-bar linkage uses a trigonometric formula for the relation of linkage lengths r_1, r_2, r_3, andr_4 and its other parameters, which is found in standard mechanics of machinery textbooks as mentioned in [1, 8]. The coupler point (P) in the global coordinate can be expressed as

$$x_P = x_{O2} + r_2\cos(\theta_2 + \theta_1) + L_1\cos(\phi_0 + \theta_3 + \theta_1) \quad (1)$$

$$y_P = y_{O2} + r_2\sin(\theta_2 + \theta_1) + L_1\sin(\phi_0 + \theta_3 + \theta_1)$$

where$x_{O2}$and y_{O2} are the coordinates position of the O_2 in the global coordinates [8].The anglesϕ_0, θ_3, θ_4, and γinside the kinematic diagram relate the known link lengths r_1, r_2, r_3, and r_4 at any crank angle (θ_2) by law of cosine.

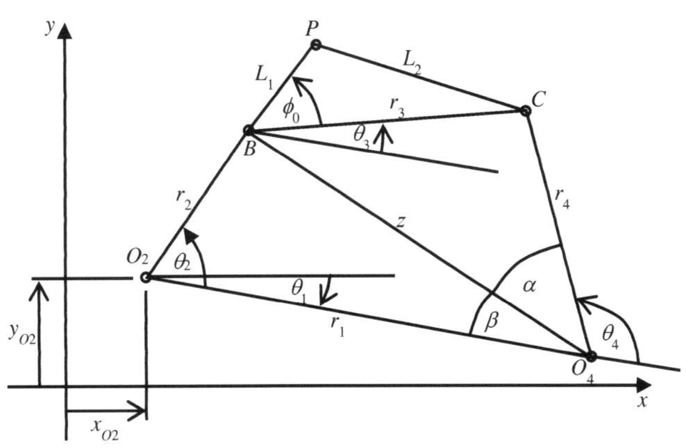

Figure 1. Four-bar linkage in the global coordinate system [1]

2.2 Optimization problem and constraint handling

The motion generation problem is different from the path generation problem as only the objective function includes the desired angles of the coupler link.As a result to the objective function of this problem composes of two parts where the first part is the position error between the desired points

$P_d(x_d, y_d)$ and the actual points $P(x_p, y_p)$. The second part of objective function is term of the angular error between desire angles (θ_{3d}) and actual angles (θ_{3p}). Then, the design variables includes r_1, r_2, r_3, r_4, L_1, L_2, and the coordinates of O_2 (x_{O2}, y_{O2}) and the angle of frame $1(\theta_1)$. In this research is focusing only the motion generation problem type, which is called synthesis without prescribed timing. Theinput set of θ_2^i values is also set as design variables. The optimization problem without prescribed timing is then written as

$$\text{Min} f(\mathrm{x}) = \sum_{i=1}^{N}[(x_{d,i}-x_{p,i})^2 + (y_{d,i}-y_{p,i})^2 +(\theta_{3d,i}-\theta_{3p,i})^2] \tag{2}$$

subject to

$$\min(r_1, r_2, r_3, r_4) = \text{crank}(r_2) \tag{3}$$

$$2\min(r_1, r_2, r_3, r_4) + 2\max(r_1, r_2, r_3, r_4) < (r_1+ r_2+ r_3+ r_4) \tag{4}$$

$$\theta_2^1 < \theta_2^2 \ldots < \theta_2^N \tag{5}$$

$$\mathbf{x}_l \leq \mathbf{x} \leq \mathbf{x}_u. \tag{6}$$

where $\mathbf{x} = \{r_1, r_2, r_3, r_4, L_1, L_2, \theta_0, x_{O2}, y_{O2}, \theta_2^i\}^T$, N is the number of points on the prescribed or desired curve, and x_l and x_u are lower and upper bounds of design vector $\mathbf{x}$,respectively.

The design constraint usually cans use an external penalty technique to handle the constraints by adding them to the objective function (2). For this case there are two parts of penalty function value. The first part is assigned to control link lengths to meet the Grashof's criteria(3-4). The second part is assigned to ensure the input crank can rotate with a complete revolution in either a clockwise or counter clockwise direction (5). The penalty function works by adding a very high value to modify the objective function when some of the constraints are violated. It was found this technique is inefficient for solving the above design problem, as a result, the unknowns θ_2^i are solved for by another technique and removed from the design variables [1, 8]. If the mechanism is a crank-rocker, this means constraint (4) and (5) are met. To tackle both constraints a new technique from [8] is used instead of the traditional penalty technique, but it has never been used for solving a motion generation problem. The constraints can be written in a common form as $g_i(\mathbf{x}) \leq g_{0,i}$.Then, a set of N input angle values (θ_2) is generated that are equally spaced from 0 to2π radian. Many more intervals indicate higher efficiency, but the disadvantage is that computationally it is more time consuming. Hence the positions of point P corresponding to all targets are calculated, the objective function is

$$f(\mathbf{x}) = \sum_{i=1}^{N} \min d_{ij}^2 \tag{7}$$

where $d_{ij}^2 = (x_{d,i} - x_{P,j})^2 + (y_{d,i} - y_{P,j})^2$ for $j = 1, \ldots, N$. The detail of this technique can be seen in [1, 8].

The usual penalty technique adds a constraint penalty term to the objective function with the constant parameter multiplied with the constraint function. If the design variables are infeasible, a large constant is added to modify the objective function value. Otherwise, the constant is 0. With this technique, the constraints can be induced and approach a feasible solution. For this reason the size of a constant is a sensitive parameter which affects the search performance of an optimizer. An alternative penalty technique was proposed by [8] for solving the path generation problem was shown to be an efficient technique for constraint handling. The focus of this research is to apply this technique in solving four-bar linkage motion generation problems. This technique creates a new equivalent objective function by using four constraint handling rules:

1) Any feasible solution has a better equivalent objective function value than any infeasible one;
2) Between two feasible solutions, the one with the better objective function value has a better equivalent objective function value;
3) Between two infeasible solutions, the one with a similar constraint violation has better equivalent objective function value; and
4) As infeasible solution with a small constraint violation ($c_i < c_0 > 0$) is considered feasible in the early stages of an optimisation run where c_0 is a small positive constraint tolerance.

The fourth rule is used to explore promising infeasible solutions which are near constraint boundaries. This "nearby rule" usually enables an algorithm to reach the real optimum faster. When constructing a new equivalent unconstrained objective function from rules 1-3, the function can be defined as [8]

$$f_p(\mathbf{x}, c_0) = \begin{cases} f' = \dfrac{f(\mathbf{x})}{\dfrac{f(\mathbf{x})+k}{1+\max|c_i(\mathbf{x})|}} & \begin{matrix} if \ \max|c_i(\mathbf{x})| \le c_0 \\ otherwise \end{matrix} \end{cases} \tag{8}$$

where $k > 0$ is a constant to be defined (for this paper $k = 10$), and c_0 starts from a positive value and is iteratively reduced to 0 at about a half iteration.

The constraint equation should be rearranged in normalized form first before being used with (8) as follows

$$c_i(\mathbf{x}) = g_i(\mathbf{x}) / g_{0,i} - 1 \le 0 \tag{9}$$

where $g_{0,i}$ is an allowable value for the constraint g_i.

3. Numerical Experiment

Three optimisation problems without prescribed timing are used to study the performance of the proposed design technique. The design problems are detailed as follows:

Case-1 : Design variables for **x** are

$\mathbf{x} = [r_1, r_2, r_3, r_4, L_1, L_2, x_{O2}, y_{O2}, \theta_1, \theta_2^1, \theta_2^2, \theta_2^3, \theta_2^4, \theta_2^5, \theta_2^6]$

Target points are (x_d, y_d) and θ_{3d}

$(x_d, y_d) = [(20,20),(20,25),(20,30),(20,35),(20,40),(20,45)]$

$\theta_{3d} = [1.9937, 1.9220, 1.8434, 1.7599, 1.6709, 1.5735]$ rad

Limits of the variables:

$5 \le r_1, r_2, r_3, r_4 \le 60$

$-60 \le L_1, L_2, x_{O2}, y_{O2}, \le 60$

$0 \le \theta_0 \theta_2^1, \ldots, \theta_2^6 \le 2\pi$

Case-2 : Design variables for **x** are

$\mathbf{x} = [r_1, r_2, r_3, r_4, L_1, L_2, x_{O2}, y_{O2}, \theta_1, \theta_2^1, \theta_2^2, \theta_2^3, \theta_2^4, \theta_2^5, \theta_2^6, \theta_2^7, \theta_2^8, \theta_2^9, \theta_2^{10}]$

Target points are (x_d, y_d) and θ_{3d}

$(x_d, y_d) = [(20, 10), (17.66, 15.142), (11.736, 17.878), (5, 16.928), (0.60307, 12.736), (0.60307, 7.2638), (5, 3.0718), (11.736, 2.1215), (17.66, 4.8577), (20, 10)]$

$\theta_{3d} = [0.4208, 0.5117, 0.7433, 0.9910, 1.1394, 1.1296, 0.9599, 0.7322, 0.5257, 0.4208]$ rad

Limits of the variables:

$5 \leq r_1, r_2, r_3, r_4 \leq 80$

$-80 \leq L_1, L_2 x_{O2}, y_{O2}, \leq 80$

$0 \leq \theta_0 \theta_2^1, \ldots, \theta_2^{10} \leq 2\pi$

Case-3 : Design variables for **x** are

$\mathbf{x} = [r_1, r_2, r_3, r_4, L_1, L_2, x_{O2}, y_{O2}, \theta_1, \theta_2^1, \theta_2^2, \theta_2^3, \theta_2^4, \theta_2^5, \theta_2^6, \theta_2^7, \theta_2^8, \theta_2^9, \theta_2^{10}, \theta_2^{11}, \theta_2^{12}, \theta_2^{13}, \theta_2^{14}, \theta_2^{15}, \theta_2^{16}, \theta_2^{17}, \theta_2^{18}, \theta_2^{19}, \theta_2^{20}, \theta_2^{21}]$

Target points are (x_d, y_d) andθ_{3d}

(x_d, y_d) = [(0, 0), (0.9356, 0.4064), (1.5983, 0.8220), (2.1704, 1.3002), (2.6621, 1.7986), (2.9957, 2.2692), (3.2614, 2.7276), (3.4941, 3.1408), (3.6635, 3.5485), (3.7798, 3.9623), (3.7992, 4.4734), (3.7170, 4.8783), (3.4675, 5.3624), (3.1242, 5.6980), (2.7158, 5.8800), (2.2517, 5.8780), (1.7997, 5.6882), (1.2732, 5.1741), (0.8336, 4.5628), (0.4003, 3.9064), (0.3, 3.38)]

θ_{3d} = [0, 0, 0, 0, 0, 0, 0.06984127, 0.13968254, 0.261904762, 0.436507937, 0.611111111, 0.838095238 1.047619048, 1.30952381, 1.571428571, 1.658730159, 1.746031746, 1.833333333, 1.746031746 1.658730159, 1.571428571] rad

Limits of the variables:

$5 \leq r_1, r_2, r_3, r_4 \leq 80$

$-80 \leq L_1, L_2, x_{O2}, y_{O2}, \leq 80$

$0 \leq \theta_0 \theta_2^1, \ldots, \theta_2^{21} \leq 2\pi$

Inorder to measure the performance of the new constraint handling technique for solving four-bar motion generation problems, SAP-TLBO is used to tackle the design problems. The parameter settings of this algorithm are defined as IReset = 20, IRange = 5, and δ = 1. The details of this optimiser can be seen in [1].

All algorithms are coded in MATLAB commercial software. In this study the population size is set n_P = 100 for all algorithms, while the maximum number of iterations is 500. The number of run times of each algorithm is set to be 30 times to study statistical performance. The value of c_0= 30 at the initial and iteratively reduced to 0 at approximately a half of an optimization run to make it more diversified for the optimizers.

4. Design Results

The design results are shown in Table 1. The table shows the mean objective function values from 30 optimization runs, , the worst result (max), the best result (min), the standard deviation (std), and the number of successful runs (a run that gives a feasible solution). Figs. 2-4 show the best path and angle traced by the coupler point and its kinematic diagram of the best linkages. In Case-1, there are 6 target points and angles. It was found that SAP-TLBO with the new penalty technique gives the best result (error = 0.002591) and the best mean objective value (error = 0.007623), which is better than the original technique. The result of Case-2 showed that SAP-TLBO with the new penalty technique gives the better result than the original one in both mean (error = 0.088245) and min (error = 0.035899). The last result of Case-3 showed that SAP-TLBO with the new penalty technique gives a better result than the original one in both mean (error = 0.407154) and min (error = 0.378623). The number of successful runs of the new penalty is quite similar to the traditional penalty technique as shown in Table 1. The results show that SAP-TLBO with the new technique is better than those using the traditional technique in all cases.

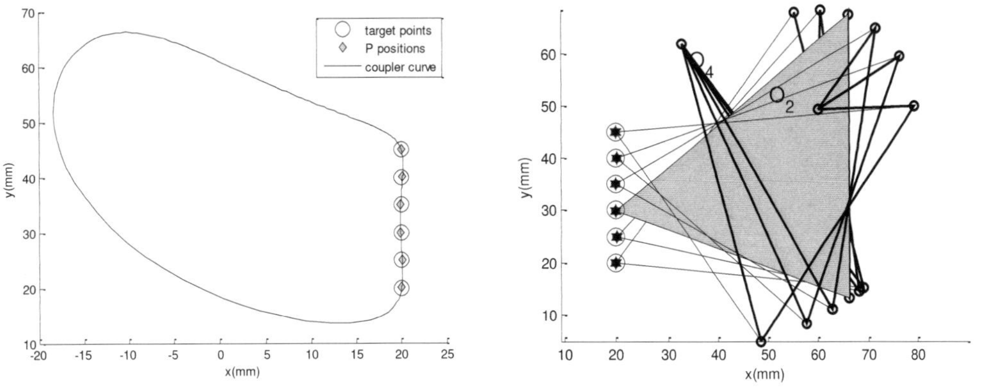

Figure 2. The best coupler curves and the best mechanism of Case-1.

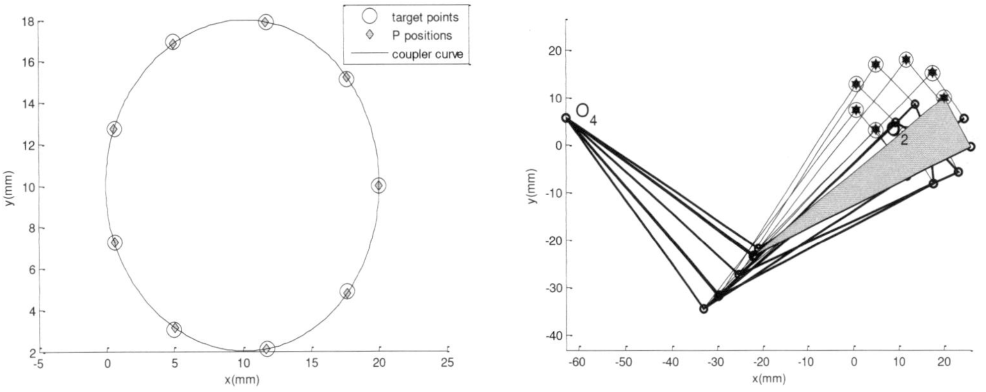

Figure 3. The best coupler curves and the best mechanism of Case-2.

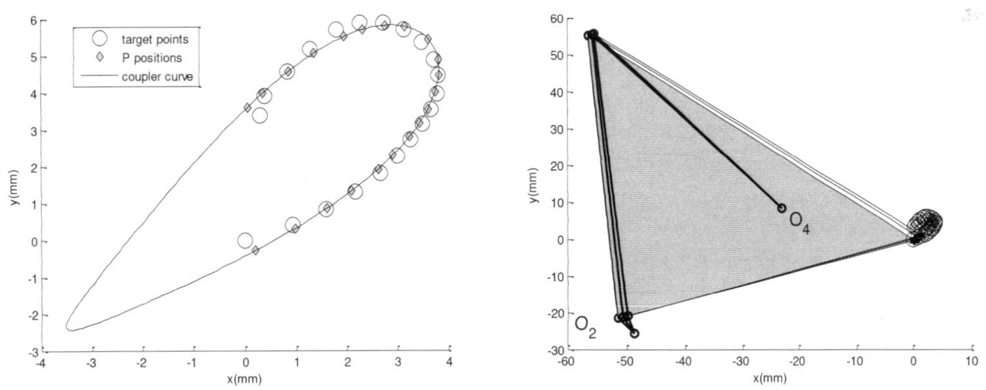

Figure 4. The best coupler curves and the best mechanism of Case-3.

Table 1. Design results of motion generation problem

Para-meters	SAP- TLBO					
	Case-1		Case-2		Case-3	
	New penalty	Traditional penalty	New penalty	Traditional penalty	New penalty	Traditional penalty
r_1	0.455280	0.747894	0.093053	0.808245	0.900515	0.997974
r_2	0.256455	0.478707	0.054969	0.061523	0	5.77E-07
r_3	0.897185	0.579066	0.638748	0.472704	0.868487	0.253568
r_4	0.982635	0.722648	0.622398	0.461557	0.694746	0.744407
L_1	0.005589	0.840120	0.502475	0.548133	0.083085	0.573494
L_2	0.909715	0.199440	0.824103	0.257639	0.513827	0.657151
x_0	1	0.983981	0.558221	0.519631	0.703590	0.522978
y_0	0.911070	0.591794	0.554886	0.542629	0.921755	0.462580
θ_0	0.431581	0.084342	0.973713	0.779455	0.551639	0.965804
mean	0.007623	0.209854	0.088245	1.036778	0.407154	8.105407
max	0.019476	1.713401	0.507658	12.39276	0.544994	18.595960
min	0.002591	0.019861	0.035899	0.375200	0.378623	6.389301
std	0.003665	0.412596	0.098038	2.391289	0.035156	2.613243
Success.*	25	22	25	25	24	30

* Success. = no. of successful runs

5. Conclusions

This paper proposed combining our previous efficient techniques for path generation of a four-bar linkage and applying them to motion generation problems. This technique is a combination of a technique to remove timing constraints from the optimization problem and a new penalty technique for handling Grashof's criterion constraint. Numerical experiments demonstrated that the new technique with SAP-TLBO outperformed the same optimiser with the traditional penalty technique. However, this is a possibility study of an alternative technique for solving motion generation problem without prescribed timing. For future work the other parameters that affect the performance of this design technique should be investigated.

Acknowledgments

The authors are grateful for the financial support provided by King Mongkut's Institute of Technology Ladkrabang, the Thailand Research Fund, and the Post-doctoral Program from Research Affairs, Graduate School, KhonKaen University (58225).

References

[1] Sleesongsom S and Bureerat S2017 *Knowledge-Based Systems* **135** 180–191
[2] Smaili A and Diab N2007 *Mechanism and Machine Theory* **42** (1) 115–130
[3] Cabrera J A, Nadal F, Muñoz J P and Simon A 2007 *Mechanism and Machine Theory* **42**(7) 791-806
[4] Acharyya S K and Mandal M 2009 *Mechanism and Machine Theory* **44**(9) 1784-1794

[5] Lin W Y 2010 *Mechanism and Machine Theory* **45**(8) 1096–1107
[6] PeñuñuriF, Peón-Escalante R, Villanueva C and Pech-Oy D 2011 *Mechanism and Machine Theory* 46(10) 1335–1349
[7] Ebrahimi S and Payvandy P 2015 *Mechanism and Machine Theory* **85** 189–204
[8] Sleesongsom S, Bureerat S 2018 *IOP Conf. Series: Materials Science and Engineering* **324** 012012
[9] Gogate G R, Mateka S B 2012 *Mechanism and Machine Theory* **52** 158–179
[10] Sleesongsom S, Bureerat S 2018 *Computational Intelligence and Neuroscience* (in press).

IOP Conf. Series: Materials Science and Engineering **501** (2019) 012043 doi:10.1088/1757-899X/501/1/012043

Numerical Investigation of In-chamber Flow inside a Wankel Rotary Engine

T Poojitganont[1,*], J Sinchai[2], B Watjatrakul[2], and HP Berg[1]

[1]Chair of Combustion Engines and Flight Propulsion, Brandenburg University of Technology Cottbus - Senftenberg, Cottbus, 03046, Germany
[2]Department of Mechanical and Aerospace Engineering, King Mongkut's University of Technology North Bangkok, Bangkok, 10800, Thailand

* Corresponding Author: Poojitth@B-Tu.De

Abstract. In this paper the in-house technique mesh moving function based on AVL-FIRE has been developed as a simulation tool to investigate the in-chamber flow phenomena of the Wankel rotary engine. The meshes from the starting rotor position at 165° BTDC until the ending rotor position at top dead center (TDC) have been created and connected together. The numerical simulations between an intake stroke and the end of a compression stroke have been successively performed. The results are validated with the selected publication on various engine speeds, at 675 rpm and 1170 rpm. In parallel three refinements of meshes have been carried out, in order to optimize the suitable meshes' elements for the calculation of this engine type. In addition, the turbulence models, which are standard k–ε and Large Eddy Simulations (LES), have been varied for detailed investigation of their predictive capabilities. The simulation results show that the flow phenomena are well corresponding to the experimental data in both engine speeds, especially with the LES. It could be also identified that the LES model performs better on predicting the flow field both in directions and characteristics. Moreover, there is no evidently difference on the results between the medium (c.a. 100,000 elements) and fine (c.a. 1,000,000 elements) meshes comparing to the experimental results.

1. Introduction

The Wankel rotary engine is categorized as one type of internal combustion engine. Based on its simple and compact design, the Wankel rotary engine provides a high power to weight ratio compared to the reciprocating engines. Its advantage is profitable for automobile with Hybrid system and several kinds of aircrafts [1]. Furthermore, it generally generates flat torque characteristics and high performance at high speed [2, 3]. The exception from those advantages of the Wankel rotary engine, there are drawbacks, which are significant to be considered for improving this engine type. Typically, the engine has lower thermodynamics efficiency due to its long combustion-chamber design increasing possibility of quenching area. This causes the incomplete combustion to occur and releases emissions of NOx, CO, HC and unburned fuel [3].

In order to improve the combustion and mixture formation of the Wankel rotary engine, it is necessary to understand and be able to predict the flow behaviors inside the chamber. Therefore, the study of fluid mechanisms in inside the combustion chamber is apparently convincing in terms of optimizing the design of its geometries, fuel atomization, and fuel-air mixing including combustion

IOP Conf. Series: Materials Science and Engineering **501** (2019) 012043 doi:10.1088/1757-899X/501/1/012043

efficiency [4-6]. Based on these ideas, the flow phenomena of the Wankel rotary engine has been investigated in this study by means of the CFD software, AVL-FIRE. The in-house mesh moving technique has been implemented to connect the group of meshes together. In order to validate various flow patterns, the numerical results has been compared with the experimental investigation from the publication on the airflow visualization of rotary engine assembly performed by F. Hamady, T. Stuecken and H. Schock [7].

2. CFD Modelling

2.1. *Geometric Models Generation*

In the investigation the flow field inside the Wankel rotary engine during an intake stroke to the end of compression stroke is focused, thus a group of meshes as same geometry as the referenced publication in small steps of rotor position between 165°R BTDC until TDC has been performed, referencing on the absolute coordinate system. Examples of these meshes could be shown in figure 1.

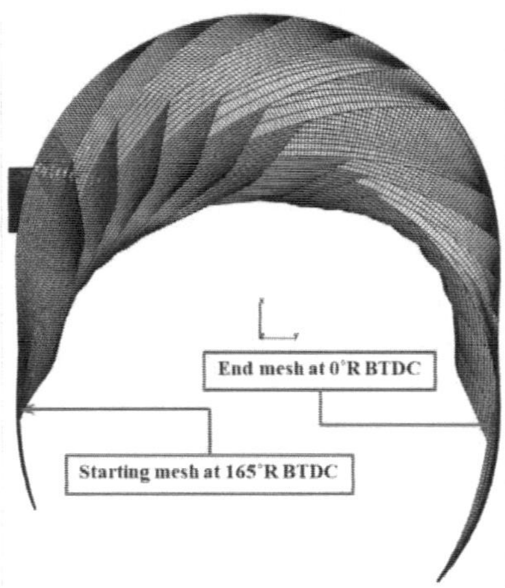

Figure 1. Example of calculating meshes.

These meshes are connected and continuously calculated during the simulation, using the in-house mesh moving technique. Conservation of mass, momentum and thermal energy are utilized as the governing equation [8]. In this study, the k–ε model and LES are chosen for capture turbulent behaviours furthered from the former research [9]. In addition, the variation of mesh quantity has been also achieved as well.

2.2. *Turbulence Models*

Regarding the flow pattern in the engine has sophisticated behaviors and high turbulence intensity, the turbulence models have to be implemented. Firstly, the k–ε model is claimed to be suitable for this kind of study [10, 11]. The defined equation of standard, high-Re-number k–ε model, integrated over a control volume is given based on AVL FIRE CFD solver [8] as follows,

$$\rho \frac{\partial k}{\partial t} + \rho U_j \frac{\partial k}{\partial x_j} = P + G - \varepsilon + \frac{\partial}{\partial x_j}(\mu + \frac{\mu_t}{\sigma_k}\frac{\partial k}{\partial x_j}) \tag{1}$$

$$\rho \frac{D\varepsilon}{Dt} = \left(C_{\varepsilon 1}P + C_{\varepsilon 3}G + C_{\varepsilon 4}k\frac{\partial k}{\partial x_j} - C_{\varepsilon 2}\varepsilon\right)\frac{\varepsilon}{k} + \frac{\partial}{\partial x_j}(\frac{\mu_t}{\sigma_\varepsilon}\frac{\partial \varepsilon}{\partial x_j}) \tag{2}$$

Where P is a production of k by mean-flow deformation, G is the production of k by body force. Both parameters are given as follows,

$$P = 2\mu_t S \cdot S - (2/3)[\mu_t(trS) + k](trS) \quad (3)$$

$$G = -\frac{\mu_t}{\rho\sigma_\rho}\nabla\rho \quad (4)$$

$$\mu_t = C_\mu \rho \frac{k^2}{\varepsilon} \quad (5)$$

In order to investigate the predictive capability, LES model is also used in the simulation since it is proved more accurate by various studies [12-14]. The definition could be illustrated as following.

$$\frac{\partial(\overline{u_i})}{\partial t} + \frac{\partial(\overline{u_i u_j})}{\partial x_j} = -\frac{1}{\rho}\frac{\partial \bar{p}}{\partial x_j} + \frac{\partial}{\partial x_j}\left[v\frac{\partial(\overline{u_i})}{\partial x_j} - \left(\overline{u_i u_j} - \overline{u_i}\overline{u_j}\right)\right] \quad (6)$$

2.3. *Mesh Quantity*
Three different grid sizes are considered in this study including 2.4 mm, 0.8 mm, and 0.4 mm of hexahedral mesh. The mesh number example of those three different sizes are shown in table 1.

Table 1. Number of meshes of three different grid sizes.

Grid size	2.4 mm	0.8 mm	0.4 mm
Mesh number at TDC	7280	58240	465920
Mesh number at BTC	17900	143200	1145600

2.4. *Initial and Boundary Condition*
The working fluid is assumed to be air throughout the simulation which is defined as unsteady flow, viscous and compressible. All the leakage goes inward and outward the combustion room is neglected. The initial condition at 165°R BTDC shown in table 2 is based on experimental condition of the referent publication.

Table 2. Initial condition.

Parameters	Value
Species:	Air
Pressure [pa]:	1×10^5
Density [kg/m^3]:	1.161
Temperature [K]:	298
Initial velocity U, V, W [m/s]:	0, 0, 0

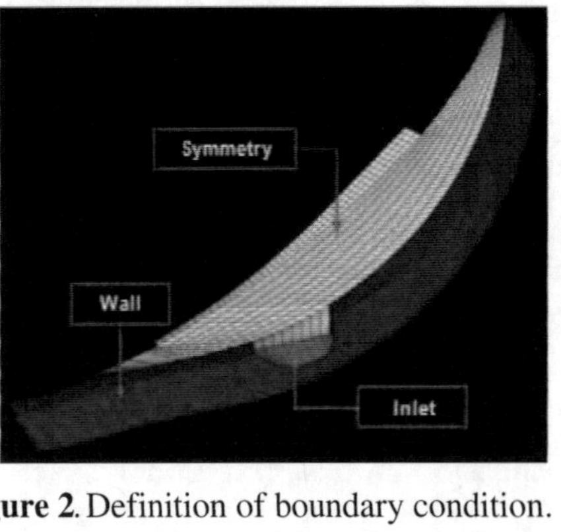

Figure 2. Definition of boundary condition.

Due to a symmetrical shape in Z-direction of the engine, the symmetry boundary condition could be applied, to reduce the global number of meshes. The inlet port has been added into and defined as inlet boundary condition shown in figure 2. The rest surfaces have been set as moving or fixed walls.

3. Simulation Results and Discussion

3.1. *Mesh Validation*

The simulation results on different grid sizes on both k–ε and LES cases are illustrated in figure 3 and 4. The results showed that the medium (0.8 mm) and fine (0.4 mm) grid size provide the closely results and capture full of details comparing to the rough one. The medium grid size has been selected to perform the further calculation since it provides adequate details and reasonable time spending.

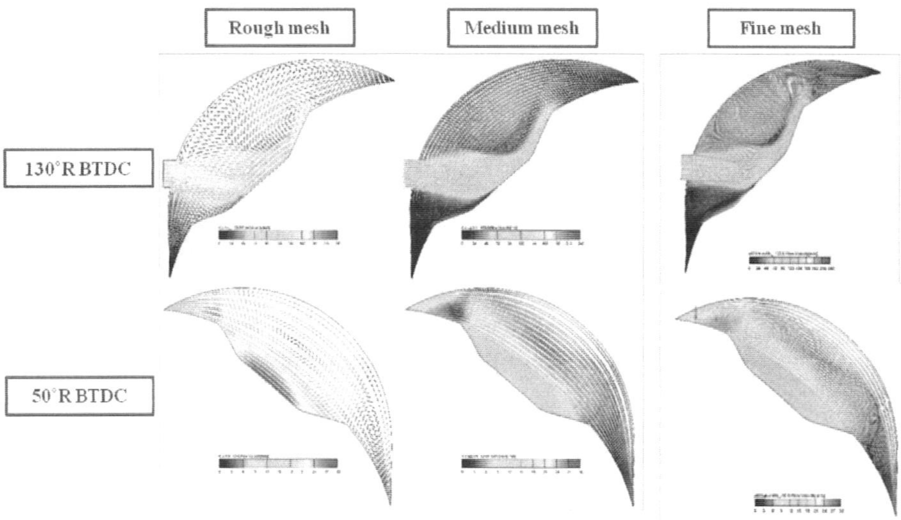

Figure 3. Velocity profiles on different grid sizes with k-e turbulence model.

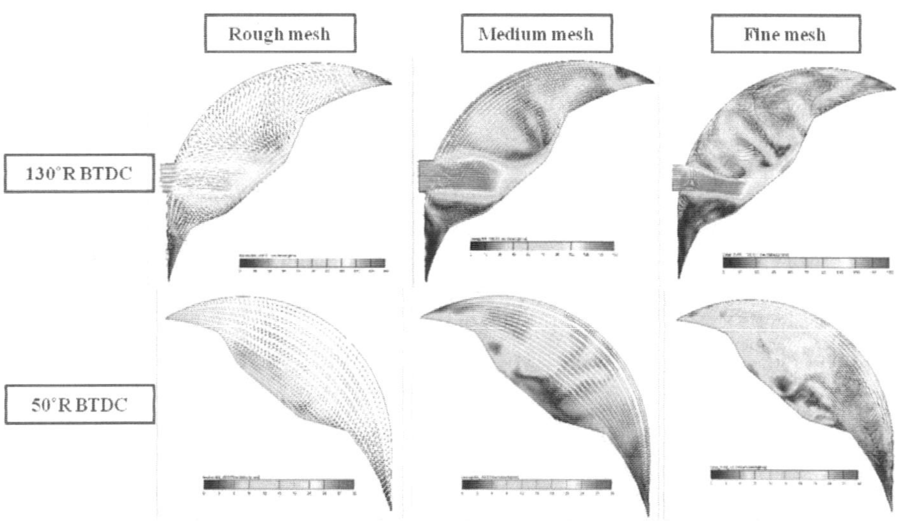

Figure 4. Velocity profiles on different grid sizes with LES turbulence model.

3.2. *Turbulence Model Variation*

In this step, the comparison on the medium mesh with various turbulence models and experimental data has been performed as shown in figure 5. It is evident that the LES provides more accurate results comparing with k- ε model. Therefore, the investigation of flow validation on different engine speeds in the next chapter has been done based on LES model.

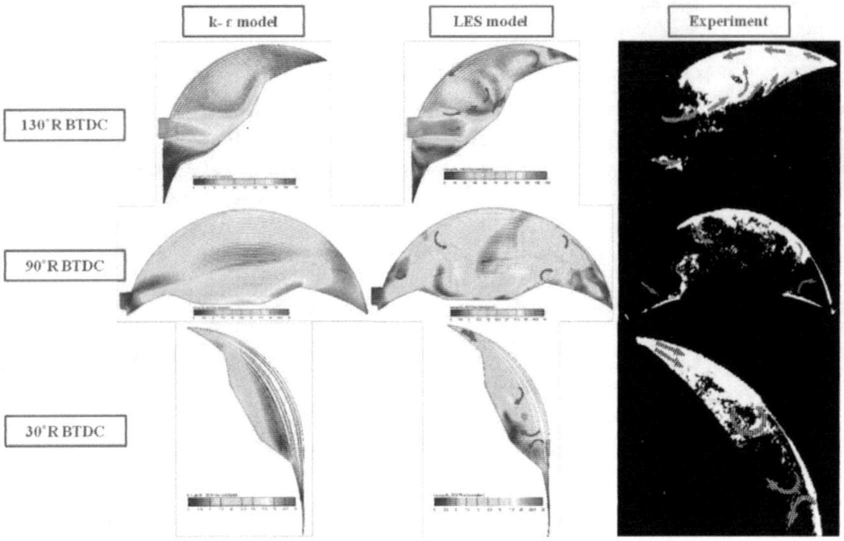

Figure 5. Velocity profiles with various turbulence models at 675 rpm.

3.3. *Flow Pattern Validation*

Starting from 165°R BTDC until the last position at TDC on the shaft speed of 675 rpm and 1170 rpm, the flow field inside the combustion chamber could be verified. During intake stroke, the flow is mainly governed by the momentum of intake and impingement on the rotor pocket, forming a counterclockwise recirculation in the main region. Since the effect of leakage at the apex seal is neglected, the simulation could not capture the counter flow along the housing adjacent to the leading edge as marked in figure 6. However, this flow seems to diminish rapidly due to the jet-like blow-by flow caused by the leakage at the apex seal [7].

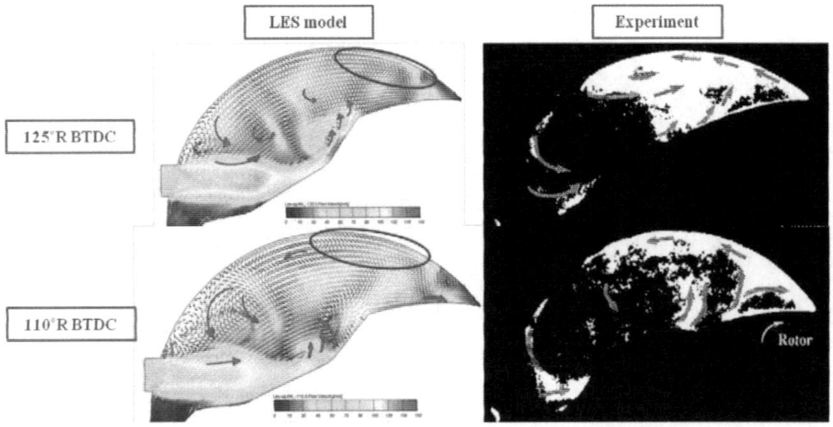

Figure 6. Comparing results during intake stroke at 675 rpm.

When the rotor moves to the bottom dead center (90°R BTDC), the intake is started to close. After this position, the fluid is mainly governed by the movement of the rotor. The flow started to change its direction forming the clockwise recirculation as illustrated in figure 7. This phenomenon induces the mixing process between the oxidizer and fuel in this type of engine.

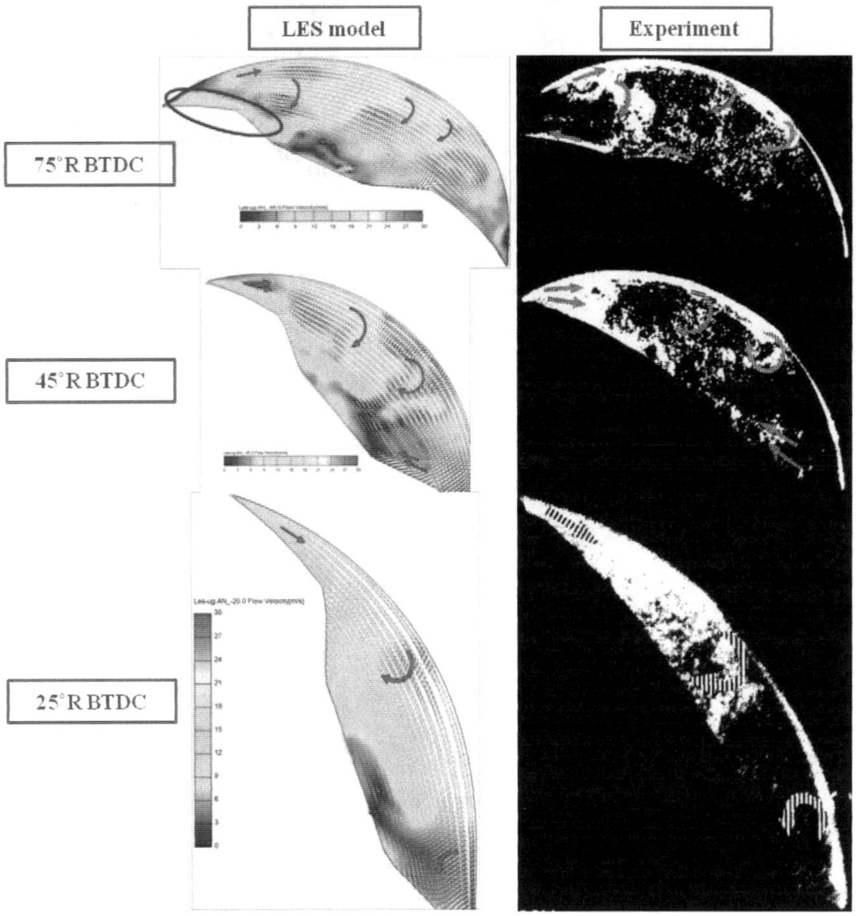

Figure 7. Comparing results during compression stroke at 675 rpm.

The reversed flow appeared in the experiment at 75°R BTDC could not be captured by the simulation. The assumption of discrepancy is the existence of the leakage of the apex seal at the tailing edge, which is ignored in the simulation and should be studied in the next paper. However, the main character of flow field could be well predicted with the LES model. In the final stage of compression, the vorticity is intensified by decreased volume.

On the higher shaft speed at 1170 rpm, the flow pattern has also the same behaviour as appeared in the slower case. However, there is a significant increase on the intensity of turbulence compared to the lower speed during intake stroke as shown in figure 8 and compression stroke in figure 9, respectively.

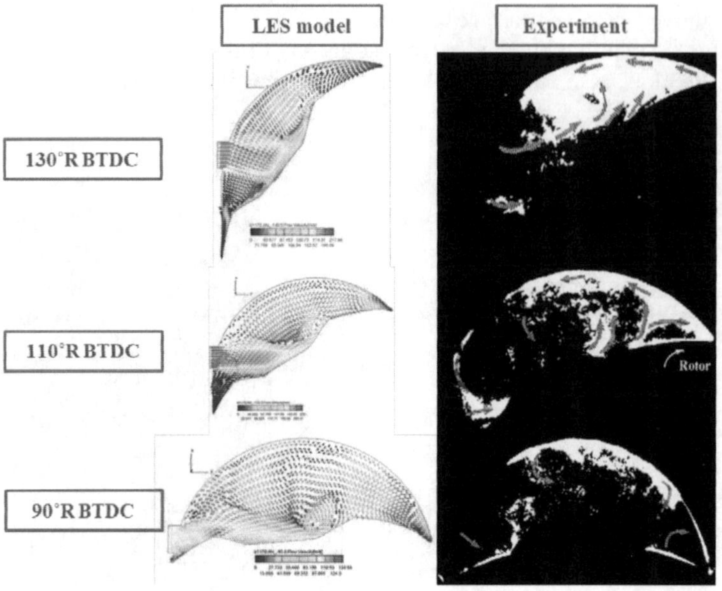

Figure 8. Comparing results during intake stroke at 1170 rpm.

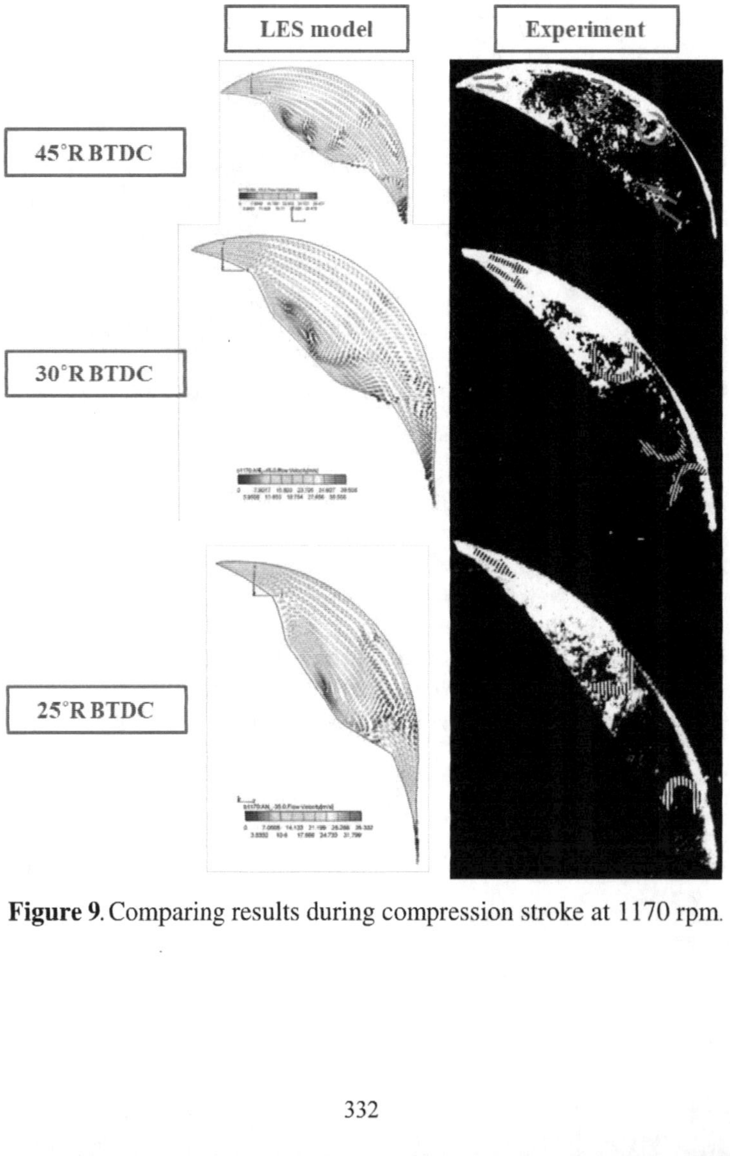

Figure 9. Comparing results during compression stroke at 1170 rpm.

4. Conclusions

In this study, the dynamic meshes and turbulence models have been gathered and simulated to capture the flow phenomena inside the combustion chamber of the Wankel rotary engine. The simulation has been validated by comparison to the experimental results. The quantity of mesh has been considered and proved. Too rough mesh could not provide satisfactory details of flow for investigation, however, too fine mesh cost time and facility for calculation. Regarding the turbulence models variation, it could be concluded that the LES can predict the flow behavior more closely to the experimental data. In addition, the flow patterns occurred in each position inside the fluid domain correspond with the experimental result. It is also found out that during the intake stroke the flow field is mainly governed by the intake charge forming the counterclockwise circulation in the domain. After the intake port is closed, the flow field is dominated by the movement of the rotor and leakage at both leading and tailing edges only. These two parameters cause the change of circulation direction to clockwise. This specific phenomenon of this engine type influences the oxidiser and fuel to mix together as well.

Acknowledgments

This research is cooperatively done under the DAAD program, called "Praxispartnerschaften zwischen Hochschulen und Unternehmen in Deutschland und in Entwicklungsländern" between the BTU C-S and KMUTNB. The AVL-FIRE software is also supported by the AVL List GmbH under the "AVL AST UNIVERSITY PARTNERSHIP PROGRAM".

References

[1] Meng P, Hady W and Barrows R 1984 An overview of the NASA rotary engine research program *NASA Tech. Memo. 83699* (Washington D.C.: National Aeronautics and Space Administration)
[2] Heywood J 1988 *Internal Combustion Engine Fundamentals* (McGraw Hill)
[3] Yamamoto K 1981 *Rotary Engine* (Hiroshima: Sankaido Co. Ltd.)
[4] Patiotis E, Kosmadakis G, Rakopoulos C 2012 Comparative analysis of three simulation models applied on a motored internal combustion engine *Energy Conversion and Management* **60** 45-55
[5] Taghavifar H, Khalilarya S and Jafarmadar S 2014 Engine structure modifications effect on the flow behavior, combustion, and performance characteristics of DI diesel engine *Energy Conversion and Management* **85** 20-32
[6] Semlitscha B, Wang and Mihăescu M 2014 Flow effects due to pulsation in an internal combustion engine exhaust port *Energy Conversion and Management* **86** 520-36
[7] Hamady F, Stuecken T and Schock H 1990 Airflow visualization and LDV measurements in a motored rotary engine assembly *J. of Engines* part 1 **99** 163-83
[8] AVL LIST GmbH 2017 *FIRE™ CFD Solver* (Graz)
[9] Poojitganont T, Berg H and Guo Z 2015 CFD simulation of flow field inside the Wankel rotary engine between intake and compression stroke *AVL AST UC 2015* Graz Austria
[10] Han Z and Reitz R 1995 Turbulence modeling of internal combustion engines using RNG-k- ε models *J. Combust. Sci. and Technol.* **106** 267-95
[11] Orszag S, Yakhot V, Flannery W, Boysan F 1993 Renormalization group modeling and turbulence simulations *Int. Conf., Near-wall Turb. Flows* Tempe Arizona
[12] Cheng Y, Lien F, Yee E and Sinclair R 2003 A comparison of large Eddy simulations with a standard k–ε Reynolds-averaged Navier–Stokes model for the prediction of a fully developed turbulent flow over a matrix of cubes *J. of Wind Engineering and Ind. Aerodynamics* **91** 1301-28
[13] Murthy B and Joshi J 2008 Assessment of standard k–ε, RSM and LES turbulence models in a baffled stirred vessel agitated by various impeller designs *Chem. Engineering Sci.* **63** 5468-95
[14] Khan Z and Joshi J 2015 Comparison of k–ε, RSM and LES models for the prediction of flow pattern in jet loop reactor *Chem. Engineering Sci.* **127** 323-33

Performance of Sixth-order Finite Surface Method in Turbulent Flow Simulations

A Hokpunna[1*]

[1*] Mechanical Engineering Department, Faculty of Engineering,
Chiang Mai University, 239 Huaykaew Rd., Chiang Mai, Thailand 50200

* Corresponding Author: E-mail arpiruk.hok@eng.cmu.ac.th

Abstract: This paper presents the finite-surface method for solving the Navier-Stokes equations (NSE). This method defines the velocities as a surface-averaged value on the surfaces of the pressure cells. Consequently, the mass conservation on the pressure cells becomes an exact equation. The only things left to approximate is the momentum equation and the pressure at the new time step. At certain conditions, the exact mass conservation enables the explicit n-th order accurate NSE solver to be used with the pressure treatment that is two or four order less accurate without losing the apparent convergence rate. This feature was not possible with finite volume of finite difference methods. The convergence rate for laminar flows are presented. In addition to these result which was already presented elsewhere, this work presents a resolution criteria needed to achieve a DNS-like solution. The turbulent channel flow with friction Reynolds number 590 is used in this study. Previously, it was found that a fourth-order scheme is effectively 10X faster than the second-order scheme, at the comparable accuracy. The newly developed sixth-order FSM is 4X faster than the fourth-order scheme and thus it is a very interesting numerical method for solving turbulent flow. This speedup is possible due to the reduction of the number of grid points.

1. Introduction

Finite Volume Method (FVM) and Finite Difference Method (FDM) have been important tools for turbulence research. In the past three decades, numerous higher-order methods have been developed. Despite having significant higher cost, they outperformed the low-order methods in both accuracy and performance in three-dimensional problems. The main reason is because higher-order methods require less number of grid points and this reduction is a quartic function (including time-step) while the cost function is a linear one. This advantage is well observed in convection and diffusion problems. Early applications of higher-order scheme to the incompressible Navier-Stokes equations (NSE) often improved the approximation of the convection and diffusion terms, but kept the approximation of pressure and the Poisson equation to second order. This was partly due to the pressure can be viewed as a Lagrange multiplier and thus it should not affect the accuracy of the momentum. Another reason is because higher-order treatment of the pressure is difficult. Later it was shown that the n-th order NSE solver must be used with the same order pressure [1,2]. This higher-order treatment of the pressure demands more computing power than the momentum equation. In fourth-order context, the Poisson's stencil is 7-point wide in 1D and the 3D Laplacian matrix is 19-banded. Comparing to the stencils of the standard second-order scheme (7-band in 3D), while the matrix of the sixth-order momentum is only tridiagonal. Thus reducing the complexity of the Poisson equation is crucial to simulation performance.

This paper presents a newly developed method called Finite-Surface Method (FSM). It fills the gap between FDM, and the FVM in which they define the data as point and volumes, respectively. FSM treats surface-averaged values as the flow unknown. The Poisson equation generated by FSM is significantly smaller than both FDM and FVM. This can be mistaken as the scheme IV of Van Leer [19] in which he proposed to improve the accuracy by using the boundary values together with the cell-averaged as the flow variables. In the same spirit, Sun and Xiao developed the multi-moment method in [3]. Hybridizible discontinuous Galerkin method also defines boundary polynomials in addition to the cell polynomial and use them to solve the instabilities in diffusion problem. These methods then compute numerical fluxes from the polynomial defined by these different types of data. At best, the mass imbalance of the cell is only exact up to the polynomial being used. The FSM presented here deliver a literally exact mass imbalance because the field variables are defined as the surface-averaged at the cell-surface. Our FSM approach, defines the velocities on these control surfaces and keep the pressure defined on the original cells. It is clear that the sum of these velocity fluxes is the exact integral of the divergence on the pressure cell. This exact mass balance can be computed at the same cost as the traditional second-order scheme. Furthermore, the positioning, of the data is still the same as they were on the staggered grid and thus half-a-cell advantage of the staggered grid in calculating the pressure gradient is maintained. Ultimately, all the advantages of the staggered grid are kept without any need to actually define the staggered cells or co-volumes at all. This approach also delivers the most compact discrete Poisson operator L under the projection method which greatly simplifies the treatment of pressure using explicit scheme. Furthermore, the simplicity of the mass conservation equation allows compact treatments of the pressure at virtually the same cost as the explicit scheme. Therefore, the finite surface method is very attractive.

2. The Navier-Stokes equations

The integral form of the incompressible Navier-Stokes equations for incompressible Newtonian fluid can be written as

$$\oint \boldsymbol{u} \cdot \boldsymbol{n}\, dA = 0 \tag{1}$$

$$\frac{\partial}{\partial t}\int_{\Omega} \boldsymbol{u}\, d\Omega + \oint_{A} (\boldsymbol{u}\cdot\boldsymbol{n})\boldsymbol{u}\, dA = \nu \oint_{A} \boldsymbol{T}\, dA - \frac{1}{\varrho}\oint_{A} p\boldsymbol{I}\cdot\boldsymbol{n}\, dA \tag{2}$$

Here, the variables and notations are velocity vector:u, pressure:p, strain rate tensor:T, identity matrix:I, density: ϱ and the kinematic viscosity: ν. The unit vector (**n**) of an infinitesimal area dA is pointing outwards of the volume $d\Omega$.

FSM discretizes the continuous domain to a discrete domain where the momentum variables are defined on the area normal to its direction while the pressure is defined as the averaged quantity over the cells. Such arrangement allows **an exact evaluation of mass imbalance over the pressure cell**. This exact imbalance can be calculated easily and efficiently. It only cost 5 floating point operations (FP) while the standard sixth-order approximation of the divergence costs 35 FP.

Such physical discretization requires a special integral form of equations (1) and (2) and the discrete form of the NSE under finite-surface method is displayed in figure 1. The square brackets represent the average and its superscript is the plane of averaging. Note that the indices *ijk* indicates the center of the pressure cells. The staggered indices e.g. *js* is the midpoint between the *j* and *j+1*. Please note that the position of *js* and *js+1/2* are the same.

IOP Conf. Series: Materials Science and Engineering **501** (2019) 012044 doi:10.1088/1757-899X/501/1/012044

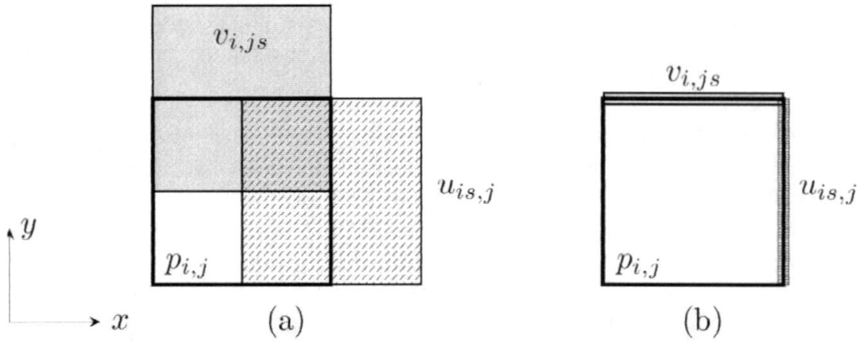

Figure 1. Spatial discretization of FSM. FSM shrinks the momentum control volume of standard FVM to the control surface of the pressure cell.

$$[div]^{xyz}_{i,j,k}\triangle x_i\triangle y_j\triangle z_k = \left([u]^{yz}_{i+\frac{1}{2},j,k} - [u]^{yz}_{i-\frac{1}{2},j,k}\right)\triangle y_j\triangle z_k + \left([v]^{xz}_{i,j+\frac{1}{2},k} - [v]^{xz}_{i,j-\frac{1}{2},k}\right)\triangle x_i\triangle z_k + \left([w]^{xy}_{i,j,k+\frac{1}{2}} - [w]^{xy}_{i,j,k-\frac{1}{2}}\right)\triangle x_i\triangle y_j$$

$$A_{j,k}\frac{\partial[u]^{yz}_{is,j,k}}{\partial t} = -\mathcal{C}_{is,j,k} + \nu\mathcal{D}_{is,j,k} - \frac{1}{\rho}\mathcal{P}_{is,j,k}$$

$$\mathcal{C}_{is,j,k} = \left[\frac{\partial uu}{\partial x}\right]^{yz}_{is,j,k}\triangle y_j\triangle z_k + \left([vu]^{z}_{is,j+\frac{1}{2},k} - [vu]^{z}_{is,j-\frac{1}{2},k}\right)\triangle z_k + \left([wu]^{y}_{is,j,k+\frac{1}{2}} - [wu]^{y}_{is,j,k-\frac{1}{2}}\right)\triangle y_j,$$

$$\mathcal{D}_{is,j,k} = \left[\frac{\partial^2 u}{\partial x^2}\right]^{yz}_{is,j,k}\triangle y_j\triangle z_k + \left(\left[\frac{\partial u}{\partial y}\right]^{z}_{is,j+\frac{1}{2},k} - \left[\frac{\partial u}{\partial y}\right]^{z}_{is,j-\frac{1}{2},k}\right)\triangle z_k + \left(\left[\frac{\partial u}{\partial z}\right]^{y}_{is,j,k+\frac{1}{2}} - \left[\frac{\partial u}{\partial z}\right]^{y}_{is,j,k-\frac{1}{2}}\right)\triangle y_j,$$

Figure 2. Discrete Navier-Stokes equation of FSM

3. Numerical Approximations

In this work we use compact sixth-order schemes from Lele [4] and Kobayashi [5] as well as a nonlinear correction concept from [1]. The numerical approximations are presented here only for *u*. The application to *v*- and *w*-velocity are straight forward.

3.1. Approximation of convective term of u in x-direction

The convective term in x-direction: $\partial[uu]^{yz}_{is,j,k}/\partial x$ can be approximated in several ways such as advective form, convective form or skew-symmetric form. In this work we use a convective form

$$\frac{\partial[uu]^{yz}_{is,j,k}}{\partial x} = 2[u]^{yz}_{is,j,k}\frac{\partial[u]^{yz}_{is,j,k}}{\partial x} + NC \quad (3)$$

The nonlinear correction (NC) in the above equation correct the linearization of the first term on the rhs. The exact form of this term is long and will not be included here due to limited space. The first term of the linearized convection $[u]^{yz}_{is,j,k}$ is the inherent variable and the second term $\partial[u]^{yz}_{is,j,k}/\partial x$ is computed using sixth-order compact scheme from [4].

3.2. Approximation of convective term of u in y- and z-directions

Along the edges of the u-momentum, the convection is a result of the line integral fluxes. In y-direction the net convective fluxes is the difference between the convective flux at the top edge and the bottom edge. At the top edge of u-control surface, the convective flux $[vu]^{z}_{is,j+1/2,k}$ need a similar approximation as equation (3). This flux requires a deconvolution of the surface-average *u* into a line-averaged. We use the sixth-order compact scheme from FVM[5]:

$$\alpha_1[u]^{z}_{is,js-1,k} + [u]^{z}_{is,js,k} + \alpha_2[u]^{z}_{is,js+1,k} = \begin{array}{c}\beta_1\left([u]^{yz}_{is,j,k} + [u]^{yz}_{is,j+1,k}\right)\\ +\beta_2\left([u]^{yz}_{is,j-1,k} + [u]^{yz}_{is,j+2,k}\right)\end{array} \tag{4}$$

The approximation of diffusive term and the pressure term is straight forward, fourth- and sixth-order approximations can be taken directly from [3] when the approximation is direct e.g. finding differentiation of a surface-averaged from the surface-averaged quantity. Otherwise, if the approximation is a deconvolution type e.g. finding a line-averaged differentiation from the surface-averaged data. The method finding sixth-order nonlinear correction (*NC*) is expanded from [2] but it will not be given here due to limited space.

4. Advantage of FSM

The projection method is used to decouple the mass conservation equation from the momentum equation and we solve NSE using the 1st and the 2nd step shown in figure 3. The second step here is equivalent to applying the projection operator **P** to the non-divergence-free velocity obtained in the 1st step. The Fourier transform of the projector is shown in step 4 and the Fourier transform of the whole mass-conservation enforcement is displayed in step 5. Effectively, the accuracy of the mass conservation depends on *i.How good did we compute the mass imbalance* and *ii.How accurate we solve the Poisson equation (the 1st equation in the second step)*. The two approximation links together as the numerator and denomination shown in step 5. Together, we call them mass conservation factor ($\mathscr{F}$). The numerator is responsible for preserving the solenoidal part of $\boldsymbol{u}^*$ while the denominator conserve its rotational part. Incorrect pressure means the final field lacks part of $\boldsymbol{\nabla}\times\boldsymbol{u}^*$.

1.Integrate the momentum equation $u^* = u^n + (\mathscr{H}(u^n) + Gp^n)\Delta t$ 2. Projection step $DG\phi = -\frac{\varrho}{\Delta t}Du^*$ $p^{n+1} = p^n + \phi$ 3.Integrate the momentum equation $P = I - G(DG)^{-1}D$	4. Fourier-Transform of Projection operator $\hat{\phi}_{kx,ky,kz} = \frac{\widehat{div}_{kx,ky,kz}}{T_L(kx)kx^2 + T_L(ky)ky^2 + T_L(kz)kz^2}$ 5. Fourier-Transform of Projection method $\hat{\phi}_{kx,ky,kz} = \frac{\left(T_{Dx}(kx) + T_{Dy}(ky) + T_{Dz}(kz)\right)\widehat{div}_{kx,ky,kz}}{T_L(kx)kx^2 + T_L(ky)ky^2 + T_L(kz)kz^2}$

Figure 3. Equations related to analysis of projection method.

IOP Conf. Series: Materials Science and Engineering **501** (2019) 012044 doi:10.1088/1757-899X/501/1/012044

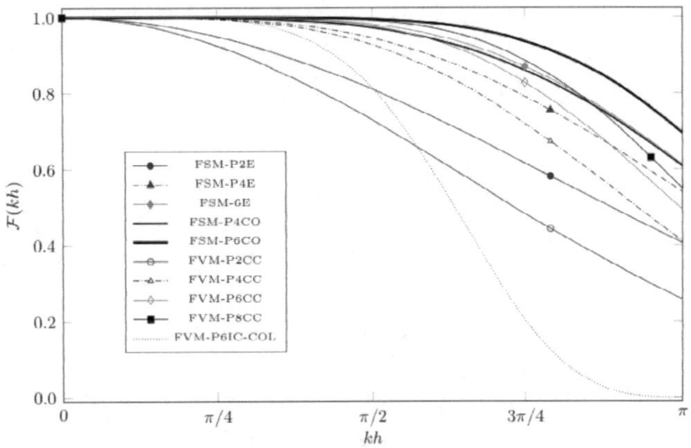

Method	Mass Conservation Efficiency (e_m)		
	$\epsilon = 0.1$	$\epsilon = 0.01$	$\epsilon = 0.001$
FVM-P2CC	0.291	0.090	0.028
FSM-P2E	0.356	0.110	0.035
FVM-P4CC	0.548	0.296	0.165
FSM-P4E	0.596	0.317	0.176
FSM-P4CO	0.688	0.391	0.222
FVM-P6CC	0.666	0.429	0.286
FSM-P6E	0.706	0.446	0.296
FSM-L6CO	0.809	0.555	0.383
FVM-P8CC	0.733	0.514	0.374

Figure 4. *Left:* Mass conservation factor under projection method; second-order cell-center method (2CC), L4E: explicit fourth-order, L6E:explicit sixth-order, L4C:compact fourth-order, L6C:compact sixth-order, L4CC:fourth-order cell-centered, L6CC:sixth-order cellcentered, L8CC:eight-order cell-centered and L6IC:sixth-order inter-cell. *Right:* Resolving efficiency $e = kh_{max}/\pi$ such that $|1 - \mathcal{F}(kh)| \geq \epsilon$.

Solving NSE on staggered grid usually more accurate than solving them on collocated grid because of the cell-center approximations which have the half-a-cell advantage. This allows a 2π Nyquist limit (the original data still limited at π). The collocated grid, on the other hand, must use inter-cell approximations whose resolution is also limited to π. Under FSM, the approximation of divergence is exact and FSM only need to approximate the Poisson equation. Therefore FSM possess an extraordinary mass-conservation property. In figure 4 we display the mass-conservation factors of the FSM in comparison with other cell-center and inter-cell approximations. We can see in the figure that the factor of the fourth-order inter-cell (IC) approximations flats out at π while those of the cell-center (CC) cross the right frame. At the same method and the same order of the truncation error, FSM is much better than FVM on staggered grid which uses cell-center approximations. At low *kh* an *n*-th order FSM and FVM-CC are comparable, but the mass conservation factor of FVM-CC soon falls down while that of the FSM is hanging around the (*n*+2)-th order of FVM-CC. This feature suggests that, when high-wave content of the flow is dominant, the mass conservation of FSM should behave like that of the FVM-CC with two order higher. In order to quantify it, we define the resolving efficiency and tabulated them on the right. The number indicates that, for a given error threshold (ϵ), how much fraction of the wave space the method can capture. Large error threshold, represents the error in high-wave components while small ϵ represents that in the low-wave components. At the lowest ϵ, the FSM-P4E is roughly 10% less accurate than FVM-P6CC. This difference shrinks to 4% at sixth- and eight-order. Interestingly, the *n*-th compact FSM outperforms (*n*+2)-th order FVM at high-wave components. The fourth-order compact FSM (FSM-P4CO) is only 6% less accurate than FVM-P8CC. The differences become 24% and 40% at the two smaller thresholds. Again, when the high-wave contents are significant like in turbulent flow, the errors will be coming from the high frequency part.

5. Performance of the Finite Surface Method

5.1 Laminar flow test case

The first test case consider here is the instability of plane channel flow. The initial parabolic laminar profile is superposed with the eigenmode obtained from Orr-Sommerfeld equations (see figure 5a). Figure 5a shows the evolution of the disturbance's energy. The slope of the curve represents the growth rate whose analytical value is $G_d = 4.46996E - 03$. Under uniform grid, FSM only needs 128 grid in the wall-normal directions to achieve 1E-5 error (0.2% relative error) in the prediction of the growth rate. In comparison, the fourth-order compact FVM[1], would need 173 points. It is found that the sixth-order FSM is 25% faster than the fourth-order compact FVM during the momentum integration. This is because, FSM do not need to interpolate the convective velocity like FVM and thus some computations were saved. Combining the faster speed and the reduction in grid requirement, the sixth-order FSM is faster than the fourth-order FVM by a factor of four, at the same accuracy.

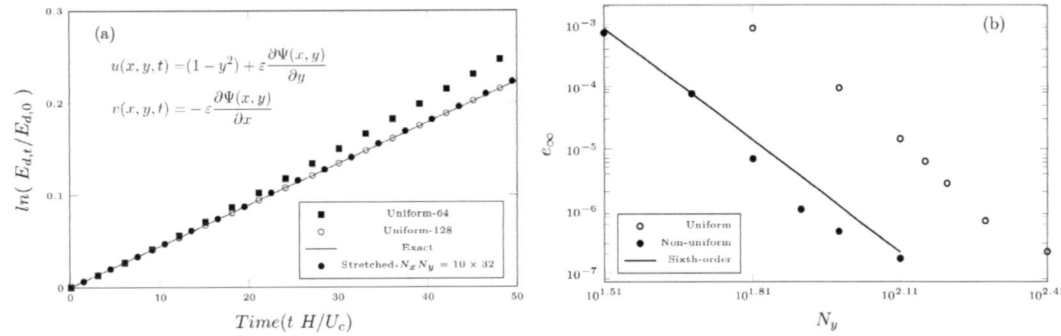

Figure 5. Evolution of the perturbations (a) and the error of the perturbations' growth rate from the analytical value.

5.2 turbulent flow test case

In the last test case we use direct numerical simulation of a turbulent channel flow with friction Reynolds number= 590 (bulk flow Reynolds number = 22,000). The computation domain is $[L_x, L_z, L_y] = [2\pi H, \pi H, 2H]$ in which x, y and z are the streamwise-, spanwise- and wall-normal directions and H is the channel's half-width. The mean streamwise velocity is plotted in figure 6. In our simulation, we fix the pressure gradient which drives the flow. The turbulent interactions generated by the direct numerical simulations will be responsible for producing shear stress. If the turbulence interactions in the numerical simulation are fewer than the reality, the flow rate will be over predicted and if they are becoming too few, the flow could continue accelerating and blow up the simulation. At the lowest resolution $L_x L_z L_y = 32 \times 32 \times 32 = 0.033M$, the numerical scheme cannot capture the shear and the turbulence interactions accurately and result in higher flow rate. Note that the plot is in outer scale. Increasing the resolution by 50% in every direction produce a better result. Surprisingly, the fourth-order FVM[1] need more grid points to match the FSM. Here FVM uses one-third more grid points in every direction, more than the FSM. This number is close to the one predicted in the laminar flow (35%). Still, FVM results is still poorer near the wall despite having more grid. It should be emphasized that the reference solution was simulated with a spectral method on 37.6M grid points which is 340 times more grid points. This does not imply that FSM is more accurate than the spectral method. It just showed that the FSM can capture main features and turbulent interactions sufficiently well, even at this rough resolution. We can predict the necessary resolution that the sixth-order FSM can deliver predictions for the mean and the second-order statistics that are on top of the spectral results. The grid resolution of $[dx, dy, d_{zmin}, d_{zmax}] = [45^+, 15^+, 3^+, 30^+]$ should be sufficient to achieve such predictions.

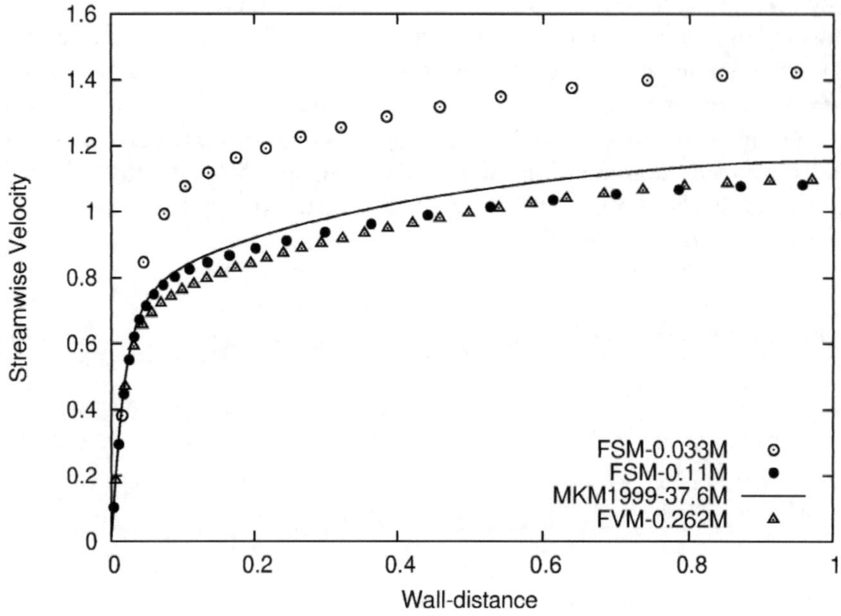

Figure 6. Mean streamwise velocity of the turbulent channel flow. The reference solution MKM1999 can be found in [6]

6. Conclusion

The finite-surface method is very attractive. This method retains all the good property of staggered grid without the need to create one. The mass-conservation is very good and the Poisson equation created by this method is very compact and can be solved efficiently. It should be noted that the FSM has two finite-volume directions and one finite-difference direction. The finite-difference direction suffers from the usual disadvantages of the FDM. Care must be given to dealiase the error and/or filter out he aliasing error.

Reference

[1] Hokpunna A and Manhart M. 2010 J. Comput. Phys. p.7545-7570

[2] Knikker R 2009 Int J Numer Methods Fluids. p.1063-1092

[3] Sun Z and Xiao F 2016 *Int. J. Numer. Meth. Fluids*, p.351–375.

[4] Lele S K 1993, J. Comput. Phys. p.16-42.

[5] Kobayashi M H 1999, J. Comput. Phys. p.137-180.

[6] Moser, Kim, Mansour 1999 Phys. of Fluids. p.943.

IOP Conf. Series: Materials Science and Engineering **501** (2019) 012045 doi:10.1088/1757-899X/501/1/012045

Stress Distribution and Displacement Analysis to Design of Patient Lifting Equipment

P. Keangin*, M. Sakunsongbunsiri, K. Auttiya and T. Kuntaraporn

Department of Mechanical Engineering, Faculty of Engineering, Mahidol University 25/25 Phutthamonthon 4 Road, Salaya, Nakhon Pathom, 73170, Thailand

* Corresponding Author: E-mail: pornthip.kea@mahidol.ac.th

Abstract. Patient lifting equipment can help the lifting of patient movement completely. It can help reduce the risk of disease that can be caused by lifting the patient and reduce the impact on the patient caused by lifting the patient in the wrong way. In addition, it can help to reduce the working time and reduces the number of personnel in moving patients. Including it can be to prevent the occupational disease on the service providers. This study aims to analyze the stress distribution and displacement to design of patient lifting equipment. The three weight conditions of 60 kg, 80 kg, and 120 kg are considered for studying about stress distribution and displacement at various points. In this test, ASTM A36 steel is used as a strength material tester. The results found that the maximum stress distribution and displacement is in the sling bar of the equipment. The obtained values will provide a basis for the guideline to design of patient lifting equipment to lift patients to suit in each situation such case from a private car go to bed or and wheelchair and from bed or and wheelchair go to a private car.

1. Introduction

Nowadays, the number of patients has been increasing every year that does not include patients who can help and move themselves. According to Ramathibodi Facilities Services (RFS) survey, there were patients average more than 350 people per day who come to healed at Ramathibodi hospital and they are necessary to use wheelchair or stretcher, as shown in Figure 1 [1]. For handling of patients who are unable to help themselves, it is necessary to have relatives or service staffs in moving patients. Especially hemiplegia case and paraplegia case, the handling in movement needs to be accurate according to the principles of physiology for increase safety of moving patients from car to wheelchair or stretcher and moving patients from wheelchair or stretcher to car. However, the handling of patients who are unable to help themselves by carrying, lifting by hand, bearing, or lifting that incorrect according to the principles of physiology may cause harm to the patients [2-3]. In addition, it may result in the relatives and the service staffs being injured by handling the patient in moving or lifting patients such as low back pain (LBP), musculoskeletal disorders (MSD) and joint disease [3]. Many researches have been done to develop and evaluate of a multifaceted ergonomics program to prevent injuries associated with patient handling tasks [4-5]. Nelson A et al. [4] designed and implemented a multifaceted program that successfully integrated evidence-based practice, technology, and safety improvement to evaluate the impact of the program on injury rate, self-reported unsafe patient handling acts, level of support for program, staff and patient acceptance, program effectiveness and costs. The program elements resulted

IOP Conf. Series: Materials Science and Engineering **501** (2019) 012045 doi:10.1088/1757-899X/501/1/012045

in a statistically significant decrease in the rate of musculoskeletal injuries of staffs as well as the number of modified duty days taken per injury. Reviewing the research related to the implementation of comprehensive patient-handling programs to improve staff quality retention and reduce time-to-injury was presented by Wardell H [5]. Efforts to reduce injuries associated with patient handling are often based on tradition and personal skills include using patient lifting and movement equipment. The utilization of patient-handling equipment and devices is a more efficient engineering control strategy within the health care for both of patients and staffs [6]. Many researchers studied about the preventing injuries using patient-handling equipment in movement. For example, Bernice D and Owen R N [7] studied the preventing of problem of nurses' back and shoulder overexertion injuries in patient-handling tasks (e.g. transferring patients on and off stretchers, repositioning patients on operating room (OR) beds) using an assistive devices include the stand-up mechanical lift. The results showed that the using the patient-handling equipment can decrease the perceived physical stress and the injury rate of nurses. The patient handling tasks with high risk for musculoskeletal disorders in health care using a powered transport device was studied by Waters T R et al. [8]. Burdorf A et al. [9] evaluated the effect of manually lifting patients on the occurrence of low back pain (LBP) among nurses and estimated the impact of lifting device use on the prevention of LBP and musculoskeletal disorder (MSD) injury claims. The results indicated that the lifting device provides a good result in significantly reducing the injuries of LBP and MSD. Therefore, several technologies were presented to improve the effectiveness of patient handling in movement tasks and reduce the risk for musculoskeletal disorder significantly for caregivers such as gait belts, transfer chairs, powered stand-assist and repositioning lifts, powered full-body sling lifts, sliding boards, gait belts and stand-assist and repositioning aids [10]. However, each device has different strengths and limitations as well as suitable for different usage situations. Traditionally, gait belts are a popular for transferring patients from a sitting to a standing position without any assistive device due to it is low-cost solution [6]. However, gait belts are not suitable for vertical transfers of weight-bearing patients. The handling of patients from bed to chair transfers using gait belts is difficult because lifts are involved. The sliding boards use for seated bed to chair or chair to toilet transfers, low-cost, rigid and smooth, low-friction material and act as a supporting bridge when seated slide transfers are done but some manual lifting is required. The stand-assist and repositioning aids can support patients that need only minimal assistance to stand up. In the operating room needs to use the moving equipment to help horizontal transfers of patient from bed to stretcher or operating room table [11]. The bed repositioning, performing tasks alone, features of a bathroom, and height-adjustable beds are required for safe patient handling and movement in home care [12]. Although there are many devices to help handling of patient, the lifting of the patient is still not complete, especially moving the patient in a restricted area such as from car to wheelchair or stretcher and moving patients from wheelchair or stretcher to car. In addition, the basis of the strength analysis of the structure design to analyze the stress distribution and displacement of the device is still minimal due to computer program limitations. A comprehensive understanding of the fundamentals of strength analysis will help the recommendation to design of patient lifting equipment to lift patients for appropriate use in each situation.

In this study focuses on the analysis of stress distribution and displacement to design of patient lifting equipment. The variations in three weight conditions i.e. 60 kg, 80 kg, and 120 kg are investigated. The ASTM A36 steel is considered as a material for strength testing. The Solidworks simulation program is used for structural analysis of patient lifting equipment design. The main components of the patient lifting equipment are built and simulated via a three-dimensional model. The results obtained from the simulation solution are examined and compared with the analytical solution using two-dimensional model to verify the accuracy of the present numerical model. The obtained results of this work will used to capture design knowledge for patient lifting equipment to lift patients and may be of assistance in determining practical guideline to improve handling of patient process in patient-care situations.

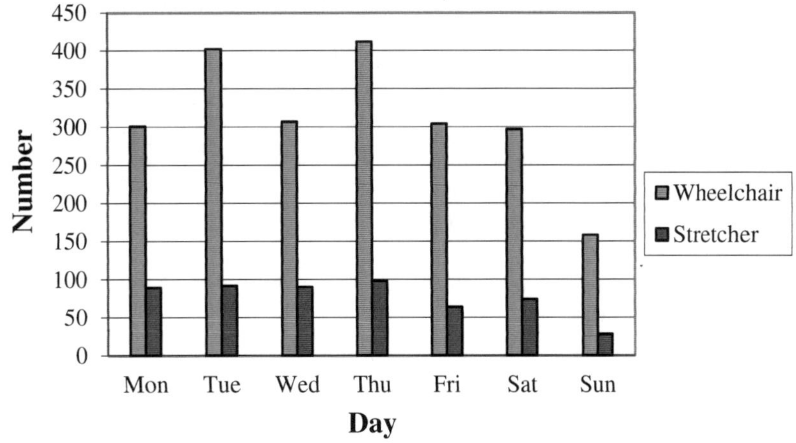

Figure 1. The number of wheelchair and stretcher using at Ramathibodi hospital in 2016 [1].

2. Designs

2.1. Structural design

The design must comply with structural calculations to refer to the height, width and weight of the patient lifting equipment. From the general car survey at the lowest altitude to the highest altitude that leads to the most appropriate level of access to design the patient lifting equipment. For a more efficient, it is necessary to measure the distance of the equipment to the point where the patient is sitting in the car to be a component of the equipment and the length and width of the equipment that the patient will be able to be balanced while lifting and moving patients. In addition, it is necessary to survey and store the height of the car door gap in each model. The design of the equipment size must be most convenient to use for smallest vehicle condition.

The design will start with a literature review and summary of information related to the general procedure to transfer patient, conceptual design, preliminary design and structural design analysis. Due to the cost, size factor and convenience, the computer program is choose to analyze the strength of the structure. The ASTM A36 steel is used as a test material for structural strength. Table 1 shows the mechanical properties of ASTM A36 steel material used in the computations. The stress distribution and displacement are analyzed to design of patient lifting equipment. The three weight conditions of 60 kg, 80 kg, and 120 kg are considered for studying about stress distribution and displacement at various points. The basic feature, part functions, advantages and limitations have been analyzed accordingly. Improving the design and mock-up inspection is prepared before proceeding to the final design concept and creating the actual equipment. The design process flowchart is shown in Figure 2. From all information, the design specification is shown as follows:

- For user with weight supports up to 120 kg on each position for lifting the patient.
- The maximum overall size of 1010 mm x 1150 mm x 1130 mm (shown in Figure 3).
- The overall weight of equipment is 75 kg.

The equipment design is shown in Figure 3. The components of the equipment can be divided into three major components, namely the sling bar, boom lift and core and base, as shown in Figures 4-6, respectively.

2.2. Computer simulation

The computer simulation of the preliminary design is performed using Solidworks simulation program. The analysis of stress distribution and displacement using finite element method on one of the most critical part, which is the sling bar at various weight conditions are performed. The three-dimensional model is discretized using triangular elements to approximate the stress distribution and displacement

IOP Conf. Series: Materials Science and Engineering **501** (2019) 012045 doi:10.1088/1757-899X/501/1/012045

variations across each element. A grid independence test is carried out to identify the appropriate number of elements required. This grid independence test leads to a mesh with approximately 59,052 elements. It is reasonable to assume that, at this element number, the accuracy of the simulation results is independent of the number of elements. To obtain a good approximation, a fine mesh is specified in the sensitive areas. This study provides a variable mesh method for solving the problem as shown in Figure 7. During the preliminary design stage, a more detailed design has been considered based from the conceptual data. For this simulation, the load of 60 kg, 80 kg, and 120 kg are applied of sling bar which represent the weight of the patient. The boundary condition for analyzing stress distribution and displacement at various points are also shown in Figure 7. To verify the accuracy of the present numerical model, the results obtained from the simulation solution are compared with the analytical solution using two-dimensional model.

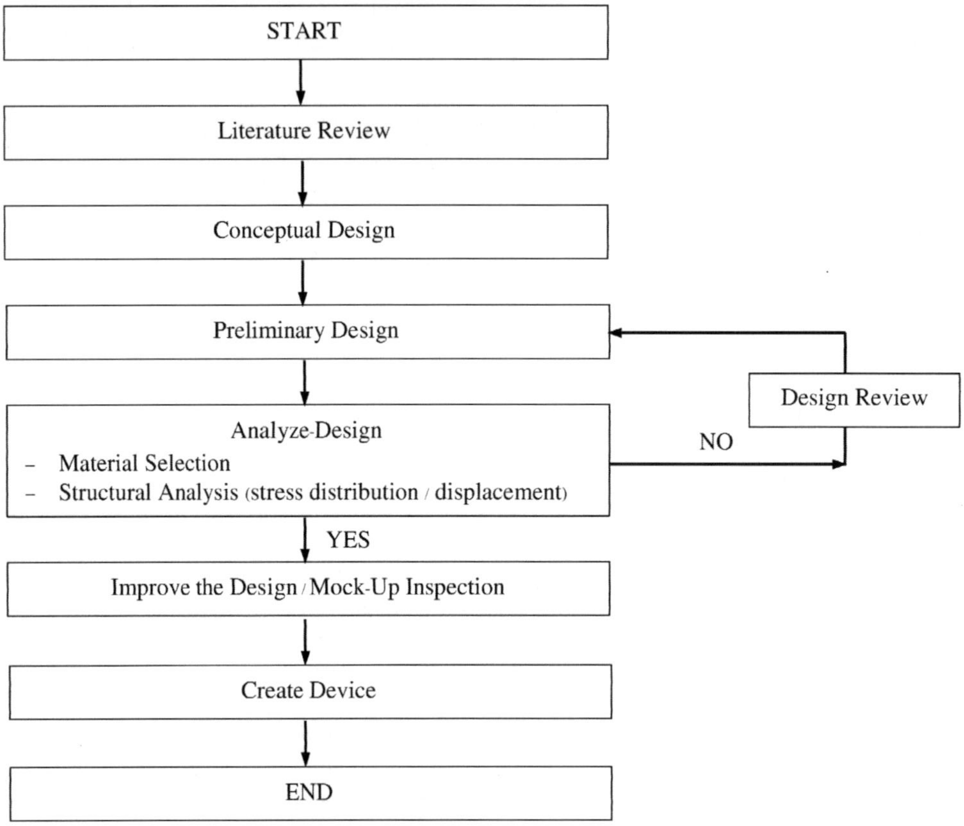

Figure 2. The design process flowchart.

Table 1. The mechanical properties of ASTM A36 steel material used in the computations used in the numerical simulation.

Property	Value	Units
Elastic Modulus	$2x10^{11}$	N/m^2
Poisson's Ratio	0.26	-
Shear Modulus	$7.93x10^{10}$	N/m^2
Mass Density	7850	kg/m^3
Tensile Strength	$4x10^8$	N/m^2
Yield Strength	$25x10^7$	N/m^2

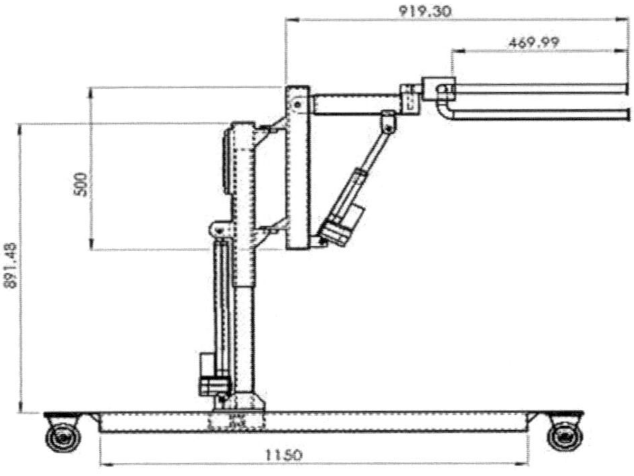

Figure 3. The equipment design.

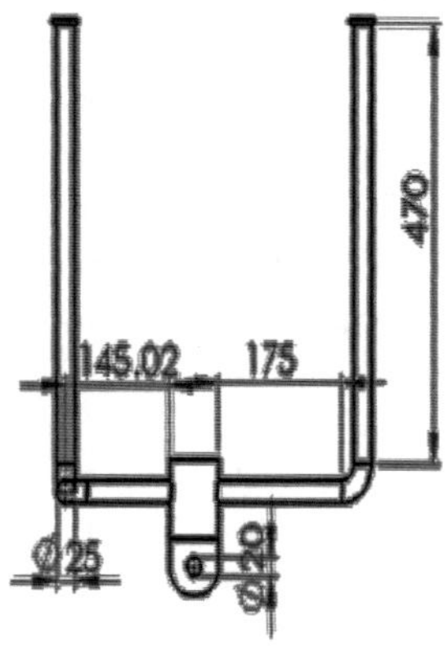

Figure 4. The sling bar component.

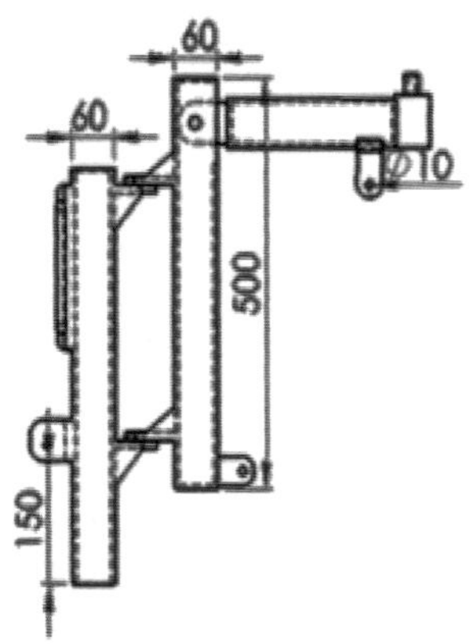

Figure 5. The boom lift and core component.

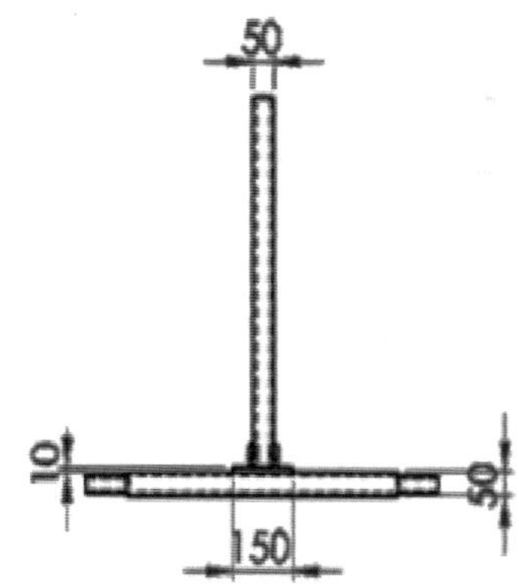

Figure 6. The base component.

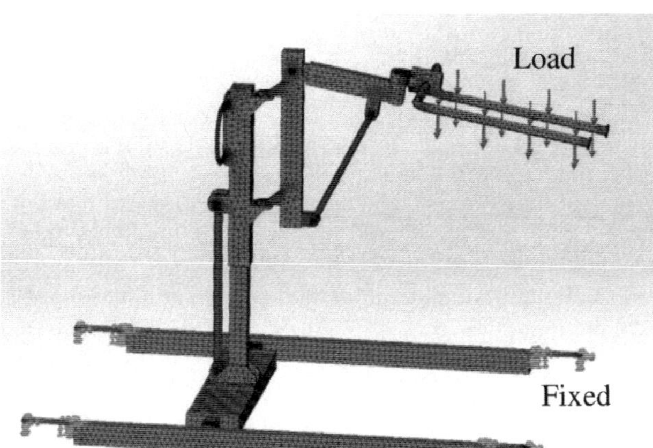

Figure 7. The three-dimensional mesh generation of equipment design and boundary condition.

3. Results and Discussion

IOP Conf. Series: Materials Science and Engineering **501** (2019) 012045 doi:10.1088/1757-899X/501/1/012045

3.1. Verification of the model

The accuracy of the present numerical model is verified with the analytical solution using two-dimensional model at various loads. The equation below is used to calculate the stress distribution of sling bar.

$$\sigma = \frac{Mc}{I} \tag{1}$$

where M is the maximum bending moment (N.m), c is the distance from the neutral plane (m) and I is the moment of inertia (m^4)

The comparison of stress distribution of simulation results with the analytical results in case of load of 60 kg, 80 kg, and 120 kg are shown in Figures 8-10, respectively. The calculation of the stress distribution using the analytical solution in case of load test of 60 kg, 80 kg, and 120 kg are shown in Equations (2)-(4), respectively. It is found that the stress distributions of sling bar of simulation results are in excellent agreement with the stress distributions of sling bar of analytical solution. Certain amounts of mismatch between the simulation results and the analytical results are caused by the numerical scheme. Figures 8-10 show that an increase in the load test results in an increase stress distribution. The comparisons of the percent error between simulation results and the analytical results are displayed in Table 2.

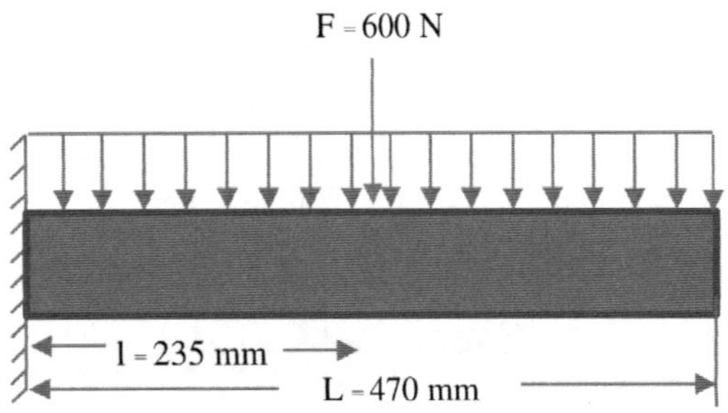

$$\sigma = \frac{Mc}{I} = \frac{32Fl}{\pi d^3} = 91.91 \text{ MPa} \tag{2}$$

(a)

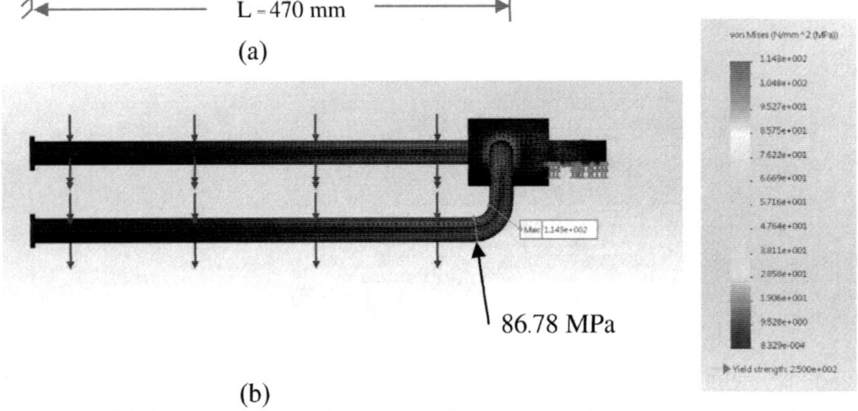

(b)

Figure 8. The comparison of stress distribution of analytical result with the simulation result in case of load of 60 kg (a) analytical method and (b) simulation method.

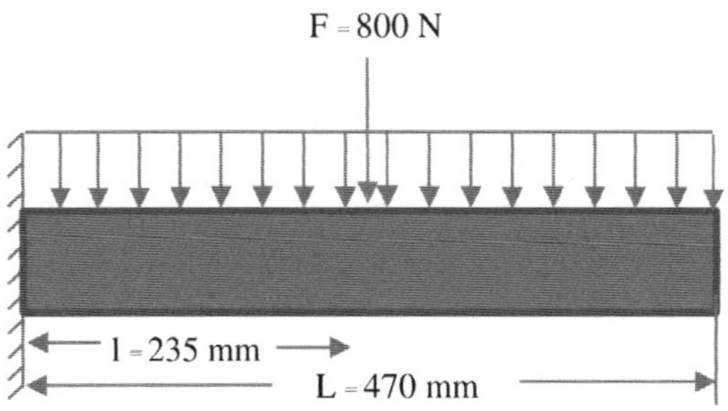

$$\sigma = \frac{Mc}{I} = \frac{32Fl}{\pi d^3} = 122.56 \text{ MPa} \quad (3)$$

(a)

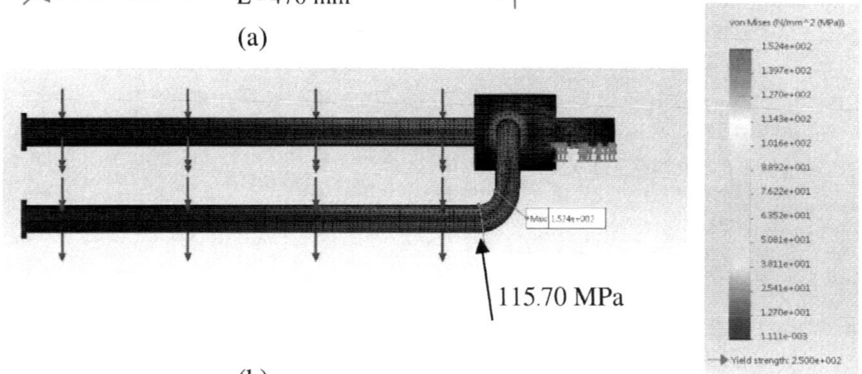

(b)

Figure 9. The comparison of stress distribution of analytical result with the simulation result in case of load of 80 kg (a) analytical method and (b) simulation method.

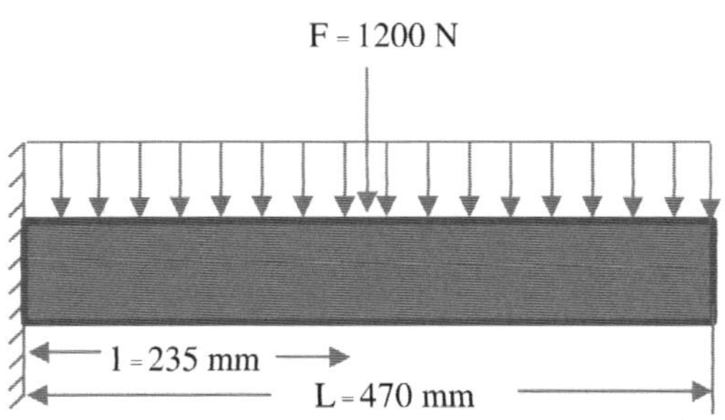

$$\sigma = \frac{Mc}{I} = \frac{32Fl}{\pi d^3} = 183.34 \text{ MPa} \quad (4)$$

(a)

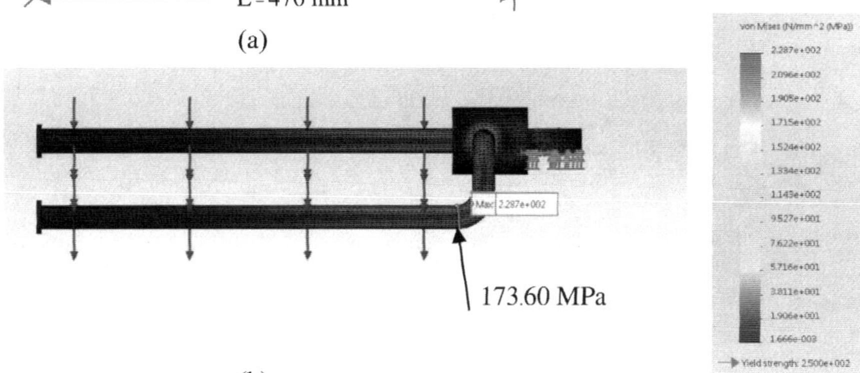

(b)

Figure 10. The comparison of stress distribution of analytical result with the simulation result in case of load of 120 kg (a) analytical method and (b) simulation method.

Table 2. The comparisons of the percent error between simulation results and the analytical results.

Weight (N)	Analytical Method (MPa)	Numerical Method (MPa)	Error (%)
600	91.91	86.78	5.58
800	122.56	115.70	5.60
1200	183.34	173.60	5.31

3.2. The effects of load test on the stress distribution

The effects of load test on the stress distribution of equipment design in case of load of 60 kg, 80 kg and 120 kg are shown in Figures 11–13, respectively. Figures 11–13 show the corresponding results, the maximum stress distribution occur at the position of sling bar because it is the part that accepts weight from patient. In addition, also found that the maximum stress distribution occur at the joints of the lower sling bar and the upper sling bar. Considering the influences of load test on the maximum stress distribution in each case, it can be seen in case of load of 120 kg has more maximum stress distribution than in case of load of 80 kg and in case of load of 60 kg, respectively. It is found that the maximum stress value in case of load of 60 kg is 76.76 MPa, in case of load of 80 kg is 105.00 MPa and in case of load of 120 kg is 157.50 MPa, respectively. Moreover, considering the maximum stress value, it is found that in each case will not exceed the yield strength (refer to the Table 1) which it shall not reach the failure state based from the design specification.

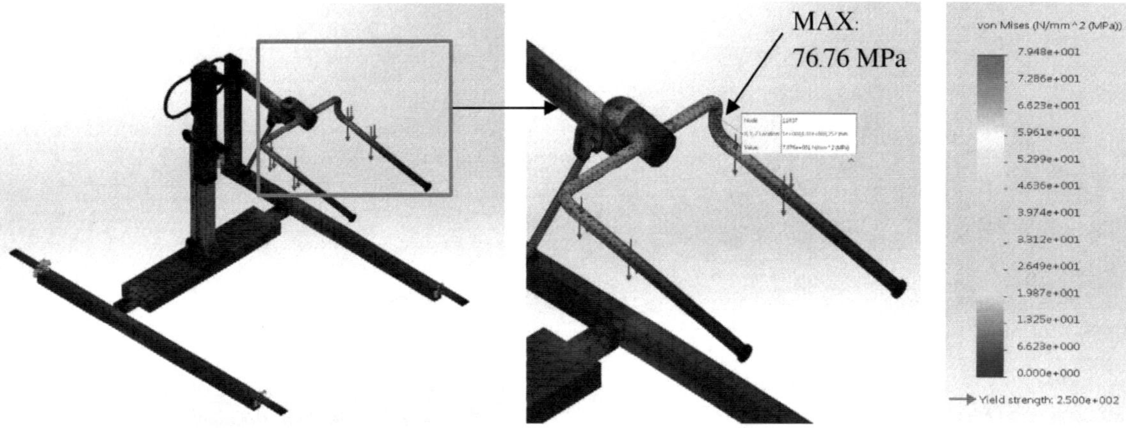

Figure 11. The stress distribution of simulation result in case of load of 60 kg.

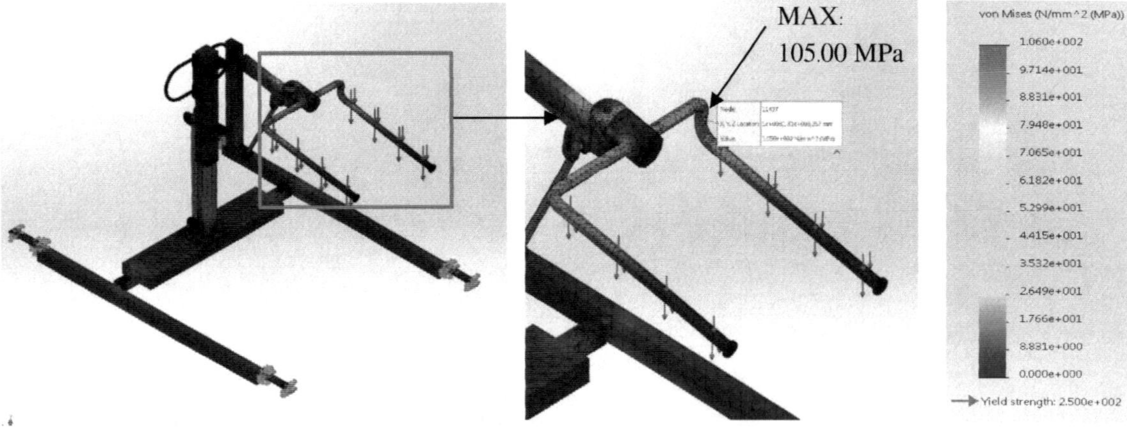

Figure 12. The stress distribution of simulation result in case of load of 80 kg.

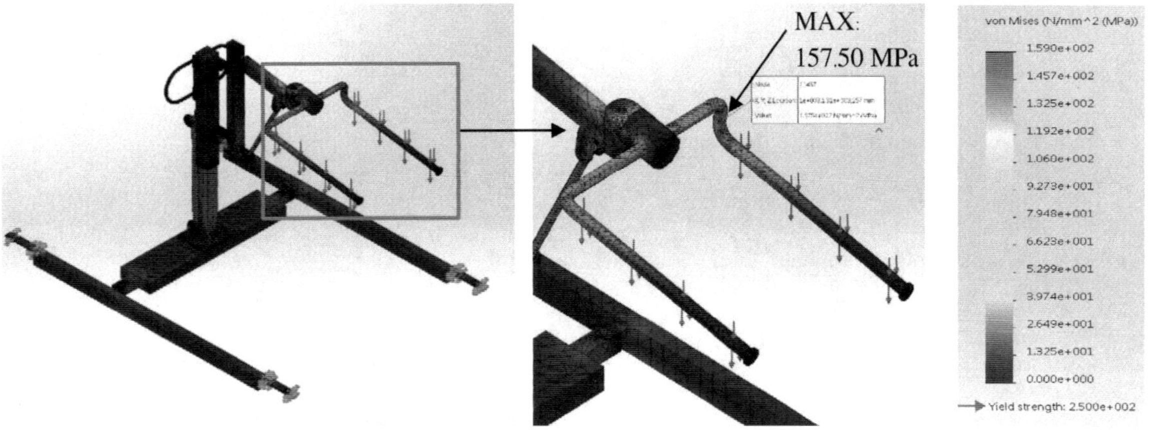

Figure 13. The stress distribution of simulation result in case of load of 120 kg.

3.3. The effects of load test on the displacement

Figures 14–16 displays the effects of load test on the displacement of equipment design in case of load of 60 kg, 80 kg and 120 kg, respectively. It is found that the displacement results follow the stress distribution results. The maximum displacement occurs at the position at the end of the sling bar. This is due to the distance between load and fulcrum point, which is accepts weight from patient. The higher load results in higher stress distribution and greater stretching displacement. It can be seen in case of load of 120 kg has more maximum displacement than in case of load of 80 and in case of load of 60, respectively. It is found that the maximum displacement in case of load of 60 kg is 8.111 mm, in case of load of 80 kg is 10.82 mm and in case of load of 120 kg is 16.22 mm, respectively.

4. Conclusions

This research is carried out to prepare the preliminary design of the equipment design (using lifting mechanism) by focusing on the engineering design process and simulation the strength of the structure through simulation software. The effects of load test on the stress distribution and displacement variations in three weight conditions i.e. 60 kg, 80 kg, and 120 kg are investigated. The ASTM A36 steel is considered as a material for strength testing. The finite element method is performed for structural analysis of patient lifting equipment design. Three-dimensional model of equipment design are considered in this work. The simulation results with computer programs are validated with analytical results using two-dimensional model. The results show that the higher load results in higher stress distribution and greater stretching displacement. It is found that the maximum stress distribution and displacement occurs at the position at the sling bar. The obtained results in this research will provide basic concept on the design in respect of the equipment to help handling of patient. In the future, further investigation and design improvement shall be made to improve the preliminary design and lead to the creation of a prototype.

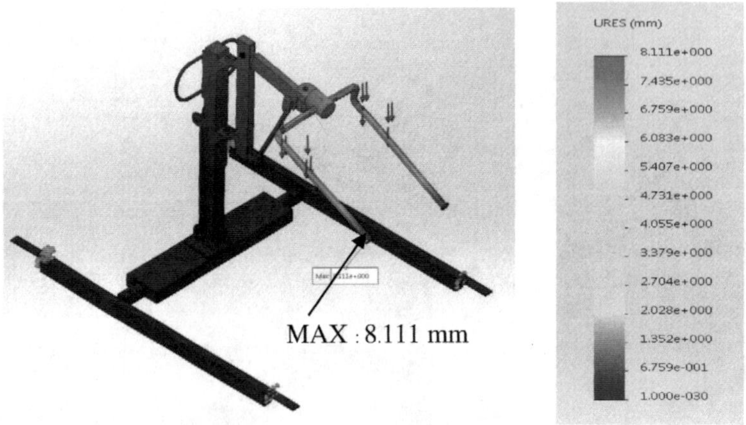

Figure 14. The displacement of simulation result in case of load of 60 kg.

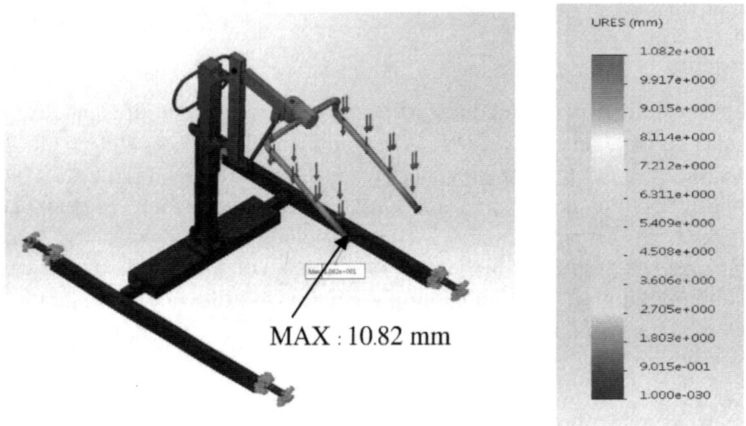

Figure 15. The displacement of simulation result in case of load of 80 kg.

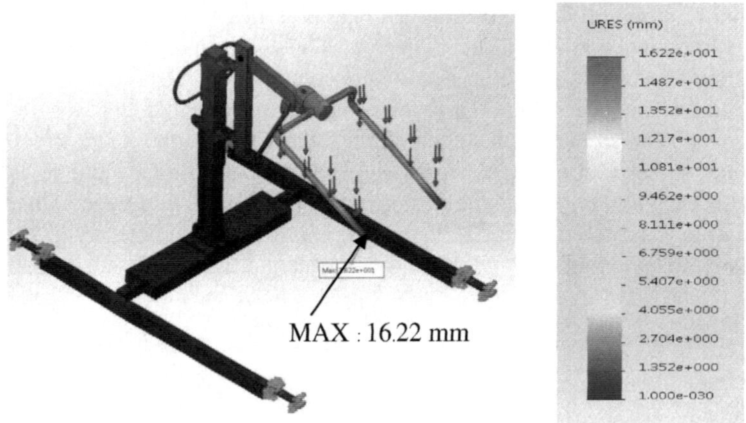

Figure 16. The displacement of simulation result in case of load of 120 kg.

5. References

[1] Ramathibodi Facilities Services, Appointment Patient Rate 2016
[2] Goubert L, Francken G, Crombez G, Vansteenwegen D and Lysens R 2002 *Behav. Res. Ther.* **40**(4) 415–429
[3] Samantha B and Mike B 2015 *Nurs. Manag. (Springhouse)* **46**(6) 46–50
[4] Nelson A, Matz M, Chen F., Siddharthan K, Lloyd J and Fragala G 2006 *Int. J. Nurs. Stud.* **43**(6) 717–733
[5] Wardel H 2007 *Workplace. Health. Th. Saf.* **55**(10) 407–412
[6] Nelson A and Baptiste A S 2006 *Clin. Rev. Bone Miner. Metab.* **4**(1) 55–69
[7] Bernice D and Owen R N 2000 *Assoc. Oper. Room. Nurs.* **72**(6) 1031–1036
[8] Waters T R, Nelson A and Proctor C 2007 *Crit. Care. Nurs. Clin.* **19**(2) 131–143
[9] Burdorf A, Koppelaar E and Evanoff B 2013 *Occup. Environ. Med.* **70**(7), 491–497
[10] Audrey N, Bernice O, John L, Guy F, Mary M, Margaret A, Judith B, Susan MC, Glenn R and Karen L 2003 *Am. J. Nurs.* **103**(3) 32–43
[11] Owen B 2000 *Assoc. Oper. Room. Nurs.* **72**(6) 1031–1036
[12] Owen B D and Staehler K 2003 *Home. Healthcare Nursing Manual* **21**(3) 180–186

IOP Conf. Series: Materials Science and Engineering **501** (2019) 012046 doi:10.1088/1757-899X/501/1/012046

Automatic Handling for Valves Trimming System

S. Vongbunyong*, P. Archeewawanich, P. Akaratham, C. Martchan, K. Thamrongaphichartkul,

Innovation and Advanced Manufacturing Research Group, Institute of Field Robotics, King Mongkut's University of Technology Thonburi.
126 Pracha Uthit Rd, Bang Mot, Thung Khru, Bangkok, Thailand 10140

Corresponding Author: supachai.von@kmutt.ac.th,
Tel: +662-47-9339, Fax: +662-470-9703

Abstract Shortage of workers in manufacturing industry has become one of the critical issues in every industry nowadays. Valves manufacturing process is involved forging and trimming process of brass workpieces which are hazardous and able to harm the workers. The aim of this project is to develop an automatic handling of valves in trimming system for replacing human workers. A 6-DOF articulated robotic arm is integrated with vision system for effectively picking the workpieces from the feeder and placing them to the rotating die for being trimmed. One of the challenging problems is the uncertainties of the shape of the forged workpieces related to flash. The specially designed impactive gripper and machine vision are developed to resolve those uncertainties. The gripper is a parallel jaws gripper designed to grip and hold an object with irregular geometries. For the selected model of valve, the system is able to work at 1,033 pieces hour at 96.67% success rate which is comparable to the capacity of the current human workers.

1. Introduction

Forging is one of oldest techniques in bulk material deformation processing. Yet, it is a very important technique used to make a variety of high-strength components nowadays. A large hydraulic press machine is generally used to generate force on a metal billet which is inserted between an upper and a lower molds. Three forging techniques, including (a) *open-die forging*, (b) *impression-die forging*, and (c) *flashless forging*, are used in metal deformation process [1]. For the impression-die forging, the shape of the final workpiece will be formed according to the surface of the molds. When the billet has been compressed within the cavity between the surfaces of two molds, some excess material will flow into the surrounding area and form flash (see Figure 1(a)-(c)). Afterwards, this flash must be trimmed off in order to obtain the forged workpiece with the desired shape. With die pressing method, impact force is used to trim off the flash by shearing at the cutting edge on the die (see Figure 1(d)).

These days, shortage of workers has become one of the most critical issues in many manufacturing industries. According to the operations performed in the factory that we study, the forged workpieces are transferred and inserted to the trimming dies manually at the rate of maximum 1,000 items/hour. Robotic system is considered to replace the human workers due to the high labor cost and safety issues. This manual operations is concerned high risk due to the size and accessibility of the trimming machine and the sharpness of the workpieces that can be harmful for the human workers while handling.

Development of automatic handling system with robot arm deals with 3 main areas of problems, including (a) motion control of robot arms, (b) design of grippers, and (c) vision system. A number of research works with respect to this application were conducted in these areas. For example, Modi *et. al.* (2015) [2] improved the performance of the CNC production line by using a robot arm picking and placing workpieces into CNC machine. Patil, G. (2013) [3] developed the visual-guided pick-and-place robot system with SIFT that is used for recognizing and locating the workpiece to be picked. SIFT is robust to work in low light condition and fast enough to operate real-time. Kumar *et. al.* (2015) [4] developed visual-guided robot system that was able to work with various types and sizes of workpieces. The workpieces were able to be classified, recognized, and localized with higher than 80% accuracy.

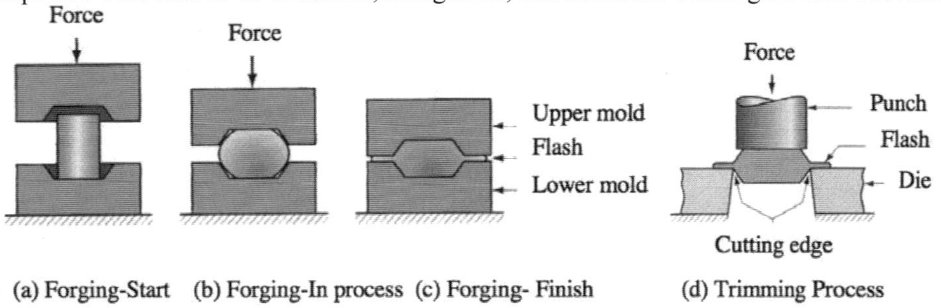

Figure 1 (a-c) Impression-die forging and (d) Trimming processes (adapted from [1])

In conclusion, in this research, a vision-guided robotic system will be implemented in workpiece handling process. The workpieces with flash will be automatically transferred and inserted to the trimming die by the robot arm. This paper is organized as follows. The function requirements of the manufacturing process and system architecture are presented in *Section 2*. Details in gripper design and vision system are presented in *Section 3*. The experiments are in *Section 4*. Conclusion is in *Section 5*.

2. Methodology

2.1 *Requirements related to forging and trimming process*

CT-15 is a tail part of the valve (see complete product, Figure 2). CT-15 is made of brass JIS C3771 with the dimension of 44.6 mm x 27.0 mm. The flash is normally formed as a result of the forging process. The dimension of CT-15 with flash is around 60 mm x 40 mm in general.

The current operation related to the forging process of CT-15 consists of 7 steps (see Figure 2). The steps (4-5) are manual and (6) is semi-automatic operations. The workers pick a forged workpiece from the bin and then manually place in the mold in the trimming machine. The machine will be manually activated. The die will be rotated inside the machine and the workpiece will be trimmed. These steps will be converted to automatic operations by using a robot arm. Only (5-6) will be focused in this paper.

2.2 *System architecture*

The system that is developed in this research mainly consists of a robot arm with gripper and vision system. The system diagram is shown in Figure 3. CFD Controller is a robot controller used as a master of this system that the main program is running. CFD connects with the gripper via I/O and the machine vision module which is acted as a slave via TCP/IP protocol. The machine vision module is a smart camera module used to detect the workpieces, in which the information will be provided to CFD. This system is connected to the trimming machine and the part feeder through a Programmable Logic Controller (PLC), so that the CFD can command them according to the designed operation sequences.

In the operation, the forged workpiece will be fed to the robot picking position which is at one end of the part feeder. The vision system will detect the shape, position, and orientation of the workpiece. The robot will use the gripper to grasp this workpiece and transfer it to the trimming machine and placed on the mold. Next, the robot will activate the trimming process by sending the command through the PLC. The workpiece that has been trimmed will be loaded out automatically by the mechanism in the trimming machine. The details of each main component are explained in the following section.

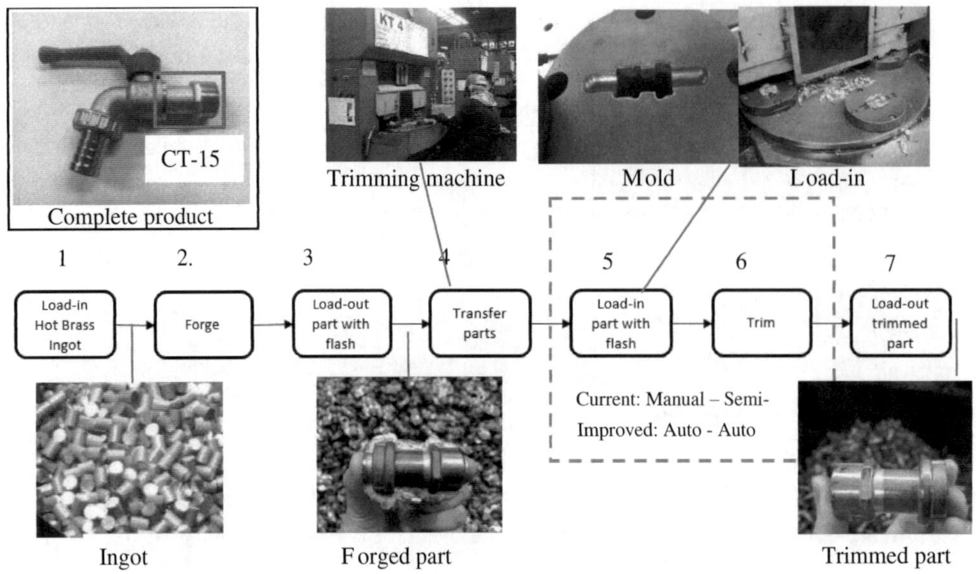

Figure 2 Process flow diagram for forging and trimming of CT-15

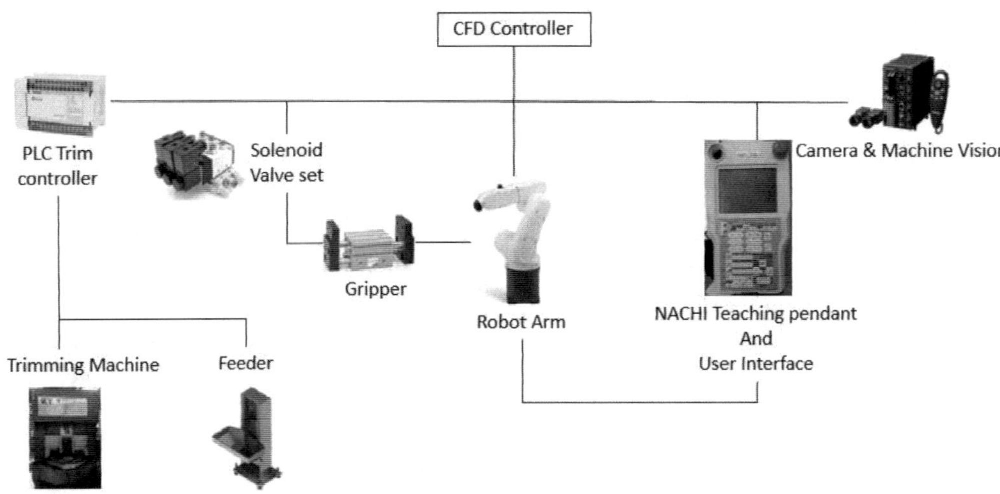

Figure 3 System diagram

3. Implementation

3.1 *Mechanical Components: Robot arm and Gripper*

3.1.1 Robot Selection

A small size 6-axis articulated robot (Nachi MZ04) [5] is used in this system. This model is selected due to specification of the high speed movement, high accuracy with ±0.02 mm position repeatability, workspace with 541mm reach, and 4 kg payload. The 6-axis robot is selected due to the position and orientation requirements of the pick-and place process. The gripper is mounted on the robot's end effector for grasping valves, CT-15 model, from the feeder to the rotating trimming die.

3.1.2 Gripper Design

For the gripper, impactive gripper type (jaw and friction grippers) [6] is used in this application. The friction grippers are impactive mechanical grippers that holding objects by using the friction force acting between active surfaces of the object and gripper's jaw. The jaw grippers use the friction force together with the jaws' geometry to support the object [7].

The impactive gripper type is chosen over the suction gripper type due to the availability of active surface to be gripped. For the suction grippers, the active surface must be the top side of the workpieces that contain numerous curve surfaces, which is challenging for the selection of proper vacuum pads. However, regarding the impactive gripper, the challenging issue is to specifically design the geometry of the jaws to match the contact surface and uncertain shape of flash (see Figure 4). From preliminary experiments, the special designed friction gripper achieved 99% success rate.

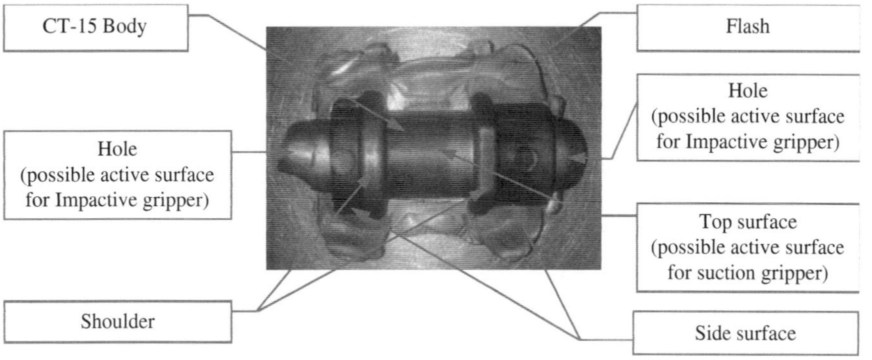

Figure 4 Possible active surface for impactive grippers and suction grippers

3.1.3 Jaw design

The shapes of possible contact surfaces are taken into account to design the active surface and the jaws' geometry. In order to secure the position and orientation of the objects, sufficient points and areas of contacts are necessary. From Figure 4, the possible contact surfaces for impactive grippers are (a) *top surface of the body*, (b) *in the holes*, (c) *shoulders*, and (d) *flash*.

Regarding the prehension strategy, firstly the gripper is at the open position (Figure 5(a)). The gripper is approaching the workpiece from the top and *body supporter* lightly presses on this area to secure the pose of the valve relative to the gripper. Next, the jaws will close (Figure 5(b), then the fingers will contact with the holes and other surfaces. As a result, the pose of the workpiece will be fully secured and ready to transfer. However, flash is not consider for being an active surface since the geometry of flash is uncertain. Therefore, there is no common geometry that the jaws can be designed to match. Two versions of jaws geometry set were designed as follows:

(a) Open position (use jaw version 1) (b) Close position (use jaw version 2)

Figure 5 Impactive Gripper: position – open and close (demonstrated with 2 versions of jaws)

For the *first version* (see Figure 6(a)), each side of the jaw is designed to match with three areas of contact surface which are 1 hole and 2 sides surface. The workpieces can be secured effectively at the desired position and orientation by the center finger, two supporting fingers on both sides, and the body supporter. However, it was occasionally found that the two supporting fingers collide with some forms of flash. As a result, this issue is resolved in the second version of design. For the *second version* (see Figure 6(b)), the two side fingers are removed to prevent the collision. Only the center finger is used to insert into the hole for supporting and the body support that pressed the valve in the secured position.

From the two versions of design, the gripper is selected according to the gripping performance which is justified with the success rate of grasping process. The experiments were done on 27 cases which are the combination of 3 position variations (-5,0,+5 mm axial offset) and 9 flash shape variations. As a result, the success rate are 77.78% and 100.00%, respectively. Hence, the gripper-2 is selected.

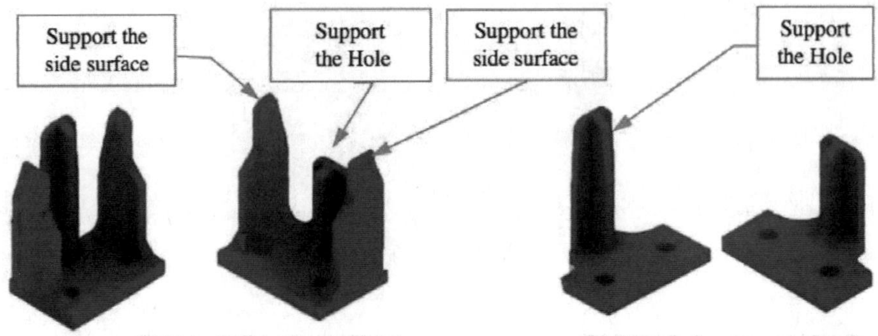

(a) Jaws design set – version 1. (b) Jaws design set - version 2

Figure 6 Impactive gripper jaws design

3.2 *Vision system*

3.2.1 Detection algorithm development

This research use the smart camera (Omron FQ Vision Sensor: FQ2-S20100F) has a built-in processor that is able to process the data and communicate with other parts of the system via I/O and TCP/IP protocol. The machine vision algorithms for detecting the workpieces are programmed by using FQ Touch Finder Simulation software. The algorithm is designed with region-based and contour-based features detection on the significant features of the CT-15. To disregard the effect of uncertainties from the flash, the features are determined from the images of the workpiece without flash. The features that are manually marked are presented in Figure 7(a). Primarily, the data in regard to position (*x, y, z*) and orientation (*yaw angle* and *up/down*) will be calculated and sent to CFD.

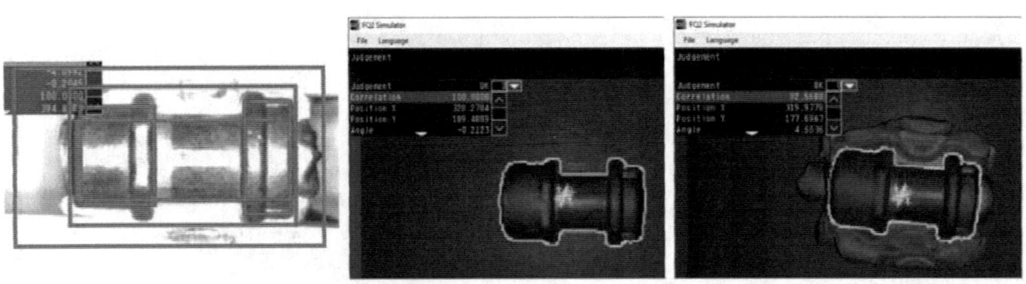

(a) Detection template (b) without flash (c) with flash and rotate

Figure 7 Examples of valves detection in various situations

The algorithms are developed based on the functions available in FQ software [8]. The functions are the integration between region-based and contour-based detection algorithm. The capability of each search algorithms are summarized as follows:

(a) *"Search"* – simple and fast features matching algorithm. However, it is ineffective to handle complex situation, for example variations in lighting condition and shape.

(b) "*Shape search 2*" – able to detect multiple objects presented in the image. It is robust to orientation, partial occlusion, and various lighting condition. It was found that this function was sensitive with the color variation in which lighter tone can be occurred in the production.

(c) "*Shape search 3*" – is similar to Shape search 2 and also robust to the object with background. It is also capable of detecting the samples in various orientations. Example results are shown in Figure 7(b-c). From these key features, this function is selected to be used for this application.

3.2.2 Verification
The experiment was conducted with 50 random samples with flash. They were tested on the feeder in the actual production line. Each sample was tested with 4 posture variations (front/back and 0°/180° yaw). For the result, the samples was able to be detected with 98.77% correlation on average.

4. Experiments

4.1 *Setup and testing procedure*
The prototype system primarily consists of a robot arm, vision system, part feeder, and trimming machine are placed in the factory shop floor. The experiment was designed to evaluated the performance of the entire system. The setup is shown in Figure 8.

Figure 8 Experimental working area setup

The robot arm movement is set to 100% speed with 2 modes of smoothness function (with / without). In general, the smoothness function modifies the motion profile by adjusting acceleration constant in order to smoothen the movement at the beginning and the end of moving paths. Consequently, enabling smoothness function should help reducing vibration of the platform due to the robot movement and increase accuracy in gripping process but the time consumption might increase.

4.2 *Experiment results*
The experiment was conducted on 100 random samples. The workpiece was fed by the feeder. The experimental result is shown in Table 1. The smoothness function was enabled in Experiment 1-3 (3.92 sec/item, 899 items/hour) and disabled in experiment 4-6 (3.33 sec/item, 1034 items/hour).

Around 4-5% of the samples were missed due to the geometrical error of the part feeder. The part feeder used in this experiment is designed for feeding numerous models of valves. Small gap between workpiece and the guide rail of the feeder is presented. This gap leads to the orientation and position error of the sample when approaching at the end position. As a result, the vision system was able to detect these errors and the system rejected the workpieces.

Table 1 Experiment result – robot pick 100 samples at 100% speed

Experiment No.	Smoothness Function	Total Time (min)	Average Time per sample (sec / item)	No. of samples that were picked (item)	Estimated No. of samples transferred to the die (items/hr.)
1	Yes	6.18	3.90	95 /100	905
2	Yes	6.34	4.18	91 /100	831
3	Yes	6.15	3.69	100 /100	960
Avg. (1-3)	**Yes**	-	**3.92**	**95.33%**	**898.67**
4	No	5.48	3.46	95 /100	983
5	No	5.26	3.22	98 /100	1,082
6	No	5.37	3.32	97 /100	1,036
Avg. (4-6)	**No**	-	**3.33**	**96.67%**	**1033.67**

5. Conclusions

In this research, the robotic system for transferring the forged workpieces to the trimming machine is developed. The 6-axis robot arm equipped with special designed gripper is used together with the vision system to precisely pick and place the workpieces. One of the challenging issues is the uncertain shape of the forged workpiece regarding the flash which is uncontrollable in the manufacturing process.

The gripper and the vision system work together to resolve this problem in pick-and-place process. (a) *Impactive gripper*, two holes in the body and top surface of CT-15 are chosen to be active surfaces, in which the shape of the jaws and other supporting parts are designed specifically. The gripper achieved 100% success rate within ±5 mm position deviation. (b) *Vision system*, "Shape search 3", working based on contour-based and area-based detection algorithm, achieved 98% detection rate. However, when the gripper and vision system components are integrated in the complete system, the overall performance little dropped from the individual tests.

About the final system configuration, the robot system without smoothness function can perform more efficiently. The system can handle the samples at the rate of 3.48 sec/item with less than 5% of unsuccessful picking. The performance is comparable with the working capacity of human workers.

Acknowledgement

We would like to thank Asahi Thai Alloy Co,. Ltd., Thailand, for technical support, financial support, and the demonstration site. Also, this project is participated in Talent Mobility program, which is partially funded by National Science Technology and Innovation Policy Office, Ministry of Science and Technology and Office of the Higher Education Commission, Ministry of Education, Thailand.

References

[1] Groover, M.P., *Bulk deformation processes in metal working*, in *Fundamentals of Modern Manufacturing: Materials, Processes, and Systems*. 2010, John Wiley & Sons. p. 405 - 420.
[2] Modi, H.A. and D.M. Patel, *Automated system design for pick and place of M/C Component of CNC.* IJIRST – International Journal for Innovative Research in Science & Technology, 2015. **1**(2).
[3] Patil, G.G., *Vision Guided pick and place robotic arm system base on SIFT.* International Journal of Scientific & Engineering Research, 2013. **4**(12) 242 - 248.
[4] Kumar, R., et al. *Object detection and recognition for a pick and place robot.* in *Asia-Pacific World Congress on Computer Science and Engineering*. 2014.
[5] Nachi-Fujikoshi-Corp., *Standard specifications MZ04-01/MZ04E-01-CFD.* 2017.
[6] Gualtiero, F., et al., *Grasping devices and methods in automated production processes.* CIRP Annals - Manufacturing Technology,, 2014. **63**(2) 679-701.
[7] Monkman , G.J., et al., *Robot Grippers*. 2017: Wiley-VCH Verlag GmbH & Co.
[8] Omron, *User's Manual*. Smart Camera FQ2-S/CH Series. 2013.

IOP Conf. Series: Materials Science and Engineering **501** (2019) 012047 doi:10.1088/1757-899X/501/1/012047

Principle and Robotic Applications of Conical Scanning Method

K Tobita[1,*], K Matsumoto[2], S Yotsuyaku[3], and C Kanamori[3]

[1]Department of Mechanical Engineering, Shizuoka Institute of Science and Technology, 2200-2 Toyosawa, Fukuroi, Shizuoka, 437-8555, Japan
[2]JGC Corporation, 2-3-1, Minatomirai, Nishi-ku, Yokohama, Kanagawa, 220-6001, Japan
[3] Department of Mechanical Engineering and Intelligent Systems, The University of Electro-Communications, 1-5-1 Chofugaoka, Chofu, Tokyo, 182-8585, Japan

* Corresponding Author: tobita.kazuteru@sist.ac.jp

Abstract. We have proposed a surrounding environmental recognition method using conical scanning distance measurement for mobile robots. The conical scanning method has a high robustness against ambient light noise and its calculation cost is low. This report describes the principle of this method and the two implementations. The first application is the quadrupedal wheeled robot. The robot can recognize a stair in sight, approach to it and climb. A second application is the indoor-outdoor wheeled mobile robot. The robot can recognize its position and obstacles on floor from information of a laser range finder and a 3D ToF camera. The validity of this method was confirmed by these robotic applications.

1. Introduction

Various methods using stereo cameras, 2D-LiDAR or 3D-LiDAR have been proposed to recognize the environment around the running area of the robot. A Kunihiro's method using stereo cameras can obtain a wide range of information at once[1], but it is difficult to accurately recognize the environment depending on the time zone and location since ambient light and the texture around the running area become noise. According to Watanabe's method using 2D LiDAR, the area ahead of the robot can be detected by sensors installed at a fixed angle with respect to the road surface[2], but there are many erroneous detections when the posture of the robot is changed due to a rough uneven road surface. Various methods have been proposed for 3D-LiDAR-based methods. Fujii suggests a method using RANSAC[3], but because of its high calculation cost, it has low real-time performance and it is not suitable for environment recognition of autonomous mobile robot. Eguchi has proposed a method of extracting a step by comparing the horizontal distance when sensing a road surface with the distance measured beforehand[4], but this is also reported that erroneous detection occurs when the robot changes its posture.

We propose a conical scanning method. In this method, the surface of the measured target is scanned conically using a distance sensor, and the shape of the measured surface is classified according to the length (measuring distance) of the generating line of the cone and the characteristic amount of the scanning curve drawn by the scanning rotation angle[5][6][7][8]. This paper describes the principle of conical scanning method and its applications as four legged wheeled robot that moves up and down stairs, and indoor / outdoor wheeled mobile robot.

IOP Conf. Series: Materials Science and Engineering **501** (2019) 012047 doi:10.1088/1757-899X/501/1/012047

2. Principle of conical scanning

When the distances $L(\alpha)$ to the target surface are measured while varying the scan angle α from 0 ° to 360 ° to the target surface (Figure 1), a graph as shown in Figure 2 is obtained (Hereinafter referred to as scan curve). Features (maxima, minimal value points, discontinuous points) appear in the scan curve as shown in Figure 3 depending on the shape of the region to be measured (plane, concave coupling, convex coupling, discontinuous surface). In other words, it is possible to recognize the shape of the measurement target surface from the characteristics of the scan curve. The parameters uniquely specifying the plane are three, distance dz, inclination angle γ, and inclination direction δ (Figure 4).

From Figure1 and Figure5, the feature amount dz, γ, δ of the plane by the following equation using the parameter $L_{max} \cdot L_{min} \cdot \alpha\,(L_{max})$ of the scanning curve can be calculated.

$$\begin{aligned} dz &= L_{max}\cos\beta - L_{max}\sin\beta\tan\gamma \\ &= L_{min}\cos\beta + L_{min}\sin\beta\tan\gamma \end{aligned} \tag{1}$$

$$\gamma = \alpha\tan\left\{\frac{(L_{max}-L_{min})\cos\beta}{(L_{max}+L_{min})\sin\beta}\right\} \tag{2}$$

$$\delta = \alpha(L_{max}) \tag{3}$$

β is the half apex angle of the cone and is a constant.

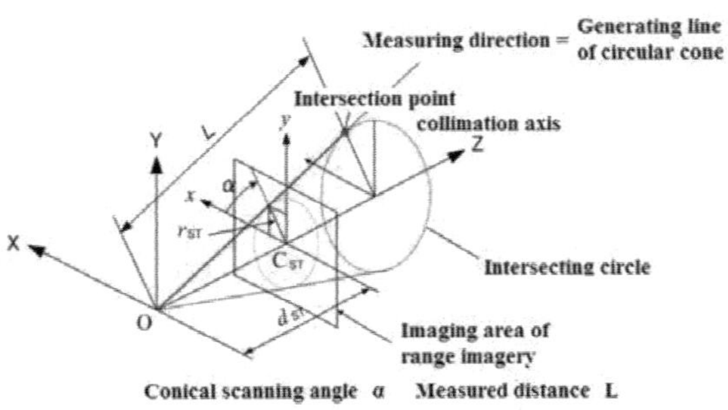

Figure 1. Principle of conical scanning

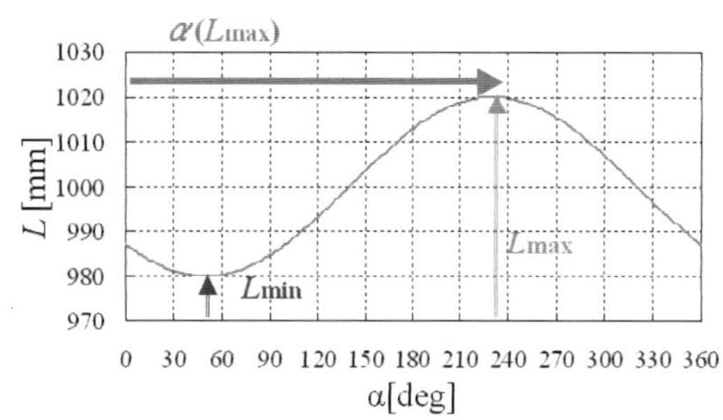

Figure 2. Example of scan curve

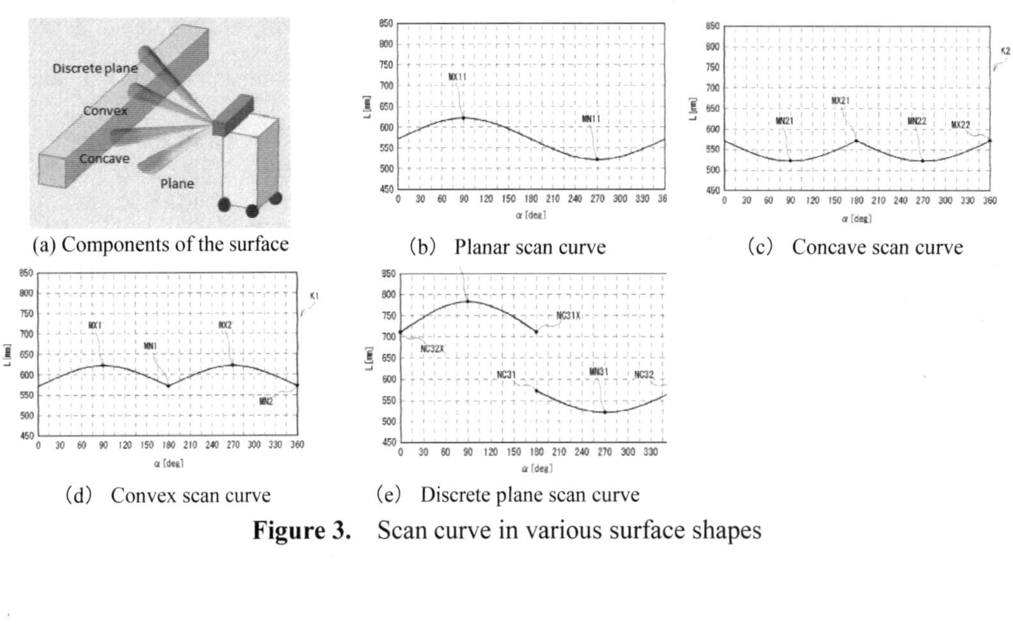

(a) Components of the surface (b) Planar scan curve (c) Concave scan curve

(d) Convex scan curve (e) Discrete plane scan curve

Figure 3. Scan curve in various surface shapes

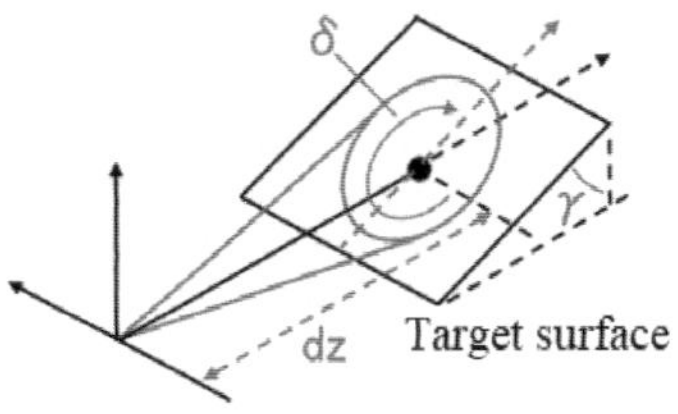

Figure 4. Parameter of plane

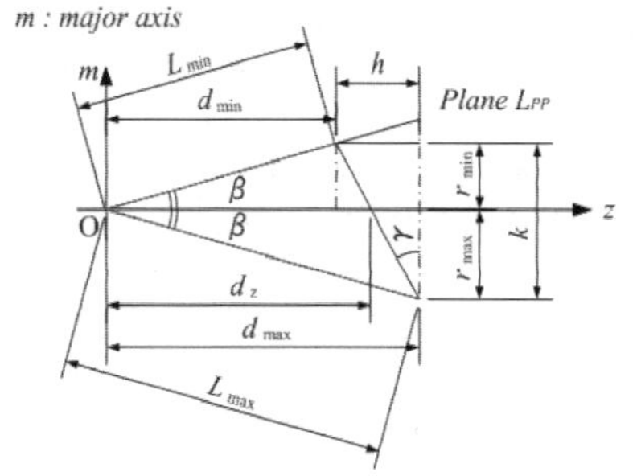

Figure 5. Parameter calculation drawing

In the following, distance image sensors are used to simultaneously measure multiple points at high speed. Conical scanning for multiple regions correspond to extracting distance on multiple circular segments within the range of the distance image sensor.

As shown in Figure 6, the center point of arbitrary i th and j th circles on the range image sensor is expressed by the following equation.

$$(x_c, y_c) = \left(\frac{w_c}{2} + \left(i - \frac{m}{2} \right) \cdot \Delta x, \frac{h_c}{2} + \left(j - \frac{n}{2} \right) \cdot \Delta y \right) \tag{4}$$

Here,

m, n: number of the circular segment in the x and y directions

Δx, Δy: spacing between the x direction and the y direction of a circular segment

w_c, h_c: number of pixels in the horizontal and vertical directions of the sensor

Based on these points, each extraction point corresponding to conical scan angle α is calculated. A scanning curve is obtained by acquiring the distance L of the pixel corresponding to the extraction point. For the intermediate position of the pixels, data smoothing interpolation processing by the bilinear method are performed. Parameter identification is performed for each segment, and the horizontal plane is extracted by residual evaluation and inclination direction evaluation.

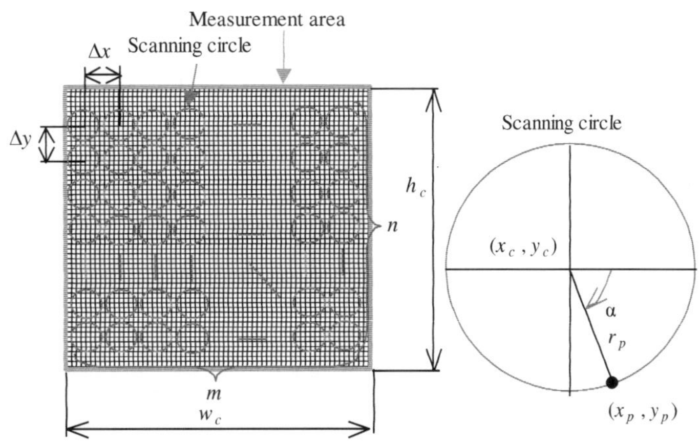

Figure 6. Image of Multiple conical scanning

3. Application for Legged wheeled robot

3.1 Overview of the legged wheeled robot.

IOP Conf. Series: Materials Science and Engineering **501** (2019) 012047 doi:10.1088/1757-899X/501/1/012047

In this section, we describe an example of applying a multiple conical scanning to staircase recognition of a four legged wheeled robot.

The appearance of the robot is shown in Figure 7. The robot has 16 DOF, with yaw, pitch axis of a hip joint, and pitch axis of a knee and a driving wheel at the toe for each leg. It supports the visually impaired person ascending and descending stairs. In the flatland, it runs with four wheel steering and four wheel driving by using driving wheels and hip joint yaw axes.

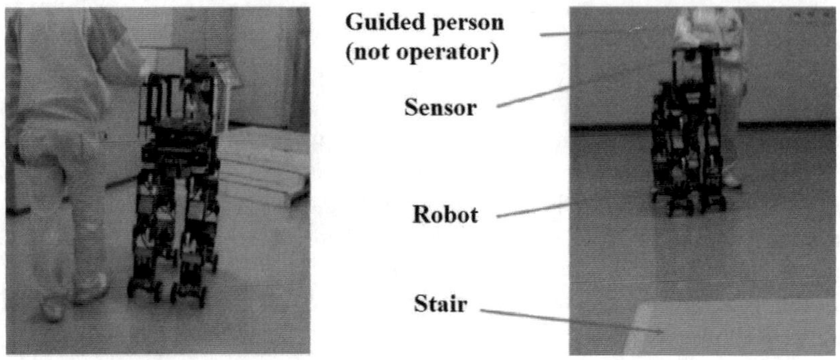

Figure 7. Appearance of leg-wheel type robot

3.2 Recognition object and recognition principle

As shown in Figure 8, the staircase information to be recognized is the stair position (S_X, S_Y), direction γ, and stair width S_W.

The staircase width is treated as the effective width within the measurement range, and the stair position is the center point of the effective width.

Figure9 shows the procedure of staircase detection. Labeling processing is executed on the image obtained by plotting the center coordinates (x_c, y_c) of the segment determined as the horizontal plane. For each extraction region, the average value Z_{ave} of the Z coordinate (height) and the variation (evaluated by the difference between the maximum value Z_{max} and the minimum value Z_{min} within the extraction region) is calculated. An extraction region in which Z_{ave} and variations are within a predetermined range is adopted as a step (first step of the stair). Line fitting is performed on the corresponding extraction region by least squares method, points on the right and left ends (x_{left}, y_{left}) , (x_{left}, y_{left}) on the approximate straight line are calculated, and converted to the left and right end points (X_{left}, Y_{left}, Z_{left}),(X_{right}, Y_{right}, Z_{right}) of the robot coordinate system.

The stair position, direction, and stair width are calculated by the following equation.

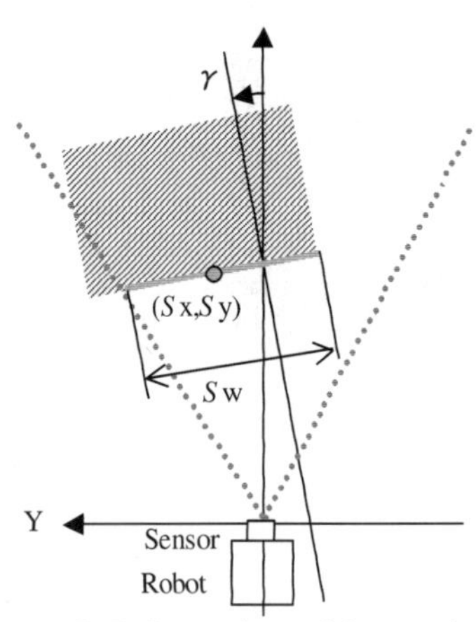

Figure 8. Information of detected stair

IOP Conf. Series: Materials Science and Engineering **501** (2019) 012047 doi:10.1088/1757-899X/501/1/012047

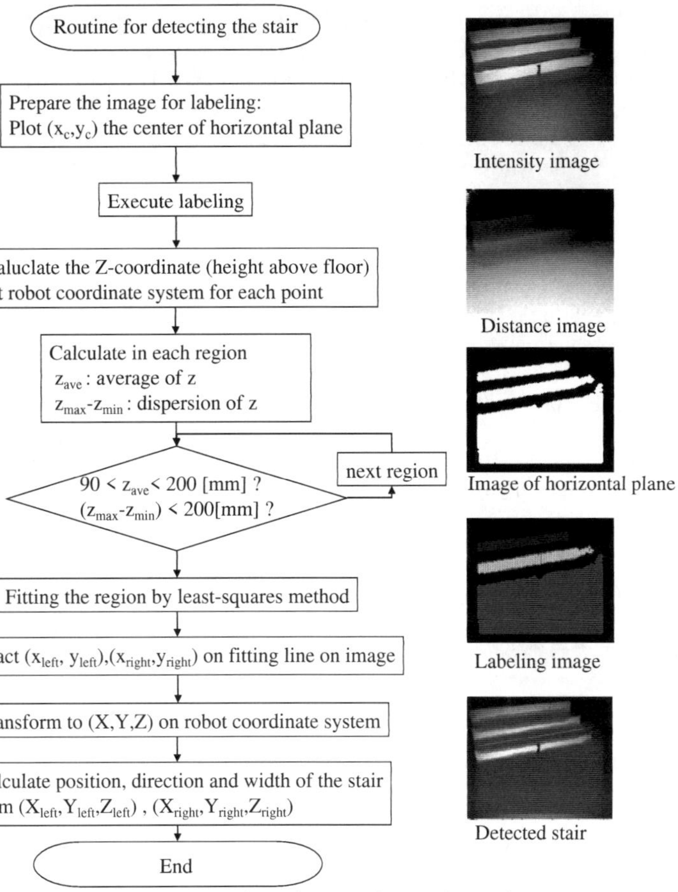

Figure 9. Flowchart to detect the stair

$$(S_X, S_Y) = \left(\frac{X_{left} + X_{right}}{2}, \frac{Y_{left} + Y_{right}}{2} \right) \tag{5}$$

$$\gamma = \tan^{-1}\left(\frac{X_{right} - X_{left}}{Y_{left} - Y_{right}} \right) \tag{6}$$

$$S_W = \sqrt{(Y_{left} - Y_{right})^2 + (X_{right} - X_{left})^2} \tag{7}$$

3.3 Recognition experiment

This algorithm is verified with TOF type distance image sensor Swiss Ranger SR-3000 (measuring range : 300 to 7500 mm, number of pixels : 176 × 144, FOV : 47.5 × 39.6 deg, frame rate : maximum 50 fps)

Experiments were carried out in both cases when facing the stairs (Figure10) and when given a predetermined angle (Figure11). Common parameters are shown in Table 1. The measurement target is a staircase with 3 steps, tread depth of 240 mm, and rise height of 120 mm.

3.3.1 Experiment when facing the stairs Table 2 shows the conditions under which recognition experiments were carried out with facing the front of the stairs. Figure12 shows recognition results of stair direction γ. Average value of 50 times and error terms of 2σ were plotted. Since it faces the front of the staircase, the ideal value of the angle is 0, but when it is 1000 mm ahead, it will be distorted as it deviates in the *Y* axis direction. It can be thought that the whole staircase can only be seen near the

IOP Conf. Series: Materials Science and Engineering **501** (2019) 012047 doi:10.1088/1757-899X/501/1/012047

center because it is too close to the staircase. At 2000 mm ahead where the entire staircase is visible, although the dispersion depending on distance measurement accuracy is large, it is within ± 5 °.

Figure13 shows the recognition results of stair position (S_X, S_Y). The part where the whole staircase is visible are surrounded by a red dashed line.

When l_X = 1000 mm, since the whole staircase can not be seen, S_Y is narrower than the actual value, but S_X can be detected over the whole area, 2 σ = 9 mm at 1000 mm and 2 σ = about 60 mm even at 2000 mm.

3.3.2 Experiment when given a predetermined angle Table 3 shows experimental conditions when a recognition experiment is performed at a predetermined angle toward the stairs. Figure14 shows the relationship between the set angle (ideal value) of the staircase direction γ and the detection angle (experimental value). Average value of 50 times and an error term of 2σ were plotted. It was able to recognize with an accuracy of about ± 1 deg at 1000 mm ahead. As the distance gets farther, the variation becomes larger under the influence of the accuracy of distance measurement, but it is about ± 5 deg even at 2000 mm ahead.

Based on the distance and angle to the staircase calculated by this algorithm, the legged wheeled robot generated the approach trajectory, approached the stairs, and could move up and down.

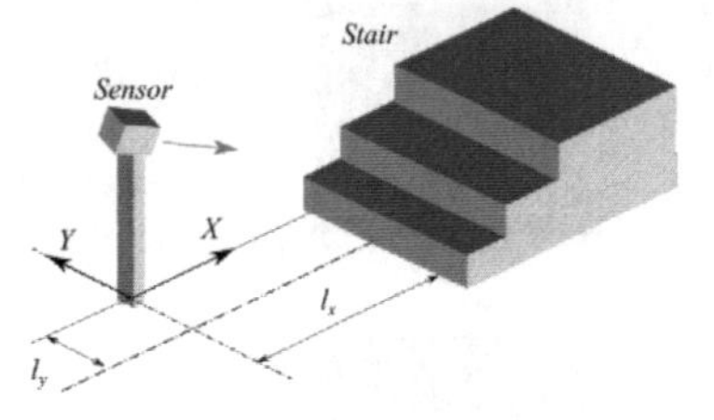

Figure 10. Experimental environment to measure the stair

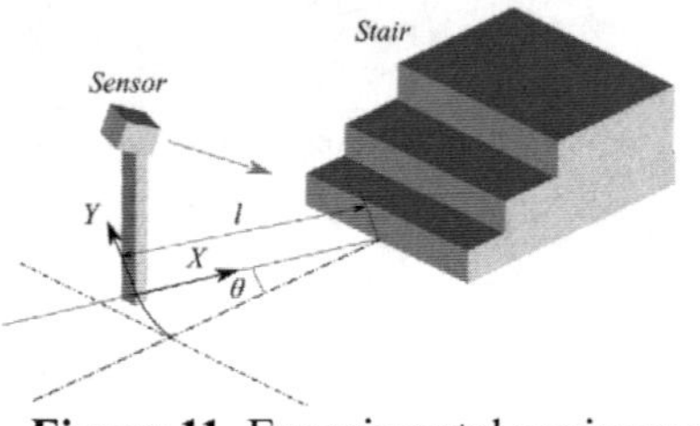

Figure 11. Experimental environment to measure the stair

Table 1. Parameter for multiple conical scanning

Radius of conical scan r_p	5 (pixel) (β=1.3625(deg))
Interval of conical scanning angle $\Delta\ \alpha$	5.625 (deg) (divided into 64)
Number of segment $m \times n$	74 × 121
X-axis Scanning interval Δx	2 (pixel)
Y-axis Scanning interval Δy	1 (pixel)
Descending vertical angle of sensor	30 (deg]
Height of the sensor	750 (mm)

Table 2. Experimental condition

X-coordinate of stair position l_x	1000, 1500, 2000 (mm)
Y-coordinate of stair position l_y	-800~+800(mm) (per 200)
Direction of stair γ	0 (deg)

Table 3. Experimental condition

Distance stair l	1000, 1500, 2000 (mm)
Direction of stair γ	-20 ~ +20 (deg) (per 5)

IOP Conf. Series: Materials Science and Engineering **501** (2019) 012047 doi:10.1088/1757-899X/501/1/012047

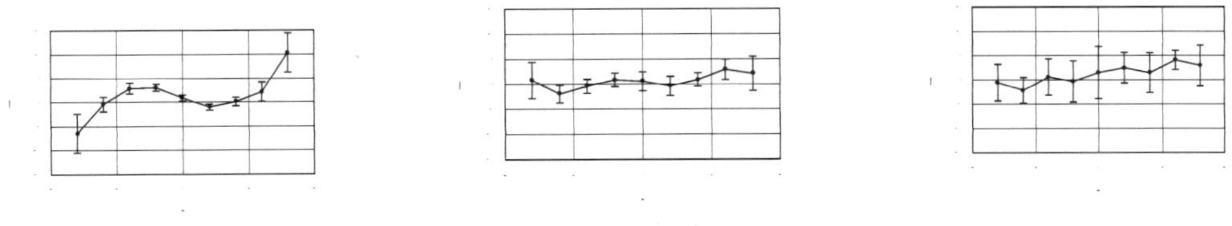

(a) 1000mm (b) 1500mm (c) 2000mm

Figure 12. Experimental results of γ

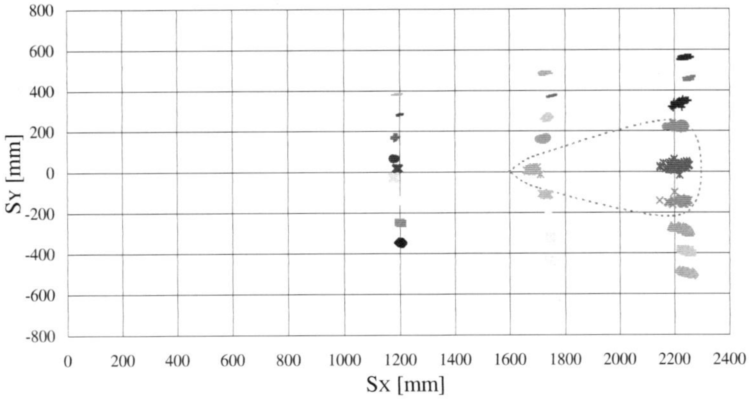

Figure 13. Experimental results of (S_X,S_Y)

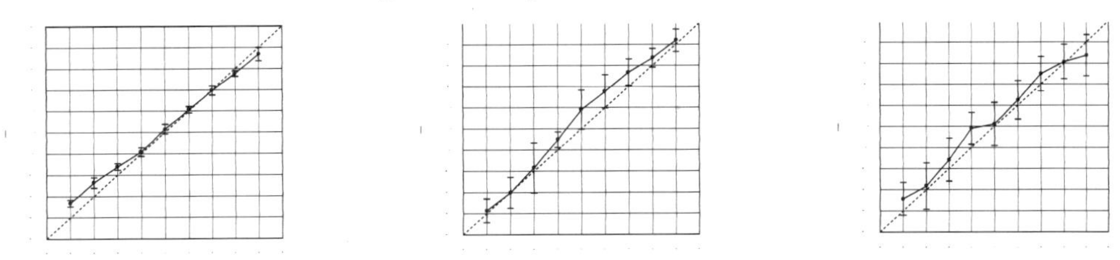

(a) 1000mm (b) 1500mm (c) 2000mm

Figure 14. Experimental results of γ

4. Application for indoor and outdoor mobile wheeled robot

4.1. Overview of the mobile robot

In this section, we describe an example of applying a conical scanning to indoor and outdoor autonomous mobile robot "KANACO (Kanamori Autonomous Navigation And COmmunication robot)" (Figure 15). This robot is an independent two-wheel drive type robot. Two air tires with a wheel diameter of 210 mm are attached to the front wheels and two casters with a wheel diameter of 125 mm are attached to the rear wheels in order to overcome small steps in the running area of the robot, such as the groove of the automatic door and the boundary between the asphalt and the side groove lid. The hardware consists of a PC, left / right motor unit, motor driver, control circuit, various sensors.

The sensor using for conical scan is Kinect v2 (color image resolution : 1920 × 1080, color frame rate : 30 fps, depth image : 512 × 424, Depth frame rate : 30 fps, Depth sensing method : TOF (Time Of Flight), Depth range : 500 mm to 8000 mm, a horizontal viewing angle of 70 degrees, and a vertical viewing angle of 60 degrees).

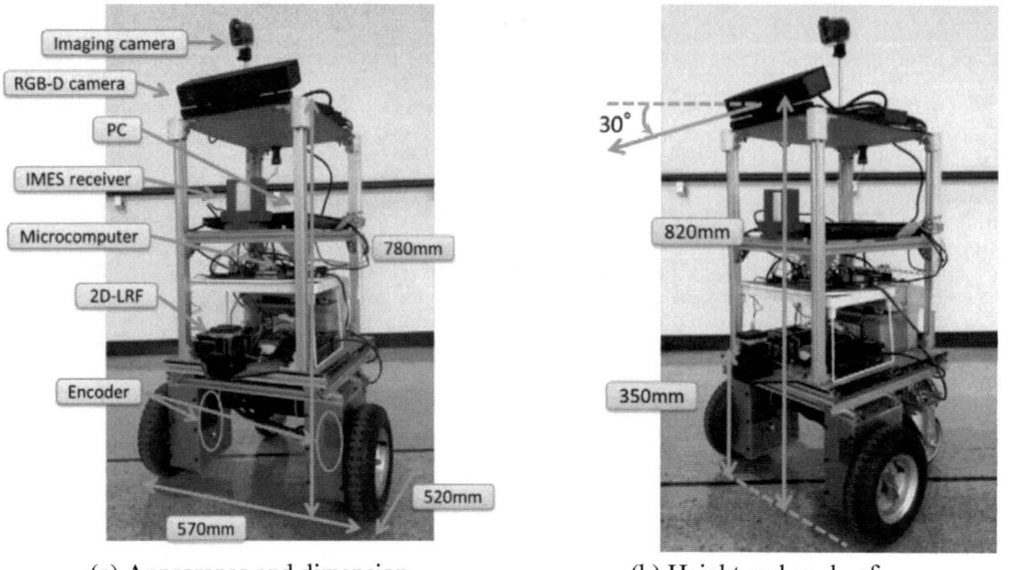

(a) Appearance and dimension (b) Height and angle of sensor

Figure 15. Overview of the robot

4.2. Extension and verification of conical scan

When the conical scanning method was used outdoors, a lot of noise was included in the measured scanning curve, and the correct shape could not be classified and the feature amount could not be calculated in some cases. Therefore, we developed a scan curve fitting method applying weighted M-estimation[9] in order to realize robust environmental recognition. Furthermore, in order to process in real time, a scan curve parallel processing algorithm using GPU (NVIDIA Geforce GT 740 M) was implemented.

A flowchart of the implemented algorithm is shown in Figure16. First, using the depth information acquired from the distance image sensor, scanning curve of each scanning circle are acquired and transmitted to the GPU. Each thread of the GPU obtains the scan curve of the designated scan circle and then performs M-estimation of the quartic equation. After M-estimation, the number of extreme values is judged. If the number of extreme values is 3, it is determined to be a region other than the plane, that is, a running dangerous region. If the number of extreme values is 2, it is determined to be a flat area, and it is judged whether it is a travel safety area or a danger area according to the inclination angle and inclination direction of the measurement surface. Thereafter, parameters of the target surface obtained by each thread are transmitted to the CPU. By repeating this series of processing, the robot consecutively recognizes the surrounding environment.

Table 4 shows the results of the processing speed executed under the four conditions. The average of 100 frames was taken as the average processing speed. Under all conditions, the parallel processing by the GPU realized about 20 to 30 times faster than sequential processing by the CPU. These are sufficient processing speeds exceeding 30 fps and real time processing can be realized.

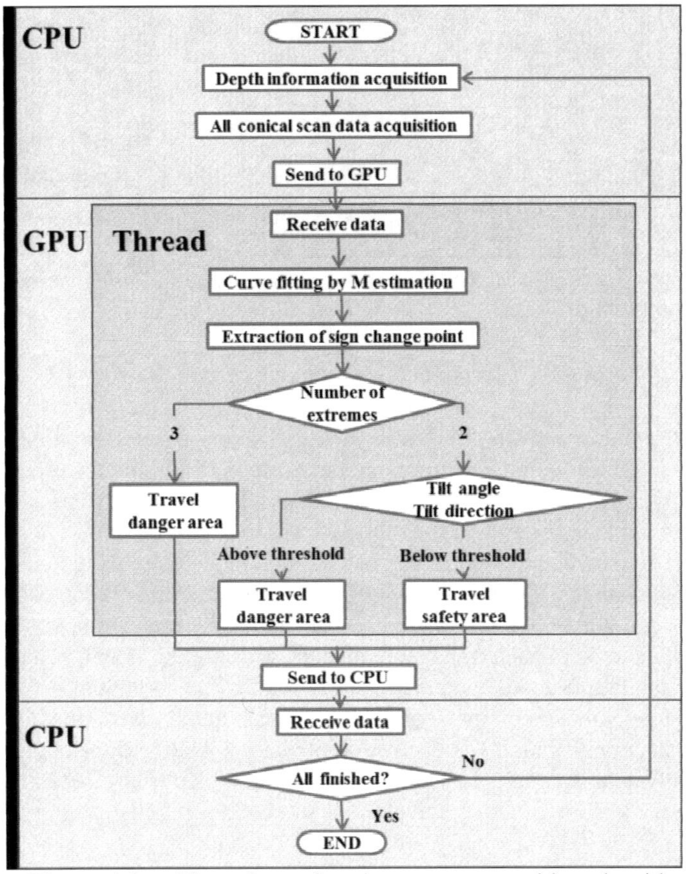

Figure 16. Flow chart of environment recognition algorithm

Table 4. Example of processing speed result

Radius of scan circle (pixel)	Average processing speed CPU only (msec/frame)	Average processing speed with GPU (msec/frame)
5 (3933 scan circle)	899.78	27.42
7 (2009 scan circle)	463.35	15.62
9 (1147 scan circle)	265.03	13.95
11 (775 scan circle)	170.79	7.43

4.3 Indoor and outdoor running experiments

Figure17 shows an example of obstacle detection indoors. In the situation (a) where a soccer ball is placed in the corridor, the running danger area is indicated by light blue in (b) according to the surrounding environment recognition result. (c) is an overlay of the traveling dangerous area on the planar map created by SLAM from the measurement data of the 2D LRF, and the dangerous area is treated as an obstacle.

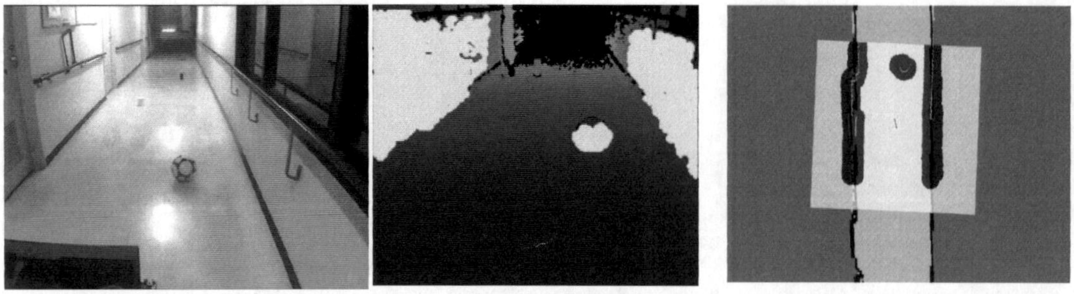

(a) Robot viewpoint image (b) Surrounding environment recognized image (c) Local map

Figure 17. Soccer ball detection during indoor driving

An outdoor running experiment was continued. Figure18 shows a state of running on a wheelchair slope. The slope was a surface with unevenness and the posture of the robot was easy to change. Also, because there are steps and fences on both sides, the road width was very narrow. The surrounding environment recognized image shows that the robot recognized steps on both sides. In addition, it was possible to stably determine whether the robot is in safe traveling area or dangerous area even when the posture of the robot changes. Figure 19 is a scene to avoid a bicycle. The spokes of the wheels with high metallic luster could not be recognized because the depth data did not return, but since tires with low gloss were able to be extracted stably, avoidance action were able to be taken safely. Figure20 is a state of running in a narrow passage surrounded by a short metal pole and a curb. There were a plurality of places where the point cloud could not be acquired due to the high metallic luster, but by performing scan curve fitting by M-estimation, it was possible to determine the travel dangerous area robustly against noise. By using our algorithm which is also robust to outdoor peculiar noise, the robot was able to autonomously travel safely about 350 m to the goal point by following the trajectory as shown in Figure 21.

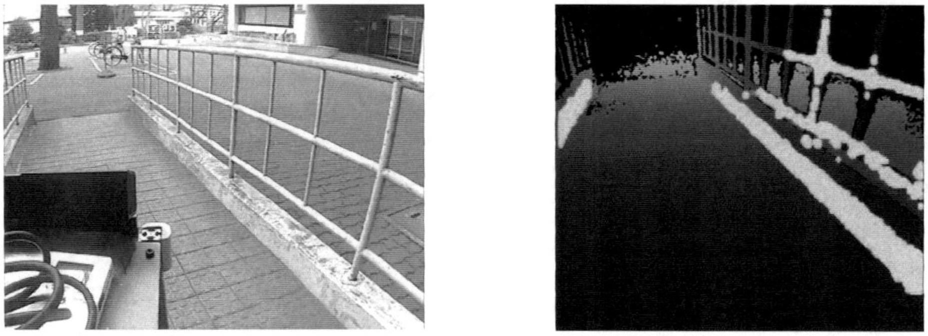

(a) RGB image (b) Recognized image

Figure18. Traveling on a slope

IOP Conf. Series: Materials Science and Engineering **501** (2019) 012047 doi:10.1088/1757-899X/501/1/012047

(a) RGB image (b) Recognized image

Figure 19. Bicycle detection

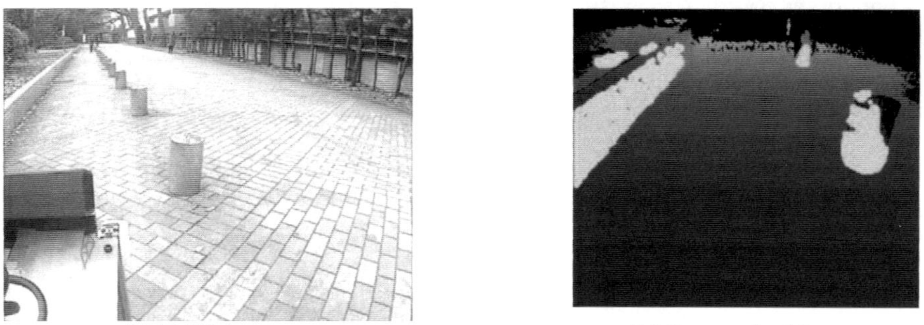

(a) RGB image (b) Recognized image

Figure 20. Driving a narrow passage

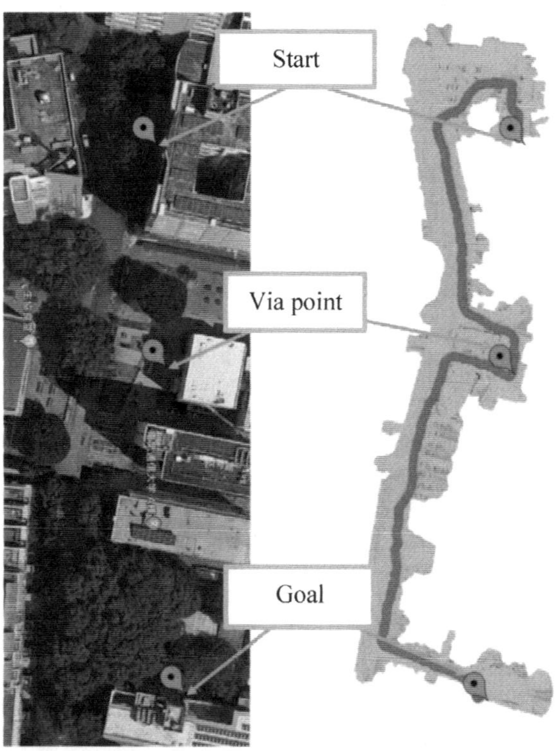

Figure 21. Example of outdoor driving test result

5. Conclusion

We discussed the basic principle of the conical scanning method for detecting the distance to the plane and the attitude of the plane and the algorithm for recognizing the surrounding environment by multiple conical scanning processes. This algorithm was applied to a quadruped wheel type robot, the robot recognized the stairs and approached, realized the ascent and descent operation. In addition, when applied to indoor and outdoor wheeled robots, the robust against ambient light noise by M-estimation and real-time peripheral environment recognition by GPU processing was implemented, the navigation to the destination was realized and the utility of this method was shown.

Acknowledgements

Conical scanning method is the result of collaborative research between NSK Ltd. and UEC. This research was promoted based on the results of the collaborative research. We appreciate the support of NSK at that time.

References

[1] Kunihiro T and Banno H, Application of Efficient Plane Detection Using Stereo Camera to Unmanned Ground Vehicle, *Journal of IHI Technologies*, **50**(1) 3-10 .

[2] Watanabe A, Bando S , Shinada K and Yuta S, Road-Following-Based Navigation in Park and Pedestrian Street with Finding Intersection and Orientation Detection, *Journal of the Robotics Society of Japan*, **30**(3), 271-279.

[3] Fujii A, Wada K, Tanaka M, Nagasaka N and Kubota F, Development of Autonomous Mobile Robot for Moving through Narrow Spaces, *Journal of the Robotics Society of Japan*, **32**(1), 74-83.

[4] Eguchi J and Ozaki K, Extraction Method of Travelable Area by Using of 3D-laser Scanner — Development of Autonomous Mobile Robot for Urban Area—, *Transactions of the Society of Instrument and Control Engineers,* **52**(3), 152-159.

[5] Japan patent JP5499355B, Device and method to recognize environment for moving

[6] Japan patent JP5742052B, Device and method to recognize environment for moving

[7] Tobita K, Kanamori C, Okubo Y, Ogawa H and Sugita S, Recognition of the stairs by multiple-region conical-scanning, *Proceedings of 2011 JSME annual Conference on Robotics and Mechatronics*, 1P1-E08.

[8] Okubo Y, Kanamori C, Tobita K and Ogawa H, Mobile environment recognition method using conical scanning distance measurement (Part3) - Recognition of the stairs and around - , *Proceedings of the 30th annual conference of the Robotics Society of Japan*, 4I2-7.

[9] Hoaglin. D. C, Mosteller. F, Tukey. J. W, “Understanding Robust and Exploratory Data Analysis”, Wiley, New York (2000).

IOP Conf. Series: Materials Science and Engineering **501** (2019) 012048 doi:10.1088/1757-899X/501/1/012048

Tube-feet Inspired Soft Gripper

Pasit Jittungboonya[1] and Thavida Maneewan[1,*]

[1]Institute of Field roBotic, King Mongkut's University of Technology Thonburi, Bangkok, 10140, Thailand

* Corresponding Author: praew@fibo.kmutt.ac.th

Abstract. Soft gripper can grasp various types of objects even deformable objects or complex shapes. In this work, the design of a soft gripper is inspired by a tube foot of a starfish. Tube foot can deform in multiple directions. The proposed gripper has three soft silicone fingers which are driven by pneumatic actuation. Each finger can be curved in 6 directions using a three sided air chamber. This gripper can be controlled by a binary command into four states: home, open, close and twist. The tube feet inspired soft gripper was developed, modelled and tested.

1. Introduction

Many robot designs have been inspired by natural mechanism in animals and plants. In this work, we are interested in a locomotion mechanism of a sea star and animal in echinoderm family such as sea cucumber and sea urchin. A sea star can move around the ocean floor using a unique organ called 'tube feet' which uses its extremity to adhere to a surface[1]. Tube foot can be considered as one kind of a leg. However, it is quite different from a so-called leg in other animals. A tube foot has no bone inside so that it can bend and create a motion in multiple degrees of freedom. A tube foot is actuated by pressure caused by ejecting water from the circulation system into each of the tube feet. This type of hydraulic actuation allows the tube foot to deform in multiple directions due to its soft structure. Furthermore, a tube foot also provides the adhesion mechanism at its end using some kind of protein-base underwater glues to attach itself to different type of surfaces. There are thousands of tube feet under the sea star's body which will have to be well coordinated in order to move its body toward the desired direction. There are many robots which try to mimic a sea star locomotion especially the motion of its quad-radial or penta-radial body [2, 3, 4,5,7]. Otake et al. created a sea star shaped robot from an electropolymer gel. Their robot can be controlled from electromagnetic field, but it can only move very slowly. Mao et al. developed the sea star robot using shape memory alloy and silicon elastomer. Their study focused on different crawling gait of the robot which can be used to transverse on different types of terrain. Mao et al. Developed the simulation model for a three legged sea star robot based on shape memory alloy actuation. The proposed forward kinematic model of this robot is based on a modified DH method which includes the bending section of a silicone elastomer structure. In 2017, Poungrat and Maneewarn also developed a quadruple soft robot using mckibben actuator as a leg module. Their soft robot has different crawling gaits which allow the robot to move in multiple directions.

However only a small number of research papers have discussed about the tube feet mechanism of a sea star. In 2007, Pei et al. [6] designsed a tube foot like mechanism based on layers of an electroactive polymer that are partitioning into two or four separately actuated sectors. This design allows a multi-degree-of-freedom motion. Recently, Bell et al.[7] has tried to replicate a tube foot using a bistable deformable dome structure with an asymmetrical rib. Their design featured the reversible adhesion to ferromagnetic surfaces using a small permanent magnet. They applied the synthetic tube feet into a wheel mechanism.

In this work, we adapt the concept of a 'hydro-stat' muscle mechanism in sea star's tube foot into a pneumatic based deformable actuator. The proposed soft actuator uses three air chambers to control its motion in five discrete states. This tube feet inspired actuator also assembled into a three fingers gripper which can be used to grasp an object as well as to manipulate an object in a twisting motion.

2. Design and Fabrication

2.1. Tube foot inspired finger design and fabrication

The structure of the proposed finger is based on three air bellow modules that are attached together in the triangular form as shown in figure 1. This soft actuator was fabricated by casting from silicone rubber based on the technique described in Marchese et al.[8]. The mold was created from PLA filament using 3D printer. The structure of each air bellow is based on the Fast Pneu-net (fPN) design proposed by Modsadegh [9]. fPN design allows faster actuation due to the increasing surface area of the extensible layer. Each air bellow module was casted from silicon rubber and cured with condensation catalyst.

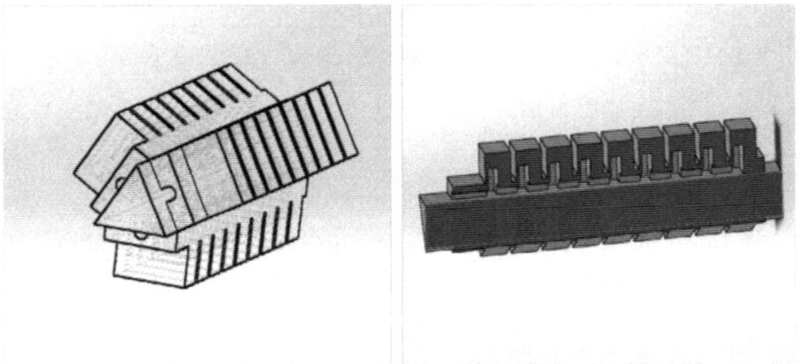

a) b)

Figure 1. a) The structure of a tube foot inspired soft finger which is created from three air bellow modules that allow the finger to bend in multiple directions b) Internal structure of the soft finger

Silicone rubber that was used for this soft finger has properties as listed: shore A = 10,viscosity = 12,000-14,000 cps, tensile strength = 24 N/mm^2, elongation at break = 620%

Each finger comprises of one core part and three air bellow parts. First, each piece of the core part and the air bellow part was molded and cured. Then, these pieces were assembled together using glue as shown in figure 2. After the core part was fully assembled, three bellow parts were attached to each side of the core part using glue as shown in figure 3.

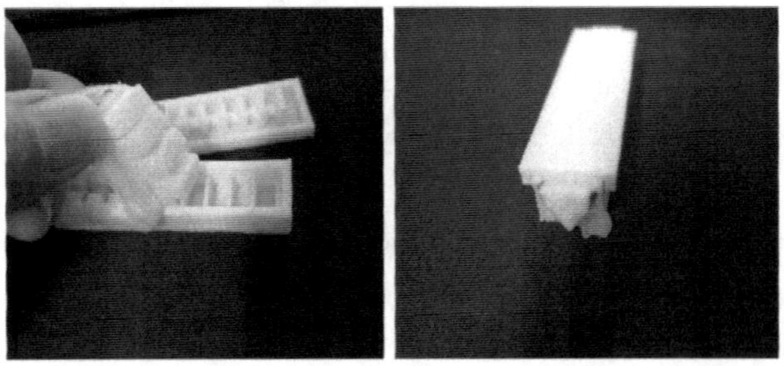

a) b)

Figure 2. The fabrication process of a soft finger a) the air bellow part was molded and cured b) the core part was molded and cured

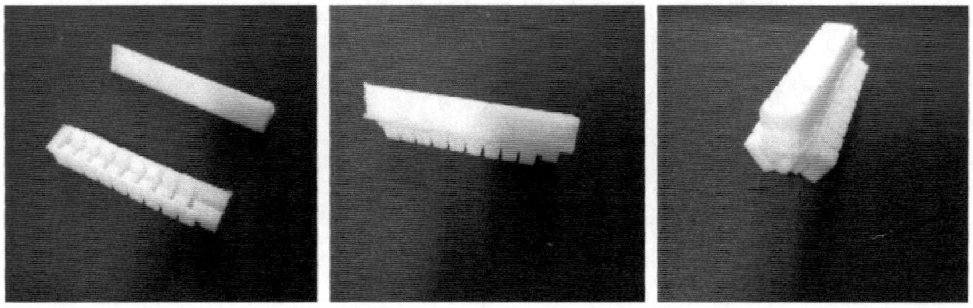

a) b) c)

Figure 3. a) core part and bellow part b) The air bellow part was glued to the core part c) The air bellow part was glued to the core part side

Each soft finger module is actuated and controlled by pumping air into each side of the air bellow. The air pressure from the pneumatic pump was reduced by a regulator to 0.5 bar. Each soft finger is controlled by three solenoid valves using a binary command which is sent from a microcontroller. Each finger can bend in 6 directions which can be selected from a 3 bit binary command. When a gripper is assembled from three soft fingers, the gripper is controlled by 9 bit binary command which is shown is figure 4.

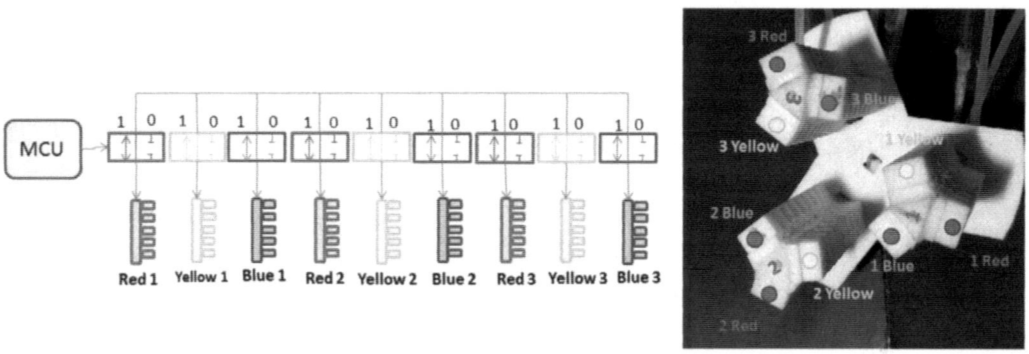

a) b)

Figure 4. A gripper with three fingers can be controlled by a 9 bit binary command. Each solenoid valve connects to the air bellow module on each side of the finger a) control diagram b) bellow name specify

2.2 *Kinematic modeling*

After the three-finger soft gripper was fabricated, the kinematic model of a tube-feet inspired soft finger is explored in this section. Since the entire structure of the soft finger is made from deformable material, the model of this finger is derived directly from experimental data. Each finger have six states which comprises of 3 states when only one air bellow is actuated and 3 states when two air bellows are actuated. In this experiment, the constant input pressure of 0.5 bar was used. For each finger, the air bellow in each side is designated by three colors: red, yellow and blue. In the first experiment, the air bellow was actuated one at a time. The shape of deformation was tracked from an image captured perpendicular to the actuated side. The deformation data was collected from three different fingers when only one side of air bellow was actuated. The discrete data from the image was curve fitted by a quadratic polynomial function and results for each finger are shown in figure 5.

In the second experiment, two air bellows were actuated at the same time. There are 3 possible pairs: red-yellow, yellow-blue and blue-red. The input pressure was set at 0.5 bar. The shape of deformation was also captured from an image as in the first experiment. The discrete data from the image was curve fitted by a quadratic polynomial function and results for each finger are shown in figure 5. An example image of a single and a double air bellow actuation is shown in figure 6.The deformation curves of the single and double bellow actuation are plotted in 3D as shown in figure 7.

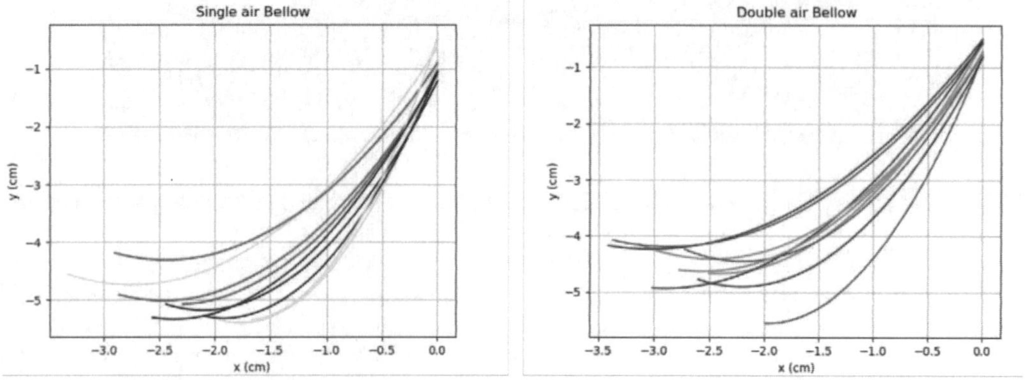

a) b)

Figure 5. Quadratic polynomial function of a deformation curve for a) a single air bellow actuation with 3 fingers b) a double air bellow actuation with 3 fingers

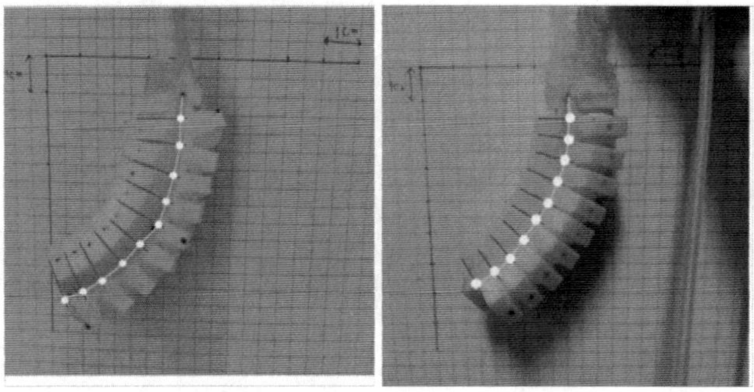

a) b)

Figure 6. An example image of the soft finger, yellow points on the air bellow were a tracking marker point for an image processing software a) a single air bellow actuation when R1 was actuated b) double air bellow actuation when Y1 and B1 were actuated

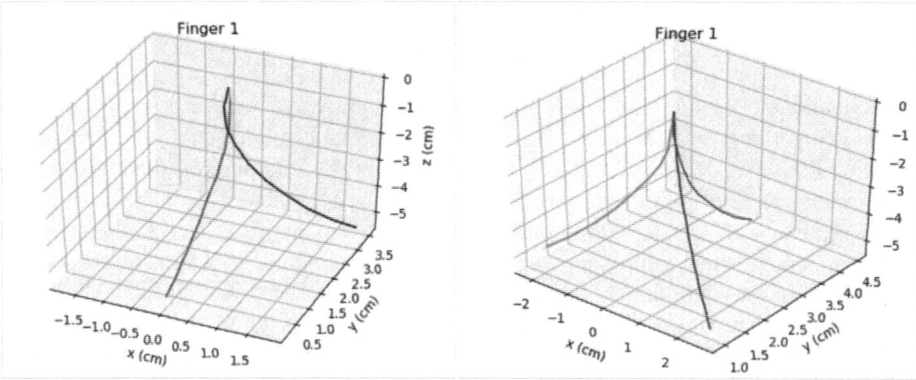

a) b)

Figure 7. The plot of a deformation curve for a) a single air bellow actuation in 3 dimensional for finger 1. Each color represents the actuated air bellow (R,Y,B). b) a double air bellow actuation. Each color represents different pair of actuation (Orange: Y-R, Purple: R-B, Green: B-Y)

3. Three-finger soft gripper

In order to demonstrate the usage of a tube feet inspired soft actuator, the proposed actuator was assembled into a three finger soft gripper. Each finger is oriented at 120 degree as shown in figure 8. Since there are 9 binary states for the three finger actuation, four types of operation can be utilized.

1. home: no actuation in all three fingers
2. open: single air bellow actuation in the inner (red) side
3. close: double air bellow actuation in the outer (yellow-blue) side
4. twist: double air bellow actuation in the side pair(blue-red or yellow-red) side

Each type of manipulation can be realized by a simple 9-bit binary command. There are 8 possible combinations for 9 bit binary commands, however only 5 commands are used. The binary command diagram for each operation is shown in figure 8. From the kinematic modeling data in the previous section, the 3D plot showing the shape of deformation for this three finger gripper when each command is applied are shown in figure 9. Figure 10 shows an actual image of the gripper when open, close and twist command were given.

Command	R1	Y1	B1	R2	Y2	B2	R3	Y3	B3
Home	0	0	0	0	0	0	0	0	0
Open	0	1	1	0	1	1	0	1	1
Close	1	0	0	1	0	0	1	0	0
Twist(CCW)	1	1	0	1	1	0	1	1	0
Twist(CW)	1	0	1	1	0	1	1	0	1

Figure 8. The 9-bit binary command diagram for each operation (home, open, close and twist)

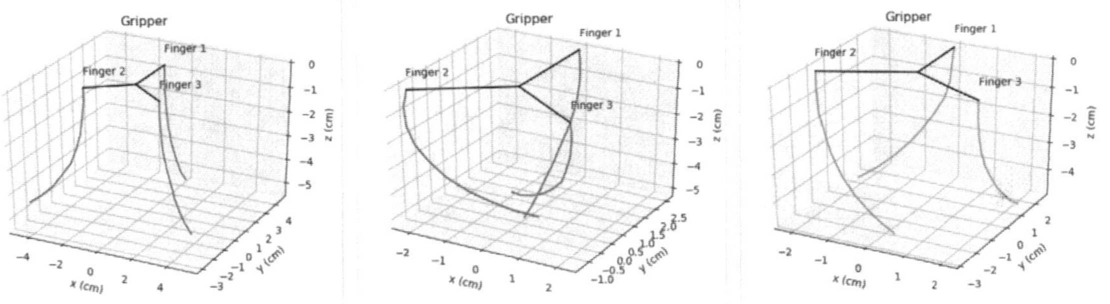

a) b) c)

Figure 9. The 3D plot showing the shape of deformation for the three finger gripper when different commands were applied a) open b) close c) twist

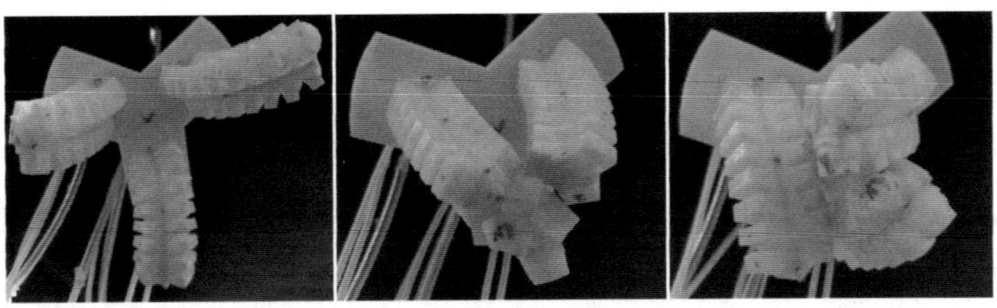

a) b) c)

Figure 10. the gripper when different commands were applied a) open b) close c) twist

When the gripper is open and close around an object, the gripper can grasp an object by either applying grasping force or caging. When the fingers apply sufficient force to an object in the force closure condition, the object can be held in place. Caging condition requires that the opening area of a gripper has to be smaller than the size of an object. In cage grasping, the gripper does not apply any force to keep an object in place. An image of a soft gripper that grasp an object successfully using cage grasping is shown in figure 11.

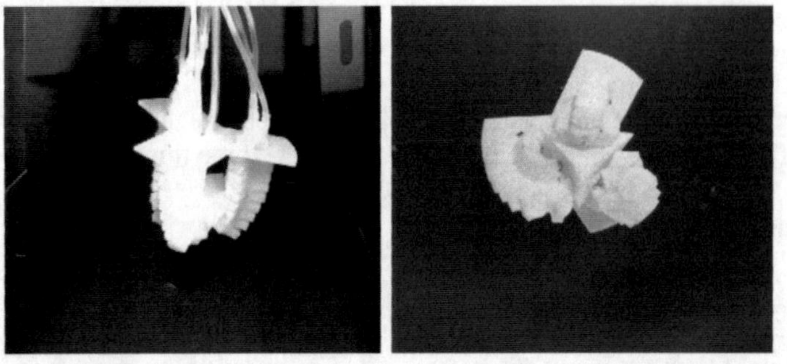

a) b)

Figure 11. the gripper grasp an object using cage grasping a) side view b) bottom view

When twisting command is performed, an object can be twisted in the specified direction as long as the object is grasped by sufficient amount of force. An image of a twisting operation is shown in figure 12.

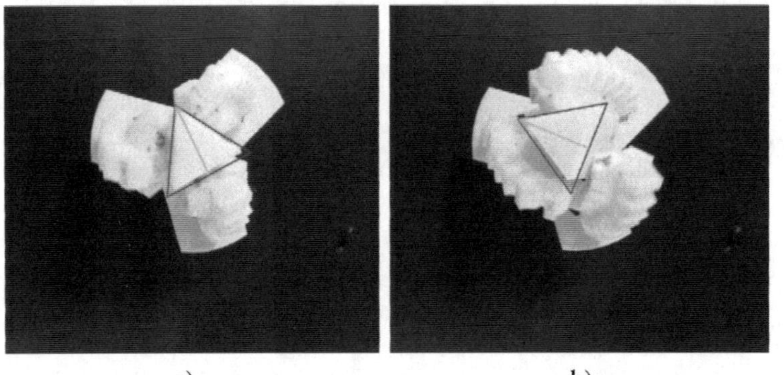

a) b)

Figure 12. the gripper rotates an object in counter clockwise direction by applying a twisting motion a) close command b) twist command

Conclusions

In this work, we have successfully developed a pneumatic based soft gripper. Each finger of this gripper can bend in multiple directions which is inspired by a tube foot mechanism of a sea star. Three air bellows on a finger allows the finger to bend in different directions which can be achieved by sending a binary command to solenoid valves that connect to each air bellow. The kinematic model of the finger is approximated by a quadratic polynomial function which derived from the experimental data. When three soft fingers were assembled into a gripper, the gripper can be controlled with 9-bit binary command in 4 states: home, open, close and twist. The soft gripper can grasp different shapes and sizes of object, as long as the size of object is larger than the area of opening when the gripper is closed. The gripper can also rotate an object in the twisting motion. In the future, more experiments will be performed in order to provide better understanding about different type of grasping and manipulation that can be utilized by this soft gripper.

IOP Conf. Series: Materials Science and Engineering **501** (2019) 012048 doi:10.1088/1757-899X/501/1/012048

References

[1] Elise H, Romana S and Patrick F 2012 *Zoosymposia* **7** 25-32.

[2] Mihoko O, Yoshiharu K, Masayuki I and Hirochika I 2002 *Robotics and Autonomous Systems* **40** 185-191

[3] Shixin M and Erbao D 2013 *Proc. Int. Conf. on Robotics and Biomimetrics (Shenzhen)* (New York: IEEE)

[4] Hu J, Erbao D, Shixin M, Min X and Jie Y 2014 Locomotion modeling of an actinomorphic soft robot actuated by SMA springs *Proc. Int. Conf. on Robotics and Biomimetics (Bali)* (New York: IEEE)

[5] Anon P and Thavida M 2017 A starfish inspired robot with multi-directional tube feet locomotion *Proc. Int. Conf. on Robotics and Biomimetics (Macau)* (New York: IEEE) pp 712-717

[6] Qibing P, Marcus R, Scott S, Harsha P, and Ron P 2004 Multiple-degrees-of-freedom electroelastomer roll actuators *Smart Material and Structures* vol 13 (IOP Publishing Ltd)

[7] Michael A B, Isabella P, William S, Kitty K, Mohammad K J, Derek A P, Carmel M, James C W and Robert J W 2018 Echinoderm-Inspired Tube Feet for Robust Robot Locomotion and Adhesion *Robotics and Automation Letters* vol 3 (New York: IEEE) Issue: 3 pp 2222-2228

[8] Andrew D M, Robert K K and Daniela R 2015 *Soft Robotic* **2** 7-25

[9] Bobak M, Panagiotis P, Christoph K, Sophia W, Robert F S, Unmukt G, Jongmin S, Katia B, Conor J W and George M W 2014 *Advanced Functional Materials* **24** pp 2163–2170

IOP Conf. Series: Materials Science and Engineering **501** (2019) 012049 doi:10.1088/1757-899X/501/1/012049

Energy Efficiency and Safety for Wall Type Air Conditioner with Propane Refrigerant

S Traivivatana[1] and W Yongchareon[2,*]

[1] Energy Research Institute, Chulalongkorn University, Phayathai Road, Pathumwan, Bangkok 10330 Thailand
[2] Department of Mechanical Engineering, Faculty of Engineering, Chulalongkorn University, Phayathai Road, Pathumwan, Bangkok 10330 Thailand

[*] Corresponding Author: yongchareon@gmail.com

Abstract. Thailand has been used the R-32 refrigerant in air conditioner since 2016, while it has high global warming potentials (GWP = 675). A propane refrigerant (R-290) is introduced as a newly alternative refrigerant, due to the global issue on refrigerant substitution for friendly environment. It has GWP of 4, which is 169 times lower than R-32. In this study, the theoretical energy efficiency was analysed under the Thailand's conditions at the saturated evaporation temperature of 10°C and the saturated condensation temperature of 47°C. However, the propane is highly flammable; therefore, the safety must be considered. The results showed that the coefficient of performance of propane was 5.54, which was 7% higher than R-32. The wall type air conditioner with propane refrigerant would be safe from fire if it was installed at a height higher than 1.8 m above the floor level and the floor area was larger than 17 and 23 m^2 for air conditioner cooling capacity of 3375 W and 5000 W, respectively.

1. Introduction

The R-32 refrigerant has commonly been used in air conditioners in Thailand since 2016. However, it has high global warming potentials (GWP = 675). Hence, the worldwide interest is focusing on alternative refrigerant substitutions for environmental friendly that reaches zero-GWP. The newly propane refrigerant (R-290) has GWP equal to 4, which is 169 times lower than R-32 and it might be one of the alternatives. Nonetheless, R-290 is highly flammable, therefore, energy and safety conditions are focused in this study.

2. R-290 Air Conditioner System

2.1. *Properties of propane*

The propane gas is heavier than air and may travel along the ground, there is a possibility of distance ignition. Moreover, due to low electric conductivity, the substance can generate electrostatic charge as a result of flow, agitation, etc. Therefore, do not use compressed air for filling, discharging or handling. Furthermore, the rapid evaporation of liquid may cause frostbite. If there is an ignition source and propane concentration is between 2-9.5%, the fire will be ignited. The properties of propane are shown in table 1.

IOP Conf. Series: Materials Science and Engineering **501** (2019) 012049 doi:10.1088/1757-899X/501/1/012049

Table 1. The properties of propane.

Formula	C_3H_8
Relative molecular mass	44.1
Boiling point, °C	-42.0
Melting point, °C	-187.0
Flash point, °C	-104.0
Autoignition temperature, °C	465.0
Critical temperature, °C	96.7
Relative density (water = 1)	0.5
Relative vapour density (air = 1)	1.6
Vapour pressure at 20°C, bar	8.5
Solubility in water	none
Explosive limit, vol.% in air	2.0-9.5
Limit (LFL), kg/m³	0.038
Minimum ignition energy (MIE), mJ	0.25
Heating value, MJ/kg	49.58

2.2. *System description*

In the vapor compression refrigeration system as shown in Figure 1, there are four major components; evaporator, compressor, condenser, and expansion valve. External work (W) is supplied to the compressor and the heat (q_{ev}) from the air condition space is added to the system in the evaporator. Whereas in the condenser, the heat rejection (q_{co}) is transferred from the system to the surrounding. Heat rejection and heat addition are dissimilar to different refrigerants, which cause a change in energy efficiency (COP) and exergy efficiency for the systems.

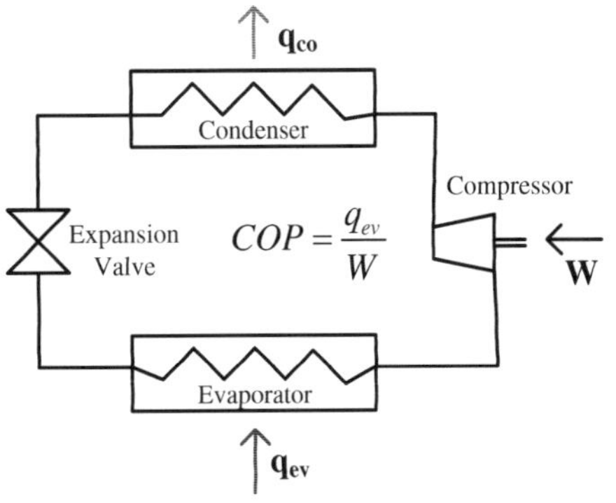

Figure 1. The vapour compression refrigeration system.

The P-h diagram of the refrigeration cycle is shown in Figure 2. The compression, condensation, expansion, and evaporation processes are shown between points 1–2, 2–3, 3–4, and 4–1, respectively. The evaporator saturated temperature was set to 7°C, while the condenser saturated temperature was taken as 47°C for comparison purpose. State 1 and 4 are assumed to be saturated vapour and saturated liquid, respectively. Whilst state 2s and 2 are obtained by the assumption that a compression process is isentropic and adiabatic with a compression efficiency (η_c) of 85%. The compression discharged enthalpy, h_2, is calculated by using equation (1),

$$h_2 = h_1 + \frac{(h_{2s} - h_1)}{\eta_c} \quad (1)$$

where h_{2s} is the discharged enthalpy for isentropic compression.

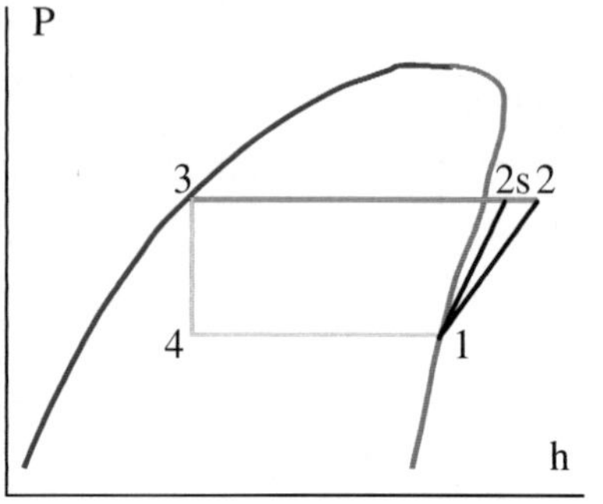

Figure 2. P-h diagram of the refrigeration cycle.

The temperature, pressure, enthalpy and entropy of each state of R-290 and R-32 are shown in Table 2 and Table 3, respectively.

Table 2. Temperature, pressure, enthalpy and entropy of each state of R-290.

State	Temperature, K	Pressure, MPa	Enthalpy, kJ/kg	Entropy, kJ/kg
1	280	0.5779	580.6	2.3594
2	325	1.5963	636.1	2.3800
3	320	1.5963	324.7	1.4125
4	280	0.5779	324.7	1.4451

Table 3. Temperature, pressure, enthalpy and entropy of each state of R-32.

State	Temperature, K	Pressure, MPa	Enthalpy, kJ/kg	Entropy, kJ/kg
1	280	1.012	516.4	2.2190
2	356	2.930	268.4	2.1650
3	320	2.930	290.7	1.2980
4	280	1.012	290.7	1.3236

3. Energy Analysis

In this study, the following assumptions are made;

1. Steady state conditions are remained in all the components
2. Pressure loses in the pipelines are neglected
3. Heat gains and heat loses from the system or to the system are not considered

4. Kinetic energy, potential energy, and pressure losses are not considered.

The energy performance of refrigeration cycle is calculated by using the following equations,

Refrigeration effect, q_{ev}; $$q_{ev} = m(h_1 - h_4) \quad (2)$$

Compression work, W; $$W = m(h_1 - h_4) \quad (3)$$

Condenser heat rejection, q_{co}; $$q_{co} = m(h_2 - h_3) \quad (4)$$

Coefficient of performance, COP; $$COP = \frac{(h_1 - h_4)}{(h_2 - h_1)} \quad (5)$$

Energy balance; $$q_{co} = q_{ev} + W \quad (6)$$

4. Safety

4.1. *Safety standard*

The fire or explosion hazard will take place if the mixture of propane and air is ignited. There are three critical elements that have to occur at the same time. Firstly, the concentration of mixture must be equal to or greater than LFL of 0.038 kg/m^3 or 2.0%. Then, the source of ignition has energy greater than MEI of 0.14 mJ, and finally, the mixed gas velocity is lower than burning velocity.

The safety standard, ANSI/ASHRAE Standard 34-2010 [1], classifies the safety class of propane as A3 which is non-toxic, LFL $\leq$ 0.1 kg/m^3, heat of combustion $\geq$ 19,000 kJ/kg, and exhibits flame propagation. Additionally, IEC 60335-40-2 [2] limits maximum refrigerant mass, M_{max}, for a given room floor area according to equation 7 or minimum floor area, A_{min}, for a given mass according to equation 8,

$$M_{\max} = 2.5(LFL)^{5/4} h_0 A^{1/2} \quad (7)$$

$$A_{\min} = \left[\frac{M}{2.5(LFL)^{5/4} h_0} \right]^2 \quad (8)$$

where LFL is the lower flammability limit of refrigerant,
A is the room floor area,
h_o is the installation height of fan coil unit.

4.2. *Risk assessment*

The flammability risk of propane in air was studied by D. Colbourne [3]. The event tree analysis (ETA) was used to identify the combination of sequence of events that may lead to a release of refrigerant and an active ignition source that could result in ignition and eventually the probability of ignition and severity of consequences. The result revealed that the total ignition frequency and the risk for maximum thermal intensity were 106 visits per million that begin servicing (3.6×10^{-4} per visit), and 3.1×10^{-3} (s(kWm^{-2})$^{4/3}$/yr, respectively. Also the ignition frequency for an industrial average worker is slightly lower than criteria 500 visits per million, where the well trained worker value is ten order of magnitude lower.

5. Results

From the theoretical refrigeration cycle analysis, both the energy of refrigerant R-290 and R-32 are shown in Table 4. The COP of R-290 and R-32 were 4.6 and 4.3, respectively. Using R-290 was increase the efficiency by 7.0% compare to R-32. Similarly, the cooling capacity of R-290 and R-32 were 256 and 226 kJ/kg, respectively, which was 13.3% increase in the cooling capacity.

The efficiency of the air conditioner is defined by energy efficiency ratio, which is the ratio of the cooling capacity and electrical power input to the unit. The actual energy efficiency ratio (EER) is

usually depended on many factors such as refrigerant type, brand, cooling capacity size, and cost. However, most of the air conditioner using R-290 is more efficient than R-32.
At present, some of the split type air conditioners with refrigerant R-290 and R-32 are available in the market as shown in Table 5.

Table 4. Energy Efficiency of R-290 and R-32.

Component	Energy, kJ/kg	
Refrigerant	R-290	R-32
Compressor Work	55.5	55.5
Condenser Heat Reject	311.4	277.7
X- Valve	0.0	0.0
Evaporator Cooling	256.0	225.7
COP	4.62	4.34

Table 5. Inverter split type air conditioners.

Refrigerant	Model	Capacity, kW	Power, kW	EER
R-290	GSC18GIG	5.28	1350	3.91
R-290	GSC12FG7	3.52	946	3.72
R-290	GSC12FIXH	3.52	684	5.15
R-32	FTRM185V25	5.19	1240	4.18
R-32	RHM50RVIN	4.86	1315	4.14
R-32	AR15MYFTAUAN	4.12	1350	3.05
R-32	AR18NYFXAWRN	5.28	1370	3.85
R-32	1V18RIN	5.28	1540	3.43
R-32	1K13RN	3.52	1030	3.42
R-32	RASDX13CJT	3.78	1182	3.20

Note: The EER based on indoor air conditions at 27°Cdb and 19°Cwb, outdoor air conditions at 35°Cdb and 24°Cwb.

The minimum room floor area according to equation 8 is shown in Table 6. There is no area requirement if the refrigerant mass is less than or equal to 0.152 kg. On the other hand, while the refrigerant mass is greater than 0.988 kg, the forced ventilation is necessary.

Table 6. Minimum room area.

Charge amount (M), kg	0.152	0.228	0.304	0.456	0.608	0.988
h_o, m	Minimum room area, m^2					
0.6		82	146	328	584	1541
1.0		30	53	118	210	555
1.8		9	16	36	65	171
2.2		6	11	24	43	115

Due to the hot and humid weather in Thailand, the typical air conditioner cooling load for building with high and light cooling load are about 12 m^2/RT and 18 m^2/RT, respectively. Likewise, the refrigerant charge in the commercial air conditioners are 0.33 and 0.41 kg for cooling capacity of 3500 W (1RT) and 5250 W, respectively. Table 7 and Table 8 illustrate the minimum room area and air condition area with high and light cooling load. According to the safety standard of International Standard IEC 60335-40-2:2018, only the 3500 W split type air conditioner with ceiling installation fan coil unit is safe. Nonetheless, the refrigerant charge in air conditioner at the same capacity depends on many factors, for example, band, model, cost, etc. Hence, if the refrigerant mass is reduced, the minimum room area or the installation height is also reduced.

Table 7. Minimum room area for air conditioner with high cooling load.

Cooling Load, W	3500	5250	7000	8750
Air Condition Area, m^2	12	18	24	30
R290 mass, kg	0.33	0.41	0.50	0.59
	Minimum Area for R-290, m^2			
Floor Installation, $h_o = 0.6$	167	269	396	547
Window Installation, $h_o = 1.0$	60	97	143	197
Wall Installation, $h_o = 1.8$	19	30	44	61
Ceiling Installation, $h_o = 2.2$	12	20	29	41

Table 8. Minimum room area for air conditioner with light cooling load.

Cooling Load, W	3500	5250	7000	8750
Air Condition Area, m^2	18	27	36	45
R290 mass, kg	0.33	0.41	0.50	0.59
	Minimum Area for R-290, m^2			
Floor Installation, $h_o = 0.6$	167	269	396	547
Window Installation, $h_o = 1.0$	60	97	143	197
Wall Installation, $h_o = 1.8$	19	30	44	61
Ceiling Installation, $h_o = 2.2$	12	20	29	41

6. Discussion and Conclusion

The results of the R-290 COP and cooling capacity, which were higher than R-32, are identical to D. Colbourne [3]. In this study, there are 7% increasing in both the COP and EER and 13.3% increasing in the cooling capacity of R-290 compare to R-32. In contrast, the R-290 compressor discharge temperature and pressure were lower than R-32 for air-cooled air conditioner. However, the refrigerant R-290 is more flammable than R-32.

The wall type air conditioner with propane refrigerant would be safe from fire if it was installed under two conditions; 1) at a height higher than 1.8 m above the floor level and 2) the floor area was larger than 17 and 23 m^2 for air conditioner cooling capacity of 3375 W and 5000 W with the propane charge of 0.31 and 0.36 kg, respectively.

To avoid fire hazard the air conditioner height and minimum room floor area should be complied to safety standard, also the service technicians should be well trained for flammable refrigerant.

Additionally, user must be aware of using flammable refrigerant and avoid any ignition source in the air condition space.

Acknowledgments
The researchers appreciate GIZ for funding this project to Chulalongkorn University.

References

[1] ASHRAE 2010 *Designation and Safety Classification Refrigerant* ANSI/ASHRAE Standard 34-2010 (Atlanta: ASHRAE)

[2] IEC 2018 *Household and Similar Electrical Appliances – Safety – Part 2-40: Particular Requirements for Electrical Heat Pumps, Air Conditioners and Dehumidifiers* International Standard IEC 60335-2-40:2018 (Geneva: IEC)

[3] Colbourne D 2011 Risk Analysis of Flammable Refrigerant Handling During Service and Maintenance Activities *Proc. 23rd IIR Int. Congress of Refrigeration* (Prague: 21–26 August 2011)

Solar Water Heater of Absorption Type using Natural Black Rock

R Aungkurabrut[1] and **P Pathike**[1,*]

[1] Department of Mechanical Engineering, School of Engineering, University of Phayao, 19 Maeka, Muang Phayao, Thailand, 56000

* Corresponding Author: pakdeeman@gmail.com, Telephone:+6654466666-3383, Fax Number:+6654466662

Abstract. This work studies the novel solar water heater using natural black rocks as the absorption material. The objective of this study is to compare the thermal performance of solar water heater between using black rocks and black copper pipes. Both options were tested at the same parameters such as 0.8 m^2 collector area, angle tilted 19° and 30 litters of water tank volume. The experimental results showed that the maximum mean temperatures of solar water heater of absorption type using natural black rock system and black copper pipes system were 54.3 ^{o}C and 50.7 ^{o}C, respectively. In addition, the efficiencies of the water heater of absorption type using natural black rock system and black copper pipes system were 44.6% and 33.5%, respectively. Therefore, the water heater of absorption type using natural black rock system is another option used in the household sector to reduce the energy consumption of the country.

1. Introduction

Nowadays, the fossil fuels are the major resources of energy to drive in many sections all around the world, especially in industrialized and developing countries. As the population continues to grow, the limited amount of fossil fuels begins to diminish. To solve this situation, one of the potential solutions is to enhance by using the renewable energy resources. Solar energy has been known as a potential renewable energy. According to use directly and indirectly from the sun, solar energy applications will increase significantly such as photovoltaics, solar drying process [1], solar absorption refrigeration [2], and solar desalination [3,9], as well as solar water heater. To get abundant, clean, and free energy from sun, solar water heater is the simple and economic way to turn solar radiation into thermal energy. In the research, the present of a solar water heater in dwelling is significantly higher in rural areas than the cities because of utility area and it can reduce the annual energy bill by ten percent for the average homeowner [4]. Because the desired hot water heater temperature is usually under 60 °C, there are many advantages of the hot water heater system such as easy to install, having a low payback period, and reducing the electricity consumption, as well as emitting negligible greenhouse gases [5]. In past few decades, the flat plate collector normally had been used to rise the water temperature and the maximum was around 50°C [6,7]. In order to improve the thermal performance of solar water heater system, the tubular collectors had been presented to enhance solar energy storage, space area saving, and decrease the fabrication cost. The system increased the water temperature up to 62°C [8]. Additionally, the concentric tube was also conducted to reduce the heating space and appropriate for typical household water usage [5,9]. However, all these absorption materials still have been made by metal or inorganic product. To be environmentally friendly, the black colored sand collector was investigated in the passive solar water heater to be the simple and low-cost compact solar water heater [11]. In the reseach, the results showed that the collector absorbed as gray material and the temperature could be reached about 90 °C. However, the solar water heater is not only limited application for the human living standard, it also can be enhanced in the aquaculture field. To illustrate, the warm water fish has been affected by the temperature in winter months. The fish weight loss occurs due to the metabolism of fish and it is directly correlated to water temperature [12]. Additionally, if the solar water heater is applied in the winter season, the fish farming has been to increase profitability and sustainability. Moreover, the economic and environmental life cycle assessments of solar water heaters had been applied to aquaculture to show the optimal operating of solar water heater for aquaculture [13]. Consequently, it will be a good choice of solar water heater, if the collector includes having a large thermal storage capacity per unit volume, a suitable operating temperature range, long life, simple, inexpensive, and natural material, as well as hygiene. The current study, the natural black rocks as the solar collector has been proposed for the solar water heater. In the experiment, the thermal performance of the system has been investigated by comparing between the natural rocks and tubular copper collector. Furthermore, other parameters such as absorptivity, emissivity, and heat capacity of collector influenced the system were also analysed.

2. Experimental setup and procedure

The experiment was carried out to investigate the thermal performance of the solar water heater. It was conducted by testing both the tubular copper absorber (Type I) and natural rocks absorber (Type II) with a specific solar collector set. Type I was made of copper tube with 0.64 cm of outside diameter and 100 cm of length. The tube was sprayed solar selective coating and closely arranged one by one. Type II was made of natural black rocks with 0.7 cm of average diameter and the amounts of rock were 6496 pieces. Similarly, system was designed in a rectangle shape of dimension 76x100 cm and 18 cm thickness and there are 3 parts of the system which are the covers, absorber, and storage tank as shown the schematic diagram in figure 1. The cover components, the double transparent glasses of thickness 5 mm each with the 3 cm airy distance between them had been used for keeping maximum amount of incident radiation. Absorber was installed on keeper plate that was attached to the down glass. The steel box was used to house all the parts of the collector. Additionally, the collector had been submerged in the storage tank. The water was allowed to flow through collector's space between the absorber and bottom surface by natural convection mechanism. In order to maintain the heat gain, the outer storage tank had been well insulated with 3 cm thickness of glass wool. The setting systems were tilted 19° (local latitude). The experiment was operated from 8.00 am to 5.00 pm. According to the temperature data, the locations were set using thermocouple type K and connected to the data logger as shown in the figure 1. In addition, the extra thermocouples were installed to measure the temperature distribution in system and ambient temperature.

Nomenclature

Q_o = optical losses (W)
Q_l = heat loss (W)
Q_{st} = heat absorber storage (W)
Q_u = useful heat (W)
$Q_{collector}$ = heat collector storage (W)
t_a = ambient temperature (K)
$t_{c1,1}$ = upper cover1 temperature (K)
$t_{c1,2}$ = lower cover1 temperature (K)
$t_{2,1}$ = upper cover2 temperature (K)
$t_{2,2}$ = lower cover2 temperature (K)
t_{ab} = absorber temperature (K)
t_w = wall temperature (K)
$h_{conv,c1-a}$ = convection heat transfer coefficient between cover1 and ambient (W/(m^2 K))
U_T= total heat loss coefficient (W/(m^2 K))
Nu= dimensionless Nusselt number
Gr= dimensionless Grashof number
Pr= dimensionless Prandtl number
Ra=dimensionless Rayleigh number
v = kinematic viscosity (m^2/s)
a = thermal diffusivity (m^2/s)
g= gravitational acceleration
I_r = solar irradiation (W/m^2)
A= collector area (m^2)
c_p= water specific heat (kJ/(kg K))
s =Stefan–Boltzmann constant (W/(m^2 K^4))
e_I = emissivity of absorber type I
e_{II} = emissivity of absorber type II

IOP Conf. Series: Materials Science and Engineering **501** (2019) 012050 doi:10.1088/1757-899X/501/1/012050

$h_{conv,c1-c2}$ = convection heat transfer coefficient between cover1 and cover 2 (W/(m^2 K))
$h_{conv,c2-ab}$ = convection heat transfer coefficient between cover2 and absorber (W/(m^2 K))
$h_{conv,ab-w}$ = convection heat transfer coefficient between absorber and wall (W/(m^2 K))
$h_{r,c1-a}$ = radiation heat transfer coefficient between covers 1 and ambient (W/(m^2 K))
$h_{r,c1-c2}$ = radiation heat transfer coefficient between cover1 and cover 2 (W/(m^2 K))
$h_{r,c2-ab}$ = radiation heat transfer coefficient between cover2 and absorber (W/(m^2 K))
$h_{r,ab-w}$ = radiation heat transfer coefficient between absorber and absorber (W/(m^2 K))
h = efficiency
b = tilt angle of water heating system
b =volumetric coefficient
L = characteristic length (m)
K_e = the extinction coefficient
l_g = the glass thickness (m)

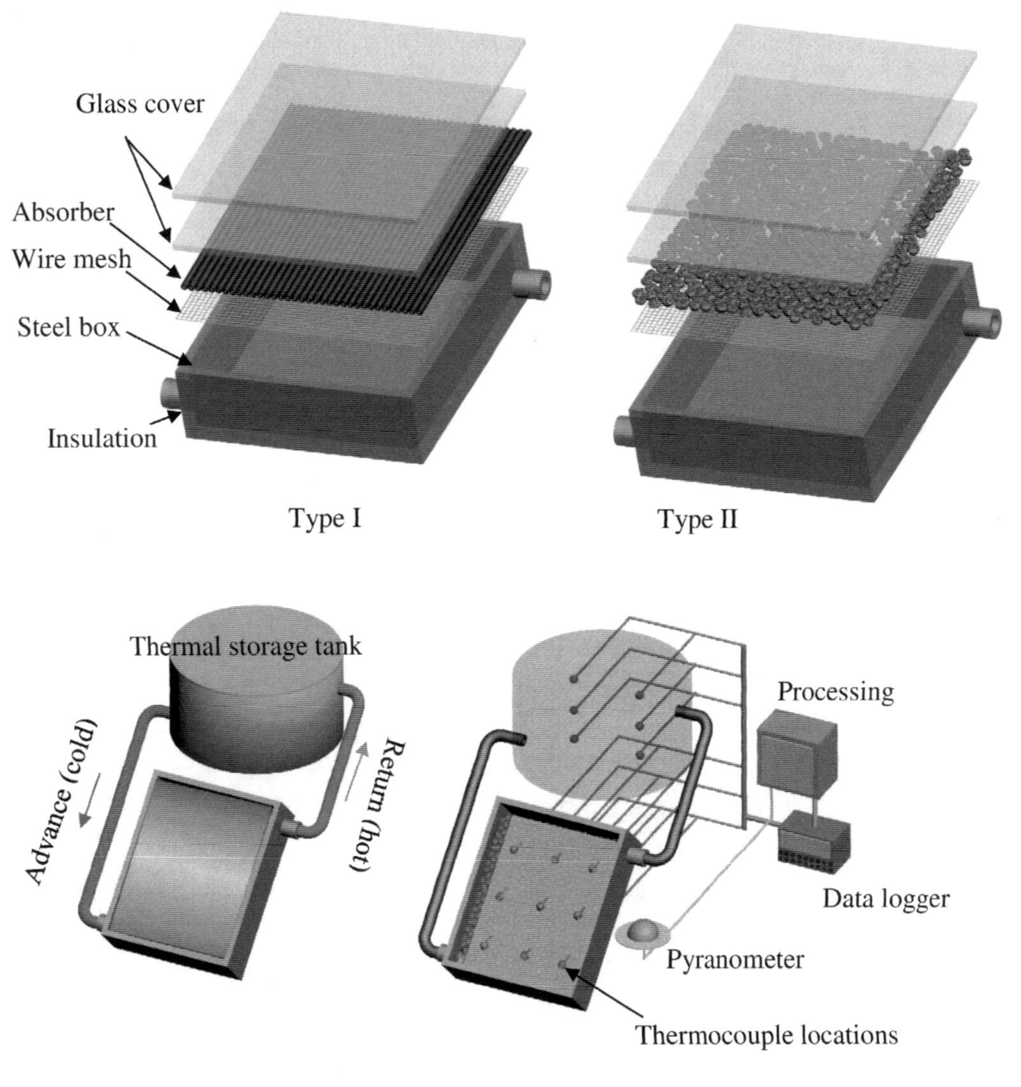

Figure 1 Schematic of solar water heater

3. Theoretical model

According to heat transfer phenomena of solar water heater, as shown on figure 2, when solar radiation incident passes through the transparent cover, the solar radiation has been converted in to heat at the absorbing surface which is transferred to a heat transfer fluid. Normally, the black colored collector is used to increase the solar radiation absorption. To achieve the maximum heat gain, the collector material has been investigated to increase the thermal performance of solar water heater such as black rock and copper tube collector.

The thermal network of the system has been shown in the figure 2. To illustrate, it begins with the solar radiation in surrounding at ambient temperature (t_a) on the upper transparent glass cover at cover$_1$ temperature ($t_{c1,1}$) and the conduction heat transfer at the upper cover increases temperature ($t_{c1,2}$) at the lower surface. When the solar radiation goes through the air gab of double glasses, it raises the cover$_2$ temperature ($t_{c2,1}$). The same conduction as the upper glass, there is the lower surface temperature ($t_{c2,2}$). At the absorber, the collector absorbs the solar radiation at absorber temperature (t_{ab}) and converts into heat transferring to the water. Last step, the heat gain of water losses through the wall surface at wall temperature (t_w). Consequently, the thermal energy from the top to the bottom of the system can be expressed in equivalent thermal network for solar collector as in figure 2.

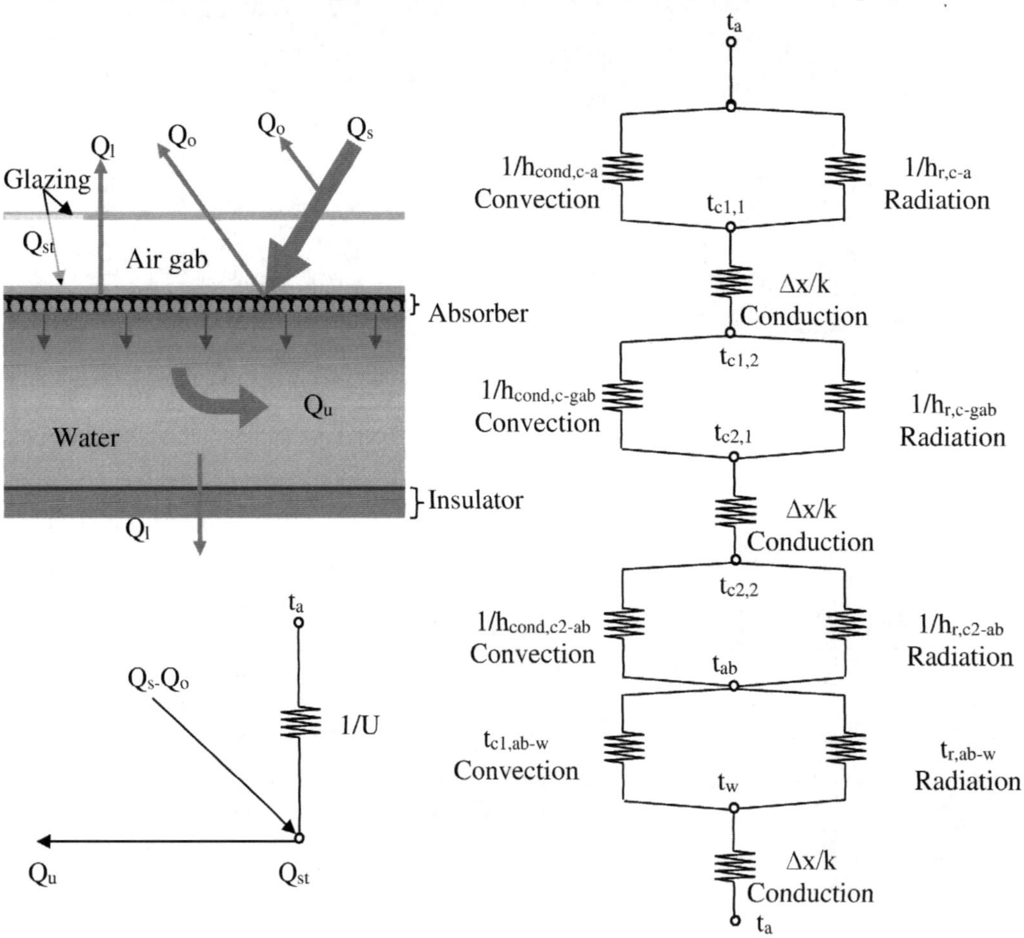

Figure 2 Heat transfer scheme and thermal network of the system

IOP Conf. Series: Materials Science and Engineering **501** (2019) 012050 doi:10.1088/1757-899X/501/1/012050

Energy balance of absorber

When the absorber receives the solar radiation, the total energy (Q_S) of the system can be presented in the energy reflection (Q_o) from the cover, energy loss (Q_l) from the system, energy storage (Q_{st}) in the absorber, and energy useful (Q_u) of warm water. The energy balance of the system can be expressed as shown.

$$Q_s = Q_o + Q_l + Q_{st} + Q_u \tag{1}$$

The collector transmittance (τ) equation is defined as the following [14]:

$$\tau = \frac{I_{transmitted}}{I_{incident}} = \exp\left(-\frac{K_e l_g}{\cos\theta_2}\right) \tag{2}$$

where θ_1 is the incidence index and θ_2 is the refraction angle after transmiting from the glass cover 1. The linked together them as follows:

$$n_1 \sin\theta_1 = n_2 \sin\theta_2 \tag{3}$$

where n is the index of refraction.

$$Q_u = Q_s - Q_s\tau \tag{4}$$

In order to obtain the efficiency of system (η), one can use the following equation as

$$\eta = \frac{\sum Q_{collector}}{A\sum I_r} \tag{5}$$

where I_r is the daily incoming radiation recorded by pyranometer (the both beam and diffuse intensity).

Furthermore, $\sum Q_{collector}$ is the summation of daily absorbed energy by the water into the storage tank and heat gain of absorber from solar radiation is calculated as the following.

$$\sum Q_{collector} = \sum Q_u + \sum Q_{st} \tag{6}$$

The heat gain ($Q_{collector}$) also can be calculated from energy useful (Q_u) of warm water and energy storage (Q_{st}) using $\sum Q = mc_p\Delta t$ where Δt is the difference of the inside collector at mean temperatures, from the beginning up to the end of the test. In addition, mean temperature is determined at 15 points by sum of terms that calculated from ratio of temperature and volume at a single point.

Balance equations of constant-flow process are used. The heat loss transfer rate of the collector can be expressed as:

$$q_T = U_T\left(t_{ab} - t_a\right) \tag{7}$$

The rate of relative heat transfer between i toward j is given as:

$$q_{i\rightarrow j} = h_{c,ij}(t_j - t_i) \tag{8}$$

The radiative heat transfer coefficient between i toward j is given as:

$$h_{r,i\rightarrow j} = \frac{\sigma\left(t_j^2 + t_i^2\right)\left(t_j + t_i\right)}{\frac{1}{\varepsilon_i} + \frac{1}{\varepsilon_j} - 1} \tag{9}$$

The conductive heat transfer coefficient between i toward j is given as:

$$h_{cond,i\rightarrow j} = \frac{\Delta x_{ij}}{k} \tag{10}$$

The convective heat transfer coefficient (air gab) between i toward j is given as:

$$h_{conv,ij} = \frac{Nuk}{L_{ij}} \tag{11}$$

Furthermore, when Nusselt number is a function of Grashof and Prandtl number, the heat transfer coefficient in natural convection shows as the following.

$$Nu = f\left(Gr, \Pr\right) \tag{12}$$

Where the Grashof number (Gr) is the dimensionless as the following equation.

$$Gr = \frac{\beta g L^3 \Delta t}{v^2} \tag{13}$$

$$\Pr = \frac{v}{\alpha} \tag{14}$$

The Rayleigh number is a function of Grashof number and Prandtl number as shown.

$$Ra = Gr.\Pr \tag{15}$$

In convection model, heat in the system transfers by natural convection. Therefore, Nu is the dimensionless, Nusselt number, $k(W/m \cdot K)$ is the thermal conductivity of air between double glasses and L is the distance between two covers. Nusselt number obtains as [14]:

$$Nu = 1 + 1.44\left[1 - \frac{1708(\sin 1.8b)^{1.6}}{Ra\cos b}\right]^{+}\left[1 - \frac{1708}{Ra\cos b}\right]^{+} + \left[\left(\frac{Ra\cos b}{5830}\right)^{1/3} - 1\right]^{+} \quad (16)$$

According to the thermal network of the system and the equations from (1) to (16) , the simulation model has been established for analyze of heat transfer at individual part of collector and evaluate the performance of absorbers based on the system efficiency by input the experimental data.

4. Results and discussion

In this section, the results of experimental and numerical data have been presented for solar water heater of type I and type II. To illustrate, the thermal performance of both systems has been compared indicating in water temperature, system efficiency and utility heat in the system as the following.

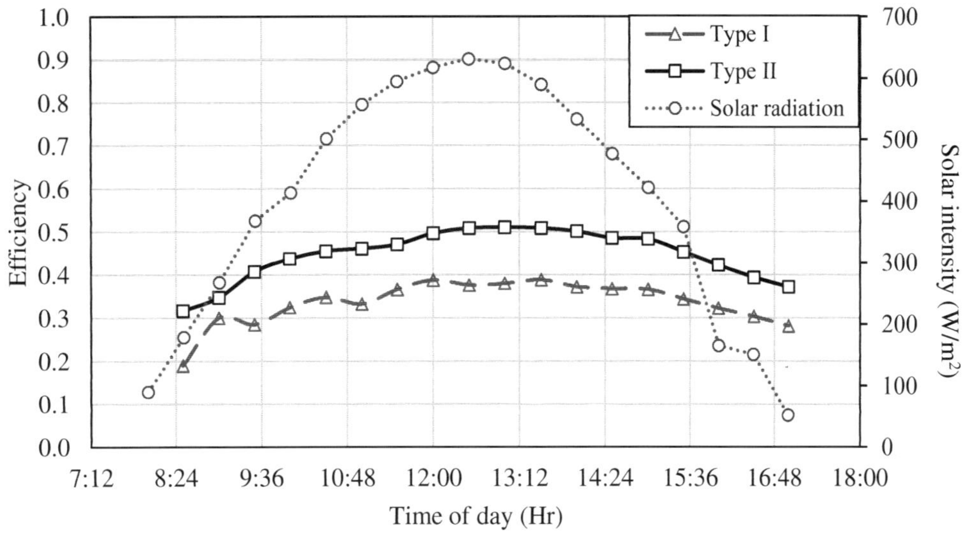

Figure 3 The hourly efficiency for natural rock and copper pipe absorber

The solar radiation is one of parameters that indicates the performance of the system. As shown in Figure 3, the system efficiency increases in the morning and reaches the maximum at the highest daily radiation. After that, when the solar radiation continuously decreases, the efficiency also slow down decreases depended on the solar radiation. In addition, as we expected, the experimental results show that system efficiency of type II absorber exceeds type I absorber, mean system efficiency of type II absorber is higher than mean system efficiency of type I about 10 % because of the significantly higher heat capacity of the natural rock indicating the higher thermal performance of the system.

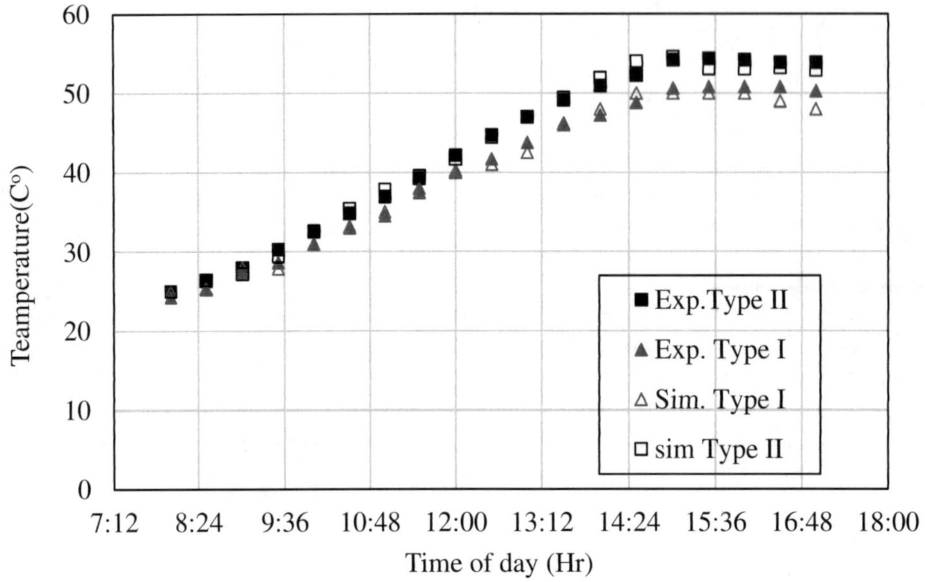

Figure 4 Temperature variation of mean water temperature along the local time

To support this reason, we plotted variation of temperature along the local time. As shown in Figure 4, the experimental results show that the mean water temperature is continuously increased in the morning and it significantly increases in the afternoon because of influenced solar radiation and ambient temperature for both absorbers. The water temperature of natural rock collector is higher than copper tube collector because the heat capacity rate of rock is much higher than copper and the heat capacity rate can be effective in the afternoon, as we can see the water temperature difference. Therefore, the more absorber has heat capacity rate the higher system gets heat gain. Moreover, comparison between mean temperatures of water obtained from the simulation and experimental data, experimental data is compared with the simulation results. There is well agreement; the standard deviation of error temperatures was 0.74. Especially, at the maximum temperature value, the simulation accurately provides performance prediction. Consequently, the simulation can be employed as implementation for further consideration of system efficiency in this study.

Figure 5 shows the simulation results, the efficiency system based on absorber temperature. As you have seen, efficiency difference between Type I and type II is about 2%. It is effect of absorber emissivity on thermal loss ($e_I = 0.3, e_{II} = 0.85$) resulting the available heat in type I more than type II. However, the higher we use absorber emissivity, the more absorption ability is obtained. In this study, the efficiency of type II exceeds type I by average 11.14% and it is much more than emissivity effect to thermal loss. Thus, the heat useful for the type II is more than type I. Furthermore, the total resistance of the type II is much lower than type I.

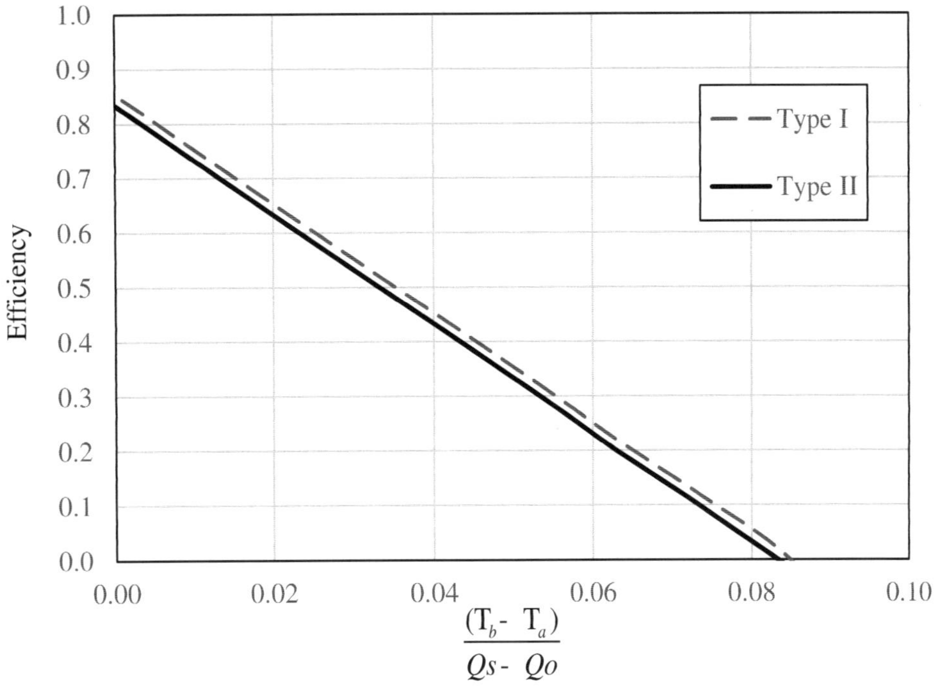

Figure 5 Efficiency curve of type I and type II absorber based on absorber temperature

5. Conclusions

The novel solar water heater using natural black rock collector is proposed in this paper. The idea of the proposed system is based on utilizing the high heat capacity of natural rock to implement the solar water heater. Such natural material leads to long life, simple, inexpensive, as well as hygiene. The thermal performance of new proposed model was experimentally investigated and analyzed experimentally by comparing with the conventional solar water heater, copper tube collector. The obtained results illustrated that the thermal performance of natural black rock collector exceed the copper tube collector by average 11.1%. Additionally, the system increased the water temperature up to 54°C, with a performance instantaneous efficiency that reached to average 44.6% which was very satisfactory. Ultimately, the proposed solar water heater system introduces an efficient practical solution utilizing solar energy for the human living standard as well as the aquaculture to reduce the energy consumption of the country.

6. References

[1] R. Ouaaboua , B. Nabila , N. Hidara , L. Lahninea , A. Idlimamc , A. Lamharrarc , H. Hanineb , M. Mahrouz, 2018, Valorization of solar drying process in the production of dried Moroccan sweet cherries, Solar Energy, Inpress, Corrected proof.

[2] M.S. Fernandes, G.J.V.N. Brites, J.J. Costa, A.R. Gaspar, V.A.F. Costa, 2014, Review and future trends of solar adsorption refrigeration systems, Renewable and Sustainable Energy Reviews, Vol.39, pp.102–123.

[3] A. Pugsley, A. Zacharopoulos, J.D. Mondol, M. Smyth, Solar Desalination PotentialAroundthe World, 2018, Renewable Energy Powered Desalination Handbook, Application and Thermodynamics, Pages 47-90.

[4] E. Aydina, P. Eichholtzb, E. Yönder, 2018, The Economics of Residential Solar Water Heaters in Emerging Economies: The Case of Turkey, Energy Economics, Vol. 75, pp. 285–299.
[5] J. Shi, K. W. Lina , Z.Q. Chena, H. Shi, 2018, Annual dynamic thermal performance of solar water heaters: A case study in China's Jiangsu Province, Energy & Buildings, Vol. 173, pp. 399–408.
[6] D. R. S. Tambunan, Y. P. Sibagariang, H. Ambarita, F. H. Napitupulu, H. Kawai, 2018, Numerical study on the effects of absorptivity on performance of flat plate solar collector of a water heater, Journal of Physics: Conference Series 978, 012099.
[7] W. M. Hashima , A. T. Shomranb , H. A. Jurmutb , T. S. Gaazb, A. A. H. Kadhumc, A. A. Al-Amieryd, 2018, Case study on solar water heating for flat plate collector, Case Studies in Thermal Engineering Vol. 12, pp. 666–671.
[8] M. M. Marmoush, H. Rezk, N. Shehata, J. Henry, M. R. Gomaa, 2018, A novel merging Tubular Daylight Device with Solar Water Heater e Experimental study, Renewable Energy, Vol. 125, pp. 947-961.
[9] R. Sathyamurthya, S.A. El-Agouzb, P.K. Nagarajana, J. Subramania, T. Arunkumarc, D. Mageshbabua, B. Madhua, R. Bharathwaaja, N. Prakashe, 2017, A Review of integrating solar collectors to solar still, Renewable and Sustainable Energy Reviews, Vol. 77, pp. 1069–1097.
[10] S. J. Mamouri, A. Bénard, 2018, New design approach and implementation of solar water heaters: A case study in Michigan, Solar Energy, Vol. 162, pp. 165–177.
[11] Y. Taheri, B. M. Ziapour, K. Alimardani, 2013, Study of an efficient compact solar water heater, Energy Conversion and Management, Vol. 70, pp.187–193.
[12] L. A. Roya, S. D. Rawlesb, A. M. Kellya, C. D. Websterb, N. Stonea , A. Haukenesa, 2017, Effect of winter feeding frequency on growth, survival, and fatty acid metabolism of juvenile bluegill (Lepomis macrochirus) and hybrid bluegill (L. cyanellus × L. macrochirus), Aquaculture, Vol. 479, pp. 780–789.
[13] Y. Kim, Q. Zhang, 2018, Economic and environmental life cycle assessments of solar water heaters applied to aquaculture in the US, Aquaculture, Vol. 495, pp. 44-54.
[14] Duffie J, Beckman W.,2006, Solar engineering of thermal processes. 3rd ed. New York: John Wiley & Sons.

IOP Conf. Series: Materials Science and Engineering **501** (2019) 012051 doi:10.1088/1757-899X/501/1/012051

Effects of Structure and Hydrophobic Treatment on Water Transport Behaviour in PEM Fuel Cell Gas Diffusion Layers

F Jinuntuya* and W Kamsanam

Department of Mechanical Engineering, School of Engineering, University of Phayao, Phayao, 56000, Thailand

* Corresponding Author: fontipj@gmail.com

Abstract. This paper examines the effects of structure and wettability on water transport behaviour in PEM fuel cell gas diffusion layers (GDLs). Different GDL materials, including carbon cloth and carbon paper with and without PTFE and MPL coatings, were used in this study. The breakthrough pressure of liquid water, water flow rate and water retention in the GDL were measured. For the GDL with and without MPL, the results indicate that the paper GDLs require higher breakthrough pressure than the cloth ones. Compared to the untreated GDL, applying PTFE on the GDL increases breakthrough pressure up to 1.5 times while it decreases water flow rate and water retention up to about 3.8 and 0.5 times respectively, over the tested conditions. The results also show that applying MPL on the GDL surface has greater impact on breakthrough pressure than applying PTFE. Liquid water requires up to about 4 times greater pressure to break through the other side of the GDL with the MPL. By applying MPL, the amount of water retention and flow through the GDL also reduced greatly, to almost zero. In addition, the results suggest that the GDLs are degraded after being exposed to water, which causes a significant decrease in breakthrough pressure and an increase in water retention.

1. Introduction

A gas diffusion layer (GDL) plays a crucial role in the overall performance of a proton exchange membrane (PEM) fuel cell by providing pathways for gaseous reactant transport from a supply channel to a catalyst layer, and product water removal from the catalyst layer to the gas channel. Excessive amounts of liquid water in the GDL hinder the reactant gases from accessing the active sites of the catalyst layer leading to decreased performance of the PEM fuel cell. Therefore, GDLs are commonly treated with a hydrophobic substance to render their fibres more hydrophobic and, thus, facilitate gas transport and water removal. Bevers et al. [1] studied the effect of PTFE (polytetrafluoroethylene) contents on water saturation in carbon paper GDL, and the results indicate a decrease in water saturation levels with an increase in PTFE content. Several works [2–6] have demonstrated numerically the phenomenon of saturation reduction with increasing hydrophobicity. Benziger et al. [7, 8] provided insight into water permeation properties under the effect of GDL wettability. They concluded that PTFE treatment of the GDL reduces the amount of water that can be trapped in the GDL and, thus, facilitates good gas transport through the GDL. In addition, a fine microporous layer (MPL) is usually applied on one side of the GDL to improve cell performance and fuel cell water management. This is an additional layer between the GDL and catalyst layer, typically consisting of carbon powder and PTFE particles. Lu et al. [9] reported that water saturation in the GDL sample with MPL is significantly lower than that of GDL without MPL. They found that MPL also restricts the number of locations where water can enter the GDL allowing a greater number of pathways for gas transport. Likewise, Gostick et al. [10] illustrated that water saturation in the GDL was considerably decreased when MPL was introduced.

Since the application of PTFE and MPL has shown great promise for water and gas transport in the GDL, the present study aims to investigate the effects of these treatments on liquid water transport, including breakthrough behaviour, water flow rate and water retention in the GDL. Accordingly, different GDL structures, including paper and cloth GDL with and without MPL, were used in this study. Wettability of the GDL was also altered so that the effect on water transport could be examined. By varying GDL structure and PTFE loading, their effects on liquid water behaviour in the GDL can be compared. This could improve the understanding of water transport behaviour in the GDLs, which is essential for the development of effective PEM fuel cell water management and the design of future GDL materials.

2. Experimental

2.1. *Materials*

The GDL samples used in this study were: a carbon paper, a carbon paper with MPL, a carbon cloth and a carbon cloth with MPL. They are commercially available and have similar thickness of about 350 μm. Carbon paper consists of several layers of carbon fibres, whilst carbon cloth is composed of woven bundles of carbon fibres. The individual carbon fibres are about 7-10 μm in diameter and the bundles are in the region of around 350-450 μm. The micrographs of the GDL samples are shown in figure 1. Paper without MPL was also examined under different PTFE loadings, from 0 wt.% to 30 wt.%, in order to evaluate its effect on water behaviour in the GDL.

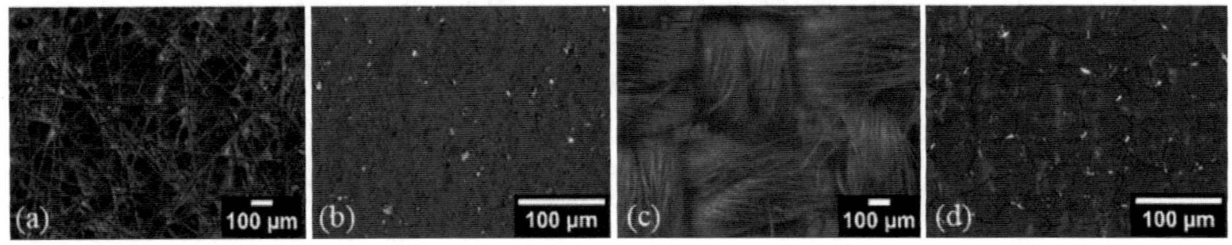

Figure 1. Micrographs of (a) carbon paper (b) carbon paper with MPL (MPL side) (c) carbon cloth (d) carbon cloth with MPL (MPL side).

2.2. *Experiments*

This work sets out to examine breakthrough pressure, water flow rate and water retention in the GDL. A circular GDL sample of 12 mm in diameter was mounted between two acrylic plates in the test apparatus. The apparatus, as illustrated in figure 2, consists of a 2 m high acrylic tube of 10 mm in diameter standing on top of the two acrylic plates where the tested sample is placed. The acrylic tube allows a water column to build up creating hydrostatic pressure for the experiments. A peristaltic pump is used to add water into the acrylic tube. Each tested sample is weighed both before and after the experiment to measure the amount of water trapped in the GDL.

For the breakthrough experiment, water was slowly added to the acrylic tube on top of the tested GDL sample. As water commenced to flow through the GDL, the height of the water column corresponding to hydrostatic pressure was recorded. The influence of wettability on breakthrough pressure was examined by altering the PTFE loadings on the GDL surface from 0 wt.% to 30 wt.%. The required minimum pressures to break through the GDLs were then compared. Each GDL sample was weighed after the first set of samples, i.e. virgin GDLs, was tested to measure the amount of water

IOP Conf. Series: Materials Science and Engineering **501** (2019) 012051 doi:10.1088/1757-899X/501/1/012051

retained in it. The samples were then dried at 70°C in a hot-air oven for 3 hours and the tests were repeated with the dried samples.

For the flow experiment, the outlet of the GDL sample was initially blocked to allow for higher hydrostatic pressure, and the peristaltic pump was used to maintain the water level during the experiment. As the GDL outlet was unblocked, water could pass through the GDL. The amount of water drainage was then measured with time.

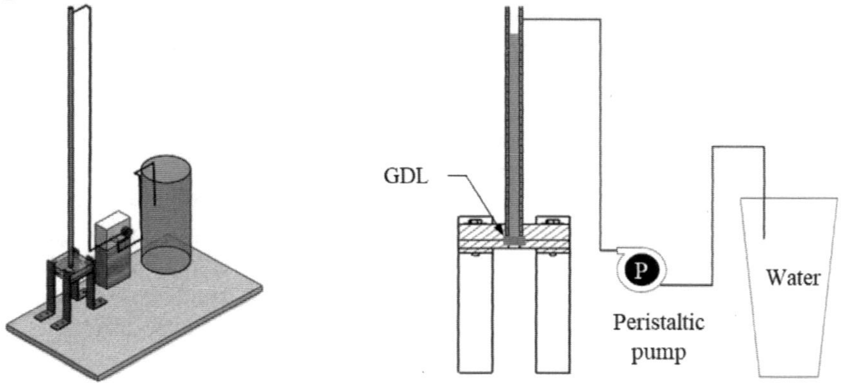

Figure 2. Schematic of the apparatus.

3. Results and discussion

3.1. *Effects of PTFE on breakthrough pressure*

The plain paper and paper with PTFE loadings of 10 wt.%, 20 wt.% and 30 wt.% were used in this study to investigate the effects on breakthrough behaviour. Figure 3 shows breakthrough pressure at varying PTFE loadings for virgin GDL, represented by a solid line, and dried GDL, represented by a dashed line. The figure indicates that PTFE loading has a significant effect on the minimum pressure required to push liquid water through the GDL samples. By applying PTFE, the required pressure has increased by about 0.5 and 1.5 times for the virgin and dried GDLs respectively. Since the wettability of the GDL was altered towards becoming more hydrophobic by applying PTFE, capillary resistance also increased according to the Young-Laplace equation. As a result, larger pressure needed to be applied in order to overcome the increased resistance.

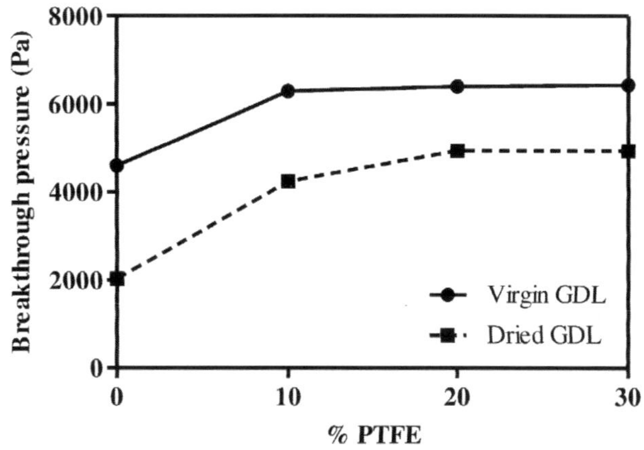

Figure 3. Effect of PTFE loading on the breakthrough pressure.

The results also show that the dried GDL, i.e. previously used GDL, requires significantly lower breakthrough pressure than that required by the virgin GDL. The reduction in breakthrough pressure observed in the dried GDL agrees with the results reported by Benziger et al. [7]. The possible reason for this is the loss of hydrophobicity of the GDL as it is exposed to water, found in the work of Ha et al. [11], which demonstrated a continuous decrease of hydrophobicity over time.

3.2. *Effects of PTFE on water flow rate*

In order to investigate the effect of PTFE loading on the flow of water through the GDL, the amount of water drainage was measured as a function of time. The outlet of the GDL sample was initially blocked to allow for higher hydrostatic pressure. Water was then freely drained through the sample by gravity and the amount of drained water was recorded. At the same time, the pressure head was maintained by adding water into the acrylic tube using the peristaltic pump. In this experiment, the GDL samples were tested over the applied hydrostatic pressures of 7.4 kPa, 9.8 kPa and 12.3 kPa.

Figure 4 shows the water flow rate according to the amount of PTFE. The dotted, dashed and solid lines represent the flow rate of drained water out of the GDL sample with varying PTFE loadings at a hydrostatic pressure of 7.4 kPa, 9.8 kPa and 12.3 kPa respectively. As expected, the largest hydrostatic pressure gives the highest flow rate, while the smallest hydrostatic pressure gives the lowest flow rate, suggesting that liquid water permeation depends on the applied hydrostatic pressure. It was observed that all hydrostatic pressures show similar trends, where the flow rate significantly decreases as the amount of PTFE increases. At the hydrostatic pressure of 12.3 kPa, the flow rate of water decreased to about a third, from 3,390 mg s^{-1} to 1,040 mg s^{-1} over the range of tested PTFE loadings, while the flow rate decreased to about a fifth, from 1,760 mg s^{-1} to 370 mg s^{-1}, and to about a quarter from 740 mg s^{-1} to 190 mg s^{-1} over the same range of PTFE loadings for the 9.8 kPa and 7.4 kPa cases respectively.

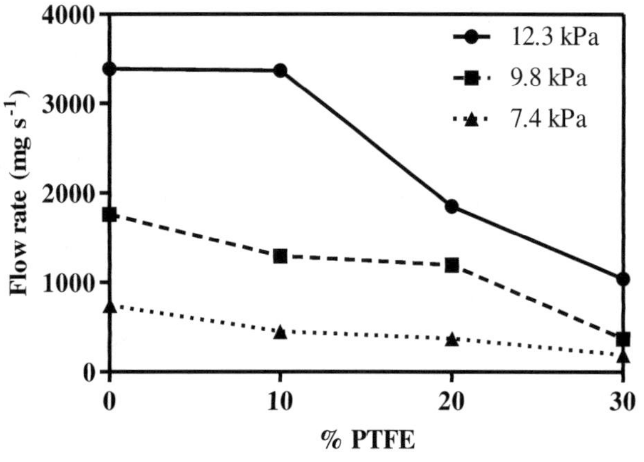

Figure 4. Effect of PTFE loading on water flow rate through the GDL.

Interestingly, for the case of 12.3 kPa, applying 10 wt.% PTFE on the GDL had only a minor effect on the rate of water flowing through the GDL compared to the plain GDL. This suggests that at a large hydrostatic pressure, a small amount of PTFE might not be able to reduce the amount of water travelling through the GDL, since the applied pressure is great enough to overcome the large capillary resistance created by small pores and is, thus, able to push liquid water through the GDL.

3.3. *Effects of PTFE on water retention*

In order to investigate the effect of PTFE loadings on water retention in the GDL, the plain paper and paper coated with 10 wt.%, 20 wt.% and 30 wt.% PTFE were used. After the virgin GDL samples were tested, all samples were dried at 70°C in a hot-air oven for 3 hours and were repeatedly tested at the same condition.

Figure 5 shows the amount of water retention in the paper GDL under different PTFE loadings. It can be observed that applying PTFE considerably reduces the amount of water retention in the GDL, which amounts to over 50% reduction in water retention compared to the untreated GDL.

This study also compares the water retention capability of the virgin GDL, represented by the solid line, with the dried GDL, represented by the dashed line. The result shows that a larger amount of water is trapped in the dried GDL than in the virgin. For example, the dried GDL with 30 wt.% PTFE could retain 60.3 g m^{-2} of water, while the virgin GDL retained 22.5% less at 46.7 g m^{-2}, indicating that the GDL tends to hold more water over time. This is again due to the loss of hydrophobicity of the dried GDL, as mentioned in [11], which allows more water to be retained in it.

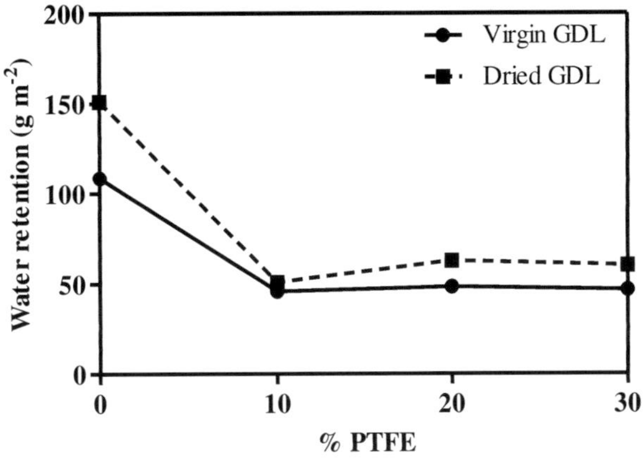

Figure 5. Effect of PTFE loading on water retention in the GDL.

3.4. *Effects of GDL structure on breakthrough pressure*

Figure 6 compares the minimum pressure required to force water through different GDL structures, including paper GDL and cloth GDL with and without MPL. The solid bar represents virgin GDL, whilst the striped bar represents dried GDL. For both GDLs with and without MPL, it is seen that paper GDL requires larger breakthrough pressure than the cloth GDL. For example, the virgin paper GDL requires 4 times larger pressure than the virgin cloth GDL. The reason is that the paper GDL has a smaller average pore size than the cloth GDL, which creates large capillary resistance, thus leading to larger pressure being required to push water through the pores of the paper GDL. This corresponds to the results in [7], which reported a significant reduction in the required breakthrough pressure for the cloth GDL. For the paper and cloth GDLs with MPL, the diminutive pore size of the MPL, as shown in figure 1 (b) and (d), creates even more resistance to water flow. As a result, considerably larger pressure is required to force water to break through the GDL with MPL, which is almost 2 and 4 times higher for the paper and cloth GDLs respectively.

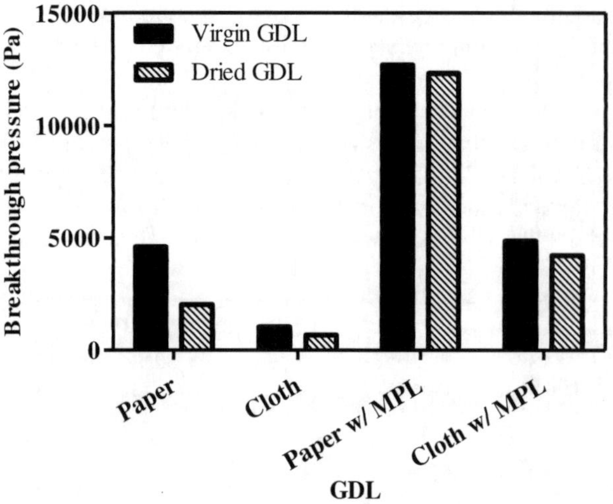

Figure 6. Effect of GDL structure on the breakthrough pressure.

As observed, the required hydrostatic pressure for the dried GDL is less than that of the virgin GDL. This implies that the required pressure would change over time, possibly due to GDL degradation. A similar phenomenon is reported by Tamayol and Bahrami [12], who observed a significant reduction in the required breakthrough pressure for the dried GDL.

3.5. *Effects of GDL structure on water retention*

The paper and cloth GDLs with and without MPL were used to examine the effect of structure on water retention. As seen in figure 7, both plain paper and cloth show a similar ability to keep liquid water within their structure. In contrast, the paper and cloth with MPL exhibit a different trend where almost no water is kept within their structure. The reason is that the MPL has very small pore sizes creating large capillary resistance, which might limit the amount of water entering the sample, thus causing very little water being retained in it. Conversely, the GDL without MPL has relatively large pore sizes creating smaller capillary resistance, which allows more water to flow through its structure, possibly increasing water retention.

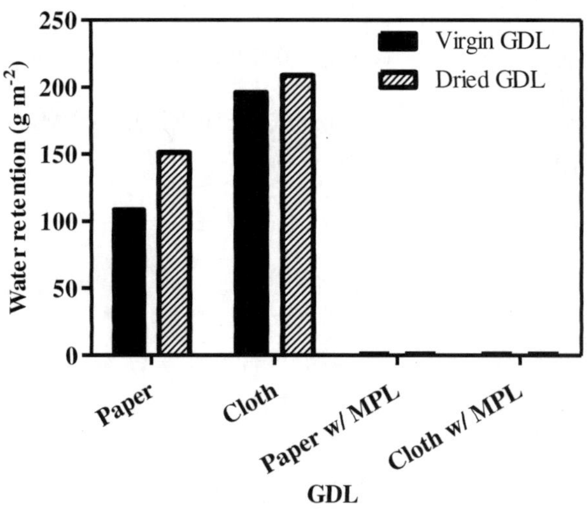

Figure 7. Effect of GDL structure on water retention in the GDL.

Considering the paper and cloth without MPL, it is seen that the cloth GDL can retain more water than the paper GDL, which is about 1 time and 0.25 times for the virgin and dried GDLs respectively. The possible reason is that the woven structure of the cloth GDL allows water accumulation within its bundle of carbon fibres, while the randomly dispersed carbon fibres of the paper GDL can retain significantly less water. For the paper and cloth GDLs with and without MPL, the dried GDL can retain a larger amount of water than the virgin GDL of the same structure, suggesting that there is a structural or surface wettability change after drying the GDL, which causes a change in the water retention properties of the GDL. This, again, corresponds to the observations by Ha et al. [11], who illustrated the loss of hydrophobicity of the GDL over the duration of the test.

4. Conclusion

In this work, the effects of GDL structure and PTFE treatment on breakthrough pressure, rate of flow and water retention were examined. The conclusions that can be drawn from the results are as follows.

- Applying PTFE on the GDL surface increases the minimum pressure required to break through the GDL. This suggests that altering GDL surface wettability to more hydrophobic creates greater capillary resistance for liquid water to pass through the pores of the GDL. As a result, the rate of water draining out of the GDL and the amount of water trapped in the GDL also decreases significantly compared to the untreated GDL.
- Applying MPL on the GDL has greater impact on breakthrough pressure than applying PTFE. Liquid water requires considerably higher pressure to travel through the GDL with the MPL due to the diminutive size of MPL pores creating greater capillary resistance, which implies that water can only travel in vapour form through the GDL with applied MPL in PEM fuel cell working conditions. In applying the MPL, the amount of water retention and flow through the GDL also reduce greatly to almost zero.
- The degradation of the GDL due to the loss of hydrophobicity can cause a significant decrease in breakthrough pressure, while at the same time cause an increase in water retention over time.

References

[1] Bevers D, Rogers R and von Bradke M 1996 *J. Power Sources* **63** 193-201
[2] Chapuis O, Prat M, Quintard M, Chane-Kane E, Guillot O and Mayer N 2008 *J. Power Sources* **178** 258-68
[3] Hao L and Cheng P 2010 *J. Power Sources* **195** 3870-81
[4] Rama P, Liu Y, Chen R, Ostadi H, Jiang K, Gao Y, Zhang X X, Fisher R and Jeschke M 2012 *AIChE J.* **58** 646-55
[5] Gao Y, Zhang X X, Rama P, Chen R, Ostadi H and Jiang K 2013 *Comput. Math Appl.* **65** 891-900
[6] Jinuntuya F, Whiteley M, Chen R and Fly A 2018 *J. Power Sources* **378** 53-65
[7] Benziger J, Nehlsen J, Blackwell D, Brennan T and Itescu J 2005 *J. Membrane Sci.* **261** 98-106
[8] Gauthier E, Duan Q, Hellstern T and Benziger J 2012 *Fuel Cells* **12** 835-47
[9] Lu Z J, Daino M M, Rath C and Kandlikar S G 2010 *Int. J. Hydrogen Energ.* **35** 4222-33
[10] Gostick J T, Ioannidis M A, Fowler M W and Pritzker M D 2009 *Electrochem Commun.* **11** 576-9
[11] Ha T, Cho J, Park J, Min K, Kim H S, Lee E and Jyoung J Y 2011 *Int. J. Hydrogen Energ.* **36** 12427-35
[12] Tamayol A and Bahrami M 2011 *J. Power Sources* **196** 6356-61

IOP Conf. Series: Materials Science and Engineering **501** (2019) 012052 doi:10.1088/1757-899X/501/1/012052

Interaction Analysis Investigations of Multiple Supply Duct Split-type Air-conditioning units in an Open Space

W Widsonkun[1], D Woradechjumroen[2*], and S Vongbunyong[1]

[1]Institute of FIeld robotics (FIBO), King Mongkut's University of Technology Thonburi, Bangkok, Thailand

[2] Department of Mechanical System Engineering and Industrial Innovation, School of Engineering, Sripatum University, Bangkok, Thailand

* Corresponding Author: denchai.wora@gmail.com, denchai.wo@spu.ac.th
(+66)948503366

Abstract. Oversizing has been seldom concerning in Thailand. However, oversizing causes short-term operations of an air-conditioning (AC) unit and excessive power consumptions. In addition, multiple operations of oversized AC units can be faulty operations when they are used to serve in an open space area because interaction among AC units are not considered before installations. This paper collects data operations of the multiple supply duct split-type air-conditioning (SAC) units used in an open space from additional sensors installed in zone and return temperature. Significant wireless sensors are added for measuring temperature effect among a interaction area between two SAC units. To measure the system variation caused by system interaction, runtime fraction (RTF) is used to quantify the compressor runtime ratio which equals to a stage-on time operation over the cycle time at an AC design condition. The results show that one of the four SAC units performs under-sizing instead of oversizing performance because it runs continuously without off cycling. To diagnose this issue, interactions between zone and return temperature are quantified in terms of correlation coefficient, which demonstrates that this SAC results in higher oversizing of the rest SAC units. This interaction analysis can be further used to design a soft-repair algorithm to reduce oversizing effect of the multiple SAC operations.

Keywords: Oversizing; Interaction; Split-type air-conditioning; Operations of multiple air-conditioning units.

1. Introduction

Nowadays, developing counties are promoting and implementing energy efficiency policies and codes to reduce unnecessary energy consumption and increase energy efficiency in both buildings and factories. Thailand is one of the developing counties in ASEAN formally issued the Energy Conservation Promotion Act B.E. 2535 (1992) which was revised in Edition B.E. 2550 (2007) [1].

The promotion states that all designated buildings and factories are required to follow energy management procedures and to comply with the energy conservation standard, criteria and methodologies that are prescribed in the ministerial regulation based on ethical standards and engineering approaches. In the method guidelines, efficient machine installations are significantly emphasized to enhance the understandings of practical engineers and technicians in terms of energy conservation measures. Most of significant machines in the designated buildings and factories are heating, ventilation and air-conditioning (HVAC) systems. Especially, split-type air-conditioning (AC) units or packaged AC units have been intensively used to provide thermal comfort in conditioned space. Although oversizing issue is intensively concerned in developed countries because the effect leads to short cycling, humidity problem and degraded life cycle before time, an oversizing understanding is seldom considered in Thailand [2].

To firstly reduce the oversizing issue, a proper selection of packaged air-conditioning units (PACs) is one of the important processes to properly control zone conditions in small commercial offices and retail stores. In previous interviews [3] and surveys [4], a number of designers estimate an rooftop units (RTU) sizing using oversizing internal load or related safety factors by software tools for commercial buildings [5]; it could incur oversized PACs resulting in waste of energy. Since the compressor of a PAC is manipulated by an on–off controller with non-synchronized individual control function such as a variable speed drive of a supply fan, the oversizing may resulting in the short-time period of a PAC cycle operation at a peak design condition, not running continuously in a hour. This frequent cycling over manufacturer's data causes over usage of the induction in the compressor motor, which results in loss cycling in terms of longer transient response before reaching a state-state value. This problem also results in higher operation cost due to the lower efficiency from the improper operation of the PAC machine [5].

In terms of energy penalty from field tests, at least 15% of the actual calculated capacity is over designed to ensure adequate cool and heat in the hottest and coldest periods. Based on real-time trended data in many field tests, it was found that the RTUs have the average oversized capacity of 84% for cooling. The lowest peak energy penalty for 12 stores with 268 RTUs was 34.66 kW in a cooling mode, while the highest peak energy penalty went to 226.41 kW in a cooling mode [5]. These RTU field tests were operated without supply ducts.

Moreover, multiple supply duct split-type air-conditioning units (SACs) have been seldom studied for field tests, especially in an open space. From the interaction engineering point of view between SACs, there are two significant points: (1) each SAC is typically controlled individually; it could lead to cooling fighting due to a non-coordinated control function between SAC units [6], and (2) heat transfer situation happens between an active control zone and the adjacent off-control SAC zones. Thus, a wrong return sensor position will cause inefficient control performance. The study of the interactions between SAC units is very useful for designing coordination control or improved performance of SAC installations without machine replacement.

2. Backgrounds

In the method analysis of this paper, significant backgrounds associated with oversizing are briefly explained for using in interaction analysis and essential effect for sensor installations.

2.1. *Typical SAC operations*

Multiple SAC units provide constant air volume to an open conditioned space. However, they are individually controlled without functions coordination, which perform inefficiency in practice:

1) Ventilation control practically associates with mixing damper and full speed of fan operations for maintaining comfort purpose which is only required the range of 0.12-0.20 cfm/ft^2. In reality, 1.00 cfm/ft^2 is approximately used (12-20% from fresh air plus 80-88% from circulating air).

2) Cooling staging control is manipulated by on-off cycling control of a thermostat sensing zone air temperatures in a conditioned space. During most times, SAC units are operated at inefficient partial load conditions.

3) Humidity control is an indirect action based on absorbing latent load on a cooling coil. During partial load conditions, humid outside air enters the zone without fully processing because compressor cycling periods are too short; it will lead to inefficiency of latent load absorption on a coil surface.

3. Sensor Installations for interaction analysis

In Figure 1, it is a layout of an example building that interaction analysis of four SAC units in an open space is investigated. Each SAC was sized 123,000 BTU/hour to provide cooling via a duct supply system. No. 1 to No. 4 are the order of SAC position with the return temperature sensors on the ceiling to provide the zones. By simply workthrough audit, the authors noticed the inefficient performance of the three duct supply patterns in Area 1, 2 and 3, respectively. The supply duct lines of these three SAC units were installed without return duct lines, so that the indoor air is impossible to circulate correctly to the return temperature sensor. As a result, the temperature sensor will sense wrong return temperature, which may be affected from temperature of the adjacent SAC. Beyond other SAC units, SAC No.1 has a straight supply duct line, and then it is possible that the supply air will circulate back to the return sensor if cooling load is suitable enough to incur practical air movement; however, SAC No.1 is used in the open space, so that heat transfer will occur between the zone and adjacent zones. These issues degraded each SAC performance due to interaction effect.

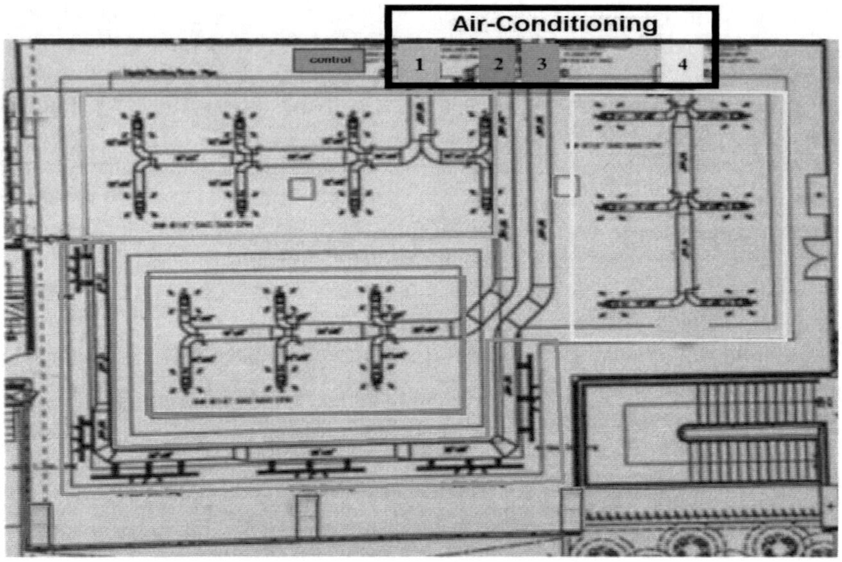

Figure 1. Layout of an example building

To investigate the interaction issue used for system improvement, sensor positions are designed to install for sensing temperature profile and system performance of each SAC unit as depicted in Figure 2. R1 to R4 stand for return temperature and humidity sensors positions of SAC No. 1 to No. 4, respectively.

Meanwhile, Z1, Z2 and Z4 are the zone temperature and humidity sensors of SAC No. 1, No. 2 and No.4. The Z3, Z4 and Z5 positions are all used to measure the zone temperature and humidity values of SAC No. 3 because this SAC unit is utilized to provide cooling load near the mirror and has too long supply duct line; many sensors are installed more than other SAC units to efficiently analyze any effect in the open space. For the rest sensor positions, Z7 and Z8 are used to measure the zone interaction between SAC No. 3 and No. 4, and zone interaction between SAC No. 2 and No. 3.

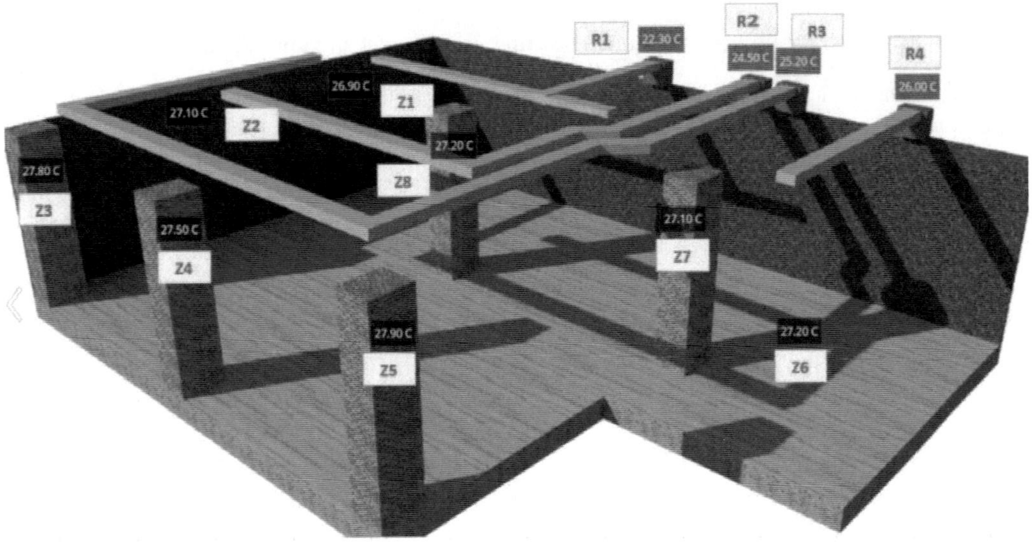

Figure 2. Sensor position installations

4. Interaction Analysis Method

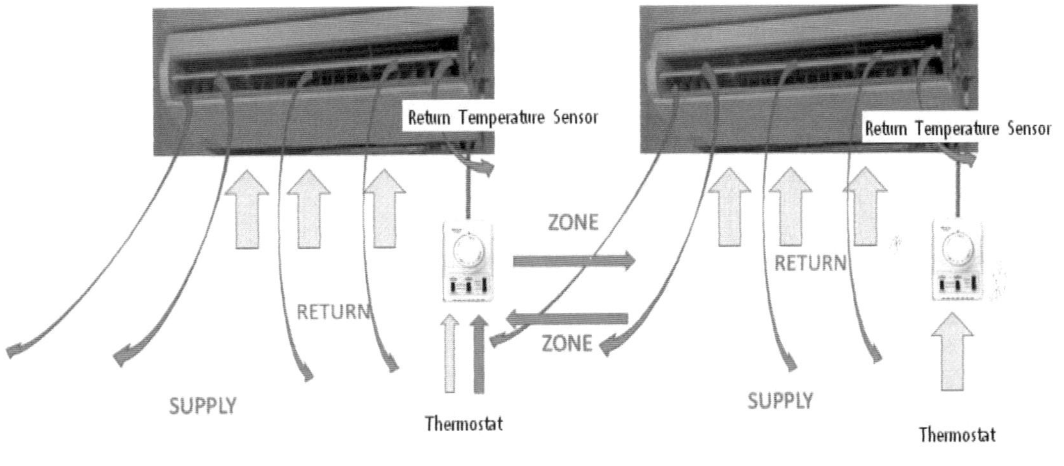

Figure 3. Interaction between two air-conditioning units

To perform the interaction analysis, Pearson correlation (r value) is used to compute significant relation between two independent variables in Equation (1).

$$r=\frac{\sum_{i=1}^{n}(X_i-\bar{X})(Y_i-\bar{Y})}{\sqrt[n-1]{\sum_{i=1}^{n}(X_i-\bar{X})^2(Y_i-\bar{Y})^2}} \tag{1}$$

Where: X_i = zone or return temperature of a zone; $\bar{X}$ = the mean value of X_i ; Y_i = zone or return temperature of the adjacent zone; $\bar{Y}$ = mean value of Y_i; n = number of data points.

Note: low level (0-0.50), medium level (0.51-0.89) and high level (0.9-1.0) [6]

In Figure 3, there are interactions between two AC units. For example, two return sensors are located closely. Thus, two possible situations can be occurred: (1) cooling fighting if the two AC stages are concurrently on; and (2) inefficient operations if one of them is activated until the return temperature of the rest is under a set-point temperature. To measure and study these occurrences, the r value of return temperature and zone temperature between the two adjacent units can be used to explain differences between a typical SAC unit in a zone and multiple SAC units in an open space.

5. Interaction results

The r values of relations are estimated for zone temperature and return temperature of each SAC unit because zone temperatures will correlate with return temperature if zone air can circulate normally from a supply to return position. In this case, the r value should be in the least medium range because there is a linear relation between two temperature positions. In contrast to the first case, the r value is low if the relation is weak or another parameter is disturbance.

This section will investigate: (1) relations between zone and return temperature to examine temperature circulation of each SAC; and (2) relations between two SAC units. The results of r values using equation (1) are tabulated in Table 1.

Table 1 r values for interaction analysis

Relation	r value	SAC investigations
R1 and Z1	0.79	No. 1
R2 and Z2	0.97	No. 2
R3 and Z3	0.85	No. 3
R4 and Z6	0.16	No. 4
R4 and Z1	0.57	Interaction between No. 1 and No. 4
R4 and Z2	0.98	Interaction between No. 2 and No. 4
R4 and Z3	0.92	Interaction between No. 3 and No. 4

From Table 1, the relations of R1 and Z1, R2 and Z2, and R3 and Z3 are at least medium level for return and zone temperatures relation of SAC No. 1, 2 and 3, respectively. This notification shows temperature profile relation between zone and return sensor positions of each SAC. However, in the testing, SAC No. 4 is turned on manually, whereas the other units are turned off. This situation confirms that the medium r values are caused by heat transfer from high to low temperature area when a SAC is off-time period. In contrast to the first notification, the r value of R4 and Z6 is low which means that the occurrence of mechanical cooling effect or other disturbances.

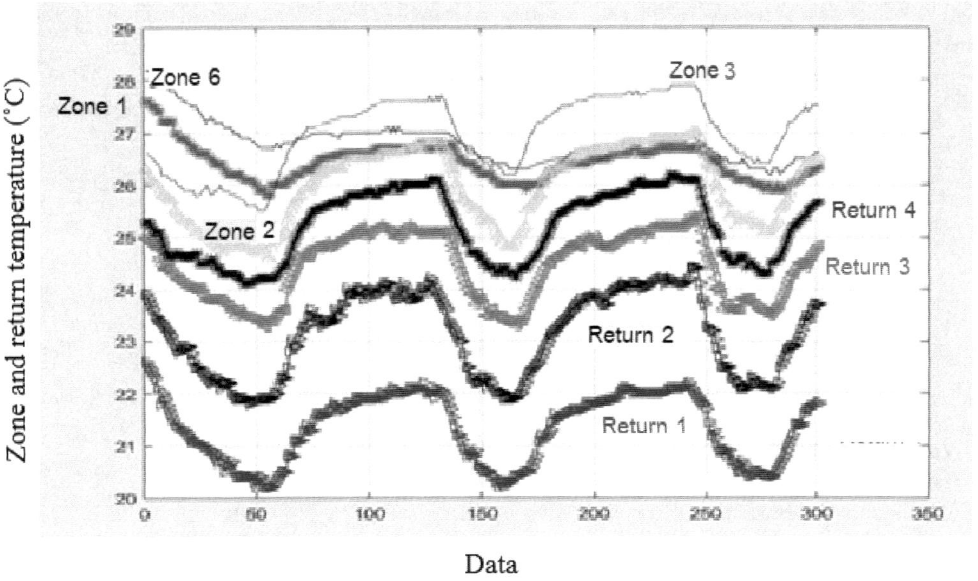

Figure 4. Temperature profile of zone and return temperature

From the field investigations, SAC No. 4 is turned on manually leading to the low r value between zone and return temperatures, whereas other zones have good relation based at least medium r value level because of heat transfer occurrence, without mechanical cooling.

The possibility of another notification is the mechanical cooling from zone 4 to other zones because the layout is an open space area. To assure this notification, the interaction of SAC. No. 4 and other SAC units can be investigated via the r value between return temperature of No. 4 and other zone temperatures. The results show the strong relation with No. 2 and No. 3 (r values are 0.98 and 0.92, respectively). It can be concluded that mechanical cooling of No. 4 moves to zone No. 2 and 3. Meanwhile, zone No. 1 is the farthest distance from zone No. 4 causing the low r value (0.57).

To further illustrate this operation, Figure 4 shows the temperature profiles between return and zone temperatures. The profiles confirm the results from the two notifications of heat transfer and mechanical cooling. In addition, by using graphical method, the third notification is analyzed based a sensor position. From Figure 4, it can be noticed that return temperature profile of No. 1, 2 and 3 follow the return temperature trend. Especially, when R4 is controlled between 24 and 26 °C (set point at 25°C with ±1°C operating differential range), other return temperatures are moved similarly with different lower and upper temperature ranges. The finding is useful for developing coordinated control or soft-repair algorithms used for this system improvement for reduced oversizing issue and energy savings.

6. Conclusions

Oversizing is seldom studied and concerned in Thailand leading to misunderstanding and wrong design and installation of multiple SAC units in an open space due to lacking of a guideline. This paper investigates the possible interaction between SAC units in the open space by installing zone and return temperature sensors to observe temperature profiles and heat transfer effect between off-time control and mechanical control. Pearson's correlation is used to quantify relations between associated variables. The findings are: (1) the different operations between off-control SAC and activated SAC; and (2) return temperature position should be considered during installations and can be further used to design coronation control of multiple SAC units to reduce oversizing effect without supply duct or SAC replacement.

Acknowledgments

The authors are deeply grateful to the Sripatum University and Myenergy Domain (Thailand) Co., Ltd for financial support to further develop innovative technologies based on this work in near future.

References

[1] DEDE. Ministry of Energy, Thailand. *Guidelines, Best Practices for Energy Conservation in Designed Building and Designed Factory*, October 2014.
[2] Woradechjumroen, W, Li, H and Tongshoob, T. 2016. *Proc. Int. Conf on ESIT*
[3] Djunaedy, E, Wymelenberg, KVD, Acker, B and Thimmanna, H. 2011. *Energy Build.* 43 (2–3) 468–475.
[4] Jacobs, P and Conlon, T. 2002. *State-of-the-art review whole building, building envelope, and HVAC component and system simulation and design tools – Part1: whole-building and building envelope simulation design tools,* Project FC05-99OR22674.
[5] Woradechjumroen D, Yu Y, Li H,Yang H and Yu D 2014 *Energy Build.* 69 131-143.
[6] Woradechjumroen D, Li H and Yu Y 2016 *Building Environ* 108 47 – 62.
[7] Woradechjumroen D 2017 *Proc. Int. Conf on Mechanical Engineering*

IOP Conf. Series: Materials Science and Engineering **501** (2019) 012053 doi:10.1088/1757-899X/501/1/012053

Heat Transfer Augmentation in Round Tube with Diamond-shaped Baffle Inserts

C. Khanoknaiyakarn[1], S. Sripattanapipat[2], P. Promvonge[3] and S. Skullong[4,*]

[1]Division of Mechanical and Automotive Engineering Technology, Faculty of Engineering and Technology, King Mongkut's University of Technology North Bangkok, Rayong Campus, 19 M.11, Nonglalok, Bankhai, Rayong, 21120, Thailand.

[2]Department of Mechanical Engineering, Faculty of Engineering, Mahanakorn University of Technology, Bangkok, 10530, Thailand.

[3]Department of Mechanical Engineering, Faculty of Engineering, King Mongkut's Institute of Technology Ladkrabang, Bangkok 10520, Thailand.

[4]Energy Systems Research Group, Department of Mechanical Engineering, Faculty of Engineering at Sriracha, Kasetsart University Sriracha Campus, 199 M.6, Sukhunvit Rd, Sriracha, Chonburi, 20230, Thailand.

*Corresponding Author: sfengsps@src.ku.ac.th, sompol@eng.src.ku.ac.th.

Abstract. Influences of diamond-shaped baffle (DSB) insertion in a heat exchanger tube on its thermal performance are experimentally presented. The DSB elements were repeatedly mounted into the tube using two straight small rods to connect the DSBs together to produce the longitudinal vortex flows inside. The experiments were carried out in a constant surface heat-fluxed tube equiped with DSBs. The DSBs were placed periodically with an attack angle of 45° and a pitch spacing of three times of tube diameter (PR=P/D=3) while four different DSB heights called blockage ratios (BR=b/D= 0.1, 0.2, 0.3 and 0.4) were offered. Air was used as the test fluid flowed into the tube to yield the Reynolds number (Re) in a range of 4190–26,000. The heat transfer rate and the pressure drop of the current work were presented in terms of Nusselt number (Nu) and friction factor (f), respectively. The experimental results showed that the DSB provides the increase in the heat transfer around 3.53–4.16 times higher than the smooth tube while friction factor increases around 16.68–29.52 times. To evaluate the gain of using the DSB, the thermal performance factor (TEF) is determined in a range of 1.25–1.55 whereas the highest TEF is found at BR= 0.1.

1. Introduction

The important part of thermal energy equipment, such as air conditioner, refrigerators, etc., is the heat exchanger, which is used to exchange thermal energy from warmer medium to colder one. Many researches were focused on the ways to obtain higher thermal performance of the heat exchangers. Several techniques have been used for heat transfer enhancement in order to meet the energy costs and to obtain the compact size. The tubular heat exchanger is widely used in most of heat exchangers [1–6]. In enhancing the heat transfer, many different types of turbulators have been developed to produce swirl/vortex flows inside by inserting these devices into the round duct/tube. All kinds of the turbulators have been used to improve the rate of heat transfer in the round tube such as twisted-tape [7–11], coiled square-wire [12], snail entry and coiled wire [13], combined non-uniform wire coil and twisted tape [14], equilateral triangle cross sectioned coiled wire [15], artificial roughness surface and wire coils [16]. The modified twisted tapes were performed to yield the thermal performance improvement higher than the typical twisted tape around 10–40%.

Apart from the turbulators mentioned earlier, there were conical rings, cross-sectional rings, V-nozzles, and conical nozzles. Promvonge [17] investigated the effect of three different conical ring diameters and its arrangement on heat transfer behaviors in a round tube. Ozeyhan et al. [18] studied the heat transfer enhancement in a tube with five different spacing circular cross-sectional rings. Thermal characteristics of a heat exchanger tube contained with V-nozzle turbulators was reported by Eiamsa-ard and Promvonge [19]. Again, Promvonge and Eiamsa-ard [20] examined the effect of three different conical-nozzle pitches and nozzle arrangements on flow friction and thermal behaviors. Also, thermo-hydraulic characteristics in a constant surface heat-flux tube with V-ring inserts were investigated by Chingtuaythong et al. [21]. They indicated that the V-ring performed the best among the other published rings.

The use of punched delta-wing/winglet tape inserts [22–24] is considered as another kind of turbulators employed to enhance the heat transfer in a round duct/tube with the slight increase in friction factor. Also, several investigations have been carried out to examine the effect of delta-wing/winglet tapes on heat transfer and pressure drop. Skullong et al. [25] studied experimentally the thermal and flow friction characteristics in a tube inserted with staggered-winglet perforated-tapes. Eiamsa-ard and Promvonge [26] investigated the influence of double-sided delta-wing tape insert with alternate-axes on heat transfer characteristics in a heat exchanger tube. Influences of quadruple perforated-delta-winglet inserts on thermal behaviors in a round tube with different winglet porosities were presented by Skullong et al. [27].

As cited above, most of the performance improvements in many heat exchanger systems have been made by considering the different geometries and arrangements of swirl/vortex flow devices with different their height/widths, pitches and angles. Convection heat behaviors for turbulent tube flow through the oblique diamond-shaped baffles have never been reported. The aim of the current work is to study the heat transfer as well as the performance augmentation in a tube equipped with diamond-shaped baffles. The experiment was carried out with air as the working fluid in a turbulent regime, Reynolds number from about 4190 to 26,000.

2. Experimental setup

A schematic diagram of the apparatus with the basic components is presented in figure 1. The experimental facility consisted of an inlet section, test section, air supply system (high-pressure blower) and heating arrangement. In the experimental apparatus, the copper tube having inner diameter (D) of 50.8 mm and 2 mm thickness (t) was 3000 mm long included the test section length (L) of 1200 mm while the inlet section was used to get a fully developed flow before entering the test section. To achieve a constant heat-fluxed tube, an electrical wire was employed and continually wound round the test tube and a variac transformer (AC power supply) was used to control the output power of the electric heater as desired. The outermost of the test section wrapped by flexible insulations was done to decrease the convection loss to air in the surrounding.

A blower with 1.9-kW motor power was used to supply the room air to the system by flowing through the orifice-plate for measuring the volumetric airflow rate before entering the test tube where the pressure drop measurement of the orifice plate was done via an inclined/U-tube manometer. The rates of airflow through the test section were varied using an inverter to control the motor speed of the blower. Two Resistance Temperature Detectors (RTD, Pt-100) were employed to measure the temperatures of air at the outlet and inlet while 24 type-T thermocouples equally located on the top and side walls of the test tube were used for measuring the surface temperatures along the test section. When outlet and surface temperature values were not changed for about half an hour indicating a steady state condition of the system, all values of temperatures were recorded via a data acquisition device (Fluke 2680A). Also, the pressure drop measurements of the test tube was conducted using a dwyer instruments, 475-1-FM 475 mark III digital manometer. The uncertainties in velocity, pressure and temperature measurements were, respectively, evaluated to be under ±5%, ±3% and ±0.4%. The maximum measurement uncertainties in various instruments were within ±1.4% for the orifice plate and ±1.2% for the digital manometer.

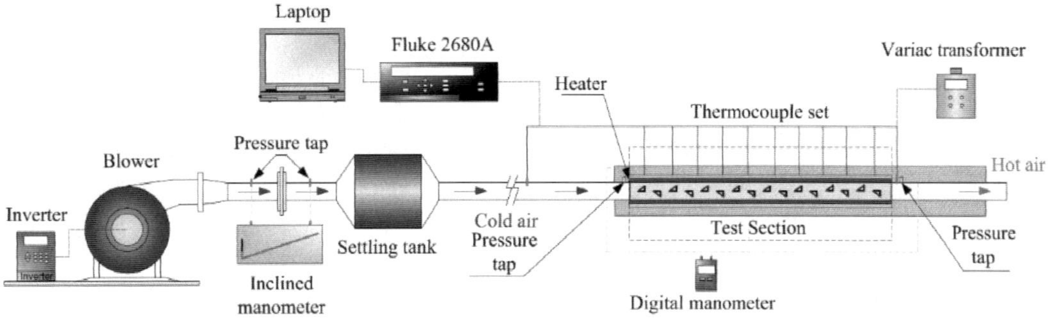

Figure 1. Experimental set-up.

The detail of 45° DSB inserted into the test tube is presented in figure 2. The DSB in the current work was made of an aluminum sheet with 0.3 mm thickness (t). The DSB was cut, drilled to form the diamond-shaped hole and placed repeatedly in the tube with four baffle blockage ratios (BR=b/D= 0.1, 0.2, 0.3 and 0.4) at a single relative baffle pitch or pitch ratios (PR=P/D=3). The DSB elements were linked to each other using two straight small-rods on two edges of the baffles. The interfaces between the DSB and the rod were tightly attached using superglue.

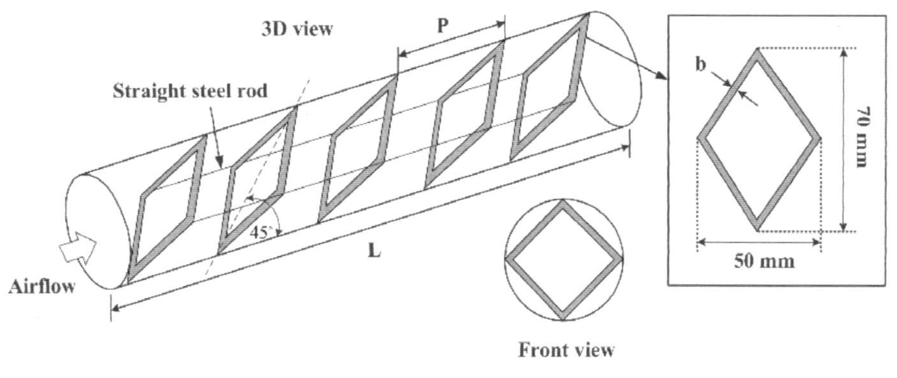

Figure 2. Detail and arrangement of 45° DSBs in test tube.

3. Data reduction

The purpose of this experiment is to examine the flow resistance and thermohydraulic behaviors in a heat exchanger tube equipped with DSBs. The parameters of interest are Nusselt number (Nu), friction factor (f) and Reynolds number (Re). The Re can be evaluated by:

$$\mathrm{Re} = \rho U D / \mu \tag{1}$$

While pressure loss (ΔP) of the test tube is used to estimated f,

$$f = \frac{2}{(L/D)} \frac{\Delta P}{\rho U^2} \tag{2}$$

where U, μ and ρ are, respectively, the average velocity, dynamic viscosity and density of air.

The heat transfer inside the tube is equal to the convection heat of air in the current work. Therefore, the average convection coefficient (h) is estimated by

$$h = mC_{p,a}(T_o - T_i)/A(\tilde{T}_w - T_b) \quad (3)$$

in which

$$\tilde{T}_w = \sum T_w / 24 \quad (4)$$

The Nusselt number (Nu) of the system is given as

$$\mathrm{Nu} = hD/k \quad (5)$$

Thermal enhancement factor (TEF) at identical pumping power, is defined by Webb [28] as

$$\mathrm{TEF} = \left.\frac{h}{h_0}\right|_{pp} = \left.\frac{\mathrm{Nu}}{\mathrm{Nu}_0}\right|_{pp} = \left(\frac{\mathrm{Nu}}{\mathrm{Nu}_0}\right)\left(\frac{f}{f_0}\right)^{-1/3} \quad (6)$$

where h_0 and h stand for convection coefficients for smooth and inserted tubes, respectively.

4. Results and discussion

4.1. Verification of plain tube

The Nu and f values of the current smooth/plain tube were compared to those of the correlations of Gnielinski and Petukhov [29].

Gnielinski's correlation,

$$\mathrm{Nu} = \frac{(f/8)(\mathrm{Re}-1000)\mathrm{Pr}}{1+12.7(f/8)^{1/2}(\mathrm{Pr}^{2/3}-1)} \quad \text{for } 3000 < \mathrm{Re} < 5 \times 10^6 \quad (7)$$

Petukhov's correlation

$$f = (0.79 \ln \mathrm{Re} - 1.64)^{-2} \quad \text{for } 3000 < \mathrm{Re} < 5 \times 10^6 \quad (8)$$

The validation of both the results is shown in figure 3a for Nu and 3b for f. A reasonably good agreement between the experimental and correlation data can be observed while the mean discrepancy is within 5.3% for Nu and 6.7% for f.

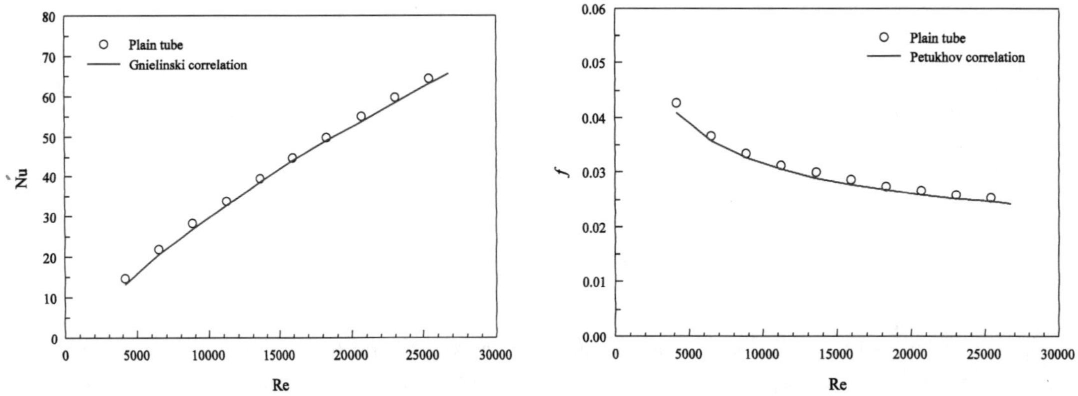

Figure 3. Nu_0 (a) and f_0 (b) against Re for smooth tube.

4.2. Influence of BR on heat transfer

The heat transfer rate in a circular tube with 45° DSB insert against Re values is clearly depicted in figure 4a. Obviously, Nu of the DSB is much higher than that of the plain/smooth tube alone and displays the rising tendency with the increase in Re. This is because the presence of DSBs can produce longitudinal vortex flows inside apart from promoting the turbulence intensity. The DSB with larger BR value can create higher flow obstruction leading to stronger vortex flow strength than the one with smaller value.

The variation of the ratio of augmented Nu to Nu of plain tube alone (called "Nusselt number ratio", Nu/Nu_0) with Re is displayed in figure 4b. It can be seen that Nu/Nu_0 shows the uptrend with the increase in BR owing to stronger degree of turbulence level. The mean values of Nu/Nu_0 for BR=0.4, 0.3, 0.2 and 0.1 are, respectively, about 4.16, 3.93, 3.71 and 3.53 times. This implies that the use of larger BR gives rise to stronger circulation of flow apart from higher turbulence intensity and thermal boundary layer destruction resulting in the considerable increase in the rate of heat transfer.

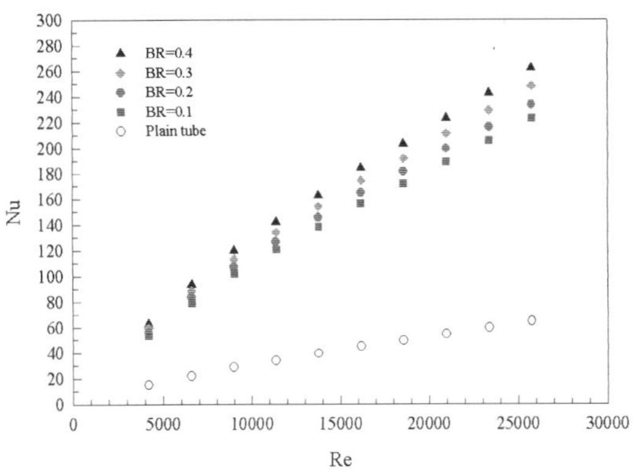

(a)

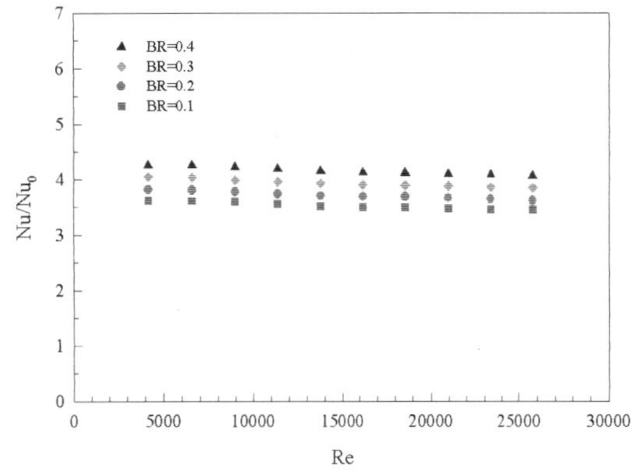

(b)

Figure 4. Nu (a) and Nu/Nu_0 (b) versus Re for DSBs.

4.3. Influence of BR on friction loss

The presence of DSBs leads to the increase in pressure drop which is presented in the form of f and friction factor ratio, (f/f_0). Figure 5a and 5b depicts, respectively, f and f/f_0 plotted against Re for different BRs. It is found in figure 5a that f for DSBs is considerably higher than the smooth/plain tube alone and it shows the slightly decreasing tendency with rising Re. f at BR = 0.4 is much higher than that at smaller BR. This is because of higher flow blockage, the dissipation of dynamic pressure of the fluid due to larger surface area and the act caused by the vortex flow.

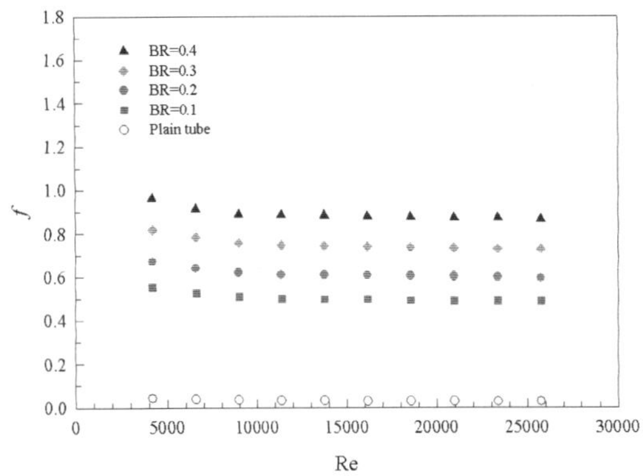

(a)

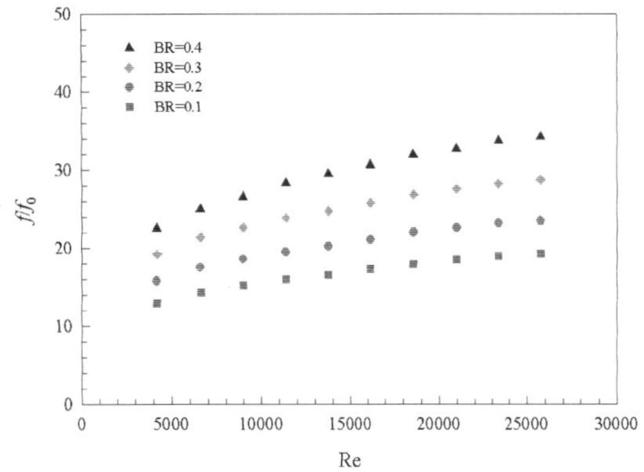

(b)

Figure 5. f (a) and f/f_0 (b) against Re for DSBs.

In figure 5b, it is visible that f/f_0 tends to rise with increasing Re. The use of DSB yields the substantial increase in f/f_0 with raising Re. f/f_0 for BR = 0.4 is much higher than that for smaller BR values at the same operating condition. It is worth noting that f/f_0 increases with the increase in BR. The mean values of f/f_0 at BR = 0.4, 0.3, 0.2 and 0.1 are found to be about 29.52, 24.89, 20.37 and 16.68 times, respectively.

4.4. Thermal performance

The potential of using DSBs in real practical is estimated by a performance indicator, namely, thermal enhancement factor (TEF) as defined in equation (6) which is plotted against Re as depicted in figure 6. In the figure, it can be noted that all of the TEF values are higher than unity. This means that employing the DSBs provides the merit over the smooth tube. For the DSBs, TEF shows the decreasing trend with the increment of Re and BR where its peak around 1.54 is seen at BR= 0.1. TEF values at BR = 0.1, 0.2, 0.3 and 0.4 are, respectively, around 1.29–1.55, 1.27–1.52, 1.26–1.51 and 1.25–1.50.

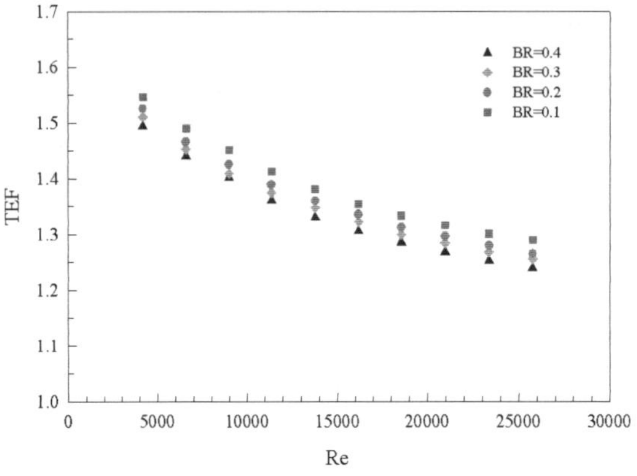

Figure 6. Effect of BR on TEF for DSBs.

5. Conclusion

An experimental investigation on thermal behaviours in a constant heat fluxed tube equipped with 45° DSBs for various BRs in turbulent flow, Re = 4190–26,000 has been carried out. The heat transfer rate in the inserted tube is enhanced at about 3.53–4.16 times over the plain/smooth tube alone while the pressure drop penalty of the tube in the form of f is increased around 16.68–29.52 times. Furthermore, the DSB at BR=0.1 provides the maximum TEF around 1.55, indicating that for practical use, this optimum value is considered as a promising device for energy saving.

References

[1] Promvonge P and Eiamsa-ard S 2007 *Int. Commun. Heat Mass Transf.* **34** 72-82
[2] Thianpong C, Eiamsa-ard P, Promvonge P and Eiamsa-ard S 2012 *Energy Procedia* **14** 1117-23
[3] García A, Solano JP, Vicente PG, Viedma A, *App. Therm. Eng.* **35** 196-201.
[4] Bhuiya MMK, Chowdhury MSU, Saha M, and Islam MT 2013 *Int. Commun. Heat Mass Transf.* **46** 49-57
[5] Nanan K, Thianpong C, Promvonge P and Eiamsa-ard S 2014 *Int. Commun. Heat Mass Transf.* **52** 106-12
[6] Gautam A, Pandey L and Singh S 2018 *Heat Mass Transf.* **54** 2009-21
[7] Chang SW Yang TL and Liou JS 2007 *Exp. Therm. Flu. Sci.* **32** 489-501.
[8] Eiamsa-ard S, Thianpong C, Eiamsa-ard P and Promvonge P 2009 *Int. Commun. Heat Mass Transf.* 36 365-71
[9] Eiamsa-ard S and Promvonge P 2010 *App. Therm. Eng.* **30** 1673-80

[10] Eiamsa-arda S, Yongsiri K, Nanan K and Thianpong C 2010 *Chem. Eng. and Process.* **60** 58-65
[11] Bas H and Ozceyhan V 2012 *Exp. Therm. and Flu. Sci.* **41** 51-8
[12] Promvonge P 2008 *Energy Convers. Manag.* **49** 980-7
[13] Promvonge P 2008 *Int. Commun. Heat Mass Transf.* **35** 623-9
[14] Eiamsa-ard S, Nivesrangsan P, Chokphoemphun S and Promvonge P 2010 *Int. Commun. Heat Mass Transf.* **37** 850-6
[15] Gunes S, Ozceyhan V and Buyukalaca O 2010 *Exp. Therm. Flu. Sci.* **34** 684-91
[16] Garcia A, Solano JP, Vicente PG and Viedma A 2012 *Appl. Therm. Eng.* **35** 196-201
[17] Promvonge P 2008 *Energ. Convers. Manag.* **49** 8-15
[18] Ozceyhan V, Gunes S, Buyukalaca O and Altuntop N 2008 *Appl. Energy* **85** 988-1001
[19] Eiamsa-ard S and Promvonge P 2006 *Int. Commun. Heat Mass Transf.* **33** 591-600
[20] Promvonge P and Eiamsa-ard S 2007 *Int. Commun. Heat and Mass Transfer.* **34** 72-82
[21] Chingtuaythong W, Promvonge P, Thianpong C and Pimsarn M 2017 *Appl. Therm. Eng.* **110** 1164-71
[22] Skullong S, Promvonge P, Jayranaiwachira N and Thianpong C 2016 *Chem. Eng. and Process.* **109** 164-77
[23] Koolnapadol N, Sripattanapipat S and Skullong S 2016 *J. Res. Appl. Mech. Eng.* **4** 166-74
[24] Skullong S Promvonge P Thianpong C Jayranaiwachira N Pimsarn M 2018 *Appl. Therm. Eng.* **129** 1197-211
[25] Eiamsa-ard S and Promvonge P 2011 *Chinese J. Chem. Eng.* **19** 410-23
[26] Skullong S, Promvonge P, Thianpong C and Pimsarn M 2016 *Int. J. Heat Mass Transf.* **95** 230-42
[27] Skullong S, Promvonge P, Thianpong C and Jayranaiwachira N 2017 *Appl. Therm. Eng.* **115** 229-43
[28] Webb RL 1981 *Int. J. Heat Mass Transf.* **24** 715-26
[29] Incropera FP, Witt PD, Bergman TL and Lavine A 2006 *Fundamentals of Heat and Mass Transfer* John-Wiley & Sons

IOP Conf. Series: Materials Science and Engineering **501** (2019) 012054 doi:10.1088/1757-899X/501/1/012054

Experimental investigation of modified Savonius wind turbines

J Priyadumkol [1,*] K Khaothong[2] and W Chaiworapuek[3]

[1]Department of Mechanical Engineering, Mahidol University, Nakhon Pathom, 73170, Thailand
[2]Department of Mechanical Engineering, Faculty of Engineering at Kamphaeng saen, Kasetsart University, Nakorn Pathom, 73140, Thailand
[3]Department of Mechanical Engineering, Kasetsart University, Bangkok, 10900, Thailand

[*] Corresponding Author: jetsadaporn.pri@mahidol.ac.th

Abstract. This aim of this study was to experimentally characterize the performance of Savonius wind turbine. The turbine rotors were modified by varying the distances, measured from blade tip to the most concave point of rotor. To achieve this objective, the investigated parameters were shape factor, including 0.5, 0.4, 0.3, and 0.2 while the Reynolds numbers were set as 136,640 and 156,160. The results showed that the Savonius rotor performed better performance at the higher Reynold number. Also, the tip speed ratio and shape factor were found to have a strong influence on the characteristic of power coefficient. The maximum power coefficient was obtained during the tip speed ratio of 0.4-0.6 while the shape factor of 0.4 provided a better performance than other rotors. This corresponds with the modified rotor 1, having the higher power coefficient compared to conventional rotor by 6.67% and 18.75% for a Reynolds number of 136,640 and 156,160, respectively.

1. Introduction

It is a well-known fact in scientific circles that wind energy is one of the largest renewable energy resources. Wind turbines are used to generate electricity from the wind energy. They can be classified into two types, which are horizontal axis wind turbine (HAWT) and vertical axis wind turbine (VAWT) based on their axis of rotation. The VAWTs have some advantages over HAWTs in that its constructions are simple with low cost, they are independent of the direction of the wind, as well as they have low noise emission and a high starting torque. Thus, these advantages make them suitable for small-scale power generation [1-2]. The Savonius wind turbine is a simple type of VAWT. The main driving force of the Savonius rotor is the difference of the drag force between the convex and the concave sides of the rotor blades [1].

Up to now, Savonius wind turbine has been continuously developed by many researchers. They have focus on studying various design parameters both experimentally and numerically [1-6]. The modified rotor shapes are one of the popular method to increase the performance of Savonius wind turbine. The changes in rotor shapes result in the changes in power coefficient. The results showed that power coefficient was in the range of 0.14 to 0.32 [7].

In this research, the experiment are carried out to study the effect of several parameters, including shape factor, tip speed ratio, and Reynolds number on the performance of modified Savonius rotors in term of power coefficient. The obtained results can be an important information, leading to the better understanding of the Savonius rotor design

2. Experimental details

2.1 Design of the Savonius wind rotor

The designed Savonius rotors were developed on CAD software. 3D printing (Wanhao duplicator 9) had been used to create Savonius wind rotor models with Fused Deposition Modeling (FDM). The nozzle was maintained a temperature of 230 °C for the extrusion of the ABS material and the bed platform was maintained at 90 °C. The layer thickness had been taken as 0.2 mm. After finalizing forming up all layers, Savonius rotor was manufactured as depicted in figure 1.

The designed Savonius rotor and its geometrical parameters are presented as shown in figure 2 and table 1, respectively. In the table, D is the rotor diameter, d is the blade chord length, e is the gap distance, H is rotor height, D_o is end plate diameter, and a is the distance from the tip of the blade. The thickness of blades and end plates is 3 mm and 5 mm, respectively. In the previous studied [8-13], they suggested that the conventional Savonius rotor with the following parameters had a good performance. Accordingly; the end plate parameter (D_o/D) = 1.1 [8]; Overlap ratio (e/d) = 0.10 - 0.15 [9-10]; Aspect ratio (H/D) =1.0 - 2.0. [11-13]

(a) (b)

Figure 1. (a) CAD model (b) Savonius rotor.

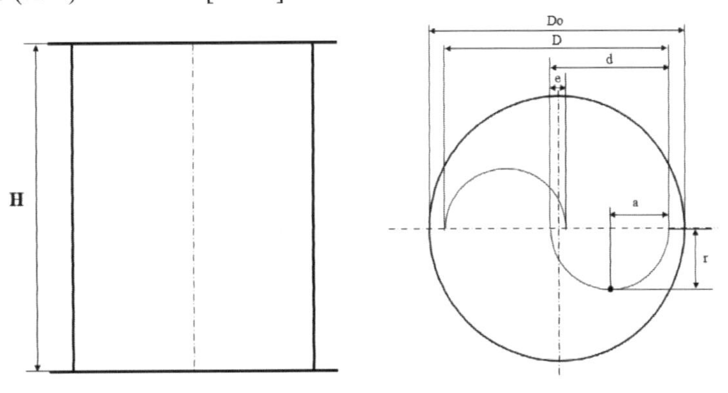

Figure 2. Schematic diagram of Savonius wind rotor.

Table 1. Geometrical parameters of the Savonius rotors.

Type	D (mm)	d (mm)	e (mm)	Do (mm)	H (mm)	r (mm)	a (mm)
Classical rotor	148	80	12	165	180	40	40
Modified rotor	148	80	12	165	180	40	32, 24, 16

The objective of this research was to study the influence of shape factor (a/d) and Reynold number (Re) on the performance of modified Savonius rotor. The ratios between the distance from the tip of the blade (a) and the blade chord length (d) have been considered in the experiments. In this paper, four different blades are given as conventional rotor, modified rotor 1, 2 and 3 for Reynolds number of 136,640 and 156,160. The conventional Savonius rotor has been made from a half ellipse with the ratio of 0.5. The modified blades have been studied by varying the ratio (a/d). The factors of blade shape have been adjusted to find the optimum factors which generate maximum power coefficient. This study focused on the shape factors less than 0.5 as in the numerical study [14]. The shape factor (a/d) of the modified rotors 1, 2, 3 were 0.4, 0.3 and 0.2, respectively. The schematic models for all blades are illustrated as shown in figure 3.

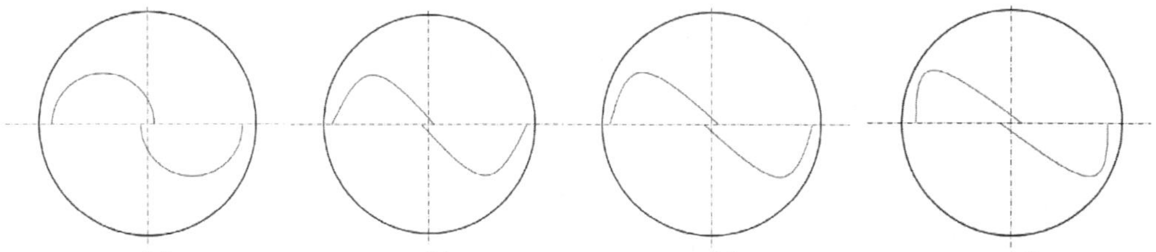

(a) (b) (c) (d)

Figure 3. Schematic models of (a) conventional rotor, (b) modified rotor 1, (c) modified rotor 2 and (d) modified rotor 3.

2.2 Experimental setup

The experimental set-up, used in the present work is shown in figure 4. It consisted mainly of a wind tunnel, the manufactured Savonius wind rotors, velocity measuring devices, and torque measuring system. In experiment, an open-circuit type wind tunnel has been used (AEROLAB Educational Wind Tunnel) and its length is 61cm while a size of cross-section area is 30.5 cm x 30.5 cm. The Savonius wind rotor has been placed vertically inside the test section. Pitot tube Anemometer (Extech HD-350) has been used to measure wind velocity. Test velocities have been conducted at two wind speeds that are v=7 and 8 m/s as in the experiments [9]. A tachometer (DIGICON DT-245P) was used to measure the rotational speed of shaft.

Figure 4 shows mechanical torque measuring system. It was used to measure the mechanical torque of the rotating rotor. As shown in Fig 4b, it consists of rotor, steel shaft, bearing housing, pulley, nylon string, spring balance, weighting pan and aluminum profile housing. Two shafts were made of steel with a diameter of 15 mm. The shafts were used to mount in bearing housings. A belt pulley was installed on the upper shaft. The pulley was assumed to be smooth. Flat leather cord and string were assumed to be massless and to move without slipping over the pulley. The string was pulled at both ends by spring balance and weighting pan. In the experiment, the weighting pan of 50 grams was loaded on the string. The tension in the string can be calculated by spring balance. If the rotor is still rotated, the mass of 10 grams is placed on the weighting pan until the rotor had stopped rotating. The experiments were conducted by repeating 3 times and averaging the results.

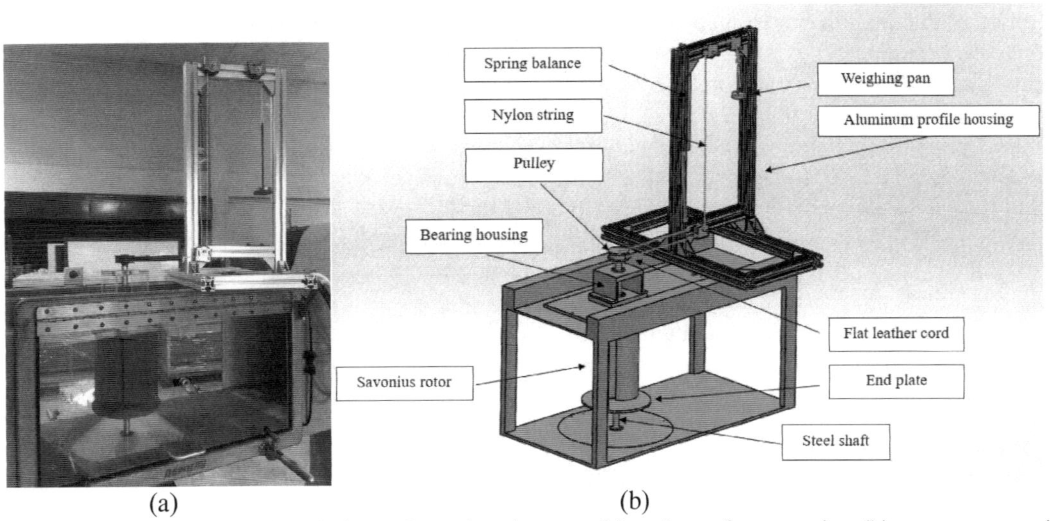

Figure 4. Three-dimensional view of rotational set-up (a) a view of test section (b) torque measuring arrangement.

3. Energy in the wind

When the mechanical torque and rotational speed are measured, the mechanical power (P_m) can be calculated at each wind speed as:

$$P_m = T\omega = T\ (2\pi n/60) \quad (1)$$

where ω is the angular speed, n is the shaft rotational speed and T is the mechanical torque.

The mechanical torque, determined from the measured load and spring balance load is given by:

$$T = Fr = (m - s)\ g\ r \quad (2)$$

where F is the force acting on the rotor shaft, r is the pulley radius, m is the mass placed on the pan, s is the spring balance and g is the gravitational acceleration.

The wind power (P_w) can be obtained as follow:

$$P_w = 0.5\rho A V^3 \quad (3)$$

where ρ is the density of air, A = the projected area for the rotor (HD), V is the wind speed.

The power coefficient (C_p) can be calculated from the following equation:

$$C_p = P_m/P_w \quad (4)$$

Tip speed ratio (TSR) means the ratio between rotor blade tip speed and freestream velocity and is defined as follows:

$$TSR = \omega d/v \quad (5)$$

The Reynold number is calculated from

$$Re = \rho V d/\mu \quad (6)$$

4. results and discussion

Conventional and modified Savonius rotors performance have been investigated experimentally to determine the optimum geometries of Savonius rotor. Experiments were conducted in a wind tunnel for two Reynolds number of 136,640 and 156,160. The rotors were fabricated to study the effect of shape factor (a/d) that are 0.5, 0.4, 0.3 and 0.2.

Table 2 gives the maximum value of power coefficient with respect to the tip speed ratio at a Reynolds number of 136,640 and 156,160. The highest power coefficients of conventional rotor were 0.15 and 0.16 for a Reynolds number of 136,640 and 156,160, respectively. These have shown good agreement with the results of Kamoji et al [6]. It was clear that the power coefficients increased according to increase in Reynolds number. The performances of the modified Savonius rotors are compared with the performances of the conventional rotor as shown in table 2. It was found that the rotor with shape factor of 0.4 was having the higher power coefficient compared to conventional rotor by 6.67% and 18.75% for a Reynolds number of 136,640 and 156,160, respectively. While, the maximum power coefficients of modified rotor 3 were decreased by about 60% and 18% from the conventional rotor for two Reynolds number. The results show that the modified rotor 2 resulted in the highest power coefficient of 0.2 at a Reynolds number of 156,160, which was higher than that of modified rotor 1 (0.19). However, at a Reynolds number of 136,640, the performance of modified rotor 2 was decreased by 6.65% from the conventional rotor.

Table 2. Comparison of main performance parameters.

Type of rotor	Re	$Cp_{,\,max}$	TSR at $Cp_{,\,max}$	Change of $Cp_{,\,max}$ [%]
Conventional rotor	136,640	0.15	0.51	-
(a/d=0.5)	156,160	0.16	0.44	-
Modified rotor 1	136,640	0.16	0.61	6.67
(a/d=0.4)	156,160	0.19	0.58	18.75
Modified rotor 2	136,640	0.14	0.59	-6.65
(a/d=0.3)	156,160	0.2	0.59	25
Modified rotor 3	136,640	0.06	0.42	-60
(a/d=0.2)	156,160	0.13	0.57	-18.75

The effect of blade shape factor on coefficient of power is shown in figure 5 and figure 6 for Reynolds number of 136,640 and 156,160, respectively. Change of power coefficient according to tip speed ratio has been showed both for conventional and modified rotors. It was clear that the power coefficient had a different behavior with varying shape factors.

In figure 5, the power coefficient of modified rotor 1 is increased over the whole tip speed ratio range, compare to the conventional rotor for a Reynolds number of 136,640. It has appeared that the power coefficient value obtained when TSR is 0.61 is the highest of 0.16. The power coefficients of conventional and modified 2 rotors were similar for tip speed ratio between 0.5 and 0.7. In contrast, it is obvious that the performance of modified rotor 3 was decreased compared to other rotors.

The variation of the power coefficients of the rotating rotor with the tip speed ratio for a Reynolds number of 156,160 is shown in figure 6. For tip speed ratio up to 0.5, the difference between the power coefficient of rotors was small. While, the performance of modified rotor 1 and modified rotor 2 were higher than the performance of conventional rotor for tip speed ratio more than 0.5. The results showed that the two peak power coefficient values obtained from modified rotor 1 and modified rotor 2 were close which are 0.19 and 0.2 at tip speed ratio of 0.58 and 0.59, respectively.

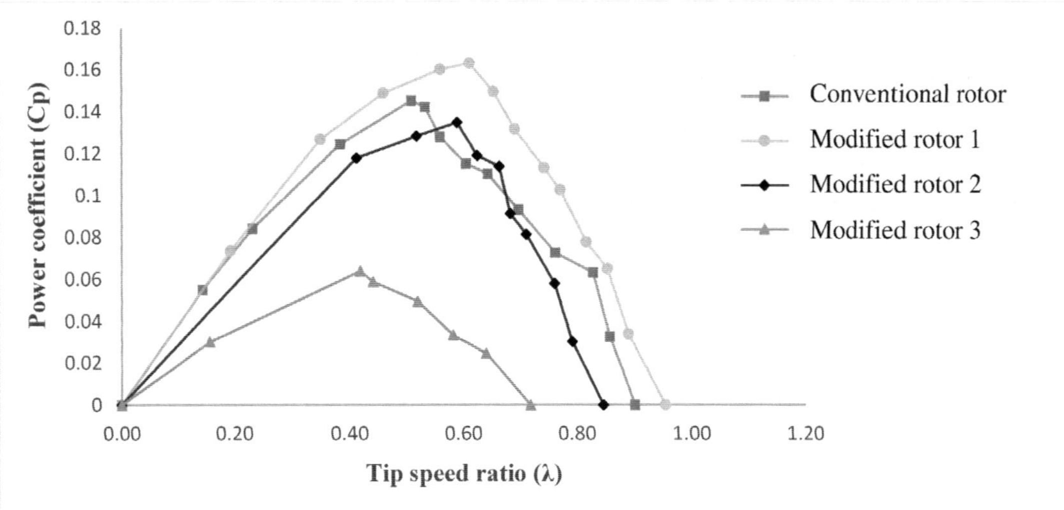

Figure 5. The effects of shape factor on the power coefficient with respect to the tip speed ratio at a Reynold number of 136,640.

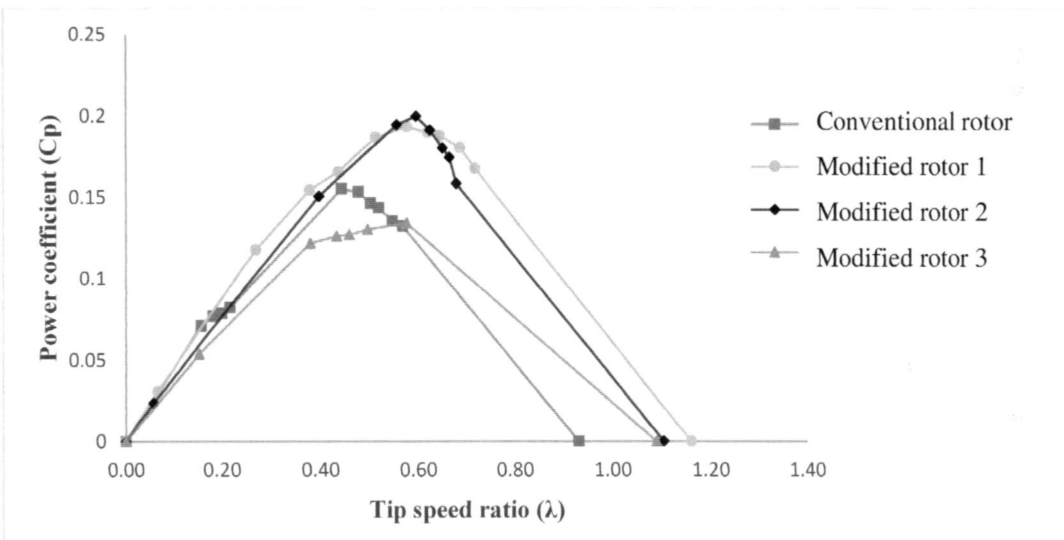

Figure 6. The effects of shape factor on the power coefficient with respect to the tip speed ratio at a Reynold number of 156,160.

5. Conclusions

In this study the performance of various rotors with shape factors of 0.5, 0.4, 0.3 and 0.2 were determined by experimental investigation. The optimum shape factor has been found as 0.4. The use of modified rotor with shape factor of 0.4 significantly increased the power coefficient by 6.64% and 18.75% compared with conventional rotor for a Reynolds number of 136,640 and 156,160, respectively. It also provides good agreement with the numerical study [14]. At a higher Reynolds number, all of the rotors exhibit better power coefficient. The obtained results can be an important information, leading to the better understanding of the developed Savonius rotors.

Acknowledgments

Conventional rotor

The author would like to thank the technical support given by T. Surisang, P. Preuklerdtrakul and A. Phanthanaphon.

References

[1] Roy S and Saha U K 2013 *Renew. Sust. Energ. Rev.* **24** 73-83
[2] Tian W, Mao Z, Zhang B and Li Y 2018 *Renew. Energy* **117** 287-299
[3] Altan B D, Atilgan M and Özdamar A 2008 *Exp. Therm. Fluid. Sci.* **32** 1673-78
[4] Altan B D, Altan G and Kovan V 2016 *Renew. Energy* **99** 584-591
[5] Kacprzak K, Liskiewicz G and Sobczak K 2013 *Renew. Energy* **60** 578-585
[6] Kamoji M A, Kedare S B and Prabhu S V 2009 *Appl. Energy.* **86** 1064-73
[7] Akwa J V, Vielomo H A and Petry A P 2012 *Renew. Sust. Energ. Rev.***16** 3054-64
[8] Ushiyama I and Nagai H 1988 *Wind Eng.* **12** 59-75
[9] Sheldahl R E, Blackwell B F and Feltz L V 1977 Wind tunnel performance data for two-and three-bucket Savonius rotors *Sandia Report SAND* 77-0131
[10] Fujisawa N 1992 *J. Wind Eng. Ind. Aerodyn.* **40** 277-292
[11] Alexander A J and Holownia B P 1978 *J. Ind. Aerod.***3** 343-352.
[12] Saha U K, Thotla S and Maity D 2008 *J. Wind Eng. Ind. Aerodyn.* **96** 1359-75
[13] Vance W 1973 *Proc. Int. Conf. on Wind Energy Conversion System* p 96
[14] Gad H E, Abd El-Hamid A A, El-Askary W A and Nasef M H 2014 *CFD Letters* vol. 6(4) December 2014

IOP Conf. Series: Materials Science and Engineering **501** (2019) 012055 doi:10.1088/1757-899X/501/1/012055

Evaluation of Contact Thermal Sensation Caused by Local Stimulation with Seating in Outdoor

A. Yoshida[1], S. Yasuda[1], D. Hayashi[1], S. Shoho[2], S. Akao[2] and S. Kinoshita[1,*]

[1]Department of Mechanical Engineering, Osaka Prefecture University, 1-1 Gakuen-cho, Naka-ku, Sakai, 599-8531, Japan

[2]Department of Technical Development, Koshii & Co.,Ltd., 1-2-158 Hirabayashi-Kita, Suminoe-ku, Osaka, 559-0026, Japan

* Corresponding Author: kinosita@me.osakafu-u.ac.jp

The experiments on contact thermal sensations felt by subjects were performed under two conditions, namely, when seated outside and in an artificial climate chamber. There was a high correlation between the contact thermal sensation and the seat surface heat flux for thermal stimulus in a hot environment. It could be quantified that the effect of the thermophysical properties of a seating surface material on the contact thermal sensation. For cold stimulation in a cold environment, the trends that were seen for hot stimulus in the hot environment could not be observed. It was concluded that this was due to the large contact thermal resistance arising from the differences in insulation performance of clothes. In an outdoor environment dominated by solar radiation, the effect of contact thermal sensation on the whole-body thermal sensation could not be comfirmed.

1. Introduction

In recent years, to improve the thermal sensation in indoor spaces, personal air conditioning systems [1] that perform local heating and cooling based on the thermal sensations of individuals and thermal control that takes into consideration the heat transfer through parts in contact with an individual, such as a floor surface and seat surface, have attracted attention. Matsumoto et al. [2] studied the influence of such contacts on the whole-body thermal sensation by assuming that floor panel heating would heat the sole of the foot. Compared to when there was no floor panel heating, there was a difference in the contact thermal sensation and whole-body thermal sensation when the sole of the foot was heated. Naka et al. [3] reported that there was a significant change in the whole-body comfort when the temperature of a car seat was adjusted. This was thought to be due to the large contact area between the seat and the body. In each of these studies, the level of skin stimulation had not been sufficiently evaluated; thus, there is no clear quantitative relationship between stimulation level and local thermal sensation.

Outdoor spaces have lately provided increasing opportunities to study thermal environment control via adaptive measures such as sun shades and window films [4-5]. The introduction of wooden chairs is being promoted, even at outdoor stadiums, for the purpose of improving user comfort [6]. Wooden materials have useful thermophysical properties, such as low thermal conductivity and low thermal effusivity. Because of these properties, wooden materials are expected to be effective insulating materials. Misaka et al. [7] performed human subject experiments to compare temperature perceptions; the subjects were seated outside during the summer either on a bench whose temperature could be

IOP Conf. Series: Materials Science and Engineering **501** (2019) 012055 doi:10.1088/1757-899X/501/1/012055

adjusted using flowing water (cooling bench) or on an ordinary bench without cooling. They observed that when subjects sat on the cooling bench for a certain duration, the whole-body thermal sensation reported when sitting on the normal bench was alleviated. The salient drawbacks of this study were that the seat surface temperature had not been set; the seat surface heat flux, namely, level of stimulation on skin was not measured; and the local thermal sensation here, namely, the thermal sensation on the buttocks was not evaluated. These constitute the future challenges for their study.

Shimazaki et al. [8] conducted a study on the local thermal sensation on different locations of the body for cold stimulation. They reported that the body trunk was more sensitive than the limbs at perceiving differences. In human subject experiments conducted in an artificial climate chamber, they indicated that there was a correlation between whole-body thermal sensation and local thermal sensation. Obata et al. [9] calculated the contact surface temperature from analysis of the heat transfer phenomenon when two objects with semi-infinite thicknesses were in contact and showed that there was a correlation between the local thermal sensation and difference between the initial and contact surface temperatures of the two objects. Moreover, they discussed the relationship between contact thermal sensation and thermophysical properties of the objects.

In this study, we installed a plastic seat surface and a wooden seat surface in a sunny place, both inside an artificial climate chamber and in an outdoor area, and conducted experiments involving human subjects sitting on these seats. Through this study, we examined the relationship between the level of local stimulation from the seat surface on the buttocks and the associated thermal sensation. The present study intended to clarify the factors controlling buttock contact thermal sensation; relationship between contact thermal sensation, whole-body thermal sensation, and comfort; difference between susceptibilities to hot and cold stimuli; and influence of contact thermal resistance.

2. Experimental Site and Methods

The experiments were conducted outdoor and in an artificial climate chamber. The outdoor experiment used a building roof in the Nakamodzu campus of the Osaka Prefecture University and the artificial climate chamber experiments used the facilities within the Technology Research Institute of Osaka Prefecture. The artificial climate chambers were set up with lighting that simulated sunlight. The outdoor summer experiments were conducted on July 27, August 3, and August 10, 2017, and the outdoor winter experiments were conducted on December 22, 2017 and January 19, 2018. The artificial climate chamber experiments were conducted on November 30 and December 1, 2017.

The outdoor experiments were conducted on days with weak wind and stable weather. The seat surface temperature was not adjusted. The summer experiment days, on average, had a temperature of 34.2 °C, relative humidity of 45.7%, and global solar radiation of 725 W/m^2. The winter experiment days, on average, had a temperature of 11.1 °C, relative humidity of 36.1%, and global solar radiation of 560 W/m^2. The artificial climate chamber simulated a hot summer environment, with a temperature of 30 °C, relative humidity of 50%, quietness, solar radiation of 800 W/m^2 using a metal halide lamp, and the seat surface temperature was adjusted to 50 °C and 55 °C.

Temperature, relative humidity, solar radiation, infra-red radiation and wind speed were measured as weather condition parameters. To measure the temperature and relative humidity, we installed a sensor inside a ventilation pipe. An ultrasonic anemometer was used to measure the wind speed. They were all installed at a height of 1 m from the ground. To evaluate the level of stimulation from the seat surface, we measured the seat surface temperature using a thermocouple, its heat flux using a heat plate, and the buttock temperature using a thermistor. The clothing temperature was also measured during winter. Skin temperatures were measured at seven sites and the rectal temperature was measured as a core temperature using a thermistor. The measurement interval for all parameters, except wind speed, was 10 s; the measurement interval for wind speed was 0.1 s. To determine the total volume of sweat during measurement, we measured the body weights of the subjects at the start and end of the experiments. As psychological declarations of the subjects, we obtained their buttock thermal sensations, whole-body thermal sensations, and levels of comfort every 3 min, according to a linear scale based on the ASHRAE thermal sensation index [10].

The subjects were seven, healthy, male undergraduate and graduate students of the university, who were between 21 to 24 years. The subjects refrained from eating since 2 h prior to the start of experiment and stayed in a waiting room set to a constant temperature of 26 °C since 1 h before moving to the experimental site. After moving from the waiting room to the experimental site, the subjects spent approximately 10 min in preparation: wearing a device to measure their physiological parameters. After data was acquired for 3 min while standing at rest, the subjects sat on the seats. The seating time was 18 min for the summer outdoor experiments and 9 min for the artificial climate chamber and winter outdoor experiments. The experiments were conducted according to procedures approved by the ethics committee of the Graduate School of Engineering, Osaka Prefecture University. On top of commercially-available wooden chairs, a plastic seat surface (44×41×4 cm) and a wooden seat surface (40×39×6 cm) were placed to obtain the chairs required for these experiments. The two types of chairs were kept side-by-side, and the experiments involved two subjects sitting on the respective chairs at the same time and were conducted were repeatedly. Their chairs are representative for ordinary use. The summer and winter outdoor experiments and the artificial climate chamber experiments involved the total number of 20, 20, and 42 subjects, respectively.

3. Measured Results and Discussion

Figure 1(a) shows an example of the change over time in the seat surface heat flux where the direction from the seat surface to the buttocks is positive for the seat surface temperature, buttocks temperature, and whole-body thermal sensation that were obtained in the summer outdoor experiment. The seat surface heat flux reached its maximum immediately after sitting and converged to 0 thereafter. The buttock temperature and the seating surface temperature gradually converged to the equilibrium temperature. There was a difference in the two temperatures even after 18 min, suggesting contact thermal resistance. There were no significant changes to the whole-body thermal sensation and comfort.

Figure 1(b) shows an example of the change over time in seat surface heat flux where the direction from the seat surface to the buttocks is positive for the seat surface temperature, buttocks temperature, buttocks clothed temperature, buttocks thermal sensation, and whole-body thermal sensation that were obtained in the winter outdoor experiment. Due to the difference in season between the summer and winter experiments, there is a difference in the type of clothing worn. The summer clothing were white T-shirt and shorts, whereas the winter clothing were gray long-sleeve sweatshirt and long sweatpants. The summer clo level was 0.3, whereas the winter clo level was 1.0. For this reason, it was predicted that the contact thermal resistance is greater during winter than in the summer, and despite it being around the same temperature difference as in the summer experiment, the seat surface heat flux was

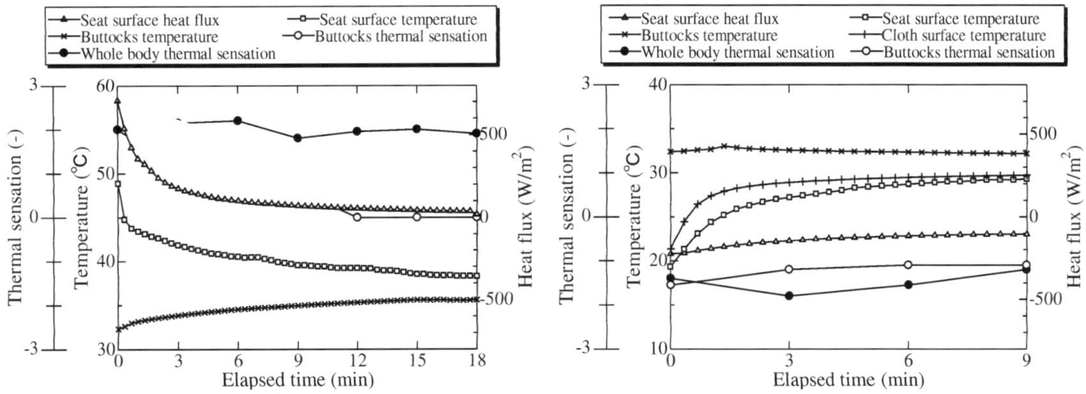

(a) summer outdoor experiment (b) winter outdoor experiment

Figure 1. Changes in the environmental factors, physical values, and psychological declaration sat on a plastic chair over time.

small, and the change in buttock thermal sensation associated with this was smaller than in summer. Even though buttock clothed temperature and seat surface temperature gradually tended toward an equilibrium temperature, there was no change in the buttock temperature.

We investigate the relationship between seat surface heat flux and buttock thermal sensation using Fechner's law [11] on sensory levels and stimulus levels, as shown in equation (1).

$$E = Klog|S| \quad (1)$$

Here, E is the sensory level, S is the stimulus level, and K is a constant. In this study, the sensory level is the buttock thermal sensation, whereas the stimulation level is the seat surface heat flux. We discuss the results obtained from summer outdoor experiments based on equation (1). We saw a correlation in the relationship between the two. The correlation coefficient was R = 0.614. However, the longer is the seating time, the smaller is the seat surface heat flux, and the buttock thermal sensation tended toward the neutral declared value. We left out the stimulus levels that could not be felt by the subjects. We will therefore use Weber's law, which is shown below.

$$\frac{\Delta S}{S} = constant \quad (2)$$

Here, let S be the magnitude of the original stimulus and ΔS be the threshold of the stimulus that can be recognized by humans. We conduct this study to evaluate the local stimulus and the contact thermal sensation. Therefore, thinking that the change in the seat surface heat flux decreases as the seating time increases and is lower than the threshold that humans can recognize, we removed data where the change in the reported buttock thermal sensation is less than 0.2, compared to the sensation reported 3 min earlier. The results obtained thereafter have been shown in Figure 2(a). The seat surface heat flux on the horizontal axis was standardized to 1 W/m^2. We believe that by using Weber's law, we could eliminate data on stimulus levels that could not be recognized by humans.

The experiment was conducted in the artificial climate chamber to maintain uniformity in experimental conditions. Figure 2(a) shows the relationship between the buttock thermal sensation and the seat surface heat flux compared with the results of the summer outdoor experiment. As with the way to process the results of the summer outdoor experiment, we used Weber's law. The data from the artificial climate chamber experiment demonstrated a strong correlation between the seat surface heat flux and the buttock thermal sensation. The correlation coefficient was R = 0.793. While variation is greater in the data measured outside, compared to those measured in the artificial climate chamber, the buttock thermal sensation and the seat surface thermal flux in the two experiments showed almost the same correlation.

Figure 2(b) shows the relationship between the buttock thermal sensation and the seat surface heat flux in the winter outdoor experiment. While the relationship between the seat surface heat flux and

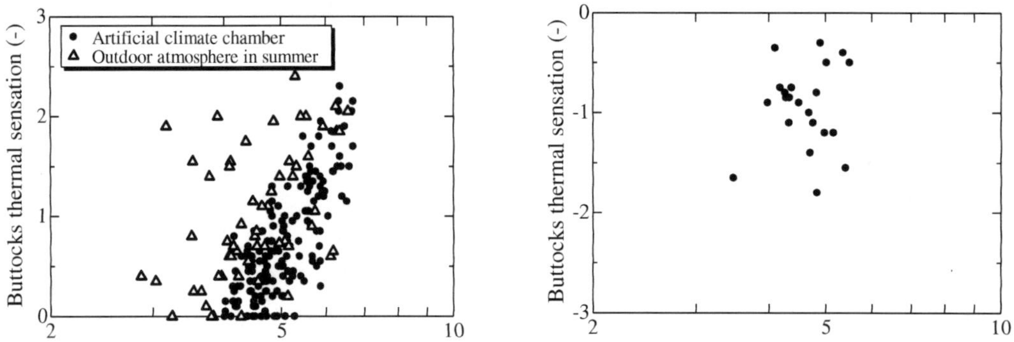

(a) summer outdoor experiment and artificial climate chamber experiment

(b) winter outdoor experiment

Figure 2. Correlation between the buttock thermal sensation and the seat surface heat flux.

the buttock thermal sensation shows a negative correlation for cold stimulus, this was not clear. The difference in the contact thermal resistance of clothes can be suggested as a reason for this.

Figure 3(a) and (b) show the results of significance testing on the seat surface heat flux and the buttock thermal sensation when the subjects sat on the plastic and wooden seat surfaces set to 50°C, which were obtained in the artificial climate chamber. The seat surface heat flux was significantly different depending on the type of seating surface; however, the buttock thermal sensation demonstrated that there was no significant difference when the seating time was long.

We examined whether the buttock thermal sensation affects the whole-body thermal sensation. Figure 4 shows the relationship between the whole-body thermal sensation and the buttock thermal sensation. Despite the change in buttock thermal sensation after seating, there was no change in the whole-body thermal sensation, which made it clear that the buttock thermal sensation does not affect the whole-body thermal sensation. The whole-body comfort levels also demonstrated similar trends. The reason for this is that solar radiation is greater than the seat surface heat flux and solar radiation is greater in the area that receives its heat, so it is conceivable that solar radiation is dominant when determining the whole-body thermal sensation and whole-body comfort levels.

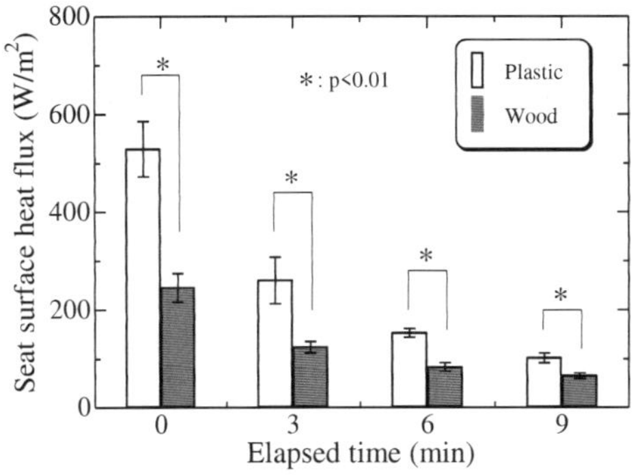

(a) Seat surface heat flux

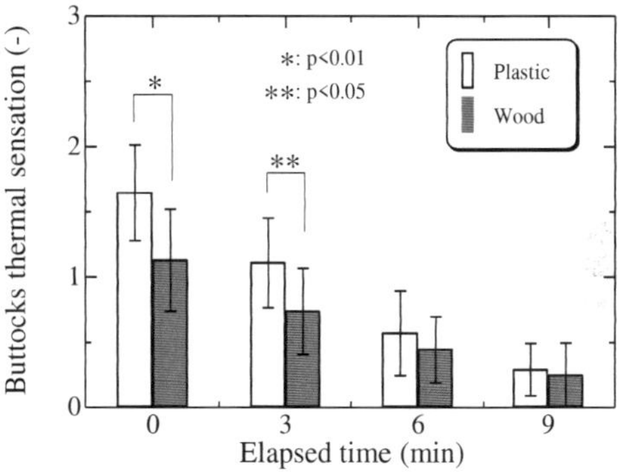

(b) Buttock thermal sensation

Figure 3. Difference in seat surface heat flux and buttock thermal sensation according to the seat surface material during the artificial climate chamber experiment (hot stimulus).

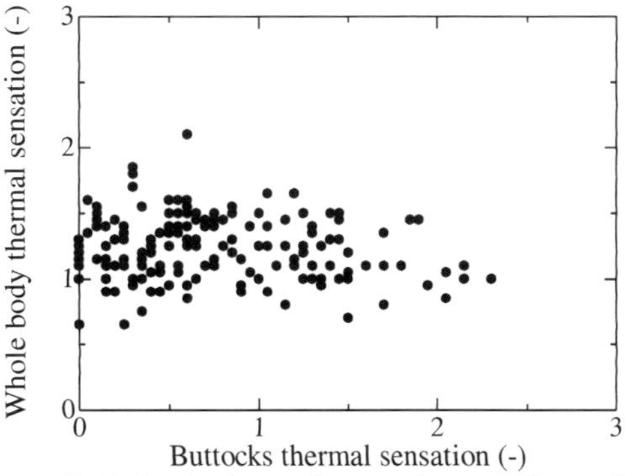

Figure 4. Correlation between whole-body thermal sensation and buttock thermal sensation in the artificial climate chamber experiment.

4. Numerical Calculations

We consider the heat transfer phenomenon when semi-infinite solids with different thermophysical properties come into contact. When the temperatures of the human body and the object material are defined as T_H °C and T_M°C, respectively, the unsteady one-dimensional heat conduction equation describing the thermal diffusion phenomena in each is shown below. The subscripts H and M represent the physical quantities of human body and object material, respectively.

$$\frac{\partial^2 T_H(t,x)}{\partial x^2} = \frac{C_H \rho_H}{\lambda_H} \frac{\partial T_H(t,x)}{\partial t} \quad (3)$$

$$\frac{\partial^2 T_M(t,x)}{\partial x^2} = \frac{C_M \rho_M}{\lambda_M} \frac{\partial T_M(t,x)}{\partial t} \quad (4)$$

Here, C is the specific heat capacity in J/(kg· K), ρ is the density in kg/m^3 and λ is the thermal conductivity in W/(m· K). Solving equation (3) and (4) by applying the boundary conditions, (a) results in equal heat flux at the contact surface between the human body and the material and (b) constant temperature (initial temperature) at a position sufficiently far away from the contact surface allows us to determine the heat flux $q(t)$ at the contact surface between the human body and the material. The initial temperature is uniform for the human body and the material. If the contact thermal resistance R is 0, then

$$q(t) = \frac{\eta_M}{\sqrt{\pi t}}(T_{iniM} - T_{CS}) = \frac{\eta_M}{\sqrt{\pi t}} \frac{T_{iniM} - T_{iniH}}{1 + \frac{\eta_M}{\eta_H}} \quad (5)$$

where η is thermal effusivity in J/s$^{1/2}$·m^2·°C and T_{CS} represents the contact surface temperature in °C. It is clear that the contact heat flux at the seat surface is dominated by the ratio between the thermal effusivity of seat surface material and thermal effusivity of the human skin.

In this experiment, contact thermal resistance due to clothing, etc., was considered non-negligible when the subjects sat on the bench. If there is contact thermal resistance R, a difference arises in the contact surface temperature between the human body and the material. The heat flux at the contact surface, that is the seat surface heat flux, is given as

$$q(t) = \frac{\frac{\eta_H \eta_M}{\sqrt{\pi t}}}{\frac{R \eta_H \eta_M}{\sqrt{\pi t}} + \eta_H + \eta_M}(T_{iniM} - T_{iniH}) \quad (6)$$

The contact thermal resistance at the seat surface is determined by changing the contact thermal resistance R and fitting it to the experimental value of the seat surface heat flux. As representative values of thermal effusivity for the wood (Radiate pine, laminated veneer lumber) and plastic (high density polyethylene) materials used in this study, we used 300 and 600 J/s$^{1/2}$·m^2·°C, respectively. The thermal effusivity of human skin [12] was set to 1590 J/s$^{1/2}$· m^2· °C.

We examined the relationship between the type of seat surface and the contact thermal resistance by comparing the results of the artificial climate chamber and winter experiments. The results are shown in Figure 5(a) and (b). While the hot stimulus significantly varied depending on the type of seat surface, there was no significant difference for the cold stimulus. There was no difference in the contact thermal resistance of the subjects. The clear relation between the buttock thermal sensation and the kind of chair as shown in Figure 3(a) and (b) was influenced by the contact thermal resistance other than the thermophysical properties of chair material for hot stimulus. The fact as shown in Figure 2(b) for cold stimulus is thought to be because of the large contact thermal resistance during the winter compared with the summer, which made it difficult to feel the buttock thermal sensation.

Since buttock thermal sensation correlates strongly with seat surface heat flux, as equation (6) that represents seat surface heat flux shows, it is affected by the thermal effusivity of the seat surface material, seat surface temperature at the time of seating, and the contact thermal resistance between the seat surface and the buttocks. In order to alleviate the buttock thermal sensation, it is desirable to use a material with a low thermal effusivity, such as wood as the seating surface, to have a seating surface with a low solar radiation absorptance, and use a seating surface with large contact thermal resistance.

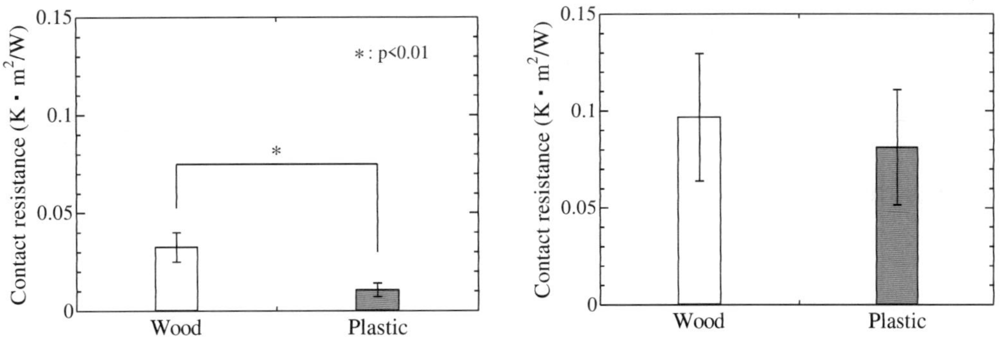

(a) Artificial climate chamber experiment (hot stimulus)

(b) Winter outdoor experiment (cold stimulus)

Figure 5. Differences in seat surface heat flux due to differences in environmental conditions and seat surface material.

5. Conclusion

(1) Based on the seat surface heat flux, which was considered as the local stimulus, we could predict the contact thermal sensation (buttock thermal sensation). We were unable to clarify the difference in susceptibilities of the buttock thermal sensations to hot and cold stimuli.

(2) We calculated the seat surface heat flux using the thermophysical properties of the seating surface material and the contact thermal resistance between the seat surface and human buttock and demonstrated that it was possible to predict the buttock thermal sensation.

(3) In the outdoor environment, we can neglect the effect of buttock local stimulation on whole-body thermal sensation and comfort.

Acknowledgments

The expense of this research was partly covered by Grants-in-Aid for Scientific Research from the Japan Society for the Promotion of Science (17H00802, representative: Atsumasa Yoshida). We would like to express our gratitude for this support.

References

[1] Sudo M, Murakami S, Kato S, Song D and Omori T 2007 *Trans. Soc. Heating Air-Cond. Sani. Eng. Japan* **32-125** 1-10
[2] Matsumoto T and Matsui I 2008 *J. Environ. Eng., AIJ* **73** 721-725
[3] Naka Y, Tochihara Y, Ohnaka T, Nagayama H and Oi H 2007 *Japanese J. Ergonomics* **43** 340-341
[4] Sakai S, Nakamura M, Furuya K, Amemura N, Omishi M, Iizawa I, Nakata J, Yamaji K, Asano R and Tamotsu K 2012 *Energy and Buildings* **55** 28-34
[5] Yoshida S, Yumino S, Uchida T and Mochida A 2016 *Procedia Eng.* **169** 384-391
[6] http://www.koshii.co.jp/product/maxelum/bw_sekourei.html
[7] Misaka I, Narita K, Ishimaru Y and Horiguchi T 2017 *Proc. 41th Symp. Human-Environ. Sys.* 1-4
[8] Shimazaki Y, Yoshida A, Sato S and Morita T 2012 *Thermal Sci. Eng.* **20** 61-67
[9] Obata Y, Takeuchi K, Imanishi H, Furuta Y and Kanayama K 2002 *Int. J. Soc. Mater. Eng. Resour.* **10** 14-19
[10] ASHRAE *ASHRAE handbook fundamentals* 2005
[11] Fechner G 1966 *Henry Holt Ed. Psychology* **1** 112-198
[12] Kagata K, Yamada T and Yoshida A 2008 *Thermal Sci. Eng.* **16** 87-93

IOP Conf. Series: Materials Science and Engineering **501** (2019) 012056 doi:10.1088/1757-899X/501/1/012056

Experimental Study of HFC-32 Split-Type Air Conditioning Systems in Different Indoor Temperatures

T Saihong[1], A Chaweram[1], P Sridara[1], A Tongkratoke[1], S Phankhoksoong[1], and A Pramuanjaroenkij[1,*]

[1]Faculty of Science and Engineering, Kasetsart University, Chalermphrakiat Sakon Nakhon Province Campus, Sakon Nakhon 47000 Thailand

*Corresponding Author: anchasa@gmail.com

Abstract. Energy management has become one of the main issues in world energy scenario, every energy consumed system should be investigated and developed to increase its performance; especially air-conditioning systems which consumes more than half of the total power required by one air-conditioned building. In agreement with "Montreal Protocol on Substances that Deplete the Ozone layer", HCFC-22 was stopped its production and its refrigerating system must take another refrigerant; HFC-32 as the working fluid. This work presented information about two 12,000 Btu/h HFC-32 split-type air-conditioning systems with the hermetically sealed swing compressor (so called Inverter compressor) and the hermetically sealed rotary compressor. Both systems were placed in two identical rooms where 3,500 Watt heaters were also set in the middle of the rooms and operated under the constant heat flux condition in two different indoor temperatures; 20 and 22 Celsius. From experiment results in both indoor temperatures, the system with the hermetically sealed swing compressor escorted better energy consumptions, Coefficient of Performances (*COPs*) and Energy efficiency ratios (*EERs*) than those of another system with the hermetically sealed rotary compressor by 29.70% and 31.86% at 20- and 22-Celsius room temperature, respectively. The performances of both HFC-32 systems in different indoor temperatures were variant. The double pipe heat exchangers (DPHXs) were also investigated in this work, the *COPs* of the system with the swing compressor and DPHX were higher than those of another system with the rotary compressor and DPHX by 65.17% and 75.88% at 20- and 22-Celsius room temperature, respectively. The information in this work was used in making decision on the A/C system selection, as well as, the A/C energy management to reduce total A/C power consumption in macro views.

1. Introduction

In countries with hot and humid weather, air-conditioning systems are preferred to adjust living space conditions to be comfortable. Such as in Thailand, the southwest monsoon influence the weather in the rainy *EER* season which increases the humidity significantly. The air-conditioning systems are operated almost throughout the year in every living space both public and private sectors; i.e. residents, offices, factories, etc. Information presented by Air Conditioning Engineering Association of Thailand (ACAT) reveals that split-type air conditioning systems; in which the condensing unit is installed separately from the fan coil unit, were the most installed system in Thailand [1]. In the energy point of views, the air conditioning systems consume about 60% to 70% of the total energy consumption in any office buildings [2]. Many air conditioning systems are older than 10 years and they have been operated regularly, however, the older the system the higher energy consumption required to produce the same refrigerating rate. One of the solutions to improve energy consumption rates of the air conditioning systems was replacing the old systems with the new systems but their investment costs were the main issue and only few system owners could effort the new systems; especially in the building with large numbers of old systems. In refrigeration theory for any refrigerators, when refrigerant enters an evaporator at lower temperatures and pressure, the system Coefficient of Performance (*COP*) is higher. To reduce the refrigerant temperatures and pressure, the refrigerant need to release its heat to another fluid or the refrigerant which enters a compressor.

Ahmadzadehtalatapeh and Yau [3] experimentally investigated the use of double heat pipe based heat exchangers (HPHXs) in a conventional fully fresh air air-conditioning (A/C) system. They found that the A/C system equipped with the double HPHX with double eight rows of HPHXs was superior in terms of energy savings. Zhang and Zhang [4] simulated and investigated the heat pipe heat exchangers (HPHXs) in the air handler to decouple dehumidification from cooling to reduce energy consumption by simulations and experiments. They claimed that the dedicated ventilation system combined with heat recovery technology could be efficiently applied to buildings; especially for laboratories in subtropical areas. Monirimanesh et al. [5] presented the effect of using nanofluid in thermosyphon-type heat pipe heat exchangers on energy conservation of an air conditioning system. They claimed to innovate two heat exchangers in-series using TiO_2/methanol nanofluids with 0–4 wt% concentrations as working fluids. They found that impacts of temperature and relative humidity on the effectiveness of 2 and 4-row heat exchangers were more than 40 % energy saving. Qian et al. [6] reviewed fundamental principles and clarified common features and major differences of regeneration methods for various typical cooling technologies by classifying regeneration methods into three categories. The first group of regeneration methods were recuperative heat exchangers, the second group of regeneration methods were regenerative heat exchangers, and the third group of regeneration methods were internal heat recovery processes, wherein fluid was applied as a regenerator to store/release thermal energy cyclically to pre-cool and pre-heat the solid-state functional materials. Devotta et al. [7] simulated and experimentally evaluated a few selected low global warming potential refrigerants; HC-290, HC-1270, HFC-161, HFC-32, and HFC-1234yf, as alternatives to HCFC-22 for room air conditioners. They evaluated each alternative refrigerant by optimizing heat exchangers, compressors and capillary to get the maximum Energy Efficiency Ratio (*EER*) Their results showed that HFC-32 gave the *EER* of 3.7 but the discharge temperatures of HFC-32 were high, they suggested that this issue needed special attention in the compressor design. Mota-Babiloni et al. [8] informed that the HFC-32 value of Global warming potential is 677, which is below the HFC regulation limit in refrigeration equipment (750) They concluded that HFC-32 has significantly good heat transfer characteristics and a level of performance that make it acceptable at low condensing temperatures and its performance is very similar to that of R410A across the entire operating range, but they also noted about overly high compressor discharge temperatures. Antunes and Bandarra Filho [9] introduced their experimental investigation of the drop-in alternative halogenated refrigerants processes; HFC-438A, HFC-404A, HFC-410A, HFC-32, HC-290 and HC-1270, for HCFC-22 in a 5-ton refrigeration system. Their experimental setup composed of a semi-hermetic reciprocating compressor, tube in tube heat exchangers and an electronic expansion valve. The main parameters were varied to verify the range and performance of each refrigerant and then compared to the HCFC-22. Their results showed that the natural refrigerants presented the best coefficient of performance and that results for HFCs. We noted that they found that the HFC-32 remained below in performance compared to HCFC-22. However, they experimentally investigated 5-ton refrigeration system with variable speed compressor.

Pramuanjaroenkij et al. [10] experimentally studied effects of a double-pipe heat exchanger which installed inside a 24,000-BTUh split-type air-conditioning (A/C) system. Refrigerant inside the suction line (the refrigerant tube between the evaporator and the compressor) received heat from high temperature refrigerant inside the liquid line (the refrigerant tube between the receiver tank and the refrigerant flow control valve). They investigated the A/C power consumptions before and after the heat exchanger installation and they also investigated the A/C power consumption when the system with the heat exchanger coupled with fabric drying application which induced the evaporative heat transfer effect at the condensing unit in 4 different weather conditions; the normal and rainy daytime and night time.

Hoonpong and Skullong [11] experimentally studied on heat transfer and flow friction characteristics in a solar air heater duct roughened artificially with V-shaped baffles. The absorber plate is mounted with V-baffle vortex generators to improve the performance of the solar thermal system for energy saving. The experiment in the test duct having the aspect ratio (AR) of 10 is conducted for Reynolds number (Re) based on the hydraulic duct diameter ranging from 5300 to 22,600. In the current work, V-baffles are placed on the absorber with three relative baffle heights (R_B=b/H = 0.1, 0.2 and 0.3) and pitches (R_P= P/H = 0.5, 1.0 and 1.5) at a single attack angle (b) of 60°. The experimental results reveal that the use of V-baffle vortex generators yields the considerable increase in heat transfer over the smooth duct around 2.32–4.3 times while the friction loss increases around 4.08–36.9 times. The heat transfer and the friction loss tend to rise for increasing R_B but show the reversing trend for increasing R_P. The maximum thermal performance for the V-baffle vortex generators around 1.57 is seen at R_B = 0.2 and R_P = 1.0.

Chingtuaythong and Chokphoemphun [12] experimental and numerical investigation on thermal performance in a uniform wall heat-fluxed tube inserted with 45° inclined oval–pentagon rings (OPR) for turbulent flow region using air as the working fluid. The ring parameters included three blockage ratios (R_B=e/D=0.05, 0.10 and 0.15) and four pitch ratios (R_P=P/D=1.0, 1.25, 1.5 and 2.0) are introduced. The results from the inserted tube are compared with those from smooth tube alone and show that the OPR can augment the heat transfer rate and friction loss around 2.25–4.86 and 14–100 times over the smooth tube, respectively, depending upon operating conditions. The maximum thermal enhancement factor for using OPR tabulators at about 1.36 is for Re=3900, R_B=0.05 and R_P=1.0. It is worth noting that the results of numerical study are in good agreement with the experimental ones. The average deviation of both results is within ±7% for Nusselt number and ±10% for friction factor.

They presented that the drying application worked as the small evaporative air cooler behind the condenser and enhanced the A/C *EER* since the fabric moisture could absorb heat to vaporize itself, this circumstance assisted for the better heat exchanging between the heated air and ambient air because applying the drying setup without installing the heat exchanger could enhance the A/C *EERs* by 30.23% and 9.09%, in the normal daytime and night time, respectively. However, the heat exchanger could enhance the A/C *EERs* in all four weather conditions as 23.26%, 13.04%, 69.70% and 63.89% increased *EERs* in normal daytime, rainy daytime, normal night time and rainy night time, respectively. When they applied the drying setup couple with the heat exchanger, the A/C *EERs* could be enhanced by 48.48% and 17.5%, only in normal and rainy daytime, respectively. From this investigation, the A/C system with the heat exchanger consumed power less than that of the system without the heat exchanger in all weather conditions. We noticed that the A/C system with the heat exchanger consumed power significantly less than that of the system without the heat exchanger in the nighttime.

If the air-conditioning system consumes less energy by installing an extra equipment; a heat exchanger, in the system while the system can operate with the same refrigerating rate, this installation should be considered as one of the energy consumption improvement techniques. In this current work, we experimentally investigated effects of the double pipe heat exchanger installation in air-conditioning systems on their energy consumption, Coefficient of Performance (*COP*) and Energy efficiency ratio *EER*. There was a limited number of literatures on HFC-32 in Thailand, therefore, this work focused on HFC-32 because HFC-32 was rising its role as the main refrigerant in the split-type air conditioning systems. The air-conditioning systems using HFC-32 as their working fluid with two compressor types were tested; one was the hermetically sealed rotary type and another was the hermetically sealed swing type (so called "Inverter type"), and installed in two identical refrigerating spaces where were controlled at 20 and 22 Celsius, respectively. This work also focused on installing double pipe heat exchangers (DPHXs) into both systems in two different room temperatures, the *COPs* and *EERs* of both systems were compared between values before and after the double pipe heat exchanger installations. These effects were important information for many sectors such as air conditioning owners, energy auditors, energy management responsible people, etc. to manage and invest on double pipe heat exchangers for lower energy consumption purpose.

2. Theoretical Study

Split-type air conditioning systems play an important role in residential applications in Thailand. Since the power consumption can be measured from the applications, one important indicator to show performance of any refrigerating systems; which can be calculated from the power consumption or power input, *EER* and *EER* can be determined as [10]

$$EER = \frac{Useful\ refrigerating\ capacity}{Power\ input} \tag{1}$$

Where the useful refrigerating capacity is in Btu/hr and the power input is in Watts. Another indicator to show performance of any refrigerating systems is *COP* and can be determined as [10]

$$COP = \frac{EER}{3.14188} \tag{2}$$

3. Experimental Study

Two refrigerated rooms were built identically in the same controllable environment and controlled at 20 and 22 Celsius, respectively. Two 12,000 Btu/h split-type A/C systems with different compressor types were installed; the hermetically sealed rotary type and the hermetically sealed swing type. In each experimental setup, there were 4 thermocouples, 2 hygrometers, and 2 power meters connected to one controller as shown in a diagram (figure 1). Four positions of the thermocouples in each setup were in the middle of the room, in front of the return air position of the A/C fancoil unit, in front of the supply air position of the A/C fancoil unit and intake air position of the condensing unit. Two positions of the hygrometers in each setup were in front of the return air position of the A/C fancoil unit and in front of the supply air position of the A/C fancoil unit while there was one separated hygrometer placed between both condensing units. Then, both systems were operated without any heat loads to test the A/C and measurement systems and their power consumptions, air temperatures and humidity inside the rooms and at the condensing unit were recorded every 10 minutes for total of 24 hours in each of 3 test periods.

After that, two 3,500-Watt heaters were installed in the middle of both rooms (figure 2) and they were set to have the maximum temperature at 37 Celsius to protect them from any overload conditions. Then, both systems were operated with 49,450 Watts of power supplied to both heaters in both rooms while their power consumptions, air temperatures and humidity inside the rooms and at the condensing unit were recorded every 10 minutes for total of 24 hours in each of 3 test periods. Finally, two copper double pipe heat exchangers (DPHXs), type HE 1.0 with the trademark; Danfoss, Denmark, were connected to both systems. Each DPHX was connected to allow hot fluid or high temperature refrigerant inside the liquid line (the refrigerant tube between the receiver tank and the refrigerant flow control valve) to release its heat to cold fluid or refrigerant inside the suction line (the refrigerant tube between the evaporator and the compressor). All mentioned parameters were recorded every 10 minutes for total of 24 hours in each of 3 test periods. All power consumption values were used to evaluate *COPs* and *EERs* of both systems in three operating conditions; refrigeration without any heat loads, refrigeration with constant heat load without DPHXs and refrigeration assisted by DPHXs with constant heat load.

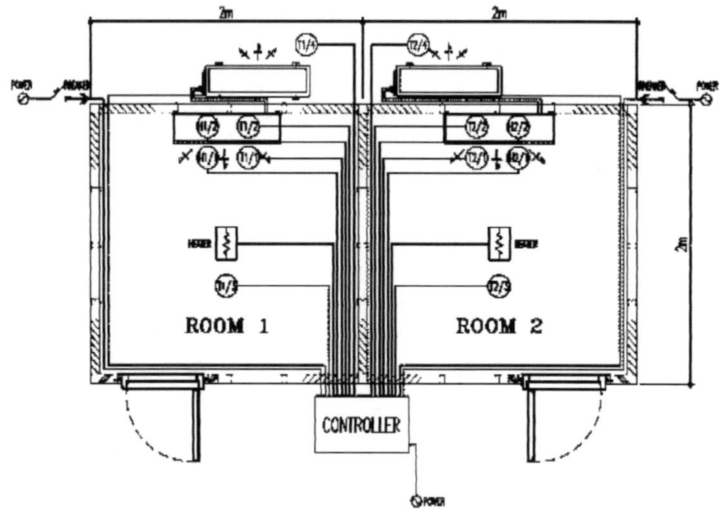

Figure 1. The diagram of positions; A/C systems, temperature and humidity measurements [11].

Figure 2. Heaters which installed inside both rooms.

4. Experimental Results and Discussion

Firstly, the experimental setups; one room with the A/C system utilizing the hermetically sealed rotary compressor and another room with the A/C system utilizing the hermetically sealed swing compressor, both room temperatures were set at 20 and 22 Celsius, respectively. Their measuring sensors were operated, the results of both air conditioning systems without any heat loads were shown in table 1. In the same condensing unit ambient without any supplementary loads, we noticed that the A/C system utilizing the hermetically sealed swing compressor at 20- and 22-Celsius room temperatures took power less than that of another system by 0.5 and 0.1 average unit different per 24 hours, respectively, the average power consumption taken from three repeating examinations. Since HCFC-22 was depleted and stopped its production, the information of the A/C systems applying HFC-32 as the refrigerant was conducted for both A/C systems; the rotary compressor and the swing compressor, this result reconfirmed that the hermetically sealed rotary compressor consumes more energy than that of the hermetically sealed swing compressor; especially during the motor starting period from the rest mode. Secondly, when the heat load was applied by the 3,500-Watt heater in each room under the constant heat flux condition for 24 hours per period in three repeating tests, the results revealed that the A/C system utilizing the swing compressor used up 7.7 and 7.2 average units per 24 hours at 20 and 22 Celsius, respectively, while the A/C system utilizing the rotary compressor used up 10 and 9.5 average units 24 hours, respectively, the latter consumed more energy than that of the former. We noted from table 1 that the heat load in the rooms made both A/C systems displaying their performances which were quite different in comparison to the rooms without any loads in the same condensing unit ambient.

Table 1. The power consumption, temperature and humidity values obtained from the A/C systems with the hermetically sealed rotary and the hermetically sealed swing compressors.

A/C systems	Conditions	Room temperature at 20°C			Room temperature at 22°C		
		Energy (kWh)	EER (BTU/hr W)	COP	Energy (kWh)	EER (BTU/hr W)	COP
The rotary compressor	- without loads and without the DPHX	2.50	-	-	0.70	-	-
	- with loads but without the DPHX	10.00	15.55	4.95	9.50	16.37	5.21
	- with loads and the DPHX	10.56	14.70	4.68	10.20	15.24	4.85
The swing compressor	- without loads and without the DPHX	2.00	-	-	0.80	-	-
	- with loads but without the DPHX	7.70	20.17	6.42	7.20	21.58	6.87
	- with loads and the DPHX	6.40	24.29	7.73	5.80	26.80	8.53

Afterwards, the copper double pipe heat exchangers (DPHXs) were installed inside both A/C systems between their liquid and suction lines, then the systems were investigated while the 3,500-Watt heaters were turned on under the same condition as before for 24 hours per period in three repeating tests. In table 1, the average power consumptions obtained from both systems at at 20- and 22-Celsius room temperatures were as; 10.56 and 10.20 units per 24 hours, respectively, for the system with the rotary compressor and 6.40 and 5.80 units per 24 hours, respectively, for the system with the swing compressor. We noticed that the system with the swing compressor performed better in the same condensing unit ambient and took noticeable power consumption than that of another compressor. We found that the DPHXs played their roles in only the A/C system with the swing compressor under the constant heat load condition as the differences of 4.16 and 4.4 units per 24 hours higher than those of the system with the rotary compressor at 20- and 22-Celsius room temperatures respectively. The A/C systems with the swing compressors installed with and without the DPHX consumed different power consumptions of 1.3 and 1.4 units per 24 hours at 20- and 22-Celsius room temperatures respectively. On the other hand, the A/C system with the rotary compressor installed with the DPHX consumed more power than that of the system without the DPHX. The results taken at 22-Celsius room temperature indicated lower power consumption than the results taken at 20-Celsius room temperature as the refrigeration theory, the higher vaporizing temperature the lower power consumption taken from the same A/C system.

We have analyzed the power consumption in terms of refrigerating performance indicators; *COP* and *EER*. From table 1 under the no load condition at 20- and 22-Celsius room temperatures, the *COP*s and *EER*s of the A/C systems with the hermetically sealed swing and hermetically sealed rotary compressors could not be calculated as shown on the table because there was no heat released from the heaters, "Useful refrigerating capacity" in equation (1) could not be substituted. When the 49,450-Watt load was applied to the A/C system with the hermetically sealed swing compressor at 20-Celsius room temperature, the *COP* and *EER* of the A/C system were higher than those of the A/C system with the hermetically sealed rotary compressor by 1.47 and 4.62, respectively, the higher percentage was 29.70% obtained from 1.47x100/4.95 and, at 22-Celsius room temperature, by 1.66 and 5.21, respectively, the higher percentage was 31.86% obtained from 1.66x100/5.21. After the DPHXs were installed in the A/C system with the hermetically sealed swing compressor at 20-Celsius room temperature under the same load condition, the *COP* and *EER* of the A/C system were higher than those of the A/C system with the hermetically sealed rotary compressor by 3.05 and 9.59, respectively, and by 3.68 and 11.56, respectively, at 22-Celsius room temperature. The A/C system with the swing compressor and DPHXs at 20- and 22-Celsius room temperatures achieved about 65.17% and 75.88% higher *COP* and *EER* than those of the system with the rotary compressor. To fulfill our purpose in introducing DPHXs for the A/C energy management, the *COPs* of the A/C systems with the swing compressor under the constant heat load condition with and without the DPHX performing were averaged as 20.40% and 24.16% at 20- and 22-Celsius room temperatures, respectively, the higher performances were found when the A/C System with the swing compressor was equipped with the DPHX. Therefore, this work found that the DPHX enhanced the performances of the A/C system with the swing compressor.

Additionally, to provide another viewpoint of energy management, the investment cost should be reviewed. When the building owner decides to replace new A/C systems with the swing compressors over the rotary compressors in this building, the saving per year in replacing the A/C systems with the swing compressors at 20- and 22-Celsius room temperatures can be calculated as 2.3 units per day x 365 days x 7 Baht per unit (assumed average energy cost) which equals to 5,876.50 Baht per year. If the owner considers total cost of the new A/C systems with the swing compressors, the owner must prepare about 20,000 Baht per system, the simple payback time without considering the future worth of the investment is 3.4 years, the amount of years may make the owner hesitates to do so. If one focuses on installing the DPHXs for the HFC-32 A/C systems with the swing compressors, the saving per year in connecting DPHXs at

IOP Conf. Series: Materials Science and Engineering **501** (2019) 012056 doi:10.1088/1757-899X/501/1/012056

20-Celsius room temperature can be calculated as 1.3 units per day x 365 days x 7 Baht per unit (assumed average energy cost) which equals to 3,321.50 Baht per year. The cost of DPHXs and their installation is about 7,000 Baht per unit, the simple payback time is 2.11 years, the amount of years can encourage the owner to consider the DPHX installation.

5. Conclusions

In the energy management, enhancing performance of refrigerating systems is one of the important issues to be considered. We would like to introduce interesting roles of compressor and heat exchangers into split-type air-conditioning systems. Two 12,000 Btu/h split-type A/C systems in this current work were connected with two different types of compressors; the hermetically sealed rotary type and the hermetically sealed swing type. The copper double pipe heat exchangers (DPHXs) were chosen to experimentally investigate in this work. There were two identical refrigerating spaces where two 3,500-Watt heaters and two A/C systems were placed, the heaters were operated under the constant heat load in both rooms. As the heat load applied, the system with the rotary compressor consumed more power than that of the system with the swing compressor by 2.3 more units per 24 hours at 20- and 22-Celsius room temperatures. The swing compressor performance expressed their potential to replace the rotary compressor. As the heat load also applied, the DPHXs only enhanced the performance of the system with the swing compressor, the DPHXs enhanced the *COPs* and *EERs* of the system with the hermetically sealed swing compressor as 20.40% and 24.16% at 20- and 22-Celsius room temperatures higher *COP* than those of the system with the rotary compressor. Noticeably, the double pipe heat exchanger has drawn our attention to apply it in the current split-type A/C system since the DPHX investment cost is acceptable, as well as, the DPHX enchantment on the system was outstanding. One more important point of views in this work was the information of the HFC-32 split-type systems with the hermetically sealed rotary compressor and the hermetically sealed swing compressor which were two choices in replacing current HCFC-22 split-type systems. According to "Montreal Protocol on Substances that Deplete the Ozone layer", production lines of HCFC-22 itself and its split-type systems were stopped, the HFC-32 split-type systems were the proper choices to replace the HCFC-22 systems. The HFC-32 split-type system with the hermetically sealed swing compressor showed its potential. This important information of the DPHX enhancement on the HFC-32 system performance was presented and emphasized the use of DPHX in the energy management in the updated systems.

Acknowledgments

The research described in this paper was financially supported by Innovation Hub project funding, Faculty of Science and Engineering and Office of Campus, Kasetsart University, Chalermphrakiat Sakon Nakhon Province Campus, and Kasetsart University Research and Development Institute, Kasetsart University

References

[1] Card D and Riddell WC 2015 *Canadian J. Econ.* **261** 149-189
[2] Pawangkarut, J 2015 *Kaset. Engin. J.* **3** 32-36
[3] Ahmadzadehtalatapeh M and Yaub YH 2014 *Int. J. Engin.* **26** 51-57
[4] Zhang L and Zhang YF 2016 Research on Heat Recovery Technology for Reducing the Energy Consumption of Dedicated Ventilation Systems: An Application to the Operating Model of a Laboratory. 2016 Research report, Tianjin University, Tianjin,

China.
[5] Monirimanesh N, Nowee SM, Khayyami S and Abrishamchi I 2015 *Heat Mass Transfer* **52** 1025-1035
[6] Qian S Yu J and Yan G 2017 *Renew. Sustain. Energy Rev.* **69** 535-550
[7] Devotta S, Padalkar AS, Mali KV 2016 *Sci. Tech. Built Environ.* **22** 1128-1135.
[8] Mota-Babiloni A, Navarro-Esbrí J, Makhnatch P and Molés F 2017 *Renew. Sustain. Energy Rev.* **80** 1031-1042.
[9] Antunes AHP and Bandarra Filho EP 2016 *Int. Nat. J. Refriger.* **70** 119-127.
[10] Kongdee K, Thongdee C and Sansuriwong N 2015 A Study of Impacts of Heat Exchanger Installation on Split Air Conditioning System, Kasetsart University, Chalerm Phrakiet Campus Sakon Nakhon Province, Thailand
[11] Hoonpong P and Skullong S 2018 *J. Res. Appl. Mechan. Engin.* **6** 29-39.
[12] Chingtuaythong W and Chokphoemphun S 2018 *J. Res. Appl. Mechan. Engin.* **6** 50-62.

IOP Conf. Series: Materials Science and Engineering **501** (2019) 012057 doi:10.1088/1757-899X/501/1/012057

Internal Resistance Computation of a Cylindrical LCO Battery for Heat Generation Model

T. Kowitkulkrai[1], A. Kaewpradap[1], S. Hirai[2], V. Lailuck[3], S. Rompho[3] and M. Masomtob[3,*]

[1] Department of Mechanical Engineering, King Mongkut's University of Technology Thonburi, Bangkok, 10140, Thailand
[2] Department of Mechanical Engineering, Tokyo Institute of Technology, Tokyo, Japan
[3] Materials for Energy Research Unit, National Metal and Materials Technology Center (MTEC), National Science and Technology Development Agency (NSTDA), Pathum Thani, 12120, Thailand
* Corresponding Author: E-mail : manopm@mtec.or.th

Abstract. The internal resistance is one of the most important parameters which directly relate to battery performance in terms of output power and heat generation. This research presents the internal resistance of the battery that is studied with using battery voltage behaviour. Lithium Cobalt Oxide (LiCoO2, LCO) battery is applied for the experiment. The voltage drop during the current discharge is calculated to obtain internal resistance by the current pulse technique (CPT). Three different terms of internal resistance are compared by the voltage drop characteristic. As the result, comparison of temperature profile between experiment and simulation was obtained by current pulse technique with discharge rate variations.

1. Introduction

The battery technology has been improved to make a new type of battery for supporting electronics. At the present, there are several battery types due to different materials. Most type of batteries include lead acid, lithium ion and Nickel batteries have different characteristics, advantages and disadvantages. Lithium Cobalt Oxide (LiCoO2, LCO) battery is a type of batteries that is used with many applications such as electronics, phones and laptops due to its high energy density but low power density [1]. However, one of the most important parameters of battery performance is the internal resistance. H. G. Schweiger *et al.* have explained that the internal resistance of battery is directly related to battery performance in terms of output power and heat generation; for example, the low internal resistance of battery provides higher output power but less among of heat generation compared to the high internal resistance [2]. The internal resistance of battery (R_{int}) includes three different terms of resistance; ohmic resistance caused by material, electrolyte and electrical connection (R_{ohm}), charge transfer resistance caused by the resistance of ion movement ($R_{ch,tr}$)and diffusion resistance caused by the concentration within electrolyte between positive and negative pole (R_d) [3]. These three internal resistances trend to increase when the depth of discharge is higher. Moreover, the increase of the internal resistance affects to the battery heat generation. The heat generation of the battery includes two heat sources which occurs within battery: joule heating caused by current flow through the internal resistance of battery, entropy change caused by chemical activation within battery [4]. Therefore, the three terms of internal resistance should be accurately measured to find the heat generation of battery. In order to obtain heat generation, the method to measure the heat generation due to entropy change using trial and error approach which can be used to model the battery heat generation [5]. The presented approach obtains the one term of internal resistance from measurement device which is not precise. Consequently, the exact heat generation ratio of those two heat sources is less accurate. This study represents a method to measure the three terms of internal resistance by battery voltage behavior. The voltage drop during the current discharge is calculated for the three terms of internal resistance. The result will be used for the heat generation simulation to compare the experimental and simulation temperature profile at 0.5C and 1C of the battery discharge rates.

2. Experiment Setup

2.1. *Operating condition of battery cell and apparatus*

As the heat generation of LCO is higher than other types of Lithium ion battery, and it is proper for the battery thermal study, 18650 Lithium Cobalt Oxide (LiCoO2, LCO) is used for the experiment [6]. The 18650 battery is 18 mm in diameter, 65 mm in the height and "0" represents the cylindrical shape as shown in figure 1. The battery cell is discharged in range of 4.2 V to 3.0 V as operating voltage and the discharge rates are determined at 1.3A (0.5C) and 2.6A (1C) as shown in table 1. using Agilent N3301A electronic load controlled via LabVIEW software which is used to determine among of current and time for the current pulse pattern. The Graphtec midi Logger GL840 data logger is used to measure battery voltage and battery surface temperature.

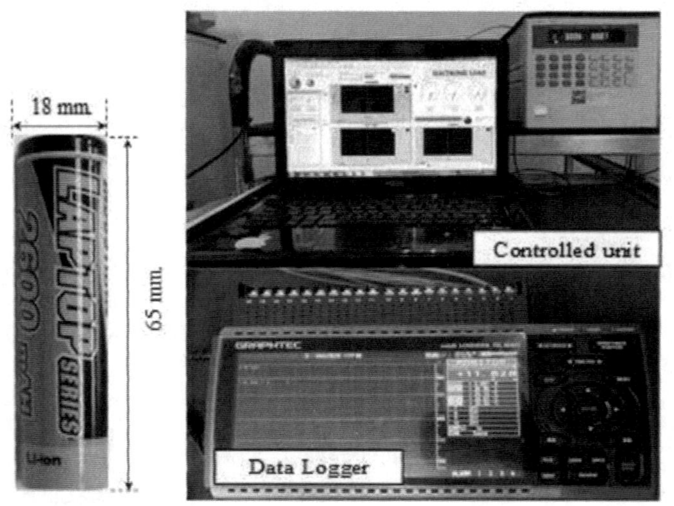

Figure 1. Experimental equipment a)18650 LCO battery cell b) Data acquisition unit.

Table 1. Specification of Lithium ion cell [7]

Cell Chemistry	Cell Geometry	Size (mm.)	Operating Voltage (V)	Nominal Capacity (mAh)
Lithium Cobalt Oxide ($LiCoO_2$)	Cylindrical	18 (D) 65 (H)	4.2 ~ 3.0	2600

3. Methodology

3.1. *Current Pulse Technique*

Lam L. explains the three terms of internal resistance can be expressed by equation (1)[2]. To investigate the three terms of internal resistance, Current Pulse Technique (CPT) is applied to obtain the internal resistance of the battery. The current pulse (see in figure 2) is used for the discharging. The battery is applied for load in one minute which decreases for 1.67% of state of charge (SOC). Then, it is rested for three minutes to restore the close circuit voltage (Vcc) of the battery to the open circuit voltage (Voc) state. In addition, Lam L. has presented the internal resistance is directly related with % SOC, it increases at high % DoD[2]. Therefore, from the current pulse pattern in figure 2, the battery capacity is removed from 0% –100% state of charge and the voltage characteristic is displayed as shown in figure 3 to obtain the three terms of internal resistance at 0–100% state of charge.

$$R_{int} = R_{ohm} + R_{ch,tr} + R_d \tag{1}$$

3.2. *Internal resistance calculation*

From section 3.1, the CPT is applied to obtain voltage characteristic for studying the three terms of internal resistance. Therefore, Thanagasundram S, Arunachala R, Makinejad K, Teutsch T, and Jossen A have presented the internal resistance calculation from the voltage drop as shown in figure 4[8]. The voltage is distinguished into two main parts. The first part is the voltage drop from u_1 to u_2 caused by the Ohmic resistance (R_{ohm}) and the diffusion resistance (R_d). This instant drop can be calculated from equation (2). The second part is the effect of charge transfer resistance ($R_{ch,tr}$) which behaves like the capacitor. Consequently, it causes the curve after instant drop from u_2 to u_3. The end of u_3 is determined by the rate of voltage change by time less than 0.001 V/s^2. This algorithm can be done using MATLAB software as the tool. The charge transfer resistance can be calculated from equation (3).

$$R_{Ohm} + R_d = \frac{(u_1 - u_2)}{i} \tag{2}$$

$$R_{ch,tr} = \frac{(u_2 - u_3)}{i} \tag{3}$$

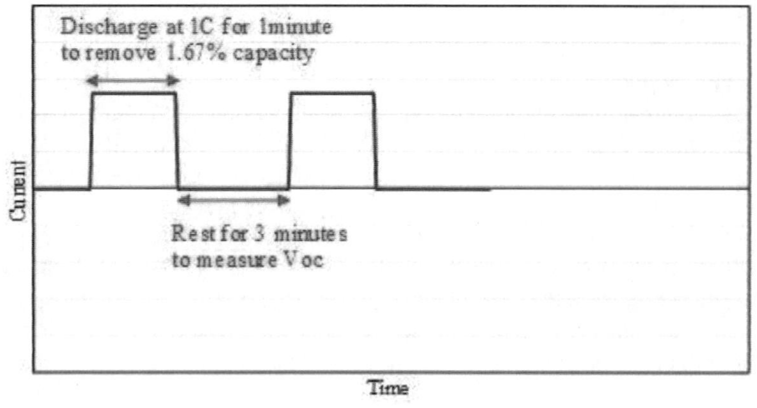

Figure 2. Current pulse technique (CPT) for discharging.

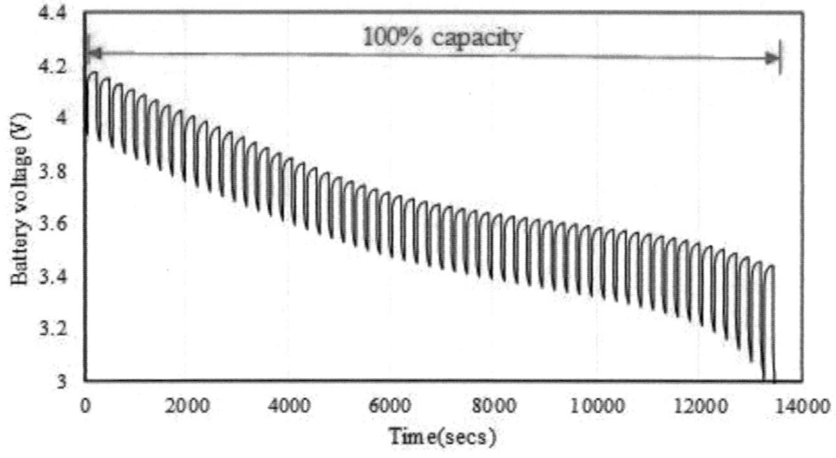

Figure 3. Battery voltage using current pulse technique.

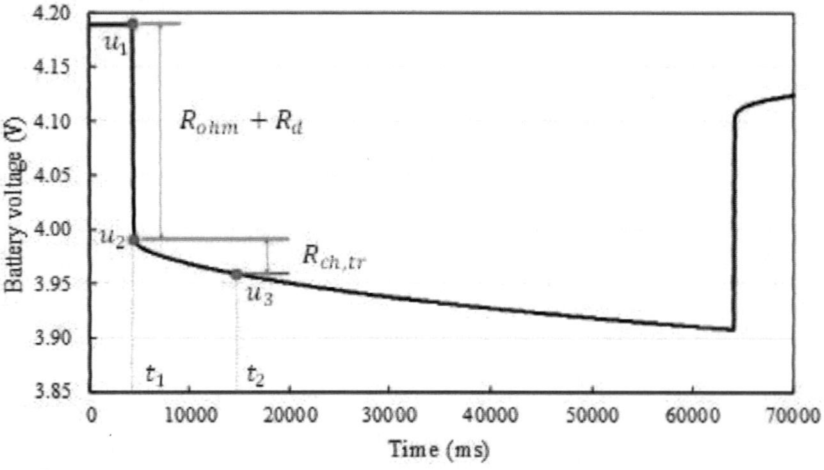

Figure 4. The voltage drop characteristic during discharging for calculation.

4. Result and Discussion

4.1. *Internal resistance*

The internal resistance of the battery is presented by two parts according to the calculation approach. At the first part as shown in figure 5, the Ohmic resistance and diffusion resistance are fairly constant at0–60% DoD, then significantly increase after 60% DoD. The increase is caused by owing to the diffusion resistance term. The concentration gradient in the electrolyte becomes higher when the battery capacity is low. Therefore, it enhanced the resistance of the ion movement which increases the diffusion resistance.

For the second part as shown in figure 6, the behaviour is similar to the first part but causes by the different reason. In this scenario, at 0 – 60% DoD, the ions can easily perform the chemical reaction at the electrode to generate to the electricity, therefore the charge transfer resistance is quite constant. At the end of discharge state, after 60% DoD, the surface area of electrode is insufficient to perform the reaction because the activated ions form at the electrode surface as the layer and resist the moving ions to reach the electrode. Thus, the charge transfer resistance is higher at the end of discharge state.

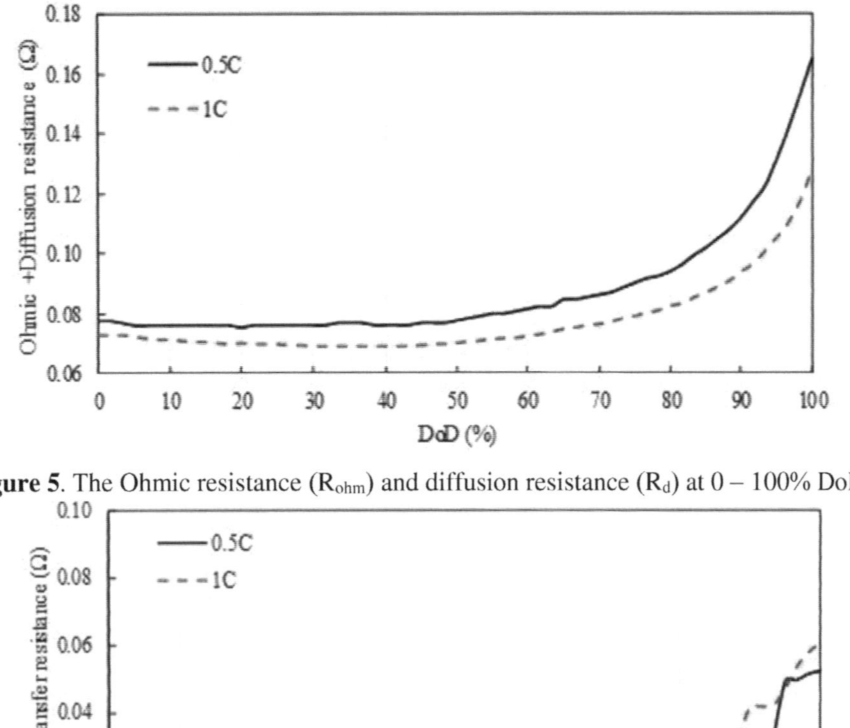

Figure 5. The Ohmic resistance (R_{ohm}) and diffusion resistance (R_d) at 0 – 100% DoD.

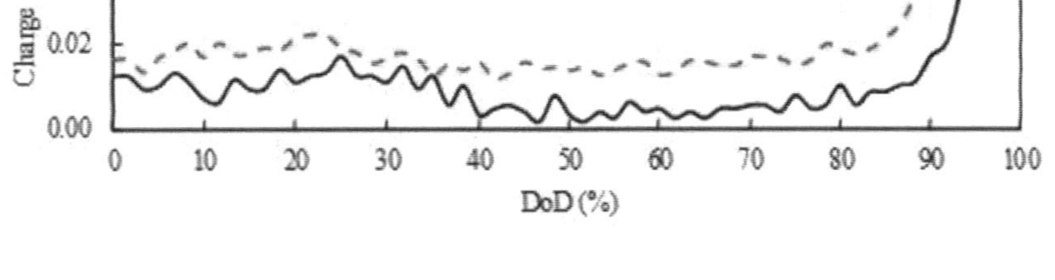

Figure 6. The charge transfer resistance ($R_{ch,tr}$) at 0 – 100% DoD.

4.2. *Thermal behavior*

The heat generation of battery includes two heat sources as joule heating and entropy change which can be expressed by equation (4). The joule heating and entropy change can be expressed by equation (5) and equation (6) respectively [3].

$$\dot{Q}_{battery} = \dot{Q}_{joule} + \dot{Q}_{entropy} \tag{4}$$

$$\dot{Q}_{joule} = I^2 \cdot R_{int} \tag{5}$$

$$\dot{Q}_{entropy} = -iT \cdot \left(\frac{\Delta S}{nF}\right) \tag{6}$$

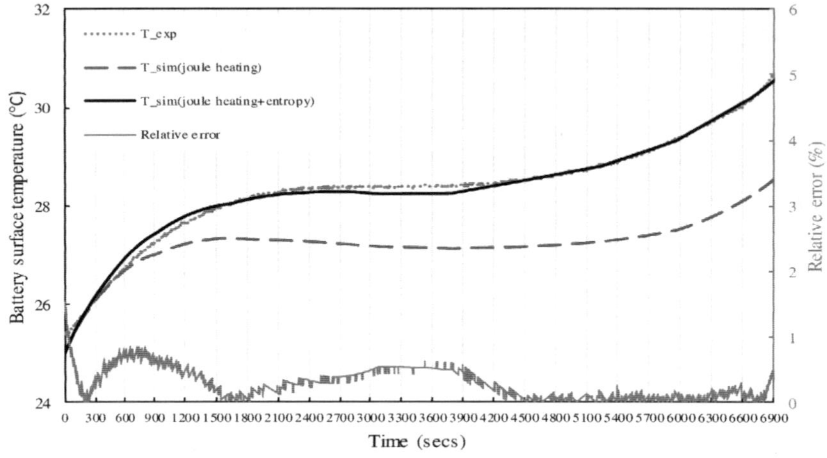

Figure 7. Battery surface temperature profile of experimental and simulation during constant current discharge at 1.3A (0.5C).

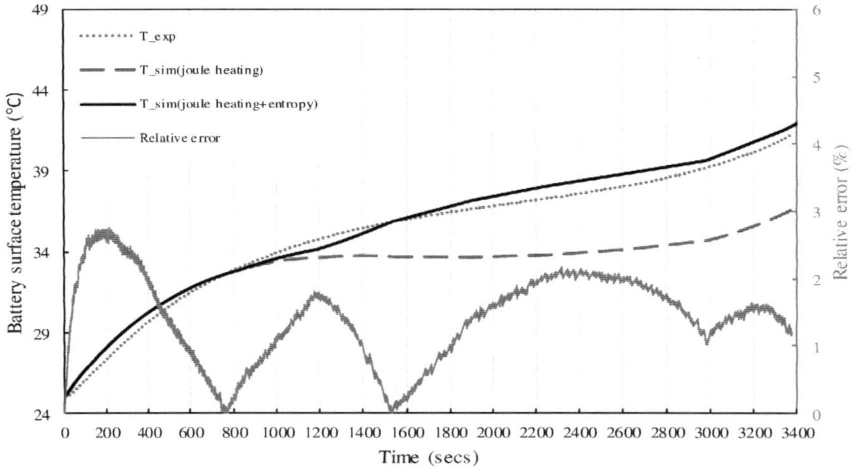

Figure 8. Battery surface temperature profile of experimental and simulation during constant current 2.6A (1C).

The experimental is discharged at 0.5C and 1C. From discharged current at 0.5C, the experimental result as shown as the dot line in figure 7. It represents that the maximum battery surface temperature which reaches 30.6°C. It comes from two heat sources. To analyse the contribution of two heat sources separately, simulation using Finite Element Method (FEM) is needed. the joule heating heat source is computed by equation (5) which use the internal resistance calculated from section 4.1. the simulation result as shown as dash line in figure 7. It is less than the experimental result because it comes from only joule heating heat source. Therefore, the temperature gap between joule heating temperature profile (dashed line) and the experimental (dot line) is considered as the heat source from the heat generation due to entropy change. To complete the temperature profile, the heat source from entropy change is used the trial and error method can be expressed by equation (6) which be similar method of Chanthevee P *et al* [4]. Thus, the simulation result which combines both joule heating heat sources and entropy change as shown as black line in figure 7. In addition, the experimental result and simulation result at 1C are represented as shown in figure 8, respectively.

5. Conclusion

5.1. *Internal resistance calculation*

This current pulse technique can be used for the internal resistance of the battery using the current pulse pattern for discharging and the voltage drop for the calculation. The result of the internal resistance calculation consists of three different internal resistance of terms which can be used for heat generation simulation. Thus, this current pulse technique (CPT) can be applied for various discharge rate using the similar current pattern and calculation.

5.2. *Thermal behaviour*

The internal resistance obtained from the current pulse technique is more precision than measuring from device because the temperature trend of joule heating is like the experimental result. Consequently, the heat source contribution of this paper is difference compared to the of Chanthevee P *et.* It is because the internal resistance measurement of this paper includes the effect of three different terms of the resistance. Thus, the temperature of joule heating is relatively higher. Thus, this is more practical for the battery heat generation modelling.

Acknowledgement

The author would like to acknowledge Materials for Energy Research Unit, National Metal and Materials Technology Center (MTEC) and King Mongkut's University of Technology Thonburi for equipment and financial support along the study.

IOP Conf. Series: Materials Science and Engineering **501** (2019) 012057 doi:10.1088/1757-899X/501/1/012057

Reference

[1] www.batteryuniversity.com

[2] Schweiger HG, Obeidi O, Komesker O, Raschke A, Schiemann M, Zehner C, Gehnen M, Keller M and Birke P 2010 Comparison of several methods for determining the internal resistance of lithium ion cells *Sensor* **10** 5604-25 .

[3] Lam L 2011 A Practical Circuit - based Model for State of Health Estimation of Li - ion Battery Cells in Electric Vehicles 174.

[4] Jeon DH and Baek SM 2011 Thermal modeling of cylindrical lithium ion battery during discharge cycle. *Energy Convers Manag* **52** 2973-81.

[5] Chanthevee P, Hirai S, Lailuck V, Laoonual Y, Sriam P, Rompho S, Chanurai N and Masomtob M 2018 A Simplified Approach for Heat Generation Due to Entropy Change in Cylindrical LCO Battery. *ITEC Asia-Pacific 2018 - 2018 IEEE Transp Electrif Conf Expo, Asia-Pacific E-Mobility A Journey from Now Beyond* 1-5.

[6] Ralph E. Williford,Vilayanur V and Viswanathan J-GZ 2009 Effects of entropy changes in anodes and cathodes on the thermal behavior of lithium ion batteries *J.Power Source* **189** 101-7

[7] www.oskabatt.com

[8] Thanagasundram S, Arunachala R, Makinejad K, Teutsch T and Jossen A 2012 A Cell Level Model for Battery Simulation *Eur Electr Veh Congr* 1-13.

IOP Conf. Series: Materials Science and Engineering **501** (2019) 012058 doi:10.1088/1757-899X/501/1/012058

Mode and Torque of Flows around a Rotating Disk in a Cylindrical Casing

T Watanabe[1,*]

[1]Graduate School of Informatics, Nagoya University, Nagoya, 464-8601, Japan

* Corresponding Author: watanabe@i.nagoya-u.ac.jp

Abstract. The flow around a rotating disk in a cylindrical casing is investigated by the direct numerical simulation. The sizes of the disk and the casing are finite and the flow field has a radial clearance and an axial clearance. The interaction of the flows in these clearances make complex three-dimensional flow structures. In the present study, the patterns and the bifurcation processes of the fully developed flows are classified and the developing process of the flow is identified. The torques acting on the disk surface and the rim of the disk are evaluated and the power laws governing the relations among the torque, the Reynolds number and the thickness ratio of the disk are determined.

1. Introduction

The flows around an axisymmetric bodies can be found in a variety of fluid machinery and bio-reactors and the investigation of the flow is important from the engineering's point of view. They also show basic three-dimensional and cross-flow phenomena and have an interest in the research area of the fluid dynamics [1]. In the present study, the direct numerical simulation is performed to study the flow around a rotating disk in a cylindrical casing. The disk has a finite thickness. It is supported at the center of the casing by a thin driving shaft. This flow configuration has a radial clearance between the sidewall of the casing and the rim of the disk and an axial clearance between the end wall of the casing and the surface of the disk. The interaction of the flows in these clearances forms a complex three-dimensional flow structures and the bifurcation processes that have not been studies well.

In the radial clearance, Taylor-Couette system is established where the sidewall of the casing and the disk rim correspond to the stationary outer cylinder wall and the rotating inner cylinder wall, respectively [2-4]. Indeed, Taylor vortices have been predicted to exist between the rotating disk rim and the stationary enclosure of the hard disk drive [5]. The axial length is limited by the stationary end walls of the casing and the effect of the Ekman pumping mechanism on the end walls makes secondary flows at the Reynolds number below that for the onset of Taylor vortices and brings the imperfect bifurcation process [6, 7]. This imperfection forms a multiple states and solutions of the flow and the flow structures includes the normal mode flow, the anomalous mode flow and the secondary mode flow in addition to the axisymmetric Taylor vortex flow, wave Taylor vortex flow and modulated flows. The appearance of these flow modes in the accelerated and the decelerated flow was investigated experimentally and numerically [8-10]. Recently, the periodically driven flow [11], the flow with passive contaminants [12] and the simulation by multiple turbulent model [13] of Taylor-Couette system have been examined.

The flow in the axial clearance can be considered as the flow in the cylindrical cavity flow between rotating and stationary disks and the flow in the annular cavity flow between disks with a hub at their center. Saric et al. [14] and Launder et al [15] reviewed the stability of these cavity flows. The detail experimental result of the cylindrical cavity flow by Schouveiler et al. [16] showed that the circular vortices and spiral vortices appeared in the wider axial clearance and the turbulent spots and solitary vortices emerged in the narrower axial clearance. Yim et al. [17] numerically represented the similarity of the flow to the von Kármán self-similar solution and examined the stability of the flow between the disk edge and the sidewall of the casing. The experimental study by Wan et al. [18] confirmed the existence of the merged and separated boundary layers on the rotating and stationary disks and estimated the effect of the superhydrophobic surface. In the annular cavity with a hub at its center, the disturbances in the Bödewadt layer on the stationary end wall propagate along the hub and reach the Ekman layer on the rotating disk. The stationary hub suppresses the propagating disturbances [19], while the hub rotating with the disk intensifies the disturbances and makes the flow turbulent [20, 21]. Other than the cylindrical and annular cavity flows, the flow between two co-rotating disks confined by a stationary enclosure has been examined. Huang and Hsieh [22] experimentally showed the polygonal flow structures around the hub and the hysteresis of the flow structures against the change of the Reynolds number. Uenishi et al. [23, 24] carried out experimental and numerical study and showed the symmetric and asymmetric flow structures in the meridional section as well as the polygonal flow structures in the axial clearance.

In this paper, we investigate the flow with a radial clearance and an axial clearance by the direct numerical simulation. For this flow, Schouveiler et al. [16] suggested the important effect of the radial clearance, though they showed no result. We clarified that various flow structure appeared in the radial clearance [25]. Following this work, we study the vortex structures and specify the scenario of their developing processes. The details of the effect of the acceleration rate of the disk from rest on the final flow structures are shown. The torques acting on the disk surface and the disk rim are evaluated and the power lows among the torques, the Reynolds numbers and the thickness ratios of the disks are presented.

2. Formulation

The schematic of the flow field and the cylindrical coordinate system $(\square, \square, \square)^{\square}$ used in the numerical calculation are shown in figure 1. The origin is at the center of the lower end wall of the casing and the axial $\square$ axis is align with the central axis. The length and the inner radius of the stationary casing are $\square_{\square}$ and $\square_{\square}$, respectively. The thickness and the radius of the disk are $\square_{\square}$ and $\square_{\square}$, respectively. The disk has a driving shaft of the radius $\square_{\square}$. The disk is supported at the center of the casing. All physical values are made dimensionless by using the disk radius as a reference length and the circumferential velocity at the disk rim as a reference velocity. The Reynolds number $\square\square$ is defined by these reference values and the kinematic viscosity of the working fluid. The size of the cylindrical casing is fixed and $\square_{\square} = 0.3150$ and $\square_{\square} = 1.118$. The radius of the driving shaft $\square_{\square} = 0.0787$ and it is small enough not to have any influence on the flow structures [26]. Eight disks with different thicknesses are used and the width of the axial clearance is settled. The thickness ratios $\square$ defined by $\square_{\square}/\square_{\square}$ are 0.50, 0.55, 0.60, 0.65, 0.70, 0.75, 0.80 and 0.85.

The equations governing the incompressible fluid flow are the equation of continuity and the three-dimensional unsteady Navier-Stokes equation

$$\nabla \cdot \boldsymbol{u} = 0\,, \tag{1}$$

$$\frac{\partial \boldsymbol{u}}{\partial t} + \nabla(\boldsymbol{u}\boldsymbol{u}^T) = -\nabla p + \frac{1}{Re}\nabla^2 \boldsymbol{u}\,. \tag{2}$$

Here, □ is time and □ is the velocity vector with its radial, circumferential and axial components of □, □ and □, and □ is the pressure. The boundary condition is the no-slip condition for the velocity components and the Neumann condition for the pressure. Initially, the fluid is at rest. At the beginning of the calculation, the disk is suddenly accelerated or it is linearly accelerated to the rotation speed

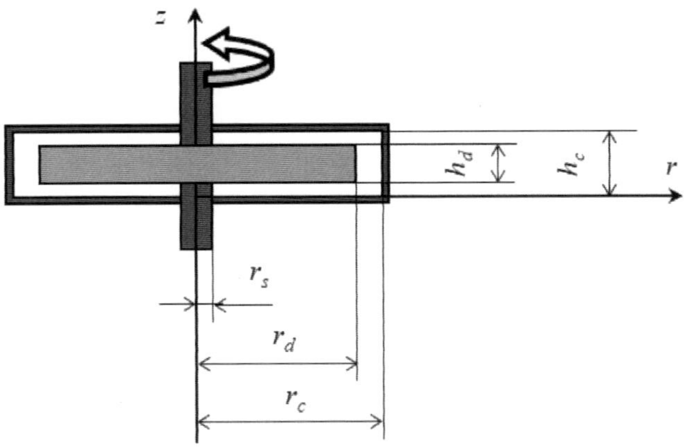

Figure 1. Schematic of the flow field.

defined by the prescribed Reynolds number. The discretization method is the finite difference method proposed by Lygren and Andersson [27]. The fully region of □ from 0.0 to 2□ is considered. The verification and the validation of the present code has been confirmed [28].

The numerical results are intended to be compared with the experimental results. The experimental apparatus has a casing with the length of 40 mm and the inner radius of 142 mm, and the disk radius is 127mm. For the dimensionless time □, the dimensional time in second is given by 641.8 □/□□ when the kinematic viscosity of the fluid is 4.0 × 10^{-6} [□^{2}/□]. The acceleration rate □ is defined by the increment of the Reynolds number per dimensional second. For convenience, we assume that the expression □ = □ means the sudden acceleration. In addition to the sudden acceleration, several values of the acceleration rate are adopted to investigate the effect on the flow development.

3. Results and discussion

3.1. *Flow structures found in the radial clearance and the axial clearance*

The modes of Taylor vortex flows between two rotating cylinders with finite lengths are determined by the number of vortices and the rotating direction of the vortices in the meridional section [29]. In the radial clearance around a rotating disk in a casing, the flows similar to Taylor vortex flows emerge. In this study, four types of flows appear as depicted in figure 2. In this result, the acceleration

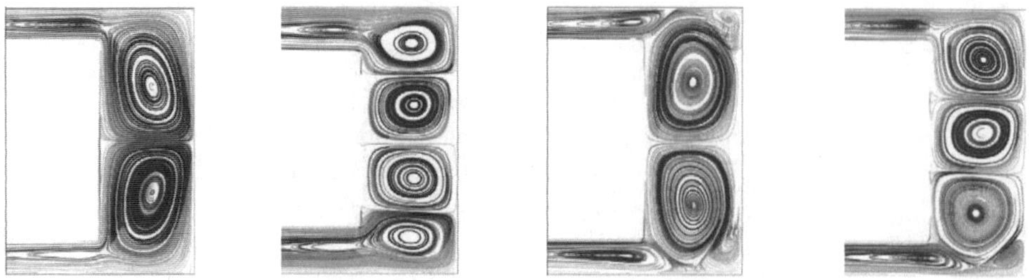

(a) Normal 2-cell mode. $Re = 5000$, □ = 0.75

(b) Normal 4-ce ll mode. $Re = 4000$, □ = 0.65

(c) Anomalous 2-cell mode. $Re = 10000$, □ = 0.75

(d) Anomalous 3-cell mode. $Re = 6000$, □ = 0.75

Figure 2. Flow structures in the radial clearance. □ = □.

is the sudden acceleration and □ = □. In each panel, the left is the rotating disk and the right is the sidewall of the casing. The flow pattern represents the streamline like lines obtained from the radial and axial velocity components. The colors designate the radial flow direction, and the red and the blue correspond to the radially outward flow and the radially inward flow, respectively. Figure 2 (a) shows the flow at $Re = 5000$ and □ = 0.75. Two large vortices appear in the radial clearance. The radial flow is outward near the axial mid-plane, and the upper and lower vortices rotate in the counterclockwise and clockwise directions, respectively. This indicates that the flows on the upper and lower end walls are radially inward. The flow structure with radial inward flows on the end walls is in the normal mode [29]. This flow has two vortices and the flow mode is the normal 2-cell mode. The flow in figure 2 (b) has four vortices. The upper most and lower most vortices rotate in the counterclockwise and clockwise directions, respectively, and the flow mode is the normal 4-cell mode. Figure 2 (c) shows the flow that includes two large vortices and the flows near the end walls are radially outward. On the stationary end wall, the flow retards and the centrifugal effect becomes small. This makes our physical intuition say that the flow on the end wall is radially inward. However, the flow in figure 2 (c) is outward near the end wall and this contradicts with our insight. Therefore, the vortex with an outward flow on the end wall is anomalous [29], and the flow mode is the anomalous 2-cell mode. The anomalous vortex accompanies an extra vortex. In figure 2 (c), the light blue regions at the upper-right and lower-right corners almost correspond to extra vortices attached to the two main anomalous vortices. The flow in figure 2 (d) has three vortices and the flow structure is axially asymmetric. The rotations of the upper most and the lower most vortices are counterclockwise and the lower most vortex is anomalous. The flow mode is the anomalous 3-cell mode. The extra vortex form at the lower right corner.

Figure 3 shows the bird's eye views of the four flow structures in the axial clearance. The disk rotates in the counterclockwise direction when it is seen from above. The vortex structures are extracted by the contour of Q that is the second invariant of the velocity gradient tensor. The color on the contour indicates the value of the circumferential velocity component. Figure 3 (a) depicts the circular vortices. The flow mode in the radial clearance is the normal 2-cell mode and two vortices surround the disk. The Reynolds number is 14000 and it is not small, small thin vortices on the sidewall of the casing attach to the two main vortices. These thin vortices have been confirmed in the study of the direct numerical simulation of the turbulent Taylor-Couette system [30]. When the Reynolds number is small, no thin vortex appears and the flow structure is axisymmetric. The flow in figure 3 (b) has a polygonal vortices of nonagon around the disk. The flow in the radial clearance is the anomalous 2-cell mode. The anomalous vortex has an extra vortex and the circumferential deformation of the extra vortex results in the polygonal vortices [25]. Following the shape change of the extra vortex, the two main vortices become wavy in the circumferential direction. In figure 3 (c), the disturbances generated in the radial clearance propagate and several tens of spiral vortices extend from polygonal vortices around the disk to the radially inward region over the disk. Because the normal direction of the front of the spiral vortex is opposite to the rotating direction of the disk, the front angle of the spiral vortex is negative. The flow in the radial clearance is the anomalous 2-cell mode and the circumferential frequency of the polygonal structure is higher than that of the flow in figure 3 (b). Figure 3 (d) represents the flow with spiral vortices with a positive front angle. The flow in the radial clearance is in the anomalous 2-cell mode but it is almost unconstructed. The Reynolds number is the high value of 40000 and the spiral vortices with a positive front angle seem to be the ones that spontaneously appear between rotating and stationary disks [31].

3.2. *Flow structures and the acceleration rate of the disk*

An unconstructed turbulent state in the axial clearance appears in addition to the four types of vortices denoted in the previous subsection. Figure 4 summarizes the flow structures in the axial clearance at the Reynolds number up to 40000. At each thickness ratio ⊓, the profile gives the ranges of the Reynolds number in which the circular vortices, polygonal vortices, spiral vortices with a negative

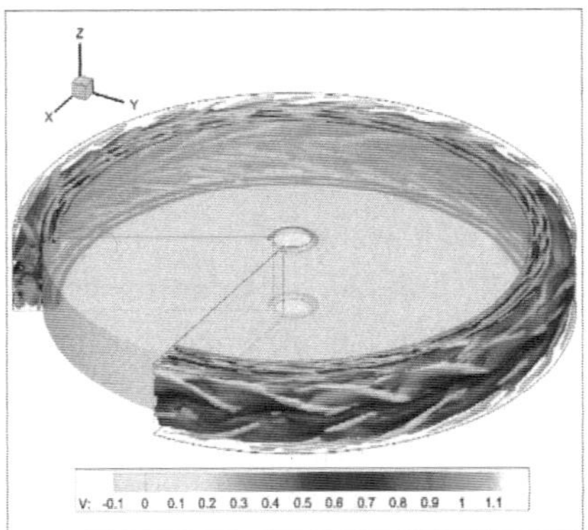

(a) Circular vortices in the normal 2-cell mode. *Re* = 14000, □ = 0.85, *Q* = 0.1.

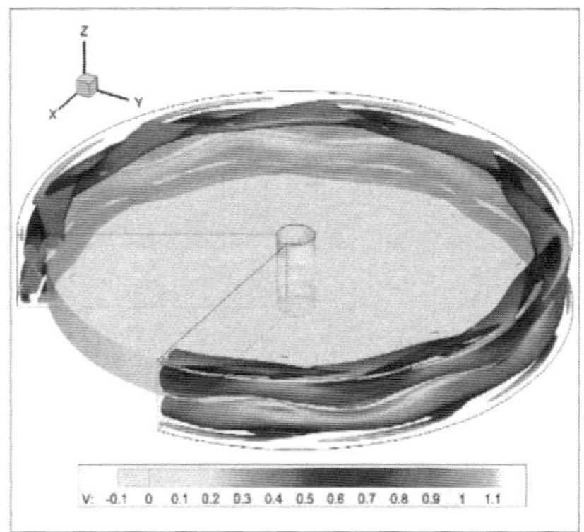

(b) Polygonal vortices in the anomalous 2-cell mode. *Re* = 8000, □ = 0.55, *Q* = 1.0.

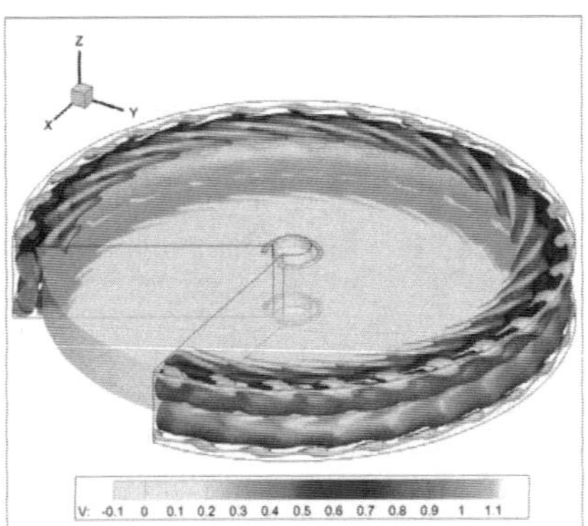

(c) Spiral vortices with a negative front angle in the anomalous 2-cell mode. *Re* = 8000, ⊓ = 0.75, *Q* = 1.0.

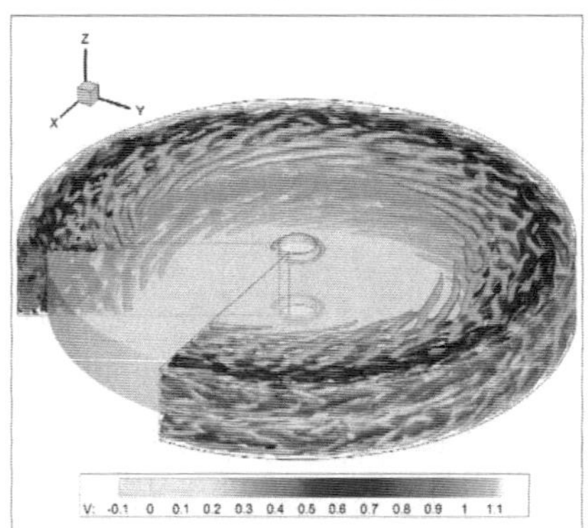

(d) Spiral vortices with a positive front angle in the anomalous 2-cell mode. *Re* = 40000, ⊓ = 0.75, *Q* = 0.1.

Figure. 3. Flow structures in the axial clearance. □ = □. The disk is rotating in the counterclockwise direction when viewed above.

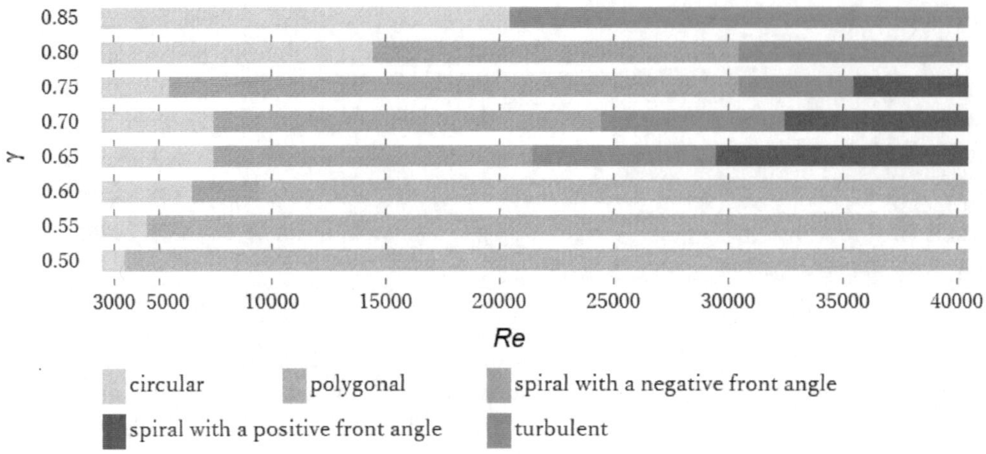

Figure. 4. Flow patterns found in the axial clearance. □ = □.

angle, spiral vortices with a positive front angle and turbulent state appear. When the thickness ratio is 0.50 and 0.55, the circular vortices arise at small Reynolds numbers and the polygonal vortices with the anomalous 2-cell mode form in a wide range of the Reynolds number. When the thickness ratio is small, the axial clearance is wide. The flow generated on the Ekman layer on the rotating disk, which has a high value of the circumferential velocity component, easily transfers. This flow tends to bring a radially outward flow on the end wall in the radial clearance and makes the flow mode anomalous. At □ = 0.60, the spiral vortices with a negative front angle follows the circular vortices. As shown in figure 3 (c), the disturbances originating from the several tens of polygonal vortices travel on the stationary end wall of the casing and spread over the disk. When the thickness ratio is from 0.65 to 0.75, the turbulent state and the spiral vortices with a positive front angle emerge after the appearance of the spiral vortices with a negative front angle. The existence of the turbulent state between the spiral vortices with negative and positive front angles supports the assumption that the spiral vortices with different front angles forms from different fluid dynamics mechanics. Schouveiler et al. [16] showed that the onset Reynolds number of the spiral vortices with a positive angle increases as the axial clearance becomes narrower. This is consistent with the current result that the region of the spiral vortices with a positive front angle found at □ = 0.75 disappears and the region of the turbulent state widens at □ = 0.80. At □ = 0.85, the convection in the axial clearance is weak and no coherent structure other than the circular vortices forms.

The acceleration rate affects the flow structures after the completion of the disk acceleration. The well-developed flow structures obtained at the acceleration rates of □ = □ and □ = 500 are shown in the Table 1 where the Reynolds number is from 3000 to 20000. In each table, the row and the column show the Reynolds number *Re* and the thickness ratio □, respectively. The colors correspond to the flow modes in the radial clearance and the characters mean the flow structures in the axial clearance. The result at □ = 1000 is in the reference [25]. In the cases with two characters of C and S, the circular vortices and the spiral vortices appear at the inner and outer part of the disk region, respectively. At □

Table 1. Final flow structures at □ = □ and 500 [a].

(a) □ = □.

	γ							
	0.50	0.55	0.60	0.65	0.70	0.75	0.80	0.85
3000	C	C	C	C	C	C	C	C
4000	P	C	C	C	C	C	C	C
5000	P	P	C	C	C	C	C	C
6000	P	P	C	C	C	C,S	C	C
7000	P	P	S	C	C	C,S	C	C
8000	P	P	S	S	S	S	C	C
9000	P	P	S	S	S	S	C	C
10000	P	P	P	S	S	S	C	C
11000	P	P	P	S	S	S	C	C
12000	P	P	P	S	S	S	C	C
13000	P	P	P	S	S	S	C	C
14000	P	P	P	S	S	S	C	C
15000	P	P	P	S	S	S	C,S	C
16000	P	P	P	S	S	S	C,S	C
17000	P	P	P	S	S	S	S	C
18000	P	P	P	S	S	S	S	C
19000	P	P	P	S	S	S	S	C
20000	P	P	P	S	S	S	S	C

(b) □ = 500.

	γ							
	0.50	0.55	0.60	0.65	0.70	0.75	0.80	0.85
3000	C	C	C	C	C	C	C	C
4000	P	C	C	C	C	C	C	C
5000	P	P	C	C	C	C	C	C
6000	P	P	C	C	C	C	C	C
7000	P	P	S	C	C	C	C	C
8000	P	P	S	S	S	C	C	C
9000	P	P	S	S	S	C	C	C
10000	P	P	S	S	S	C	C	C
11000	P	P	P	S	S	C	C	C
12000	P	P	P	S	S	C	C	C
13000	P	P	P	S	S	S	C	C
14000	P	P	P	S	S	S	C	C
15000	P	P	P	S	S	S	C	C
16000	P	P	P	S	S	S	C	C
17000	P	P	P	T	S	S	C	C
18000	P	P	P	T	S	S	C	C
19000	P	P	P	T	S	S	C	C
20000	P	P	P	T	S	S	C	C

a. yellow: normal 2-cell mode, orange: normal 4-cell mode, green: anomalous 2-cell mode, blue : anomalous 3-cell mode. C : circular vortices, P : polygonal vortices, S : Spiral vortices with a negative front angle, T: turbulent flow.

= □ and □ = 500 and □ up to 0.55, the flow has the anomalous 2-cell mode in all cases and the polygonal vortices in cases except those at the lowest Reynolds number. When the thickness ratio is from 0.60 to 0.70, the spiral vortices tend to emerge instead of the polygonal vortices. At □ = □ and □ = 0.75 and 0.8, the number of cases of the circular vortices in the normal 2-cell mode increases. When □ is 500, the region of the normal 4-cell mode expands at □ from 0.70 to 0.80 and the incoherent turbulent state appear at □ = 0.65 and $Re \geq 17000$. At both acceleration rates, all cases have the normal 2-cell mode at □ = 0.85.

Table 2 represents the flow structures obtained at lower acceleration rates of □ = 333, 250, 220 and 200. The Reynolds number is from 3000 to 7000 and the thickness ratio is from 0.50 to 0.70. The flow at □ = 0.70 has the circular vortices in the normal 2-cell mode at the Reynolds number up to 7000. The cases of this flow structure occupy wider region as the decrease of the acceleration rate and the all cases are the circular vortices in the normal 2-cell mode at □ = 200. As shown above, the flow on the rotating disk makes the radially outward flow on the end wall in the radial clearance. At higher acceleration rate, the strong radially outward flow appears just after the beginning of the rotation and its effect is large. When the acceleration rate is low, the radially outward flow is not strong enough to suppress the radially inward flow generated by the Ekman pumping that is consistent with the physical intuition, and the normal mode tends to appear frequently.

Table 2. Final flow structures at lower acceleration rates.

□ = 333.

□ = 250.

γ

	0.50	0.55	0.60	0.65	0.70
3000	C	C	C	C	C
4000	P	C	C	C	C
5000	P	C	C	C	C
6000	P	P	C	C	C
7000	P	P	S	C	C

□ = 230.

γ

	0.50	0.55	0.60	0.65	0.70
3000	C	C	C	C	C
4000	P	C	C	C	C
5000	P	C	C	C	C
6000	P	C	C	C	C
7000	P	P	C	C	C

□ = 200.

γ

	0.50	0.55	0.60	0.65	0.70
3000	C	C	C	C	C
4000	P	C	C	C	C
5000	P	C	C	C	C
6000	P	C	C	C	C
7000	P	C	C	C	C

γ

	0.50	0.55	0.60	0.65	0.70
3000	C	C	C	C	C
4000	C	C	C	C	C
5000	C	C	C	C	C
6000	C	C	C	C	C
7000	C	C	C	C	C

3.3. *Process of the flow development*

The finial flow modes appear some developing processes from the rest. Figure 5 shows one of the processes. The acceleration is sudden and the final flow includes the circular vortices in the normal 2-cell mode. First, small vortices appear at the tips of the rotating disk ($t = 2.0$, 0.257 (s) in the dimensional time). Then, as these vortices grow, a pair of another vortices form on the rim of the disk ($t = 4.0$, 0.513 (s) in the dimensional time), and the flow have four vortices in the radial clearance. This 4-cell mode is transient. The central two vortices collapse ($t = 60$, 7.70 (s) in the dimensional time), and then the flow stabilizes in the final normal 2-cell mode state. The appearance and disappearance of vortices and the mode competition as shown in figure 5 have been studied in Taylor-Couette system [32, 33].

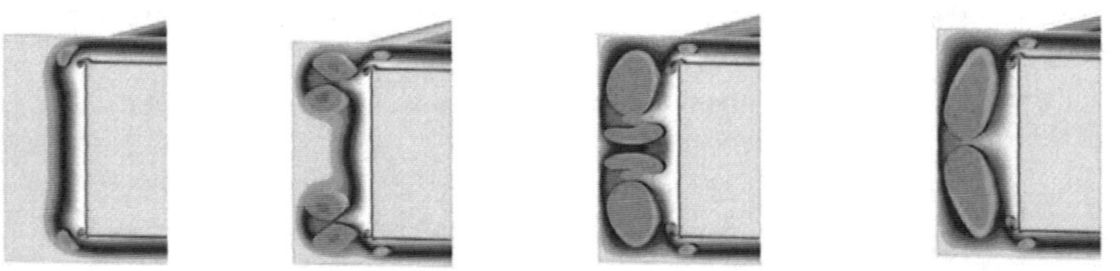

$t = 2.0$ $t = 4.0$ $t = 60.0$ $t = 76.0$

Figure 5. Formation process of the normal 2-cell mode. □ = □, $Re = 5000$, □ = 0.75.

3.4. *Torque of the disk and its power law*

The evaluation of the torque acting on the disk is important to improve the efficiency of fluid machinery. In this subsection, we estimate the effects of the Reynolds number and the thickness ratio on the torques on the surface and the rim of the rotating disk in the anomalous 2-cell mode.

Figure 6 shows the relations among the torque on one side of the disk □, the Reynolds number and the thickness ratio. The profiles in figure 6 (a) give the variations of the torque with respect to the Reynolds number for the thickness ratios from 0.50 to 0.85. In the logarithmic scales, the torque □ is almost linearly proportional to the Reynolds number. Thought the effect of the thickness ratio on the torque is not large, the torque decreases as the thickness ratio increases and the axial clearance becomes narrower. In order to figure out the effect of the thickness ratio, we assume the torque □□ has a power law with the Reynolds number *Re* and the thickness ratio □ and drive the following equation

$$T_s = 0.05 Re^{1.58} \gamma^{-0.26}. \tag{3}$$

The profile of the normalized torque is in figure 6 (b). The normalization properly reduces the effect of the thickness ratio and the all plots are almost on the power law given by the yellow line. The figure includes the empirical result by Daily and Nece [34], which evaluates the torque of a rotating plane disk enclosed by a cylindrical chamber. The empirical profile gives lower values when compared with the law of equation (3). This may be because that the existence of the radial clearance in the present study gives an increase of the torque on the disk.

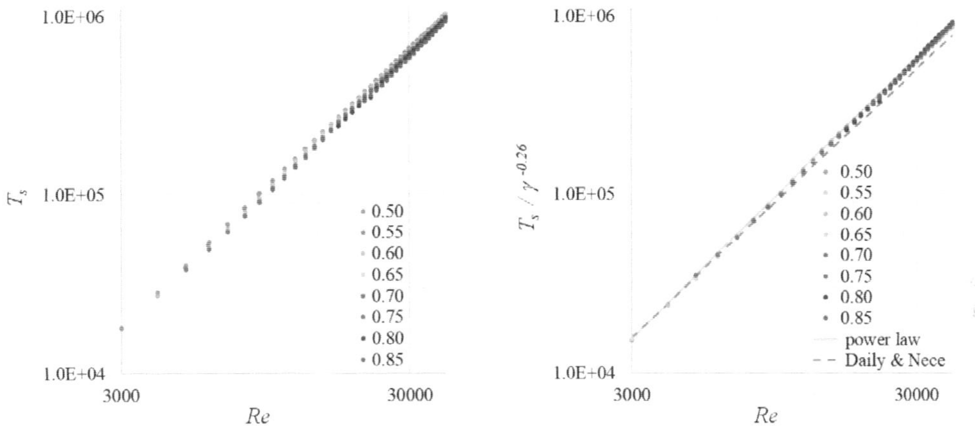

(a) Raw profiles. (b) Normalized profiles.

Figure 6. Torque on a disk surface in the anomalous 2-cell mode.

Figure 7 presents the torque acting on the disk rim □□. Thought the area causing the torque □□ is smaller than that of the torque □□, the radius is maximum at the disk rim and the value of □□ is comparable with that of □□. The thickness ratio directly affects the area of the rim, and each thickness ratio gives its own straight profile of the torque as shown in figure 7 (a). The power law among the torque, the Reynolds number and the thickness ratio gives the equation

$$T_r = 0.32 Re^{1.46} \gamma^{1.65}. \tag{4}$$

The normalized torque in figure 7 (b) gives the unique profile regardless the thickness ratio. We may reasonably consider that the torque □□ is linearly proportional to the thickness ratio. However, the exponent of □ in equation (4) is not unity. This nonlinearity may depends on the interaction of the flows in the radial and axial clearances, and the large interaction at the lower thickness ratio decreases the torque on the disk rim.

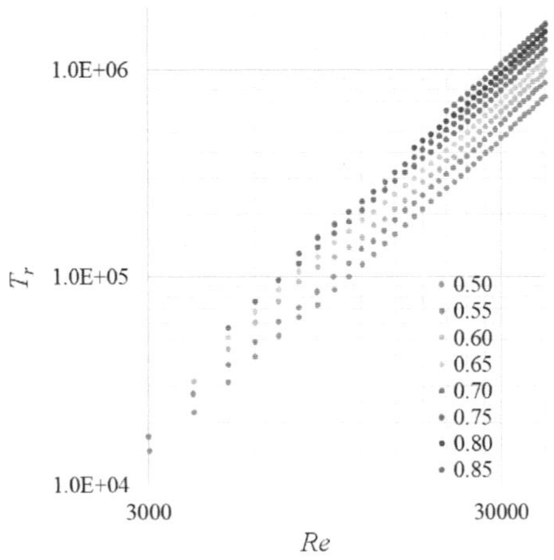

(a) Raw profiles.

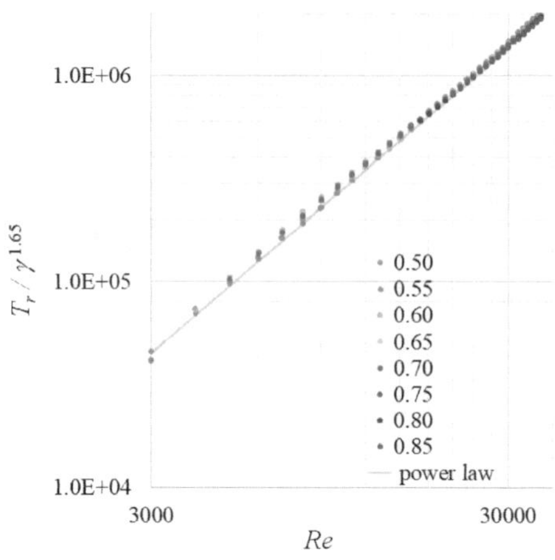

(b) Normalized profiles.

Figure 7. Torque on disk rim in the anomalous 2-cell mode.

4. Conclusion

The flow around a rotating disk in a cylindrical casing is numerically investigated. The flow field has a radial clearance and an axial clearance and characteristic flows develop in these clearances. The Reynolds number based on the disk radius and the circumferential velocity at the disk rim is from 3000 to 20000 or 40000. The ratio of disk thickness to the length of the cylinder is from 0.50 to 0.85. In the radial clearance, the modes of the normal 2-cell, normal 4-cell, anomalous 2-cell and the anomalous 3-cell appear. In the axial clearance, the circular vortices, polygonal vortices, spiral vortices with a negative and a positive front angle and the turbulent state exist. The diagram of these flow structures in the space of the Reynolds number and the thickness ratio is clarified for the case of the sudden acceleration. The anomalous mode and the polygonal vortices tend to appear at the smaller thickness ratios where the interaction of the flows between the radial and the axial clearances is rather strong. When the thickness ratio is large, the normal 2-cell mode usually found in Taylor-Couette system frequently forms. In case that the acceleration rate is small and the flow development is quasi-steady, the circular vortices in the normal 2-cellmode appear. It is shown that the flow develops to the final state via several structures. The torques acting on the disk surface and the disk rim are evaluated and the torques' laws are introduced, which represent the relations with the Reynolds number and the thickness ratio.

References

[1] Childs P R N 2011 *Rotating Flow* (Amsterdam: Butterworth-Heinemann/Elsevier).
[2] Tagg R 1994 *Nonlinear Science Today* **4** 1-24
[3] Fardin M A, Perge C and Taberlet N 2014 *Soft Matter* **10** 3523-35
[4] Grossmann S, Lohse D and Sun C 2016 *Annu. Rev. Fluid Mech.* **48** 53-80
[5] Hendriks F 2010 *Microsyst. Technol.* **16** 93-101
[6] Benjamin T B 1978 *Proc. Royal. Soc. Lnd. A* **359** 1-26
[7] Benjamin T B 1978 *Proc. Royal. Soc. Lnd. A* **359** 27-43
[8] Nakamura I and Toya Y 1996 *Acta Mechanica* **117** 33-46

[9] Watanabe T, Furukawa H and Nakamura I 2002 *Phys. Fluids* **14** 333-41
[10] Furukawa H, Watanabe T, Toya Y and Nakamura I 2002 *Phys. Rev. E* **65** 036306
[11] Verschoff R A, te Nijenhuis A K, Huisman S G, Sun C and Lohse D 2018 *J. Fluid Mech.* **846** 834-45
[12] Majji M V, Morris J F 2018 *Phys. Fluids* **30** 033303
[13] Luo G, Yao Z and Shen H 2018 *Phys. Fluids* **30** 065103
[14] Saric W S, Reed H L and White E B 2003 *Annu. Rev Fluid Mech.* **35** 413-40
[15] Launder B, Poncet S and Serre E 2010 Annu. *Rev Fluid Mech.* **40** 229-48
[16] Shouveiler L, Le Gal P and Chauve M P 2001 *J. Fluid Mech.* **443** 329-50
[17] Yim E, Chomaz J-M, Martinand D and Serre E 2018 *J. Fluid Mech.* **848** 631-47
[18] Wang C, Tang F, Li Q and Wang X 2018 *Exp. Fluids* **59** 47
[19] Serre E, Crespo del Arco E and Bontoux P 2001 *J. Fluid Mech.* **434** 65-100
[20] Séverac É, Poncet S, Serre É and Chauve M-P 2007 *Phys. Fluids* **19** 085113
[21] Makino S, Inagaki M and Nakagawa M 2015 *Flow Turbulence Combust.* **95** 339-413
[22] Huang R F, Hsieh M K 2011 *Exp. Fluids* **51** 1529-47
[23] Uenishi R, Tsugawa S, Maeda T, Uesaka M, Watanabe T, Tanigawa H and Hirata K 2016 *Proc. the 7th TSME Int. Conf. on Mech. Engng. (Chaing Mai)* TSF0025
[24] Uenishi R, Watanabe T, Tanigawa H and Hirata K 2017 *Proc. the 8th TSME Int. Conf. on Mech. Engng. (Bangkok)* TSF0005
[25] Watanabe T, Endo S 2017 *Proc. the 8th TSME Int. Conf. on Mech. Engng. (Bangkok)* TSF0002
[26] Poncet S, Serre É and Le Gal P 2009 *Phys. Fluids* **21** 064106
[27] Lygren M and Andersson H I 2001 *J. Fluid Mech.* **426** 293-326
[28] Watanabe T and Furukawa H 2010 *Phys. Fluids* **22** 114107
[29] Nakamura I, Toya Y, Yamashita S and Ueki Y 1990 *JSME Int. J. Ser. II* **33** 685-91
[30] Dong S 2008 *Phys. Rev. E* **77** 035301
[31] Cros A and Le Gal P 2002 *Phys. Fluids* **14** 3755-65
[32] Wei T, Kline E M, Lee S H-K and Woodruff S 1992 *J. Fluid Mech.* **245** 47-68
[33] Park J, Billant P, Baik J-J and Seo J M 2018 *J. Fluid Mech.* **840** 5-24
[34] Daily J W and Nece R E 1960 *J. Basic Engng.* **82** 217-30

IOP Conf. Series: Materials Science and Engineering **501** (2019) 012059 doi:10.1088/1757-899X/501/1/012059

Flow Analysis of the Symmetric Airfoil NACA0015 Blade at Different Angles of Attack and Mixing Velocities in a Continuous Stirred Tank Reactor (CSTR)

W Manosroi[1,*], C Phapisarn[1], N Moonmuang[1] and N Yasan[1]

[1]Department of Mechanical Engineering, Faculty of Engineering, Chiang Mai University, Chiang Mai, Thailand

[*]Corresponding Author: E-mail: woradej.manosroi@gmail.com

Abstract. Biogas produces from the degradation of organic compounds of garbage and water wastes from various sources such as ranch and agricultural industries, usually by the fermentation process with anaerobic microorganisms. Layers of sediment are often occurred that can block the flow and longer mixing time is required for a more homogenous mixing. The objective of this study is to solve the non-uniform and non-homogenous as well as to save the mixing time in the fermentation process of biogas production using a model-mixing reactor. Flow analysis using an image processing technique of the symmetric airfoil NACA0015 blade at different angles of attack and mixing velocities in a Continuous Stirred Tank Reactor (CSTR) was examined. The CSTR was equipped with two and three airfoil blades at the angles of attack 0, 10, 16(stall angle) and 20° operated at the mixing velocities at 80, 110, 140 and 190 rpm. The mixing efficiency was evaluated from the homogenous appearance of plastic particles (5 mm diameter) dispersed in water by an image processing technique. The results showed that the mixing efficiencies of CSTR with three blades were higher than that of the CSTR with two blades of about 1.3 folds. The mixing efficiency increased with increasing mixing velocities and angles of attack, and was almost constant when the blade angles of attack increased from 16° to 20° and the mixing velocity increased from 140 to 190 rpm. This may be due to the airfoil blade stall and the saturation of the mixing. A new blade design for the CSTR system from this study can give a more efficient stir and mixing flow which will not only be beneficial for biogas production in the laboratory scale, but also a model design for the industrial biogas production as well.

1. Introduction

Continuous Stirred Tank Reactor or CSTR is a reactor widely used for producing biogas from energy crop, manure, municipal waste, sewage, green waste or food waste due to its high mixing capability. CSTR gives higher stir and mixing efficiency than the traditional biogas digester tank. A more homogenous system will be obtain due to the higher mixing and mass transfer rate. With effective mixing, the reactive materials homogeneously distributed with negligible mass transfer resistance resulting a more biogas generation [1]. The degradation time is also shorter and higher biogas production rate achieve. A parametric study on the design variable and flow distribution of CSTR by the numerical method using the Lattice Boltzmann Technique conducted by Satjaritanun et al. [1]. CSTR for investigation of the anaerobic co-digestion of *Pennisetum purpureum* cv. Pakchong1 grass and chicken layer has been performed [2]. Watanabe et al. [3] have examined the continuous production of 6-*O*-

decanoyl, dodecanoyl, or tetradecanoyl L-ascorbate through the immobilized lipase-catalyzed condensation of L-ascorbic acid and decanoic, dodecanoic, or tetradecanoic acid in acetone using a CSTR. The dynamic behavior and thermal stability of an autothermal CSTR with two stable operation units have investigated by Furusawa et al. [4]. They found that both experimental and computational results were in good agreement. Mavros et al. [5] have used the Laser Doppler Velocimetry (LDV) to investigate the flow pattern in a stirred tank by an axial-flow impeller, while Bakker et al. [6] have investigated the flow pattern of a pitched blade turbine using Laser Doppler Velocimetry (LDV) and Digital Particle Image Velocimetry (DPIV). The flow was generated by an axial flow impeller in batch has been simulated by Khopkar et al. [7]. The effects of shaft eccentricity on the hydrodynamics of unbaffled stirred tank examined by Montante et al. [8]. There are also some studies on the flow on the hydrofoil and airfoil blade. The flow characteristics of the novel hydrofoil OK-2003 comparing with NACA0015 in the water tunnel investigated by Amromin et al. [9]. They have also investigated the lift and drag characteristics of hydrofoil shape comparing with NACA0015 in the water tunnel [10]. Weetman et al. [11] have examined the fluid forces of impeller and airfoil blade in tank.

Airfoil can generate different flows in the CSTR. However, there is no study on flow analysis of symmetric airfoil NACA0015 blades at different angles of attack and mixing velocities in the CSTR system, also, there is still no data of mixing efficiency comparison in the CSTR between the airfoil blades and other blade types. The objective of this study is to analyze and compare the flow characteristics and mixing efficiency of particles suspended in water using the two and three airfoil NACA0015 blades at different angles of attack and mixing velocities. This study can obtain a new blade design of the CSTR system for a more efficient stir and mixing flow of substances suspended in water, which will be applicable for biogas production in both the laboratory and industrial production scale.

2. Materials and Methods

2.1. Equipment and Materials

The experimental equipment was consisted of the CSTR system and a camera.

2.1.1 The CSTR system

The CSTR system consisted of an acrylic tank, a motor and the blades. The height, diameter and thickness of the tank were 0.4, 0.3 and 0.02 m respectively. **Figure 1** presented the CSTR system (b) and an acrylic tank (a). The motor was the Mitsubishi Super Line Series Induction Motor SF-JR 1/2 HP. **Figure 2** showed the plastic particles for imitating the organic substance for biogas production. The weight, density and diameter of each plastic particle were 0.11 g, 1680 kg/m^3 and 5 mm respectively. Five hundred plastic particles used in this study. The blades were the symmetric airfoil NACA0015 with two and three blades as shown in **Figure 3**.

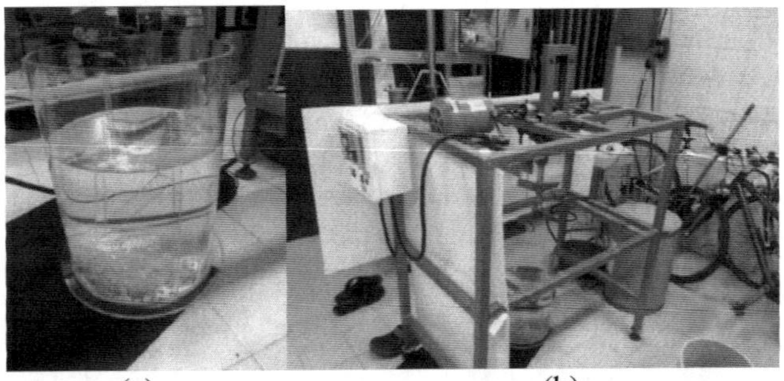

(a) (b)

Figure 1. (a) The acrylic tank (b) The CSTR system

Figure 2. The plastic particles used in the experiment

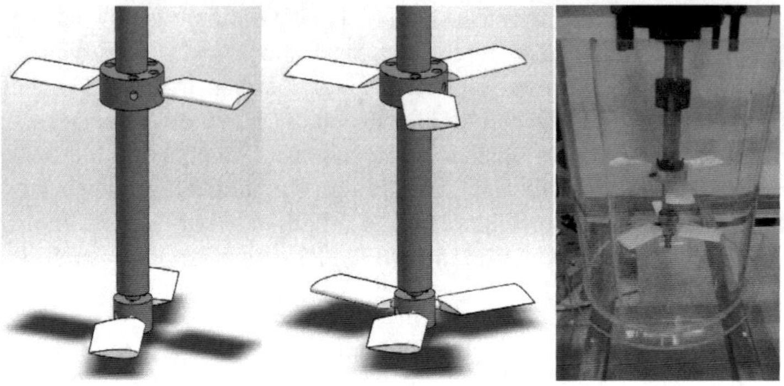

(a) (b) (c)

Figure 3. (a) The airfoil NACA0015, (b) two blades, (c) three blades, (d) The airfoil NACA0015 with three blades in the CSTR system

2.1.2 The Camera

The Canon 600D camera used for taking pictures to investigate the mixing efficiency.

2.1.3 The Mixing Efficiency

The Mixing Efficiency analyses from the distribution of the particles using the ImageJ program and determines by the cross section area of plastic particles that detect by the camera divide by 500 plastic particles cross section. Cross-section area (A_c) of the 500 sphere plastic particles (diameter 0.5 cm each) determine by:

$$A_c = \frac{\pi(0.005)^2}{4} x500 = 9.817x10^{-3}\ m^2$$

The mixing efficiency calculates by cross section area that inject by the camera divide by cross-section area (A_c) of the 500 sphere plastic particles. For example the inject area of the plastic particles by the camera is $5.788 \mathrm{x} 10^{-3}$ m^2:

$$Mixing\ efficiency = \frac{5.788x10^{-3}}{9.817x10^{-3}} = 0.59$$

2.2 *Experimental*

Figure 4 shows the experimental setup. The camera mount in front of the tank. The white paper put behind the tank for good image quality. The core in the CSTR system rotated by the motor.

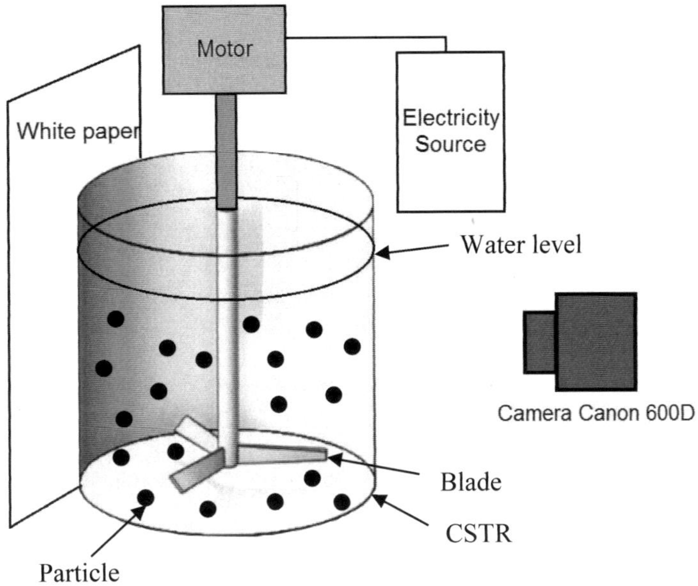

Figure 4. Experimental setup

The experiment started by five hundred plastic particles were suspended in 23 litres of water in the CSTR system as shown in with two or three blades at different angles of attacks (0, 10, 16 and 20°). Then, the motor turned on. The pictures took by the camera every second as shown in **Figure 5**. The mixing efficiency was then analysed. The mixing efficiency was increased with times and approximately constant at three second as presented in **Figure 6**. The constant mixing efficiency was determined at each combination of mixing velocities and blade angles of attack.

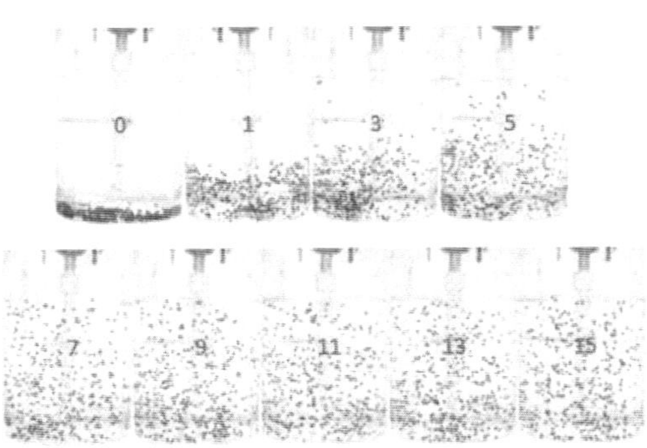

Figure 5. The CSTR pictures taken during the experiment at 0-15 s

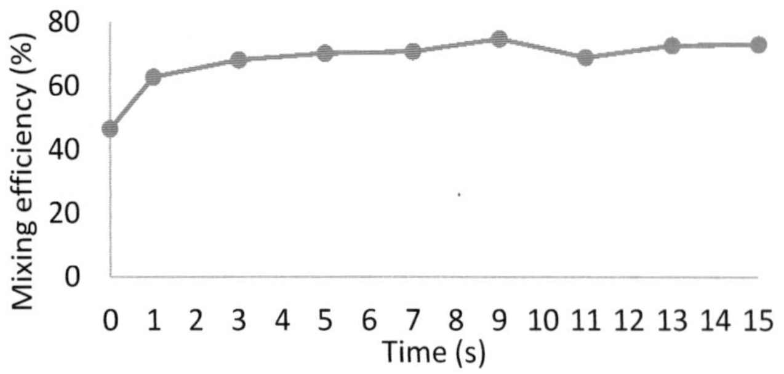

Figure 6. The mixing efficiency at each time interval (in seconds) at the angle of attack 10° and velocity at 80 rpm

3. Results and Discussion

3.1 *Effects of angles of attack on mixing efficiencies*

Figures 7, **8**, **9** and **10** displayed the effects of angles of attack at the mixing velocities of 80, 110, 140, 190 rpm respectively. The highest and lowest mixing efficiencies were 90.89 and 25.55% respectively. In almost every case, mixing efficiencies increased with increasing angles of attack. This may be due to the increase of the performing force between the particles and the blade, which increased with increasing angles of attack. The mixing efficiencies in every case of the three blades were higher than that of the two blades due to the increase of the contact area between the particles and the blades. For the two-blade system at the low mixing velocity (80 rpm) and low angles of attack (0 and 10°), mixing efficiencies were constant when the angles of attack increased from 0 to 10° owing to the not sufficient force to distribute particles in the water as shown in **Figure 7**. At high mixing velocities (140 and 190 rpm) and high angles of attack (16 and 20°), the mixing efficiencies were almost constant when the angles of attack increased from 10 to 20° because of the reaching of mixing saturation and the airfoil blade stall (16°) [12] as shown in **Figures 9** and **10**. This phenomenon agree with the work of Weetman et al. [11]. A stall is a reduction in lift coefficient generated by as Airfoil angles of attack increase. It caused by the separation of the flow behind the airfoil blade at higher angles of attack.

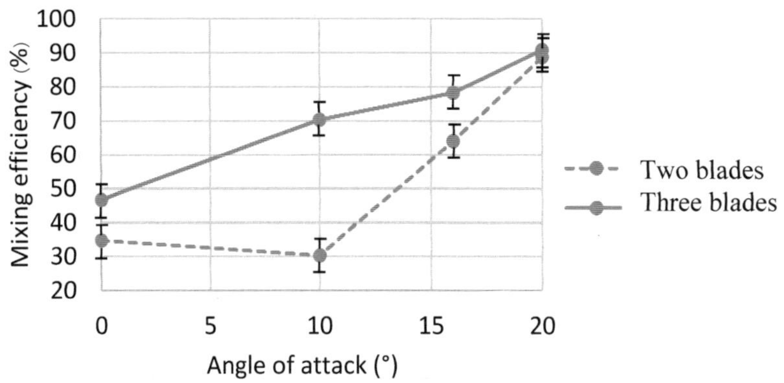

Figure 7. The mixing efficiency at each angle of attack of the mixing velocity at 80 rpm

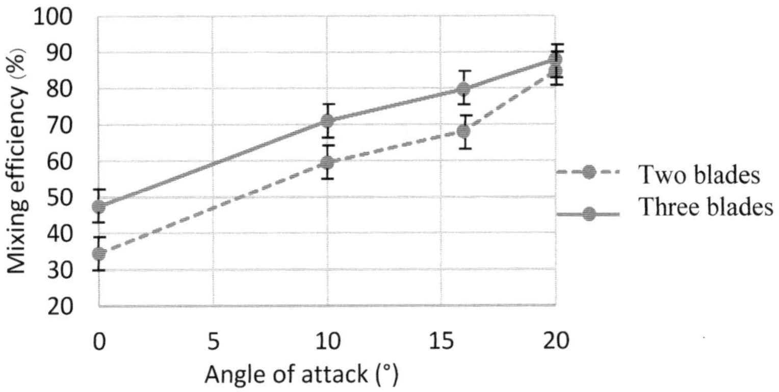

Figure 8. The mixing efficiency at each angle of attack of the mixing velocity at 110 rpm

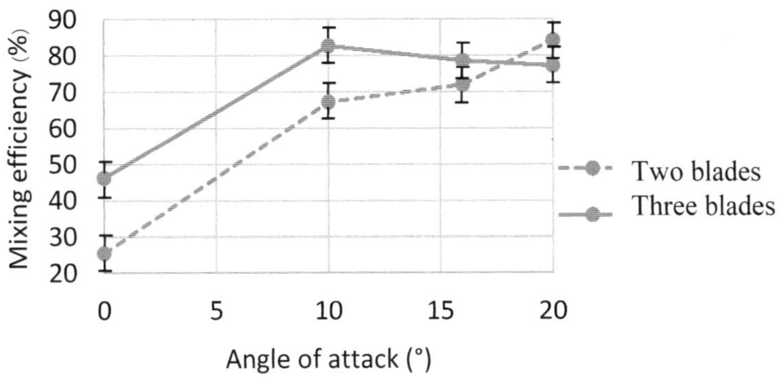

Figure 9. The mixing efficiency at each angle of attack of the mixing velocity at 140 rpm

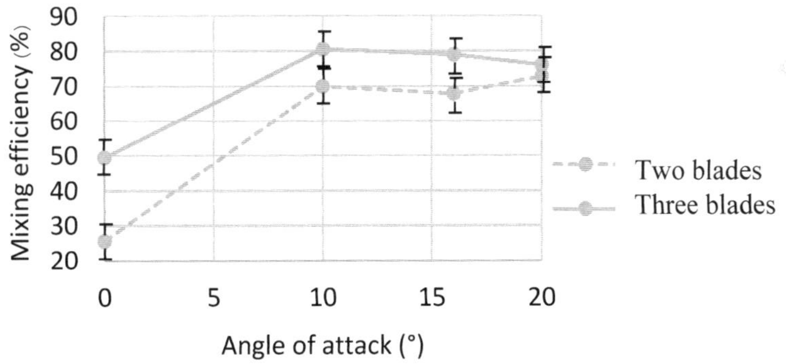

Figure 10. The mixing efficiency at each angle of attack of the mixing velocity at 190 rpm

3.2 Effects of mixing velocities on mixing efficiencies

Figures 11, **12**, **13** and **14** exhibited the effects of mixing velocities at angles of attack 0, 10, 16 and 20° on mixing efficiencies respectively. In **Figure 11** at the angle of attack at 0°, the mixing efficiency of the two blades was slightly decreased and almost constant in the case of the three blades system with increasing mixing velocity. This can be explained that at the angle of attack 0°, the force was very weak and was not enough to distribute the particles in water. In **Figure 12**, at the angle of attack 10°, the mixing efficiency increased with increasing mixing velocity due to the increasing force also the turbulent intensity of the flow that can drive the particles to distribute into the water. This result appeared to agree with the study of Khopkar et al. [7]. They reported that turbulent intensities increased with increasing mixing velocities. At high angles of attack (16° and 20°), mixing efficiencies were almost constant at the angle of attack 16° and a slight decrease at the angle of attack 20° was observed due to the airfoil blade stall (16°) as shown in **Figures 13** and **14**.

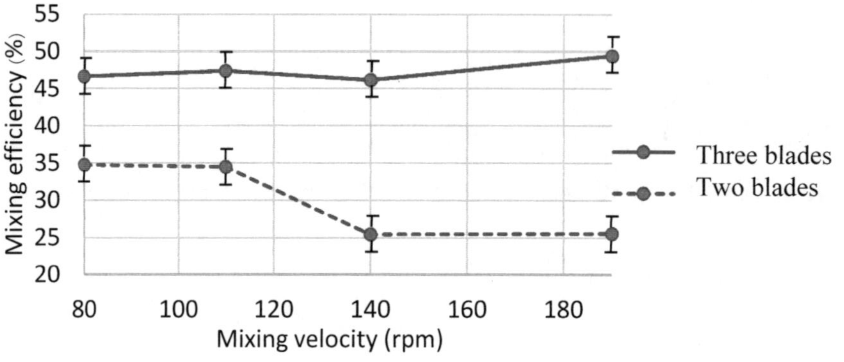

Figure 11. The mixing efficiency of each mixing velocity at the angle of attack 0°

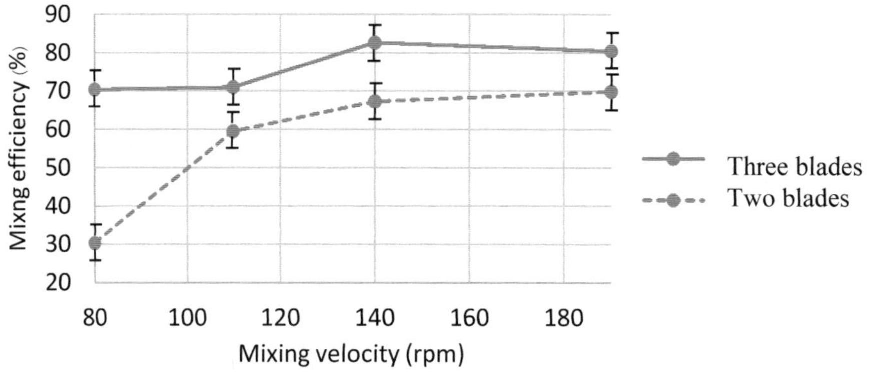

Figure 12. The mixing efficiency of each mixing velocity at the angle of attack 10°

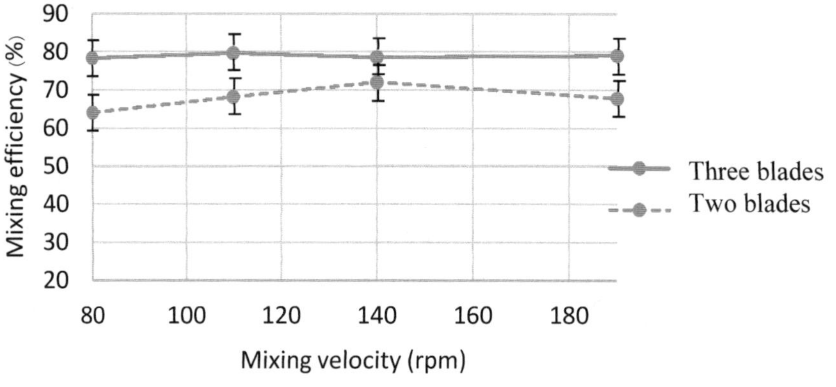

Figure 13. The mixing efficiency of each mixing velocity at the angle of attack 16°

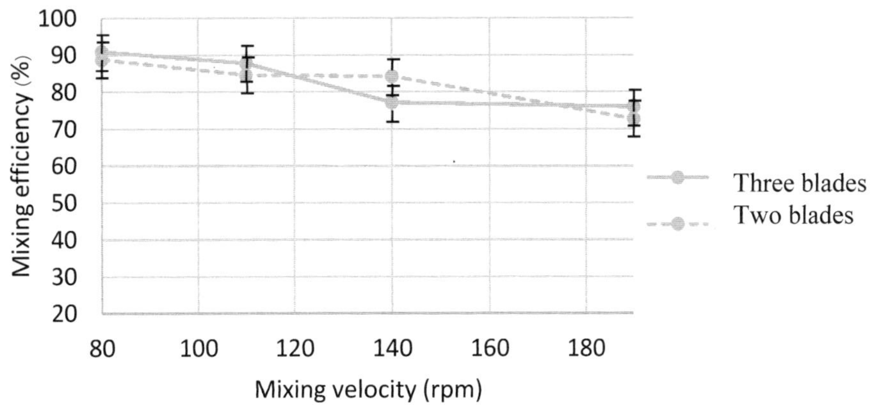

Figure 14. The mixing efficiency of each mixing velocity at the angle of attack 20°

4. Conclusion

This study has presented the flow analysis in the CSTR system using the airfoil NACA0015 of two and three blades. Effects of angles of attack and mixing velocities on mixing efficiencies were demonstrated. The mixing efficiency of the three-blade system was higher than that of the two-blade system of about 1.3 folds due to the more contact area between the particles and the blades of the three-blade system than that of the two-blade system. The mixing efficiency increased with increasing mixing velocity and angles of attack. The mixing efficiency was almost constant when the blade angles of attack increased from 16° to 20° owing to the airfoil blade stall. The mixing velocity was also almost constant when the mixing velocity increased from 140 to 190 rpm because of the reaching of mixing saturation. The information from this study can applied for a novel airfoil blade design for more mixing efficiency in the CSTR system that can be applied for not only in the biogas production, but also for other mixing operation in the CSTR system as well.

References

[1] Satjaritanun P, Khunatorn Y, Vorayos N, Shimpalee S, Bringley E 2016 *Biomass and Bioenergy* **86** (2016) 53-64.

[2] Wilawan W, Pholchan P, and Aggarangsi P 2014 *Energy Procedia* **52** (2014) 216 – 222.

[3] Watanabe Y, Kuwabara K, Adachi S, Nakanishi K, and Matsuno R 2003 *J. Agric. Food Chem.* 2003, **51**, 4628-4632

[4] Furusawa T, Nishimura H and Mizauchi T 1969 *Journal of Chemical Engineering of Japan* Vol.**2** No.1

[5] Mavros P, Xuereb C, Fort I and J. Bertrand 2002 *Chem Eng Sci*, **57**:3939-3952.

[6] Bakker A, Myers K.J, Ward R.W. and Lee C.K. 1996 *Trans IChemE, Part A, Chemical Engineering Research and Design*, Vol **74**:485-491.

[7] Khopkar A.R, Mavros P, Ranade V.V and Bertrand J 2004 *Trans IChemE, Part A, Chemical Engineering Research and Design*, **82**(A6): 737-751.

[8] Montantea G, Bakkerb A and Paglianta A, Magellia F 2006 *Chemical Engineering Science* **61**, 2807-2814.

[9] Amromin E, Kopriva J, Arndt R and Wosnik M 2006 *Journal of Fluids Engineering* Vol **128**/931.

[10] Amromin E, Hansberger J, Wang H, Wosnik M and Arndt R 2003 *Fifth International Symposim on Cavitation*, Osaka, Japan.

[11] R. Weetman and B. Gigas "Mixer Mechanical Design- Fluid Forces" Proceeding of the 19th International Pump Symposium.

[12] İ. Şahin and A. Acir 2015 "Numerical and Experimental Investigations of Lift and Drag Performances of NACA 0015 Wind Turbine Airfoil" *International Journal of Materials, Mechanics and Manufacturing,* Vol. **3**, No. 1.

Heat Transfer Enhancement in a Heat Exchanger Tube with 45° Rectangular-Winglet Pairs

P. Tongyote[1], S. Sripattanapipat[2] and P. Promvonge[1,*]

[1]Department of Mechanical Engineering, Faculty of Engineering, King Mongkut's Institute of Technology Ladkrabang, Bangkok, 10520, Thailand.
[2]Department of Mechanical Engineering, Faculty of Engineering, Mahanakorn University of Technology, Bangkok, 10530, Thailand.

* Corresponding Author: E-mail address: kpongjet@gmail.com

Abstract. An experimental investigation on thermal characteristics in a heat exchanger equipped with 45° rectangular winglet pairs (RWP) is conducted using air flowing in a constant heat flux tube with Reynolds number (Re) in a range of 5,200 to 23,000. Parameters of RWP included three relative height or blockage ratios (BR = 0.06, 0.08 and 0.1) and two pitch-spacing ratios (PR = 1.0 and 2.0). The heat transfer and the pressure drop in the test tube are displayed in the form of Nusselt number (Nu) and friction factor (f), respectively. The investigation shows that Nu and f values of the tube with RWP insert are, respectively, in the range of about 1.55–3.2 and 3.76–6.44 times above the plain tube alone, depending on the operating condition. The maximum thermal performance for using the RWP is found to be 1.7 at the lowest Re.

Keywords: heat transfer, pressure loss, thermal enhancement, rectangular-winglet pairs

1. Introduction

In several engineering applications and industrials, various techniques are devised to enhance the convection coefficient between fluid flow and heated walls of thermal systems. Vortex flow devices are among the techniques that are commonly used to increase thermal performance of those systems. The employ of vortex flow in thermal systems has been extensively found in the literature [1–4]. For decades, many investigations have been emphasized on heat transfer augmentation by passive techniques. A tube insertion with vortex-flow devices is one of the passive techniques that have been widely employed for augmenting the heat transfer in a heat exchanger tube [5–9].

The aim of heat transfer enhancement in a heating/cooling system is to reduce the exchanger size and its cost apart from the reduction of pumping/blowing power required in a prescribed thermal system for operating cost saving. Thus, several studies have been conducted to examine the influence of insert devices for increasing the heat exchanger performance. For example, vortex-flow devices to produce the vortex flows included wire coils [10, 11], nozzles [12], twisted tapes [13, 14], fin tapes [15, 16], baffles [17, 18], winglets [19, 20], for augmenting the convection heat transfer are the commonly known in the literature. Ozceyhan [21] studied numerically the heat transfer and thermal stress in a constant heat fluxed tube with wire-coil insert. The turbulent flow friction and thermal behaviors in a circular tube equipped with winglets was examined experimentally by Chokphoemphun et al. [22]. Recently, Tamna [23] investigated numerically the heat transfer behaviors and flow structure in a tube inserted with winglet pairs. They reported that the use of winglet pairs can induce two main counter-rotating vortices inside the tube leading to greater increase in thermal performance of the heating system

In the literature survey cited earlier, the vortex devices with various configurations have been introduced for augmenting thermal performance of a heat exchanger system.

The heat transfer augmentation is strongly dependent of the device geometry and the proper device geometry can provide higher performance of thermal systems. Among the mentioned devices, the winglet is one of the best efficient vortex-flow devices because of its lowest pressure drop penalty and ease of use by placing on flat surfaces of ducts/channels. In the current study, the 45° winglet pairs mounted inside the test section with two different pitch-spacing (P) ratios (PR=P/D=1.0 and 2.0) and three relative winglet heights (b), called blockage ratio (BR=b/D = 0.06, 0.08 and 0.1) were offered. Owing to curved wall of the tube, the winglets were alternately arranged and mounted repeatedly on a straight tape employed for supporting the winglets before insertion. In the present experiment, air as the working fluid entered the test tube for the Re ranging from 5,200 to 23,000.

2. Experimental Setup

The experimental apparatus is schematically sketched as depicted in figure 1. As can be seen, air induced by a blower was directed through the orifice-type flowmeter, settling tank and then, the test section. The flowmeter used to measure the volumetric airflow rate in the tube was built as per ASME standard [24] where the pressure drop of the orifice plate was measured by an inclined manometer filled by manometric fluid with specific gravity of 0.826. The varied airflow rate in the tube was made using an inverter for controlling the blower-motor speed.

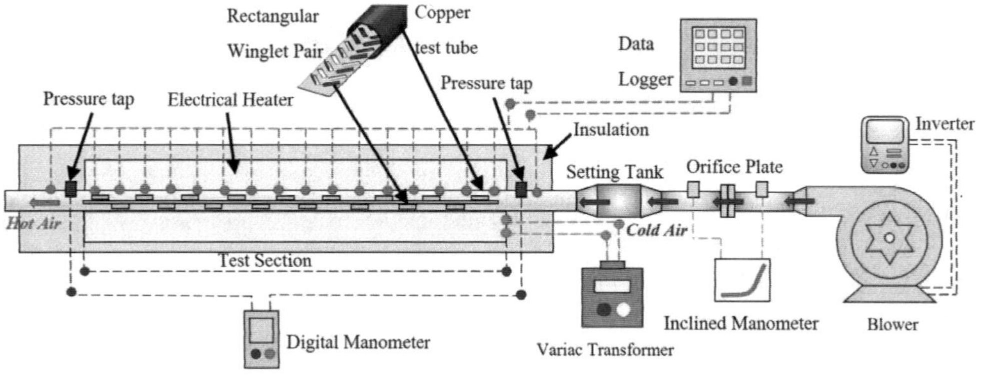

Figure 1. Schematic of experimental setup.

The system tube made of copper had a 50-mm inner diameter (D) and 3000 mm in total length that included a 1600-mm length (L) of the test tube as shown in figure 1. To provide a constant wall heat-flux condition, a continually winding flexible electrical wire was utilized to wrap around the tube. Also, a duct insulation was applied and covered the outermost wall of the test tube to reduce the convection heat loss to the surrounding air. The bulk temperatures of air at the inlet and the outlet (T_i and T_o) were measured by a data acquisition unit in common with the RTD-type PT-100 thermocouples while wall temperature measurements (T_w) along the test section were made by 16 thermocouples (T-type) attached on the outer wall tube and equally positioned along the tube wall. The measurement of pressure drop of the test tube was made by using a digital manometer coupled to two static pressure taps across the test tube.

IOP Conf. Series: Materials Science and Engineering **501** (2019) 012060 doi:10.1088/1757-899X/501/1/012060

Figure 2 depicts the arrangements of rectangular-winglet pairs (RWP) employed in the current work. The formation of RWP was done by cutting three sides of a preset rectangle marked on a 0.5-mm thick aluminum tape before punching it to become the rectangular winglet as displayed in figure 2. Each of the winglet had a length (l) of 19.8 mm whereas the tape had 1600-mm in length. The RWPs were mounted alternately on double sides of the tape as can be seen in figure 2. The RWPs were arranged with a fixed angle of attack (α) of 45°, two relative winglet pitches or pitch ratios (PR=P/D) and three relative winglet heights (b) or blockage ratios (BR=b/D). The insertion of the RWP-attached perforated-tape into the test tube is shown in figure 2.

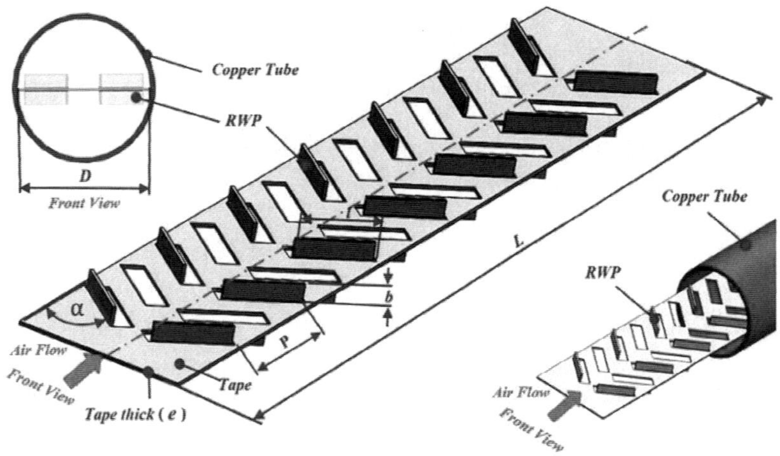

Figure 2. Rectangular Winglet Pair (RWP) arrangements.

3. Data Processing

In this section, the measured data are rearranged to be in terms of Reynolds number, Re, Nusselt number (Nu), pressure loss in the test tube or friction factor (f), and thermal enhancement factor (TEF). The Re is defined as

$$\mathrm{Re} = UD/\nu \tag{1}$$

where ν and U are, respectively, the kinematic viscosity and mean velocity in the tube of air. The f calculated from the pressure drop (Δp) of the tube length (L) is obtained by

$$f = \frac{\Delta p}{(L/D)\rho U^2/2} \tag{2}$$

In the current work, air entered the insulated tube and thus, the steady-state heat transfer rate is assumed to be equal to the heat absorbed from the tube which can be expressed as:

$$Q_{\mathrm{air}} = Q_{\mathrm{conv}} \tag{3}$$

where

$$Q_{\mathrm{air}} = \dot{m}C_{\mathrm{p,air}}(T_{\mathrm{o}} - T_{\mathrm{i}}) \tag{4}$$

in which $\dot{m}$ and $C_{\mathrm{p,air}}$ are the mass flow rate and the specific heat of air, respectively. The convection heat absorbed from the test section is estimated by

$$Q_{\mathrm{conv}} = hA(\tilde{T}_{\mathrm{w}} - T_{\mathrm{b}}) \tag{5}$$

where A is the heat transfer inner surface area of the test tube and

$$T_b = (T_o + T_i)/2 \tag{6}$$

$$\tilde{T}_w = \sum T_w /16 \tag{7}$$

T_w represents the local wall temperature located equally on the top of the test section and the mean wall temperature, $\tilde{T}_w$ is evaluated by using 16 points of the wall temperatures. The average convection coefficient (h) and Nusselt number (Nu) are determined via

$$h = \dot{m}C_{p,air}(T_o - T_i)/ A(\tilde{T}_w - T_b) \tag{8}$$

and

$$\mathrm{Nu} = hD/k \tag{9}$$

Thermal-physical properties of fluid are evaluated at the bulk air temperature (T_b) as given in equation (6).

TEF is defined as the ratio of h of the tube with RWP insert to h of the smooth/plain tube alone at an identical pumping/blowing power (pp) and is presented by

$$\mathrm{TEF} = \left.\frac{h}{h_0}\right|_{pp} = \left.\frac{\mathrm{Nu}}{\mathrm{Nu}_0}\right|_{pp} = \left(\frac{\mathrm{Nu}}{\mathrm{Nu}_0}\right)\left(\frac{f_0}{f}\right)^{1/3} \tag{10}$$

where the subscript "0" stands for the smooth tube acting alone.

4. Results and Discussion

4.1. Verification of smooth tube

Nu and f data of the present smooth tube are verified with those from Dittus-Boelter and Blasius correlations [25] as prescribed in equations (11) and (12), respectively, as displayed in figure 3. It can be observed in the figure that the present results agree well within ± 6% with both Nu and f from the correlations.

Dittus-Boelter correlation,

$$\mathrm{Nu} = 0.023\,\mathrm{Re}^{4/5}\,\mathrm{Pr}^{0.4} \tag{11}$$

Blasius correlation,

$$f = 0.316\mathrm{Re}^{-0.25} \tag{12}$$

4.2. Effect on heat transfer

Figure 4 shows the relationship between Nusselt number ratio, $\mathrm{Nu}/\mathrm{Nu}_0$ and Re for using RWPs. In the figure, $\mathrm{Nu}/\mathrm{Nu}_0$ tends to decrease slightly with the rise of Re for all cases. It is noted that the case of BR = 0.1 and PR = 1.0 gives the highest $\mathrm{Nu}/\mathrm{Nu}_0$. This is because of higher blockage of flow at BR = 0.1, more flow interruption resulting in higher vortex-flow strength and, then, more efficient convection coefficients between air and the tube surface. The $\mathrm{Nu}/\mathrm{Nu}_0$ values are varied in the range of 1.55–3.2 depending on Re, BR and PR, and the mean increases in $\mathrm{Nu}/\mathrm{Nu}_0$ for BR=0.06, 0.08 and 0.1 are about 2.15, 2.49, and 2.68; and 1.8, 1.98 and 2.31 times for PR=1.0 and 2.0, respectively.

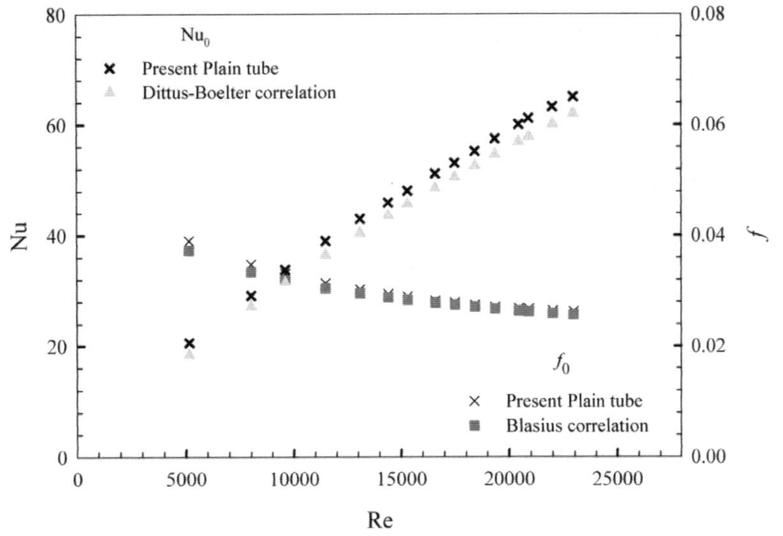

Figure 3. Nu_0 and f_0 of the present work with those of published correlations [25].

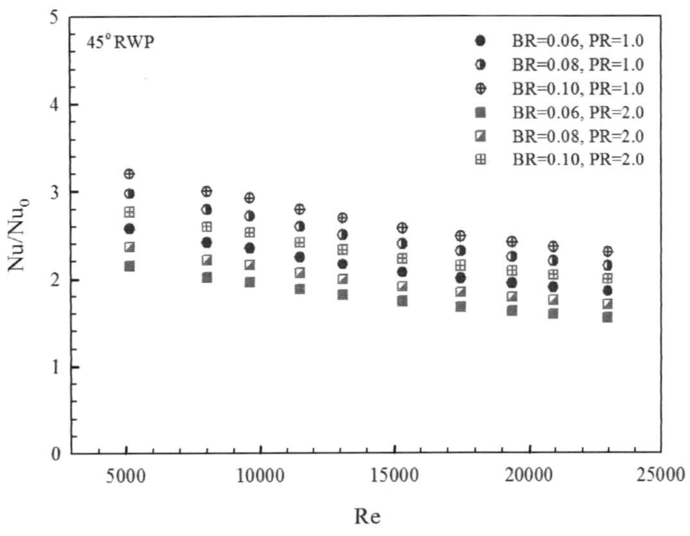

Figure 4. Nu/Nu_0 versus Re for RWP insert.

4.3. Effect on friction factor

Figure 5 depicts the variation of friction factor ratio, f/f_0 with Re for RWP inserts. It is clearly observed that the use of RWP leads to the extreme increase of f higher than the smooth tube acting alone, owing to the fluid dissipation of kinetic energy and larger surface friction from the presence of RWP including the appearance of reversing flow. The increase of f/f_0 for using the RWP is in a range of 3.76–6.44 times and its values are found to be nearly free from Re at a given BR and PR. The RWP with PR=1.0 provides considerably higher f/f_0 than that with PR=2.0 and also f/f_0 shows the increasing tendency with rising Re for all RWPs used. The maximum f/f_0 around 6.44 times can be seen for the RWP with BR=0.1, PR=1.0. The mean f/f_0 for the RWP with PR=1.0 is about 12 % higher than the one with PR= 2.0, at a prescribed BR.

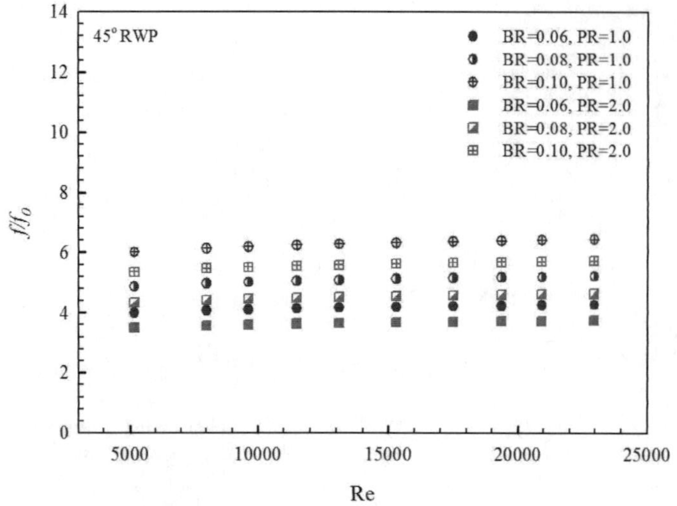

Figure 5. f/f_0 versus Re for RWP insert.

4.4. Assessment of thermal performance

The TEF variation with Re values for the tube inserted with RWPs is displayed in figure 6. TEF was evaluated from the Nu and f of the present tube with RWP and those of the smooth tube at a similar pumping/blowing power condition as defined by equation (10). It is interesting to note that the TEF shows the decreasing tendency with the increment in Re for all RWP cases. The maximum TEF of 1.7 can be observed at BR = 0.1, PR = 1.0 at lower Re whereas the lowest one of 0.96 is seen at BR = 0.06, PR = 2.0.

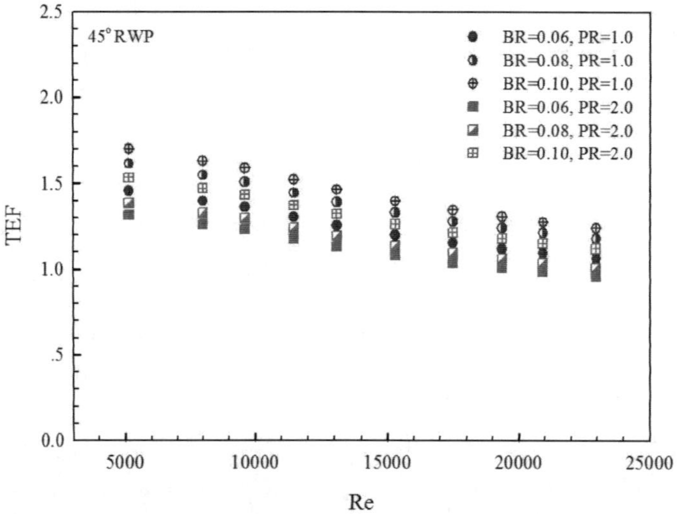

Figure 6. Variation of TEF with Re for RWP insert.

5. Conclusion

Thermal characteristics in a heat exchanger tube equipped with 45° RWP tapes have been examined experimentally and the current experiment has been carried out for the turbulent tube flow, Re = 5,200–23,000 for a uniform tube-wall heat-flux. The study shows that Nu/Nu_0 displays the decreasing tendency while f/f_0 exhibits the increasing tend with rising Re. The application of 45° RWP insert results in a considerable increase in the heat transfer rate above the smooth tube alone around 1.55–3.2 times depending on PR and BR. Also, the increase of f/f_0 for the RWP is seen around 3.76–6.44 times. As a performance indicator, the highest TEF around 1.7 can be observed at BR=0.1 and PR=1.0.

References

[1] Eiamsa-ard S, Wongcharee K and Promvonge P 2010 *Int. Commum. Heat Mass Transf.* **37** 156-62
[2] Promvonge P and Eiamsa-ard S 2007 *Int. Commun. Heat Mass Transf.* **34** 838-48
[3] Promvonge P 2010 *In. Commun. Heat Mass Transf.* **37** 835-40
[4] Skullong S, Thianpong C and Promvonge P 2015 *Heat Mass Transf.* **51** 1475-85
[5] Liu S and Sakr M 2013 *Renew. Sustain. Energy Rev.* **19** 64-81
[6] Promvonge P and Eiamsa-ard S 2007 *Int. Commun. Heat Mass Transf.* **34** 72-82
[7] Thianpong C, Eiamsa-ard P, Promvonge P and Eiamsa-ard S 2012 *Energy Procedia* **14** 1117-23
[8] Nanan K, Thianpong C, Promvonge P and Eiamsa-ard S 2014 *Int. Commun. Heat Mass Transf.* **52** 106-12
[9] Koolnapadol N, Sripattanapipat S and Skullong S 2016 *J. Res. Appl. Mech. Eng.* **41** 166-74
[10] Promvonge P 2008 *Int. Commun. Heat Mass Transf.* **35** 623-29
[11] Promvonge P 2008 *Energy Covers. Mange.* **49** 980-87
[12] Promvonge P and Eiamsa-ard S 2007 *Int. Commun. Heat Mass Transf.* **34** 72-82
[13] Eiamsa-ard S, Thianpong C and Promvonge P 2006 *Int. Commun. Heat Mass Transf.* **33** 1225-33
[14] Promvonge P, Pethkool S, Pimsarn M and Thianpong C 2012 *Int. Commun. Heat Mass Transf.* **39** 953-59
[15] Promvonge P, Skullong S, Kwankaomeng S and Thianpong C 2012 *Int. Commun. Heat Mass Transf.* **39** 617-24
[16] Promvonge P, Skullong S, Kwankaomeng S and Thianpong C 2012 *Int. Commun. Heat Mass Transf.* **39** 625-33
[17] Tandiroglu A 2006 *Int. J. Heat Mass Transf.* **49** 1559-67
[18] Skullong S, Thianpong C, Jayranaiwachira N and Promvonge P 2016 *Chem. Eng. Process Intensif.* **99** 58-71
[19] Eiamsa-ard S, Wongcharee K, Promvonge P and Thianpong C 2010 *Appl. Therm. Eng.* **30** 310-18
[20] Thianpong C, Eiamsa-ard P, Promvonge P and Eiamsa-ard S 2010 *Energy Procedia* **14** 1117-23
[21] Ozceyhan V 2010 *Energy. Cnver. Manag.* **46** 1543-59
[22] Chokphoemphun S, Pimsarn M, Thianpong C and Promvonge P 2015 *Chin. J. Chem. Eng.* **23** 605-14
[23] Tamna S, Sripattanapipat S and Promvonge P 2016 *Energy Procedia* **100** 518-21
[24] ASME 1984 *Standard Measurement of fluid flow in pipes using orifice, nozzle and venturi* (United Engineering Center)
[25] Frank P, Theodore L and Adrienne S 2006 *Fundamentals of heat and mass transfer* (John-Wiley & Sons)

IOP Conf. Series: Materials Science and Engineering **501** (2019) 012061 doi:10.1088/1757-899X/501/1/012061

Application of Computational Fluid Dynamics to Design of Polymer Extrusion Dies

N Hagioka[1], H Fukuoka[1,*], Y Minegishi[2] and K Kodama[2]

[1] National Institute of Technology, Nara College, 22 yata-cho, yamato koriyama-shi, Nara, 639-1058, Japan
[2] ASKA Corporation, 458-1 Minosho-cho, Yamato Koriyama-shi, Nara, 639-1103, Japan

* Corresponding Author: fukuoka@mech.nara-k.ac.jp, TEL/FAX +81-743-55-6076

Abstract. Extrusion is one of the most widely used processes for polymer processing, e.g., producing film, pipe, sheet, cable and pellet. The shape of extrusion die has an important part for the manufacturing products. The purpose of this study has been to confirm the influence of nozzle shape on polymer flow. We calculated the die flow simulations in the various parameters. The calculation results are expected to improve the efficiency of pelletization. A commercial computer fluid dynamics program Ansys Fluent was used to calculate the die passage. The parameters for the calculations were taper angle, outlet diameter, properties of polymer and two models. The governing equations were the incompressible momentum equation and the continuity equation. As results, the increase in flow rate was confirmed by increasing taper angle and outlet diameter in the specific dimensional range. The average outlet velocity of single and multi-nozzle models almost corresponded in each same parameter.

1. Introduction

Extrusion is one of the most widely used processes for polymer processing, e.g., producing film, pipe, sheet, cable and pellet. In each type of extrusion, the polymer is forced through a die to create the desired shape. The shape of extrusion die is an important part for the manufacturing products. It determines the cross-sectional shape of products and makes the flow distribution more uniform. The die design of extrusion is needed knowledge about material characteristics, flow, heat transfer phenomena, and extensive experience with extrusion processing [1]. The predicting these effect is very complex task. Design of the complex extrusion die is still an art rather than a science.

Plastic pellet production process known as pelletization is combining the polymer strand fabrication by extrusion and cutting techniques. In pelletization, the properties of polymer pellet change slightly depending on the processing time. The reduction of processing time will stabilize the quality of the pellet and increase the production rate. One of the method to reduce the processing time is to increase of extrusion pressure. The adjustment of extrusion pressure is difficult task due to constraint of equipment.

Therefore, we proposed on optimization of die. The die that is easy to pass through can shorten processing time and reduce quality irregularity of the pellet. The die for the pelletization has several types. This many type is based on taper nozzle. In this study, we focused on the die in which taper nozzles were arranged in double rows of staggered types.

Numerical simulation is expected one of the method for trial and error on die design. Extrusion die design for a new product and product improvement is developed on the basis by experiments and in-plant trials. In many cases costly experiments can be replaced by numerical simulation. The numerical simulation has the potential to uncover important interior details of the extrusion process, such as velocity, shear stress, pressure, and temperature fields in the region of interest, which is not possible to do experimentally [1,2].

Various studies are conducted on numerical simulation of extrusion die design. However, most studies of numerical simulation for extrusion die design have focused on profile extrusion [1-9]. Very few studies have been reported on die design for pelletization. Therefore, little is known about typical optimized shape parameters in the die design for pelletization.

The purpose of this study is to confirm the influence of nozzle shape on polymer flow. In this paper, the parameters of calculations were the outlet diameter D_{out}, the taper angle θ, the properties of polymer, two kinds of models. The models were prepared simple two-dimensional models and three-dimensional models with reference to the actual die. The evaluation of this study focused on the velocity distribution of the outlet.

2. Numerical method

Ansys Fluent is a finite element computational program for heat and fluid problems. The Ansys Fluent 17.2 was used in this study.

The single nozzle model for the simulation is shown in Figure 1. The single nozzle models were made by DesignModeler. The polymer flows through the large inlet to the small outlet via tapered wall part. The drawing of the multi-nozzle models is shown in Figure 2. The multi-nozzle models were made by SolidWorks. The dimensions of the single nozzle model were determined with reference to Figure 2. The influence on the flow field by varying the parameters were calculated.

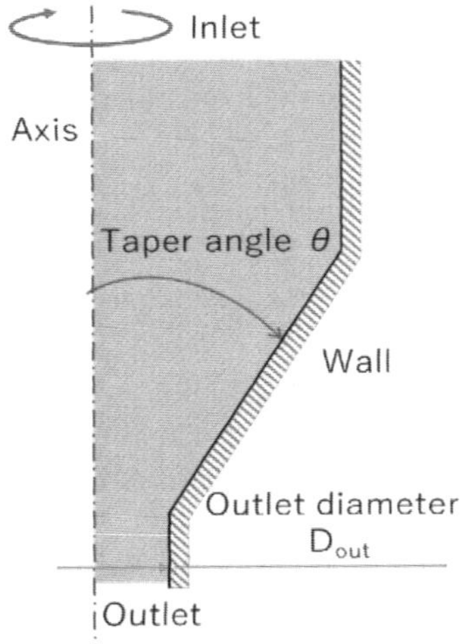

Figure 1. Single nozzle model.

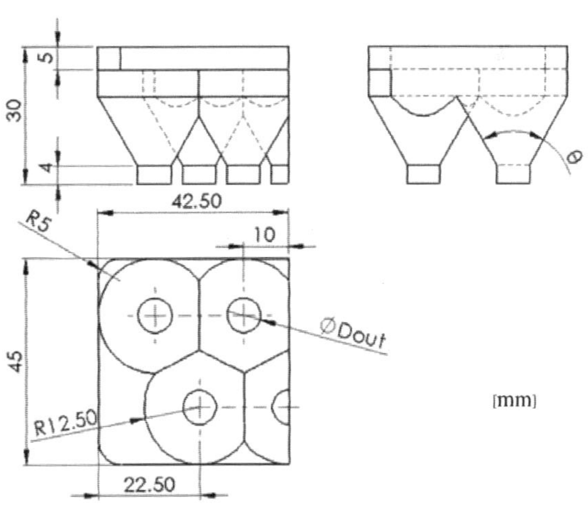

Figure 2. Drawing of multi-nozzle model.

The governing equations of the calculation were the incompressible momentum equation (1) and the continuity equation (2). Gravitational forces are neglected. The simulation was discretized by finite volume method and solved by SIMPLE method.

$$\frac{\partial v}{\partial t} = -\nabla \cdot p + \nabla \cdot \tau_{ij} \qquad (1)$$

$$\nabla \cdot v = 0 \qquad (2)$$

In boundary conditions, the inlet was set pressure-inlet and imposed gage pressure 115000 Pa. The outlet was set pressure-outlet and imposed gage pressure 0 Pa. The wall was set no slip condition. The simulations were performed time-transient, time step size 0.01 s and number of time step 300. The simulation was confirmed that the flow becomes almost steady-state within the calculation time.

The simulations were performing by varying outlet diameter, taper angle, property of polymer and simulation model. The outlet diameter D_{out} was 7.5, 9.5 and 11.5 mm. The taper angle θ were 60° and 90°. The materials of extrusion were two kinds of polymers. The polymers were used a power-low viscosity model.

$$\eta = k\,\gamma^{n-1} \qquad (3)$$

The properties of polymer are showed in Figure 3. In order to prevent divergence in calculation, the viscosity was set upper and lower limit. The viscosity of the polymer A is lower than the polymer B. The viscosity upper and lower limits of the polymer A were set 800 and 70 Pa·s. The viscosity upper and lower limits of the polymer B were set 1400 and 100 Pa·s.

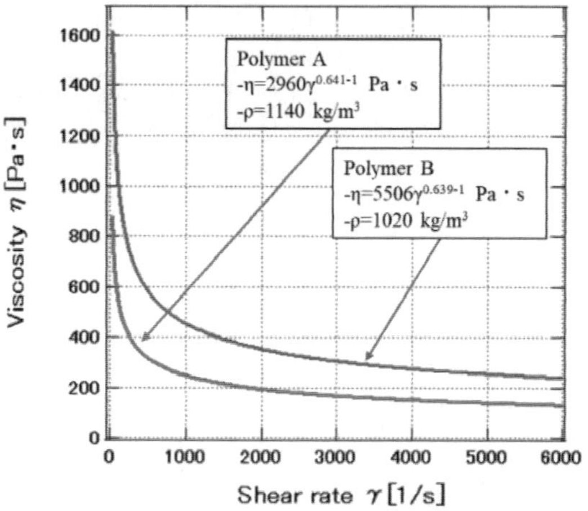

Figure 3. Properties of polymer.

Mesh density strongly affects on calculation accuracy. It is necessary to confirm the effect of the mesh size on calculation accuracy. The numerical simulations were performed for θ = 60°, D_{out} = 7.5 mm, polymer A, single nozzle model by varying the element size of mesh. The element sizes of the mesh were D/5,10,15,20 and 100. Since these results, the velocity difference of each element sizes was little. The element size $D_{out}/10$ was used for the calculations of this paper. The element size of the mesh is one tenth of the outlet diameter D_{out} in each simulation.

3. Results and Discussion

3.1. Single nozzle models

Figure 4 shows the velocity contour for the taper angle θ = 60°, the outlet dimeter D_{out} = 9.5 mm, the polymer A and single nozzle model. The velocity of the upper domain is smaller than lower domain. The velocity of center domain is larger than nearly wall. The maximum velocity appears at the center of outlet.

In above, the polymer accelerates in the taper domain and obtains the maximum velocity at the center of outlet. Therefore, the taper domain affects on the outlet velocity. In order to find the effect of flow field on the taper angle, the relationship between the average outlet velocity and the taper angle are shown in Figure 5. Figure 5 shows the average outlet velocity for θ = 60°, 90°, D_{out} = 9.5 mm, polymer A, and single nozzle model. The average outlet velocity was calculated to compare with another parameter condition. The average outlet velocity at θ = 90° is 1.09 times as large as θ = 60°.

The outlet diameter is one that affects on the outlet velocity. For the confirmation for the effect of the outlet diameter, the relationship between the average outlet velocity and the outlet diameter are shown in Figure 6. Figure 6 shows the average outlet velocity for θ = 60°, 90°, D_{out} = 7.5-11.5 mm, polymer A, and single nozzle model. The average outlet velocity is increasing with increasing the outlet diameter. The average velocity at θ = 90° is larger than θ = 60° for each outlet diameter.

In practical production, a common pelletizing die is used for different kind of polymer. It is important to confirm the effects of properties of polymer. The two kinds of polymers were compared. Figure 7 shows the average outlet velocity for θ = 60°, 90°, D_{out} = 7.5-11.5 mm, polymer A, B and single nozzle model. The average velocity of the polymer A is larger than the polymer B at same condition. For the slope of plot owing to the outlet diameter, the polymer A is also lager than the polymer B. Therefore, the effect of the outlet diameter is greater as low viscosity. The effect of the taper angle is relatively smaller than another parameter.

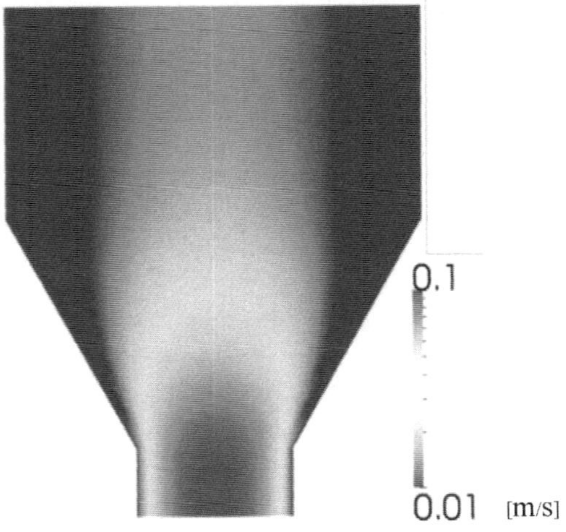

Figure 4. Velocity contour for θ = 60°, D_{out} = 9.5 mm, polymer A and single nozzle model.

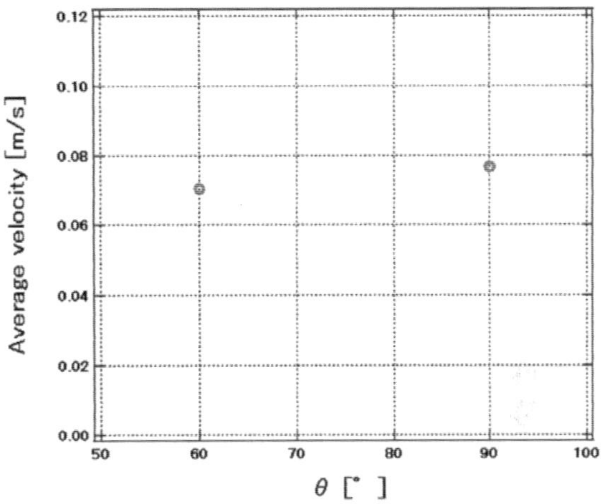

Figure 5. Average outlet velocity for θ = 60°, 90°, D_{out} = 9.5 mm, polymer A and single nozzle model.

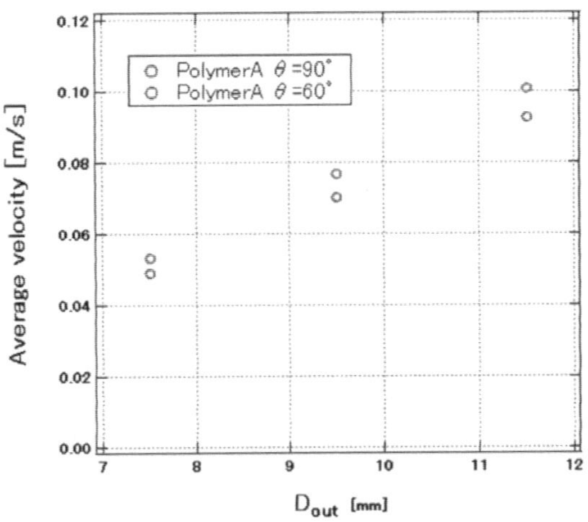

Figure 6. Average outlet velocity for θ = 60°, 90°, D_{out} = 7.5-11.5 mm, polymer A, and single nozzle model.

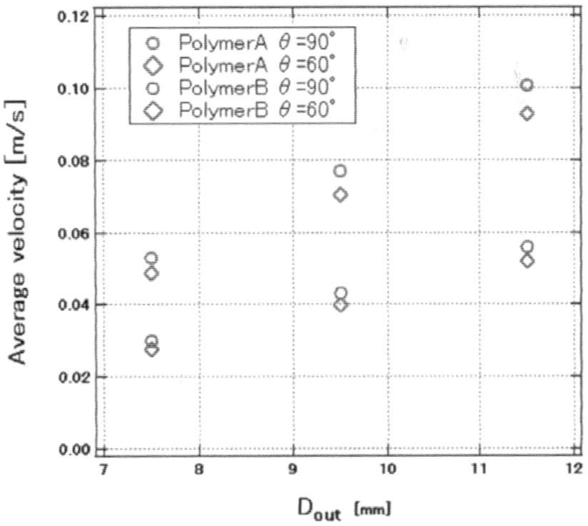

Figure 7. Average outlet velocity for θ = 60°, 90°, D_{out} = 7.5-11.5 mm, polymer A, B and single nozzle model.

IOP Conf. Series: Materials Science and Engineering **501** (2019) 012061 doi:10.1088/1757-899X/501/1/012061

3.2. Multi-nozzle models

The die for pelletization is usually arranged with the tapered nozzles. The upper taper shapes of the tapered nozzles are slightly lack due to overlapping nozzles as shown in Figure 8. The effect of overlapping nozzles on the polymer flow is still unclearly since the complex flow might be occurring. Figure 8 shows the velocity contour for $\theta = 60°$, $D_{out} = 9.5$ mm, the polymer A and multi-nozzle model. The contour seems to be almost same with single nozzle model shown in Figure 4. The maximum velocity of each holes appears at center of the hole.

The comparison of the average outlet velocity with single nozzle models and multi-nozzle models is shown in Figure 9. Figure 9 shows the average outlet velocity for $\theta = 90°$, $D_{out} = 7.5\text{-}11.5$ mm, polymer A, single and multi-nozzle model. The results of single and multi-nozzle model almost corresponded. Since the results, the effect of overlapping nozzles is little small. From the point of calculation time, the simulation of single nozzle model is valid for die design than the multi-nozzle model. However, if the model is set lager the outlet diameter and the taper angle, the upper taper shape become short and the effect of overlapping nozzles may appear.

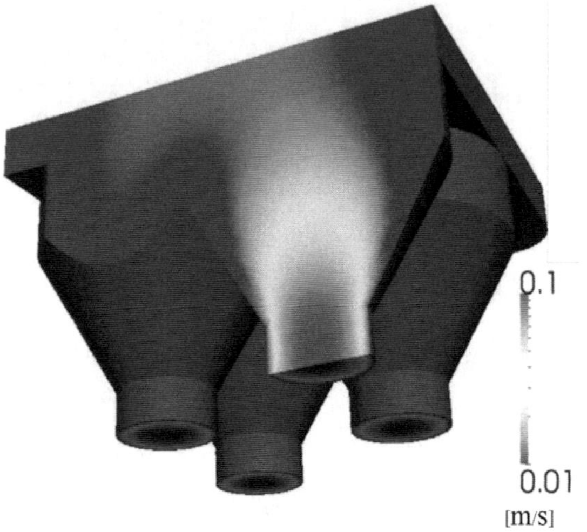

Figure 8. Velocity contour for $\theta = 60°$, $D_{out} = 9.5$ mm, polymer A and multi-nozzle model.

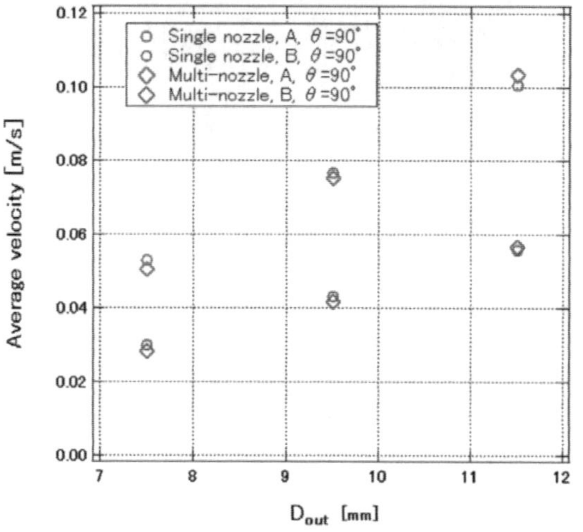

Figure 9. Average outlet velocity for $\theta = 90°$, $D_{out} = 7.5\text{-}11.5$ mm, polymer A, single and multi-nozzle model.

4. Conclusion

In this paper, the purpose is to confirm the influence of nozzle shape on polymer flow. We calculated the die flow simulations in the various parameters. The parameters of calculations were the outlet diameter D_{out}, the taper angle θ, the properties of polymer, single and multi-nozzle models. The largest outlet velocity was for $D_{out} = 11.5$ mm, $\theta = 90°$ and polymer A. The smallest outlet velocity was for $D_{out} = 7.5$ mm, $\theta = 60°$ and polymer B. The average outlet velocity of single and multi-nozzle models almost corresponded in each same parameter. The calculation time of the single nozzle model is shorter than the multi-nozzle model. Therefore, the simulation of single nozzle model is valid for pelletizing die design in the specific dimensional range.

References

[1] Kostic M M and Reifschneider L G 2006 *Encyclopedia of Chemical Proc.* (Oxford, Taylor & Francis Group) pp. 633-649

[2] Sienz J Goublomme A and Luege M 2010 *Computers and Structures* (Amsterdam, Elsevier) 88(9-10) pp. 610-624

[3] Karol P and Arkadiusz M 2011 *Journal of Polish CIMAC* (Gdańsk)

[4] G N Kouzilos, G V Seretis, C G Provatidis and D.E. Manolakos 2018 *Extrusion of Metals, Polymers, and Food Products* (London, IntechOpen) chapter10

[5] Reifschneider L G, Kostic M, and Vaddiraju S R 2004 *In ANTEC-Conf. Proc.- Vol. 1* (Pennsylvania, Society of Plastics Engineers) pp 246-250

[6] Lin Z, Juchen X, Xinyun W and Guoan H 2003 *Journal of Materials Processing Technology* (Amsterdam, Elsevier) **142**(3) pp 659-664

[7] Hatzikiriakos S G and Mitsoulis E 2009 *Polymer Engineering & Science* (Wiley Online Library) *49*(10) pp 1960-1969

[8] Ahmed A M A 2014 Computer Aided Design for Polymer Extrusion and its Dies (Doctoral dissertation, Sudan University of Science and Technology)

[9] Pandey V P and Gahir G S 2017 *International Journal of Engineering and Management Research* (*IJEMR*) (Lucknow, Bandana Publications) *7*(3) pp 494-500

Optimal height ratio of Y-shape pin fin to plate fin of PPFHS

Thicomporn Lekaphol, Pensiri Sompong and Sungkom Srisomporn*
Faculty of Science and Engineering, Kasetsart University Chalermphrakiat Sakon Nakhon Province Campus, 59/5 moo1 Chiangkrua Muang Sakon Nakhon, Sakon Nakhon 47000, Thailand
* Corresponding Author: sungkom.s@ku.th, 66-04-272-5033, 66-04-272-5034

Abstract. The purpose of this study is to determine the optimal ratio of Y-shape pin fin height to plate fin height in order to maximize the cooling performance of the plate pin fin heat sinks. The three-dimensional model of plate pin fin heat sink is considered in this study. Computational fluid dynamic method is used as a tool for solving the model. In this study, four values of height ratio, 0.2, 0.4, 0.6 and 0.8, and five values of air velocity, 0.5, 1.0, 1.5, 2.0 and 2.5 m/s, are analysed. The results of the study showed that the optimal fin height ratio is 0.6 resulting in the lowest thermal resistance of the plate pin fin heat sink.
Keywords: Computational fluid dynamics, plate pin fin heat sink, Y-shape pin fin

1. Introduction

One of the most significant problems associated with heat inside a computer is occurrs while the computer is working. For example, more heat is caused inside the computer due to the increase in temperature if the central processing unit (CPU) of a computer is working with high data processing. A heat sink is a piece of equipment that reduces the operating temperature of computers and electronics devices. The heat sink is designed to increase the surface area linked with the thermal convection and radiation from the heat sink to the surrounding environment, such as the air. A heat sink is essential for releasing heat because it can prevent damage or unusual work by other components in the computer when in use. Various designs for the heat sink have been developed, including plate fin heat sink, pin fin heat sink and plate pin fin heat sink resulting from the combination of the plate fin heat sink and pin fin heat sink. The plate pin fin heat sink can improve the efficiency of cooling better than plate fin and pin fin heat sinks. Mostly, heat sinks are made from highly conductive materials such as aluminium or copper. A heat sink can be connected to other components in the computer, such as a display card or central processing unit (CPU). The solder material is bonded between the heat sink and components. In addition, a fan or pump can be installed to enhance the rate of convection heat transfer from the heat sink to surrounding air. A plate pin fin heat sink (PPFHS) has seen less development. Many studies are related to the application of splitter plates to modify the hydro-thermal behaviour of plate pin fin heat sinks and the heat flow characteristics of compact plate-pin fin heat sinks [1-2]. The development of the plate fin heat sink and pin fin heat sink can be found in [3-8]. There are also studies associated with the creation of tools for testing heat transfer [9-10]. Applying computational fluid dynamics to determine heat transfer performance and optimal estimation can be found in [11-13]. Thus, the heat transfer behaviour of Y-shape pin fin heat sink resulting from the influence of air velocity and the height of pin is investigated in this study. It is expected that the results of the study will be used to make decisions concerning types of heat sinks.

2. The analytical model

The 3-D physical model of plate pin fin heat sink (PPFHS) in the present study is shown in Figure 1. The bases width is 50 mm (D) and length is 55 mm (L). The height of the base plate pin fin heat sink is 3 mm. The height of plate fin heat sink (A) is 50 mm and the distance between pin fin heat sink is 5 mm. The plate pin heat sink thickness and length are 1 mm and 4 mm, respectively. Air enters the heat sink through an inlet of win. The Y-shape pin fins heat sink thickness is 2 mm and the angle between two wings Y-shape pin fin heat sink is 6 degrees. The height of the pin fin heat sink (H) is measured from the base to the position changed to Y-shape. The fluid domain width, height and length are 50 mm, 53 mm and 105 mm, respectively. The surface around the plate pin fin heat sink surface is symmetric and the width of all wins is uniform.

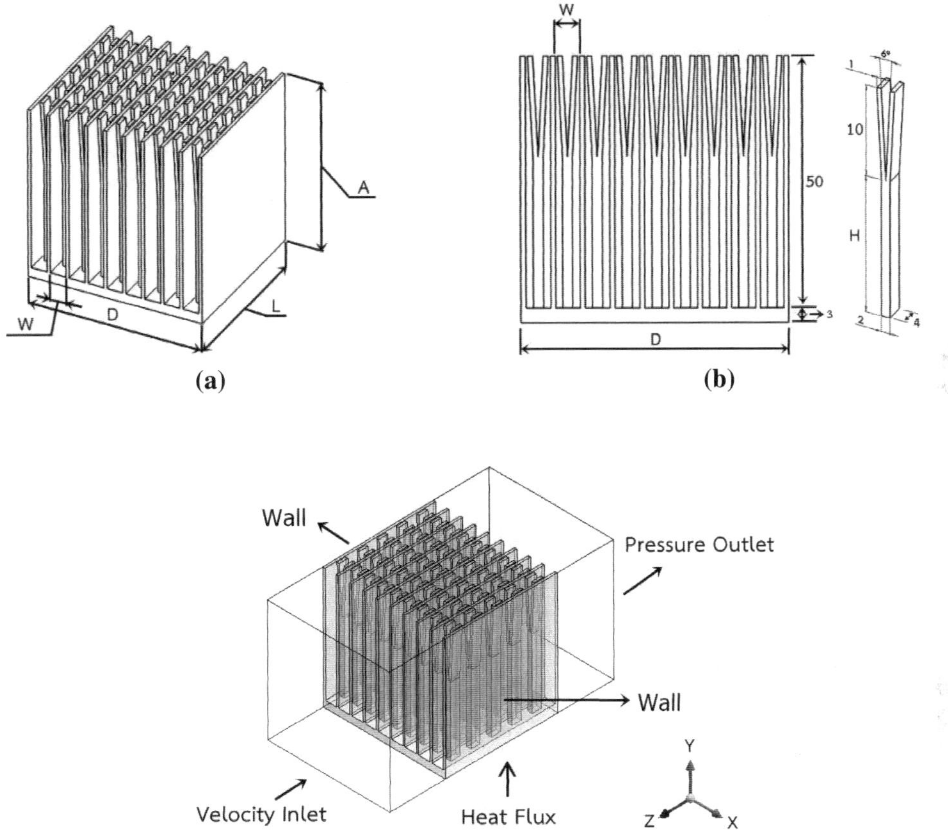

Figure 1. (a) Illustration of Y-shape plat pin fin heat sink (b) Illustration of dimensions for Y-shape plate pin fin heat sink (c) Illustration of 3D domain used for Y-shape plate pin fin heat sink simulation.

3. Assumptions and properties

In this study, heat distribution throughout the base area and the velocity of air flow passing through the inlet are uniform. The air flow is assumed to be steady and turbulent due to obstruction of the Y-shape pin fin. Radiation heat transfer from the heat sink is not considered. The temperature of air is 25 °C. The heat sink surfaces used in this study is made from 6061 aluminium alloy ASTM B 221 which is manufactured for extruding, welding, or wrapping a thin metal sheet on surface. The thermal properties of air and the aluminium heat sink are shown in Table 1.

Table 1. Air and Aluminium properties.

Properties	Values			
	Density (ρ) (kg/m^3)	Thermal conductivity (k) $(W/m\cdot K)$	Specific heat (c_p) $(J/kg\cdot K)$	Viscosity $(kg/m\cdot s)$
Air[a]	1.225	0.0242	1006.43	1.789e-5
Aluminium[b]	2719	202.4	871	-

[a]Air properties in this study are constant and not dependent on ambient temperature.
[b]ASM Handbook, Volume 2: Properties and Selection: Nonferrous Alloys and Special-Purpose Materials ASM Handbook Committee, p 102 DOI: 10.1361/asmhba0001060

4. Theory

In order to study the influence of the Y-shape pin fin heat sink on heat transfer, computational fluid dynamics is used to obtain the solution for analysis. The equations used in this study are continuity, momentum and energy equations, as shown from Equations: (1) to (3):

Continuity equations:

$$\nabla\cdot(\rho u)=0, \tag{1}$$

Momentum equations:

$$(u\cdot\nabla)\rho u=-\nabla p+\mu\nabla^2 u, \tag{2}$$

Energy equations:

$$u\cdot\nabla T=\frac{k}{\rho C_P}\nabla^2 T, \tag{3}$$

where $\nabla=\left(\frac{\partial}{\partial X},\frac{\partial}{\partial Y},\frac{\partial}{\partial Z}\right)$, ρ is the density of air, u is the velocity of air, p is the pressure of air, μ is the viscosity, k is the thermal conductivity and T is temperature. The results obtained from the analysis of the equation (1) to (3) can be used to calculate the value of the heat sink by using Equations (4) and (5) as follows:

Thermal resistance equations:

$$R_{HS}=\frac{\Delta T}{Q}, \tag{4}$$

Fan pumping power equations:

$$P_F=\left(\frac{\Delta p}{\rho}\right)\left(\frac{dm}{dt}\right), \tag{5}$$

where R_{HS} is thermal resistance of the heat sink, $\Delta T=T_{jc}-T_a$ is the difference of temperature across the heat sink, T_{jc} is the junction temperature between the base plate and the electronic device, T_a is ambient air temperature, Δp is pressure drop across the heat sink $\left(N/m^2\right)$, $\frac{dm}{dt}$ is air flow rate per time $\left(kg/s\right)$, P_F is fan pumping power.

5. Computational Method

To study the influence of the Y-shape pin fin heat sink on heat transfer by using the 3-D physical model detailed in the previous section, computational fluid dynamic is used to solve for the solutions. In the present study, air flow is assumed to be forced convection. The five values of air velocity that flow through the inlet that are studied are 0.5 m/s, 1.0 m/s, 1.5 m/s, 2.0 m/s, and 2.5 m/s, respectively. The ratio H/A is set as 0.2, 0.4, 0.6, and 0.8, respectively. The temperature of ambient air is 298 K. The heat flux on the base of the plate pin fin heat sink is kept constant at 18,200 W/m^2, while gravitational acceleration is set at 9.81 m/s^2 in a negative y direction. A 3-D tetrahedral mesh type of 0.001 m is used for mesh dependency in the experiment. The geometry parameters for the Y-shape plate pin fin heat sink are shown in Table 2.

Table 2. Geometry parameters of heat sinks.

Fin height ratio	Node	Element	Number of fins
0.2	23,195	84,561	45
0.4	24,729	88,365	45
0.6	26,611	93,123	45
0.8	27,980	95,542	45

6. Results

The influence of the Y-shape plate pin fin heat sink on heat transfer performance can be measured by using the value of thermal resistance. The results of the study are obtained by changing four values of pin height at 10, 20, 30, and 40 mm, which are written in the form of ratio of pin fin height to plate fin height at 0.2, 0.4, 0.6, and 0.8, respectively. The values for pressure drop and thermal resistance of the Y-shape plate pin heat sink are shown in Figure 2.

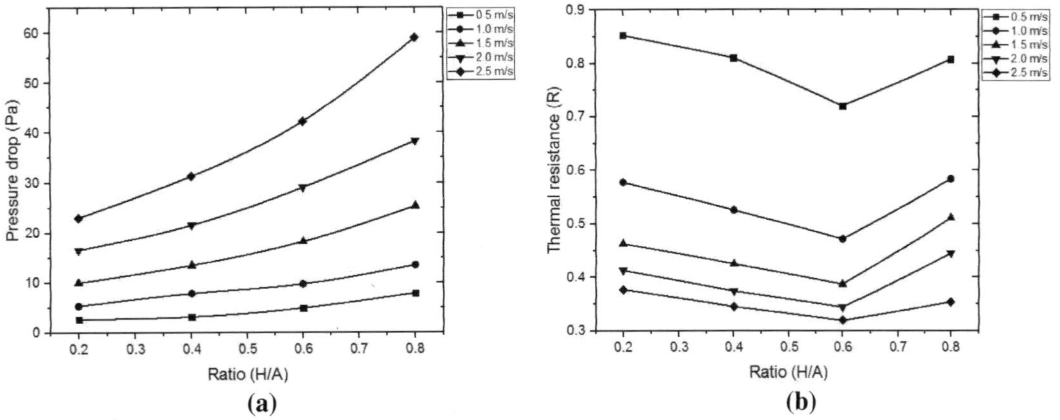

(a) **(b)**

Figure 2. (a) Effect of pressure drop on the PPFHS and (b) Thermal resistance of the PPFHS.

Figure 2(a) displays the effect of the cross-section area on pressure drop for the pin fin heat sink. It can be seen that the increase in cross-section area results in an increase in the value of pressure drop. When pressure drop is increased, it is necessary to increase the energy of the pumping power, causing noise and energy consumption. Therefore, other factors have to be considered for future applications.

The values of thermal resistance obtained from the experiments are shown in Figure 2(b). The highest value for thermal resistance can be seen at the lowest value of air velocity, while lower values of thermal resistance can be seen when the velocity of air is increased. As the ratio is increased from 0.2 to 0.4, thermal resistance values tend to decrease. They are increased as the ratio is increased from 0.6 to 0.8. Although the surface area of heat sink is large, the increase in fin height causes the obstruction of air flow associated with heat transfer performance. From the study results, it can be concluded that the optimal fin height ratio is 0.6 for all air velocity values.

7. Conclusions

In the present study, it was found that the effect of thermal resistance was reduced when the air flow velocity was increased resulting in more energy consumption to force the air. The pressure drop tends to increase when air velocity and the ratio of Y-shape pin fins height are increased. The optimal ratio for the plate pin fin heat sink is 0.6 for this study causing the lowest thermal resistance of the plate pin fin heat sink (PPFHS). It was observed that the graphs of thermal resistance and pressure drop are similar to the parabolic curve and linear.

Acknowledgments

This research was supported by the Faculty of Science and Engineering, Kasetsart University, Chalermphrakiat Sakon Nakhon Province Campus

References

[1] S.E. Razavi, B. Osanloo, R. Sajedi. Application of splitter plate on the modification of hydro-thermalbehavior of PPFHS. Applied Thermal Engineering. 2015;80:97-108.

[2] Amer Al-damook, FahadSalehAlkasmoul. Heat transfer and air flow characteristics enhancement of compact plate-pin fins heat sinks. Propulsion and Power Research. 2018;7(2):138-46.

[3] Daechan Jeon, Chan Byon. Thermal performance of plate fin heat sinks with dual-height fins subject to natural convection. International Journal of Heat and Mass Transfer. 2017;113:1086-92.

[4] Tae Hoon Kim, Kyu Hyung Do, Sung Jin Kim. Closed-form correlations of pressure drop and thermal resistance for a plate fin heat sink with uniform air jet impingement. Energy Conversion and Management. 2017;136:340-9.

[5] Waleed Al-Sallami, Amer Al-Damook, H.M. Thompson. A numerical investigation of the thermal-hydraulic characteristics of perforated plate fin heat sinks. International Journal of Thermal Sciences. 2017;121:266-77.

[6] Lingen Chen, Aibo Yang, Zhihui Xie, Fengrui Sun. Constructal entropy generation rate minimization for cylindrical pin fin heat sinks. International Journal of Thermal Sciences. 2017;111:168-74.

[7] Assel Sakanova, King Jet Tseng. Comparison of pin-fin and finned shape heat sink for power electronics in future aircraft. Applied Thermal Engineering. 2018;136:364-74.

[8] Sanchai Ramphueiphad, Sujin Bureerat. Synthesis of multiple cross-section pin fin heat sinks using multiobjective evolutionary algorithms. International Journal of Heat and Mass Transfer. 2018:462-70.

[9] Shi Zeng, Bugra Kanargi, Poh Seng Lee. Experimental and numerical investigation of a mini channel forced air heat sink designed by topology optimization. International Journal of Heat and Mass Transfer. 2018;121:663-79.

[10] Hamdy Hassana, Souad Harmand. An experimental work on the effect of the eccentric rotation of heat sink on the convective heat transfer. International Journal of Thermal Sciences. 2018;124:68-75.

[11] Kaj Lampio, Reijo Karvinen. Optimization of convectively cooled heat sinks. Microelectronics Reliability. 2017;79:473-9.

[12] Kumru Güres, çi, Faruk Yes, et al. Numerical analysis with experimental comparison in duct flow using optimized heat sinks. Journal of Radiation Research and Applied Sciences. 2018;11(2):116-23.

[13] Jan H. K. Haertel, Kurt Engelbrecht, Boyan S. Lazarov, Ole Sigmund. Topology optimization of a pseudo 3D thermofluid heat sink model. International Journal of Heat and Mass Transfer. 2018;121:1073-88.

IOP Conf. Series: Materials Science and Engineering **501** (2019) 012063 doi:10.1088/1757-899X/501/1/012063

Study of Interaction between Unsteady Supersonic Jet and Vortex Rings Discharged from Elliptical Cell

K Ueda[1], H Fukuoka[1*], N Kuniyoshi[2], M Yaga[3], EUeno[4], T Takiya [4] and N Fukuda[4]

[1]National Institute of Technology, Nara college, 22 Yata-cho, Yamatokoriyama-shi, Nara, 639-1080, Japan
[2]Tokyo University of Marine Science and Technology, 2-1-6 Etchujima, Koto-ku, Tokyo, 135-8533, Japan
[3]University of the Ryukyus, 1 Senbaru, Nishihara-cho, Okinawa, 903-0213, Japan
[4] Hitachi Zosen Corporation, 7-89 Nakano-kita 1 chome, Suminoe-ku, Osaka, 559-8559, Japan

* Corresponding Author: fukuoka@mech.nara-k.ac.jp

Abstract. The unsteady supersonic jet formed by a shock tube with a small high-pressure chamber was used as a simple alternative model for pulsed laser ablation. Understanding the vortex ring formed by the shock wave is crucial in clarifying behavior of unsteady supersonic jet discharged from elliptical cell. The purpose of this study has been to reveal behavior of the supersonic jet and the vortex rings. The unsteady behavior of a flow is investigated numerically by solving the axisymmetric two-dimensional compressible Navier-Stokes equations. The system of the calculation is a model of laser ablation of a certain duration followed by discharge through the cell exit. The parameters for the calculations are the pressure ratio of the shock tube and the cell diameter. As a result, it was found that the diameter of the elliptical cell had the effect on the interaction between vortex rings and the jet.

1. Introduction

Pulsed laser ablation (PLA) is a promising technique that has been applied to grow high-quality thin films. This technique has recently been used for the preparation of nano-clusters in gas phase [1]. Subsequently, thin films can be formed by layering the clusters discharged through pulsed lasing, and the optimal size cluster can then be selected through conventional pulsed laser deposition [2][3].

Iwata et al. [4] proposed a method for producing monodispersed clusters, which is expected to allow the direct generation of monodispersed clusters. This method uses an interaction phenomenon between the plume and shock wave arising in an elliptical cell following a laser ablation in inert gas. For simplifying the numerical analysis, the laser-ablated plume is substituted with a supersonic jet formed by a shock tube with a small high-pressure chamber can describe the pulsed expansion dynamics of the gas flow. This approach is a good alternative to describe gas ejection from the target during PLA. The high-pressure gas region formed by the laser irradiation, the laser-ablated plume, and the substrate correspond to the shock wave with a small high-pressure section, the supersonic jet injected from the shock tube, and the impingement plate, respectively.

Sakamoto et al. proposed vortex ring formation due to shock wave discharge from the exit of the elliptical cell [5]. Kitazono et al. performed a parametric study to demonstrate the effect of pressure ratio on the interaction among vortex rings and the supersonic jet [6]. However, it is not clear the effect of diameter of the elliptical cell on the interaction among vortex rings and the supersonic jet. Therefore, we investigated the behavior of vortex ring and jet in the present study.

2. Numerical Method and Boundary Conditions

The axisymmetric two-dimensional compressible Navier-Stokes equations, which can be derived from the continuity, momentum and energy equation, are solved using the finite-volume method with the total variation diminishing scheme in curvilinear generalized coordinates [7]. The inviscid flux is evaluated by third-order Roe 's approximate Riemann solver [8]. The grid number is 1259 x 713. In later work by Kitazono et al., the present calculation has already proven to accord with the experimental result [6].

Figure 1 shows the elliptical contours. Three contours indicate the maximum and minimum diameters of the elliptical cell, and shape in the middle shows the standard elliptical cell. The flat plate is installed downstream of the elliptical cell. The symbols represent the focal points for D_m/D = 3.04-6.00, where D_m is the cell diameter at x/D = 4.00 and D is the diameter of shock tube, respectively, and x is the horizontal distance from the exit of the shock tube. The flow field is shown in Figure 2. For the boundary conditions, non-slip conditions are applied to the cell wall. The parameters for the calculations are the cell diameter D_m/D and the pressure ratio P_h/P_b, where P_h and P_b are the pressure of the high-pressure area and backpressure. The downstream area of the elliptical cell is focused on to investigate the jet and the vortex rings generated by the shock wave.

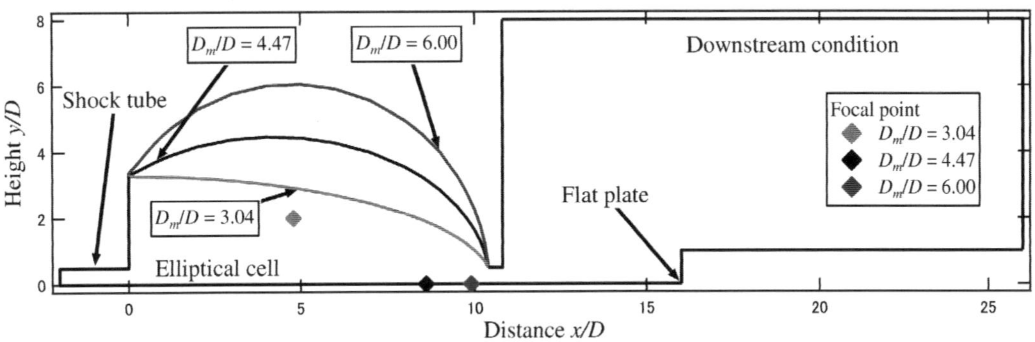

Figure 1. Calculation shape and focal point.

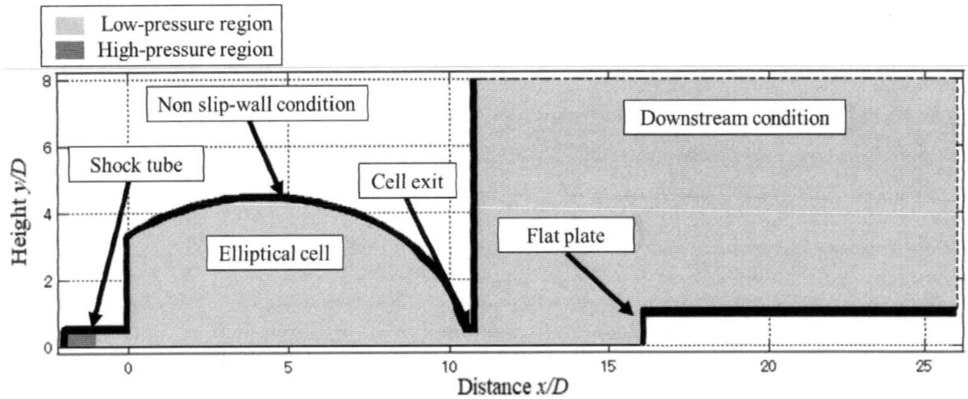

Figure 2. Flow field for computation and boundary conditions.

3. Results and Discussion

3.1. *Interaction between jet and vortex ring*

Figures 3(a)-(h) show the calculated vorticity contours and computer schlieren images for $P_h/P_b = 22.6$ and $D_m/D = 4.47$, where t denotes the time elapsed from the moment of the jet injection. Figure 3(a) shows the first vortex ring. Figure 3(b) shows the second generated vortex ring immediately downstream of the cell exit. We refer to first generated vortex ring and second generated vortex ring as "VR_1" and "VR_2", respectively. VR_2 is formed by VR_1 because of Kelvin-Helmholtz instability [6]. Figure 3(c) shows the jet discharged from the cell exit. In Figure 3(d), the distance among VR_1, VR_2, and the jet decreased as compared with that in Figure 3(c). Figure 3(e) shows the interaction among VR_1, VR_2 and the jet, and Figure 3(g) depicts VR_1 and VR_2 merging into the merging vortex ring (VR_{12}). At this time, the jet is closer to the flat plate than VR_{12}, which shows that the jet passes through the vortex ring because the velocity of the jets is higher than those of the vortex rings. Figure 3(h) shows the jet impingement with the flat plate.

For the interaction between the jet and the vortex ring, it is important to determine the effect of the vortex rings on the jet velocity. The time history of wall pressure on the center of the flat plate and the jet velocity for $P_h/P_b = 22.6$ and $D_m/D = 4.47$ are shown in Figure 4. In figure 4, the times of the indicated by (c)-(h) correspond to those of figures 3(c)-(h). As shown in figure 4, the jet velocity increase during the period of $t = 624$ μs and $t = 688$ μs. The jet acceleration during the period of 624 μs and 688 μs in figure 4 is attributed to the interaction with the vortex ring in figure. 3(d)-(f).

For the quality of thin film, it is important to discuss the static pressure at the substrate [9]. As shown in figure 4, there is a peak at $t = 864$ μs. This peak is caused by the jet impingement on the flat plate.

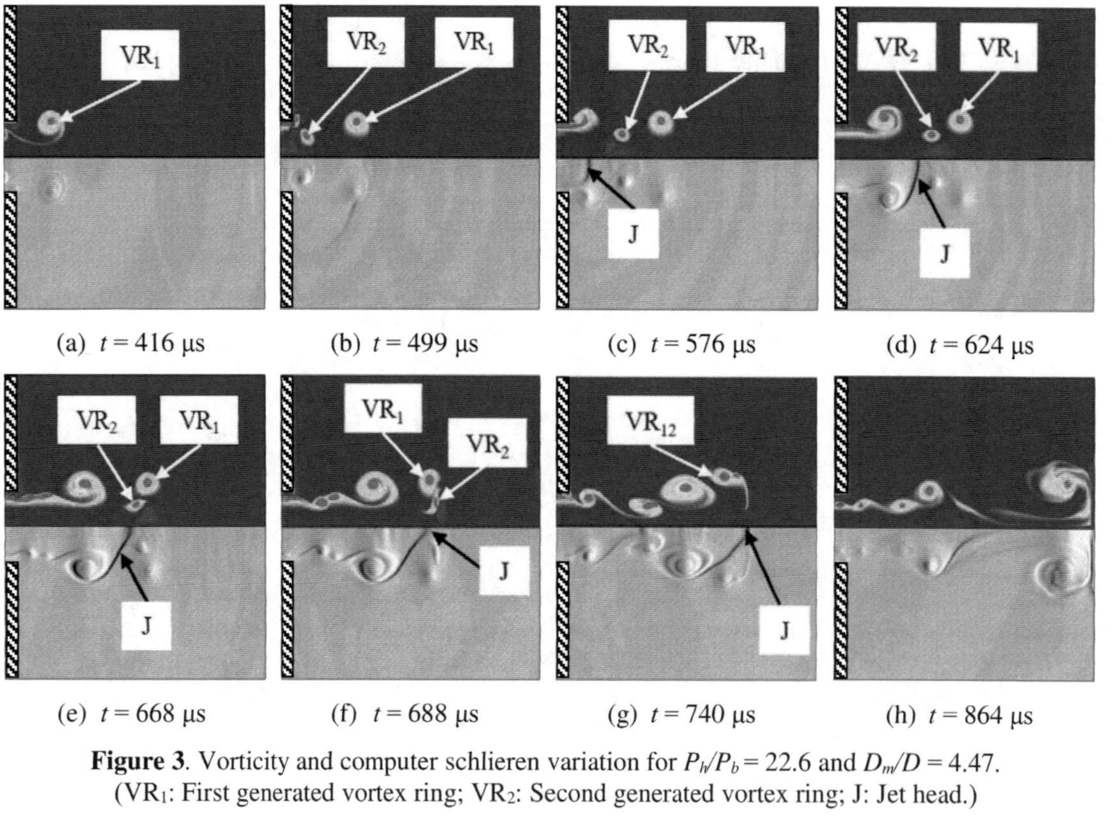

Figure 3. Vorticity and computer schlieren variation for $P_h/P_b = 22.6$ and $D_m/D = 4.47$.
(VR_1: First generated vortex ring; VR_2: Second generated vortex ring; J: Jet head.)

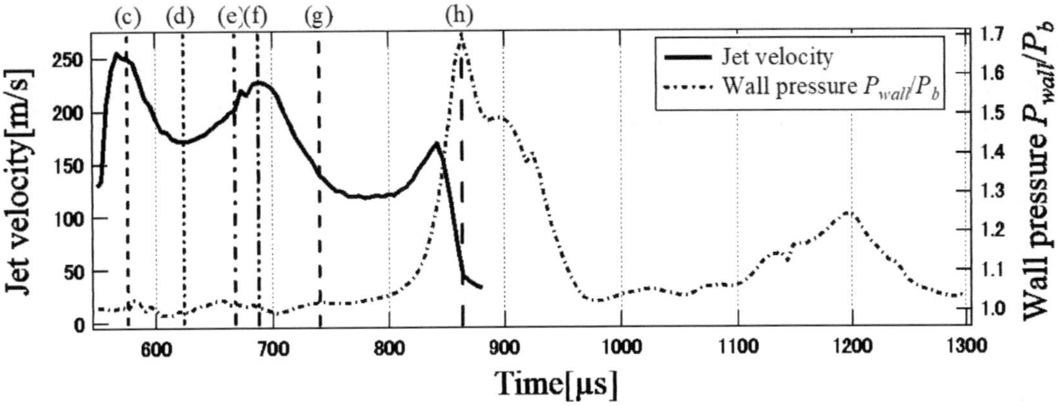

Figure 4. Time history of wall pressure on the center of the flat plate and jet velocity for P_h/P_b = 22.6 and D_m/D = 4.47.

3.2. *Effect of diameter of elliptical cell*

Figures 5(a)-(h) show the calculated vorticity contours and computer schlieren images for two different diameters of the elliptical cell D_m/D = 3.04 and 6.00, where the rest of conditions are the same as for Figures 2. In Figure 5(a) shows VR_1, VR_2 and the jet are observed. This tendency is the same as that in the case of D_m/D = 4.47. Figure 5(b) shows the interaction between VR_1 and VR_2. Figure 5(c) shows the impingement of VR_1 and VR_2 on the flat plate. Figure 5(d) shows the jet impingement on the flat plate. It is found that the interaction between the jet and the vortex ring does not occur for D_m/D = 3.04 differently to D_m/D = 4.47. The reason for no interaction is that the cell diameter is smaller than that of the standard elliptical cell. The smaller diameter makes the shock wave reach in the arrival time at the cell exit earlier than that for the standard elliptical [10]. Therefore, the vortex rings for D_m/D = 3.04 are generated earlier than those for the standard elliptical cell.

Figure 5(e) shows the jet discharged from the cell exit, where VR_2 stagnate. Figure 5(f) shows the interaction between the jet and VR_1. Figure 5(g) shows VR_1, VR_2 and the jet merging. Figure 5(h) shows and the jet impingement on the flat plate. It is found that VR_1, VR_2 and the jet for D_m/D = 6.00 merge at the same time downstream of the cell exit differently to D_m/D = 4.47.

Figure 6 shows the time history of the jet velocity for D_m/D = 3.04, 4.47 and 6.00, the pressure ratio P_h/P_b = 22.6, respectively. In figure 6, the times of the indicated by (c)-(h) correspond to those of figures 5(a)-(h). As shown in figure 6, the jet velocity for D_m/D = 3.04 decreases during the period of t = 648 μs and t = 837 μs due to the decline of the jet momentum. On the other hand, the jet velocity increases for D_m/D = 4.47 and 6.00 during the period of t = 624-688 μs and 726-763 μs, respectively, due to the interaction between the jet and the vortex ring.

The jet velocity at x/D = 11.0 for D_m/D = 3.04, 4.47 and 6.00 is plotted against P_h/P_b in figure 7. As shown in figure 7, the jet velocity first increases for D_m/D = 3.04, 4.47 and 6.00 at P_h/P_b = 10.7-22.6, 10.7-20.0 and 22.6-47.4, respectively, due to the fact that the initial jet momentum is proportional to P_h/P_b. Furthermore, the jet velocity decreases for D_m/D = 3.04 and 4.47 at P_h/P_b = 22.6-29.0 and 20.0-47.4, respectively. This is suggestive of the effect of the vortex ring at the cell exit. However, the detail reason is not clear at this moment. Moreover, the jet velocity remains practically constant for D_m/D = 3.04 at P_h/P_b = 29.0-47.4.

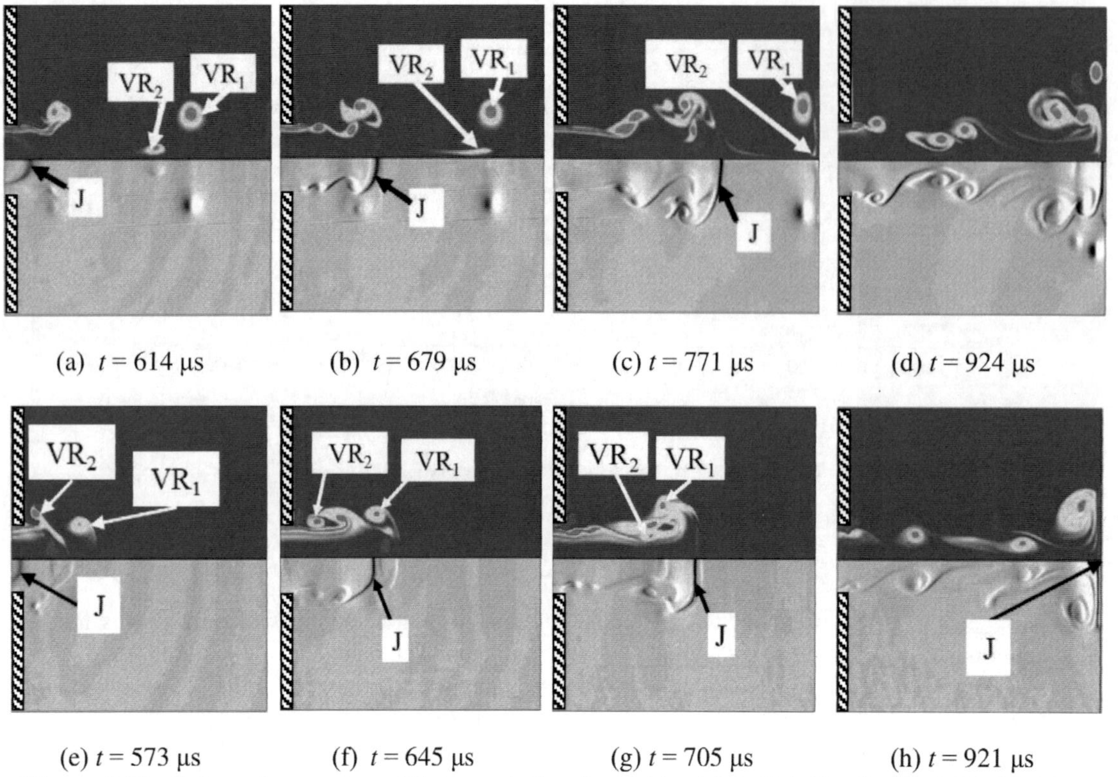

Figure 5. Vorticity and computer schlieren variation for $P_h/P_b = 22.6$. ((a)-(d) : D_m/D =3.04, (e)-(h): D_m/D =6.00 VR_1: First generated vortex ring; VR_2: Second generated vortex ring; J: Jet head.)

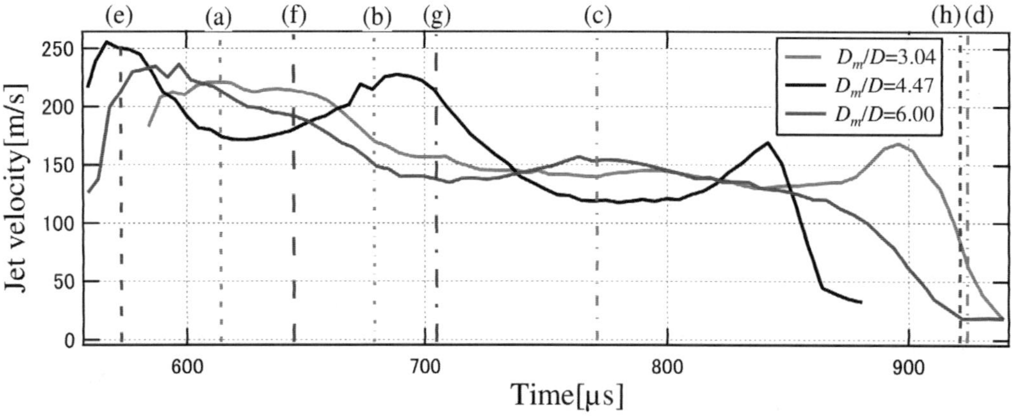

Figure 6. Time history of jet velocity for D_m/D = 3.04, 4.47 and 6.00.

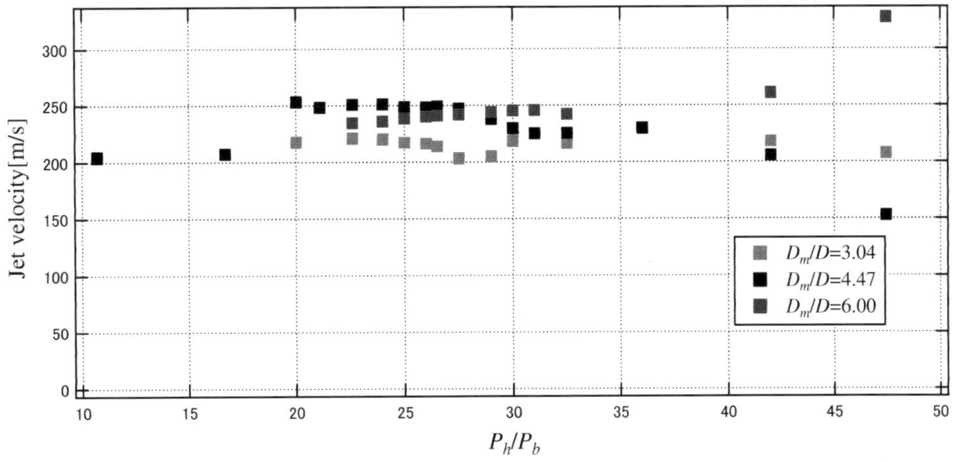

Figure 7. Jet velocity versus P_h/P_b for D_m/D = 3.04, 4.47 and 6.00.

Conclusion

The purpose of the present study is to investigate the influence of the cell diameter and the pressure ratio on the jet and the vortex ring downstream of the elliptical cell. The parameter of the calculations are the cell diameter and the pressure ratio. The conclusions can be summarized as follows:

1) The jet velocity for P_h/P_b = 22.6 and D_m/D = 4.47 greatly increases during the period of t = 624 μs and t = 688 μs. This is due to the interaction between the jet and the vortex ring.
2) The peak wall pressure on the center of the flat plate occurs at t = 864 μs for P_h/P_b = 22.6 and D_m/D = 4.47. This is due to the jet impingement on the flat plate.
3) The interaction between the jet and the vortex ring does not occur for D_m/D = 3.04. At the same time, the jet velocity decreases during the period of t = 648 and t = 837 μs as a result of the decline of the jet momentum.
4) The interaction between the jet and the vortex ring occurs for D_m/D = 4.47 and 6.00. Accordingly, the jet velocity increases as a result of the interaction between the jet and the vortex ring.
5) The jet velocity first increases for D_m/D = 3.04, 4.47 and 6.00 at P_h/P_b = 10.7-22.6, 10.7-20.0 and 22.6-47.4, respectively, due to the fact that the jet momentum is proportional to P_h/P_b.

Reference

[1] Umezu I, Yamamoto S and Sugimura A 2010 *Appl. Phys.* A **101** 131-136
[2] Seto T, Orii T, Hirasawa M and Aya N 2003 *Thin Solid Film* **437** 230-234
[3] Naono Y, Kawabata S, Huh S H and Nakajima A 2006 *Sci. Tech. Adv. Master. Int. Conf. on Nuclear Phys* **7** 209-215
[4] Iwata Y, Kishida M, Muto M, Yu S, Sawada T, Fukuda A, Takiya T, Komura A and Nakajima K 2002 *Chem. Phys. Lett.* **358** 36-42
[5] Sakamoto M, Matsui M, Fukuoka H, Yaga M and Takiya T 2013 *Mater. Sci. Forum.* **767** 80-85
[6] Kitazono K, Fukuoka H, Kuniyoshi N, Yaga M, Ueno E, Fukuda N and Takiya T 2016 *Int. J. Mech., Aerosp., Industrial, Mechatron. and Manuf. Eng.* **10** 1801-1805
[7] Hatta N 2010 *Flow calculation* (Tokyo: Morikita) pp 193-216
[8] Tanahashi T 1993 *CFD* (Tokyo: IPC) p 925
[9] Anders C, Meβlinger S and Urbassek H M 2006 *Surface Sci.* **600** 2587-2593
[10] Fukuoka H, Yaga M and Takiya T 2008 *J. Fluid Sci. and Tech.* **3** 881-889

IOP Conf. Series: Materials Science and Engineering **501** (2019) 012064 doi:10.1088/1757-899X/501/1/012064

The Experimental Investigation of Double Pipe Heat Exchangers Prepared from Two Techniques

A Tongkratoke[1,*], A Pramuanjaroenkij[1], S Phankhoksoong[1] and S Kakac[2]

[1]Department of Mechanical and Manufacturing Engineering, Faculty of Science and Engineering, Kasetsart University, Chalermphrakiat Sakon Nakhon Province Campus, Sakon Nakhon 47000 Thailand

[1]Department of Mechanical Engineering, TOBB University of Economics and Technology, Ankara, 06560 Turkey

* Corresponding Author: E-mail ammarin.to@ku.th

Abstract. Heat exchangers are the equipment used in variety of industries at all levels related to heat and thermal systems. Using heat exchangers at maximum efficiency are to reduce the amount of energy lost that impacts on environment and manufacturing cost in the industry. This research focused on comparing performances of two heat exchangers; two double-pipe heat exchangers were fabricated domestically with two different welding techniques (flare and Argon welding techniques). Both heat exchangers were tested on the same experimental setup and investigated for their thermal efficiency. Their working fluid was water, hot water and cold water flowed inside the heat exchangers were arranged as parallel (co-current) and counter flow configurations. Thermal efficiency when hot water and cold water flow rates varied in the range from 0.5 to 2 liters per minute, were also investigated. The results showed that the second heat exchanger fabricated with the Argon-welding technique showed the better thermal performance than that of the first heat exchanger fabricated with the flare technique. The thermal performances of the second heat exchanger were varied from 92% to 99% while the performances of another heat exchanger were varied from 32% to 38%. Therefore, the heat exchanger manufacturing played an important role to the heat exchanger performance. The same heat exchanger design but different fabricating technique caused lost in heat transfer process which affected the thermal performance and operating cost of the industry.

1. Introduction

The double pipe heat exchanger uses various inserts inside tube so as to enhance heat transfer and hence increase heat transfer coefficient [1]. These types of heat exchangers found their applications in heat recovery processes, air conditioning and refrigeration systems, chemical reactors, and food and dairy processes [2]. The double pipe heat exchanger would normally be used for many continuous systems having small to medium headuties. The heat exchanger is a device which transferred the heat from hot medium to cold medium without mixed both of medium since both mediums are separated with a solid wall generally. There are many types of heat exchanger that used based on the application. For example, double pipe heat exchanger is used in chemical process like condensing the vapors to the liquid. When to construct this type of heat exchanger, the size of material that want to uses must be considered since it affected the overall heat transfer coefficient. For this type of heat exchanger, the outlet temperature for both hot and cold fluids that produced is estimated by using the best design of this type of heat exchanger [3].

Prabhat et al. [4] studied on a performance evaluation of counter flow heat exchanger considering for low temperature application. He found those Counter flow heat exchangers are commonly used in cryogenic systems because of their high effectiveness. They observed that losses such longitudinal conduction through wall, heat in leak from surrounding, flow misdistribution etc. They extended their study to understand quantitative effect of heat in leak and axial conduction parameters on degradation of heat exchanger performance for 300-80,000 and 80-20,000 temperature range.

Swapnil et al. [5] has discussed about fabrication and analysis of counter flow helical coil heat exchanger. The heat transfer coefficient increases with increase in Reynolds number as Nusselt number increases. Greater will be the turbulence higher will be the heat exchange. As the Dean number is increasing with increase in Reynolds number so, the heat transfer rate will also increase with the Reynolds number.

Christian et al. [6] has discussed about experimental investigation on the effect of fluid flow rate on the performance of a parallel flow heat exchanger. The effect of fluid flow rate on the performance of parallel flow heat exchangers in an extended plate with regard to thermal efficiency, overall heat transfer coefficient, convective heat transfer coefficient, flow rate, and Reynolds number. Physical characteristics and thermal performance of a real heat exchanger were studied in this work. Also the convective heat transfer coefficient increases with both Reynolds and Nusselt numbers, which increases the overall heat transfer coefficient.

Tapre et al. [7] has discussed about Review on Heat Transfer in Spiral Heat Exchanger Energy saving is major matter in our global world, and heat exchanger is very useful for energy saving. Study on the various effects of feed flow rate and the coil diameter was done and concluded that on increasing the feed flow rate the pressure drop increases and vice versa. A relation between the pressure drop and the feed flow rate for the steady state Newtonian fluid into the Archimedean spiral tubes was developed.

Magadum et.al [8] has discussed about Review of Experimental analysis of parallel and counter flow heat exchanger. In this paper, we have discussed about various literature review of various paper and we are now going to calculate LMTD by varying flow rate and temperature of hot water and cold fluid.

Hoonpong and Skullong [9] experimentally studied on heat transfer and flow friction characteristics in a solar air heater duct roughened artificially with V-shaped baffles. The absorber plate is mounted with V-baffle vortex generators to improve the performance of the solar thermal system for energy saving. The experiment in the test duct having the aspect ratio (AR) of 10 is conducted for Reynolds number (Re) based on the hydraulic duct diameter ranging from 5300 to 22,600. In the current work, V-baffles are placed on the absorber with three relative baffle heights (R_B=b/H = 0.1, 0.2 and 0.3) and pitches (R_P= P/H = 0.5, 1.0 and 1.5) at a single attack angle (b) of 60°. The experimental results reveal that the use of V-baffle vortex generators yields the considerable increase in heat transfer over the smooth duct around 2.32–4.3 times while the friction loss increases around 4.08–36.9 times. The heat transfer and the friction loss tend to rise for increasing R_B but show the reversing trend for increasing R_P. The maximum thermal performance for the V-baffle vortex generators around 1.57 is seen at R_B = 0.2 and R_P = 1.0.

Chingtuaythong and Chokphoemphun [10] experimental and numerical investigation on thermal performance in a uniform wall heat-fluxed tube inserted with 45° inclined oval–pentagon rings (OPR) for turbulent flow region using air as the working fluid. The ring parameters included three blockage ratios (R_B=e/D=0.05, 0.10 and 0.15) and four pitch ratios (R_P=P/D=1.0, 1.25, 1.5 and 2.0) are introduced. The results from the inserted tube are compared with those from smooth tube alone and show that the OPR can augment the heat transfer rate and friction loss around 2.25–4.86 and 14–100 times over the smooth tube, respectively, depending upon operating conditions. The maximum thermal enhancement factor for using OPR tabulators at about 1.36 is for Re=3900, R_B=0.05 and R_P=1.0. It is worth noting that the results of numerical study are in good agreement with the experimental ones. The average deviation of both results is within ±7% for Nusselt number and ±10% for friction factor.

In this work, we focused on the performance of heat exchangers; two double pipe heat exchanger were built domestically with two different welding techniques. Both heat exchangers were tested on the same experimental set and investigated thermal efficiency. The hot water and cold water flow inside the heat exchangers were arranged as parallel and counter flow configurations. The effects of hot water and cold water flow rates; in the range of 0.5 to 2 liters per minute.

2. Experimental Setup

2.1 Double-pipe heat exchanger

A double-pipe heat exchanger is considered as the simplest type of heat exchangers. It consists of two concentric pipes having different diameters (Fig. 1). One fluid flows through the smaller pipe while another fluid flows through the annular space between both pipes. Two flow arrangements in any double pipe heat exchangers could be parallel flow (both hot and cold fluids move in the same direction, enter and exit the heat exchanger at the same side of the heat exchanger) and the counter flow (hot and cold fluids enter and exit the heat exchanger at the opposite ends and flow in the opposite directions). In this work, the double pipe heat exchanger dimensions were as 300 mm in length, the inside-pipe diameter of 3.9 mm, the outside-pipe diameter of 15.4 mm, and 5 mm in thickness. The cold fluid or cold water flowed through the inside pipe while the hot fluid or hot water flowed simultaneously through the annular space between both pipes.

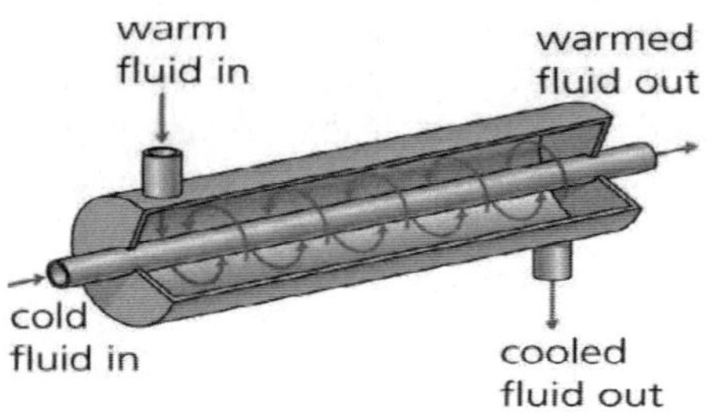

Figure 1. A double pipe heat exchanger [9].

The two double pipe heat exchanger were built domestically with two different welding techniques. Firstly, the flare technique was a process that is used to join metal with another metal by using electricity to create enough heat to melt metals. When the melted metals are cooled, their result is a binding of the metals. It is a type of welding that applies a welding power supply to create an electric arc between an electrode and the base material to melt the metals at the welding point. We can apply alternating (AC) current, and consumable or non-consumable electrodes. The welding region is usually protected by some type of shielding gas, vapor, or slag (Fig. 2 (a)). Secondly, the argon-welding technique is a welding process that uses a non-consumable tungsten electrode to produce the weld (Fig. 2 (b)). The weld area and electrode are protected from oxidation or other atmospheric contamination by an inert gas (argon or helium).

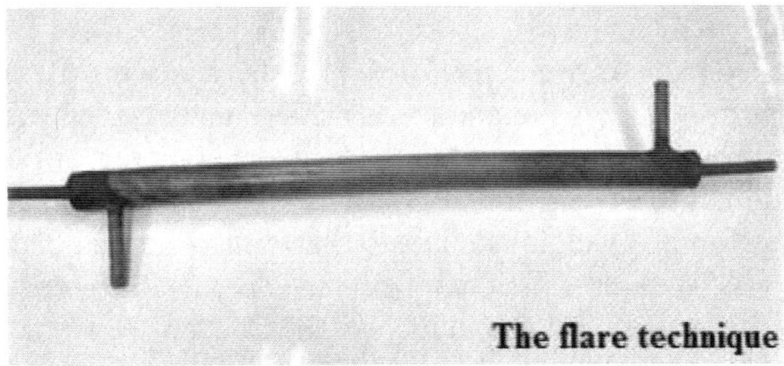

(a) The flare technique

(b) The argon-welding technique

Figure 2. The heat exchanger fabricated with
(a) the flare technique and (b) the argon-welding technique.

2.2 Theoretical considerations

The Nusselt number is calculated as follows:

$$Nu = \frac{hD_h}{k} \tag{1}$$

Heat absorbed by hot fluid from the annulus, Q can be written by:

$$Q = \rho VAC_p (T_{out} - T_{in}) \tag{2}$$

The convection heat transfer coefficient of fluid respectively:

$$U = \frac{Q}{A_{in} \Delta T_{lm}} \tag{3}$$

Where A_i and ΔT_{lm} are heat transfer area and logarithmic mean temperature difference (LMTD).

The LMTD calculated from

$$\Delta T_{lm} = \frac{(T_{hot,in} - T_{cool,out}) - (T_{hot,out} - T_{cool,in})}{\ln[(T_{hot,in} - T_{cool,out}) - (T_{hot,out} - T_{cool,in})]} \tag{4}$$

3. Results and Discussion

The two double pipe heat exchangers were built domestically by two different welding techniques. From the fabrication results, we found splinters of the melted metal on the surface of the double pipe heat exchanger fabricated by the flare technique which implied that there were some metals were moved out of their positions. Additionally, we also found that there were leakages on this heat exchanger, as on Fig. 3, because of the flakes from the welding. While the surface of the heat exchanger fabricated by the argon-welding technique was smooth which implied that all metals were melted with each other regularly. We found that there was no leakage in this heat exchanger. Noticeably, the double pipe heat exchanger fabricated by the argon-welding technique contained the area around the joints of the copper pipe which was smoother than that of the flare technique.

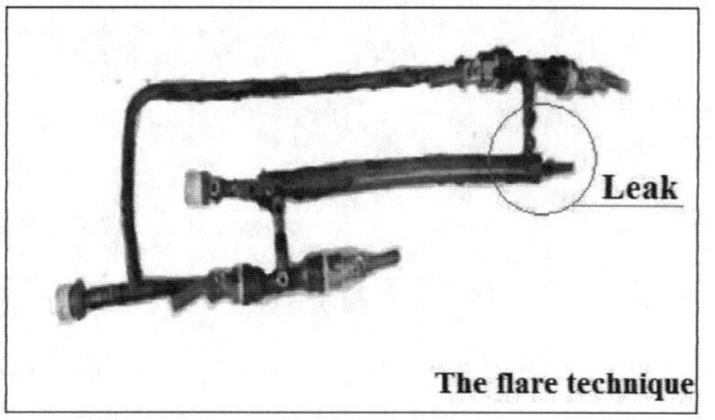

Figure 3. The fabrication result of the heat exchanger with the flare technique

Fig. 4 and 5 showed the variations of thermal effectiveness with different mass flow rates of the cold fluid for parallel and counter flows respectively. The plot clearly indicated that the effectiveness was increased with decreasing the cold-water mass flow rates for the counter flow (Fig. 4). While the plot in Fig. 5 clearly revealed that the effectiveness was decreased with increasing the cold-water mass flow rates for the counter flow. We noticed that the effectiveness of the counter flow heat exchanger was greater than that of the parallel flow heat exchanger.

Both heat exchangers were also investigated when the hot water and cold water flow rates; in the range of 0.5 to 2 liters per minute. The results showed that the second heat exchanger fabricated with the argon-welding technique showed the better thermal performance than that of the first heat exchanger fabricated with the flare technique. The thermal performances of the second heat exchanger were varied from 92% to 99% while the performances of another heat exchanger were varied from 32% to 38%. Therefore, the heat exchanger manufacturing played an important role to the heat exchanger performance. The same heat exchanger design but different fabricating technique caused lost in heat transfer process which affected the thermal performance and operating cost of the industry.

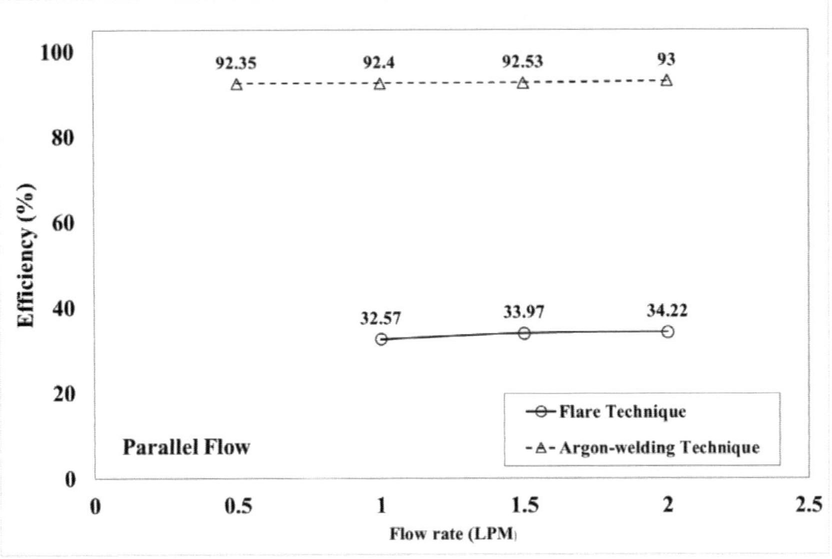

Figure 4. Efficiency of the heat exchangers were arranged as parallel flow configurations.

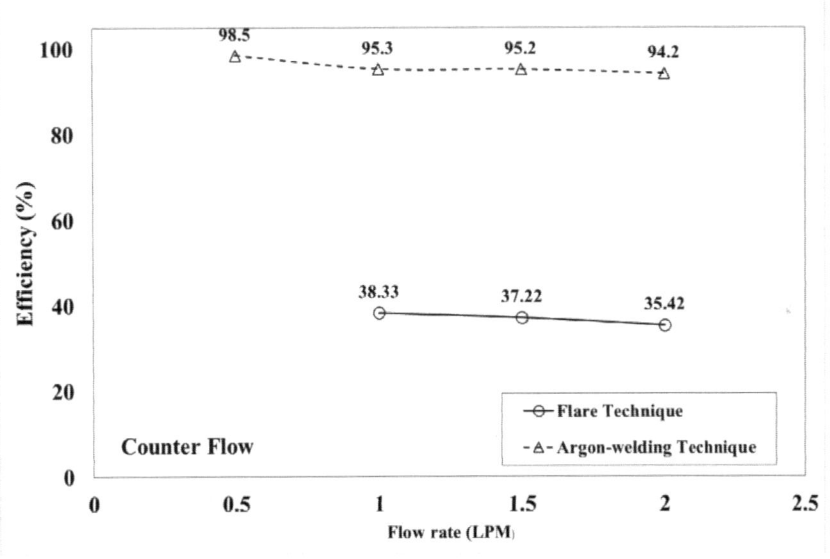

Figure 5. Efficiency of the heat exchangers were arranged as counter flow configurations.

4. Conclusions

This research focused on the performance of heat exchangers; two double pipe heat exchanger were built domestically with two different welding techniques. Both heat exchangers were tested on the same experimental set and investigated thermal efficiency. The hot water and cold water flow inside the heat exchangers were arranged as parallel and counter flow configurations. The effects of hot water and cold water flow rates; in the range of 0.5 to 2 liters per minute, were also investigated. The results showed that the second heat exchanger fabricated with the argon-welding technique showed the better thermal performance than that of the first heat exchanger fabricated with the flare technique. The thermal performances of the second heat exchanger were varied from 92% to 99% while the performances of another heat exchanger were varied from 32% to 38%. Therefore, the heat exchanger manufacturing played an important role to the heat exchanger performance. The same heat exchanger design but different fabricating technique caused lost in heat transfer process which affected the thermal performance and operating cost of the industry.

Acknowledgments

We would like to thank our students; Patpong WONGTHAM, Phathai KOONCHONNABOT, Apichat KEAWHON., The research described in this paper was financially supported by Faculty of Science and Engineering and Office of Campus, Kasetsart University, Chalermphrakiat Sakon Nakhon Province Campus, and Kasetsart University Research and Development Institute, Kasetsart University.

References

[1] Naphon P and Suchana T 2011 *Int. Communications in Heat and Mass transfer* **38** 236-241.
[2] Sarma KV, Sunder LS and Sarma PK 2009 *Int. Communications in Heat and Mass transfer* **36** 503-507.
[3] Sunder LS and Sarma KV 2010 *Int. Communications in Heat and Mass transfer* **53** 1409-1416.
[4] Gupta P and Atrey MD 2000 *Elsevier Sci.* **40** 469-474.
[5] Mushtaq IH, Abdul MA and Rageb MY 2012 *J. of Electronics Cooling and Thermal Control* **2** 35-43.
[6] Ahire S, Shelke P, Shinde B and Totala N 2014 *Int. J. of Engineering Trends and Technology* **15** 2231-5381.
[7] Christian O, Anthony O, and Adeyinka A 2015 *Int. J. of Innovative Research in Advanced Engineering* **2** 2349-2163.
[8] Tapre RW and Kaware P 2015 *Int. J. of Scientific and Research Publications* **5(6)** 2250-3153.
[9] Zare KB, Kanchan D and Patel N 2016 *Int. J. of Science Technology and Management* **5(12)** 161-174.
[10] Hoonpong P and Skullong S 2018 *J. of Research and Applications in Mechanical Engineering* **6(1)** 29-39.
[11] Chingtuaythong W and Chokphoemphun S 2018 *J. of Research and Applications in Mechanical Engineering* **6(1)** 50-62.

Main Aspects of Improvements for Centrifugal Soot Removal from Exhaust Gas

T Poojitganont*, M Kolur, O Antoshkiv, and HP Berg

[1]Chair of Combustion Engines and Flight Propulsion, Brandenburg University of Technology Cottbus - Senftenberg, Cottbus, 03046, Germany

* Corresponding Author: E-mail poojitth@b-tu.de

Abstract. In order to study the horizontal particle separator, equipped in the External-Fired Gas Turbine system (EFGT), the model experimental investigation and numerical simulation have been set up. The particle separator is used for separate small particles from gas by passing the flow through a spiral, which creates a centrifugal force. The particles will be trapped to a collector before reaching the gas exist. This study aims to analyze the efficiency of separation based on various designs of screw and cone. The experimental results show that decreasing of the cone length, the system pressure loss could be significantly reduced and the capacity of particles capture could be increased. However, the deepest cone position provides the higher loss comparing to the others. Moreover, it is evidently that the reducing number of screw's turn and it's pitch, the better capture capacity of the separator could be obtained. In parallel, the numerical simulation has been setup to support the experimental studies. The conditional parameters, including number of mesh, air mass flow rate and outlet condition, have been varied. In addition, the validation of flow simulation has been accomplished. The results show that there is the compatibility between the numerical simulation and experimental investigation as well.

1. Introduction

Large amount of particle-laden flow appears in many conventional engineering processes. Gas cyclones have been widely used as gas-solids separators for more than a century because of their simple construction and high reliability, which results from the lack of moving parts and the ability to withstand harsh operating conditions, such as high pressure and temperature applications [1]. Many researches have been done in the past to study various aspects of cyclone, for example, geometrical design [2, 3], its efficiency and performance [4, 5], optimization [6, 7] and flow characteristics [8]. Both on experimental investigation and also numerical simulation [9, 10]. However, comparing with the cyclone separator, the horizontal particle separator could provide less pressure drop and it is more suitable for a design of the EFGT system. The general flow pattern on a spiral can be deduced from basic considerations of fluid dynamics. It consists of a primary (down trough) flow component with a secondary (transverse) circulation superimposed [11].

The separator is equipped between the burner and recuperator in the External-Fired Gas Turbine system (EFGT) to split the burned ash out from the hot air. In principle, it uses the centrifugal force to separate the particles, produced by combustion of alternative fuel. The flue gas, produced from the burners, flows along the exhaust pipe. Before it reaches the heat exchanger, the exhaust gas has to be washed out. Inside the separator, the screw profile is constructed to govern the flow direction. The centrifugal force is applied to the flow. The particles, mixed in flow, will be filtered out and conveyed to the collector, whereas the cleaned air flows out on another outlet into the heat exchanger.

It is evident that the performance of a spiral would depend strongly on its design parameters, including diameter, height, number of turns, pitch and slope [12]. In order to study the effects from spiral and cone geometries, the experimental model has been built up from transparent acrylic. The test on variation of spiral's span, cone length and position has been achieved. Moreover, the different air mass flow rates from 0.025 kg/s to 0.150 kg/s have been also altered. In addition, the numerical simulation has been also setup. Their results are compared together to validate the simulation results.

2. Experimental Research

2.1. *Experimental setup*

The investigations of the developing particle separator have been carried out on a test bench at the chair's laboratory. The doser is used for supplying metal powders by rotating its feeder arm. The amount of particles is weighed before and feeding time is also counted during the test. The particle mass flow rate is achieved at the end. The structure of the test bench could be shown in figure 1.

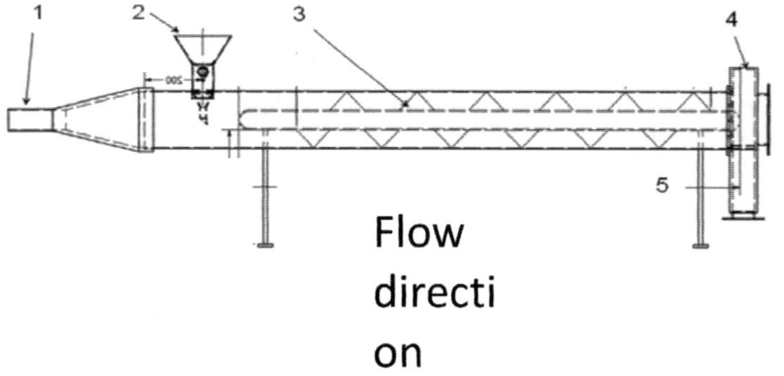

Figure 1. Principle structure of the test bench, including air inlet diffuser (1), doser (2), screw (3), collector (4) and drainage (5).

The air is supplied via the air inlet diffuser to the particle doser. There, the metal particles are added to the air flow. The rotation of the doser axis was synchronized with a pendulum, in order to have a constant mass flow during the test. The mixture of the air and particles is circulated along the screw and finally a majority of particles are collected by the collector. The low particulate air is then directed into the environment through the exit port. Figure 2 illustrates the test stand from inlet side.

IOP Conf. Series: Materials Science and Engineering **501** (2019) 012065 doi:10.1088/1757-899X/501/1/012065

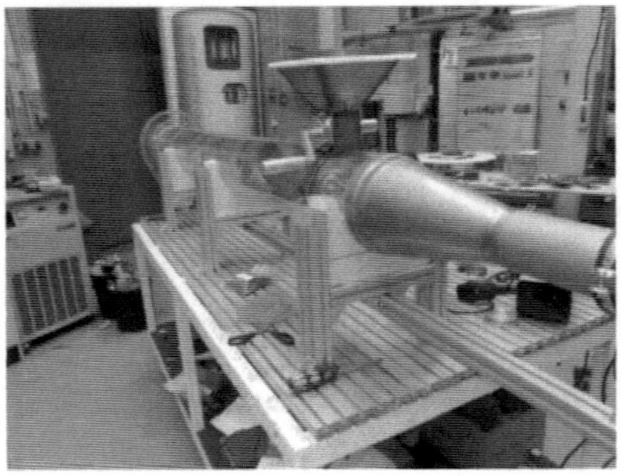

Figure 2. Test bench (view from the inlet side).

The effect of the centrifugal forces causes the separation of solid particles before the flow reaches the collector, which is shown in figure 3. The particles are collected in the lower part of the container and at the end of the experiment they are fed via flange 6 (figure 3) into the measuring vessel. Purified air is discharged via the flange 7 into the environment through the flexible tubes.

The mass flow measurement is carried out with a standardized measuring orifice according to DIN EN ISO 5167-1. The control of the air mass flow via the bypass and the separators were organized by two pneumatically controllable valves. These valves have a regulator that places the air valve piston and throttle in the appropriate position. The air is led into a separator. Here again the observation of pressure and temperature takes place. At the front of the separator is the mounting plate to which temperature probes, pressure measurement and the tested modules can be attached.

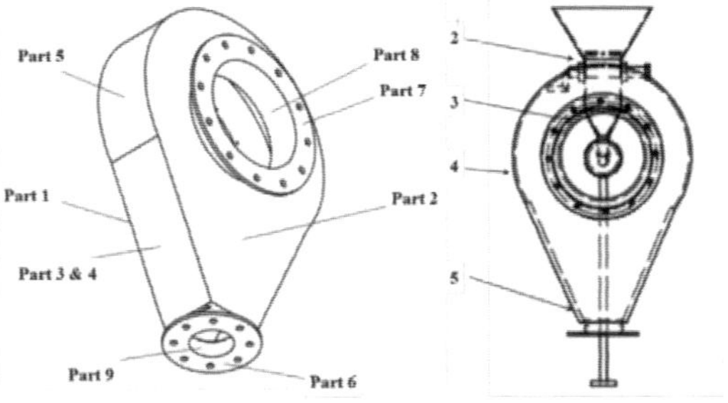

Figure 3. Basic structure of the collector, including doser (2), screw (3), collector (4) and drainage (5).

2.2. *Measurement technique*

The data acquisition on the test bench was carried out for pressure and mass flow measurements. The air mass flow was measured by means of a metering orifice. The DSA 3018 from Scanivalve Inc. with 16 parallel channels has been used for pressure measurement. The temperature measurements were carried out by means of temperature sensors at the exist of the compressor. The evaluation of the measurement signals took place in LabView. Application of the pressure sensors at the test bench is shown in figure 4.

Figure 4. Application of the measuring on the test bench.

2.3. *Design principal of the test bench geometry variation*

In order to determine optimal configuration of the separator, various geometric parameters of the particle separator have been investigated during testing. First of all, the length and span (S_{sn}) of the spiral have been considered to be varied (figure 5). The geometry variation of the screw could be concluded as in the table 1.

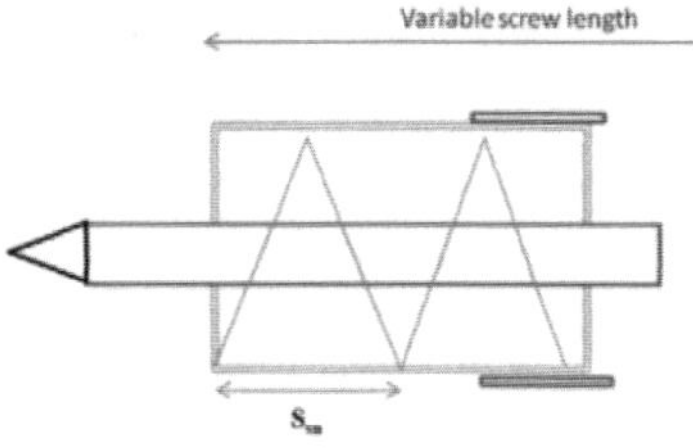

Figure 5. Variation of the screw parameters.

Table 1. Geometry parameters of screw.

Screw number	S_{sn} [mm]	Number of strand
Scw11	235	6
Scw12	235	5
Scw13	235	4
Scw21	165	10
Scw22	165	9
Scw23	165	8

In addition, the shape of dip tube, i.e. cylindrical and conical shown in figure 6, have been investigated. In both cases, the installation distance (s) was varied in 3 steps and the position of cone was also changed in 5 different steps.

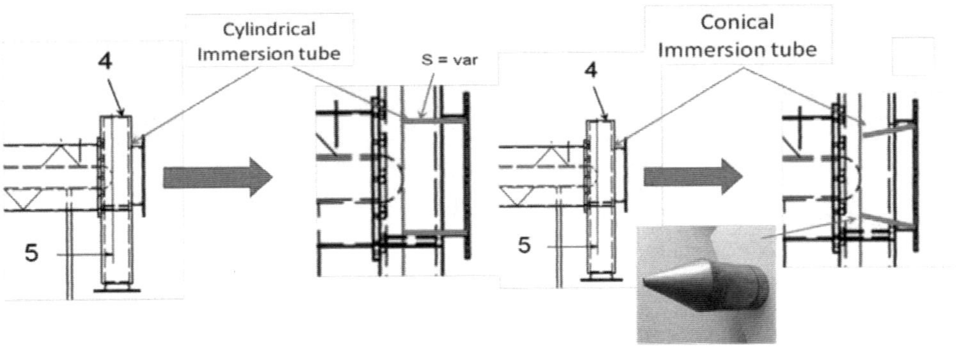

Figure 6. Positioning of the cylindrical (left) and conical (right) immersion tube.

The test bench is designed based on the similarity law [13]. The equations of fluid mechanics describing a class of flows, i.e. equations for frictional incompressible flows are necessary to know.

If the variables of these equations could be made dimensionless with constant values of velocity, length, time, etc. that are characteristic of the flow problem, then dimensionless measures like Re as factors to the terms of the equations automatically result. Problems with the same Re are described with identical equations and therefore lead to identical solutions in the dimensionless variables.

Based on existing standard part sizes of Plexiglas tubes in the market, the 200 mm diameter is selected for the housing pipe. For the model (mo) and original (or) their relationship could be written down as following:

$$\left(\frac{w \cdot l \cdot \rho}{\mu}\right)_{mo} = \left(\frac{w \cdot l \cdot \rho}{\mu}\right)_{or} \rightarrow \mathrm{Re}_{mo} = \mathrm{Re}_{or} \quad (1)$$

Where w is speed of flow, l is the characteristic length, ρ is the air density and μ is the air viscosity.

3. Simulation Setup

3.1. *Geometry model*

In parallel, the numerical simulation to study the fluid flow inside the separator has been also prepared. The calculation volume and meshes have been created. Two different quantities of mesh, i.e. 1.3 million and 2.6 million elements, on Scw22 have been varied in this study. Basically, the grid consists of air inlet, particle inlet, spiral, air outlet and outlet portion for particles as same as the model on the test rig.

3.2. *Boundary conditions and numerical setup*

In order to perform the simulation, the commercial program, ANSYS Fluent, has been used. The air inlet condition has been set as mass flow inlet, which it value is taken from the experimental data. The pressure at air outlet has been also specified. At the beginning, the particles have not been loaded into the calculation. Their particle in-/outlet have been set as no-slip wall. The realizable k-ε turbulence model has been implemented in this case. The model value could be shown in the table 1. The pressure at air outlet has been evaluated to calculate the pressure drop and compare with the experimental results achieving the validation of the simulation.

Table 2. Turbulence model values.

Parameters	Value
C_2- Epsilon	1.9
TKE Prandtl number	1.0
TDR Prandtl number	1.2

4. Experimental Results

The test results on different configurations of cones, i.e. lengths and immerged position, could be shown significant variation on the percentage of pressure loss over the particle separator with screw 11 in figure 8. In the picture the letter "C" refers to the cone configuration, which C1 is the longest one, C2 and C3 is 20 mm shorter respectively. The letter "Pos" refers to the position of each cone, where "Pos 5" is the deepest position immerged into the separator. In addition, the percent pressure loss could be defined as the equation 2.

$$\% \, pressure \; loss = \left(\frac{P_{t,in} - P_{t,out}}{P_{atm}}\right) \times 100\% \qquad (2)$$

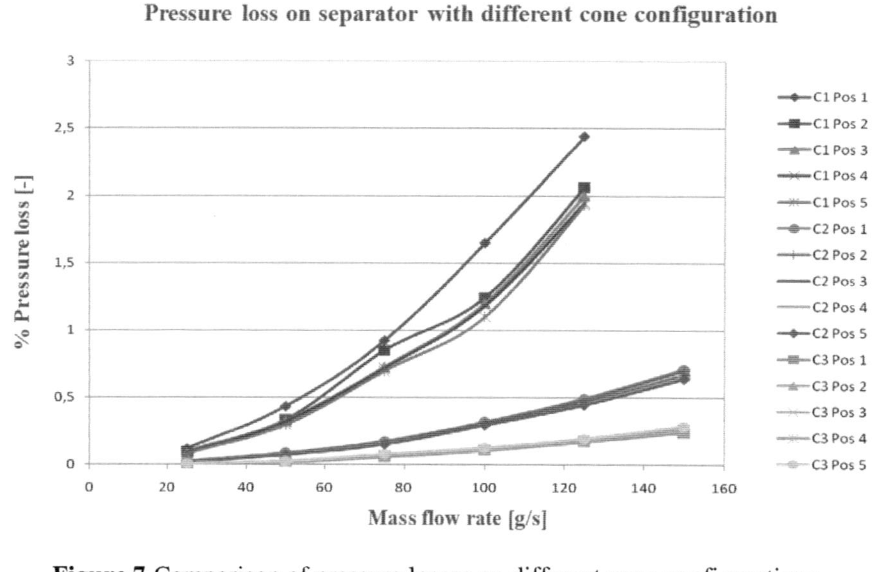

Figure 7. Comparison of pressure losses on different cone configurations.

Figure 7 illustrates evidently that the cone 1 creates highest drop on the system. Based on this cone configuration, the shallower position, i.e. position 1, provides also the higher loss comparing to other positions. However, the cone's position does not provide much effect on the cone 2 and 3, which create lower pressure loss on the system.

The deposit mass after separation as a function of cone position could be illustrated in figure 8. Considering between the different cone's length cases, i.e. "Scw11 - C1" and "Scw11 - C2", it is obviously that the shorter cone configuration can capture more particles in the system. In addition, on the most cases it seems that the position of cone has no effect on the capture capability of the separator. Moreover, this diagram shows that maximal deposit of particles in separator can be achieved by the case of screw with higher step, i.e. screw 22 or screw 23. However, this screw design causes also higher pressure loss over the separator as well.

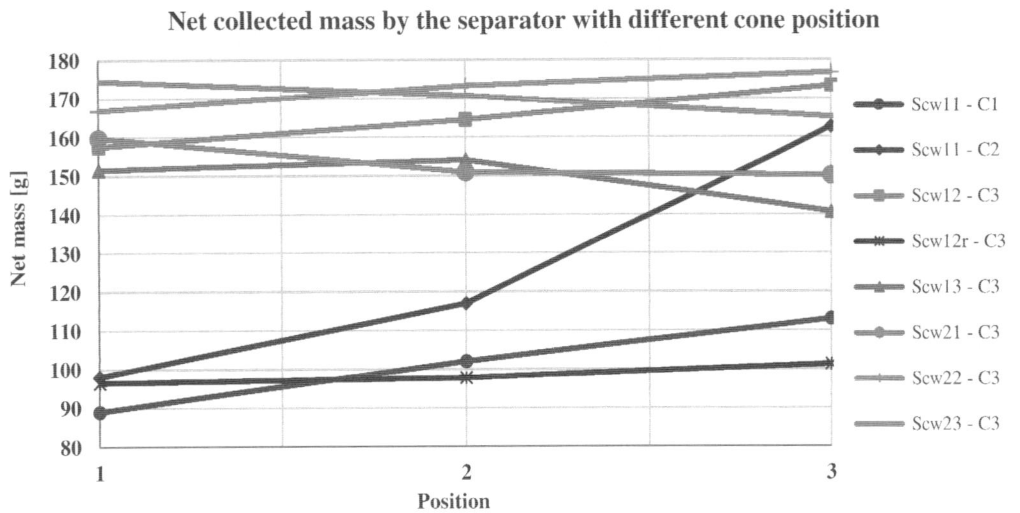

Figure 8. Comparison deposit mass function of cone type and cone position.

5. Simulation Results

For validation of simulation, the experimental result on screw 22 with cone 3 at position 3 has been selected. The results on mesh variation between medium (1.3 million elements) and finer (2.6 million elements) meshes comparing with the experimental data could be illustrated in figure 9. In these cases the simulation is finished at 2000 iterations. It is evidently that the medium mesh can provide more corresponding results with the test. This mesh has been performed further study on variation of air mass flow rate.

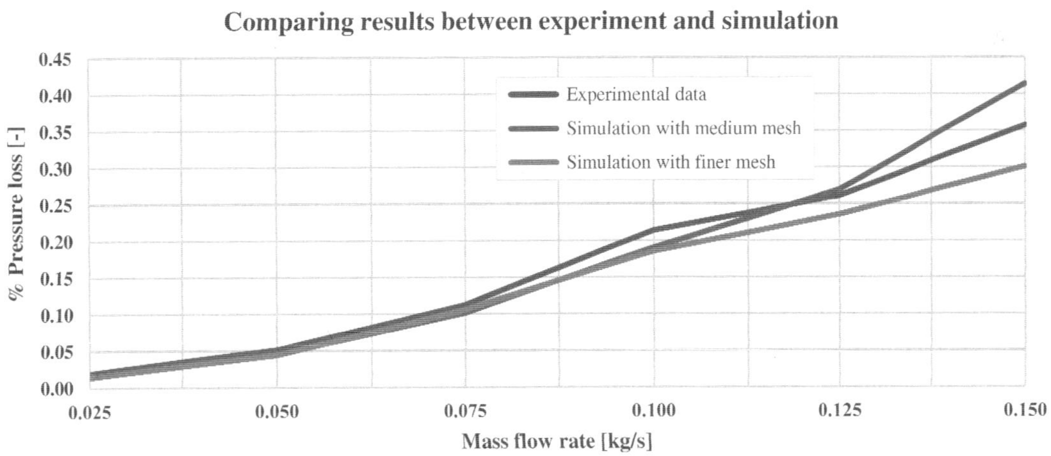

Figure 9. Comparing results on different number of meshes.

The outlet condition, i.e. target mass flow, has been further tested to achieve the better conformity with the experimental data. The results until 5000 iteration comparing on this parameter could be presented in figure 10. It is clearly that with this option the correspondent comparison with the experiment is obtained.

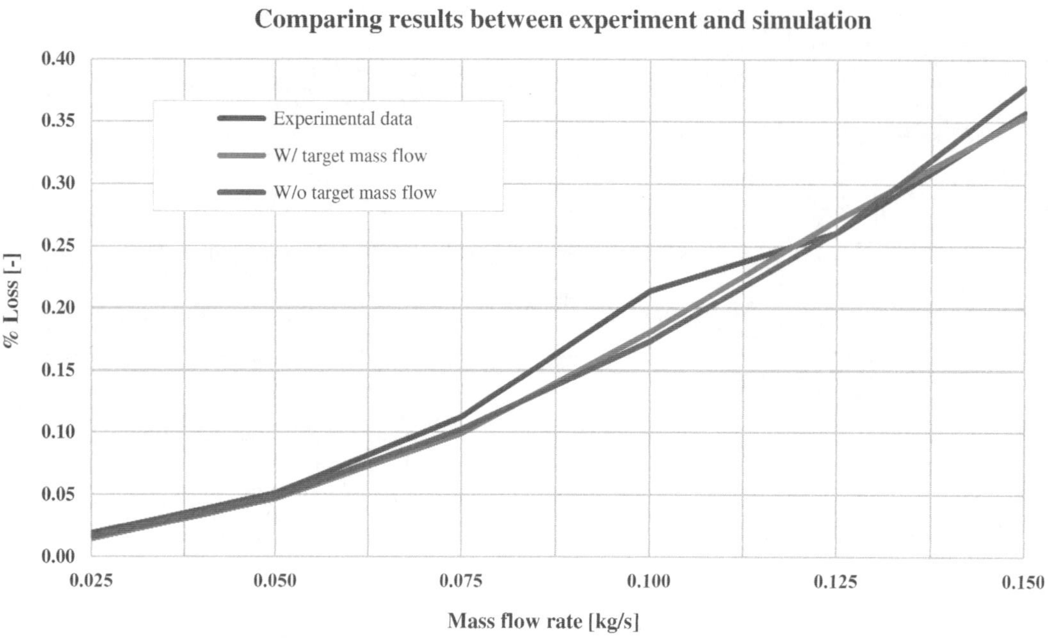

Figure 10. Comparing various mass flow rates on pressure drop.

6. Conclusion

It has to be mentioned, that any changes of screw geometry or immersion tube arrangement may increase the pressure losses over the separator. Therefore, the optimal solution for particle separator is based on estimation of both parameters - particle deposit and pressure losses as function of geometric characteristics of the separator screw and immersion tube.

The experiments show that the number of steps of a screw has a direct effect on the deposit mass trapped in the collector. The maximal deposit of particles in separator can be achieved by the case of screw with higher step, however, this screw design causes also higher pressure loss over the separator.

It was also shown, that the configuration of the immersion tube at the bunker inlet is the main tool improving the rate of deposit. The shorter cone configuration can capture more particles from the system. Moreover, it creates also lower pressure drop over the system. Based on the same cone configuration, the shallower position of the cone at the inlet collector provides higher loss comparing to the others, however, its position has not so much effect on the capture capability of the separator.

The simulation results show that the medium mesh quantity is good enough to predict the flow phenomena of the separator. The results show well correspondence comparing with the experimental data. For further investigation, the simulation could be benefited to capture more details of separation mechanism of the device. To achieve the optimal solution for particle separator, a large optimization work using modern simulation tools and sophisticated experimental set-up are required.

IOP Conf. Series: Materials Science and Engineering **501** (2019) 012065 doi:10.1088/1757-899X/501/1/012065

Acknowledgments
Authors would like to thank Mr. Philipp Sicker from our industry partner for technical supports and Ms. Kanyanat Pothijan, the exchange student in the DAAD cooperation project with KMUTNB, Thailand, for supporting the numerical simulation.

References
[1] Misiulia D, Andersson A and Lundström T 2017 Effects of the inlet angle on the collection efficiency of a cyclone with helical-roof inlet *Powder Technology* **305** 48–55
[2] Souza F, de Vasconcelos Salvo R and de Moro Martins D 2015 Effects of the gas outlet duct length and shape on the performance of cyclone separators *Sep. and Pur. Tech.* **142** 90–100
[3] Demir S, Karadeniz A and Aksel M 2016 Effects of cylindrical and conical heights on pressure and velocity fields in cyclones *Powder Tech.* **295** 209–17
[4] Balestrin E, Decker R, Noriler D, Bastos J and Meier H 2017 An alternative for the collection of small particles in cyclones: Experimental analysis and CFD modeling *Sep. and Pur. Tech.* **184** 54-65.
[5] Li Q, Xu W, Wanga J and Jin Y 2015 Performance evaluation of a new cyclone separator – Part I experimental results *Sep. and Pur. Tech.* **141** 53–8
[6] Wasilewski M and Brar L 2017 Optimization of the geometry of cyclone separators used in clinker burning process: A case study *Powder Tech.* **313** 293–302
[7] Xiong Z, Ji Z and Wu X 2014 Development of a cyclone separator with high efficiency and low pressure drop in axial inlet cyclones *Powder Tech.* **253** 644–49
[8] Wei J, Zhanga H, Wang Y, Wena Z, Yao B and Dong J 2017 The gas-solid flow characteristics of cyclones *Powder Tech.* **308** 178–92
[9] Safikhani H and Mehrabian P 2016 Numerical study of flow field in new cyclone separators *Adv. Powder Tech.* **27** 379–87
[10] Doheim M, Gawad A, Mahran G, Abu-Ali M and Rizk A 2013 Numerical simulation of particulate-flow in spiral separators: Part I. Low solids concentration (0.3% & 3% solids) *App. Math. Modelling* **37** 198–215
[11] Holland-Batt A 1995 Some design considerations for spiral separators *Miner. Engineering* **8** (11) 1381-95
[12] Kapur P and Meloy T 1998 Spirals observed *Int. J. Miner. Process.* **53** 15-28
[13] Munson B, Young D and Okiishi T 2002 Fundamentals of fluid mechanics (New York: John Wiley & Sons, Inc.)

IOP Conf. Series: Materials Science and Engineering **501** (2019) 012066 doi:10.1088/1757-899X/501/1/012066

Experiment on Influent Parameters on Wood Vinegar Burning Process

Phineath Than, and Keerati Suluksna*

School of Mechanical Engineering, Institute of Engineering, Suranaree University of Technology, 111 University Avenue, Muang District, Nakhon Ratchasima 30000, Thailand

* Corresponding Author: Keerati@sut.ac.th

Abstract This research aims to develop a burning tank for making charcoal and experiment for the wood vinegar burning process using wood as raw material. To design, and operation of such a burning tank for making charcoal require a detailed understanding of parameters that affect the system performance. In the research, the device designed for testing and developing a burning system that can improve the distillation rate of wood vinegar. The developed system consists of two tanks, a burning tank for making charcoal, and a distrillation tank for collecting wood vinegar. The burning tank for making charcoal made of steel with 400mm diameter and 600mm long for containing 15-20 kg of wood provides tight airflow and has two exhaust ducts attached to top and bottom of the tank side. The distillation tank for collecting wood vinegar has 24 copper pipes with a diameter of 12mm, and 400mm long, resulting in a condensing surface area of 0.36 m^2. The condensing pipes installed in 65 litres tank contains cool water with a temperature of 30 ˚C to 40 ˚C, used as media for removing the heat from those pipes surface. In the experiment, four influent parameters are investigated; two kinds of woods, three fire conditions of airflow rate, three moisture contents in woods are varied and two positions of exhaust duct are observed. The developed system tested by using 15 kg of Eucalyptus and River tamarind wood within 8 hours, producing 6.39L of yield wood vinegar. The experiment results show that the interaction between the new model designed and parameters significantly effect on the yield percent of wood vinegar collected. That is the effect of the prototype and those parameters can increase the wood vinegar 16 times compared to the conventional system.

1. Introduction

In a rural area, charcoal was made from wood by using an earthen kiln. The woods are often deadwood collected from the forest or sometimes bamboo is used. In addition, the charcoal then sells or use for cooking. Traditional kilns release very harmful smoke and pollution into the environment [1]. Recently, newer styles of the burning tank for making a charcoal have been studied to make a good quality charcoal and are suitable for adaption wood vinegar during a gasification process.

Wood vinegar is a liquid smoke, generated from the gas and combustion of fresh wood burning under airless condition. It is a dark liquid, contains more than 200 natural compounds primarily acetic acid, methanol, phenols, and formaldehyde [2]. Its principal components are acetic acid and methanol:

Acetic acid is used as a soil disinfectant, and methanol is used to increase crop yield utilizing less water, whereas phenols are known as a natural source for antioxidants [3]. Such attributes of its ingredients make it a better choice for sustainable farming thereby increasing the crop yield. Food and Fertilizer Technology Center [2] also shown that fruit trees produce increased amounts of fruit after applying wood vinegar in an orchard. It is also safe in the food chain living matters, especially insects that help pollinate plants. There are many rseachers have been studied on types of wood vinegar. Most of the research has been taken in China and Japan. Wessapanet et al, [4] who investigated undertaken to develop a pyrolysis oven for the production of wood vinegar by using charcoal residual as fuel. The oven was designed with an inside chamber for the heat to be evenly distributed throughout the oven with a lid covered while Somsuk, et al. [5] who also investigated pyrolysis oven in wood vinegar production by developing a heat exchanger using copper tubes and uncondensable wood gas to reheat by passing to burning chamber. Yang, et al. [6] who studied influent of airflow rate and the moisture of fuel on biomass burning behaviors. The result from this research shows that primary air has a significant effect on the moisture evaporation, devolatilisation, and char burning. Increasing primary airflow initially increasing each of the process rates but it causes a decrease in the rates beyond a certain point (the critical airflow rate). Moreover, higher moisture in the fuel results in higher moisture evaporation rate. Shen et al, [7] who observed on the influent of oxygen supply and burning rate on emission factor as a result state that a fast burning or the burning with a low air supply amount could result in oxygen deficient atmosphere. In addition, increased air supply may cool the combustion temperature, both of which could lead to the change in the formation and emission of carbonaceous particulate matter. Chalermsan, et al. [1] who produced wood vinegar from a rural charcoal kiln and it used as plant protection. Jiang, et al. [8], also carried out an experiment of wood vinegar produced from pine nut shell continuous pyrolysis. Other researchers have focused on improving animal health by using a compound of wood vinegar mixing in animal food. Suzilawati [9] investigated the design and development of a pyrolysis system to produce wood vinegar by using rice husk as a raw material. Other researchers include Yamauchi; et al. [10] who investigated wood vinegar made of broadleaf tree bark carbonized has a testing effect on red mites by injection into their bodies. Now, this natural fertilizer is receiving much attention from researchers, organizations, and institutions in the region.

However, there is still a lack of detailed or no supporting documents on the experimental study of parameters for wood vinegar burning process and develop the distillation rate of wood vinegar using wood as raw material. The advantage of an experimental study can improve the distillation rate of wood vinegar and lies in its ability to reveal the detailed behavior of parameters in the burning process inside. The research investigates two kinds of woods, three fire conditions of airflow rate, three moisture contents in woods are varied and two positions of the exhaust duct. River tamarind and Eucalyptus wood are used as a raw material for the production process. This research aims to contribute a better understanding, controlling of wood vinegar burning process.

2. Methodology

2.1. Design concepts

Schematic of wood vinegar processing system shown in Figure 1 The system consists of five parts: (1) burning gate, (2) burning kiln, (3) pipe connecting between burning tank to distillation tank, (4) distillation tank, and (5) condensing tubes. The wood vinegar burning device designed to contain 15-20 kg of wood. It provides a gate for flowing the air into the tank for burning process. The burning tank uses Liquid Petroleum Gas (LPG) as fuel. The required temperature to heat the wood is about 120-430°C to give the good quality of wood vinegar under the amount of airless condition. The limit-air known as the incomplete combustion gases that are a mixture of volatile organic compounds pass through the connection pipe to the distillation tank. This has 24 copper pipes used as a condensing tube to be condensed into wood vinegar resulting in a condensing surface area of 0.362 m^2. The condensing pipes installed in a 65 liter-tank contains with cool water, which is used as media for removing the heat from those pipe surfaces.

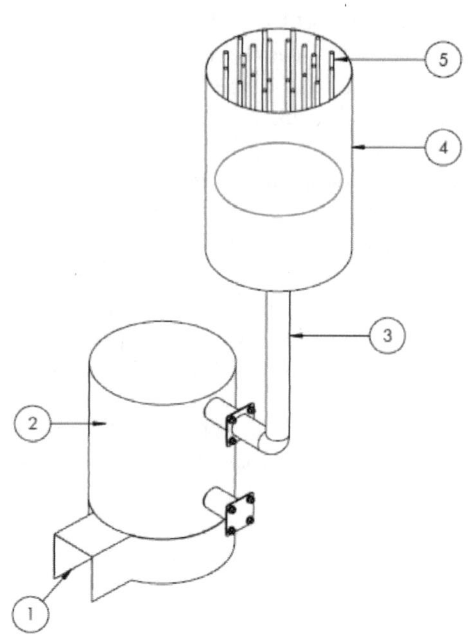

Figure 1. Schematic of the wood vinegar production system.

2.2. Operation and structure design

The burning system consists of two main components: burning tank as shown in Figure 2 and distillation tank Figure 3 the burning tank has 400 mm diameter, 600 mm tall and 1.5 mm thickness. The sand residual is used as isolation to protect the heat loss from inside, shown in Figure 11. It is designed for wood burning under the airless condition and obtains the heat from the burning gate. The tank made of sheet metal. The kiln has a kiln hood at the outside tank used for LPG fuel heating as shown in Figure 12. The capacity of the tank is 0.075 m^3 for wood. Two exhaust ducts with a 60 mm diameter and 150 mm long, attached to the tank side for more smooth ventilation and experiment investigated.

Figure 2. Burning tank.

Figure 3. Distillation tank.

The distillation system used in the extraction process for raw wood vinegar collection. The burning tank designed for the concept; the experiment of parameters for a wood vinegar burning process using wood as raw material and thermal decomposition of organic substances harmful gases pass through the copper tube. Cool water used as a media for removing the heat from those pipe surfaces. The vapor condense to be wood vinegar Figure 15. The concept helps to increase performance and decrease the greenhouse effect.

3. Experiment

In this research, the experiment process has been analized base on factorial design. The seven replicating have been employed in order to study the influence parameters; two kinds of woods, three fire conditions of airflow rate, three moisture contents and two positions of exhaust duct. Testing conditions are set based on yield percent of obtained wood vinegar. In order to determine the optimal conditions for producing the wood vinegar. The influences of exhaust duct positions are investigated by testing two replicates, fire conditions of airflow rate are varied free air mass flow, 4.2L/s and 6.4L/s will test with 2 replicates, the moisture of woods inside are removed 0kg, 1kg, 2kg will test with 2 replicates and two kinds of woods are used to test with 2 replicates. River tamarind and Eucalyptus wood use to test in the experiment as raw materials. In each experiment, 15 kg of wood is used, within a duration 5 hours, and 0.5kg of LPG. Finally, the yield percent of wood vinegar production performed.

3.1 Factorial diagram design for the experiment

1. Experiment on exhaust duct positions: factorial diagram design for exhaust duct position experiments is revealed in Figure 4. The experiment is conducted by test on the top 500mm from bottom and bottom 100mm from the bottom positions of the exhaust duct. The position which can maximize the producing of wood is able to produce wood vinegar maximum will be selected to be the setup condition for two kinds of wood experiment.

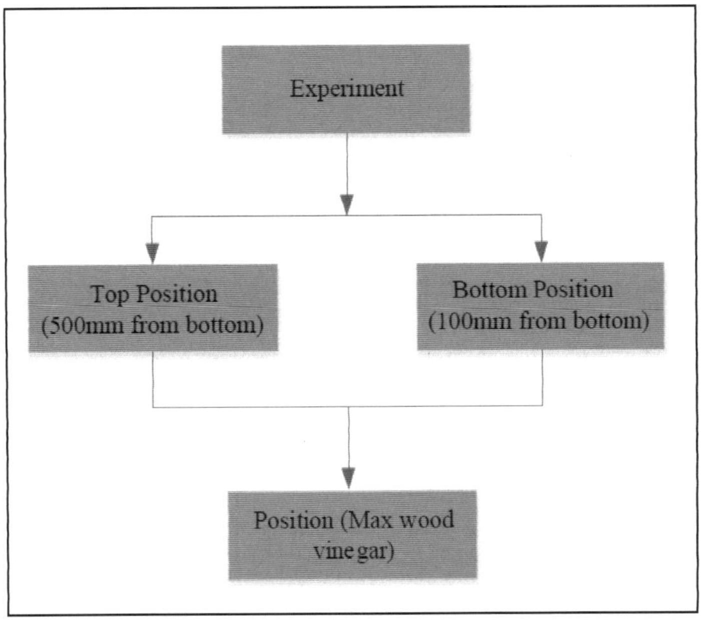

Figure 4. Factorial diagram experiment for exhaust duct position.

2. Experiment on Eucalyptus and Revier tamarind wood: factorial diagram design for Eucalyptus and Revier tamarind wood experiments is displayed in Figure 5. The experiment is conducted by test on the kind of woods. The wood is able to produce wood vinegar maximum will be selected to be the setup conditions for wood moisture experiment in the next step.

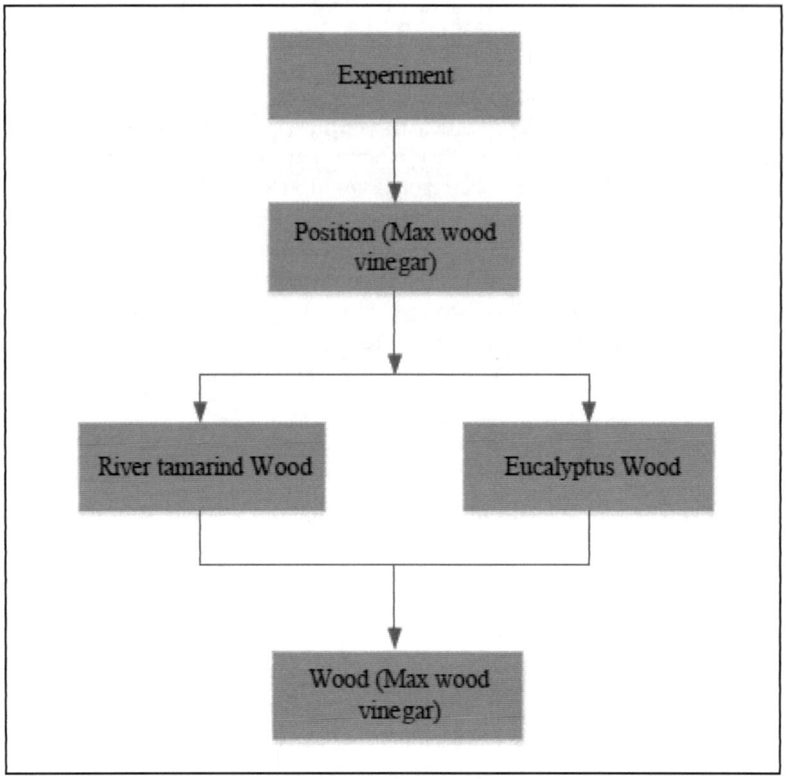

Figure 5. Factorial diagram experiment for kind of wood.

3. Experiment on wood moistures: factorial diagram design for wood moistures experiments is shown in Figure 6. The experiment is conducted by test on three kinds of wood moisture in sense of moisture removed from wood, no-moisture removed, 1kg moisture-removed, and 2kg moisture-removed. The wood moisture weight can produce a maximum wood vinegar will be selected to be the test condition for the airflow rate experiment.

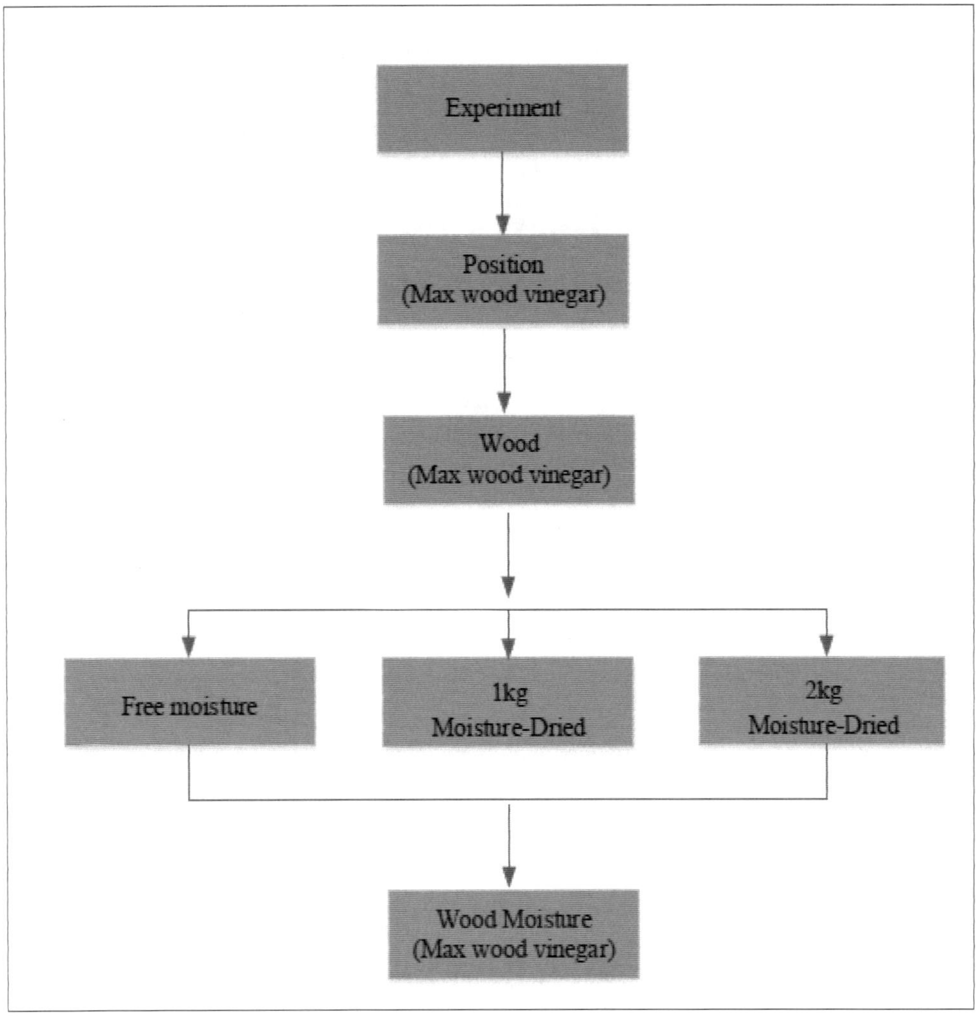

Figure 6. Factorial diagram experiment for air mass flow rate experiments

4. Experiment on air mass flow rate: factorial diagram design for wood airflow rate experiments as illustrated in Figure 7. The experiment is conducted by test on three kinds of airflow rate; free flow rate, 4.2L/s flow rate, 6.4 L/s flow rate. The wood moisture which can be produce the maximum wood vinegar will be point as the optimize condition as shown in Figure 22.

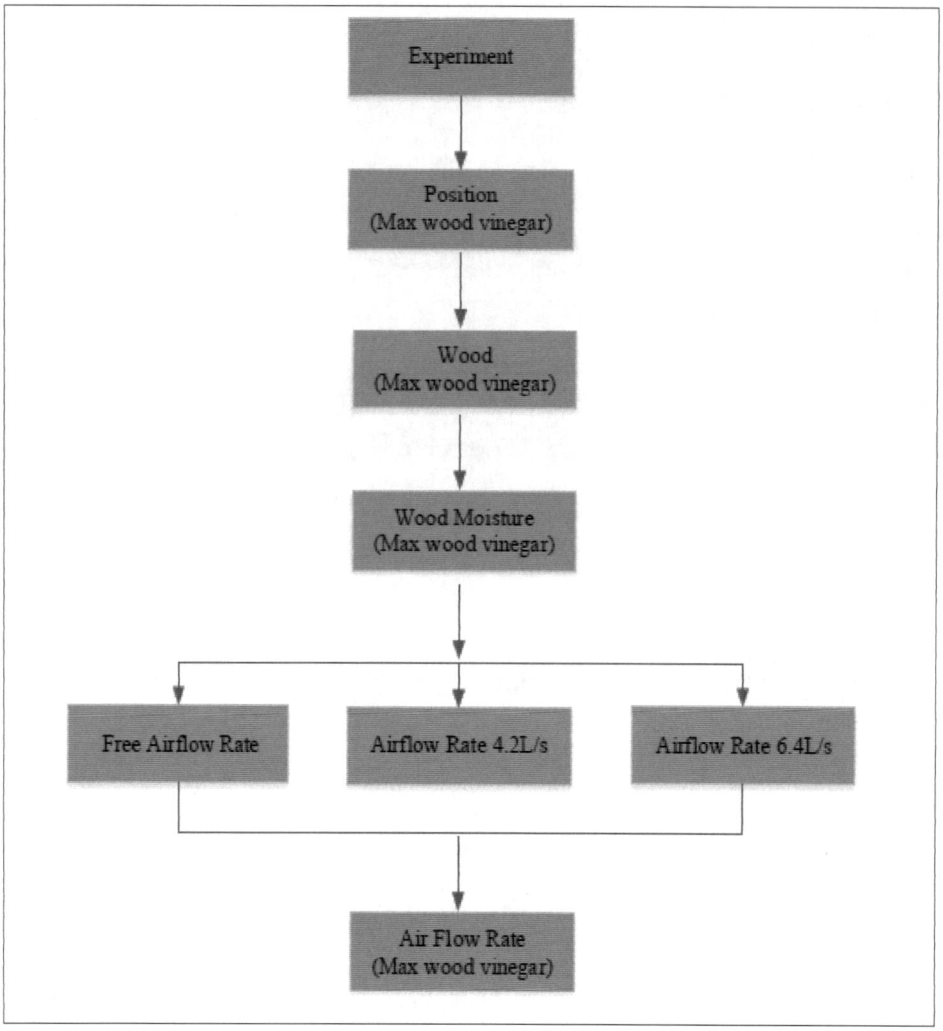

Figure 7. Factorial diagram experiment for airflow rate.

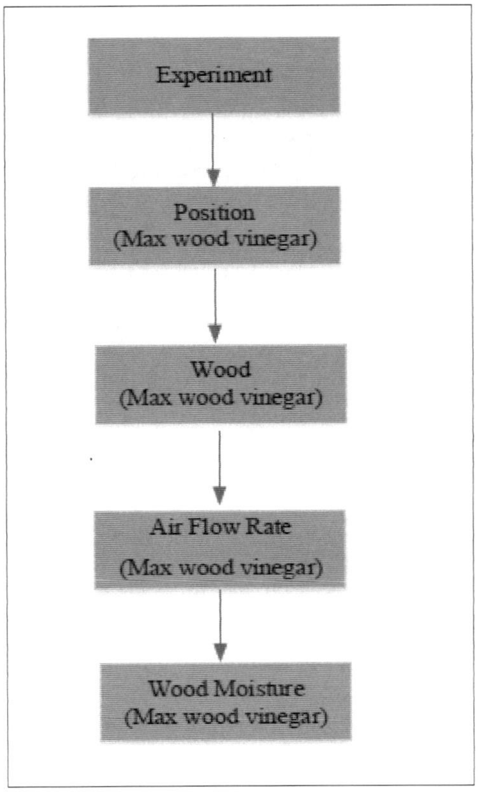

Figure 8. Factorial diagram designs for the testing.

3.2 *The steps of the experiment:*

1. Leave the fresh River tamarind wood for 25-30 days (depend on condition of moisture to be removed) before conducting an experiment step, (step 1).

Figure 9. Eucalyptus wood.

Figure 10. River tamarind wood.

2. Set up the device and fill wood river tamarind 15 kg to the burning tank, and set up LPG in the burning gate, (steps 2, 3 and 4).

Figure 11. Fill wood in the burning tank and set up the device.

3. Burn the LPG. Start to time, measure the temperature every 1 hour in the buring tank, (steps 4 and 5)

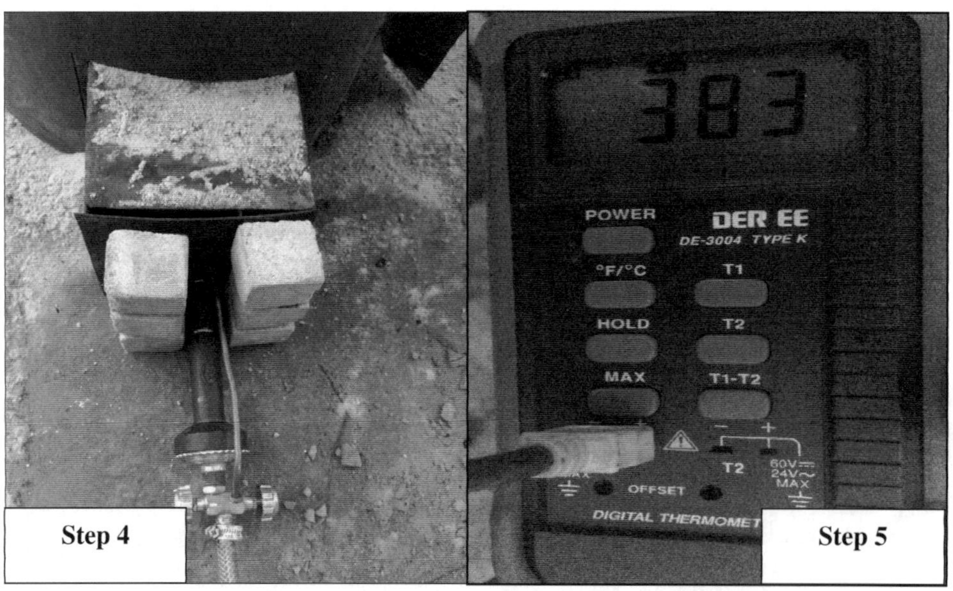

Figure 12. Liquid petroleum gas set up and temperature measurement.

4. Notice the color of the wood vinegar product from the distillation tank. If the color turns to dark-brown liquid, which means it is becoming raw wood vinegar. Collection the vinegar drops from the wood vinegar drain (step7).

Figure 13. Wood Vinegar collecting.

5. If the color of wood vinegar turns to dark brown color, which means the moisture substance in those wood are all evaporated and the wood becomes good charcoals.
6. Stop to time after five hours. The period of timing since the beginning is the process time for producing wood vinegar. Leave the burning tank and the charcoal cool down for 6 hours.
7. Measure the amount of obtained charcoal, wood vinegar, and ash.
8. Repeat the step 1-7, with another replicating again.
9. Close the raw wood vinegar in the bottles or tanks for 3 months in order to separate light oil, sour vinegar, and tar. Remove the light oil, and tar out by using sourwood vinegar only.

4. Result and discussion

The result of the experiments influent parameters are investigated and different cases are discussed according to the behavior and characteristics of wood vinegar performance as following sections; two positions of exhaust ducts are varied, two kind of woods, three moisture contents in woods, and three fire condition of airflow rates. Those experiments were tested by using 15 kg of River tamarind and Eucalyptus wood within duration 5 hours. Amount 0.5kg LPG is used to burn for each cases. The result of the interaction between those parameters is significant to show the behaviour and effect on the yield percentage of obtained wood vinegar collected.

4.1 *Experiment results on Exhaust duct position*

The burning system constructed and tested its operation on positions of exhaust ducts by using River tamarind wood. The result from the operation tested is shown in (Table 1):

Table 1. Effect of exhaust duct position on amount of products.

Testing	Unit	Exhaust Duct Positions		Percentage (%)
		Bottom Position	Top Position	
Wood	kg	15	15	0%
Time	h	5	5	0%
Fuel	Kg	0.5	0.5	0%
Wood vinegar	mL	3,392	4,215	24%
Charcoal	kg	8.1	6.2	31%
Ash	kg	0.1	0.2	50%

The average interval temperature in the burning tank process is 68-410°C that meets the desired temperature.

Figure 14. The obtained charcoal.

Figure 15. The obtained wood vinegar.

The average interval temperature in the burning tank process is about 68-410°C that meets the desired temperature. Exhaust duct on the yield of wood vinegar with respect to time collected, with two replicates. The experiment results are shown in (Table 2).

Table 2. Amount of wood vinegar with respect to time on each exhaust duct positions.

Time (h)	Burning tank (°C)				Wood Vinegar (mL)			
	Bottom		Top		Bottom		Top	
	Exp.1	Exp.2	Exp.1	Exp.2	Exp.1	Exp.2	Exp.1	Exp.2
1	124	136	71	65	8	12	0	0
2	176	154	89	99	287	253	321	349
3	228	252	144	132	840	910	1,242	1,198
4	336	316	332	348	1,240	1,180	1,409	1,431
5	419	401	337	339	1,087	967	1,273	1,207

From Figure 16, the top exhaust duct position gives the bigger yield of raw wood vinegar for the production of wood vinegar by using the burning system. The distillation rates of wood vinegar of these two position are shown in Figure 15. At initial, the bottom position gives wood vinegar 10 mL in the first hour while the top position gives no condensed wood vinegar. It is seen that the bottom position started to evaporate moisture from the wood in the first stage. From second to third hour, the temperature of the bottom position and top position were increased rapidly, but the top position was able to produce wood vinegar much more than the bottom position. During the third to fourth hour, the temperature of the bottom position and top position reached to 326 °C and 340 °C respectively. In this stage, it is found that the wood was strongly self-burning which could produce wood vinegar maximum. The wood vinegar began slightly fall distillation. After the tank was cold down, the bottom position was fill with many wood vinegar stuck to the burning surface while the top position there was lesser. This is because the hot gas in the bottom zone was hard to flow out the exhaust duct. This leads to some of the wood vinegar condensed in the burning tank and remain the temperature increasingly. The top position allowed more smooth the gas flow at the exhaust duct to the distillation tank to condense into wood vinegar.

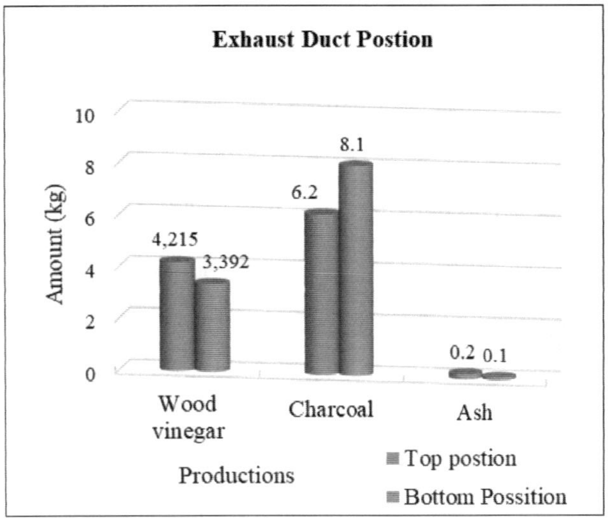

Figure 16. Comparison of product amount obtains from different position of exhaust ducts.

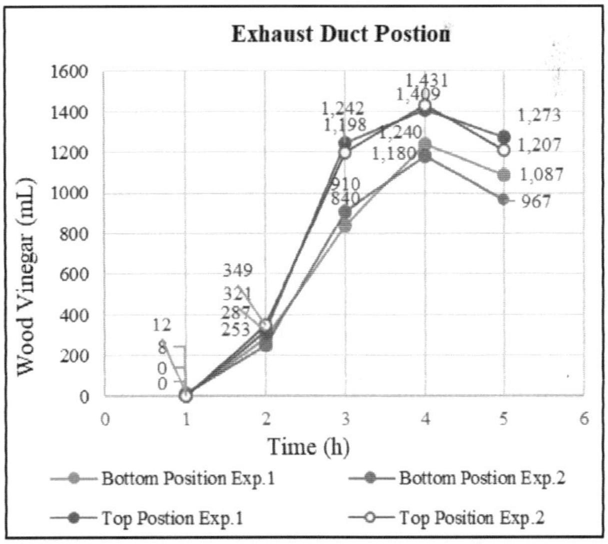

Figure 17. Wood vinegar distillation rate based on position of exhaust ducts.

4.2 *Woods experiment result and discussion*

The burning system has been tested on two kind of woods; Eucalyptus, and River tamarind wood, to find the yield percentage of obtained wood vinegar. The results from the experiment are shown in (Table 3).

Table 3. Kind of wood with respect to amount of obtained wood vinegar.

Testing	Unit	Kind of Wood		Percentage (%
		Eucalyptus	River tamarind	
Wood vinegar	mL	4,505	4,215	6.44%
Charcoal	kg	6	6.2	3%
Ash	kg	0.2	0.2	0%

The average interval of the measured temperature at the burning tank is 64-445°C that meets the desired temperature. Results from the effect of kind of woods with two replicates shown in (Table 4).

Table 4. Amount of wood vinegar with respect to time based on kind of wood experiments.

Time (h)	Burning tank (°C)				Wood Vinegar (mL)			
	Eucalyptus		River tamarind		Eucalyptus		River tamarind	
	Exp.1	Exp.2	Exp.1	Exp.2	Exp.1	Exp.2	Exp.1	Exp.2
1	61	67	71	65	0	0	0	0
2	219	231	89	99	755	683	321	349
3	359	381	144	132	1,248	1,272	1,242	1,198
4	398	418	332	348	1,331	1,309	1,409	1,431
5	439	451	337	339	1,218	1,200	1,273	1,207

Figure 18 shows the comparison of product amount between Eucalyptus and River tamarind woods. The by-production got from this experiment are charcoal and ash. The the amount of charcoal and ash obtain from both wood experiments are quite similar. Nevertheless, the observation on the results in (Table 4) revealed that the interaction of temperature between the Eucalyptus and River tamarind woods, in which the temperature in the burning tank of Eucalyptus case is higher than the River tamarind case in every hour. This is mean that the Eucalyptus wood has high heating power than the River tamarind wood. The heating power increase the temperature in the burning tank meets the temperature of wood vinegar desired (120-430°C) faster. In this case, the wet smokes happen in the burning tank are flow into the distillation

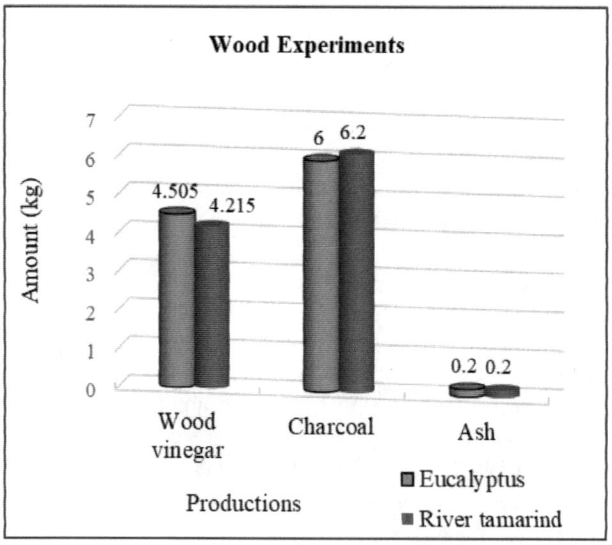

Figure 18. Comparison of product amount obtains from two kind of woods experiment.

tank. The smokes go up through condensing tube and then condense to be wood vinegar. In the second hour, 721 mL of wood vinegar was distillated from this wood. This obtained amount is higher than the use of River tamarind. However, in the fourth hour, the amount of wood vinegar condense turns to slightly drop down even the temperature in the burning tank of the wood was remain increasing. On this condition, It cause from the number of liquid substances in those woods are slightly evaporated. The temperature of River tamarind wood is 68-340°C, which is lower than the temperature of Eucalyptus case. In initial time, the wood produced wood vinegar is slower than the Eucalyptus wood case. Until the third-fourth hour, the temperature in the burning tank rises up and meets the temperature of wood vinegar desired. The wood vinegar distillation rate trend to produce faster and higher than Eucalyptus wood case. That it may cause from the number of liquid substances in those woods remain evaporating higher than the Eucalyptus wood case. Within 5 hours, the amounts of wood vinegar obtained from both of the wood were different. That is the Eucalyptus wood was able to produce wood vinegar more than River tamarind wood about 6.44% see the (Table 3). The experiment shows the type of wood is significantly affect the yield percentage of raw wood vinegar. That is, the Eucalyptus wood case can increase the wood vinegar about 6.44 % compared to the River tamarind wood case.

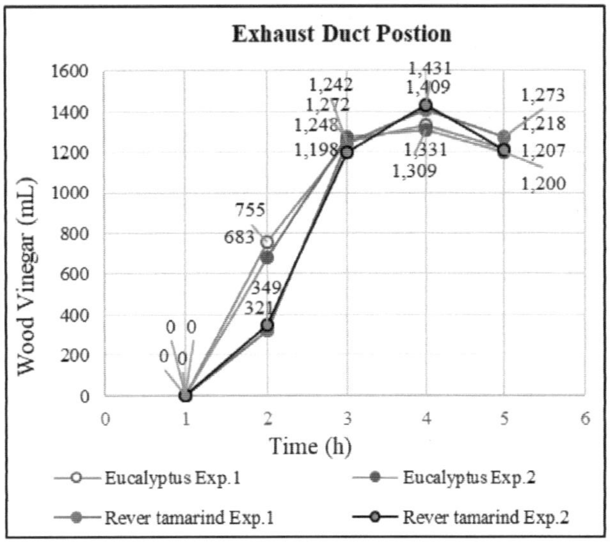

Figure 19. Wood vinegar distillation rate.

4.3 *Wood moistures experiment result and discussion*

In this research, the affect of wood moisture contents on distillation rate have been tested based on the conditions; freed moisture, 1kg and 2kg removed moisture respectively. The experiment results shown in (Table 5).

Table 5. Productions from testing.

Testing	Unit	Moistures			Percentage (%)
		0kg Removed-moisture	1kg Removed-moisture	2kg Removed-moisture	
Wood vinegar	mL	4,215	3,178	2,915	44.6%
Charcoal	kg	6.2	6	5.4	13%
Ash	kg	0.15	0.2	0.3	50%

The average interval of the measured temperature at the burning tank is around 64-452°C that meets the desired temperature. Results from the study the influence on wood moistures on the yield percent of wood vinegar collected vs. time, with two replicates. The experiment result is shown in (Table 6).

Table 6. Amount of wood vinegar with respect to wood moisture contents.

Time (h)	Burning tank (°C)						Wood Vinegar (mL)					
	0kg Removed - moisture		1kg Removed - moisture		2kg Removed - moisture		0kg Removed - moisture		1kg Removed - moisture		2kg Removed - moisture	
	Exp.1	Exp.2	Exp.1	Exp.2	Exp.1	Exp.2	Exp.1	Exp.2	Exp.1	Exp.2	Exp.1	Exp.2
1	71	65	108	112	118	130	0	0	0	0	0	0
2	89	99	174	188	201	187	321	349	496	556	690	750
3	144	132	354	352	365	359	1,242	1,198	921	949	778	822
4	332	348	407	422	406	400	1,409	1,431	1,079	1,097	856	834
5	337	339	440	464	449	449	1,273	1,207	624	636	568	532

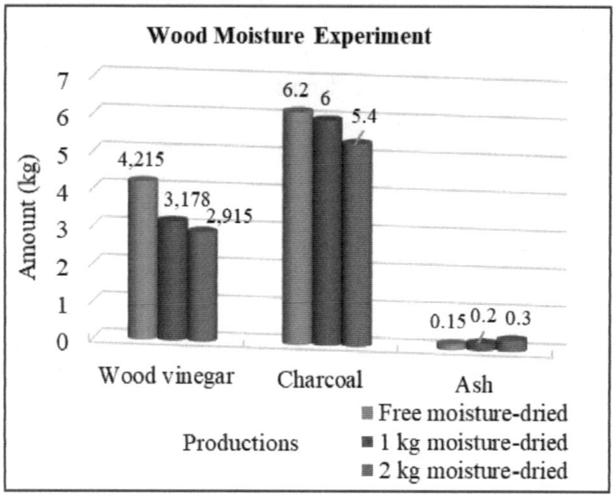

Figure 20. Comparison of product amount obtains based on wood moisture experiment.

From Figure 20, it is found that the lesser moisture removed, the higher wood vinegar condensed. The by-product got from this experiment is charcoal and ash. The charcoal obtained from none removed-moisture condition. The result in the (Table 6) shown that the temperature of none removed- moisture is lower than the temperature of high removed-moisture condition. It is also found that the wood with none removed-moisture is not good for making higher temperature and its effect on the charcoal quality. However, the temperature of the wood meets the temperature of wood vinegar desired (120-430°C). Thus, the moisture substance in those woods started to evaporate to the liquid smoke and then pass through to condense tank to condense into raw wood vinegar in the third-fifth hour see (Table 6). At the fourth-fifth hour, the amount of wood vinegar just started to fall slightly condense. This cause from the number of liquid substances in those woods is still evaporating. Moreover, the distillation rate of wood vinegar is increased when the temperature of the burning tank reach to the temperature of wood vinegar need. But in the period of last third hour the distillation rate of wood vinegar was drop down quickly even the wet smoke temperature in the burning tank remain increasing. This is because the moisture content inside those woods is mostly diffused out as wet smoke. As a result, amount of wood vinegar from burning process, it is necessary to use the raw wood with more moisture content. However, the most moisture content it reduces of wood vinegar quality.

Besides, the moisture content in the wood is also prevent the combustion turn to completed combustion. The complete combustion condition is not good for the wood vinegar as well as the charcoal production process. As ending of experiment it is found that the charcoal obtained from the wood with non-removed moisture content are almost completed. The quality of the wood vinegar is pH=3.6. While the charcoal obtained from the wood with higher moisture content, the charcoal produce from the process are also perfect charcoals as well. The quality of the wood vinegar is pH=3.1. In the experiment, it is seen that the moisture content in the woods is strongly effects to the amount of raw wood vinegar and wood vinegar quality obtained from the distillation process.

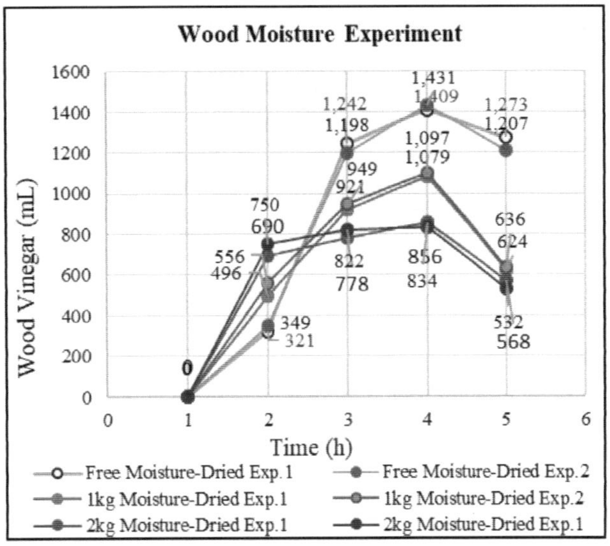

Figure 21. Wood vinegar distillation rate based on moisture content condition of woods.

4.4 *Airflow rate result and discussion*

As previous mentioned the airflow rate is one of important parameter that effect to the wood vinegar burning process. In this research, the experiments on airflow rate are on free airflow rate, 4.2L/s of airflow rate and 6.4L/s airflow rate. Results from the experiment shown in the following (Table 7):

Table 7. Productions from testing.

Testing	Unit	Airflow rate			Percentage (%)
		Freed airflow rate	4.2L/s	6.4L/s	
Wood vinegar	mL	4,235	4,215	4,175	1.42%
Charcoal	kg	4.7	6.2	6.3	24.2%
Ash	kg	0.4	0.2	0.2	50%

During perform experiment, the temperature of wet smoke inside the burning tank is varies within the range 68-430°C in order to meet the desired condition of temperature with two replicates. The experiment results are shown in (Table 8):

Table 8. Amount of distillated wood vinegar with respect to airflow rate.

Time (h)	Burning tank (˚C)						Wood Vinegar (mL)					
	Freed airflow		4.2L/s		6.4L/s		Freed airflow		4.2L/s		6.4L/s	
	Exp.1	Exp.2	Exp.1	Exp.2	Exp.1	Exp.2	Exp.1	Exp.2	Exp.1	Exp.2	Exp.1	Exp.2
1	69	71	71	65	72	80	192	268	0	0	0	0
2	268	280	89	99	98	106	1026	974	321	349	455	529
3	374	354	144	132	132	120	1345	1,335	1,242	1,198	1,298	1,372
4	380	412	332	348	305	329	918	932	1,409	1,431	1,250	1,300
5	422	438	337	339	324	332	722	758	1,273	1,207	1018	1,125

Figure 22 shows that all tested conditions of airflow rate give the quite similar amount of wood vinegar. The products obtained from the experiment are remarkable. The charcoal and ash from a freed airflow rate experiment are higher than the other rest cases. Base on the (Table 8), the temperature of freed airflow rate experiment is higher than the other rest cases. It is seen that with freed airflow rate condition, the burning tank is able to remain the temperature increase rapidly. The temperature in the burning tank meets the temperature of wood vinegar desired. Thus, the evaporation of the liquid smoke happen in the burning tank and flow to the distillation tank bypass through the condensing tube. The temperature of those tubes which surrounded by cool water is too low compared to the temperature of the vapor. That causes the temperature of the wet smoke drop down and distillated as liquid in early hour of the experiment see the Figure 23. The temperature keeps remaining at the third-fifth hour, but the amount of wood vinegar start to drop down. This is because the moisture content of those woods are almost diffused out. Therefore, the airflow rate into burning process is significant effect to increase the heat transfer between wet smoke and condensing tube. It helps to push the wet smoke from inside burning tank to the distillation. Furthermore, this can avoid the condensed smoke within the burning tank during the process. However, the heat in the burning tank will transferred out the wet smoke flow through distillation tank.

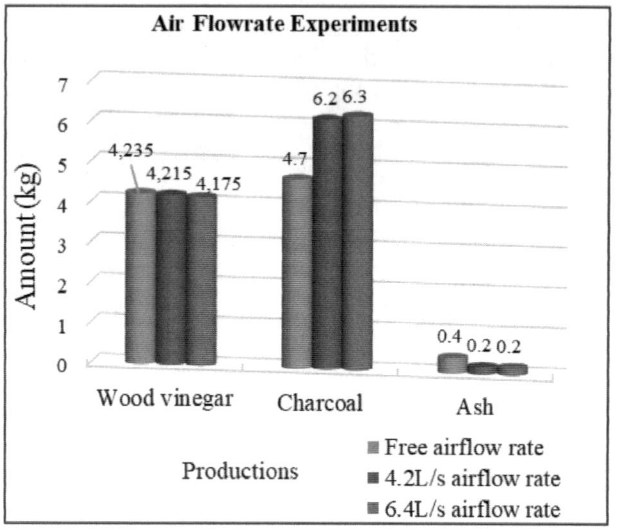

Figure 22. Comparison of product amount obtains based on airflow rate experiment.

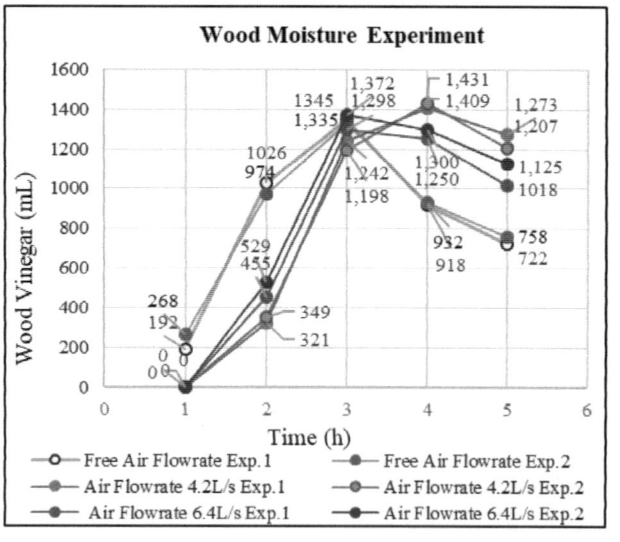

Figure 23. Wood vinegar distillation rate with various airflow.

IOP Conf. Series: Materials Science and Engineering **501** (2019) 012066 doi:10.1088/1757-899X/501/1/012066

Therefore, the temperature of airflow obtained from experiments is lower than the freed-airflow condition. That is the temperature did not match with the desired temperature of the wood vinegar. Thus, the amount of distillated wood vinegar in this period is less as well see (Table 8). As ending of the experiment, it is seen that the freed airflow experiment give a lot of wood vinegar distillate on the wall of the burning tank than other cases. The more woods closed to the fire the more charcoals and ashes become. From this experiment, in order to increase the yield of raw wood vinegar and to obtain a good charcoal, the airflow should be supplied when the temperature in the burning tank meet the wood vinegar desired temperature (120-430°C).

4.5 *Overall Result and discussion*

The burning system has been constructed and tested its operation on hold experiment. The tests are performed with various moisture contents, airflow supplied as the temperature in the burning tank meet the wood vinegar desired temperature. The experiment has been investigate on 15 kg of River tamarind wood and the position of exhaust duct has been tested. The duration of this experiment is about 8 hours. The average interval measured temperature of the burning tank is varied within the range of 65-419°C that meets the desired temperature. Results from the experiment based on the amount of distillate wood vinegar collected are shown in (Tables 9, and 10).

Table 9. Conditions and output of the test based on hold experiment condition.

Wood (kg)	15
Time (h)	8
Fuel (kg)	0.5
Wood vinegar (mL)	6,390
Charcoal(kg)	3.3
Ash (kg)	0.4

Table 10. Amount of wood vinegar with respect to time based on hold experiment.

Time (h)	Burning Tank Temperature (˚C)	Distillate Wood Vinegar (mL)
1	65	181
2	293	1,163
3	326	1,244
4	420	1,085
5	426	744
6	432	715
7	426	665
8	419	593

The proposed model design is able to produce wood vinegar 6.39 liters with the quality of pH=3.4. The measured temperature at burning tank while the wood vinegar burning process is about 65-452°C that meets the desired temperature. If the temperature below about 120°C the wood vinegar will consist mainly of water. In the other hand, if the temperature is above 400°C, it will consist mainly of tar. The product of the designed burning tank is wood vinegar and the by-products are charcoals as shown in Figure 15 and Figure 14, respectively. The wood vinegar is very valuable for agriculture. The charcoal can be used as solid fuel for cooking or can be sold. The first investment of the burning tank is about 5000 Baht. It is a simple burning tank structure, low investment but makes a high gain.

5 Conclusion

In this research, experiment has been investigated on parameters for wood vinegar burning process. The applicability of the proposed system in replacing the conventional system for wood vinegar production is determined. The experiment based on four parameters studies as explained in the case study overview have been examined. The effect of the parameters as following conclusion can be drawn:

(a) The interaction between the positions of the exhaust duct significantly affects the yield of the obtained wood vinegar. The results shown that the top position of exhaust duct can increase the wood vinegar 24 % compared to the bottom position.
(b) The type of wood significantly affects the yield of the obtained wood vinegar. From experiment, the Eucalyptus wood can increase the wood vinegar about 6.44 % more than the River tamarind wood.
(c) The moisture content in the woods is strongly affect on the yield of obtained wood vinegar. The wood with moisture content is able to produce wood vinegar much more than the 2kg removed-moisture 44.6%.
(d) The rate of airflow rate affects the amount of obtained wood vinegar when the burning tank meets the desired temperature within the range of 120-430°C.

Finally, the proposed methodology of the experiment is suitable for wood vinegar production. To eliminate or minimize the total wet smoke lost caused by a burning tank. The developed design new model can be used in wood vinegar production.

Acknowledgment
The author would like to thank the Suranaree University of Technology (SUT) for the financial support of this research.

References

[1] Chalermsan Y, Peerapan S 2009 Wood vinegar: by-product from the rural charcoal kiln and its role in a plant, *Conf., Asian Journal of Food and Agro-Industry*, S189-S195.
[2] Wood vinegar 2005 Thailand. Food & Fertilizer Technology Center. Wood Vinegar, Available: http://www.agnet.org/library/pt/2005025
[3] Burnette R 2013 an Introduction to wood vinegar. Research reports of the ECHO Technical Note #76. Retrieved from: https://www.echocommunity.Org/en/resources/fb152384-0355-4d2abcdb-dd896790a67b.
[4] Wessapan T, Sutthisong S, Somsuk N, Hussaro K, and Teekasapa S 2006. A Development of pyrolysis oven for wood vinegar production, *Conf., Eau.*
[5] Nisakorn S, Teerapot W, & Sutthisong S, Hussaro K, Teekasap S 2009 Development of a Heat Exchanger Used With Pyrolysis Oven in Wood Vinegar Production.
[6] Yang Y B, Sharifi V N, and Swithenbank J 2004 Effect of air flow rate and fuel moisture on the burning behaviors of biomass and simulated municipal solid wastes in packed beds, *Conf., Fuel*, Vol 83.
[7] Shen G, Xue M, Wei S, et al 2013 Influence of fuel mass load, oxygen supply and burning rate on emission factor and size distribution of carbonaceous particulate matter from indoor corn straw burning, *Conf., J Environ Sci (China)* 25(3):511-9.
[8] Jiang E, Chenxi Z, Liyuan Q, and Chen A 2014 *China* Experiment of wood vinegar produced from pine nut shell continuous pyrolysis, *Conf.*, *Transactions of the Chinese Society of Agricultural Engineering*, Volume 30, and pp. 262-269 (8).
[9] Suzilawati I 2013 Design and Development of Pyrolysis System to Produce Wood Vinegar. Dissertation, Universiti tun hussein onn Malaysia
[10] Yamauchi K, Matsumoto Y 2016 Japan Wood vinegar from broadleaf tree bark carbonized at low temperature has an exterminating effect on red mites by invading into their bodies, *Conf., Academia Journal of Agricultural Research*, Vol 145-159.